Springer Natural Hazards

The Springer Natural Hazards series seeks to publish a broad portfolio of scientific books, aiming at researchers, students, and everyone interested in Natural Hazard research. The series includes peer-reviewed monographs, edited volumes, textbooks, and conference proceedings. It covers all categories of hazards such as atmospheric/climatological/oceanographic hazards, storms, tsunamis, floods, avalanches, landslides, erosion, earthquakes, volcanoes, and welcomes book proposals on topics like risk assessment, risk management, and mitigation of hazards, and related subjects.

More information about this series at http://www.springer.com/series/10179

Arun Kumar Shandilya •
Vinod Kumar Singh •
Suresh Chandra Bhatt •
Chandra Shekhar Dubey
Editors

Geological and Geo-Environmental Processes on Earth

Springer

Editors
Arun Kumar Shandilya
Department of Applied Geology
Dr. Hari Singh Gour Vishwavidyalaya
Saugor, India

Vinod Kumar Singh
Department of Geology
Bundelkhand University
Jhansi, India

Suresh Chandra Bhatt
Department of Geology
Bundelkhand University
Jhansi, India

Chandra Shekhar Dubey
Professor Emeritus in Science Faculty
Sri Sri University
Cuttack, India

ISSN 2365-0656 ISSN 2365-0664 (electronic)
Springer Natural Hazards
ISBN 978-981-16-4121-3 ISBN 978-981-16-4122-0 (eBook)
https://doi.org/10.1007/978-981-16-4122-0

This Springer imprint is published by the registered company Springer Nature Singapore Pte Ltd.
The registered company address is: 152 Beach Road, #21-01/04 Gateway East, Singapore 189721, Singapore

Dedication

Professor Prem Swarup Saklani (1941–2017)

Eminent Himalayan Geoscientist

Himalayan geologist who made legendary contributions and excellent research on Geology and Tectonics of Himalaya

Former Vice Chancellor of HNB Garhwal University, Srinagar, India, Former Chairman CSTT, HRD Ministry, Government of India and Former Professor, Department of Geology, University of Delhi, India

This commemorative book having a large number of research papers pertaining to **Geological and Geo-Environmental Processes on Earth** is dedicated to Prof. Prem Swarup Saklani, who died on 24 April 2017 in Dwarka, New Delhi, at the age of 76. Professor Saklani is recognized as an eminent Himalayan geoscientist and established himself as a legendary and veteran scientist in the field of **Tectonics and Geological research of Himalaya** and has served in several academic and administrative positions at the University of Delhi and in other organizations.

Professor P. S. Saklani (1941–2017) was born on 23 August 1941 in Payar village in Tehri District of Uttrakhand. He got his school education from Inter College Srinagar Garhwal, in 1958, and obtained bachelor and master degrees in Geology from Banaras Hindu University (BHU), Varanasi, in 1960 and 1962, respectively. He has started his academic career as Assistant Professor in Geology at Punjab University, Chandigarh (1965–1968), and completed his doctoral degree (Ph.D.) under the stewardship of Late Prof. A. G. Jhingran at the University of Delhi in 1972. He brilliantly pursued his doctoral work on the '*Geological studies of the area south of the Mukhem, Garhwal Himalaya'*. Professor Saklani served his service for more than three decades in the Department of Geology, University of Delhi (1968–2006).

Professor Saklani was a visiting Fellow of the German Academic Exchange programme (DAAD) from 1973 to 1975 and worked for two years with Prof. Werner Schwan in Erlangen University, Schlossgarten in Germany. He effectively pursued research on geological and tectonic aspects of the Himalayas and Alps and recognized himself through his research contributions as a leading researcher in the field of structural and tectonic geology. To promote his academic and research career, Prof. Saklani visited the University of Tokyo, Japan, in 1984 and Germany (1989–1990) as a visiting fellow. Professor Saklani supervised one German student for dissertation work on Himalayan geology.

He has a distinguished record of service in the academic and administrative fields. During his long academic career at University of Delhi, he served as the Head of Department (1982 and 1984–1987), Dean of Science and chairman of Research and Development Committee (1985–1987), Chairman Publication Advisory Committee, University of Delhi (1986–1988). He has served in many other prestigious positions, such as the Chairman of the Commission for Scientific and Technical Terminology (CSTT), Government of India (1994–1998). Professor Saklani was elected as the President of Earth System Science of the 85th Session of Indian Science Congress in 1997–1998. In 1999, he was appointed as a Vice-Chancellor of Hemwati Nandan Bahuguna (HNB) Garhwal University, Srinagar, Garhwal, Uttarakhand, where he served until 2001. Professor Saklani had a passion for teaching and living amidst his students and has supervised several students working for their M.Sc. thesis. He has the recognition of mentoring 11 Ph.D. scholars and 11 M.Phil students, who are also well placed in different positions in academia as well as the administrative realm.

He was the recipient of several awards, viz. Consulting Editor of American Biographical Institute, Raleigh, U.S.A.; 21st Century Award: U.P. Ratna Award 1999, All India Intellectuals, New Delhi; Special Achievement Award-2001 (Education), JANYOG, New Delhi, India; Honorary Colonel Commandant of National Cadet Core (2000), Govt. of India, New Delhi; etc. He was the Founder Secretary-General of the Geoscience Foundation, India, Member of Geological Society of India and Member of Indian Council of World Affairs. He continued to render his academic services as Professor Emeritus in the Netaji Subhash Chandra Bose National Institute of Technology (NIT) for 3 years (2007–2009) after his superannuation.

He has been superannuated as Professor at the Department of Geology, University of Delhi, in August 2006. Professor Saklani authored many books, viz. Tectonic Geology, Tectonics of Garhwal Himalaya, Geology of Lower Himalaya (Garhwal), Structure and Tectonics of Himalaya, Metamorphism and Tectonics of Himalaya, Glossary of Structural Geology & Tectonics and edited many books on Geological aspects of Himalaya (volumes 1–5). His books are very popular among the students of geology in India and abroad. Professor Saklani published more than 56 research papers of national and international repute on the Himalayan tectonics. Professor Saklani served as an expert in many selection committees of DST, UGC, CSIR, UPSC and various universities in India and in Project Assessment Committees.

The unfortunate demise of Prof. Saklani has created a gap among the geo-fraternity, and the academia has lost a dedicated teacher in Earth Science and society a popular science writer and educator. His contributions to the Geological and Tectonic aspects of Himalayan and other themes will forever be remembered. He has left behind his legacy through a number of his students serving at different academic and research institutes. These students are doing significant research on tectonics and other aspects of the Himalayas and Indian peninsula. Professor Saklani was very much supported by his inspiring wife Mrs. Sushila Saklani and son Mukul and daughters Rajani (deceased) and Anjuli. His irreparable loss will be felt by all his students and admirers globally.

Research Contributions: The significant research contributions of Prof. Saklani are listed below:

Authored Books

1. 2011 Recent Geology of the Himalaya Selective Scientific Books Pandav Nagar, Delhi, 293 p.
2. 2008 Glossary of Structural Geology and Tectonics, Satish Serial Publishing House, Delhi, 191p.
3. 2006 Tectonic Geology, Satish Serial Publishing House, Delhi, 161 p.
4. 2004 Geology: an Introduction, Selective & Scientific Books, New Delhi, 93 p.
5. 1993 Geology of the Lower Himalaya (Garhwal), 254 p., International Books and Periodicals, New Delhi.
6. 1991 Elementary Geology, 82 p., Today and Tomorrow's Printers and Publishers, New Delhi.

Edited Books

7. 2008 Himalaya (Geological Aspects). Vol. 5, Satish Serial Publishing House, Azadpur, Delhi, 320p. (with an introduction by P. S. Saklani).
8. 2006 Himalaya (Geological Aspects), Vol. 4, Satish Serial Publishing House, Azadpur, Delhi, 355 p. (with an introduction by P. S. Saklani).
9. 2005 Himalaya (Geological Aspects), Vol. 3, Satish Serial Publishing House, Azadpur, Delhi, 341 p. (with an introduction by P. S. Saklani).
10. 2005 Himalaya (Geological Aspects), Vol. 2, Satish Serial Publishing House, Azadpur, Delhi, 660 p. (with an introduction by P. S. Saklani).
11. 2005 Himalaya (Geological Aspects), Vol. 1, Satish Serial Publishing House, Azadpur, Delhi, 503 p. (with an introduction by P. S. Saklani).
12. 2002 Geoindicators and related environmental studies, 248 p. Pilgrims Publishing, Varanasi-Kathamandu (jointly with C. S. Dubey).
13. 1991 Tectonic and Metamorphic investigations of Kumaun—Garhwal-Himachal—Lesser —Himalaya, 216 p., Today and Tomorrow's Printers and Publishers, New Delhi. (with an introduction by P. S. Saklani).
14. 1990 Metamorphism Ophiolites and Orogenic Belts. Today and Tomorrow's Printers and Publishers, New Delhi. (with an introduction by P. S. Saklani).
15. 1989 Himalayan Mountain Building, 198 p., Today and Tomorrow's Printers and Publishers, New Delhi. (with an introduction by P. S. Saklani).
16. 1988 Geological Society of India (Special issue), Seminar on Himalayan Crystallines, Metamorphics and Structures, Geol. Soc. Ind., Vol. 31, No. 2, Bangalore, 185–251. (with an introduction by P. S. Saklani).
17. 1986 Himalayan Thrusts and Associated Rocks, 290 p., Today and Tomorrow's Printer and Publishers, Delhi (with an introduction by P. S. Saklani).
18. 1983 Himalayan Shears, p. 113. Himalayan Books, New Delhi.
19. 1982 Himalaya: Landforms, and Processes, 116 p. Today and Tomorrow's Printers and Publishers, New Delhi (with an introduction by P. S. Saklani).
20. 1981 Metamorphic Tectonites of the Himalaya, 370 p. Today and Tomorrow's Printers and Publishers, Delhi (with an introduction by P. S. Saklani).
21. 1980 Shortening Structures in Eastern and North-Western Himalayan Rocks 62 p. Today and Tomorrow's Printers and Publishers, Delhi (Authored by W. Schwan; with an introduction by P. S. Saklani).

22. 1979 Structural Geology of the Himalaya, 394 p. Today and Tomorrow's Printers and Publishers, Delhi (with an introduction by P. S. Saklani).
23. 1978 Tectonic Geology of the Himalaya, 340 p. Today and Tomorrow's Printers and Publishers, Delhi (with an introduction by P. S. Saklani).

Published Papers and Articles

24. 2008 Saklani, P. S.* Vertical Folds and Mesoscopic Fabrics, Srinagar Area, Garhwal Himalaya. In: Himalaya (Geological Aspects), Vol. V, Satish Serial Publishing House, Delhi.
25. 2008 Bhatt, S. C. Saklani, P. S., Tectonic significance of Shear Indicators and Petrofabrics of Quartz Mylonites, Garhwal Himalaya, "Himalaya (Geological Aspects), V. 5, edited by P. S. Saklani, pp. 159–181.
26. 2006 Saklani, P. S.* Geology and Structure of Srinagar Garhwal-Himalaya In: "Himalaya (Geological Aspects), Vol. IV, Satish Serial Publishing House, Delhi 153–159.
27. 2006 Saklani, P. S., Geology of Garhwal Himalaya: Retrospect and Prospect. In: "Himalaya (Geological Aspects), Vol. IV, Satish Serial Publishing House, Delhi, 129–152.
28. 2005 Saklani, P. S., Tectonics of Central Crystallines Nappes, Garhwal Himalaya In: "Himalaya (Geological Aspects), Vol. II, Satish Serial Publishing House, Delhi, 265–280.
29. 2005 Saklani, P. S., Summary of Himalayan and Alpine structure In: "Himalaya (Geological Aspects), Vol. I, Satish Serial Publishing House, Delhi, 117–131.
30. 2005 Saklani, P. S., and Bahuguna V. K., Metamorphics of the Bhagirathi Valley, Garhwal Himalaya In: "Himalaya (Geological Aspects), Vol. II, Satish Serial Publishing House, Delhi, 207–218.
31. 2005 Saklani, P. S.* Metamorphic and Deformation History of the Central Crystallines of the Upper Alaknanda Valley, Garhwal Himalaya and its bearing on inverted metamorphism In : "Himalaya (Geological Aspects), Vol. III, Satish Serial Publishing House, Delhi, 237–255.
32. 2005 Saklani, P. S.* Geology and Morphotectonic Studies using satellite images of a part of Tehri-Uttarkashi district, Lesser Garhwal Himalaya. In: "Himalaya (Geological Aspects), Vol. III, Satish Serial Publishing House, Delhi, 153–171.
33. 2005 Saklani, P. S., Tectonic Geology of the Main Central Thrust, Garhwal, Uttaranchal In: "Himalaya (Geological Aspects), Vol. III, Satish Serial Publishing House, Delhi, 179–189.
34. 1998 Singh, V. K., Singh, S. P., Saklani, P. S., and Dubey, C. S., Mesostructure and deformational history of the Central Crystallines: an example from Garhwal Himalaya, India. J. Nepal Geol. Soc., V. 17: pp. 59–69.
35. 1998 Singh, S. P., Singh, V. K., and Saklani, P. S., Metamorphism and Thrust Tectonics in the Munsiari Group of Central Crystallines Zone of Garhwal Himalaya, India, J. Nepal Geol. Soc., V. 18: pp. 85–95.
36. 1998 Saklani, P. S., Geology of the Garhwal Himalaya: Retrospect and Prospect, Presidential Address, Earth System Sciences, Ind. Sci. Cong., Calcutta, 32p.
37. 1997 Singh, S. P., Singh, V. K., and Saklani, P. S., Metamorphic evolution of the Central Crystallines of the Higher Himalaya along Dhauli Ganga Valley, Garhwal Himalaya. N. Jb. Geol. Palaont. Abh., V. 206: pp. 249–275.
38. 1997 Singh, S. P., Singh, V. K. and Saklani, P. S., Metamorphism in the Central Crystallines of Higher Himalaya in Kedarnath valley, Garhwal Himalaya, Him. Geol. V. 18: 119–133.

39. 1996 Singh, V. K., Singh, S. P. and Saklani, P. S., Metamorphic and geodynamic evolution of the Central Crystallines of the Garhwal Himalaya, India, N. Jb. Geol. Palaeont. Abh., 199(1), 89–109.
40. 1996 Singh, V. K. and Saklani, P. S., Metamorphic and Tectonic history of the Budhakedar area, Garhwal Himalaya, Uttar Pradesh, G. S. I. Sp. Publ. No. 21 (1): pp. 199–203.
41. 1994 Singh, V. K., Singh, S. P. and Saklani, P. S., Kyanite-Sillimanite-Andalusite-bearing schist from Central Crystallines of Budhakedar area Garhwal Himalaya. Bull. Ind. Geo. Assoc. Chandigarh, V. 27: pp. 143–153.
42. 1994 Bhatt, S. C. and Saklani, P. S., Strain Transition and Microstructural Fabric analysis of Quartz-Mylonites exposed within Pratapnagar Nappe, Garhwal Himalaya, India. J. Geol. Soc. India 43(4): pp. 381–394.
43. 1993 Saklani, P. S., Geological formation and structure of the Central Himalaya Mountain system—A study in Nature's Equilibrium, The Central Himalaya Panorama, Inst. Soc. Res. & App. Anthro., Calcutta, 70–80.
44. 1992 Saklani, P. S.,* Strain estimation in the deformed Hafla (Jutogh) quartzite of Jutogh thrust sheet, Garhwal Himalaya, Jour. Him. Geol., 3(2): 149–154.
45. 1991 Saklani, P. S., Nainwal, D. C. and Singh, V. K., Geometry of the composite Main Central Thrust (MCT) in Yamuna Valley, Garhwal Himalaya, India, N. Jb. Geol. Palaeont. Abh., V. 1991: 364–380.
46. 1991 Saklani, P. S., Mesoscopic structures along the Bhatwari-Tehri-Narendranagar traverse (Garhwal, Uttar Pradesh, India)—and their chronologic and kinematic integration in the deformation plan of the Himalaya, Z. dt. Geol. Ges., 142: 97–113.
47. 1991 Saklani, P. S., Bhatt, S. C., Singh, V.K. and Satendra, Deformation and kinematic analysis of folds along Krol Thrust of Garhwal Himalaya. Jour. Scientific Research B. H. U., Platinum Jubilee Year, 41B: 141–150.
48. 1990 Bhatt, S. C. and Saklani, P. S., Kinematic framework of heterogenous deformation within Pratapnagar thrust sheet, Bhagirathi Valley, in Lesser Garhwal Himalaya, U.P., Jour. Geol. Soc. India, v. 36(3), pp. 247–261.
49. 1989 Saklani, P. S. and Bahuguna, V. K., Geothermometry of Central Crystallines of the Bhagirathi Valley, Garhwal Himalaya, Metamorphism, Ophiolites and Orogenic Belts, In : Saklani, P. S. (Ed), 1–16, Today and tomorrow's Publishers, N. Delhi.
50. 1989 Bhatt, S. C. and Saklani, P. S., Analysis of Folds in Dharkot Thrust Sheet. In: Saklani, P. S. (editor), Himalayan Mountain Building, Today and Tomorrow's publishers, New Delhi, pp. 63–90.
51. 1988 Bahuguna, V. K. and Saklani, P. S., Tectonics of the Main Central Thrust in Garhwal Himalaya. J. Geol. Soc. Ind., v. 31, 197–209.
52. 1986 Saklani, P. S. and Bahuguna, V. K., Thrust tectonics of Garhwal Himalaya. In: Saklani P. S. (Ed), Himalayan Thrusts and Associated Rocks, 1–25, Today and tomorrow's Publishers, New Delhi.
53. 1983 Saklani, P. S. and Bahuguna, V. K., Main Central Thrust Zone and associated imbricate structures in Chhatera area, Garhwal Himalaya, In: Saklani, P. S. (Ed) Himalayan Shears, Himalaya books, N. Delhi, 1–9.
54. 1981 Saklani, P. S. and Doval, S. C. Ghuttu Garhwal Himalaya: Geologic and Metamorphic Characteristics, Metamorphic Tectonites of the Himalaya In: Saklani, P. S. (Ed.), 15–26, Today and Tomorrows Publishers, New Delhi.
55. 1979 Saklani, P. S., Folded rocks of Northern Tehri Garhwal, In: Saklani. P. S. (Ed.), Structural Geology of the Himalaya. Today and Tomorrow's Printers and Publishers, New Delhi, 101–112.
56. 1979 Saklani, P. S., An outline of the Geology of the Lesser Himalaya of Garhwal with special reference to its Ecology and Development JOSHARD, Garhwal University, 178–45.
57. 1979 Saklani, P. S., A note on the study of Minor Structures in rocks of Srinagar area, Garhwal Himalaya, U.P., 10th Seminar on Himalayan Geology, W. I. H. G., 178–190.

58. 1978 Singh, S and Saklani, P. S., Some geomorphologic observations in the Ghansyali area, Garhwal Himalaya Him. Geol. 8, 813–821.
59. 1978 Saklani, P. S., Uttarakhand—its development and economic geology, Himalaya: Man and Nature, Delhi, 1(10): 10–17.
60. 1978 Saklani, P. S., Metamorphic events in relation to Deformation in the Bhilangna and Balganga Valleys, Ghansyali Area. Garhwal Himalaya, Uttar Pradesh, 9th Seminar on Himalayan Geology, 9(1): 199–210.
61. 1978 Saklani, P. S., Geology and Natural Wealth of Garhwal Himalaya, Himalaya: Man and Nature, Delhi, (5 & 6): 26–30.
62. 1978 Saklani, P. S., Deformation and Tectonism of Mukhem Area, Lesser Himalaya, In: Saklani, P. S. (Ed.), Tectonic Geology of the Himalaya. Today and Tomorrow's Printers and Publishers, New Delhi, 15–42.
63. 1977 Saklani, P. S., Khanduri, H. C. and Singh, S., A reappraisal of the geology of a part of Mussoorie Syncline around Satengal, Garhwal Himalaya. Chayanika Geologica 3, 95–103
64. 1976 Saklani, P. S., Some remarks on the Geology of the Himalaya and adjoining mountains chains of Iran, Pakistan and USSR, International Himalayan Geology Seminar, New Delhi, Structure, Tectonics, Seismicity and Evolution, Geol., Surv., Ind. Misc. Publ., 41: 165–174.
65. 1975 Saklani, P. S., Outline of the Structure of Himalaya and Alps: a comparison, Acta Universitatis Carolinae, Czechoslovakia, Prague, 3: 221–236.
66. 1975 Saklani, P. S., Metamorphic Petrology of the Area, South of Mukhem Garhwal Himalaya, Recent Researches in Geology, Hind. Publ. Corp. India, Delhi, 82–106.
67. 1975 Saklani, P. S., Deformation of Quartz and Mica in tectonites of Tehri Garhwal Himalaya, 5th Himalayan Geology Seminar, Panjab University, 1969, Pub. Cent. Adv. Stud. Geol., Panjab University, Chandigarh, 10: 34–44.
68. 1975 Saklani, P. S., A review of Himalayan Orogenesis, Cas. Min. Geol., Czechoslovakia, Prague, 18: 193–198.
69. 1974 Saklani, P. S., Geologischer Aufbau Des Mukhem Gebiets, Garhwal Himalaya-Eine Zusammenfassung, Nachr. Deuch. Geol. Ges., Hannover, 10: 20–21.
70. 1973 Saklani, P. S., Occurrence of Stromatolites in the Bhelunta limestone formation of Pratapnagar area of Garhwal Himalaya, Sci. and Cult., 37: 215–216.
71. 1973 Saklani, P. S., Geological studies of the area, South of Mukhem, Garhwal Himalaya, (Summ. Doc. Thesis, Univ. Delhi), Geomag., 1: 11–13.
72. 1973 Saklani, P. S., Cataclasis of Allochtohon of the Pratapnagar Quartzite Group, Tehri Garhwal, Sci. and Cult., 87: 438–439.
73. 1972 Saklani, P. S., Metamorphism in rocks of par autochthonous zone of Mukhem area, Garhwal Himalaya. Indian Mineralogist, 13, 69–73.
74. 1972 Saklani, P. S., Tectonics of the Main Central Thrust Zone, Lesser Garhwal Himalaya, 27th International Geological Congress, Moscow, USSR, III 1–9, Sections 06–07 : 387–388.
75. 1972 Saklani, P. S., Metamorphism in the rocks of Pratapnagar area, Tehri Garhwal, Uttar Pradesh, Bull. Ind. Geol., Asso., Punjab University, 7: 115–128.
76. 1972 Saklani, P. S., Lithostratigraphy and Structure of the area between the Bhagirathi and Bhilangna rivers, Garhwal Himalaya, Himalayan Geology, 2 : 342–355.
77. 1972 Saklani, P. S., and Pande, I. C. Geology of the Pratapnagar area, Tehri Garhwal, Uttar Pradesh, Bull. Ind. Geol. Asso., 3(1 & 2): 12–18.
78. 1971 Saklani, P. S., Structure and tectonics of Pratapnagar area, Garhwal Himalaya. Him. Geol., 1, 75–91.
79. 1970 Saklani, P. S., Metamorphism in rocks of Garhwal Nappe of Garhwal Himalaya. Pub. Centre of Advanced Study in Geology, Punjab University, Chandigarh Vol. 5, pp. 115–120.

* Co-authored

Editors:
A. K. Shandilya, Department of Applied Geology, Dr. HS Gour University, Sagar India
#Vinod K. Singh, Department of Geology, Bundelkhand University, Jhansi, India
S. C. Bhatt, Department of Geology, Bundelkhand University, Jhansi, India
C. S. Dubey, Department of Geology, University of Delhi, Delhi, India
 # Corresponding editor email: vinodksingh@bujhansi.ac.in

Preface

The earth is a restless planet and has always been influenced by geological and geo-environmental processes. The interaction of magmatic, tectonometamorphic and metal-logenic processes are responsible to produce crustal blocks. These blocks are represented by cratons, shields, and orogenic mobile belts. The studies done by the pioneers on geodynamics of lithospheric shields are adequate but the new insights are still needed to understand the detail geological processes. The geo-environmental events attributed to cause natural calamities and loss of life, land and properties. Although the natural disasters cannot be prevented, but their impact can be minimize by applying scientific remedial measures. Therefore, the researchers are more exited to know about the geochemical, geotectonic and geomorphological changes occurring in the exterior and interior domains of the earth.

Keeping in view the above facts the idea has been immerged to edit a book on 'Geological and Geo-Environmental Processes on Earth' which is dedicated to an eminent Himalayan geoscientist late Prof. Prem Swarup Saklani. This book is structured into two sections; (I) Geological Processes, (II) Geo-Environmental Processes. The chapters pertaining to section Geological Processes are focusing on the geology, structure and tectonics, geodynamics, crust-mantle interactions, mineral chemistry, rare-earth deposits, metamorphic processes, geochemical and geochronological aspects of different parts of the continental crust.

The chapters included in the section Geo-Environmental Processes are highlighting on seismic hazards, role of geotectonic controls in groundwater contamination and geothermal systems. It also explores the applications of remote sensing, Geographical Information System and SRTM data for evaluation of the morphometric and morphotectonics of different water-sheds and vulnerability of earth towards landslides and natural hazards.

This book will be useful to students, scientists, researchers, academician and planners working in various organizations. It is a noteworthy collection of articles written by eminent scholars from different countries of the world.

Sagar, India — Arun Kumar Shandilya
Jhansi, India — Vinod Kumar Singh
Jhansi, India — Suresh Chandra Bhatt
Cuttack, Odisha — Chandra Shekhar Dubey

Contents

About the Editors

Prof. Arun Kumar Shandilya is a faculty at Department of Applied Geology, Dr. Hari Singh Gour University, Sagar, India. He has 41 years of Research and Teaching experience. He has published more than 80 Research papers in national and International journals. He has received Dr. V. S. Prabhakar Rao Gold Medal at graduation level (1974) and Dr. Radha Krishana Award in 1993, by Government of Madhya Pradesh. He served as the Head Department of Applied Geology, Dean, Student's Welfare, Chief Coordinator (DST Instrumentation Centre -SIC) and Director, UGC Remedial Coaching Centre, Professor In Charge University Store and Purchase. He has served as the Subject Expert in the UPSC, state Public Service Commission (PSC), Scientific Terminology Commission, Govt. of India, and in the National Working Group for International Geological Correlation Programme (IGCP) of Geological Survey of India, Kolkata. He has completed 06 Research projects and supervised 06 Ph.D. students. Presently, he is working as member of editorial board of many journals, expert in Research Degree Committee (RDC) and Board of Studies (BOS) in various Universities.

Prof. Vinod Kumar Singh is working on Precambrian geodynamics of Bundelkhand Craton and teaching at Bundelkhand University, Jhansi, India since November, 1999. He did Ph.D. in Geology from University of Delhi and worked on Structure and Tectonics on Himalayan orogenic belts. He worked on Plaeaomagnetism at Tubingen University, Germany as scientists and co-researcher with Prof. Erwin Appel in a collaborative project funded by German Research Council. He is working in collaboration with Institute of Geology, KRC, Petrozavodsk, Russia on geodynamic evolution of earth crust. Prof. Singh using integrated approach involving new and innovative methods, geological fields indeed represent both Archean/Proterozoic and Himalayan (and related) orogens, to understand the geodynamics of crustal evolution. He received ILTP and DST-RFBR projects to understand the global geodynamics by comparing the lithospheric structures and geodynamics of Bundelkhand Craton and Karelian Craton with Russian collaborators Prof. A. I. Slabunov and Prof. DrSc Sergei A. Svetov. Prof. Singh has published about sixty contributions in reputed journals, book chapters and also edited five books and journals. He is a member of editorial board of several journals.

Prof. Suresh Chandra Bhatt a faculty in geology at Bundelkhand University Jhansi, India has specialization in Structural and Tectonic geology, Sedimentology, and Environmental Geology. He has distinguished record of services in academic and administrative field and has been working at several academic and administrative positions since last two decades. He is a recipient of Young Scientist project (1990), USSR fellowship (1987), CSIR pool scientist (1994) and did innovative research on tectonics of Himalaya and Bundelkhand craton and on palaeoclimate and chronosratigraphic status of Thar Desert. He authored two text books, three edited volumes and published more than 55 research papers. He guided five Ph.D. scholars under his stewardship. Professor Bhatt is a life member of Geological Society of India, Indian Society for Rock Mechanics and Tunnelling Technology, and Indian Society for Construction and Materials. He is working as a member of editorial board of several journals.

Prof. Chandra Shekhar Dubey is presently working as a Professor Emeritus in the Faculty of Science, Sri Sri University, Cuttack, Odisha, India. He has worked in the field of Earth and Environmental Science for the last three decades. He worked as Vice Chancellor in Sanskriti University Mathura and Director of Institute of Lifelong Learning and Campus of Open Learning besides being Head, Centre of Advanced Studies, Department of Geology and Dean of Science in University of Delhi, India. He has published 57 papers in peer reviewed International and National Journals, one patent for removal of Arsenic and has supervised 15 Ph.D. students. He has been a visiting Professor to Joseph Fourier University, Grenoble, France and recipient of International UNESCO CNR-Indo- Italian, NSF-DST Indo-US and B. P. Koirala India-Nepal Fellowship grants. He has worked as a Chairman/Expert member in several National and International committees in Ministry of Environment and Forests, Ministry of Science and Technology, Ministry of Water resources, University Grants Commission, DRDO and ICDE, Norway, HKT etc.

Geological and Geo-Environmental Processes on Earth: Introduction

A. K. Shandilya, V. K. Singh, S. C. Bhatt, and C. S. Dubey

The geological (endogenic and exogenic) processes operating in interior and exterior domains of earth are bringing dynamic changes in the lithospheric and geomorphic terrains of this planet since its origin. The interplay of magmatic, tectonometamorphic and metallogenic processes is continuously evolving new continental and oceanic plates (blocks) enriched with magmatic plume and economic minerals. The diverging and converging margins of these blocks were stressed and formed orogenic mobile belts and mountains. The undeformed stable blocks are referred to as shields and cratons. These small nuclear cratonic blocks were welded together by troughs and rifts and became a good site for sedimentary deposits through Precambrian. The geo-environmental events manifesting as seismic, volcanic, floods, cyclones and drought are poising great threat to the environment and socio-economic status of the world. These calamities are attributed to causing loss of life, land and depletion of resources and are continuously reshaping the dimension of various landscapes.

Despite the detailed study (based on petrological, geochemical, geophysical and geochronological data) done on geodynamics of lithospheric shield and geo-environmental events, the innovative research is still needed with more scientific ideas to understand these processes. The present days' studies focussed on magmatic and geodynamic systems of various shields are concentrated on U-Pb SHRIMP age, Lu-Hf isotope, trace elements studies, detrital zircons investigations, and metallogeny of the different parts of the earth's crust. These research investigations will be helpful to understand the mechanism of geological processes involved in crustal evolution.

This edited volume dedicated to late Prof. Prem Swarup Saklani addresses the multidisciplinary themes pertaining to the role of tectonism and magmatism in crustal evolution and global distribution of metallic and non-metallic mineral deposits. It gives valuable information on geodynamic evolution, structural, petrological, isotopic, metamorphic, geochemical and geochronological attributes of continental and oceanic crust and is challenging reassessments of the existing paradigms. This book also advocates the role of tectonics in contamination of groundwater, and control of drainage patterns and geothermal systems. It explores the vulnerability of earth towards natural hazards, viz. earthquakes, floods, cyclones, tsunami, volcanism, cyclones and drought. This volume throws light on the applications of remote sensing, Geographical Information System (GIS) and SRTM data for evaluation of the morphometric and morphotectonic parameters and exploring the susceptibility of river basins toward erosion and flood.

The book is organized into two sections; (I) Geological Processes, (II) Geo-Environmental Processes. The main findings of these chapters are summarized in the following paragraphs.

N.B. Kuznetsov et al. discussed the structural, tectonic and geochronological status of the composite Bashkir anticlinorium of the Southern Urals consisting of two parts: the Bashkir (western part) and Uraltau (eastern part) Uplifts. The detrital zircon from Ordovician sandstones of the Uraltau and Bashkir Uplifts stated that the amalgamation of the Precambrian Bashkir and Uraltau basements occurred before Late Ordovician. Detritus of post-Upper Ordovician sedimentary complexes of the Bashkir and Uraltau Uplifts originated from identical feeding provinces and was deposited in the same basin, sealing its pre-Ordovician composite heterogeneous basement.

T.V. Kaulina et al. studied the Precambrian structures of the Kola-Karelian region of East Baltic Shield, which can be impact related. Impact signs were found in two structures of

A. K. Shandilya
Department of Applied Geology, Dr. Hari Singh Gour University, Sagar, India

V. K. Singh (✉) · S. C. Bhatt
Department of Geology, Institute of Earth Sciences, Bundelkhand University, Jhansi, India
e-mail: vinodksingh@bujhansi.ac.in

C. S. Dubey
Sri Sri University, Cuttack, India

A. K. Shandilya et al. (eds.), *Geological and Geo-Environmental Processes on Earth*, Springer Natural Hazards,
https://doi.org/10.1007/978-981-16-4122-0_1

the Kola region: the Jarva-varaka massif of 2498 ± 6 Ma and Javrozero structure of ~1.9 Ga. They described the tectonic and metamorphic history of these two probable astroblemes and established their geochemical and geochronological status.

A.I. Slabunov and V.K. Singh pointed out that Indian Shield consists of Bundelkhand, Aravalli, Singhbhum, Bastar, Western and Eastern Dharwar Cratons. The crustal evolutions of these cratons (except Singhbhum Craton) show that geodynamic mechanisms, similar to modern plate-tectonic and mantle-plume mechanisms, were active during Paleo–Neoarchean time. However, all cratons exhibit their own crustal evolution pattern in Mesoarchean. Crustal evolution in Neoarchean of Bundelkhand and Aravalli Cratons, Western and Eastern Dharwar Cratons have many similar features, therefore, it is assumed as a part in Meso–Neoarchean time elements of the Kenorland Supercontinent, but not in a single block.

V. Vijaya Rao and Damodara Nara focussed on the role of geological processes (external and internal) on the evolution of crustal structures of Indian shield. They pointed out that the seismic images are providing key evidence of the mechanism of plate tectonics, subduction process, and interaction of crust and mantle plume and evolution of sedimentary basins from Archean to Recent. The crust–mantle interaction played a vital role in forming rift-basins and post-collisional extension in the form of mafic dyke swarms and alkaline complexes. Mantle convection process is bringing the earth's terranes together to form supercontinents while scattering is caused by the accumulation of heat above the rising convection cells due to mantle plume activity. The production of huge sediments from Himalayan mountain ranges by erosion and weathering are responsible for the formation of the world's largest Bengal fan.

Prabhakar Sangurmath contributed to the geological characteristics of Archean Orogenic Hutti gold deposits of India. He remarked that the Dharwar gold deposits are confined to subparallel reefs hosted within sheared metabasalts and are processing 2500 tons of ore body per day in six reefs. The detailed underground geological studies on individual lodes are essential to obtain ore shoots extension along the strike as well as at depth, for the Hutti Gold Mine, which seems to be a suitable area.

Yamuna Singh has presented a synthesis of rare-earth deposits hosted in various geological domains of India. According to him, the rare earth in beach placer deposits occur in the eastern and western coast of India and the Chhotanagpur granitic complex is found rich in heavy rare-earth-bearing stream placer deposits in Jharkhand and Chhattisgarh. Amba Dongar and Kamthai are rich in carbonatite-hosted rare-earth deposits in Western India, whereas low-grade Y ore was found in the Samchampi carbonatite complex. The concentrations of REE-bearing apatite-magnetite hydrothermal veins are significantly appearing along the Singhbhum Shear Zone and South Purulia Shear Zones. The vast stretches of deposits may be operated in the large-scale mining operations, whereas small-scale mining would be preferred in small-scale deposits.

V.P. Malviya et al. reported the whole rock's major, trace, REE and PGE geochemistry and mineral chemistry of Cr-spinel and other minerals of the serpentinite associated talc-tremolite-chlorite and chromitite rock from western Bastar Craton, in Central India, which occurr as enclaves within Amgaon granite gneiss. They suggested that serpentinite and talc-tremolite schist were formed from dunite, hurzburgitic and lherzolitic protolith, respectively. High Cr # indicate, high-degree of partial melting of mantle peridotite. Cr-spinel crystallized from boninitic melt at a high degree of partial melting. The IPGE (Ir, Ru) enrichment relative to PPGE (Pd, Pt, Rh) and the relatively low Pd/Ir values suggest that their parental melts were produced by melting of a moderately depleted peridotite similar to the ones described for other IPGE-rich chromitites originated in back-arc basins in the supra-subduction tectonic setting.

O.S. Sibelev et al. documented at least three stages of metamorphic processes in the Central Bundelkhand terrane of the Bundelkhand Craton: (1) ca 2.8 Ga eclogite-facies; (2) ca. 2.7 Ga amphibolites-facies and metasomatism associated with its retrograde branch; (3) ca 1.9–1.8 Ga prehnite-pumpellyite-facies presumably associated with reef formation giant two-phase hydrothermal quartz veins. The Mesoarchean (ca 2.8 Ga) eclogite-facies metamorphism is associated with subduction processes; Neoarchean (ca. 2.7 Ga) metamorphism is predetermined by accretion-collision events; metamorphic processes associated with 2.55–2.50 Ga subduction-accretion events are extremely poorly defined; the latest Paleoproterozoic (1.9–1.8 Ga) prehnite-pumpellyite-facies metamorphism is associated with rifting in the Bundelkhand Craton.

V.K. Singh et al. confirmed that the Bundelkhand Craton consists of three Tectonostratigraphic terranes: Central, Northern and Southern Bundelkhand terrane. This tectonic division is in good agreement with available geodynamic models showing the evolution of the craton. Individual terrane displays its own Paleo–Mesoarchean evolution pattern. In the Neoarchean, that they amalgamated during an accretion-collision event and became one crustal block, a craton was built.

S.C. Bhatt and V.K. Singh have remarked that the central Bundelkhand tectonic unit delineated by E-W sub-vertical Neoarchean crustal shear zone was subjected to three phases of deformations (D_1-D_3), which were followed by D_4 magmatic and NE-SW trending D_5 Riedel shears (quartz reefs). The central Babina-Prithvipur shear zone (~500 m wide) are characterized by mylonitic foliation, asymmetrical rotated porphyroblasts (σa and σb mantled clasts), asymmetrical pressure shadows and deformation lamellae. The sinistral top-to-SW sense of shear movement was dominant and the presence of undulose extinction and deformation lamellae infers that the crystal plastic process was dominant during the evolution of mylonite.

D.C. Banerjee presented a detailed synthesis of field geology combined with imagery analysis from WNW-ESE trending Dudhmania Shear Zone. It is unique in its characteristics, having cross-cut relationship with the ENE-WSW trending Paleoproterozoic rocks of the Mahakoshal Group. He stated that the WNW-ESE Dudhmania shear zone is evolved by mylonitization of rocks of Agori and Parsoi Formations of Mahakoshal Group, and was affected by a non-coaxial sinistral movement between the northern Obra-Amsi-Jiawan fault and the southern Songarh-Kasar-Dudhi fault.

A.K. Rai and R.K. Nayak discussed methods to determine the shallow crustal structure, and its importance in understanding the site characteristics and the seismic hazards. They used ambient noise records to compute Horizontal to vertical spectral ratio (HVSR) and estimate the peak frequency, whereas multi-channel surface wave (MASW) dispersion data were used to estimate shallow subsurface velocities. They conclude that shallow seismic parameters could provide critical information which can be used for assessment of impacts of natural hazards, and particularly the seismic hazard in a region. Nearly 40% population which resides within 100 km of the coastline are vulnerable to various types of natural hazards such as earthquakes, tsunamis, cyclones, etc.

M.A. Alam et al. focused on delineation and segmentation of Arsenic enriched portions of the lithosphere, based on some insightful works and representative examples from across the world, form the basis for discussing the role of geotectonics in Arsenic contamination of aquifers in this chapter. Apart from the origin and development of the magmatic arc, which is the primary source of Arsenic, the chapter deals with the geotectonic control for other primary Arsenic sources as well, viz. Arsenic-enriched hydrothermal and magmatic fluids that contaminate surface and groundwater, natural leaching of ore (especially metal sulfide) deposits that releases Arsenic to surrounding environments, which is accelerated by metal mining and processing activities, and coal deposits and hydrocarbon reservoirs, together with the exploitation processes of both these fuel resources.

M.A. Alam discussed the role of regional (tectonic) and local (structural) controls on the geothermal systems, with representative examples from across the world. The discussion takes a break from the traditional description of the geothermal systems as high-medium-low enthalpy, liquid or vapor dominated, etc., and focuses on what defines and drives the geothermal systems. It follows the concept of treating the geothermal systems as "play types", similar to the oil and gas industry, based on how geothermal resources can be found and developed effectively. An array and combination of tectonic and structural conditions that favor geothermal systems' existence and sustenance have been presented with examples.

S.C. Bhatt et al. carried out the quantitative morphometric analysis of the Jamini watershed by using advanced techniques of remote sensing and GIS. They inferred that the mean bifurcation ratio of Jamini and its all sub-watersheds are suggestive of strong structural control. The drainage density reveals that the impermeable soil is indicative of low groundwater storage capacity in hard rock granitic terrain. The form factor, elongation and circulatory ratios imply that the Jamini basin and its sub-watersheds are significantly elongated and show a low peak flow of longer duration. The Ruggedness and Melton ruggedness number infers that the terrain is slightly rugged and less prone to erosion and sediment transport. The S-shaped hypsometric curves are indicating moderate susceptibility of the basin towards erosion whereas the hypsometric index values are suggestive of occurrence of mature geomorphic terrain with moderately eroded landscapes.

S.C. Bhatt et al. have presented the morphometric and morphotectonic attributes of the Sindh basin of central India. They concluded that the low density and frequency of the watershed is indicative of the occurrence of coarse permeable soil in the low relief terrain of the basin. The elongation ratio and form factor reveal that the basin is elongated and less prone to erosion. The high density of lineaments represented by fractures, faults, foliations corresponds the orientations of low-order streams. The high-order streams following major NE-SW trending lineaments are controlled by major tectonic features.

Further, S.C. Bhatt et al. stated that the Rohni watershed of Bundelkhand of central India represented by dendritic drainage pattern is structurally controlled. The five sub-watersheds of this watershed showing high Rho Coefficient value is suggestive of high hydrologic storage and the form factor and elongation ratio indicates that the basin is elongated and is less prone to erosion and flood. The youth and mature stages are dominant in the basin with semipermeable soil and less erosion exposure. The drainage orientation is represented by polymodal distribution (NE-SW, E-W and NW-SE) and the ENE-WSW trending high order streams show major tectonic control.

S.C. Bhatt et al. attempted the morphometric evaluation of the Upper Pahuj basin of the Bundelkhand region of Central India. Five orders of streams showing dendritic drainage pattern are inherently controlled by geological structures. The moderate drainage density and low frequency are suggestive of low infiltration rate and have sparse vegetation on less permeable rocky terrain. The low elongation ratio and form factor infer that the basin catchment is elongated and is showing low peak flows of greater intervals. The circulatory ratio and ruggedness number indicate that the youth stage was dominant in the basin and the terrain is showing moderate slope and slight morphology.

B.B. Ghute and S. Md. Babar have attempted the morphotectonic and morphometric analysis of the Kayadhu river basin in Washim-Hingoli-Nanded districts, Maharashtra. They inferred that the geomorphic indices of the river show a positive marker of the tectonic inscription on the drainage network. According to them, the moderate hypsometric integral value indicates that the basin is still under a mature stage of erosion. The presence of Knick points implies that southwestward tilting of the drainage basins with asymmetry exhibits the lineament control on smaller tributaries and the area is inherently influenced by tectonics.

In the last chapter of this dedicated volume, Negi et al. have carried out landslide investigations between the Nandprayag and Gopeshwar corridor in the Chamoli district of Uttarakhand, India. They studied six landslides along this corridor for a detailed study and found that the rainfall, slope, lithology, geological structures, drainage, cutting of mountain slope for road widening, toe cutting by river play an important role in increasing landslides in this region. Therefore, it is essential to identify the safe and unsafe zones and the geological and geotechnical investigations are considered important factors to be studied before carrying out any major construction in such a vulnerable hilly terrain.

This book has wide perspectives of understanding the geodynamic, tectonic and petrogenetic history of mineralized domains of earth's crust and geoenvironmental sensitivity of earth. It would be beneficial to researchers, planners, graduate and post-graduate students at the global level.

We are extremely thankful to the Department of Geology, Bundelkhand University, Jhansi for providing facilities to complete the editorial work of this book. We express our gratitude to Mr. Mukul Saklani for providing the bio-data of his father late Prof. P.S. Saklani. We are highly grateful to Prof. Prosun Bhattacharya for helping us in correcting the relevant part of the 'Dedication' of this volume. We gratefully acknowledge Mr. Aninda Bose and Ms. Silky Abhay Sinha of Springer Nature for their immense help and support during several stages of editing of the book. We extend our sincere thanks to all the authors for their scientific contribution to this volume. We are deeply grateful and extend our thanks to all the following reviewers for giving their constructive and fruitful reviews on the manuscripts within the timelines for this dedicated book: Alper Baba, Victor Balagansky, Suresh C. Bhatt, Tapas K. Biswal, A.K. Biyani, Ramon Carbonell, Gautam K. Dinkar, M.L. Dora, Syed Hilal Farooq, Adam A. Garde, Bhagwan Ghute, Dinesh Gupta, Varun Joshi, J.R. Kayal, Niteshkumar N. Khonde, Alok Kumar, Niraj Kumar, Rajinder Kumar, Manoj Limaye, Jyotirmoy Mallik, Vivek P. Malviya, S.N. Mohapatra, Walter D. Mooney, R.C. Patel, B.C. Prabhakar, Divya Prakash, Ritesh Purohit, Abhishek K. Rai, D.S. Rao, D.V. Subba Rao, Jyotisankar Ray, Ajit K. Sahoo, Uma Shankar, Dhruv Sen Singh, Sandeep Singh, Padum K. Singh (departed), A.I. Slabunov, Vaibhava Srivastava, Sergei Svetov, Joseline Soledad Tapia, Sanjeet K. Verma and K.S. Vignesh.

Geological Processes

A Composite Structure of the Bashkir Anticlinorium: Insights from Detrital Zircons Search in Ordovician Sandstones of the Uraltau Uplift, Southern Urals

N. B. Kuznetsov, T. V. Romanyuk, and E. A. Belousova

Abstract

The Bashkir anticlinorium is an extensive outcrop of Precambrian rocks within the Southern Urals, located near the south-eastern edge of the East European Platform (EEP). The Bashkir anticlinorium is subdivided by the Zyuratkul fault into two parts: the Bashkir (western part) and Uraltau (eastern part) Uplifts. The Late Precambrian strata of the Bashkir Uplift were formed at a passive margin of the Volga-Uralian part of Baltica. In contrast, the Late Precambrian strata of the Uraltau Uplift were formed far from its present-day location. Later, the Uraltau Uplift block moved along the Zyuratkul fault to its present-day position with a large-amplitude displacement. This study presents the first results of the integrated (U-Pb age, Hf-isotope and trace-elements contents) study of detrital zircons (dZr) from the Upper Ordovician sandstones of the northern part of the Uraltau Uplift. The integrated characteristics of the studied dZr provide new constraints for their primary sources. A comparison of obtained data with rock types, U-Pb and Hf model ages of the crystalline complexes of the of Volga-Uralia basement, as well as characteristics of dZr from Late Precambrian strata of the Bashkir Uplift and Kazakhstan have revealed that the Upper Ordovician sandstones of the northern part of the Uraltau Uplift contain dZr "alien" to crystalline complexes of Volga-Uralia, Kazakhstan, and Late Precambrian strata of the Bashkir Uplift. The sources of these "alien" dZr were other crustal blocks. A very high similarity of the age spectra of dZr from the Upper Ordovician sandstones, which overlain the Late Precambrian rocks units of the Uraltau and Bashkir Uplifts allows concluding that the spatial conjunction of the Bashkir and Uraltau Uplifts had occurred before the Late Ordovician time. Post-Upper Ordovician sedimentary complexes of the Bashkir and Uraltau Uplifts were sourced from identical feeding provinces in the same sedimentary basin, sealing its composite pre-Upper Ordorvician heterogeneous basement.

Keywords

Southern Urals • Bashkir anticlinorium • Bashkir and Uraltau Uplifts • Ordovician • Detrital zircons • U-Pb age • Hf systematic • Trace-elements • Provenance • Sources

N. B. Kuznetsov
Geological Institute, Russian Academy of Sciences, Pygevsky 7, Moscow, 119017, Russian Federation

N. B. Kuznetsov (✉)
Institute of the Earth's Crust, Siberian Branch, Russian Academy of Sciences, Lermontova St. 128, Irkutsk, 664033, Russian Federation

N. B. Kuznetsov · E. A. Belousova
Australian Research Council Centre of Excellence for Core to Crust Fluid Systems/GEMOC, Macquarie University, Sydney, NSW 2019, Australia

N. B. Kuznetsov
Peoples' Friendship University of Russia (RUDN University), 6 Miklukho-Maklaya St, Moscow, 117198, Russian Federation

T. V. Romanyuk
Schmidt Institute of Physics of the Earth, Russian Academy of Science, B. Gruzinskaya 10, Moscow, 123242, Russian Federation

1 Introduction and Geological Settings

The Bashkir anticlinorium (BA) is located in the west of the Southern Urals (Fig. 1a). The BA is an extensive outcrop of Precambrian rocks, traditionally interpreted as a relic of a Late Precambrian passive margin of Baltica (Precambrian basement of the East European Platform) and its Early Precambrian basement (Puchkov 2010), or a relic of an inland rift-like basin, transformed at the beginning of the Paleozoic into a passive margin of Baltica (Ivanov 1998). The BA is divided into two parts by the Zyuratkul fault: Bashkir Uplift and the northern Uraltau Uplift (Figs. 1b, c).

A. K. Shandilya et al. (eds.), *Geological and Geo-Environmental Processes on Earth*, Springer Natural Hazards,
https://doi.org/10.1007/978-981-16-4122-0_2

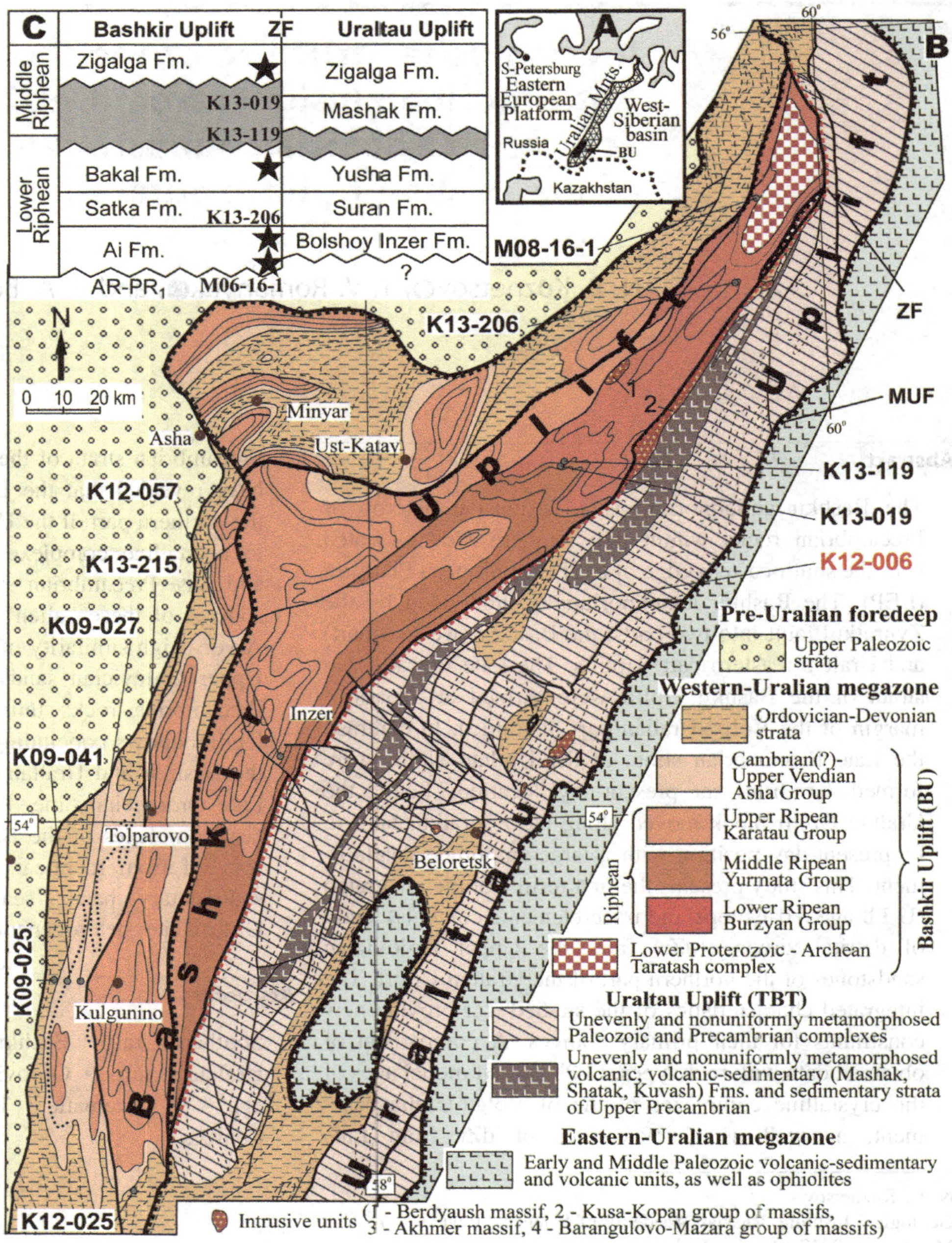

Fig. 1 Scheme of the geological structure of the northern part of the Western-Uralian megazone of the Southern Urals (the Bashkir anticlinorium) and the positions of the samples discussed in the text (**B**). Compiled using data from V.I. Kozlov (small-scale geological map, Ufa sheet), as well as authors' own materials. Faults: ZF = Zyuratkul, MUF = Main Uralian. Insert in the upper left corner: stratigraphic charts of the Bashkir and Uraltau Uplifts for Lower and Middle Riphean (**C**). **A** Index scheme

To the west of the Zyuratkul fault within the Bashkir Uplift, the Upper Precambrian Formations occur and did not experience Pre-Ordovician deformations and metamorphism (Puchkov 2010). Their lower part is composed of predominantly sedimentary (including the bottom level of the Early Riphean) non-metamorphosed rocks interpreted as a strato-typical Riphean section (Stratotype 1983; Ivanov 1998; Maslov 2004; Puchkov 2010). The upper part is composed of terrigenous formations of the Upper Vendian Asha Group (Maslov 2004; Puchkov 2010) or Upper Vendian-Cambrian in age (Kuznetsov and Shazillo 2011). Late Precambrian Formations that are unevenly metamorphosed and experienced Pre-Ordovician deformations (Golionko and Artemova 2016; Puchkov 2010) are widespread eastward the Zyuratkul fault within the northern Uraltau Uplift (Uraltau Uplift hereafter). Middle Riphean volcanic and intrusive rocks (Mashak, Shatak, and Kuvash Formations) are widely represented here (Ivanov 1998; Maslov 2004; Puchkov 2010).

In the western and south-western Bashkir Uplift, the Upper Precambrian Formations are covered by the Paleozoic strata, the section of which begins with low-thickness Upper Emsian quartzous sandstones (Takaty Formation) (Kuznetsov et al. 2014a, b; Puchkov 2010). At the same time, in the southern Bashkir Uplift and within the Uraltau Uplift, the Paleozoic section begins with a thin-thickness of Upper Ordovician quartzous sandstones (Kuznetsov et al. 2016; Puchkov 2010). The oldest sedimentary units overlying the Asha Group are Lower Devonian in age (Takaty Formation) in the western part of the Bashkir Uplift and Middle Ordovician age in the southern one. The most striking difference between the Paleozoic Formations that overlay the Upper Precambrian complexes within the Bashkir and Uraltau Uplifts is the nature of parallel and angular unconformities between these rocks and the underlying formations.

Differences in the structure of the Precambrian Formations of the western and eastern parts of the BA (see Fig. 1C) made it possible to suggest its composite structure (Kuznetsov 2009). In accordance with this, the Bashkir Uplift is a relic of a Late Precambrian passive margin of Baltica (Kuznetsov et al. 2013; Romanyuk et al. 2013), and the Uraltau Uplift is a relic of an alien structure in relation to the contiguous part of Baltica. The Uraltau Uplift block moved along the Zyuratkul fault to its present-day position with a large-amplitude dextral displacement (Kuznetsov 2009).

To develop and to test the idea of a composite structure of the BA (i.e. to unravel the ultimate origin and nature of the Uraltau Uplift block and details of its evolution, as well as to constrain the time of conjugation of the Bashkir and Uraltau Uplifts), we have studied detrital zircons (dZr) from different stratigraphic units of the Bashkir and Uraltau Uplifts (Romanyuk et al. 2013, 2014, 2017, 2018, 2019a, b, 2000; Kuznetsov et al. 2012, 2013, 2014a, b, 2016, 2017a, b, 2018). This paper deals with the first results of the integrated detrital zircons study of the Upper Ordovician sandstones (sample K12-006), which overlap unevenly metamorphosed Late Precambrian rocks of the Uraltau Uplift with an angular unconformity. The integrated characteristics of individual detrital zircon grains aim to better identify the provenance of dZr and to try to reveal relationships of affinity or alien relationships between studied strata and those of Volga-Uralia, Kazakhstan and the Western Urals. This paper focuses on a comparison of age spectra of dZr from Ordovician sandstones of the Uraltau Uplift (sample K12-006) and age spectra of dZr from Ordovician sandstones of the Bashkir Uplift (sample K12-025), which overlap non-metamorphosed Late Precambrian rocks with a parallel unconformity.

2 Methodology

The study was carried out using the TerraneChron®, analytical approach (Griffin et al. 2000, 2002, 2004, 2006, 2007; Belousova et al. 2002, 2006) developed at the CCSF/GEMOC Center (Macquarie University, Sydney). The methodology integrates in situ U-Pb age, trace-element and Lu-Hf-isotope analyses on zircons. Such an integrated approach makes it more reliable to identify the source rocks of detrital zircons and to reconstruct the evolution of the supplying provinces than can be done based on U-Pb ages of detrital zircons only (Veevers et al. 2005, 2006; Belousova et al. 2015; Romanyuk et al. 2018; Kuznetsov et al. 2018, 2019). The study of dZr was carried out using LA-ICP-MS technique, a description of which and detailed methodology are given in previous publications (Griffin et al. 2000; Jackson et al. 2004).

U-Pb ages with discordance D: $10\% > D > -5\%$ are used to plot the histograms and probability density plots (*PDP*) of the ages (Ludwig 2012; Vermeesch 2012, 2018). For zircons with age over 1 Ga, the used age is calculated based on $^{206}Pb/^{207}Pb$ ratio, for zircons with age younger than 1 Ga, $^{206}Pb/^{238}U$ ratio was used.

Attempts to determine such characteristics of individual zircon grains as indicators of magmatic, metamorphic or hydrothermal (sometimes called «metasomatic») nature of a crystal (or even separate its core/rim), a forecast of the type of parental rocks of zircon, assessment of crystallization temperatures, zoning, etc., by the morphology of the crystal and metamict zones features in it, optical, CL-and BSE-images, contents of trace and REE for zircon, the composition of gas/fluid and inherited inclusions and other data, have a very long history (Heaman et al. 1990; Hoskin et al. 2000; Hoskin and Ireland 2000; Griffin et al. 2000, 2004; Liu et al. 2001; Belousova et al. 2002, 2006, 2010, 2015; Corfu et al. 2003; Hoskin and Schaltegger 2003; Liu and Xu 2004; Watson et al. 2006; Hawkesworth and Kemp 2006; Harrison et al. 2007; Grimes et al. 2007, 2015; Ferry and Watson 2007; Fedotova et al. 2008; Fu et al. 2008; Skublov et al. 2012; Fornelli et al. 2014; Chapman et al. 2016 and many others). But unlike the study of U-Pb and Lu-Hf isotopic systems of zircons, where there have been developed common adopted technologies, the interpretation of concentrations of trace and REE elements in zircons has not been yet developed up to a generally accepted technology.

There were many attempts to develop some criteria to distinguish magmatic, metamorphic and hydrothermal zircons (e.g. Rubin et al. 1989, 1993; Corfu and Davis 1991; Claoue-Long et al. 1990; Kerrich and King 1993; Ramezani et al. 2000; Hu et al. 2004; Hoskin 2005; Schaltegger et al. 2005; Pettke et al. 2005; Kebede et al. 2007; Pelleter et al. 2007; Fu et al. 2009; Rubatto 2017 and many others). For example, Hoskin and Schaltegger (2003) had tried to summarize textural and compositional characteristics of hydrothermal zircons, but they had to conclude that the characteristics are not definitive. The hydrothermal zircons may be zoned or unzoned on cathodoluminescence images (CLI); spongy in texture; anhedral or faceted in morphology; and either high or low in common-Pb. By now, there are no doubtless criteria to distinguish between metamorphic and magmatic zircons.

Nevertheless, it has been reliably established that certain statistical trends and appreciable differences in the trace-elements contents are recorded in zircons from certain types of rocks and different origins (Hoskin and Ireland 2000; Hoskin 2005; Belousova et al. 2002, 2006, 2015; Schulz et al. 2006; Fedotova et al. 2008; Kaczmarek et al. 2008; Kostitsyn et al. 2015; Fershtater et al. 2012; Grimes et al. 2015, and others). Thus, in general, Rare Earth Elements (REE) contents in zircons increase from basic rocks (gabbroids and basaltoids), in which the total content of REE is (5–10) × 100 ppm up to (1–2) × 1000 ppm, to felsic rocks (granitoids and their effusive analogs), in which the total content of REE is (2–5) × 1000 ppm. Furthermore, the REE contents in zircons from pegmatites and nepheline syenites can reach even 1–3 mass %. For such rocks as kimberlites, the typical total content of REE in zircons is usually less than 50 ppm. In zircons from carbonatites and lamproites the total REE content is larger and can increase up to 500 ppm (Belousova et al. 2002, 2015), but its average value is 100–250 ppm (Hoskin and Ireland, 2000).

An important indicator is the Th/U ratio (see the review by Kirkland et al. 2015; Rubatto 2017), which varies from 0.1 to 1 in most zircons. Low Th/U ratios are statistically considered to be characteristic of zircon crystals of metamorphic origin, unlike zircons of magmatic origin. However, there is still no consensus on the threshold Th/U value: in different works, values Th/U from 0.5 to 0.1 are established. Thus, value 0.5 in (Kirkland et al. 2015), 0.2 in (Hoskin and Schaltegger 2003), 0.1 in (Teipel et al. 2004). In general, the accumulated data (see Rubatto 2017) indicate that Th/U values from 0.5 to 0.1 are fixed in zircons of both magmatic and metamorphic origin. For example, zircon rims from the Sulu UHP mafic and felsic rocks show higher values of Th/U up to 0.4 (Zhang et al. 2009). On the other hand, zircons crystallized in granites at low temperatures are characterized by higher U contents and lower Th, which, as a result, leads to lower Th/U values in low-temperature crystals (Harrison et al. 2007). Therefore, in the diagrams, we mark the interval Th/U from 0.5 to 0.1 as «Zircons magmatic or metamorphic origin» and believe that only zircons with a ratio Th/U < 0.1 can have a metamorphic origin with a high degree of probability.

High ratios of Th/U > 1.5, together with other characteristics, are statistically inherent in zircons from mafic rocks (Heaman et al. 1990; Kaczmarek et al. 2008; Linnemann et al. 2011). It can't be ignored that in the zircons with high Th/U are sometimes formed in environments of high degrees of metamorphism (Wanless et al. 2011).

Very low U/Yb ratios are recorded for zircons from oceanic basalts NMORB (U/Yb < 0.1) (Grimes et al. 2015). Higher values U/Yb are not unique and can be inherent in a wide range of rocks. However, in general, an increase in the ratio of U/Yb and the Hf content in zircons indicate the fractionation of their parent magmas and reflect the trends in their composition in the direction from the primitive crust towards a more enriched crust. The diagrams «U/Yb versus Hf» and «U/Yb versus Nb/Yb» are most statistically effective for distinguishing between zircons from oceanic basalts and zircons from the continental crust (Grimes et al. 2015).

Important information on zircons is carried by the REE spectra, which are represented in the form of values normalized to chondrite (we use the values for CI-chondrite from (McDonough and Sun 1995)). A monotonic increase in the content of elements from light to heavy is typical for the normalized REE spectra of zircons, which is complicated by two anomalies: positive Ce and negative Eu.

The REE spectra of zircons from granitoids are highly ordered for the heavy REEs: in the interval from Dy to Lu, the slope of the zircon spider diagrams is rather stable. Its value is characterized by the Lu/Dy (sometimes Yb/Sm or Lu/Gd) ratio, and the contents of the heaviest REEs (Yb and Lu) are used as a marker for classifications. Statistically, the zircon that is crystallized in the igneous rock has a steeper slope for heavy REEs (larger Yb/Sm values) than the crystal formed in rocks of high degrees of metamorphism (Rubatto and Hermann 2007; Rubatto 2017).

For zircons of high-temperature metamorphic origin, which are crystallized in the presence of garnet, the lower heavy REE and Y are described (Rubatto 2002; Rubatto and Hermann 2007; Fedotova et al. 2008; Skublov et al. 2012; Fornelli et al. 2014), because the garnet competes with zircon on these items. For eclogites, also a low Th concentration (no higher than 3 ppm on average) and a significant decrease in the concentrations of all REE (to 22 ppm) and particularly LREE (<2 ppm), and relatively low concentrations of Y (34 ppm), U (100 ppm), and P (41 ppm) at an elevated Hf concentration (11 400 ppm on average) have been revealed (Skublov et al. 2012).

The discriminative «(Sm/La)N versus La» and «Ce/Ce* versus (Sm/La)N» diagrams based on distinct REE patterns from magmatic and hydrothermal zircons were presented in (Hoskin 2005). Further researches revealed that not all data points reported for zircons which are thought to be hydrothermal ones fill in the «hydrothermal» area in this discriminate diagram (Fu et al. 2009). However, the proposed in (Hoskin, 2005) signs of «uplifted» LREEs spectra (a higher content of La (La > ~ 2 ppm) and a lower (Sm/La)N ((Sm/La)N < 10))) and weak Ce/Ce* anomaly (Ce/Ce* < 10) are the effective indicators to suppose if not a pure hydrothermal origin of zircon, but at least a hydrothermal imprint on the zircon. Metamictic zircons also often show «uplifted» LREE spectra.

The above information was summarized in the scheme of classification shown in Fig. 2 to distinguish between a magmatic/metamorphic/hydrothermal origin of zircon. For zircons whose magmatic origin is supposed, the CART

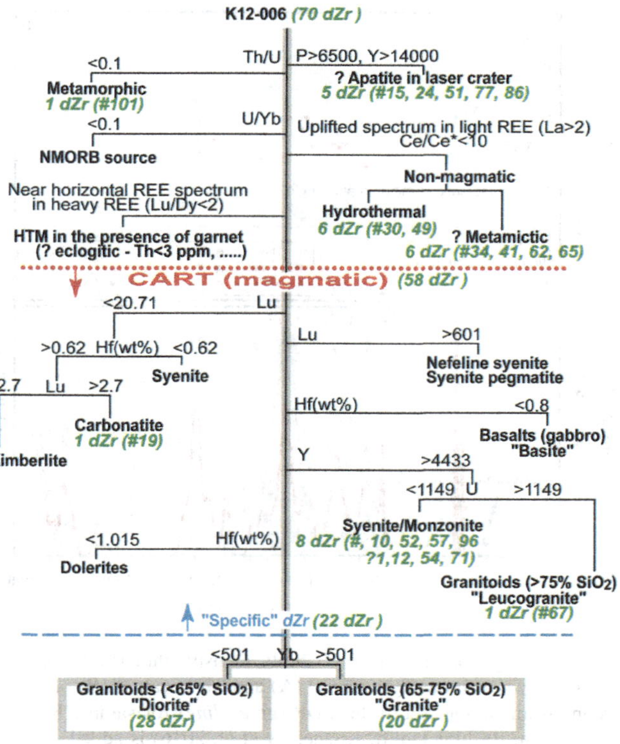

Fig. 2 The scheme of classification of detrital zircons (dZr) from sample K12-006 based on the CART algorithm (Belousova et al. 2002); a number of dZr of each type in studied samples are marked by green bold italic letters

algorithm (Belousova et al. 2002) was applied to predict a type of parental zircon rocks. The main purpose of using the CART algorithm is the primary separation of zircons, the parental rocks for which there could be rare rock complexes («specific» zircons), from zircons from granitoids and their effusive material analogues, which are the main sources of zircon. The latter, as a result of the classification, are subdivided by the percentage of SiO_2 in the rock into three groups —rocks with reduced ($SiO_2 < 65\%$), normal ($SiO_2 = 65$–75%) and increased ($SiO_2 > 75\%$) SiO_2 contents. For the sake of brevity, these groups are called, respectively, «diorite», «granite» and «leucogranite» and corresponding zircons as «dioritic», «granitic», «leucogranitic».

Unlike zircons from «granites» or «diorites», which form large groups, are determined and interpreted statistically, the identification of «specific» zircons requires «manual» testing and debugging, checking the occurrence of measured trace-elements concentrations at boundary marker values for the concentrations of the elements the CART algorithm operates on. For example, a content of Y = 4433 ppm is the marker to distinguish between «granite» and «syenite/monzonite». If content Y in zircon is a little bit less than 4433 ppm, for example, Y = 4400 ppm, then this zircon is classified as «?syenite/monzonite» and is additionally checked in other classification diagrams (Summary REE versus Ti and REE spectra). The reliability of this approach and the percentage of misclassification are discussed in Belousova et al. (2002).

3 Sampling Rocks; Separation, Imaging, Morphology and Analytical Results of Detrital Zircons Study

3.1 Sampling Rocks, Imaging and Morphology Zircons, Measurements, Standards

Sample K12-006 (~1.5 kg) was collected from the light-yellow fine-middle-grained quartzous sandstones in the basal level of the Paleozoic section (see sample location in Fig. 1) on the south-western slope of the Yuryuzan syncline (53°36′21.15″ N and 58°46′57.30″ E) on the left side of the Tyulyuk spring (a right tributary of the Yuryuzan River). Approximately 300 zircon grains were separated from the sample (technology is described in Romanyuk et al. 2018), of which 202 randomly selected grains were mounted into an epoxy disk.

Zircons were studied with a microscope and in cathode rays and backscattered electrons (BSE). All grains are rounded, mostly small (<100 μm). Almost all dZr contain inherited inclusions, metamict zones, and in some cases are broken by cracks. These are indications that zircons participated in magmatic/metamorphic processing (Corfu et al. 2003). For dating, we selected grains in which we managed to map out areas (Ø ≤ 40 μm) using CLI without obvious metamictization, violations, inclusions and cracks.

At the first stage, the simultaneous measurement of the U-Pb-isotope system and trace-elements contents of dZr was executed, at the second stage, a separated study of the Lu-Hf-isotope system of the selected dZr were executed. Measurements of the parameters of the U-Pb-isotope system in zircons were carried out with the Red-JG-1 zircon standard for calibration (Jackson et al. 2004; Elhlou et al. 2006), and Mud-Tank and 91,500 as measured zircons for monitoring. During the measurements, the mean ages were Mud-Tank = 737 ± 5 Ma (n = 7) and 91,500 = 1059 ± 8 Ma (n = 7), which agrees with the ages of these standards (Wienedbeck et al. 1995, 2004; Jackson et al. 2004; Black et al. 2004; Yuan et al. 2008; Horstwood et al. 2016).

Measurements of the Lu–Hf-isotope system were monitored by Mud-Tank and Temora II zircon standards. During the measurements, the average ratio of $^{176}Hf/^{177}Hf$ was 0.282551 ± 0.0000086 (n = 4) for Mud-Tank and 0.282619 ± 0.000025 (n = 4) for Temora II, which is in agreement with the values for these zircon standards (Jackson et al. 2004; Yuan et al. 2008). Measurement of the content of trace elements in dZr was carried out for 22 elements, the NIST standard was used for external control. Data processing was performed using the commercial program "GLITTER" (Griffin et al. 2008) and the programs (Ludwig 2012; Vermeesch 2012, 2018) available in the public free access.

3.2 U–Pb Age Results

A total of 70 analyses were performed. For 12 grains, a large analytical error (>50 Ma, #51) or strongly discordant analyses (#17, 24, 51, 77 and 41) have been obtained (Fig. 3). These analyses were excluded from consideration. The remaining 58 analyses were used to plot the age histogram and the PDP (Fig. 4a). The youngest age is 530 ± 4 Ma (D = 6.8%, #30) and the oldest is 2885 ± 31 Ma (D = −0.3%, #50).

3.3 Zircon Trace-Element Content and Parental Rock Type Classification

When studying the trace-elements content (Figs. 4b, 5, 6, 7, 8, 9, 10, 11 and 12), the La content was not determined for

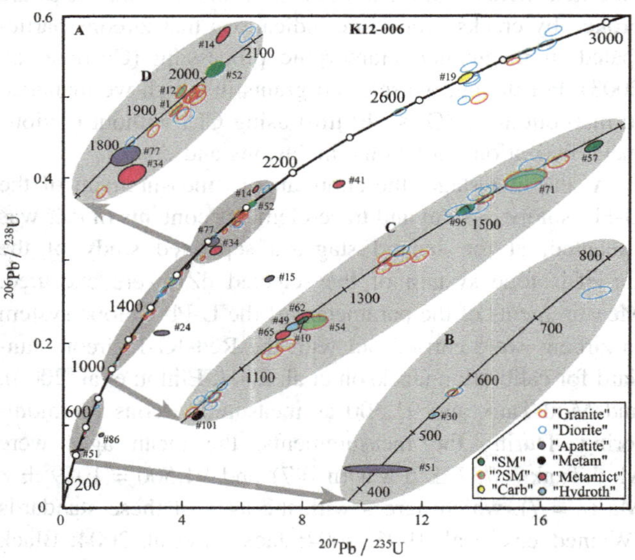

Fig. 3 Results of the U-Pb dating of detrital zircons from the K12-006 sample. A—Concordia and ellipses (some with analysis numbers), showing a 68.3% confidence interval for measurements. B, C and D (at the grey background)—the enlarged fragments of Concordia. Classification of the zircons (see Table 1 and Fig. 2): "SM" = «syenite/monzonite», "?SM" = «granite», but probably «syenite/monzonite», "Carb" = «carbonatite», "Metamict" = metamictic zircon, "Hydroth" = hydrothermal zircon, "Apatite" = apatite inclusion in the LA sampling, "Metam" = metamorphic zircon

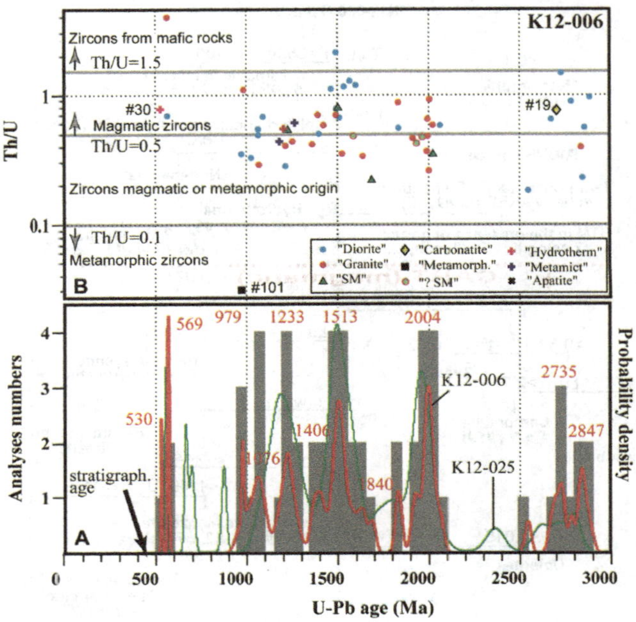

Fig. 4 Histograms and the Probability Density Plot (PDP) of U-Pb ages («*conditioned*» analyses only) (**A**) and Th/U values (**B**) of detrital zircons from sample K12-006. PDP (*green line*) for sample K12-025 (Ordovician, Bashkir Uplift, Kuznetsov et al. 2016) is added in (**A**)

grains # 19, 28, 43, 44, 45, 46, 56, 84, 95 and 101 (the content is below the detection limit). Further, for calculating Ce anomalies, the La content for these zircons was adopted at 0.01 ppm. Five grains have yielded anomaly high P and Y contents (Table 1), which may suggest an apatite inclusion in the LA crater. They also show very high contents of total REE and Ti (Fig. 7). They are marked as «apatite» in Table 1.

Zircon #101 showed very low Th/U = 0.03 (Fig. 4b) and it was classified as «metamorphic». It also differs from other zircons in a high content of Hf (Fig. 9) and steep REE spectra in heavy REE (Figs. 6 and 8).

Seven zircons (#14, 30, 34, 41, 49, 62 and 65) showed elevated La content of more than 2 ppm and weak Ce/Ce* anomalies less than 10. Their data points on discriminant diagrams in Fig. 5 fall in a field far besides «Magmatic zircons fields», so they were classified as «non-magmatic» zircons. Two zircons (#30 and 49) from «non-magmatic» zircons fit well to fields of «Hydrothermal zircons» in discriminant diagrams of Fig. 5 and were classified as «Hydrothermal zircons». The rest «non-magmatic» zircons are marked as «Metamictic». There were revealed no zircons with flat REE spectra in heavy REE (Figs. 6 and 8) that would suggest their «HTM-Gr» origin, neither zircons with the signature of NMORB (Fig. 12).

For the 58 detrital zircons having the most probable magmatic origin, the CART classification (Fig. 2) indicates that the parental rocks were most likely «granites» (24) and

Table 1 Some characteristics of the studied zircons with specific trace-element contents and results of their classification

#	Analysis #	Th/U	P (ppm)	Y (ppm)	La (ppm)	Yb (ppm)	Lu (ppm)	U (ppm)	Total REE (ppm)	Lu_{CI} /Dy_{CI}	Hf %	U–Pb Age ± 1S (Ma)	D%	ε_{Hf} (Epsilon Hf)	T_{DM}^{C} (Ga)	Classification
1	1	0.43	1826	4353	0.38	1245	219	464	2835	5.42	2.13	1968 ± 18	2.7	−4.2 ± 0.7	2.9	?SM
2	10	0.52	525	5249	0.28	1344	227	672	3260	4.25	1.47	1230 ± 17	0.8	5.5 ± 0.7	1.7	SM
3	12	0.42	1820	4334	0.16	1336	239	367	2837	6.60	2.09	1937 ± 15	−2.0	0.4 ± 0.6	2.6	?SM
4	14	0.89	165	1930	2.66	446	89	154	1063	3.27	1.36	1969 ± 19	−8.8			Metc
5	15	0.55	12,698	28,252	47.88	1721	230	1184	11,035	7.52	1.05	2651 ± 22	45.2			Apatite
6	19	0.741	271	429	<0.008	116	20.5	98	287	4.82	1.90	2703 ± 17	−0.8	−2.5 ± 0.5	3.3	Carbon
7	24	0.465	20,163	36,674	9.23	5851	798	1018	20,244	1.66	2.11	*1842 ± 83	35.7			Apatite
8	30	0.784	3080	1768	68.32	587	107	685	1626	6.72	1.71	530 ± 4	6.8	7.3 ± 0.5	1.1	Hydroth
9	34	0.497	789	1647	2	431	77	576	1212	3.53	1.90	1941 ± 47	11.0			Metc
10	41	0.65	1349	4202	2.48	636	105	741	2701	1.44	1.96	*2556 ± 22	19.1			Metc
11	49	0.546	3302	4402	45.52	1153	196	230	3127	4.13	1.60	1269 ± 31	4.6			Hydroth
12	51	0.372	12,081	16,806	74.72	3322	509	3106	10,750	2.50	1.56	435 ± 4	−2.4			Apatite
13	52	0.35	537	5058	0.26	1317	219	785	3194	4.33	1.89	2027 ± 26	0.03	4.2 ± 0.6	2.4	SM
14	54	0.474	1568	4207	0.95	972	163	806	2800	3.27	1.90	1292 ± 52	6.4			?SM
15	57	0.221	2567	5241	0.84	2365	439	499	4384	11.27	2.21	1692 ± 18	2.9	-1.1 ± 1.0	2.5	SM
16	62	0.555	261	1725	2.08	401	69	112	1127	3.39	1.44	1208 ± 30	0.1	5.5 ± 0.4	1.7	Metc
17	65	0.44	600	3245	2.97	840	145	414	2050	4.53	1.96	1185 ± 29	−0.1			Metc
18	71	0.48	302	3940	0.15	1064	163	483	2559	3.93	1.65	1593 ± 44	1.4			?SM
19	77	0.124	6947	15,024	7.46	2220	368	2659	8302	1.67	2.70	*1850 ± 50	3.0			Apatite
20	86	2.034	10,818	14,468	111.37	3082	520	2537	11,678	2.39	1.31	549 ± 71	71.7			Apatite
21	96	0.81	611	8632	0.20	2323	394	926	5392	4.87	1.53	1506 ± 24	0.9	3.2 ± 0.5	2.1	SM
22	101	0.022	145	1151	<0.012	668	159	700	1140	22.95	2.82	978 ± 7	7.0	19.6 ± 0.7	0.5	Metamor

Notes D = 100% *(age($^{206}Pb/^{238}U$)/age($^{207}Pb/^{206}Pb$) – 1). Correction for non-radiogenic lead was executed in according to (Andersen 2002). The model ages of the T_{DM} mantle and the crustal substrate T_{DM}^{C} (two-stage model) were estimated. T_{DM}^{C} assumes that zircon parental magma originated from an average continental crust ($^{176}Lu/^{177}Hf$ for the averaged crust = 0.015) that was produced from the depleted mantle (Griffin et al., 2000, 2004, 2006; Belousova et al., 2006, 2010; Liu et al. 2013). The decay-constant λ for ^{176}Lu was adopted as $\lambda^{176}Lu$ 1.867 × 10^{-11} yr^{-1} (Scherer et al. 2001). The current ratios in chondrite $(^{176}Hf/^{177}Hf)_{CH}$ = 0.282785 ± 0.000011 and $(^{176}Lu/^{177}Hf)_{CH}$ = 0.0336 ± 0.0001 were taken from (Bouvier et al. 2008), in the depleted mantle $(^{176}Hf/^{177}Hf)_{DM}$ = 0.28325 and $(^{176}Lu/^{177}Hf)_{DM}$ = 0.0384 from (Griffin et al. 2000)

$T_{DM} = 1/\lambda \times \ln\{1 + ((^{176}Hf/^{177}Hf)_Z - (^{176}Hf/^{177}Hf)_{DM})/((^{176}Lu/^{177}Hf)_Z - (^{176}Lu/^{177}Hf)_{DM})\}$, $T_{DM}^{C} = T_{DM} - (T_{DM} - t) \times ((F_C - F_Z)/(F_C - F_{DM}))$, $F_{C,Z,DM} = \{((^{176}Lu/^{177}Hf)_{C,Z,DM})/((^{176}Lu/^{177}Hf)_{CH}) - 1\}$

Classification: SM = «syenite/monzonite», ?SM = «granite», but probably «syenite/monzonite», because a small deviation down from CART-criteria (Y = 4433 ppm); Carbon = «Carbonatite»; Metamor = metamorphic zircon (Th/U < 0.1); "Hydroth" = probable hydrothermal origin, "Metc" = probable metamictic alteration of zircon; Apatite = probable apatite inclusion in the studied zircon due to high P (>6500 ppm) and Y (>14,000 ppm); * near U-Pb Age—correction of age for common-Pb is > 20 Ma

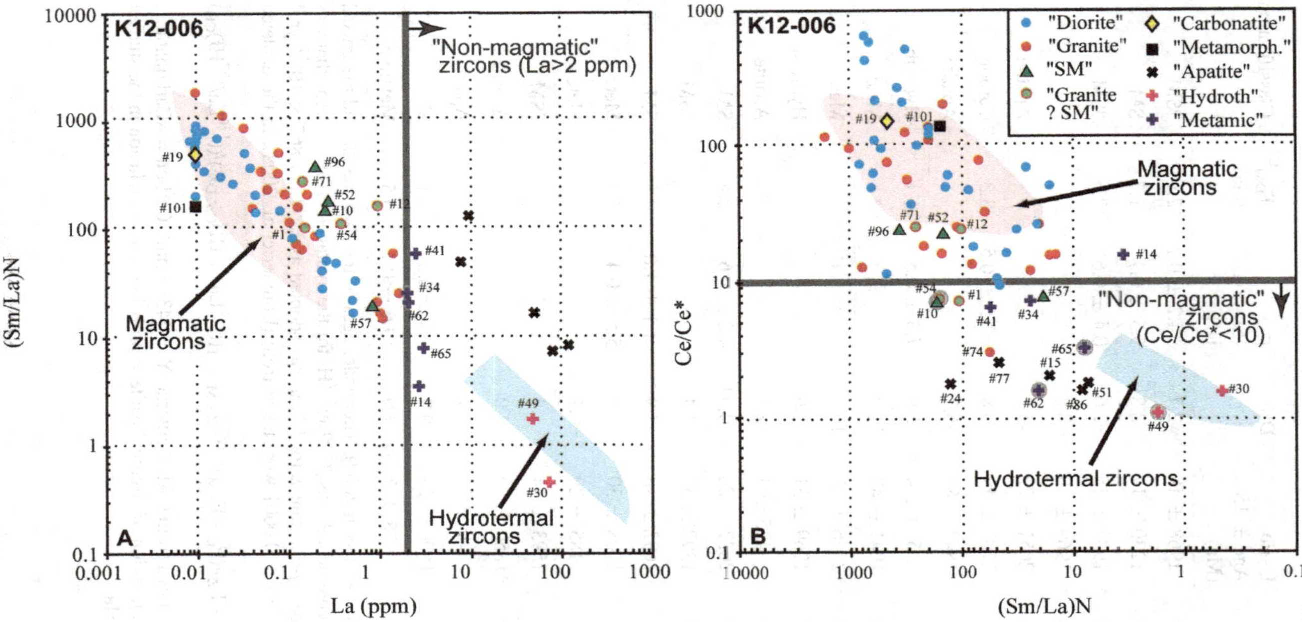

Fig. 5 Discriminant diagrams of «$(Sm/La)_N$ versus La (ppm)» (**A**) and «Ce/Ce* versus $(Sm/La)_N$» (**B**) for studied zircons from sample K12-006. The two outlined pink and blue areas are defined by magmatic and «hydrothermal» zircons from the Boggy Plain Zoned Pluton (Hoskin 2005)

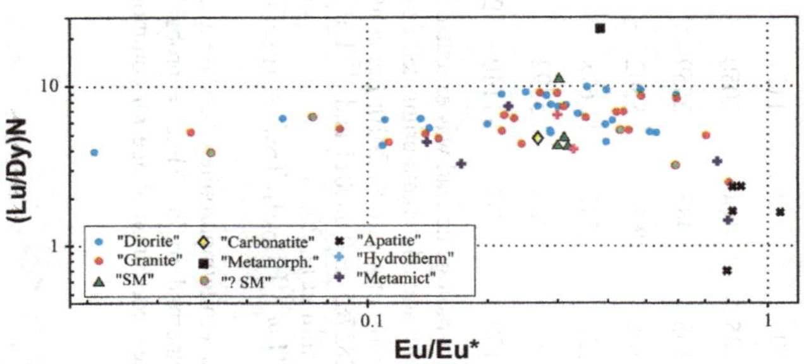

Fig. 6 «Lu/Dy versus Eu/Eu*» diagram for detrital zircons from sample K12-006

«diorites» (28) for the majority of zircons. In the classification diagrams, zircons from «diorites» are characterized by higher contents of Yb, Y, total REE and heavy REE compared to zircons from «granites». As a specific, the parental rocks were assigned «carbonatite» (1) and «syenites/monzonite» (4). Characteristics for four «granitic» zircons are very close to the thresholds values between «granites» and «syenites/monzonite», so they were marked as «?syenites/monzonite».

«Carbonatitic» zircon (#19) is clearly distinguished from other zircons by low total REE (Fig. 7), low Lu (Fig. 8) and Y (Fig. 9) contents. In contrast, «syenites/monzonite» zircons have the upper total REE (Fig. 7) and Y (Fig. 9) contents. No other clear signs of zircons were noted in other discriminant diagrams such as «Ta versus Nb» (Fig. 10) and others.

3.4 Hf Isotope Analysis of the Detrital Zircons

For dZr, the size of which allowed a second crater of 40 μm size and the U-Pb age estimates yielded an acceptable concordance, the Lu-Hf isotope zircon system was also studied (Fig. 13a). Zircon #101 («metamorphic», age of 978 Ma) has yielded a very high value of $\varepsilon_{Hf} = 19.6 \pm 0.7$. This value is essential higher ε_{Hf} of the depleted mantle of this age and $T_{DM}{}^{C} = 0.6$ Ga which is less than the U-Pb age of the zircon. Together with other atypical characteristics of this zircon (very steep spectra in Heavy REE (Figs. 6 and 8)

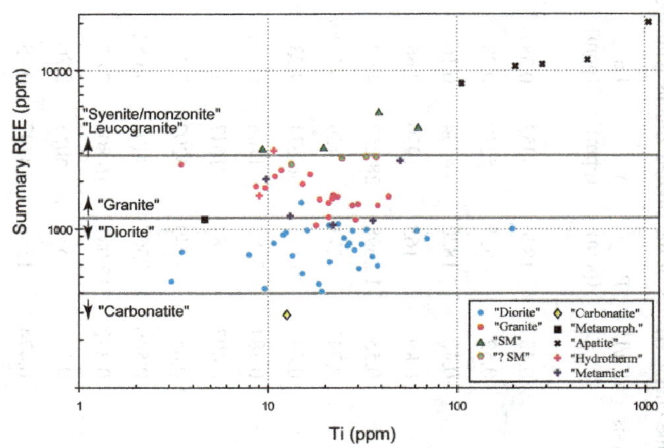

Fig. 7 «Sum or total REE versus Ti» diagram for detrital zircons from sample K12-006

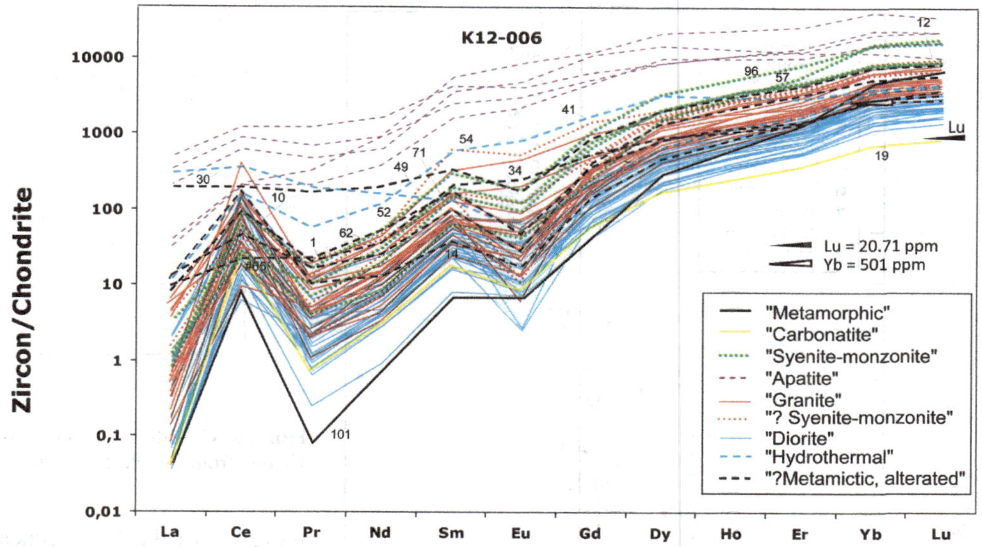

Fig. 8 Chondrite-normalized spider-diagram of REE contents in detrital zircons from sample K12-006

and very high contents of Hf (Fig. 9)), all these supposes a complicated history or/and structure of this zircon. Although CLI of this zircon did not show any inclusion within the LA crater area, it is possible that a deeply located inclusion does get in the LA sampling material, so we prevent giving any meaningful interpretation to this zircon. One more zircon has yielded strongly positive $\varepsilon_{Hf} = 13.7 \pm 3.0$ (#47) which is close to ε_{Hf} of a depleted mantle of this age. All others ε_{Hf} are within interval (-9.0 + 8.8). No zircons with strongly negative ε_{Hf} were revealed that would yield unreasonably very old T_{DM}^{C}, and obtained estimates T_{DM}^{C} vary in the interval 1.1–3.4 Ga.

4 Discussion and Interpretation of Obtained Analytical Data

The integrated isotope-geochemical characteristics of the studied dZr were compared with known data on the composition, age and model ages of the crystalline complexes of the Western Urals and Volga-Uralia, as well as with characteristics of dZr from Riphean strata of the Northern Kazakhstan (Fig. 13) and Bashkir Uplift (Fig. 14).

All data points of Archean dZr (10 zircons) from sample K12-006 form a compact area in «Diagrams ε_{Hf}» (Fig. 13a) within values U-Pb ages 2.55–2.90 Ga and ε_{Hf} −2.5 to +2.5. This compact area does not match the known fields of the Archean complexes Volga-Uralia: the «Tashlyar», «Aktanysh», «Bak-2» and «Kolyvan enderbites». The only 3 «dioritic» points are located back to back to the «Bak-1» field and the «carbonanitic» point fits well to the «Bak-1» field. The «Bak-1» field characterizes the quartz diorite of the Bakal block. Note that the predicted type of parent rocks of most of the Archean dZr is «diorites», which coincides with the type of rocks of the Bakal block. However, we think that available data does not allow us to identify quartz diorites of the Bakal block as a local source. If the basement of the Volga-Uralia has been the primary source of Archean zircons for Upper Ordovician sandstones of Uraltau Uplift, we would have seen a much wider variety of ages and Hf isotope signs of dZr, including dZr with ages >2.9 Ga and dZr with Hf isotopic signs of the oldest continental crust (T_{DM}^{C} > 3.5 Ga) that typical for the Volgo-Uralian basement. Note that such zircons have been found in sandstones of all Riphean stratigraphic levels in the Bashkir Uplift (Fig. 14). So, the obtained data do not allow us to allocate the basement of the Volga-Uralia and Riphean strata of the Western Urals as the most probable primary source of Archean dZr from sample K12-006. It is highly likely that it was a Neoarchean block beyond the Volga-Uralia, composed of mostly juvenile rocks of intermediate to mafic in composition and including some alkaline rocks.

The data points of the Paleoproterozoic dZr form area in «Diagrams ε_{Hf}» (Fig. 13a) with a range of the values of ε_{Hf} from the maximum positive («DM» composition) to substantially negative values ($\varepsilon_{Hf} = -10$) corresponding to the Hf-model ages of the magma-generating substrate with T_{DM}^{C} = 3.0 Ga. Such arrays are a sign of mixing juvenile and isotope-mature materials, which can occur in long-acting volcanic arcs on continental margins or in collision orogeny tectonic settings. A large body of geochronological data on crystalline rocks occurring in Paleoproterozoic orogens has been accumulated (Zhang et al. 2012). Most of these orogens are composed of complexes with ages <2.0 Ga. Ages of 2.1–2.0 Ga have been obtained within the EEP only for the Volga-Sarmatian orogen (Bibikova et al., 2009; Terentiev and Santosh, 2016; Terentiev et al., 2016a, b, 2017, 2018) and the Taratash Orogen (Sindern et al. 2005; Tevelev et al.

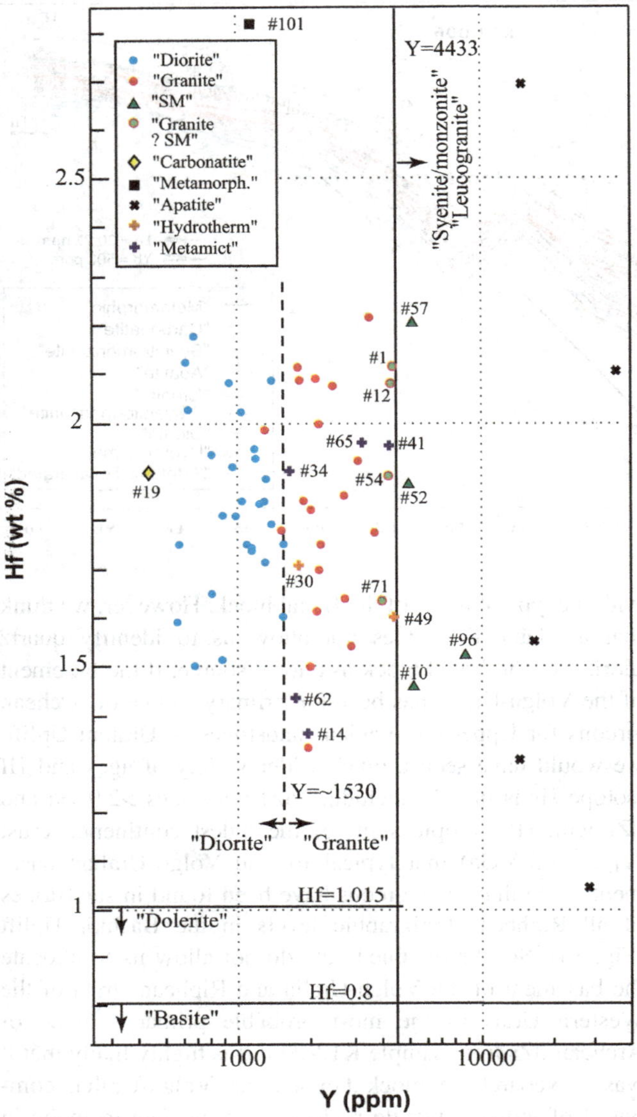

Fig. 9 Classification diagrams of Hf and Y contents in detrital zircons from sample K12-006

2014, 2015, 2017; Khotylev and Tevelev 2017), which are oldest among these orogens and closest to the Southern Urals. Note that several data points fall directly into the field of isotopic parameters of the rocks of the southern Volga-Sarmatian collisional orogen ("VSO-field"). The age interval of the Paleoproterozoic dZr of 1.85–2.15 Ga fit well to a time interval of magmatic activity within the Volga-Sarmatian and Taratash orogens. The obtained data provide a very strong argument to consider the Volga-Sarmatian and Taratash orogens as the most probable source of Paleoproterozoic zircons in sample K12-006. However, it is needed to note that Hf isotopes of the Paleoproterozoic dZr from sample K12-006 have not recorded

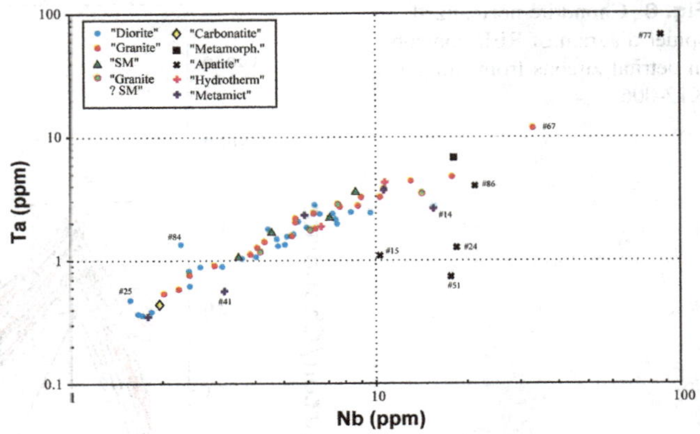

Fig. 10 Classification diagrams of Ta and Nb contents in detrital zircons from sample K12-006

recycling of the Mesoarchean and older crust. This is difficult to explain because numerous dZr with signs of recycling of the Mesoarchean, Paleoarchean and even Hadean (single dZr) have been found in the Riphean strata of the Bashkir Uplift and they are considered as sourced from the Volga-Sarmatian and Taratash orogens (Fig. 14).

Numerous specific «carbonatite» zircons with ages of about 2.0, 2.5, 2.85 and 3.6 Ga were revealed in Riphean strata of the Bashkir Uplift (Fig. 14). However, in sample K12-006, only a single «carbonatite» zircon was found with an age of about ~2.7 Ga. Among the dZr from samples K12-006, numerous zircons with ages younger than 1.65 Ga have been found. These ages are absolutely atypical for the crystalline complexes of Volgo-Uralia. The primary sources of these dZr might be located either within the parts of the EEP that are situated very far from the Southern Urals, or outside the EEP, for example, in Kazakhstan. Erosional products of all these rocks might have been recycled in the Upper Ordovician Uraltau basin by well-developed long river systems.

For example, within Fennoscandia, the Mesoproterozoic rapakivi granites and paragenetically associated migmatites are widely developed (Sharkov 2010). The Sveconorvegian domain is composed mainly of the Mesoproterozoic (Danopolonian) and Early Neoproterozoic granite-metamorphic complexes (Bogdanova et al. 2008). The Volyn volcanic province in the southwestern part of Sarmatia composes of the Late Neoproterozoic (Ediacaran) basaltoids, volcanogenic-tuffogenic formations and felsic rocks (Shumlyanskyy et al. 2016b). Relicts of the Proto-Uralide-Timanide orogen, in the structure of which Late Neoproterozoic-Cambrian granitoids and metamorphites are widely represented (Kuznetsov et al. 2014a), are located in the northeastern margin of the EEP.

Small-volume crystalline complexes with ages <1.65 Ga are known in Western Urals, as well as in some of the Mesoproterozoic–Early Neoproterozoic (Riphean)

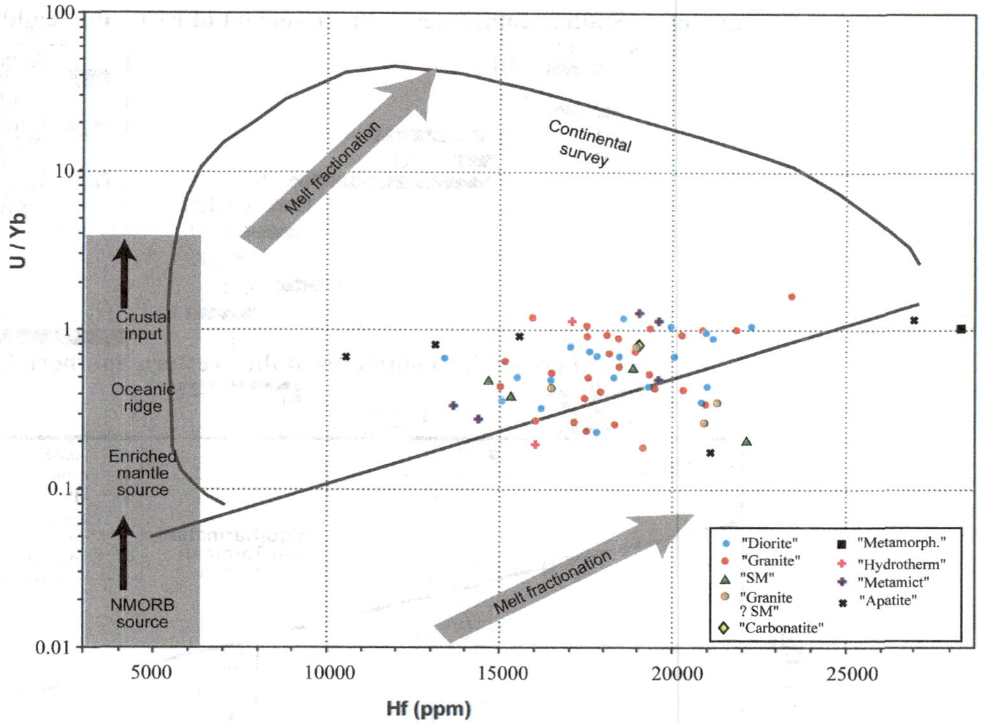

Fig. 11 Diagram «U/Yb versus Hf» in dZr. Field "Continental survey", arrows and inscriptions after (Grimes et al. 2015, Fig. 2)

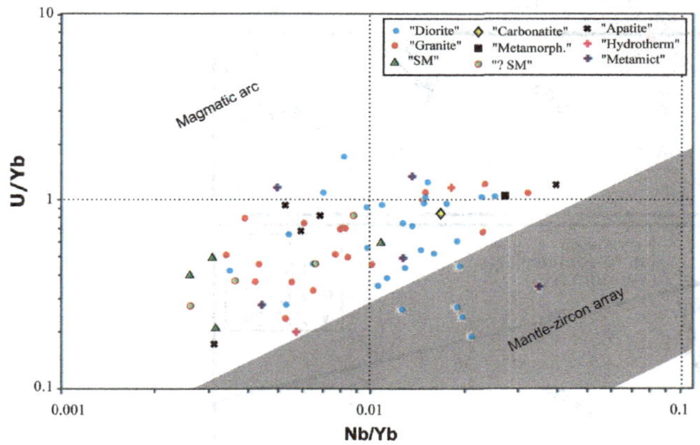

Fig. 12 Diagram of the ratios "U/Yb versus Nb/Yb" in dZr. Fields "Magmatic arc" and "Mantle-zircon array" after (Grimes et al. 2015, Fig. 6c)

aulacogenes in the southeastern EEP (Gusikhin complex in the Pachelma aulacogene, etc.). However, ages that are available from those crystalline complexes form a very narrow time interval of 1.2–1.45 Ga and these complexes cannot be the only sources for numerous Mesoproterozoic 1.0–1.6 Ga dZr available in sample K12-006 (Fig. 13). Three ages of dZr from sample K12-006 coincide well with the age of intrusive rocks of the Berdyaush massif. However, their Hf isotope signatures differ considerably from those obtained for zircons from gabbro ($\varepsilon_{Hf} = 4.6 \pm 1.0$) and from nepheline syenites, granite-rapakivi and quartz syenite-diorites (ε_{Hf} of -5.3 ± 0.7, -6.0 ± 0.9 and -7.6 ± 1.4 respectively) (Ronkin et al. 2015a; b). Therefore, it is unlikely that the Berdyaush complex was a source of dZr in the K12-006 sample.

Other potential primary sources of the Mesoproterozoic zircons might have been crustal blocks, which form now the basement of the Scythian-Turanian plate or the Paleozoides of Central Kazakhstan (the basement of the Epi-Paleozoic Kazakhstan continent). A representative body of geochronological data has already been accumulated for the basement of the Epi-Paleozoic Kazakhstan continent, which allows us to reconstruct the main stages of the formation of its heterogeneous continental crust (Degtyarev et al. 2016, 2017). In accordance with these data (Fig. 13b), the basement of the Kokchetau microcontinent located in a core of the basement of the Epi-Paleozoic Kazakhstan continent is composed of juvenile material of Late Proterozoic and Mesoproterozoic intra-oceanic arcs. The earliest magmatic episode (K1) is dated at 1.85–1.65 Ga, the next episode (K2) lasted at least from 1.5 to 1.0 Ga. Its final phases occurred at 1.2–1.0 Ga and are known as the Tarim orogeny, relicts of which are preserved in Northern Kazakhstan (the North-Kazakhstan tectonic zone). The assembly of the Kazakhstan continent as a large heterogeneous crustal block and consolidation of its basement most likely took place during this orogeny.

The study of dZr from the Mesoproterozoic–Early Neoproterozoic (Riphean) cover of the Kokchetau Massif (4 samples) (Kovach et al. 2017) revealed several Archean dZr,

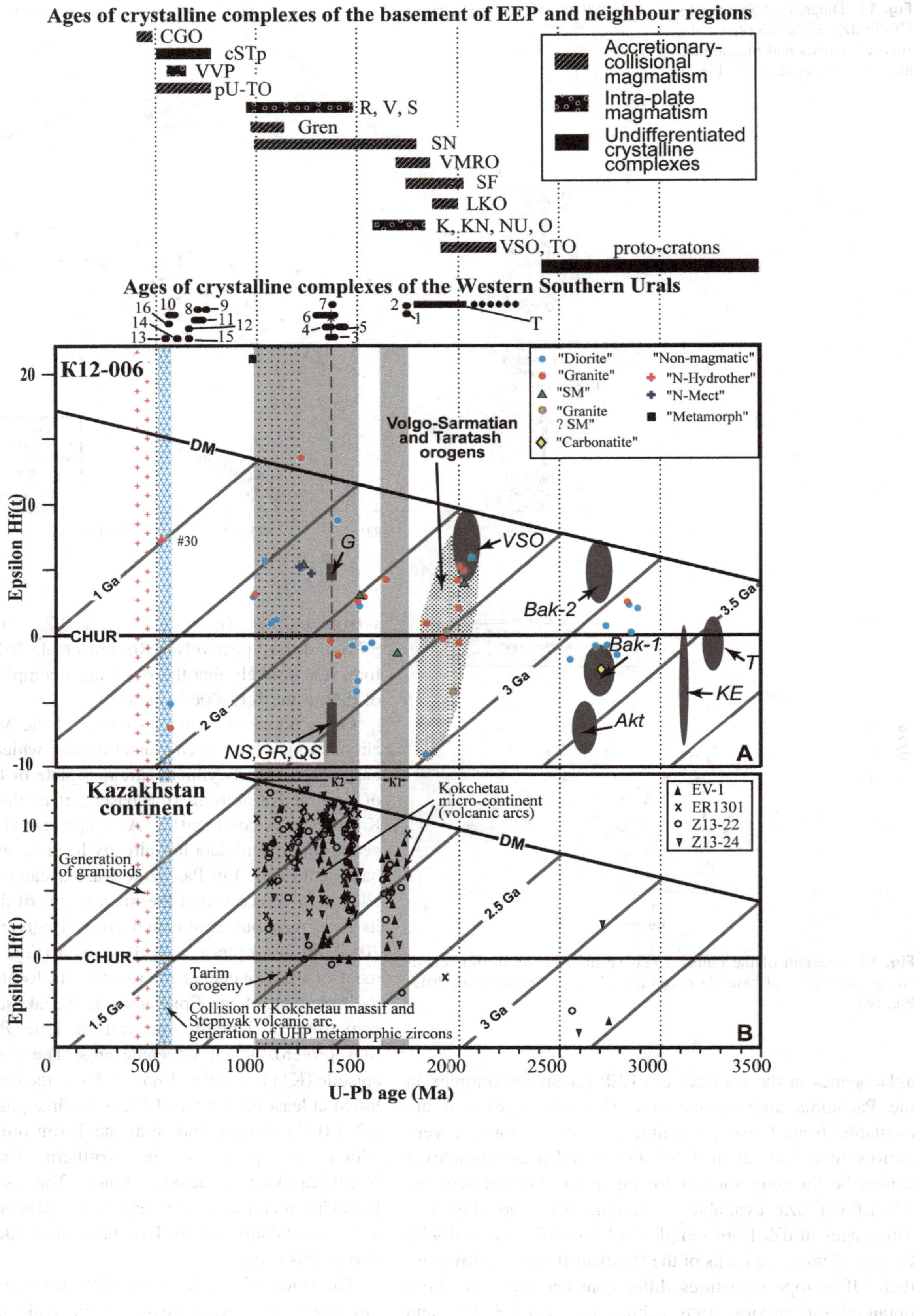

Ages of crystalline complexes of the basement of EEP and neighbour regions
CGO
cSTp
VVP
pU-TO
R, V, S
Gren
SN
VMRO
SF
LKO
K, KN, NU, O
VSO, TO
proto-cratons
Accretionary-collisional magmatism
Intra-plate magmatism
Undifferentiated crystalline complexes
Ages of crystalline complexes of the Western Southern Urals
T
K12-006
"Diorite"
"Granite"
"SM"
"Granite ? SM"
"Carbonatite"
"Non-magmatic"
"N-Hydrother"
"N-Mect"
"Metamorph"
Volgo-Sarmatian and Taratash orogens
DM
CHUR
1 Ga
2 Ga
3 Ga
3.5 Ga
#30
G
VSO
Bak-2
Bak-1
T
KE
Akt
NS,GR,QS
A
Epsilon Hf(t)
Kazakhstan continent
Kokchetau micro-continent (volcanic arcs)
EV-1
ER1301
Z13-22
Z13-24
K2
K1
Generation of granitoids
1.5 Ga
2.5 Ga
Tarim orogeny
Collision of Kokchetau massif and Stepnyak volcanic arc, generation of UHP metamorphic zircons
B
U-Pb age (Ma)

◀ **Fig. 13** A comparison of «Diagrams ε_{Hf}» for detrital zircons from sample K12-006 from Upper Ordovician strata of the Uraltau Uplift (**A**) with «Diagrams ε_{Hf}» for detrital zircons from Riphean strata of Northern Kazakhstan (**B**), as well as with known ages of EEP basement and Western Urals complexes (upper part of the figure), which are possible sources of detrital zircons. **A** «Diagram εHf» for dZr from sample K12-006. Grey oblique lines = lines of the Hf crustal model age of the protolith $T_{DM}{}^{C}$. Gray ellipses of the field of data points of ages and T_{DM}, estimated by the Sm-Nd isotopic system (complexes: T = Tashlyar, Akt = Aktanysh, Bak-1 and Bak-2—quartz diorites of Bakaly block) from (Bogdanova et al., 2010), for the southern part of the Volga-Sarmatian orogen (VSO-field) from (Bibikova et al. 2009), for Kolyvan enderbites (KE) from (Bogdanova et al. 2013). Gray rectangles are the fields of data points of ages and ε_{Hf} for rocks of the Berdyaush complex of the Western Urals (G-gabbro, NS, GR, QS-nepheline syenites, granites rapakivi and quartz syeno-diorites, correspondently) from (Ronkin et al. 2015a; b). For other complexes of the Western Urals, the age intervals are shown only in the upper part of the figure. The dotted pattern shows the area correlated with the zircons generated by the Taratash and Volga-Sarmatian collisional orogen. **B** «Diagram ε_{Hf}» for dZr from Riphean Group sandstones of the Kokchetau Massif (samples EV1, ER1, Z13-22 and Z13-24) from (Kovach et al., 2017), which characterize the basement of the Kokchetau microcontinent. In the upper part of the figure, the bars show the age intervals for the EEP and Western Urals complexes. EEP: VSO, TO = complexes of the Volga-Sarmatian and Taratash orogens (Bibikova et al. 2009; Tevelev et al. 2014, 2015, 2017; Terentiev and Santosh 2016; Terentiev et al. 2016a, b, 2017, 2018; Khotylev and Tevelev 2017); K, KN, NU, O = plutons and rapakivi-like granites: Korosten, Korsun-Novomirgorod, Novo-Ukrainian, October plutons of the Ukraine Shield (Bogdanova et al. 2004; Shumlyanskyy et al., 2006, 2015, 2016a, 2017), granites of the Voronezh Crystalline Massif (Savko et al. 2014); LKO = complexes of the Lapland-Kola orogen, sutured Kola and Karelia proto-cratons (Lahtinen and Huhma 2019); SF = complexes of the Svecofennian domain (Kahkonen 2005); VMRO = complexes of the Volyn-Middle-Russian orogen, sutured Fennoscandia and Volgo-Sarmatia (Bogdanova et al. 2008); SN = complexes of the Sveco-Norwegian domain (Bingen et al. 2008a; b); Gren = complexes of the Sveconorwegian (Grenville) orogen, which sutured Proto-Baltica, Proto-Laurentia and Amazonia in the process of assembling Rodinia; R, V, S = plutons and rapakivi-like granites: Riga, Vyborg and Salmi plutons (Bogdanova et al. 2008); pU-TO = complexes of the Proto-Uralide-Timanide orogen, which sutured Baltica and Arctida (Kuznetsov et al. 2009); VVP = complexes of the Volyn volcanic province (Shumlyanskyy et al., 2016b); cSTp = the basement of the Scythian-Turanian plate (Cadomides); CGO = Greenland Caledonian orogen, which sutured Arctic Europe and Laurentia. Western Urals: T = granitoids and metamorphites of the Taratash complex; 1—alkaline basaltoids of the Navysh complex (Ai Fm, Lower Riphean Burzyan Group, stratotypical location, Southern Urals) 1752 ± 18 Ma (Krasnobaev et al. 2013a); 2—Saran gabbro-ultrobasic intrusive complex (Middle and Northern Urals), related to magma chambers of depleted mantle over a mantle plum, ~1750 Ma (Petrov, 2017); 3—Berdyaush intrusive massif 1395 ± 20, 1373 ± 21, 1372 ± 12, 1368.4 ± 6.2 and 1369 ± 13 Ma (Puchkov 2010) (G—gabbro, NS—nepheline syenites, GR—granites-rapakivi, QS—quartz syeno-diorites); 4 and 5—granites of the Akhmerovo massif 1413 ± 45 Ma (I-generation) and 1381 ± 23 Ma (II-generation) (Krasnobaev et al., 2008); 6—intrusive rocks of the Kusa-Kopan and Ryabinov complexes (marking the northern side of the Zyuratkul fault), gabbro of the Kopan massif 1385 ± 25 Ma, granites of Ryabinov 1386 ± 40 Ma and Guben 1330 ± 16, 1330 ± 27 Ma massifs (Puchkov 2010); 7—bimodal volcanics of the Mashak Fm and its age analogous—Shatak and Kuvash Fms ~ 1385 Ma (Puchkov, 2010); 8—volcanics of the Arsha Group –709.9 ± 7.3 Ma (Puchkov, 2010); 9—granitoids of the Barangulovo massif 725 ± 5 Ma (Puchkov 2010); 10—metamorphic rocks (including HP and UHP rocks) of the Beloretsk block 600–550 Ma (Puchkov 2010); 11—rocks of the Mazara magmatic areal: granites 681 ± 14, 667 ± 9.6 Ma and gabbroids 709 ± 10, 704.2 ± 8.3 Ma (Kuznetsov 2009); 12—pyroxenites of the Kiryabin massif 680 ± 3.4 Ma (Krasnobaev et al. 2013b); 13—trachybasalts and hyalonephelinites of Dvorets and granitoids of Europa complexes 570 ÷ 550 Ma (Petrov 2017); 14—weakly alkaline rocks of the Kusa complex and basalts of basal levels of the Tanin Fm 625–600 Ma (Maslov et al. 2013; Petrov 2017); 15—trachyandesites of the Shegra Fm, granosyenitic Troitsk and verlitic-gabbro-granodioritic Zhuravlik complexes 680 ÷ 670 Ma (Maslov et al. 2013; Petrov 2017); 16—tuffs layers of the Sylvitsa Group 563.5 ± 3.5 (Kuznetsov et al. 2017c)

a small group of Late Proterozoic dZr (correlated with the **K1** episode) and a dominant population of the Mesoproterozoic zircons (correlated with the **K2** episode) (Fig. 13b). In the Late Neoproterozoic, the Kokchetau Massif collided with the Stepnyak volcanic arc, then the UHP complexes of a subduction zone were exhumed and became a source of specific metamorphic zircons with anomalously low Th/U ratios (Glorie et al. 2015). Later, extensive areas of the Kazakhstan continent were intruded by Paleozoic granitoids (Degtyarev et al. 2016, 2017).

Mesoproterozoic U-Pb ages and Hf isotope characteristics of dZr from sample K12-006 are well consistent with those from the Mesoproterozoic–Early Neoproterozoic (Riphean) cover of the Kokchetau Massif.

The only exception is that no 1.5–1.65 Ga dZr have been found in the Kokchetau Massif (a magmatic gap between **K1** and **K1** episodes), whereas several dZr with these ages have been found in the K12-006 sample. The three youngest ages of dZr from the K12-006 sample agree well with the time of the collision between the Kokchetau Massif and the Stepnyak volcanic arc. The zircons with extremely low Th/U ratios sourced from the UHP complexes related to this event have been reported by Glorie et al. (2015). There have been no similar zircons found in the K12-006 sample. Note, however, that there has been revealed one «specific» dZr #30, classified as a zircon of «Hydrothermal» origin, which may be related to the same collisional event.

There are no geochronological data available on the deep-buried complexes of the basement of the Scythian-Turanian plate. This plate is interpreted as a belt of the Cadomian terranes, which extends from Europe along the southern margin of the EEP into the Central Asian Orogenic Belt as suggested by geophysical data and indirect correlations (Kuznetsov and Romanyuk 2021).

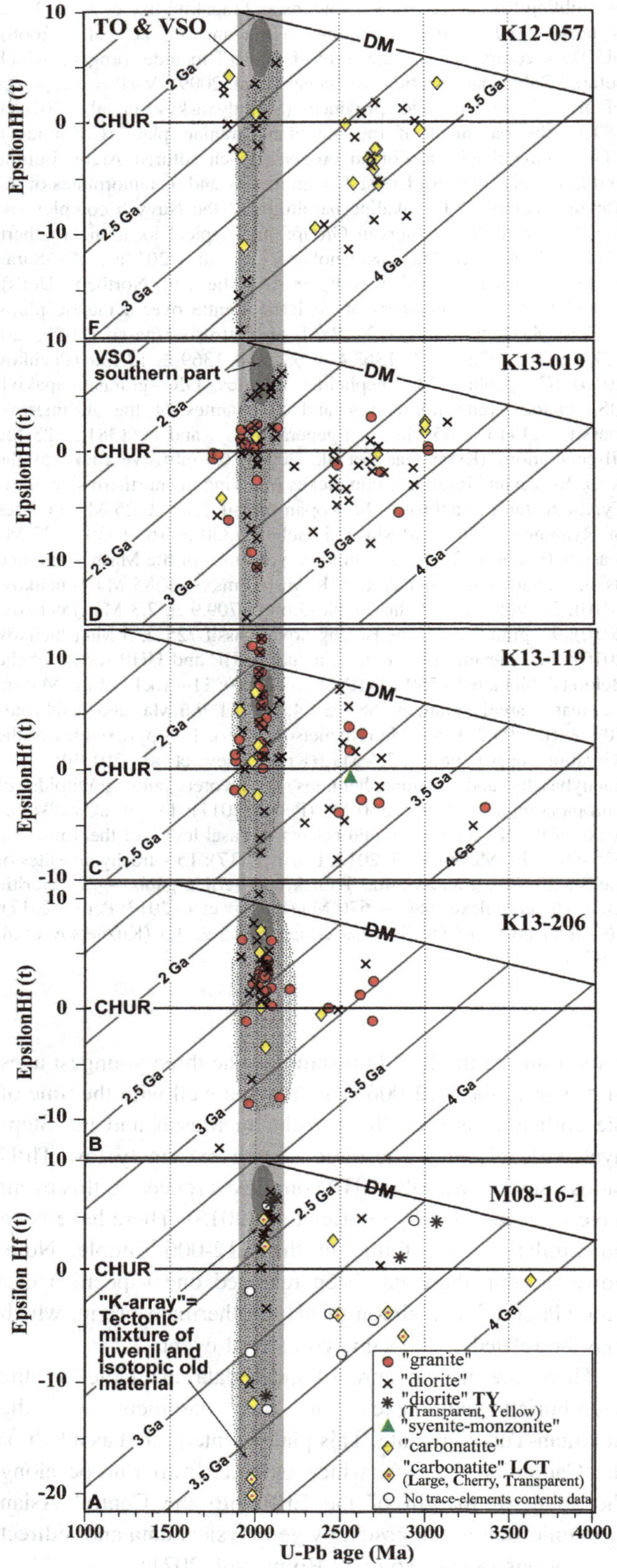

Fig. 14 «Diagrams εHf» For detrital zircons from Riphean strata of the Bashkir Uplift. VSO and TO = the Volgo-Sarmatian and Taratash orogens. Sample locations see in Fig. 1. Data after Romanyuk et al. (2013, 2014, 2017, 2018, 2019a)

5 Conclusions

Integral characteristics of detrital zircons from Upper Ordovician sandstones of the Uraltau Uplift studied in the sample K12-006 show little similarities with U-Pb and Lu-Hf model ages of rocks, from the basement of Volga-Uralia (the southeastern part of the EEP) and with the same characteristics of detrital zircons from the Late Precambrian strata of the Bashkir Uplift and Kazakhstan. In contrast, the Mesoproterozoic detrital zircons from the sample K12-006 are well consistent with U-Pb ages and Hf isotope signatures of detrital zircons from the Mesoproterozoic–Early Neoproterozoic (Riphean) cover of the Kokchetau Massif. However, studied detrital zircons also revealed that they have alien relation to Volga-Uralia, Kazakhstan and the Western Urals and thus must have originated from other crustal blocks.

A very high similarity of the age spectra of dZr from Ordovician sandstones of the Uraltau and Bashkir Uplifts (samples K12-025 and K12-006, respectively; see Fig. 4a) allows concluding that the amalgamation of the Precambrian Bashkir and Uraltau basements occurred before Late Ordovician. Detritus of post-Upper Ordovician sedimentary complexes of the Bashkir and Uraltau Uplifts originated from identical feeding provinces and was deposited in the same basin, sealing its pre-Ordovician composite heterogeneous basement.

Acknowledgements We thank Vinod Singh for his help with preparing this publication and two anonymous reviewers for the valuable comments and recommendations, which improved the manuscript considerably. This study has been carried out by following the plans of the scientific research of the Geological Institute and Schmidt Institute of Physics of the Earth (both Institutes of Russian Academy of Science). The analytical data were obtained using instrumentation funded by DEST Systemic Infrastructure Grants, ARC LIEF, NCRIS/AuScope, industry partners and Macquarie University, and ARC FT110100685 grant support (Belousova EA). This is contribution 1568 from the ARC Centre of Excellence for Core to Crust Fluid Systems (http://www.ccfs.mq.edu.au) and 1432 in the GEMOC Key Centre (http://www.gemoc.mq.edu.au). Synthesis of materials and article preparation were supported by the grant of RF government № 075-15-2019-1883 (Orogenesis: Formation and Growth of Continents and Supercontinents).

References

Andersen T (2002) Correction of common lead in U-Pb analyses that do not report ^{204}Pb. Chem Geol 192:59–79

Belousova EA, Walters S, Griffin WL et al (2002) Igneous zircon: trace element compositions as indicators of source rock type. Contrib Miner Petrol 143(5):602–622

Belousova EA, Griffin WL, O'Reilly SY (2006) Zircon crystal morphology, trace element signatures and Hf isotope composition as a tool for petrogenetic modeling: examples from eastern Australian granitoids. J of Petrology 47(2):329–353

Belousova EA, Kostitsyn YA, Griffin WL et al (2010) The growth of the continental crust: constraints from zircon Hf-isotope data. Lithos 119(3–4):457–466

Belousova EA, Gonzalez JM, Graham JI et al (2015) The enigma of crustal zircons in upper-mantle rocks: clues from the Tumut ophiolite, southeast Australia. Geology 43(2):119–122

Bibikova EV, Kirnozova TI, Fugzan MM et al (2009) Sarmatia-Volgo-Uralia junction zone: isotopic-geochronologic characteristic of supracrustal rocks and granitoids. Stratigr Geol Correl 17(6):561–573

Bingen B, Nordgulen O, Viola G (2008) A four-phase model for the Sveconorwegian orogeny, SW Scandinavia. Norwegian J of Geology 88:43–72

Bingen B, Andersson J, Soderlund U et al (2008) The Mesoproterozoic in the Nordic countries. Episodes 31(1):29–34

Black LP, Kamo SL, Allen CM et al (2004) Improved $^{206}Pb/^{238}U$ microprobe geochronology by monitoring of a trace-element-related matrix effect: SHRIMP, ID-TIMS, ELA-ICP-MS and oxygen isotope documentation for a series of zircon standards. Chem Geol 205:115–140

Bogdanova SV, Bingen B, Gorbatschev R et al (2008) The East European Craton (Baltica) before and during the assembly of Rodinia. Precambr Res 160:23–45

Bogdanova SV, De Waele B, Bibikova EV et al (2010) Volgo-Uralia: the first U-Pb, Lu-Hf and Sm-Nd isotopic evidence of preserved Paleoarchean crust. Am J Sci 310:1345–1383

Bogdanova SV, Belousova EA, De Waele B et al (2013) Zircon from Mesoarchean enderbites of Volgo-Uralia: U-Pb age, REE Hf and O-isotope compositions. Mineral Mag 77(5):727

Bogdanova SV, Pashkevich IK, Buryanov VB et al (2004) The 1.80–1.74-Ga gabbro–anorthosite–rapakivi Korosten Pluton in the Ukrainian Shield: a 3-D geophysical reconstruction of deep structure. Tectonophysics 381(1–4):5–27

Bouvier A, Vervoort JD, Patchett PJ (2008) The Lu–Hf and Sm–Nd isotopic composition of CHUR: constraints from unequilibrated chondrites and implications for the bulk composition of terrestrial planets. Earth Planet Sci Lett 273:48–57

Chapman JB, Gehrels GE, Ducea MN et al (2016) A new method for estimating parent rock trace element concentrations from zircon. Chem Geol 439:59–70

Claoue-Long JC, King RW, Kerrich R (1990) Archean hydrothermal zircon in the Abitibi Greenstone Belt: constraints on the timing of gold mineralization—reply. Earth Planet Sci Lett 109:601–609. https://doi.org/10.1016/0012-821X(90)90091-B

Corfu F, Hanchar JM, Hoskin P-WO et al (2003) Atlas of zircon textures. Rev Mineral Geochem 53(1):469–500

Corfu F, Davis DW (1991) Comment on "Archean hydrothermal zircon in the Abitibi greenstone belt: constraints on the timing of gold mineralization". In: Claoué-Long JC, King RW, Kerrich R (eds). Earth Planet Sci Lett 104(2–4):545–552

Degtyarev KE, Shatagin KN, Tret'yakov AA, (2016) Sources of Paleozoic granitic rocks and isotopic heterogeneity of the continental crust of the Aktau-Dzhungar microcontinent Central Kazakhstan. Doklady Earth Sci 470(2):1010–1013

Degtyarev K, Yakubchuk A, Tretyakov A et al (2017) Precambrian geology of the Kazakh uplands and Tien Shan: an overview. Gondwana Res 47:44–75

Elhlou S, Belousova E, Griffin WL et al (2006) Trace element and isotopic composition of GJ-red zircon standard by laser ablation. Geochmica Et Cosmochimica Acta 70(18):158

Fedotova AA, Bibikova EV, Simakin SG (2008) Ion-microprobe zircon geochemistry as an indicator of mineral genesis during geochronological studies. Geochem Int 46(9):912–927

Ferry JM, Watson EB (2007) New thermodynamic models and revised calibrations for the Ti-in-zircon and Zr-in-rutile thermometers. Contrib Miner Petrol 154:429–437

Fershtater GB, Krasnobaev AA, Bea F et al (2012) The geochemistry of zircon from the Urals igneous and metamorphic rocks. Litosfera (lithosphere) 4:13–29 (in Russian)

Fornelli A, Langone A, Micheletti F et al (2014) The role of trace element partitioning between garnet, zircon and orthopyroxene on the interpretation of zircon U-Pb ages: an example from high-grade basement in Calabria (Southern Italy). Int J Earth Sci 103(2):487–507

Fu B, Page FZ, Cavosie AJ et al (2008) Ti-in-zircon thermometry: applications and limitations. Contrib Miner Petrol 156:197–215

Fu B, Mernagh TP, Kita NT et al (2009) Distinguishing magmatic zircon from hydrothermal zircon: a case study from the Gidginbung high-sulphidation Au–Ag–(Cu) deposit, SE Australia. Chem Geol 259:131–142

Glorie S, Zhimulev FI, Buslov MM et al (2015) Formation of the Kokchetau subduction–collision zone (northern Kazakhstan): insights from zircon. Gondwana Res 27:424–438

Golionko BG, Artemova OA (2016) Late Precambrian and Paleozoic deformations of the eastern part of the Bashkir anticlinorium (Southern Urals). MOIP Bull Serie Geol 91(6):3–10 (in Russian)

Griffin WL, Pearson NJ, Belousova EA et al (2000) The Hf isotope composition of cratonic mantle: LAM-MC-ICPMS analysis of zircon megacrysts in kimberlites. Geochim Cosmochim Acta 64:133–147

Griffin WL, Wang X, Jackson SE et al (2002) Zircon chemistry and magma genesis, SE China: in-situ analysis of Hf isotopes, Pingtan and Tonglu igneous complexes. Lithos 61:237–269

Griffin WL, Belousova EA, Shee SR et al (2004) Archean crustal evolution in the northern Yilgarn Craton: U-Pb and Hf-isotope evidence from detrital zircons. Precambr Res 131:231–282

Griffin WL, Belousova EA, Walters SG et al (2006) Archean and Proterozoic crustal evolution in the Eastern Succession of the Mt Isa district, Australia: U-Pb and Hf-isotope studies of detrital zircons. Aust J Earth Sci 53:125–149

Griffin WL, Belousova EA, O'Reilly SY (2007) TerraneChron analysis of zircons from Western Australian samples record. Geol Surv West Aust 4:1–52

Griffin WL, Powell WJ, Pearson NJ et al (2008) GLITTER: data reduction software for laser ablation ICP-MS / Ed. P.J. Sylvester. Laser ablation ICP-MS in the Earth sciences: current practices and outstanding issues. Mineral Assoc Canada Short Course 40:308–311

Grimes CB, John BE, Kelemen PB et al (2007) Trace element chemistry of zircons from oceanic crust: a method for distinguishing detrital zircon provenance. Geology 35:643–646

Grimes CB, Wooden JL, Cheadle MJ et al (2015) "Fingerprinting" tectono-magmatic provenance using trace elements in igneous zircon. Contrib Mineral Petrol 170(5–6):Article 46

Harrison TM, Watson EB, Aikman AB (2007) Temperature spectra of zircon crystallization in plutonic rocks. Geology 35(7):635–638

Hawkesworth CJ, Kemp AIS (2006) Using hafnium and oxygen isotopes in zircons to unravel the record of crustal evolution. Chem Geol 226:144–162

Heaman LM, Bowins R, Crocket J (1990) The chemical composition of igneous zircon suites: implications for geochemical tracer studies. Geochim Cosmochim Acta 54(6):1597–1607

Horstwood MSA, Kosler J, Gehrels G et al (2016) Community-derived standards for LA-ICP-MS U-(Th-)Pb geochronology—uncertainty propagation, age interpretation and data reporting. Geostand Geoanal Res 40(3):311–332

Hoskin PWO (2005) Trace-element composition of hydrothermal zircon and the alteration of Hadean zircon from the Jack Hills, Australia. Geochim Cosmochim Acta 69:637–648

Hoskin PO, Ireland TR (2000) Rare earth element chemistry of zircon and its use as aprovenance indicator. Geology 28(7):627–630

Hoskin PWO, Kinny PD, Wyborn D et al (2000) Identifying accessory mineral saturation during differentiation in granitoid magmas: an integrated approach. J Petrol 41:1365–1396

Hoskin PWO, Schaltegger U (2003) The composition of zircon and igneous and metamorphic petrogenesis. In: Hanchar JM, Hoskin PWO (Eds) Zircon. Rev Mineral Geochem 53:27–62

Hu FF, Fan HR, Yang JH et al (2004) Mineralizing age of the Rushan lode gold deposit in the Jiaodong Peninsula: SHRIMP U-Pb dating on hydrothermal zircon. Chin Sci Bull 49:1629–1636

Ivanov KS (1998) Main Features of Geological History (1.6–0.2 Ga) and Structure of the Urals. Nauka, Yekaterinburg (in Russian)

Jackson SE, Pearson NJ, Griffin WL et al (2004) The application of laser ablation-inductively coupled plasma-mass spectrometry to in situ U-Pb zircon geochronology. Chem Geol 211:47–69

Kaczmarek MA, Müntener O, Rubatto D (2008) Trace element chemistry and U-Pb dating of zircons from oceanic gabbros and their relationship with whole rock composition (Lanzo, Italian Alps). Contrib Miner Petrol 155(3):295–312

Kostitsyn YA, Belousova EA, Silant'ev SA, Bortnikov NS, Anosova MO (2015) Modern problems of geochemical and U-Pb geochronological studies of zircon in oceanic rocks. Geochem Int 53:759–785. https://doi.org/10.1134/S0016702915090025

Kahkonen Y (2005) Svecofennian supracrustal rocks. In: Lehtinen M, Nurmi JA, Ramo OT (eds) Precambrian geology of finland—key to the evolution of the fennoscandian shield. Elsevier B.V, Amsterdam, pp 343–406

Kebede T, Horie K, Hidaka H et al (2007) Zircon 'microvein' in peralkaline granitic gneiss, western Ethiopia: origin, SHRIMP U-Pb geochronology and trace element investigations. Chem Geol 242:76–102

Kerrich R, King R (1993) Hydrothermal zircon and baddeleyite in Val-Dor Archean mesothermal gold deposits: characteristics, compositions, and fluid-inclusion properties, with implications for timing of primary gold mineralization. Can J Earth Sci 30:2334–2351

Khotylev AO, Tevelev AV (2017) Geochemical characteristics of the early Riphean Navysh volcanic complex (Southern Urals). Mosc Univ Geol Bull 73(1):24–30

Kirkland CL, Smithies RH, Taylor RJM et al (2015) Zircon Th/U ratios in magmatic environs. Lithos 212–215:397–414

Kovach V, Kotov A, Tolmacheva E, Degtyarev K, Tretyakov A, Wang K-L, Chung S-L, Lee H-Y, Jahn B-M (2017) Sources and provenance of the Neoproterozoic placer deposits of the Northern Kazakhstan: Implication for continental growth of the western Central Asian Orogenic Belt. Gondwana Res 47:28–43

Krasnobaev AA, Puchkov VN, Kozlov VI et al (2008) The Akhmerovo granite massif: a proxy of Mesoproterosoic intrusive magmatism in the Southern Urals. Dokl Earth Sci 418(1):103–108

Krasnobaev AA, Puchkov VN, Sergeeva ND et al (2013) Zirconology of pyroxenites from the Kiryabinka pyroxenite-gabbro complex (Southern Urals). Dokl Earth Sci 450(1):531–535

Krasnobaev AA, Puchkov VN, Kozlov VI et al (2013) Zirconology of Navysh volcanic rocks of the Ai Suite and the problem of the age of the Lower Riphean boundary in the Southern Urals. Dokl Earth Sci 448(2):185–190

Kuznetsov NB, Romanyuk TV (2021) Peri-Gondwanan Blocks in the Structure of the Southern and Southeastern Framing of the East European Platform. Geotectonics 55:439–472. https://doi.org/10.1134/S0016852121040105

Kuznetsov NB, Shazillo AV (2011) The first finds of skeletal fossils in the Kuk-Karauk Formation of the Asha Group (Southern Urals) and their significance for determining the beginning of the Pre-Uralian-Timanian orogeny. Dokl Earth Sci 440(1):1239–1244

Kuznetsov NB, Romanyuk TV, Shatsillo AV et al (2012) The first results of the bulk U/Pb-dating of detrital zircons (LA-ICP-MS) from Asha Formation, the Southern Urals—the stratigraphic, paleogeographic and paleo-tectonic aspects. Dokl Earth Sci 447 (1):1240–1246

Kuznetsov NB, Maslov AV, Belousova EA et al (2013) The first U-Pb (LA-ICP-MS) isotope data of detrital zircons from the basal levels of the Riphean Stratotype. Dokl Earth Sci 451(1):724–728

Kuznetsov NB, Meert JG, Romanyuk TV (2014) Ages of the detrital zircons (U/Pb, La-ICP-MS) from latest Neoproterozoic—Middle Cambrian(?) Asha group and early devonian Takaty formation, the south-western urals: a testing of an Australia-Baltica connection within the Rodinia. Precambr Res 244:288–305

Kuznetsov NB, Romanyuk TV, Shatsillo AV et al (2014) The first results of the U/Pb-dating (LA-ICP-MS) of the detrital zircons from sandstones of the Upper Emsian Takata Formation, the Western Urals (with a problem of an ultimate sources of the Uralian diamond placers). Dokl Earth Sci 455(2):370–375

Kuznetsov NB, Belousova EA, Degtyarev KE et al (2016) The first results of the U/Pb dating of detrital zircons from the later ordovician sandstones of the Bashkir uplifts (the Southern Urals). Dokl Earth Sci 467(2):325–330

Kuznetsov NB, Belousova EA, Romanyuk TV et al (2017) The first results of U/Pb dating detrital zircons from sandstones of Zigalga Formation (Middle Riphean, the South Urals). Dokl Earth Sci 475 (2):862–866

Kuznetsov NB, Gorozhanin VM, Belousova EA et al (2017) The first results of U/Pb dating of detrital zircons from Ordovician terrigenous sequences of Sol-Iletsk arch of the East-European Platform. Dokl Earth Sci 473(2):381–385

Kuznetsov NB, Belousova EA, Krupenin MT et al (2017) The results of geochronological and isotope-geochemical study of zircon from tuffs Sylvitsa Group (Western slope of the Middle Urals): the origin of ash layers in Vendian strata of the East European platform. Dokl Earth Sci 473(1):360–363

Kuznetsov NB, Belousova EA, Romanyuk TV (2018) Geochemical and Lu-Hf isotope (LA-ICP-MS) systematic of detrital zircons from the Ordovician sandstones of the Sol-Iletsk arch (Russia, Northern Caspian, borehole Ordovician-2). Archaeol Anthropol 3(1):31–58

Kuznetsov NB, Belousova EA, Griffin WL et al (2019) Pre-mesozoic crimea as a continuation of the Dobrogea platform: insights from detrital zircons in upper jurassic conglomerates Mountainous Crimea. Int J Earth Sci 108(7):2407–2428

Kuznetsov NB (2009) Protouralide-Timanide complexes and Late Cambrian-Early Paleozoic evolution of eastern and north-eastern margins of East-European platform (Doctoral thesis). Geological Institute RAS, Moscow (in Russian). https://www.dlib.rsl.ru/viewer/01003464837#?page=1

Lahtinen R, Huhma H (2019) A revised geodynamic model for the Lapland-Kola Orogen. Precambr Res 330:1–19

Linnemann U, Ouzegane K, Drareni A et al (2011) Sands of West Gondwana: an archive of secular magmatism and plate interactions—a case study from the Cambro-Ordovician section of the Tassili Ouan Ahaggar (Algerian Sahara) using U-Pb-LA-ICP-MS detrital zircon ages. Lithos 123:188–203

Liu F-L, Xu Z-Q (2004) Fluid inclusions hidden in coesite-bearing zircons in ultrahighpressure metamorphic rocks from southwestern Sulu terrane in eastern China. Chin Sci Bull 49:396–404

Liu J-B, Ye K, Maruyama S, Cong B, Fan H (2001) Mineral inclusions in zircon from gneisses in the ultrahigh-pressure zone of the Dabie Mountains, China. J Geol 109:523–535

Liu J, Liu F, Ding Z et al (2013) U-Pb dating and Hf isotope study of detrital zircons from the Zhifu Group, Jiaobei Terrane, North China Craton: provenance and implications for precambrian crustal growth and recycling. Precambr Res 235:230–250

Ludwig KR (2012) User's manual for Isoplot 3.75. A geochronological toolkit for Microsoft Excel. Berkeley Geochronology Center. Spec. Publ, No 5. 75 p

Maslov AV, Meert J, Levashova NM et al (2013) New data about age of the Vendian glacial deposits (Middle Urals). Dokl Earth Sci 449 (1):303–308

Maslov AV (2004) Riphean and Vendian sedimentary sequences of the Timanides and Uralides, the eastern periphery of the East European Craton. In: The Neoproterozoic Timanide Orogen of Eastern Baltica. Gee DG, Pease V (Eds). Geological Society of London, Memoirs. 30:19–35

McDonough WF, Sun SS (1995) The composition of the Earth. Chem Geol 120:223–253

Pelleter E, Cheilletz A, Gasquet D et al (2007) Hydrothermal zircons: a tool for ion microprobe U-Pb dating of gold mineralization (Tamlalt-Menhouhou gold deposit—Morocco). Chem Geol 245:135–161

Petrov GA (2017) Geology of the Pre-Palaeozoic complexes in the middle part of the Uralian mobile belt. (Doctoral thesis). Geological Institute RAS, S-Petersburg. University (in Russian). https://www.docplayer.ru/45643414-Petrov-georg-askoldovich-geologiya-dopaleozoyskih-kompleksov-sredney-chasti-uralskogo-podvizhnogo-poyasa.html

Pettke T, Audetat A, Schaltegger U et al (2005) Magmatic-to-hydrothermal crystallization in the W-Sn mineralized Mole Granite (NSW, Australia)—Part II: evolving zircon and thorite trace element chemistry. Chem Geol 220:191–213

Puchkov VN (2010) Geology of the Urals and Cis-Urals (actual problems of stratigraphy, tectonics, geodynamics and metallogeny). Design Poligraph Service, Ufa, 280 p (in Russian)

Ramezani J, Dunning GR, Wilson MR (2000) Geologic setting, geochemistry of alteration, and U-Pb age of hydrothermal zircon from the Silurian Stog'er Tight gold prospect, Newfoundland Appalachians, Canada. Explor Min Geol 9:171–188

Romanyuk TV, Kuznetsov NB, Maslov AV et al (2014) Geochemical and Lu/Hf Isotopic (LA-ICP-MS) signature of detrital zircons from sandstones of the basal levels of the Riphean stratotype. Dokl Earth Sci 459(1):1356–1360

Romanyuk TV, Kuznetsov NB, Belousova EA et al (2017) Geochemical and Lu/Hf-isotopic (LA-ICP-MS) systematics of detrital zircons from the Upper Ordovician sandstones of the Bashkir uplift (Southern Urals). Dokl Earth Sci 472(2):134–137

Romanyuk TV, Kuznetsov NB, Belousova EA et al (2018) Paleotectonic and paleo-geographic conditions for the accumulation of the lower Riphean Ai Formation in the Bashkir uplift (Southern Urals): the Terrane Chrone® detrital zircon study. Geodyn Tectonophys 9 (1):1–37

Romanyuk TV, Kuznetsov NB, Maslov AV et al (2013) Geochemical and Lu/Hf (LA-ICP-MS) systematics of detrital zircons from Upper Riphean Lemeza Fm., Southern Urals. Dokl Earth Sci 453(2):1200–1204

Romanyuk TV, Belousova EA, Kuznetsov NB et al (2019a) A search for sources of the detritus of Ordovician sandstones from the Sol-Iletsk block (Ordovician-2 borehole) based on the first data of the geochemical and Lu/Hf isotopic systematics of zircons. Doklady Earth Sci 487(1):795–799. https://doi.org/10.1134/S1028334X19070067

Romanyuk TV, Kuznetsov NB, Puchkov VN et al (2019b) Local source of detritus for rocks of the Ai Formation (basal level of the Lower Riphean stratotype, the Bashkir Uplift, Southern Urals): evidence from U-Pb (LA-ICP-MS) dating of detrital zircons. Doklady Earth Sci 484(1):53–57. https://doi.org/10.1134/S1028334X19010069

Romanyuk TV, Kuznetsov NB, Puchkov VN et al (2000) Age and Stratigraphic Position of Sedimentary Sequences of the Bagrush Mountains (Southern Urals) Based on the Results of U–Pb Dating (LA–ICP–MS) of Detrital Zircons. Doklady Earth Sciences 493:593–599 https://doi.org/10.1134/S1028334X20080188

Ronkin YuL, Maslov AV, Gerdes A (2015) REE and Lu-Hf systematics of zircons from rapakivi granites and associated rocks of supercontinent Nuna (Columbia). Dokl Earth Sci 461(1):277–282

Ronkin YuL, Gerdes A, Nesbit R (2015) Zircon of the granites rapakivi and associating rocks of the Southern Urals: REE and Lu-Hf isotopic constraints. Proc Inst Geol Geochem Ural Branch Russ Acad Sci 162:222–228 (in Russian)

Rubatto D (2002) Zircon trace element geochemistry: partitioning with garnet and the link between U-Pb ages and metamorphism. Chem Geol 184(1–2):123–138

Rubatto D (2017) Zircon: the Metamorphic Mineral. Rev Mineral Geochem 83(1):261–295

Rubatto D, Hermann J (2007) Experimental zircon/melt and zircon/garnet trace element partitioning and implication for the geochronology of crustal rocks. Chem Geol 241(1–2):38–61

Rubin JN, Henry CD, Price JG (1989) Hydrothermal zircons and zircon overgrowths, Sierra-Blanca Peaks, Texas. Am Miner 74:865–869

Rubin JN, Henry CD, Price JG (1993) The mobility of zirconium and other "immobile" elements during hydrothermal alteration. Chem Geol 110(1–3):29–47

Savko KA, Samsonov AV, Larionov AN et al (2014) Paleo-Proterozoic A and S-granites in the eastern Voronezh crystalline massif: geochronology, petrogenesis, and tectonic setting of origin. Petrology 22(3):205–233

Schaltegger U, Pettke T, Audétat A et al (2005) Magmatic-to-hydrothermal crystallization in the W-Sn mineralized mole granite (NSW, Australia)—Part I: crystallization of zircon and REE-phosphates over three million years—a geochemical and U–Pb geochronological study. Chem Geol 220:215–235

Schaltegger U (2003) The composition of zircon and igneous and metamorphic petrogenesis. In: Hanchar JM, Hoskin PWO (eds) Zircon, Reviews in Mineralogy and Geochemistry, 53, pp 27–62. Mineralogical Society of America

Scherer E, Münker C, Mezger K (2001) Calibration of the Lu-Hf clock. Science 293:683–687

Schulz B, Klemd R, Braetz H (2006) Host rock compositional controls on zircon trace element signatures in metabasites from the Austroalpine basement. Geochim Cosmochim Acta 70:697–710

Sharkov EV (2010) Middle-proterozoic anorthosite–rapakivi granite complexes: an example of within-plate magmatism in abnormally thick crust: evidence from the East European Craton. Precambr Res 183(4):689–700

Shumlyanskyy L, Ellam RM, Mitrokhin O (2006) The origin of basic rocks of the Korosten AMCG complex, Ukrainian shield: implication of Nd and Sr isotope data. Lithos 90(3–4):214–222

Shumlyanskyy L, Ernst R, Billstrom K (2015) A U-Pb Baddeleyite age of the Davydky gabbro-syenite massif of the Korosten plutonic complex. Geochem Ore Form 35:37–42

Shumlyanskyy L, Nosova A, Billstrom K et al (2016) The U-Pb zircon and baddeleyite ages of the Neoproterozoic Volyn Large Igneous Province: implication for the age of the magmatism and the nature of a crustal contaminant. GFF 138(1):1–14

Shumlyanskyy L, Hawkesworth C, Billström K et al (2017) The origin of the Palaeoproterozoic AMCG complexes in the Ukrainian shield: new U-Pb ages and Hf isotopes in zircon. Precambr Res 292:216–239

Shumlyanskyy L, Mitrokhin O, Billstrom K et al (2016a) The ca. 1.8 Ga mantle plume related magmatism of the central part of the Ukrainian shield. GFF 138(1):86–101

Sindern S, Hetzel R, Schulte BA et al (2005) Proterozoic magmatic and tectonometamorphic evolution of the Taratash complex, Central Urals, Russia. Earth and enviromental science. Int J Earth Sci 94 (3):319–335

Skublov SG, Berezin AV, Berezhnaya NG (2012) General relations in the trace-element composition of zircons from eclogites with implications for the age of eclogites in the Belomorian mobile belt. Petrology 20(5):427–449

Stratotype of Riphean. Stratigraphy. Geochronology (1983) Nauka, Moscow (in Russian)

Teipel U, Eichhorn R, Loth G et al (2004) A U-Pb SHRIMP and Nd isotopic data from the western Bohemian Massif (Bayerischer Wald, Germany): implications for Upper Vendian and Lower Ordovician magmatism. Int J Earth Sci (geol Rundsch) 93:782–801

Terentiev RA, Santosh M (2016) Detrital zircon geochronology and geochemistry of metasediments from the Vorontsovka terrane: implications for microcontinent tectonics. Int Geol Rev 58 (99):1108–1126

Terentiev RA, Skryabin VY, Santosh M (2016) U-Pb zircon geochronology and geochemistry of Paleoproterozoic magmatic suite from East Sarmatian Orogen: tectonic implications on Columbia supercontinent. Precambr Res 273:165–184

Terentiev RA, Savko KA, Santosh M (2016) Paleoproterozoic crustal evolution in the East Sarmatian Orogen: petrology, geochemistry, Sr–Nd isotopes and zircon U-Pb geochronology of andesites from the Voronezh Massif, Western Russia. Lithos 246–247:61–80

Terentiev RA, Savko KA, Santosh M (2017) Paleoproterozoic evolution of the back-arc-arc system in the East Sarmatian orogen (East European craton): zircon SHRIMP Geochronology & geochemistry of the Losevo volcanic suite. Am J Sci 317(5):707–753

Terentiev RA, Savko KA, Santosh M (2018) Post-collisional two-stage magmatism in the East Sarmatian ogogen, East European craton: evidence from the Olkhovsky ring complex. J Geol Soc 175(1):86–99

Tevelev AV, Kosheleva IA, Khotylev AO et al (2014) Peculiarities of the structure and evolution of the Riphean Ai volcanic complex South Urals. Mosc Univ Geol Bull 69(5):289–298

Tevelev AV, Kosheleva IA, Tevelev AV et al (2015) New data on the isotope ages of the Taratash and Aleksandrovka metamorphic complexes. Mosc Univ Geol Bull 70(1):24–40

Tevelev AV, Mosejchuk VM, Tevelev AV et al (2017) The zircon-age distribution in metamorphic rocks of the Taratash block, Southern Urals (an initial provenance signal). Mosc Univ Geol Bull 72 (5):314–319

Veevers JJ, Saeed A, Belousova EA et al (2005) U-Pb ages and source composition by Hf-isotope and trace-element analysis of detrital zircons in Permian sandstone and modern sand from southwestern Australia and a review of the paleogeographical and denudational history of the Yilgarn Craton. Earth Sci Rev 68:245–279

Veevers JJ, Belousova EA, Saeed A et al (2006) Pan-Gondwanaland detrital zircons from Australia analysed for Hf-isotopes and trace elements reflect an ice-covered Antarctic provenance of 700–500 Ma age, TDM of 2.0–1.0 Ga, and alkaline affinity. Earth Sci Rev 76:135–174

Vermeesch P (2012) On the visualisation of detrital age distributions. Chem Geol 312–313:190–194

Vermeesch P (2018) IsoplotR: a free and open toolbox for geochronology. Geosci Front 9:1479–1493

Wanless VD, Perfit MR, Ridley WI et al (2011) Volatile abundances and oxygen isotopes in basaltic to dacitic lavas on mid-ocean ridges: the role of assimilation at spreading centers. Chem Geol 287:54–65

Watson EB, Wark DA, Thomas JB (2006) Crystallization thermometers for zircon and rutile. Contrib Miner Petrol 151:413–433

Wiedenbeck M, Alle P, Corfu F et al (1995) Three natural zircon standards for U-Th-Pb, Lu-Hf, trace-element and REE analyses. Geostand Newsl 19:1–23

Wiedenbeck M, Hanchar JM, Peck WH et al (2004) Further characterization of the 91500 zircon crystal. Geostand Geoanal Res 28:9–39

Yuan H-L, Gao S, Dai M-N et al (2008) Simultaneous determinations of U-Pb age, Hf isotopes and trace element compositions of zircon by excimer laser-ablation quadrupole and multiple-collector ICP-MS. Chem Geol 247:100–118

Zhang Z-M, Shen K, Wang JL et al (2009) Petrological and geochronological constraints on the formation, subduction and exhumation of the continental crust in the southern Sulu orogen, eastern-central China. Tectonophysics 475(2):291–307

Zhang Sh, Li Z-X, Evans DAD et al (2012) Pre-Rodinia supercontinent Nuna shaping up: a global synthesis with new paleomagnetic results from North China. Earth Planet Sci Lett 353–354:145–155

Astroblems in the Early Earth History: Precambrian Impact Structures of the Kola-Karelian Region (East Baltic Shield)

Tatiana V. Kaulina, Lyudmila I. Nerovich, Vadim L. Il'chenko, Lyudmila M. Lialina, Evgeny L. Kunakkuzin, Maria A. Ganninbal, Sergey V. Mudruk, Dmitry V. Elizarov, and Elena S. Borisenko

Abstract

Studying early Precambrian astroblems is complicated by their localization in structural and compositional complexes that underwent superimposed metamorphism and tectonic processing. Back in the 1980s, seven sites with occurrences of breccias and other rocks resembling impact structures were identified as potentially diamondiferous impact structures within the Karelian-Kola region. But at present, only two astroblems are known in the region: Janisjärvi of 725 ± 5 Ma and Suavjärvi of ~2400 Ma. In the Kola part, impact origin is assumed for two objects: the Javrozero circle structure in the Tanaelv belt and the Järva-varaka layered massif in the Monchegorsk ore area. The most promising structure to be an astrobleme with serious signs of impact origin is the Paleoproterozoic Järva-varaka massif. According to geological, petrochemical, and geochemical characteristics of rocks the Järva-varaka massif is most similar to the 1.85 Ga Sudbury structure (Canada), for which an impact origin was assumed. Shock metamorphism of the Järva-varaka massif was manifested in amorphization of zircon from the country rocks with formation of plagioclase and sillimanite glasses in inclusions, as well as planar deformations in quartz.

1 Introduction

Widespread meteorite (impact) craters on Moon, Mercury, Mars, and Venus indicate the important role of the impacts in the evolution of solid cosmic bodies. It is obvious that the meteorite bombardment contributed significantly to the formation of the early Earth's crust. The role of meteorites in the history of the Earth has been the subject of serious scientific research since the 1980s. Many researchers believe that the impact events have significantly affected the dynamics of intra-planetary processes (Masaitis et al. 1980; Melosh 1989; French 2004; Reimold and Koeberl 2008; Glukhovsky and Kuzmin 2015; Glikson 2013, 2014). Studies focused on the external factors of the Earth evolution show that the mantle plume activity as well as the formation of large igneous provinces were associated with large asteroid impacts (for example, Jones et al. 2005; Glikson and Vickers 2010).

By now, it has also been shown that impact events can create both syngenetic mineral deposits and favorable structural and lithological conditions for subsequent formation of such deposits. In the identified impact structures, PGE-Cu-Ni, iron, polymetal ores, zeolites, bauxites, oil shales, oil and gas, etc. are concentrated (Grieve and Masaitis 1994; Reimold et al. 2004). The most pronounced is the manifestations of Cu-Ni and Cu–Zn-Pb mineralization, which, apart from the largest deposits of the Sudbury structure (Lightfoot et al. 2001; Naldrett 2004), also include smaller manifestations in the astroblems of Morokweng, Siljan, Serpent Mound, Decaturville, Suavjärvi (for references see Grieve and Masaitis 1994; Reimold et al. 2004; Naumov 2005; Mashchak and Naumov 2012 etc.), as well as impact diamonds found in huge amounts in the Popigai astrobleme, and also known in other impact structures, such as Kara, Zapadnaya, Lappajärvi and et al. (Masaitis et al. 2018; Naumov 2002). On the early stages of the Earth's crust formation, however, impact events should have produced more significant effect on the distribution of ores. However, this issue is still at the stage of posing the problem and is limited to speculative assumptions, e.g., those regarding the relation between the gold ore objects of the Witwatersrand field and the Wredefort impact structure (Gibson and Reimold 1999) or between large ferruginous quartzite deposits and iron meteorites (Glikson and Vicker 2007).

T. V. Kaulina (✉) · L. I. Nerovich · V. L. Il'chenko · L. M. Lialina · E. L. Kunakkuzin · M. A. Ganninbal · S. V. Mudruk · D. V. Elizarov · E. S. Borisenko
Geological Institute of Kola Science Centre of Russian Academy of Sciences, Fersman Str. 14, Apatity, 184209, Russia
e-mail: kaulina@geoksc.apatity.ru

A. K. Shandilya et al. (eds.), *Geological and Geo-Environmental Processes on Earth*, Springer Natural Hazards,
https://doi.org/10.1007/978-981-16-4122-0_3

Studying early Precambrian astroblems is complicated by their localization in structural and compositional complexes that underwent superimposed dynamo-thermal metamorphism and tectonic processing, as well as long-term degradation due to the denudation of the Earth's surface. The poor preservation of meteorite impact traces has to be addressed by developing a set of diagnostic criteria for identifying astroblems on the Earth's surface and distinguishing them from similar structures created by endogenous geological processes (French and Koeberl 2010).

According to Melosh (1989) the meteorite itself in the main destroyed in an impact event, thus diagnostic indicators of this event can be found only in the target rocks. Among these indicators are: shatter cones, high-pressure (diaplectic) mineral glasses, high-pressure minerals, high-temperature glasses and melts, planar fractures (PFs), and planar deformation features (PDFs) in quartz (French and Koeberl 2010). Another useful indicator of shock metamorphism is zircon, which is a widespread and weathering-resistant mineral and can be used in the study of old, deeply eroded, and metamorphically overprinted impact structures (Gucsik et al. 2002, 2004). It is considered that zircon retains the signs of shock metamorphism, even under the granulite facies metamorphism (Reimold et al. 2002). Impact zircon is characterized by several features that occur together or separately in zircon of the same sample: the formation of planar microstructures; high-pressure polymorphic modification of $ZrSiO_4$—reidite; granular texture - crystallite-grains of zircon in a glassy $ZrSiO_4$ matrix; baddeleyite inclusions; microtwins (e.g. Wittman et al. 2006; Garde et al. 2012; Cavosie et al. 2015; Erickson et al. 2017).

Up to the present, about 180 astroblems have been reliably established on Earth (Earth Impact database, 2012), while only four among them are older than 1.6 Ga: Wredefort (2023 ± 4 Ma) in South Africa (Gibson et al. 1997), Sudbury (1850 ± 3 Ma) in Canada (Davis 2008), Yarrabubba (supposedly Paleoproterozoic) in Australia (McDonald et al. 2003) and Suavjärvi (~2.4 Ga) in Karelia (Mashchak and Naumov 2012). In addition, a number of ring structures of the Early Precambrian are assumed to have the impact origin, e.g., Maniitsoq in western Greenland (Garde et al. 2012), and some mineralogical objects of cosmogenic origin were found in the Upper Archean complexes of South Africa and Australia (Lowe et al. 2014).

2 Astroblems of the Kola-Karelian Region (NE Baltic Shield)

Back in the 1980s, the Russian Geological Research Institute (VSEGEI) carried out a geological survey of sites with occurrences of breccias and other rocks resembling impact structures in the Karelian-Kola region in the frame of the project “Preliminary assessment of the cosmogenic structures of the Baltic Shield for mineral deposites”. As a result, some sites were identified as potentially diamondiferous impact structures: Arvarench, Ust-Ponoi, Panajärvi, Kumsa and Girvas, Motko, Keljärvi (Fig. 1).

At present, two impact structures are known in Karelia: Janisjärvi of 725 ± 5 Ma (Vishnevsky et al. 2002) and Suavjärvi of ~2400 Ma (Mashchak and Naumov 2012). In the Kola region, impact origin is assumed for two objects: the Javrozero circle structure in the Tanaelv belt (Kaulina et al. 2015) and the Järva-varaka layered massif in the Monchegorsk ore area (Nerovich et al. 2015, 2019), discussed below (Fig. 1).

3 Janisjärvi Structure

Janisjärvi crater with a diameter of 14 km is located in the western part of the Karelian Region (lat. 61°58′ N, long. 30° 55′ E) (Fig. 1) and is filled with the same name Lake (Masaitis and Shadenkov 1980; Feldman et al. 1981). An age of the impact event was determined as 725 ± 5 Ma (Vishnevsky et al. 2002). The target rocks are represented by the Lower and Middle Proterozoic metamorphosed sandstones, quartzites, siltstones, as well as biotite and andalusite schists. The outcrops of impactites can be seen on the islands in the center of the Lake, and on the western shore of the Lake. Impactites are represented by allogeneic breccia, zuvites, and overlapping tagamites. Zuvites contain fragments of schists and microschists, sometimes with well-formed shatter cones, glass fragments, and fragments of shock-metamorphosed quartz and feldspar-quartz veins. Tagamites are crystallized and consist of plagioclase grains surrounded by rims of potassium feldspar, quartz, cordierite with an insignificant amount of hypersthene, biotite, ilmenite, and magnetite. The matrix has a granophyre texture and consists of potassium feldspar and quartz aggregates. The internal structure of the Janisjärvi crater is assumed simple without central uplift (Feldman et al. 1981), while other researchers suggest the presence of a central uplift (Masaitis and Shadenkov 1980). The discovery of microdiamonds in impactites was reported (Vishnevsky and Palchik 2002).

4 Suavjärvi Structure

Suavjärvi structure is located in the central part of the Karelian Region (lat. 63°07′ N, long. 33°23′ E) (Fig. 1) and is considered to be a remnant of a deeply eroded and metamorphosed impact structure, which has a diameter of 16 km and was formed during the Paleoproterozoic (~2.4 Ga) time due to the age of the overlying Jatulian volcanic-sedimentary sequence. The Suavjärvi structure is

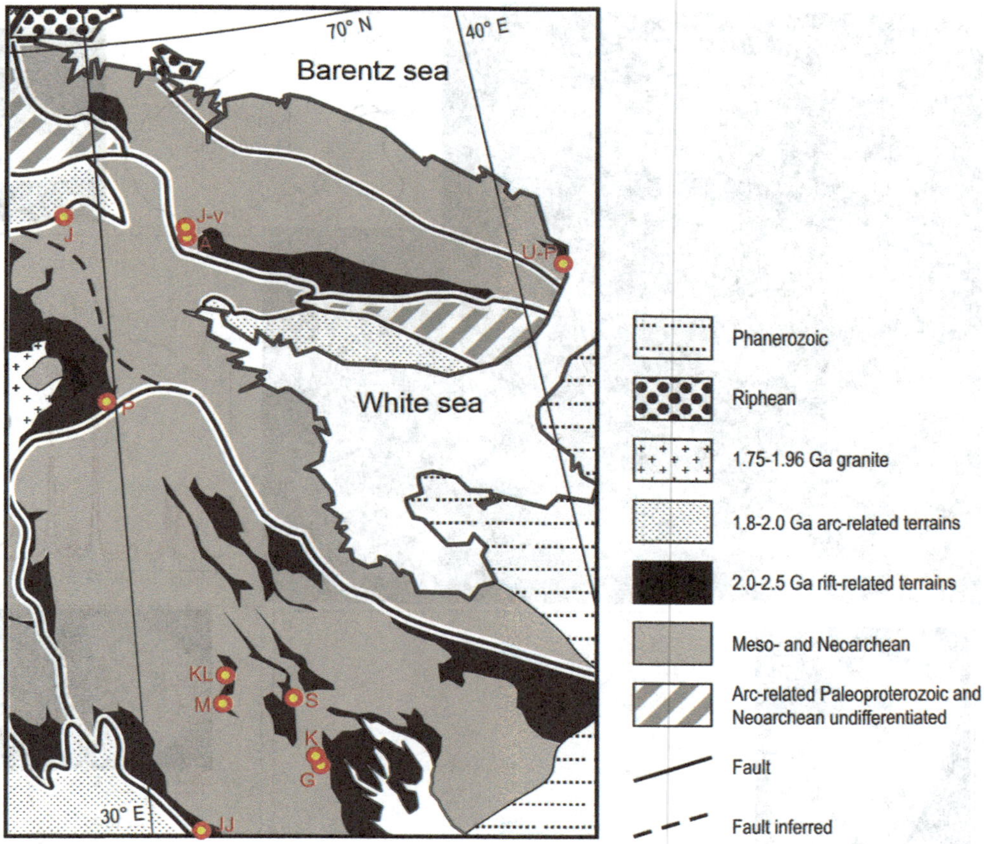

Fig. 1 Map of the Kola-Karelian region with locations of possible impact structures (A—Arvarench, J-v—Järva-varaka, J—Javrozero, JJ—Janisjärvi, G—Girvas, Kl—Keljärvi, K—Kumsa, M—Motko, P—Panajärvi, S—Suavjärvi, U-P—Ust-Ponoi). Terrains are shown after Daly et al. (2006)

described in details in (Maschak and Naumov 2012). The structure is characterized by gravity and magnetic lows. In the northeastern and southwestern parts of the structure, monument and polymict megabreccia were described. These comprise blocks of both basement granitoids and supracrustal greenstone rocks.

Shock metamorphic features in Suavjarvi are rare. The impact origin of polymict megabreccia and therefore of the Suavjärvi structure is confirmed by observations of deformation bands and mosaic structure in quartz, and deformation features in feldspars and biotite. The distinctive planar microstructures in quartz are now generally accepted as unique criteria for shock waves and meteorite impact events (Grieve et al. 1996; French and Koeberl 2010).

Massive sulfides occur within megabreccia; originating probably from postimpact redeposition of pre-existing mineralization due to hydrothermal activity (Naumov 2005, 2010). Sulfide lodes are confined to the contact zone between polymict greenstone-dominated megabreccia and monomict granitic megabreccia.

Suavjärvi underwent regional metamorphism that resulted in obliteration or transformation of shock metamorphic effects. At present time we got the first Rb-Sr data for minerals from microcline granite with "griess" structure sampled from the polymict breccia outcrop in the Suavjärvy structure (Mashchak and Naumov 2012). A whole rock-biotite isochrone defined an age of 2329 ± 23 Ga, confirming the Proterozoic age of an astrobleme. Rb-Sr data for whole rock+albite+calcite defines younger ages of 1790–1700 Ga, reflecting last low temperature metamorphic transformation of granites and cement matter in the megabreccias to epidote-actinolite, epidote-chlorite-albite, and other schists and possibly, for recrystallization of impact glass to quartz-albite-biotite-epidote aggregates.

5 Javrozero Circle Structure—A Possible Astrobleme

The Javrozero circle structure was found in the Javrozero part of the Tanaelv belt (Fig. 1). It is distinctly manifested in the relief (Fig. 2a) and in aerial photographs. Graphite ores were found within the structure, represented in zones of graphite-bearing metasomatites in the Javrozero anorthosite 2.10–1.95 Ga complex. It is believed that the formation of graphite is associated with a high-carbon fluid impulse at the boundary of about 1970 Ma (Astaf'ev et al. 2006). This boundary correlates with the time of the Proterozoic Kola plume according to the isotopic composition of Os in the ferropicrites of the Pechenga structure (Walker et al. 1997),

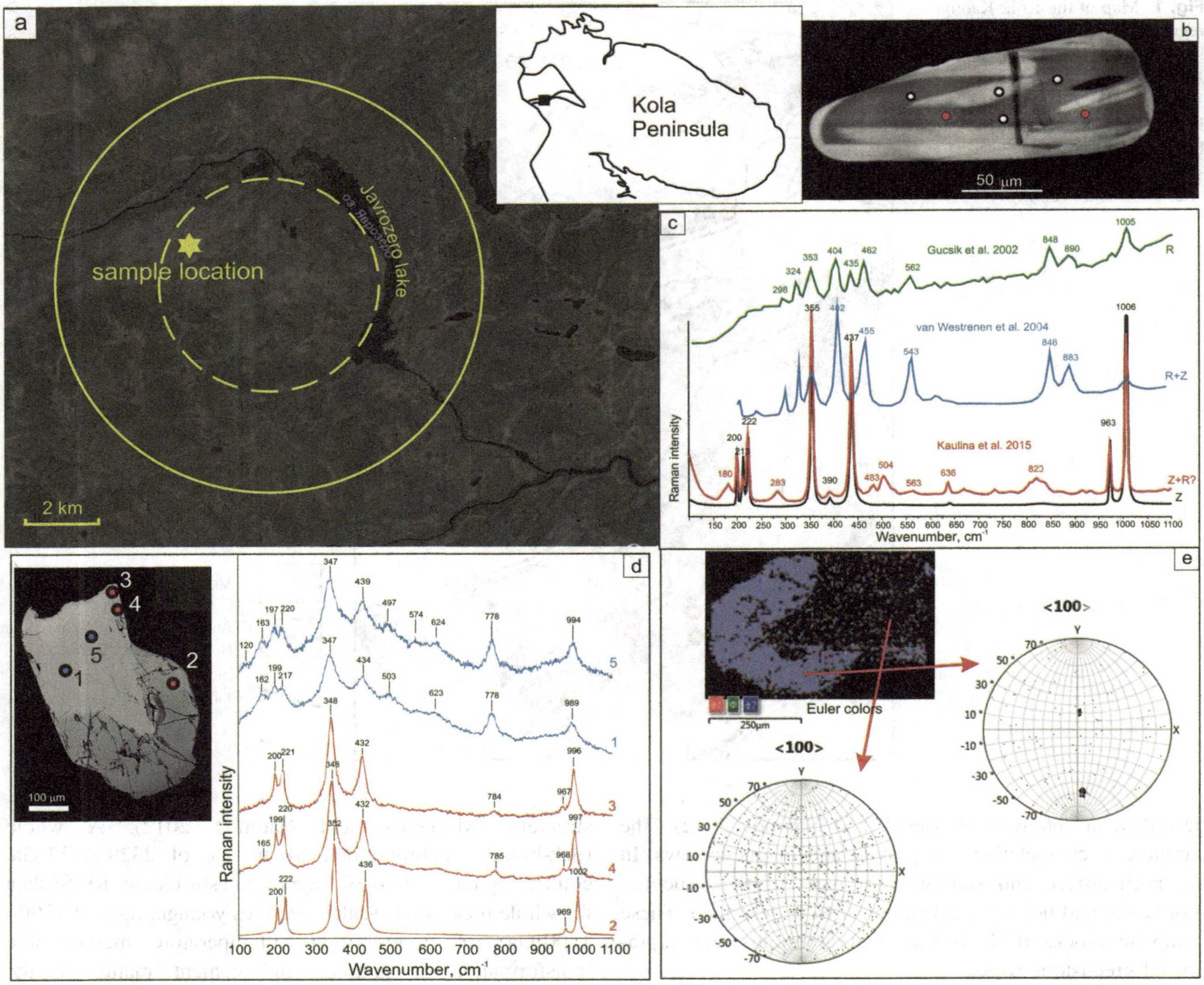

Fig. 2 **a** Circle structure in Javrozero area: satellite image with probable boundaries of the impact crater and central uplift; **b** CL image for zircon grain from gabbro-anorthosite with points of Raman analyzes; **c** Raman spectra for zircon domains—red and black lines—our data, blue and green—examples of reidite spectra from Gucsik et al. (2002) and van Westrenen et al. (2004); **d** BSE image of another zircon grain with from gabbro-anorthosite with points of Raman analyzes in core and rim parts; **e** EBSD map and polar figures for the same zircon grain showing complete distortion of crystal structure of zircon core

which, in turn, could be associated with intense meteorite bombardment at this time.

We studied zircon from gabbro-anorthosite of the Javrozero massif by means of Raman spectroscopy as an express tool to study zircon on a micro scale. Zircon grains were separated from rock samples, embedded in epoxy mounts, ground to about half of thickness and polished. The back-scattered electron (BSE) and cathodoluminescence (CL) images, as well as the qualitative composition of inclusions, were studied using a LEO-1450 Scanning Electron Microscope equipped with an XFlash-5010 Bruker Nano GmbH energy dispersive spectrometer at the Geological Institute, Kola Science Center of RAS in Apatity, Russia.

Raman spectroscopy of zircon was conducted in the St. Petersburg Resource "Geomodel" centre (Russia) at Raman spectrometer HORIBA Jobin-Yvon LabRAM HR 800, Ar-laser with wavelength of 514.5 and 488 nm, exposure time 3c, the number of repetitions—5, the laser power of 50 mW.

To determine the deformed and undeformed domains in zircon crystals method of the electron back-scattering diffraction (EBSD) mapping was used. The EBSD mapping reveals distorted domains invisible in CL and BSE. EBSD mapping also was carried out in the "Geomodel" centre. Zircon mounts were additionally polished with an ending stage of Ar ion etching for 10 min (Oxford IonFab 300) in order to obtain the flattest surface required for EBSD analysis. EBSD measurements were made using Hitachi S-3400 N scanning electron microscope equipped with Oxford HKLNordlys EBSD detector. Operating conditions were the following: accelerating voltage—20 kV, beam current—1 nA, acquisition time—16 s, and a 2 µm step during EBSP mapping. The used Oxford Instruments AZtecHKL analysis software identified the crystal orientation from the Kikuchi pattern using the crystal structures of zircon from ICSD database.

Zircon in anorthosites is represented by short prismatic pinkish transparent zircon crystals (100–300 µm), sometimes fractured, faceted with prisms {100}, {110} and dipyramids {111}, {311}. Some grains have metamorphic rims. The studied zircons showed wavy extinction and diffuse "aurora-like" cathodoluminescence (Fig. 2b), which is considered characteristic of high-pressure zircon (Corfu et al. 2003).

Raman spectrum of crystalline zircon consists of several bands: 202, 214, 225, 357, 439, 974, and 1008 cm^{-1} with the most intense bands at 1008, 439, and 357 cm^{-1}, which correspond to the internal vibration of SiO_4 tetrahedrons, while the three bands at 202, 214, and 225 cm^{-1} are external modes, corresponding to vibrations involving movements of SiO_4 tetrahedrons and Zr ions (Dawson et al. 1971; Kolesov et al. 2001).

In most of the crystals examined, these lines are manifested, and the obtained spectrum (Fig. 2c) is an example of typical crystalline zircon Raman spectrum with sharp and well resolved Raman modes. At two points the spectrum has additional Raman bands (180, 283, 390, 483, 504, 563, 636, 820 cm^{-1}) which is rather difficult to interpret. Potentially the signals at 390, 483, and 563 cm^{-1} can correspond to 402–404, 455–462, and 543–562 cm^{-1} reidite bands (Gucsik et al. 2002; VanWestrenen et al. 2004) but with certain distortion (Fig. 2c). We do not obtain reidite signals at 846–848 and 883–890 cm^{-1} but it can be masked by wide epoxy band at 820 cm^{-1}. The high crystallinity of the enclosing zircon may contribute to the reidite formation, since it is known that "moderate metamictization can inhibit reidite formation, thereby highlighting that the transformation is controlled by zircon crystallinity" (Erickson et al. 2017). In any case the results obtained demands much more detailed further study.

Another zircon grain, on the contrary, shows the disturbed structure with reduced intensity of Raman peaks, which are shifted toward shorter wavelengths (Fig. 2d). Imaging of the zircon by BSE reveals a bright core with darker fractured rim. The appearance of additional bands in the core at about 162–163, 497–503, 623–624, 778–785 cm^{-1} may be due to the appearance of enriched Zr clusters or submicron noncrystalline Zr-O phases (Titorenkova et al. 2006). The peak at 778 cm^{-1} is close to the most intense peak of silicates with quasi-isolated SiO_4 tetrahedra, i.e., this signal can arise from Si-enriched phases, which, together with ZrOn clusters, are formed when the zircon crystal lattice is strongly disturbed (Titorenkova et al. 2006).

The EBSD data for this zircon crystal also show a strong disturbance of the central (lighter in BSE) part of the crystal (Fig. 2e). Since the total content of uranium and thorium in the studied zircon does not exceed 600 ppm, its structure cannot be disturbed as a result of radioactive decay, therefore, a change in the structure of zircon can be associated with shock metamorphism.

An age of the impact event is limited by 1.95 Ga age of the central disturbed part of zircon and 1.92 Ga age of the rim, which was not subjected to shock metamorphism. The presence of a circle structure in the Javrozero area and possibly features of impact metamorphism in zircon can serve as a reason for further study and confirmation of an impact event.

6 Järva-Varaka Structure (Layered Massif) of the Monchegorsk Ore District—A Probable Astrobleme

Layered intrusions of the Baltic Shield have long been studied as a source of Pt-Pd and Cu-Ni mineralization and have been interpreted to be related to mantle plume activity (Amelin and Semenov 1996; Puchtel et al. 1997; Hansky et al. 2001). Recent study of Os isotope composition of the Monchepluton intrusion has shown a mantle plume source with subchondritic Re/Os and Os isotope composition (Yang et al. 2016).

Järva-varaka layered massif is located in the Monchegorks ore district of the Kola Peninsula (the eastern part of the Baltic Shield) and represents an intrusive body of irregular shape 1.7 × 2.2 km in size with thickness up to 2 km among gneisses of the Kola series (Fig. 3).

Differentiation of the Järva-varaka massif has resulted in a sequence of rocks including, norites, hypersthene diorites, pigeonite-augite diorites, quartz diorites, and granodiorites. The age of quartz diorites is 2496 ± 9 Ma (Layered intrusions 2004), and the age of mafic norites is 2550 ± 9 Ma (Nerovich et al. 2019).

All rock varieties are characterized by massive structure and good preservation of magmatic textures with sections of micropegmatite or granophyre textures. Granophyre is often observed in layered complexes, where it is interpreted as a derivate of residual magma crystallization, or the result of the roof acid rocks melting under the thermal influence of the mafic magma of intrusions (Naldrett 2004; Layered intrusions 2004). Major and incompatible elements show that the most probable process, contributing to the formation of the Järva-varaka initial melts is an assimilation of crustal

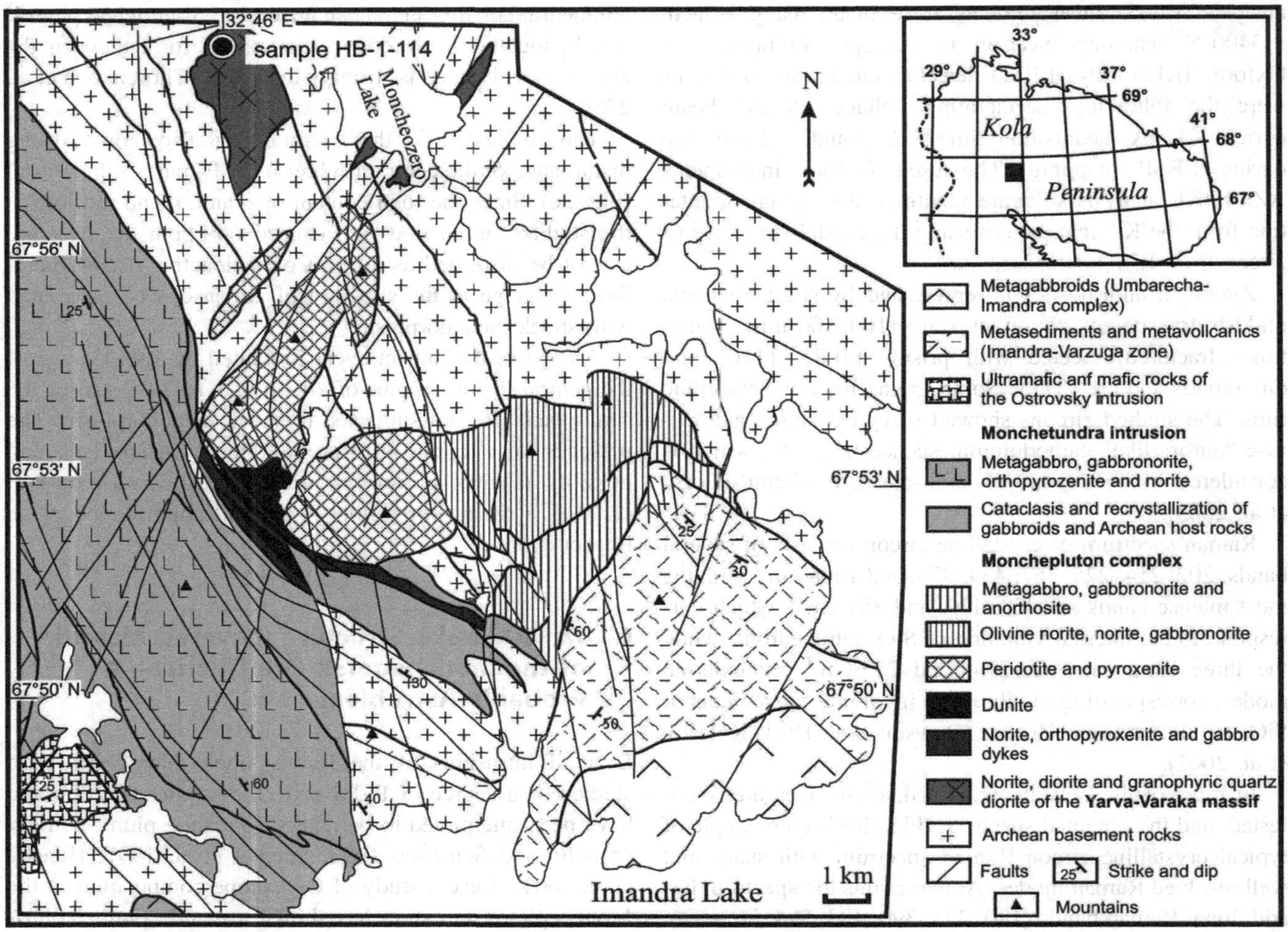

Fig. 3 Location of the Järva-varaka massif in the Monchegorsk ore region (geological map after Smolkin et al. 2004)

material by magnesia mantle magma. The assimilation nature of initial melt is confirmed by the high content of trace elements, the prevalence of LILE (K, Ba, Rb, LREE) relative to heavy REE, and Nb–Ta minima comparable with the continental crust (Nerovich et al. 2015).

According to the significant proportion of granophyres, petrochemical, and geochemical characteristics of rocks, greater role of crustal contamination, the Järva-varaka massif is most similar to the 1.85 Ga Sudbury structure, for which an impact origin was assumed (Grieve et al. 1994; Naldrett 2004). Moreover, in the country rocks of the Järva-varaka massif there were found pseudotachylite breccia and planar deformations in quartz. Pseudotachylite breccia consists of angular fragments of granite and gneiss composition, as well as quartz and plagioclase crystalloclasts, immersed in glassy cement. Breccia cement at high magnification reveals a cryptocrystalline structure and in chemical composition is close to that of the host alumina gneisses. Apparently, the formation of pseudotachylite breccia is associated with a shock metamorphism of gneisses during an impact event. Findings of pseudotachylite breccia in the host rocks, as well as planar quartz and the "kink banding" structure in biotite and clinozoisite in combination with similarity of petrographic-geochemical characteristics of the Järva-varaka and Sudbury massifs supports an impact origin of the Järva-varaka massif (Nerovich et al. 2020).

Negative and subchondritic εNd values ranging −11.5 to +0.4 for rocks of the Jarva-varaka massif (Nerovich et al. 2019) reflect variable degrees of crustal assimilation, as might occur when the mafic melt incorporates crustal rocks in the complex environment of the newly formed crater.

These data stimulated an investigation of zircon from the Javra-varaka massif. Here, we present the first results of spectroscopic Raman study of zircon from norites of the massif with back-scattered electron (BSE) imaging and single-crystal X-ray diffraction analysis.

6.1 Zircons Textures and Chemical Composition

Granophyric norites (sample HB-1/114—Fig. 3) are composed of plagioclase (40–65%), orthopyroxene (10–40%), granophyre (Pl+Kfs+Qtz, 10–17%) (Nerovich et al. 2015). Granophyre is a mineral aggregate consisting of mutual, often graphic, intergrowths of quartz, K-feldspar, and plagioclase, located in the interstices of early euhedral plagioclase grains and orthopyroxene.

Zircon separated from granophyre norites is represented by transparent light brown to yellow and colorless short prismatic grains and their fragments. Zircon grains contain numerous inclusions of Al, Si, O composition, indicating polymorphs: andalusite, sillimanite, and kyanite; and of Si, Al, Na, Ca, O composition, suggesting feldspar-group minerals. This mineral association is common for aluminous gneisses of the Kola series—host for the Järva-varaka massif. More probable that studied zircon grains are xenogenic and derived from aluminous gneisses.

BSE images reveal complicated internal texture—darker central domains (cores) with fine zoning and light rims without texture (Fig. 4). Some cores are obviously xenogenic with clearly seen boundaries (Fig. 4c), others have diffusive transition to rims. Few grains show the later alteration process– dark domains with curved boundaries along fractures and around inclusions, for example, grain 13 (Fig. 4d).

6.2 Raman Spectra and X-Ray Diffraction

Raman spectroscopy of zircon was conducted in the St. Petersburg Resource "Geomodel" as described earlier for Javrozero zircon. Ten grains of zircon were studied by means of Raman spectroscopy; 7 of them are represented by brown crystals because of their better preservation, yellow and transparent grains are represented by fragments only.

The central domains of the Järva-varaka zircon exhibit Raman spectra with typical zircon Raman modes, though with broader bands, with decreased total Raman intensity and bands shifted toward smaller wavenumbers (Fig. 4), corresponding to zircon with damaged structure (Presser and Glotzbach 2009). Xenogenic cores show less distorted Raman spectrum: points 22, 23 (Fig. 4 c). Other cores display distorted Raman spectra with strong decrease of peaks intensity, while rims demonstrate total absence of spectral bands. Amid an absence of zircon bands epoxy peaks often appear near 638 and 821 cm^{-1}.

Raman spectra of plagioclase and sillimanite inclusions have total absence of peaks also (Fig. 4a), though characteristic Raman bands of these minerals, as for all silicates, lie in the same range of wavenumbers as the Raman spectrum of zircon (Freeman et al. 2008; Mernagh and Liu 1991).

Some zircon grains, for example, grain 15 exhibit additional bands at 161–166, 503–506, and 780–782 cm^{-1} (Fig. 4e). The occurrence of extra Raman bands near 162, 509, 635 cm^{-1} is attributed to Zr-rich clusters of linked ZrO_n polyhedra or to the presence of a submicrometric noncrystalline Zr-O phase. In our case the peak near 635 cm^{-1} can be masked by the close 637 cm^{-1} epoxy peak. The position of the extra Raman band at 785 cm^{-1} is considered to arise from Si-rich phases that are formed together with ZrO_n clusters as a result of the strong violation of zircon crystal lattice (Titorenkova et al. 2006).

X-rays diffraction method was used for one zircon grain only due to small size of cores and rims. X-Ray diffractograms were taken with a Bruker D8 Discover X-Ray diffractometer (St. Petersburg Resource "X-rays diffraction" centre) using a filtered point-focus parallel-beam CuKa radiation. The orientation of the crystal with respect to the sample surface normal was determined as close to [210]. The beam was collimated to 100 mkm and focused on different parts of the crystal. The other parts have been masked with a Pb foil. Figure 5 compares the profiles of the 420 reflections taken from the area 1 (core) and area 2 (rim) of the brown crystal No 6 shown in Fig. 4b. It is seen that the intensity of the peak is considerably different in the core and rim domains of the crystal: It is strong in the core (Area 1) and weak in the rim (Area 2). This result confirms partial amorphization of the crystal rim.

Analysis of Raman spectra has shown that zircon cores are mainly represented by crystalline domains with moderate to strong degree of metamictization while the total absence of peaks in the spectrum of rims can indicate amorphization of the media. The impact of metamictization on the position and width of Raman bands for zircon is similar to the impact of shock pressure in the same material (Gucsik et al. 2004). The amorphization of zircon was obtained in experiments at 60 GPa (Gucsik et al. 2002). Due to radioactive decay of U and to lesser extend of Th, the structure of zircon can be heavily damaged by α-decay and as a result, natural zircon is commonly found to be metamict. Moderate U concentration (<600 ppm) in studied zircon prevents this conclusion since metamictization is commonly characteristic for zircons with U concentrations above 1000 ppm (Nasdala et al. 1995). At the same time metamict zircon shows the same band pattern as crystalline zircon, with bands lower in intensity, significantly broader and shifted toward lower wavenumbers (Nasdala et al. 1995, 2003). Results of Zhang et al. (2000) show that radiation-damaged zircon gives two Raman signals: one from distorted crystalline zircon and the other from amorphized zircon. Therefore diaplectic $ZrSiO_4$ glass due to shock metamorphism is more probably in our case than heavily damaged zircon due to radioactive decay. Moreover mineral inclusions of Pl and Sill no Raman spectra too, and they cannot be metamict. Plagioclase is a sensitive indicator

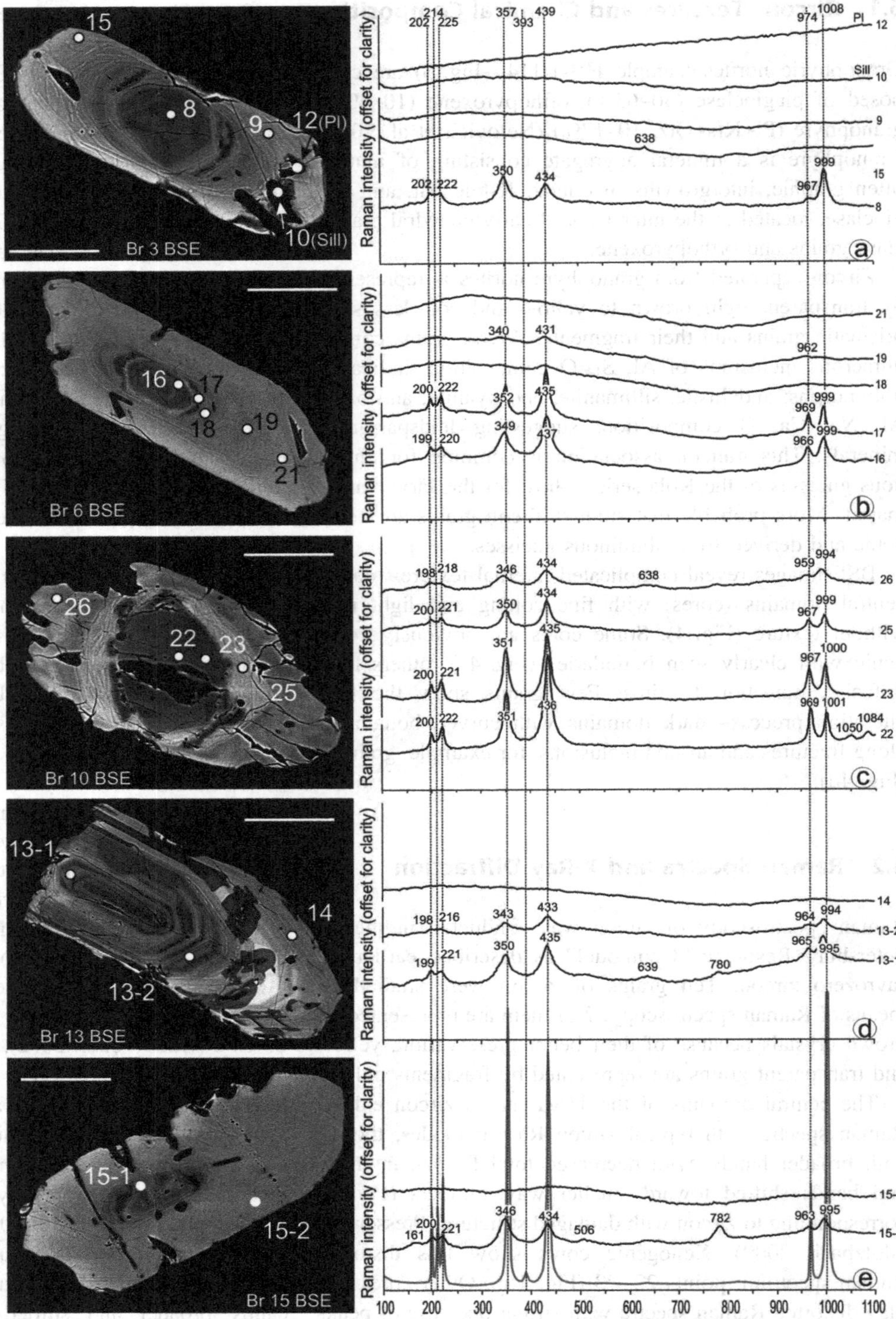

Fig. 4 BSE images of brown zircon grains separated from granophyre norites of the Yarva-varaka massif (scale bars —50 μm) with points of RAMAN analysis and corresponding RAMAN spectrum for each point. Abbreviations at BSE images, for example "Br 3 BSE" means brown grain, number 3, BSE image. Gray line at the bottom of RAMAN spectra shows spectrum of crystalline zircon

of shock conditions and at shock pressures above ~29 GPa, plagioclase transforms to a diaplectic glass, which has been known as maskelynite (Stöffler et al. 1986).

The fact of zircon crystal structure more distorted further from the centre supports an impact effect when crystal inner parts are screened from the shock influence. A large number of zircon grains captured from aluminous gneisses confirm a strong melting of the target rocks. Measurements of the experimentally shocked zircon show that zircon can transform to a diaplectic glass at temperature of 1300–1500 °C and pressure of 60 GPa (Gucsik et al. 2002; Wittmann et al. 2006).

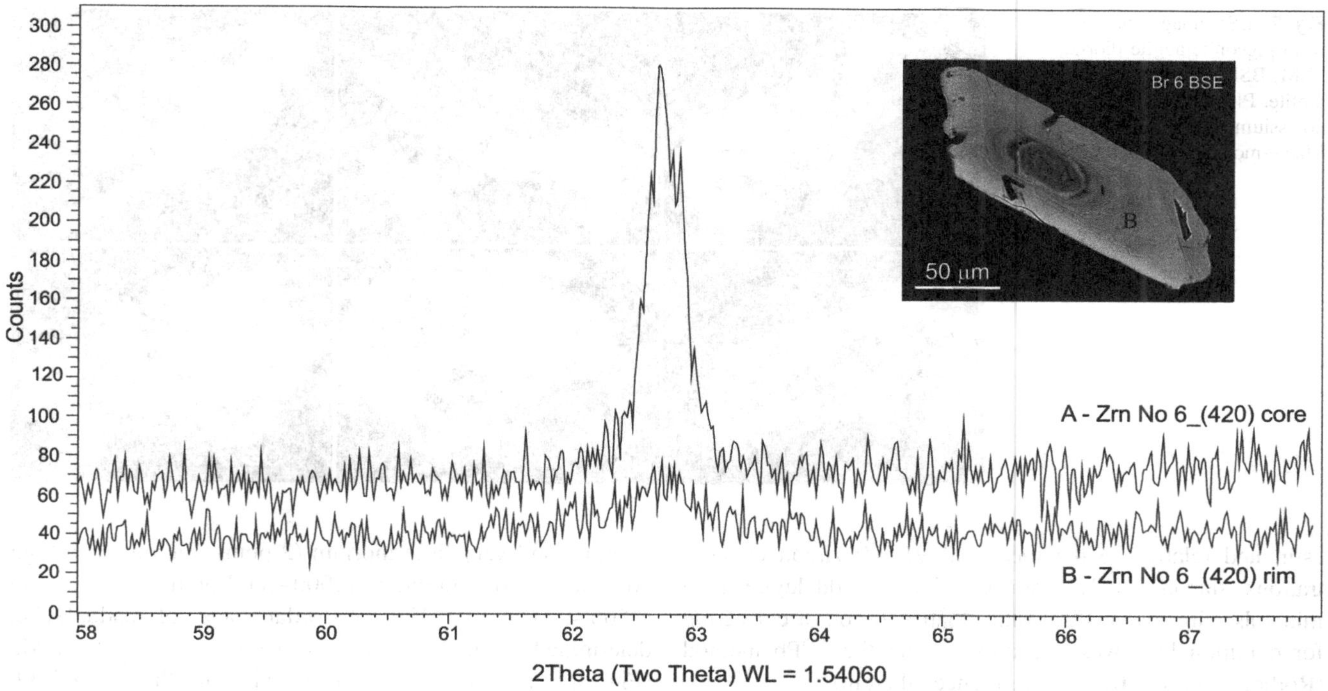

Fig. 5 X-Ray diffractograms for zoning core and textureless rim of the crystal 6. The intensity of the peak is strong in the core and weak in the rim, reflecting poor crystallinity of zircon structure in rims

6.3 U–Pb Age of Baddeleyite and Zircon

During field work, samples of the Järva-varaka massif rocks (felsic norites, pigeonite-augite diorites and granodiorites), as well as host biotite gneisses were collected. Zircon for isotopic dating was extracted from all samples. The biotite gneiss sample, apart from zircon provided monazite separate. Baddeleyite (Fig. 6) was only found in a sample of pigeonite-augite diorites; therefore, dating was started with this sample.

The dating of zircon and baddeleyite was carried using a SHRIMP-II mass spectrometer at the Center for Isotope Research of the Karpinsky All-Russian Geological Institute (St. Petersburg). The procedure of SHRIMP-II U-Pb analysis of baddeleyite is described in detail in (Rodionov et al. 2012). Baddeleyite and zircon grains were implanted into the epoxy resin along with Phalaborwa baddeleyite and 91,500 zircon standard grains. Since currently there is no generally accepted concentration standard for baddeleyite, the ^{238}U and ^{232}Th concentrations in the baddeleyite grains were

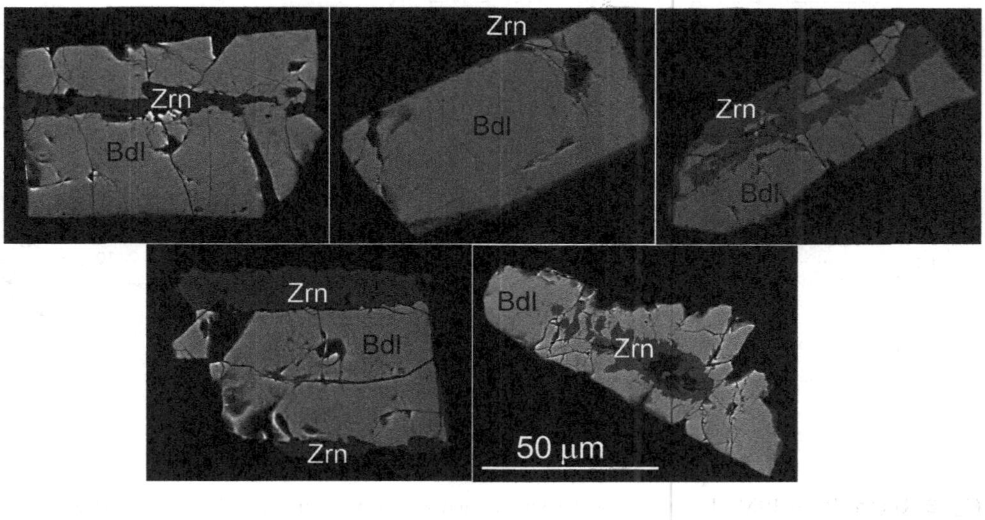

Fig. 6 Images of baddeleyite (Bdl) crystals with zircon (Zrn) inclusions and rims from pigeonite-augite diorite. Photographs are in BSE image

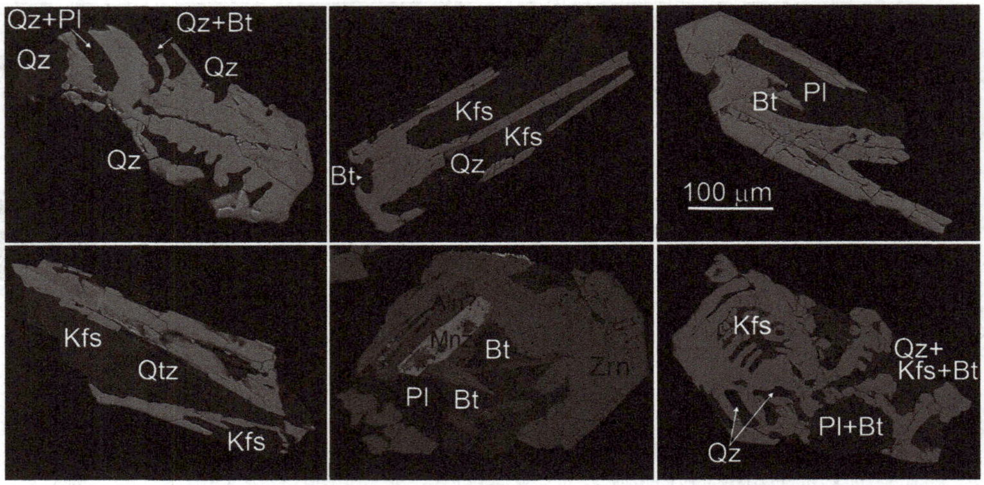

Fig. 7 BSE images of zircon from pigeonite-augite diorite. SEM, BSE. Qz—quartz, Bt—biotite, Pl—plagioclase, Kfs—potassium feldspar, Zrn—zircon, Mnz—monazite, Aln—allanite

estimated relative to the measured 91,500 zircon concentrations standard (81.2 ppm U). Since baddeleyite is a minerals with lower Th/U ratio, than a zircon, a correction for common lead was introduced using the ^{208}Pb method (Rodionov et al. 2012 and references therein).

Baddeleyite from the sample of pigeonite-augite diorites is represented by small (60–100 μm) dark brown grains. In all grains, zircon overgrowths, ingrowths, or rims along the border with biotite and other minerals were found (Fig. 6). Zircon of the pigeonite-augite diorite sample is represented by long prismatic crystals with inclusions or by skeletal crystals in intergrowths with rock-forming minerals: quartz, feldspars, and biotite (Fig. 7). Such a composition of inclusions and intergrowths suggests the simultaneous formation of zircon crystals with granophyre (Pl+Kfs+Qtz). Skeletal growth indicates rapid crystallization of the melt.

SHRIMP dating of baddeleyite and zircon showed concordant ages (Fig. 8). Baddeleyite contains up to 200 ppm of uranium and very little thorium (2 ppm). Zircon has a high content of both uranium (1000–1800 ppm) and thorium (2500–4100 ppm). The concordant age of baddeleyite, determined from 15 analytical points, is 2502 ± 11 Ma (Fig. 8a), the average age calculated using the ^{207}Pb/^{206}Pb ratio is 2498 ± 6 Ma (Fig. 7b). Since the probability of Concordance is zero (Fig. 8a), it is more reliable to calculate the average age from the ^{207}Pb/^{206}Pb ratio. Thus, the value of 2498 ± 6 Ma was assumed as baddeleyite crystallization age. The upper intercept zircon age, obtained by 10 nearly concordant data points, is 2492 ± 3 Ma (Fig. 8b) and coincides with the average age obtained using the ratio of ^{207}Pb/^{206}Pb.

The baddeleyite and zircon ages overlap within the analytical error, but given that biotite was formed after pyroxene, which together with plagioclase No. 47 forms the early paragenesis of the rock, while the association of quartz, potassium feldspar, and oligoclase-andesine No. 30 forms a

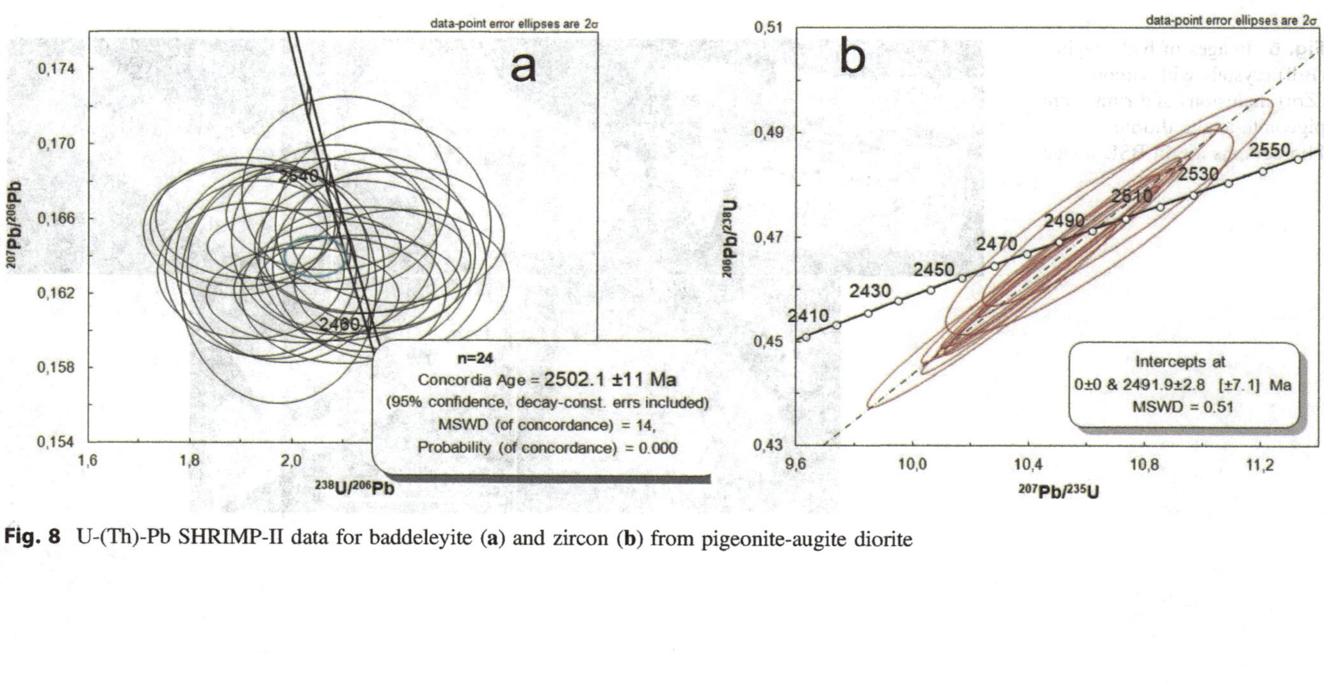

Fig. 8 U-(Th)-Pb SHRIMP-II data for baddeleyite (**a**) and zircon (**b**) from pigeonite-augite diorite

later granophyre paragenesis, an earlier crystallization of baddeleyite compared to zircon in a narrow age range could be assumed.

The obtained age of 2498 ± 6 Ma of pigeonite-augite diorites of the Järva-varaka massif completely coincides with the age of quartz diorites of the massif (2496 ± 9 Ma—Layered intrusions 2004). The crystallization of granophyre occurred at the last stage of the massif solidification, 2492 ± 3 Ma ago. The data obtained, along with geological evidence of the impact origin of the Järva-varaka massif, suggest that today, the Järva-varaka astrobleme is the oldest known on the Baltic shield.

7 Concluding Remarks

We considered three Proterozoic astroblems in the Karelian-Kola region, two of which were newly identified. The difference between the Javrozero and Suavjärvi astroblems and the Järva-varaka massif is that the first two structures show no indication of large volumes of melt formed by impact, while in the Järva-varaka structure both mantle magmatism and significant melting of the target rocks seem to have taken place. The sizes of all three astroblems are 11, 16 and 2 km in diameter, respectively, which means that most powerful melting took place in the smallest structure. Comparison of the Järva-varaka and the Sudbury massifs reveals close resemblance of geological and petrochemical features, which makes the difference in the size of these massifs even more striking. We suggested that the Järva--varaka massif is a marginal part of the impact structure, which is possibly the entire Monchegorsk ore area. This assumption is confirmed by the location of the Järva-varaka massif at the edge of the ring structure with a diameter of about 40 km, which is manifested in the gneisses and amphibolites of the Kola complex. Perhaps it should be worthwhile to look for traces of shock exposure in the gneisses framing the Monchegorsk ore area.

In the Javrozero and Suavjärvi massifs, there was no impact-related magmatism, however there was hydrothermal activity associated with shock metamorphism, which resulted in the formation of graphite and the concentration of sulfide ores, that is, in these structures impact event created favorable structural conditions for subsequent localization of ore occurrences.

Thus complex investigations (petrographic, mineralogical, geochemical, and isotope-geochemical studies) could help to diagnose new astroblems in the Baltic shield and to work out the role of impact event in the mantle plume triggering processes. Since impact events can create both syngenetic deposits of mineral raw materials and favorable structural and lithological conditions for localization of deposits in subsequent stages, it is logical to assume that the analysis of structural heterogeneities of the Earth's crust caused by large-scale impact events is relevant in order to identify certain factors controlling the localization of ore occurrences.

Acknowledgements Authors are grateful to Bocharov V. and Shilovsky V. for Raman and EBSD study of zircon, to for Rodionov N. and L'vov P. for SHRIMP-II zircon and baddeleyite dating, to Naumov M., Vinogradova N., Konstantinova L., and Sosnovskaya M. for their help during field trip and to the two anonymous reviewers for all the constructive comments they provided to improve this manuscript. The work was performed in the framework of the State Contract No. 0226-2019-0052 of GI KSC RAS and conducted using an equipment of the "Geomodel' and "X-rays diffraction methods" resource centers of the St. Petersburg State University Science Park.

References

Amelin YV, Semenov VS (1996) Nd and Sr isotopic geochemistry of mafic layered intrusions in the eastern Baltic Shield: implications for the evolution of Paleoproterozoic continental mafic magmas. Contrib Mineral Petrol 124:255–272

Astaf'ev BYu, Voinova OA, Lokhov KI, et al (2006) Age and genesis of early precambrian graphite mineralization of the lapland Belt (Kola Peninsula). Russ Geol 4:75–82 (in Russian)

Cavosie AJ, Erickson TM, Timms NE (2015) Nanoscale records of ancient shock deformation: Reidite (ZrSiO4) in sandstone at the Ordovician Rock Elm impact crater. Geology 43(4):315–318. https://doi.org/10.1130/G36489.1

Corfu F, Hanchar JM, Hoskin PWO et al (2003) Atlas of zircon textures. In: Hanchar JM, Hoskin PWO (eds) Zircon. Rev Mineral Geochem 53:469–500

Daly JS, Balagansky VV, Timmerman MJ (2006) European lithosphere dynamics. Geol Soc Lond Memoir 32:579–598

Davis DW (2008) Sub-million year age resolution of Precambrian igneousevents by thermal extraction (TE-TIMS) Pb dating of zircon: application to crystallization of the Sudbury impact melt sheet. Geology 36:383–386

Dawson P, Hargreave MM, Wilkinson GR (1971) The vibrational spectrum of zircon ($ZrSiO_4$). J Phys C Solid St Phys 4:240–256

Earth Impact Database (2012) Earth Impact Database. https://www.passc/. Accessed 30 Mar 2012

Erickson TM, Pearce MA, Reddy SM et al (2017) Microstructural constraints on the mechanisms of the transformation to reidite in naturally shocked zircon. Contrib Mineral Petrol 172(6). https://doi.org/10.1007/s00410-016-1322-0

Fel'dman VI, Granovsky LB, Korotaeva NN et al (1981) Meteorite crater Janis-järvi. In: Impactites. Moscow

Freeman JJ, Wang A, Kuebler KE et al (2008) Characterization of natural feldspars by Raman spectroscopy for future planetary exploration. Can Mineral 46:1477–1500. https://doi.org/10.3749/canmin.46.6.1477

French BM (2004) The importance of being cratered: the new role of meteorite impact as a normal geological process. Meteorit Planet Sci 39:169–197

French BM, Koeberl C (2010) The convincing identification of terrestrial meteorite impact structures: What works, what doesn't, and why. Earth Sci Rev 98:123–170

Garde AA, McDonald I, Dyck B, Keulen N (2012) Searching for giant, ancient impact structures on Earth: the Mesoarchaean Maniitsoq structure, West Greenland. Earth Planet Sci Lett 337–338:197–210

Gibson RL, Reimold WU (1999) Significance of the Vredefort Dome for metamorphic-mineralization studies in the Witwatersrand Basin. Mineral Petrol 66:25–53

Gibson RL, Armstrong RA, Reimold WU (1997) The age and thermal evolution of the Vredefort impact structure: a single-grain U-Pb zircon study. Geochim Cosmochim Acta 61:1531–1540

Glikson AY (2013) The asteroid impact connection of planetary evolution. Springer-Briefs, Dordrecht

Glikson AY, Vickers J (2007) Asteroid mega-impacts and Precambrian banded iron formations: 2.63 Ga and 2.56 Ga impact ejecta/fallout at the base of BIF/argillite units, Hamersley Basin, Pilbara Craton Western Australia. Earth Planet Sci Lett 254:214–226

Glikson AY, Vickers J (2010) Asteroid impact connections of crustal evolution. Aust J Earth Sci 57:79–95

Glikson AY (2014) The archaean: geological and geochemical windows into the early earth. Springer

Glukhovsky MZ, Kuzmin MI (2015) Extraterrestrial factors and their role in the tectonic evolution of the earth in early Precambrian. Tecton Geodyn 56:1225–1249

Grieve RAF, Masaitis VL (1994) The economic potential of terrestrial impact craters. Int Geol Rev 36:105–151

Grieve RAF (1994) An impact model for the Sudbury structure. In: Lightfoot PC, Naldrett AJ (eds) Proceedings of the Sudbury-Noril'sk Symposium: Ontario Geological Survey Special, vol 5, pp 119–132

Gucsik A, Koeberl C, Brandstätter F et al (2002) Cathodoluminescence, electron microscopy, and Raman spectroscopy of experimentally shock-metamorphosed zircon. Earth Planet Sci Lett 202:495–509

Gucsik A, Zhang M, Koeberl C et al (2004) Infrared and Raman spectra of $ZrSiO_4$ experimentally shocked at high pressures. Mineral Mag 68:801–811

Hanski E, Walker RJ, Huhma H et al (2001) The Os and Nd isotopic systematics of the c. 2.44 Ga Akanvaara and Koitelainen mafic layered intrusions in northern Finland. Precam Res 109:73–102

Jones AP (2005) Meteorite impacts as triggers to large igneous provinces. Elements 1:277–281

Kaulina TV, Il'chenko VL, Lyalina LM et al (2015) Yavrozerskaya koltsevaya struktura—vosmozhnaya astroblema (The Yavrozersky circle structure—a possible astrobleme). Proc Fersman Sci Session GI KSC RAS 12:214–217 (in Russian)

Kolesov BA, Geiger CA, Armbruster T (2001) The dynamic properties of zircon studied by single-crystal X-ray diffraction and Raman spectroscopy. Eur J Mineral 13:939–948

Lightfoot PC, Keys RR, Doherty W (2001) Chemical evolution and origin of nickel sulfide mineralization in the Sudbury igneous complex, Ontario, Canada. Econ Geol 96:1855–1875

Lowe DR, Byerly GR, Kyte FT (2014) Recently discovered 3.42-3.23 Ga impact layers, Barberton Belt, South Africa: 3.8 Ga detrital zircons, Archean impact history, and tectonic implications. Geology 42:747–750

Masaitis VL, Danilin AN, Mashchak MC (1980) Geologiya astroblem (Geology of astroblems). Nedra, Leningrad (in Russian)

Masaitis VL, Mashchak MS, Raikhlin AI et al (2018) Popigai impact structure and its diamond-bearing rocks. Springer

Masaitis VL, Shadenkov EM (1980) Janis-jarvi astrobleme. In: Geology of astroblems, p 27–32. Leningrad (in Russian)

Mashchak MS, Naumov MV (2012) The Suavjärvi impact structure NW Russia. Meteorit Planet Sci 47(10):1644–1658

McDonald FA, Bunring JA, Cina CE (2003) Yarrabubba—a large, deeply eroded impact structure in the Yilgarn Craton, Western Australia. Earth Planet Sci Lett 213:235–247

Melosh HJ (1989) Impact cratering: a geological process. Oxford University Press, New York

Mernagh TP, Liu L (1991) Raman spectra from the Al_2SiO_5 polymorphs at high pressure and room temperature. Phys Chem Miner 18:126–130

Mitrofanov FP, Smolkin VF (eds) (2004) Layered intrusions of the Monchegorsk ore region: petrology, mineralization, isotopy, deep structure. KSC RAS, Apatity

Naldrett AJ (2004) Magmatic sulfide deposits: geology, geochemistry, and exploration. Springer

Nasdala L, Irmer G, Wolf D (1995) The degree of metamictization in zircon: a Raman spectroscopic study. Eur J Mineral 7:471–478

Nasdala L, Zhang M, Kempe U et al (2003) Spectroscopic methods applied to zircon. In: Hanchar JM, Hoskin PWO (eds) Zircon. Rev Mineral Geochem 53:427–467

Naumov MV (2002) Impact-generated hydrothermal systems: data from Popigai, Kara, and Puchezh-Katunki impact structures. In: Plado J, Pesonen LJ (eds) Impacts in precambrian shields. Springer-Verlag, Berlin Heidelberg New York, pp 117–171

Naumov MV (2005) Principle features of impact-generated hydrothermal circulation systems: mineralogical and geochemical evidence. Geofluids 3:165–184

Naumov MV (2010) Impact-generated sulfide ore from the Suavjärvi impact structure, Russian Karelia (abstract # 1117). In: 41st Lunarand Planetary Science Conference, CD-ROM

Nerovich LI, Bayanova TB, Kunakkuzin EL et al (2015) Novye rezultaty geologo-petrograficheskogo I geochimicheskigo izucheniya rassloennogo massiva Yarva-varaka, Monchegorsky rudnyi raion (New results of geological-petrographic and geochemical study of the layered massif of Järva-varaka, Monchegorsk ore area). Proc Fersman Sci Session GI KSC RAS 12:141–146 (in Russian)

Nerovich LI, Bayanova TB, Kunakkuzin EL et al (2019) O problemah genezica massiva Yarva-varaka, Monchegorsky rudnyi raion (On the Genesis Problem of the Järva-varaka Massif, Monchegorsk Ore Area). Proc Fersman Sci Session GI KSC RAS 16:413–417. https://doi.org/10.31241/FNS.2019.16.083 (in Russian)

Nerovich LI., Il'chenko VL, Kaulina NV et al (2020) Signs of impact origin of the Paleoproterozoic layered Järva-varaka massif, Kola region. Proc Fersman Sci Session GI KSC RAS 17:384–389 (in Russian). https://doi.org/10.31241/FNS.2020.17.073

Presser V, Glotzbach C (2009) Metamictization in zircon: Raman investigation following a Rietveld approach. Part II: sampling depth implication and experimental data. J Raman Spectrosc 40:499–508

Puchtel IS, Haase KM, Hofmann AW et al (1997) Petrology and geochemistry of crustally contaminated komatiitic basalts from the Vetreny Belt, southeastern Baltic Shield: evidence for an early Proterozoic mantle plume beneath rifted Archean continental lithosphere. Geochim Cosmochim Acta 61:1205–1222

Reimold WU, Leroux H, Gibson RL (2002) Shocked and thermally-metamorphosed zircon from the Vredefort impact structure, South Africa. Eur J Mineral 14:859–868

Reimold WU, Koeberl C, Gibson RL et al (2004) Economic mineral deposits in impact structures: a review. In: Koeberl C, Henkel H (eds) Impact tectonics. Springer-Verlag, Berlin Heidelberg New York, pp 479–552

Reimold WU, Koeberl C (2008) Catastrophes, extinctions, and evolution: 50 years of impact cratering studies. In: Golden jubilee memoir of the geological society of India, vol 66, p 69–110

Rodionov NV, Belyatsky BV, Antonov AV et al (2012) Comparative in-situ U-Th-Pb geochronology and trace element composition of baddeleyite and low-U zircon from carbonatites of the Palaeozoic Kovdor alkaline-ultramafic complex, Kola Peninsula Russia. Gondwana Res 21(4):728–744. https://doi.org/10.1016/j.gr.2011.10.005

Smolkin VF, Fedotov ZA, Neradovsky YN et al (2004) Layered intrusions of the Monchegorsk ore region: petrology,

mineralization, isotope features and deep structure. Kola Science Centre RAS, Apatity (in Russian)
Stoffler D, Langenhorst F (1994) Shock metamorphism of quartz in nature and experiment: 1 Basic observation and theory. Meteoritics 29:155–181
Stöffler D, Ostertag R, Jammes C, Pfannschmidt G (1986) Shock metamorphism and petrology of the Shergottyachondrite. Geochim Cosmochim Acta 50:889–903
Titorenkova R, Mihailova B, Konstantinov L (2006) Raman spectroscopic study of variably recrystallized metamict zircon from amphibolite-facies metagranites, Serbo-Macedonian massif, Bulgaria. Can Mineral 44:1357–1366
VanWestrenen W, Frank MR, Hanchar JM et al (2004) In situ determination of the compressibility of synthetic pure zircon (ZrSiO4) and the onset of the zircon-reidite phase transition. Am Mineral 89:197–203
Vishnevsky SA, Pal'chik NA, Moroz TN, et al (2002) Shock Metamorphism of Carbonaceous Matter in Impactites of the Janisjärvi Astrobleme Karelia. Dokl Earth Sci 387A(9):1024–1027
Vishnevsky SA, Palchik NA (2002) Carbon matter in impactites of the Janisjärvi astrobleme, Russia: high-pressure shock transformations (abs.1676). In: LPSC XXXIII, 11–15 March, 2002, Houston, USA, LPI (CD-ROM)
Walker RJ, Morgan JW, Hanski EJ et al (1997) Re-Os systematics of early ptoterozoic ferropicrites, Pechenga complex, northwestern Russia: evidence for ancient Os-enriched plumes. Geochim Cosmochim Acta 61(15):3145–3160
Wittmann A, Kenkmann T, Schmitt RT et al (2006) Shock-metamorphosed zircon in terrestrial impact craters. Meteorit Planet Sci 41(3):433–454
Yang S-H, Hanski E, Li C et al (2016) Mantle source of the 2.44–2.50-Ga mantle plume-related magmatism in the Fennoscandian Shield: evidence from Os, Nd, and Sr isotope compositions of the Monchepluton and Kemi intrusions. Miner Deposita 51:1055–1073. https://doi.org/10.1007/s00126-016-0673-9
Zhang M, Salje EKH, Farnan I, Graeme-Barber A, Daniel P, Ewing RC, Clark AM, Leroux H (2000) Metamictizationof zircon: Raman spectroscopic study. J Phys 12:1915–1925

Crustal Evolution of Bundelkhand Craton in Archean and Comparison with Other Indian Cratons

A. I. Slabunov and Vinod K. Singh

Abstract

Indian Shield consists of Bundelkhand, Aravalli, Singhbhum, Bastar, Western and Eastern Dharwar Cratons. The crustal evolutions of these cratons (except Singhbhum Craton) show that geodynamic mechanisms, similar to modern plate-tectonic and mantle-plume mechanisms, were active during Paleo-Neoarchean time. The Mesoarchean crustal evolution of the Bundelkhand Craton show subduction–accretion processes, which is different from other cratons of Indian Shield, whereas in other cratons, plume processes were more active during this period. During the Neoarchean period (2.7–2.5 Ga), all the cratons exhibit subduction–accretion processes. Each of the cratons demonstrates its own crust formation model. It gives the impression that the cratons of Indian Shield were parts of the Kenorland Supercontinent in Mesoarchean time, rather than one block.

Keywords

Archean • Earth crustal evolution • Geodynamics • Indian shield • Bundelkhand craton • Western Dharwar craton • Eastern Dharwar craton • Aravalli craton • Singhbhum craton • Bastar craton

A. I. Slabunov
Institute of Geology, Karelian Research Centre, Russian Academy of Sciences, Petrozavodsk, Russia
e-mail: slabunov@krc.karelia.ru

A. I. Slabunov
Petrozavodsk State University, Petrozavodsk, Russia

V. K. Singh (✉)
Department of Geology, Bundelkhand University, Jhansi, India
e-mail: vinodksingh@bujhansi.ac.in

1 Introduction

Archean cratons are continuously studied for understanding geodynamics of continents around the world, which were formed prior to 2500 Ma ago (de Wit 1998, 2001; Condie 2004; Brown 2007; Windley et al. 2021). It generally consists of tonalite–trondhjemite–granodiorite (TTG); greenstone belts; and K-rich granitoid rocks. The distinctive structures of Archean cratons contain sedimentary and volcanogenic rocks consider as greenstone belts. Now, it has been recognized that plate-tectonic and mantle plume-tectonic processes were operational since ca. 3.0 Ga or earlier (e.g. Cawood et al. 2006; Witze 2006; Windley et al. 2021). While some workers (Hamilton 1998; Sharkov et al. 2000; Bedard 2018) believe that they had not been active until the Paleoproterozoic (ca. 2.0 Ga).

The Indian Shield also reveals early Earth's crust formation which consists of two groups of Archean cratons: a northern (Bundelkhand and Aravalli) and a southern (Dharwar, Bastar and Singhbhum) group separated by the Central Indian tectonic zone (Fig. 1; Ramakrishnan and Vaidyanadhan 2010; Radhakrishna et al. 2013; Jain et al. 2020). The important Bundelkhand Craton contains the fragmented evidence of geological events from the early Archean ca. 3.5–2.7 Ga (TTG-associated granitoids) up to the Paleoproterozoic (2.5–2.4 Ga; for several phases of granite) (Sarkar et al. 1995; Mondal et al. 2002; Kaur et al. 2016; Verma et al. 2016; Singh et al. 2021a). The existence of Banded Iron Formation (BIF) along greenstone belts (GB) and 3.3 Ga TTG rocks in the craton support it as a distinctive Archean Bundelkhand Craton. The oldest TTG-gneissic rocks and basic–ultrabasic rocks in the central part of the Bundelkhand Craton are possibly occur at 3.6–3.4 Ga and resemble other cratonic rocks of Indian Shield. The well-established crustal structural ensue of Western and Eastern Dharwar, Singhbhum and Bastar Cratons are suitable to compare the geological associations globally. Thus, we present this document through comparative analysis of the Archean crustal evolution of the Bundelkhand Craton with southern group of Indian Shield.

A. K. Shandilya et al. (eds.), *Geological and Geo-Environmental Processes on Earth*, Springer Natural Hazards,
https://doi.org/10.1007/978-981-16-4122-0_4

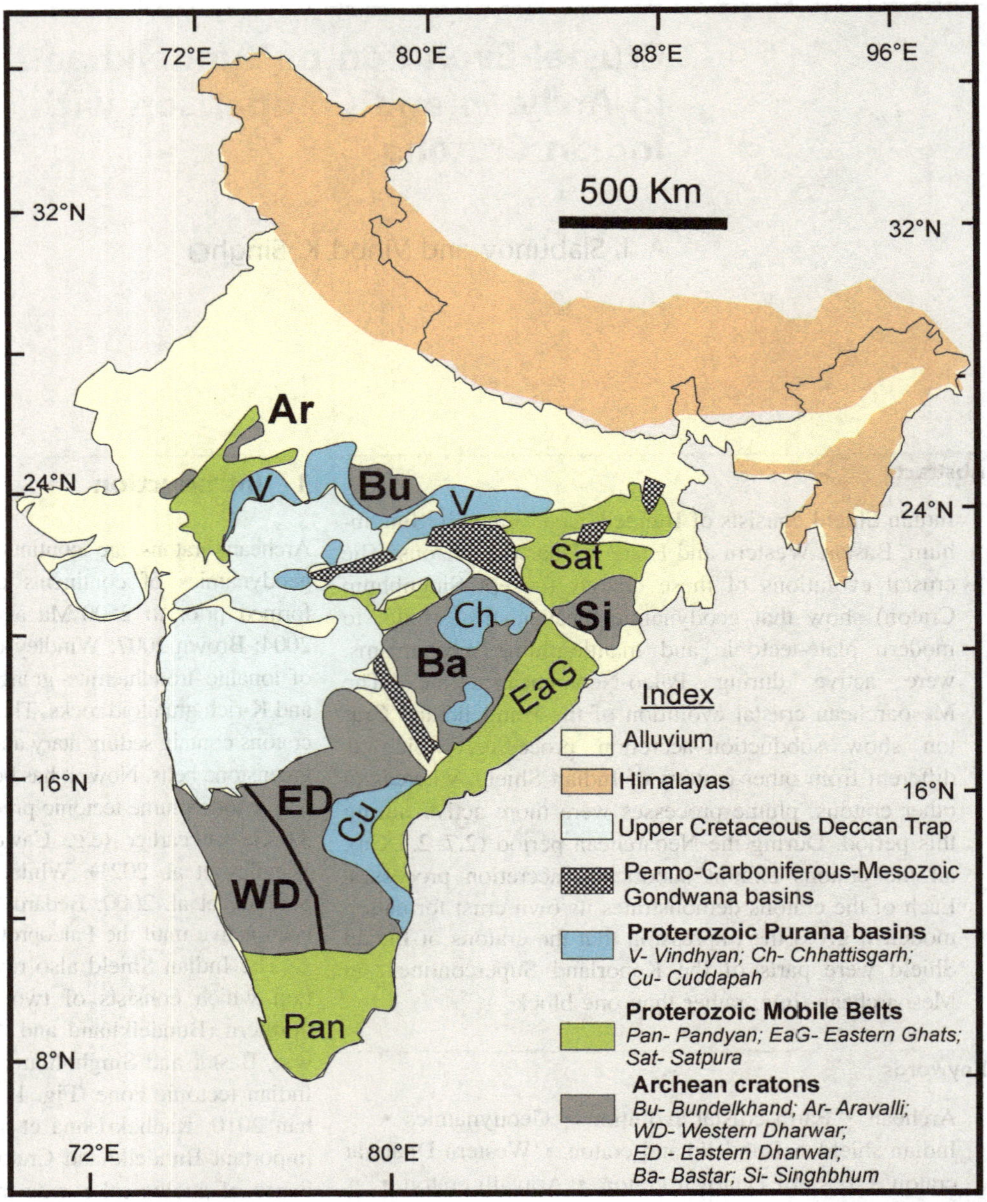

Fig. 1 Main Archean cratonic blocks of Indian Shield (after Ramakrishnan and Vaidyanadhan 2010)

2 Geology of Bundelkhand Craton

The typical TTG gneisses, supracrustal (greenstone and schist), sanukitoids, mafic–ultramafic layered and gabbroids intrusions, and K-rich granites are recognized in the Bundelkhand Craton (Mondal et al. 2002; Malviya et al. 2006; Singh and Slabunov 2013, 2015a, 2015b, 2016; Verma et al. 2016; Slabunov et al. 2017; Joshi et al. 2017; Singh et al. 2019a, 2019b, 2020, 2021a; Slabunov and Singh 2019a, 2019b; Pati 2020). The oldest TTG complexes in the craton are dated at 3.6–3.2 Ga (Mondal et al. 2002; Kaur et al. 2014, 2016; Saha et al. 2016; Singh et al. 2021a). These granitoids are associated with amphibolites. In the Babina and Mauranipur greenstone belts, the amphibolites inferred as the earliest mafic–ultramafic association of the Central Bundelkhand greenstone complex and its Sm–Nd isochron age is estimated at 3435 ± 161 Ma (Singh et al. 2019a); hence, it is similar in age to early TTG. The Bundelkhand craton notable into Central, Northern and Southern Bundelkhand terranes (Slabunov and Singh 2019b; Singh et al. 2021b).

The Central Bundelkhand greenstone complex consists association of two major assemblages of rocks, i.e. (i) an early (Mesoarchean) assemblage that contains basic–ultrabasic, felsic volcanic (2810 ± 13 Ma) and BIF rocks; and (ii) a late (Neoarchean—ca. 2.54 Ga) assemblage composed of felsic volcanic rocks. The Archean polymetamorphic evolution pattern amphibolite/granulite facies, eclogite facies metamorphism revealed and possibly associated with metasomatic events in the craton. The BIFs of the Mauranipur belt formed in a back-arc basin and BIFs of the Babina belt occur in a fore-arc basin (Slabunov and Singh 2019a). Neoarchean ca. (ca 2.7 Ga) metamorphic events in Central Bundelkhand greenstone complex under amphibolite facies and metasomatism associated are result of accretional tectonic (Sibelev et al. 2021).

The Southern Bundelkhand schist/metasedimentary complex formed the Girar belt which consist from two groups of rocks, i.e. (i) quartzite, (ii) BIFs and traces chlorite schist lenses near the quartzite/BIF boundary are of Archean age (Singh and Slabunov 2016; Slabunov et al. 2017). The foliated rocks of the Girar metasedimentary belt with 3.43–3.25 Ga detrital zircon are overlain by gently dipping undeformed sediments of Paleoproterozoic Bijawar Group (Slabunov et al. 2017). This is indirect evidence for older, most probably Meso/Neoarchean age for quartzite and BIF rocks of the Girar belt. Archean age has inferred for Ikauna peridotite–gabbro–diorite layered intrusive rocks which located near Girar belt too (Slabunov et al. 2017, 2018; Ramiz et al. 2019). The Girar belt and Ikauna layered intrusive formed under mantle plume activity at Meso/Neoarchean time apparently (Slabunov et al. 2017, 2018).

Neoarchean (2.58–2.56 Ga) sanukitoid massifs was revealed in craton as one of indicator of subduction style tectonic in this time (Joshi et al. 2017; Joshi and Slabunov 2019; Singh et al. 2019b, 2020). Late- to post-kinematic Neoarchean (2.53–2.51 Ga) granites are most common and the melting of granitoids is associated with post-accretionary processes in the crust (Verma et al. 2016; Singh et al. 2019b; Slabunov et al. 2020).

3 Aravalli Craton

Aravalli Craton is the north-western part of the Indian Shield represents the Archean granitiods mostly. They are exposed among the Paleoproterozoic rocks of the Aravalli Supergroup, which experienced tectonothermal events during ca 1.7 Ga orogeny (Buick et al. 2006; Jain et al. 2020). So this Archean rock is a part of the Paleoproterozoic Aravalli Fold Belt. Archean rock of this belt (which is called craton too) is separated from the Bundelkhand Craton by the Paleo-Mesoproterozoic Vindhyan basin. Archean rocks in the craton are represented by migmatized TTG granitoids (known as the Banded Gneissic Complex (BGC)), fragments of greenstone complexes and granites (Ramakrishnan and Vaidyanadhan 2010; Roy and Purohit 2018; Kaur et al. 2019). The BGC (TTG granitoids) display oldest rocks dated ca 3.31–3.28 Ga mostly, but have protolith with age up to 3.7 Ga (Kaur et al. 2013). Meso–Neoarchean TTG granitoids of 2.88 Ga and 2.56–2.54 Ga were also noticed in the region (Roy and Kröner 1996; Kaur et al. 2019) and conclude new continental crust formation events.

The fragments of greenstone complexes occur among BGC (e.g. the Rakhiawal greenstone belt) (Roy and Jakhar 2002; Roy and Purohit 2018). The greenstone belts have not been dated, but Mesoarchean metadykes (2828 ± 46 Ma; Gopalan et al. 1990) cut quartzites, which are a fragment of a greenstone belt and can thus be used to date the upper age boundary of the greenstone belts. This greenstone complex and dykes indicated Mesoarchean mantle plume activity in this area. Neoarchean (2562–2450 Ma) granodiorite–granite–leucogranites and K-rich granites are interpreted as having been derived by melting of basement rocks during accretion–collision processes.

4 Singhbhum Craton

The Singhbhum Craton is typical granite–greenstone terrane which consisting of TTG, greenstone belts and granite (Saha 1994; Mukhopadhyay et al. 2012; Das et al. 2017; Olierook et al. 2019). Singhbhum mobile belt occurs in the north side of craton with minor amount of supracrustal rocks (Mukhopadhyay et al. 2006; Mazumder et al. 2015; De et al. 2016). Singhbhum Craton commonly show association of the Older Metamorphic Tonalite Gneiss (OMTG) include biotite–hornblende tonalite, trondhjemite and granodiorite gneisses with enclaves of amphibolite-grade pelitic schists, arenites and calc-silicates (Older Metamorphic Group) and Iron Ore Group comprise low metamorphic-grade mafic volcanic, banded iron formation and argillaceous sedimentary rocks (Mukhopadhyay et al. 2008; Basu et al. 2008; Acharyya et al. 2010; Upadhyay et al. 2014; Dey et al. 2017; Olierook et al. 2019).

The age of the Older Metamorphic Tonalite Gneiss estimated as 3.52–3.3 Ga, but tonalite gneiss of from the Champua area have 4.24–4.03 Ga xenocrystic zircons (Chaudhuri et al. 2018), indicated the oldest (Hadean age) matter in the Indian Shield. The Iron Ore Group contains predominantly BIF with pillow basalt, dacite, pyroclastics and ultramafics in the lower parts. This group was deposited under deep marine conditions in spreading centres in arc systems (Mukhopadhyay et al. 2012). There is no precise age estimation of this group, but it brackets between 3.5 and 3.1 Ga.

The Singhbhum granite (SG) occurs as major part of the craton and divided for the three groups of pluton, i.e. 3.35 Ga (SG I), 3.1 Ga (SG II) and 2.9 Ga (SG III) (Dey et al. 2017; Nelson et al. 2014). Mesoarchean gabbro-anorthosite units (ca 3.12 Ga) and Mayurbhanj Granite intrudes (ca 3.09 Ga) are existing in the eastern and western parts of the Singhbhum Craton (Saha 1994; Misra et al. 2002; Augé et al. 2003; Mondal et al. 2007; Nelson et al. 2014; Sunder-Raju et al. 2015). Mesoarchean (ca 2.8 Ga) metabasalts, komatiites or picrites and fluvial–marginal marine sedimentary successions are exposed as the Simlipal, Dhanjori, Dalma, Ongarbira, Jagannathpur, Malangtoli Volcanics. The Simlipal basin, for example, undeformed and unmetamorphosed remains unaffected by any granitoid intrusion. They formed most probably under mantle plume activity.

Significant Neoarchean (ca. 2758 Ma) magmatism occurred along the southern margin of the Singhbhum Craton, including the Pal Lahara Gneiss and other Rengali Province granitoids (Das et al. 2017; Olierook et al. 2019). There are three Neoarchean mafic dykes Swarm as root a large igneous province exists in craton, i.e. Keshargaria swarm with age 2800 Ma and Ghatgoan Swarm of 2764–2760 Ma and ca. 2750 Ma (Kumar et al. 2017; Olierook et al. 2019).

Kolhan Basin unconformably overlies OMTG, Iron Ore Group, Singhbhum granite and consists of conglomerate, sandstones, shale and limestone. They formed in half graben-type intracratonic basin and its age estimated as 2.6–2.5 Ga (Roy and Purohit 2018), but it should be Proterozoic.

5 Bastar Craton

The Bastar Craton located between Singhbhum and Eastern Dharwar Cratons and separated from Bundelkhand Craton by Proterozoic Satpura Mobile Belt. It consists of TTG gneisses, granitoids, ca. four generations of greenstone belt and mafic dykes rocks occurring from Archean to Paleoproterozoic (Mondal et al. 2006; Roy and Purohit 2018). The TTG granitoids (Sukma I, II, Amgaon Complexes) are exposed in the south and west parts of craton. These gneisses have trondhjemitic character and dated at 3.6–3.5 Ga (Sarkar et al. 1993; Mohanty 2013, 2015; Dora et al. 2019; Asokan et al. 2020; Santosh et al. 2020; Meshram et al. 2021). Ca 3.0 Ga Granitoid plutons (Sukma granite III) are more common also which intruded into the gneisses and metasupracrustals rocks throughout the craton (Sarkar et al. 1993; Mohanty 2015).

The NW–SE trending mafic dykes and dyke swarms have cross-cut the older rocks in the craton. Mondal et al. (2006) conclude that the felsic magmatism is dominant during Archean in the Bastar Craton which changed to bimodal (acidic–basic) magmatism in the Proterozoic time. The Neoarchaean (Ca 2.6 Ga) intrusive leucocratic granite includes the Malanjkhand Granite hosting Andean-type Porphyry Copper–Molybdenum deposits. One of the oldest granulites in the craton are the Kondagaon complex indicates Neoarchaean (Ca 2.6 Ga; Roy and Purohit 2018).

The four generations of greenstone complexes are noted, i.e. Sukma, Bengpal, Bailadila and Kotri–Dungargarh–Sakoli–Sonakhans in the craton (Ramakrishnan and Vaidyanadhan 2010; Roy and Purohit 2018). The former consist mostly from quartzite, paragneiss and amphibolite, schist (included cordierite–garnet), ferruginous quartz and BIF interlayered with basalt and tuff. Kotri–Dungargarh–Sakoli–Sonakhans greenstone belts formed with volcano-sedimentary, meta-basalts with meta-chert, conglomerate and BIF. The most part of these greenstone complexes formed under mantle plume activity. Although there are basalt–andesite–dacite–rhyolite (BADR) series, rocks in Sonakhan greenstone belt which marked subduction setting in Neoarchean (Mondal and Raza 2009; Jain et al. 2020). There is very poor geochronology reported from these complexes, but Sukma, Bengpal, Bailadila complexes estimated as Meso–Neoarchean, and Kotri–Dungargarh–Sakoli–Sonakhans as Neoarchean (2530 Ma age of felsic from Dungargarh belt).

6 Dharwar Craton

The Dharwar Craton has been considered now into two cratons: Western and Eastern, based on their evolution patterns and crustal structures (Fig. 1). Each of them is 3–4 times the size of the Bundelkhand Craton.

6.1 Western Dharwar Craton

The Western Dharwar Craton consists of mainly Paleoarchean (3.36–3.2 Ga) TTG complex (Peninsular gneisses). But there are the oldest detrital zircons (3.58 Ga) in quartzites in Dharwar Group of rocks also noted. Three generations of greenstone complexes (Sargur, Bababudan and Chitradurga, agreeably) and several granitoid massifs are reported from Western Dharwar Craton (Radhakrishna and Ramakrishnan 1990; Jayananda et al. 2013). The Sargur greenstone belt is composed of mafic–ultamafic rocks (metabasalts, komatiites and their intrusive comagmates and metaanorthosites) which often predominate, and metasediments (kyanite/sillimanite–staurolite–biotite gneisses, quartzites, BIF, local marble, calc-silicate rocks, bedded barite); with limited exposures of felsic volcanics. The age of the complex is estimated at 3.1–3.3 Ga, based on the Sm–Nd whole-rock isochron age of komatiites of 3352 ± 110 Ma (Jayananda et al. 2008) and the U–Pb age of zircon from felsic volcanics of 3298 ± 7 Ma (Peucat et al. 1995). It forms small greenstone belts where dominated by mafic–ultramafics (e.g. Ghatti Hosahalli, Krishnarajapet and Nagamangala) and those with abundant sediments (e.g.

Sargur and Hole Narasipura). The Sargur greenstone complex was formed presumably in both rift-related structures on an early continental crust (3.58–3.23 Ga zircons in quartzites have been found, suggesting the existence of an older crust) and an oceanic plateau-type setting. The occurrence of 3.2 Ga TTGs in the region suggests subduction processes of that age.

A second generation of greenstone complexes in the Western Dharwar Craton is the Meso–Neoarchean Bababudan Group at the base of the Dharwar Supergroup in Bababudan, Chitradurga schist belts. The base of the Bababudan Group sequence consists of cross-bedded quartz conglomerates with ripple marks (Kalasapura Formation). These sediments rest with angular unconformity on Peninsular gneisses and Sargur Group rocks. In addition to quartz conglomerates, the Bababudan Group comprises phyllites and BIF. Mafic (metabasalts and gabbroic rocks) and ultramafic bodies seem to occur among them as sills. Felsic volcanics, occurring as part of the Santaveri Formation, are scarce. The mafic–ultramafic are dated at 2.9–2.85 Ga (Sm–Nd whole rock isochron ages are 2911 $\pm$ 49 and 2848 $\pm$ 70 Ma, (Kumar et al. 1996). The formation of the Bababudan greenstone complex was associated with plume activity and took place in an intracontinental basin.

A third greenstone/schist complex of the craton corresponds with the Chitradurga Group of the Dharwar Supergroup, which makes up the largest exposures around Shimoga and Chitradurga area. This Group consists dominantly of sediments (quartz and polymictic conglomerates containing TTG and Bababudan Group rock fragments, chert-phyllite, manganese and iron formation and stromatolitic carbonates) with pillow basalt and lesser felsic volcanic intercalations. The complex is dated at 2.75–2.58 Ga result as the Sm–Nd whole rock isochron age is 2747 $\pm$ 15 Ma (Kumar et al. 1996); the U–Pb age of zircon from the felsic volcanic is 2677 $\pm$ 2 to 2576 $\pm$ 20 Ma (Jayananda et al. 2013). The sedimentation basin of the Chitradurga Group seems to have been controlled by mantle plume activity (Hokada et al. 2013).

6.2 Eastern Dharwar Craton

The Eastern Dharwar Craton is separated from the Western Dharwar Craton by a large fault, the Chitradurga shear zone, and differs from the latter in deep geophysical structure (a thinner earth crust (Gupta et al. 2003) and the compositions and ages of Archean granitoid and greenstone complexes (Ram Mohan et al. 2013; Yang and Santosh 2015). The Eastern and Western Dharwar Cratons consists dominantly of commonly migmatized TTG granitoids, although in contrast to the Eastern Dharwar Craton, they are dominated by 2.7–2.55 Ga rocks with minor fragments of 3.0–3.38 Ga crust (Jayananda et al. 2013 and references therein). Moreover, the contribution of older crustal material to granitoid composition decreases markedly Nd T_{DM} up to 2.8–3.0 Ga in the eastern part (Dey 2013), but 2.56–2.5 Ga juvenile (εNd = +3.3) calc-alkaline to potassic granitoids are widespread here. The 2.51–2.53 Ga sanukitoid-like Closepet Granite batholiths N–S-trending occur in the western part of the Eastern Dharwar Craton, which cross-cuts the entire craton.

The greenstone belts of the Eastern Dharwar Craton are small narrow N–S and NW–SE trending linear structures, e.g. Kolar, Hutti, Kushtagi, etc. They consist mainly of metabasalts (often pillowed) associated with komatiites and BIF; felsic volcanics, associated with greywacke and polymictic conglomerates (Kolar GB), are more common; and metasediments, occurring as schists, are less common. An early association (beginning probably with 2.75 Ga, but mainly arise at ca. 2.7 Ga) of basalts and komatiites was formed in an oceanic setting under the influence of plumes, i.e. oceanic plateaus), but this stage was also terminated by subduction processes (Sangur GB). However, the main episode in the subduction processes, which gave rise to continental crust, occurred at 2.58–2.52 Ga, when felsic volcanics and various granitoids (including sanukitoids) originated. Ca. 2.5 Ga granulite facies metamorphism, widespread in the southern part of the Eastern and Western Dharwar Cratons, was associated with accretion–collision processes (Slabunov and Singh 2018, 2020).

7 Discussions and Conclusions

All cratons of Indian Shield have old (Paleoarchean) core (Fig. 2). The Singhbhum Craton has extreme older (Hadean; up to 4.2 Ga) protolith. While the old core of Eastern Dharwar Craton have Paleo-Mesoarchean age. The oldest rocks on all cratons are TTG gratitoids but with enclaves of amphibolites and gneisses.

Plate-tectonic and mantle plume mechanisms are operating for lithospheric formation during Paleo-Neoarchean in the Bundelkhand, Aravalli, Singhbhum, Bastar, Western and Eastern Dharwar Cratons, as common for World (Windley et al. 2021). The subduction–accretion processes are more responsible for crustal evolution of the Bundelkhand Craton during Mesoarchean. While other Indian cratons considerably different for crustal evolution in general, dominated by plume processes (Fig. 2). But at this time, the crustal evolution of the southern part of Bundelkhand Craton, i.e. formation of a sedimentary basin consisting quartzites and BIFs as metasedimentary complex (Singh and Slabunov 2016; Slabunov et al. 2017) and peridotite–gabbro–diorite layered intrusive rocks (Slabunov et al. 2018) might have affected by mantle plume. It should be noted that other point of view on

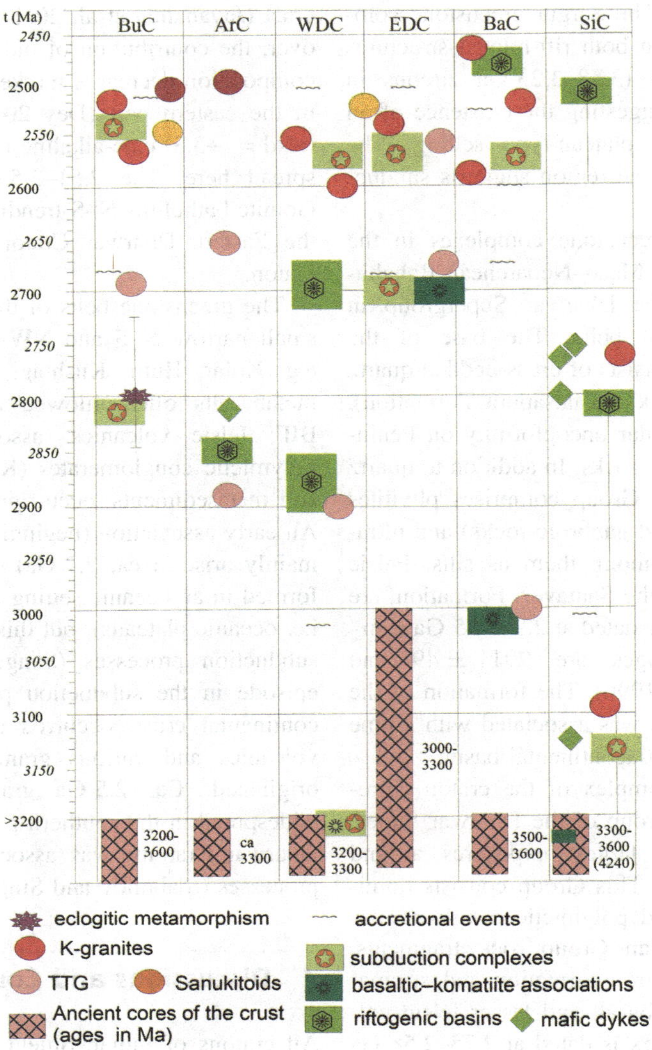

Fig. 2 Correlation of Meso–Neoarchean crustal evolution in the cratons of Indian Shield (BuC: Bundelkhand; ArC: Aravalli; WDC: Western Dharwar; EDC: Eastern Dharwar; BaC: Bastar; SiC: Singhbhum Craton) (data used as interpretation after Hokada et al. 2013; Jayananda et al. 2013, 2015; Kaur et al. 2014, 2016; Kumar et al. 2017; Mondal et al. 2002; Ramakrishnan and Vaidyanadhan 2010; Ram Mohan et al. 2013; Saha et al. 2016; Slabunov and Singh 2019a, 2020; Singh et al. 2021a)

formation of the southern part of a craton exists; it provides the leading role of subduction processes at this time (Ramiz et al. 2019). This complex has the certain features of similarity with Chitradurga Group of Dharwar Supergroup, Western Dharwar Craton. Mafic magmatisms marks mantle plume activity exists in Aravalli, Western Dharwar and Singhbhum Cratons in Mesoarchean (Fig. 2). At ca. 2.7–2.6 Ga the subduction processes are noted in the Bundelkhand, Eastern Dharwar and Bastar Cratons (Fig. 2).

In Bundelkhand Craton subduction process noted in the Babina belt, as signified by Neoarchean (2542 Ma; Singh and Slabunov 2015a), felsic volcanic formed in an active continental margin, sanukitoid massif similar in age (2577–2559 Ma; Joshi et al. 2016; Singh et al. 2019b) and meta-dacites (2557 Ma; Slabunov and Singh 2019a) in the Mauranipur belt. It means an accretion stage in the evolution of the greenstone in Bundelkhand Craton took place at about 2.53 Ga, after the youngest 2542 Ma felsic volcanics and prior to the formation of the earliest post-kinematic granites (2531 Ma; Verma et al. 2016). At this stage its Meso- and Neoarchean constituents are combined to form one greenstone complex. The melting of large volumes of granitoids in the period 2.53–2.51 Ga is associated with post-accretionary processes in the crust (Fig. 2).

Therefore, the Paleo-Neoarchean crustal evolution of the Bundelkhand Craton provides a basis for comparing with other Indian cratons (Fig. 2). It noted that the mafic–ultramafic rocks of Central Bundelkhand greenstone complex have derived from thick oceanic crust in a subduction processes in Paleoarchean (3.44 Ga; Singh et al. 2019a) and first arc-forming felsic volcanics in Mauranipur greenstone belt proceeded during the Mesoarchean (ca. 2.81 Ga; Slabunov and Singh 2019a) time. Similar processes also involved, i.e. the interaction of plumes with old cores in other cratons in the Indian Shield at that time. Thus, the formation of sedimentary rock complexes with bimodal magmatism occur (Fig. 2), in the Bababudan Group (Western Dharwar craton), the Koira (ca 3.1 Ga) and Simlipal (ca. 2.8 Ga) Groups (Singhbhum Craton) and the Sukma Group (ca 3.0 Ga; Bastar Craton). The old supracrustal enclaves in Aravalli Craton granitoids look like to be part of similar characters. The existences of metamorphosed Mesoarchean mafic dykes in Indian Shield point out the relic of the old continental core (Fig. 2). Therefore, subduction–accretion processes took place only in the Bundelkhand Craton for the formation of a new continental crust during Mesoarchean in Indian Shield. While other cratons of the shield more favourable for the transformation of the old cores by plumes mechanism at that time. During the Neoarchean the formation of the continental crust of the Bundelkhand Craton is very similar to other cratons of the Indian Shield and is different from that in the Karelian Craton and in the Superior Province (Lubnina and Slabunov 2011, 2017; Slabunov and Singh 2020). The subduction–accretion processes were more common in the Western and Eastern Dharwar and in the Bastar and Aravalli Cratons at 2.6–2.5 Ga (Fig. 2), while the situation in the Laurasian group of cratons, preceding a split-up which began at ca. 2.5 Ga, had stabilized at that time.

The observed correlation of the crustal evolution of the Bundelkhand Craton can be explained, assuming that in Mesoarchean time the craton was probably in the northern part of the Neoarchean Kenorland Supercontinent near the Karelian Craton and the Superior Province (Lubnina and Slabunov 2011, 2017; Slabunov and Singh 2020). As the model projected is based on only geological evidence, it should also be tested by paleomagnetic data.

During Neoarchean (2.7–2.5 Ga), most part of the Indian cratons (except Singhbhum) displays subduction–accretion processes. However all cratons exhibit its own crustal evolution pattern in Mesoarchean (Fig. 2). Crustal evolution in Neoarchean of Bundelkhand and Aravalli Cratons, Western and Eastern Dharwar Cratons have many similar features therefore it assumed as part in Meso–Neoarchean time elements of the Kenorland Supercontinent, but not in a single block.

Acknowledgements The work was supported by DST-RFBR (INT/RUS/RFBR/P-279 and RFBR-17-55-45005 IND) grants, it continues under MOU between Bundelkhand University, Jhansi, India and the Institute of Geology, Karelian Research Centre, RAS, Petrozavodsk, Russia, and is a contribution to Project AAAA-A18-118020290085-4. We thank Dr. Sanjeet Verma and Vivek Malviya for their valuable comments and criticism which seriously helped in improving this paper.

References

Acharyya SK, Gupta A, Orihashi Y (2010) New U-Pb zircon ages from Paleo-Mesoarchean TTG gneisses of the Singhbhum Craton. Eastern India. Geochemical J 44(2):81–88

Asokan AD, Elangovan R, Vishwakarma N et al (2020) Petrogenesis of the Kanker granites from the Bastar craton: implications for crustal growth and evolution during the Archean-Proterozoic transition. Front Earth Sci https://doi.org/10.3389/feart.2020.00212

Augé T, Cocherie A, Genna A et al (2003) Age of the Baula PGE mineralization (Orissa, India) and its implications concerning the Singhbhum Archaean nucleus. Precambr Res 121(1):85–101

Basu A, Bandyopadhyay P, Chakrabarti R et al (2008) Large 3.4 Ga Algoma-type BIF in the Eastern Indian craton. Geochimica et Cosmochimica Acta Supplement 72:A59

Bedard JH (2018) Stagnant lids and mantle overturns:Implications for Archaean tectonics, magmagenesis, crustal growth, mantle evolution, and the start of plate tectonics. Geosci Front 9:19–49

Brown M (2007) Metamorphic conditions in orogenic belts: a record of secular change. Int Geol Rev 49:193–234

Buick IS, Allen C, Pandit M et al (2006) The Proterozoic magmatic and metamorphic history of the BGC, central Rajasthan, India: LA-ICP-MS U-Pb zircon constraints. Precambrian Res 151:119–142

Cawood PA, Kroner A, Pisarevsky S (2006) Precambrian plate tectonics: criteria and evidence. GSA Today 16:4–11

Chaudhuri T, Wan Y, Mazumder R et al (2018) Evidence of enriched, Hadean mantle reservoir from 4.2–4.0 Ga zircon xenocrysts from Paleoarchean TTGs of the Singhbhum craton, Eastern India. Scientific Reports 8(1): 7069. https://doi.org/10.1038/s41598-018-25494-6

Condie KC (2004) Supercontinents and superplume events: distinguishing signals in the geologic record. Phys Earth Planet Inter 146:319–332

Das K, Bose S, Ghosh G (2017) The Neoarchean-Paleoproterozoic basin development and growth of the Singhbhum craton, Eastern India and its global implications: insights from detrital zircon U-Pb data. Precambr Res 298:123–145

De S, Mallik L, Mazumder R et al (2016) Sedimentation history of the Paleoproterozoic Singhbhum group of rocks, Eastern India and its implications. Earth Sci Rev 163:141–161

de Wit MJ (1998) On Archean granites, greenstones, cratons and tectonics: does the evidence demand a verdict? Precambr Res 91:181–226

de Wit MJ (2001) Archaean tectonics: wading through a mine-field of controversies. In: Cassidy KF, Dunphy JM, Van Kranendonk MJ (eds) 4th International Archaean symposium, extended abstracts, Perth, Australia, p 4–6

Dey S (2013) Evolution of Archaean crust in the Dharwar craton: the Nd isotope record. Precambr Res 227:227–246

Dey S, Topno A, Liu Y, Zong K (2017) Generation and evolution of Palaeoarchaean continental crust in the central part of the Singhbhum craton, eastern India. Precambr Res 298:268–291

Dora ML, Singh Y, Joshi S et al (2019) The first report of CHIME monazite age of Mul granite pluton, Western Bastar Craton and its metallogenic significance. J Earth Sys Sci 128:124. https://doi.org/10.1007/s12040-019-1165-y

Gopalan K, Macdaugall JD, Roy AB et al (1990) Sm-Nd evidence for 3.3 Ga old rock in Rajasthan, north-western India. Precambr Res 48:287–297

Gupta S, Rai SS, Prakasam KS et al (2003) The nature of the crust in southern India: implications for Precambrian crustal evolution. Geophys Res Lett 30:419

Hamilton WB (1998) Archean magmatism and deformation were not products of Plate Tectonics. Precambr Res 91:143–179

Hokada T, Horie K, Satish-Kumar M et al (2013) An appraisal of Archaean supracrustal sequences in Chitradurga Schist Belt, western Dharwar craton, southern India. Precambr Res 227:99–119

Jain AK, Banerjee DM, Kale VS (2020) Tectonics of the Indian subcontinent. Springer, Cham, p 576

Jayananda M, Kano T, Peucat J-J et al (2008) 3.35 Ga komatiite volcanism in the western Dharwar craton, southern India: constraints from Nd isotopes and whole rock geochemistry. Precambr Res 162:160–179

Jayananda M, Peucat J-J, Chardon D et al (2013) Neoarchean greenstone volcanism and continental growth, Dharwar craton, southern India: constraints from SIMS U-Pb zircon geochronology and Nd isotopes. Precambr Res 227:55–76

Jayananda M, Chardon D, Peucat J-J et al (2015) Paleo- to Mesoarchean TTG accretion and continental growth in the western Dharwar craton, southern India: constraints from SHRIMP U–Pb zircon geochronology, whole-rock geochemistry and Nd–Sr isotopes. Precambr Res 268:295–322

Joshi KB, Slabunov A (2019) Neoarchean sanukitoids from the Karelian and Bundelkhand cratons: comparison of composition, regional distribution and geodynamic setting. Transactions of the Karelian Research Centre of the Russian Academy of Sciences. 2:5–25. https://doi.org/10.17076/geo841

Joshi KB, Bhattacharjee J, Rai G et al (2017) The diversification of granitoids and plate tectonic implications at the Archaean–Proterozoic boundary in the Bundelkhand craton, central India. In: Halla J, Whitehouse MJ, Ahmad T et al (eds) Crust–Mantle interactions and granitoid diversification: insights from Archaean cratons. Geological Society, London, Special Publications, 449:123–157. https://doi.org/10.1144/SP449.8

Kaur P, Zeh A, Chaudhri N et al (2013) Nature of magmatism and sedimentation at a Columbia active margin: insights from combined U-Pb and Lu–Hf isotope data of detrital zircons from NW India. Gondwana Res 23:1040–1052

Kaur P, Zeh A, Chaudhri N (2014) Characterisation and U-Pb-Hf isotope record of the 3.55 Ga felsic crust from the Bundelkhand Craton, northern India. Precambr Res 255:236–244

Kaur P, Zeh A, Chaudhri N et al (2016) Unravelling the record of Archaean crustal evolution of the Bundelkhand craton, northern India using U-Pb zircon–monazite ages, Lu–Hf isotope systematics, and whole-rock geochemistry of granitoids. Precambr Res 281:384–413

Kaur P, Zeh A, Chaudhri N (2019) Archean crustal evolution of the Aravalli Banded Gneissic complex, NW India: constraints from zircon U-Pb ages, Lu-Hf isotope systematics, and whole-rock geochemistry of granitoids. Precambr Res 327:81–102

Kumar A, Bhaskar Rao YJ, Sivaraman TV et al (1996) Sm–Nd ages of Archaean metavolcanic of the Dharwar craton, south India. Precambr Res 80:206–215

Kumar A, Parashuramulu V, Shankar R et al (2017) Evidence for a Neoarchean LIP in the Singhbhum craton, eastern India: implications to Vaalbara supercontinent. Precambr Res 292:163–174

Lubnina NV, Slabunov AI (2011) Reconstruction of the Kenorland supercontinent in the Neoarchean based on Paleomagnetic and geological data. Mosc Univ Geol Bull 66(4):242–249

Lubnina NV, Slabunov AI (2017) The Karelian craton in the structure of the Kenorland supercontinent in the neoarchean: new paleomagnetic and isotope geochronology data on granulites of the Onega complex. Mosc Univ Geol Bull 72(6):377–390

Mazumder R, De S, Ohta T et al (2015) Palaeo-Mesoproterozoic sedimentation and tectonics of the Singhbhum craton, eastern India, and implications for global and craton-specific geological events. Geol Soc London Mem 43(1):139–149

Malviya VP, Arima M, Pati JK et al (2006) Petrology and geochemistry of metamorphosed basaltic pillow lava and basaltic komatiite in the Mauranipur area: subduction related volcanism in the Archean Bundelkhand craton, central India. J Min Petrol Sci 101:199–217

Meshram T, Dora ML, Baswani SR et al (2021) Petrogenesis and U Pb geochronology of charnockites flanking the Pranhita Godavari rift in peninsular India-link between the Bastar and eastern Dharwar cratons. Gondwana Res 92:113–132. https://doi.org/10.1016/j.gr.2020.12.024

Misra S, Sarkar SS, Ghosh S (2002) Evolution of Mayurbhanj granite pluton, eastern Singhbhum, India: a case study of petrogenesis of an A-type granite in bimodal association. J Asian Earth Sci 20(8):965–989

Mohanty SP (2013) Spatio-temporal evolution of the Central Indian Shield and its correlation with south African and western Australian cratons. In: Singh VK, Chandra R (eds) International association for Gondwana research conference series 16. 3rd International Conference precambrian continental growth and tectonism, Jhansi, India, p 109–112

Mohanty SP (2015) Palaeoproterozoic supracrustals of the Bastar craton: Dongargarh supergroup and sausar group. In: Precambrian basins of India: stratigraphic and tectonic context. Mazumder R, Eriksson PG (eds) Geol Soc London Mem 43:151–164. https://doi.org/10.1144/M43.11

Mondal MEA, Raza M (2009) Tectonomagmatic evolution of the Bastar craton of Indian shield through plume-arc interaction: evidence from geochemistry of the mafic and felsic volcanic rocks of Sonakhan greenstone belt. In: Ahmad T, Hirsch F, Charusiri P (eds) J Virtual Expl 32, Paper 7. https://doi.org/10.3809/jvirtex.2009.0024

Mondal MEA, Goswami JN, Deomurari MP et al (2002) Ion microprobe 207Pb/206Pb ages of zircon from the Bundelkhand massif, northern India: implication for crustal evolution of Bundelkhand—Aravalli protocontinent. Precambr Res 117:85–100

Mondal MEA, Hussain M, Ahmad T (2006) Continental growth of Bastar craton, central Indian shield during Precambrian via multiphase subduction and Lithospheric extension/rifting: evidence from geochemistry of gneisses, granitoids and mafic dykes. J Geosci Osaka City Univ 49(8):137–151

Mondal SK, Frei R, Ripley EM (2007) Os isotope systematics of Mesoarchean chromitite-PGE deposits in the Singhbhum craton (India): implications for the evolution of lithospheric mantle. Chem Geol 244(3):391–408

Mukhopadhyay J, Ghosh G, Nandi AK et al (2006) Depositional setting of the Kolhan group: its implications for the development of a Meso to Neoproterozoic deep-water basin on the south Indian craton. South African J Geol 109(1–2):183–192

Mukhopadhyay J, Beukes NJ, Armstrong RA et al (2008) Dating the oldest greenstone in India: A 3.51-Ga Precise U-Pb SHRIMP Zircon age for dacitic Lava of the southern iron ore group, Singhbhum craton. J Geol 116(5):449–461

Mukhopadhyay J, Ghosh G, Zimmermann U et al (2012) A 3.51 Ga bimodal volcanic-BIF-ultramafic succession from Singhbhum craton: implications for Palaeoarchaean geodynamic processes from the oldest greenstone succession of the Indian subcontinent. Geol J 47(2–3):284–311

Pati JK (2020) Evolution of Bundelkhand craton. Episodes 43:69–87

Nelson DR, Bhattacharya HN, Thern ER et al (2014) Geochemical and ion-microprobe U Pb zircon constraints on the Archaean evolution of Singhbhum craton, eastern India. Precambr Res 255:412–432

Olierook HKH, Clark C, Reddy SM et al (2019) Evolution of the Singhbhum craton and supracrustal provinces from age, isotopic and chemical constraints. Earth Sci Rev 193:237–259

Peucat JJ, Bouhallier H, Fanning CM et al (1995) Age of the Holenarsipur greenstone belt, relationships with the surrounding gneisses (Karnataka, South India). J Geol 103:701–710

Radhakrishna BP, Ramakrishnan M (eds) (1990) Archaean greenstone belts of south India. Geological Society of India, Bangalore, p 497

Radhakrishna T, Chandra R, Shrivastava AK et al (2013) central/eastern Indian bundelkhand and bastar cratons in the palaeoproterozoic supercontinental reconstructions: a palaeomagnetic perspective. Precambr Res 226:91–104

Ramakrishnan M, Vaidyanadhan R (2010) Geology of India, vol. 1: Geological society of India, Bangalore, p 556

Ramiz MM, Mondal MEA, Farooq SH (2019) Geochemistry of ultramafic–mafic rocks of the Madawara ultramafic complex in the southern part of the Bundelkhand craton, central Indian shield: implications for mantle sources and geodynamic setting. Geol J 54:2185–2207. https://doi.org/10.1002/gj.3290

Ram Mohan M, Piercey SJ, Kamber BS et al (2013) Subduction related tectonic evolution of the Neoarchean eastern Dharwar craton, southern India: new geochemical and isotopic constraints. Precambr Res 227:204–226

Roy AB, Jakhar SR (2002) Geology of Rajasthan: precambrian to recent. Scientific Publishers (India), Jodhpur, p 421

Roy AB, Kröner A (1996) Single zircon evaporation ages constraining the growth of the Archaean Aravalli craton, northwestern Indian shield. Geol Mag 133:333–342

Roy AB, Purohit R (2018) Indian shield: precambrian evolution and phanerozoic reconstitution. Elsevier, p 375

Santosh M, Tsunogae T, Yang C et al (2020) The Bastar craton, central India: a window to Archean -Paleoproterozoic crustal evolution. Gondwana Res 79:157–184

Saha AK (1994) Crustal evolution of Singhbhum-North Orissa. eastern India. Geol Soc India Mem 27:339

Saha L, Frei D, Gerdes A et al (2016) Crustal geodynamics from the Archaean Bundelkhand craton, India: constraints from zircon U-Pb–Hf isotope studies. Geol Mag 153:79–192

Sarkar G, Corfu F, Paul DK et al (1993) Early Archean crust in Bastar craton, central India—a geochemical and isotopic study. Precambrian Res 62:127–137

Sarkar A, Paul DK, Potts PJ (1995) Geochronology and geochemistry of the Mid–Archean, Trondhjemitic gneisses from the Bundelkhand craton, central India. In: Saha AK (ed) Recent researches in geology, 16:76–92

Sharkov EV, Bogatikov OA, Krasivskaya IS (2000) The role of mantle plumes in the early precambrian tectonics of the eastern baltic shield. Geotectonics 34(2):85–105

Sibelev OS, Slabunov AI, Singh VK, Mishra S (2021) Metamorphism of the central Bundelkhand greenstone complex of the Bundelkhand craton, Indian shield and its geodynamic setting. In: Shandilya AK, Singh VK, Bhatt SC, Dubey CS (eds) Geological and Geo-environmental Processes on Earth, pp 143–154 (this volume)

Singh VK, Slabunov A (2013) The greenstone belts of the Bundelkhand craton, central India: new geochronological data and geodynamic setting. In: Singh VK, Chandra R (eds) International association for Gondwana research conference series 16. 3rd International Conference Precambrian continental growth and tectonism, Jhansi, India, p 170–171

Singh VK, Slabunov A (2015a) The Central Bundelkhand Archaean greenstone complex, Bundelkhand craton, central India: geology, composition, and geochronology of supracrustal rocks. Int Geol Rev 57(11–12):1349–1364

Singh VK, Slabunov A (2015b) Geochemical characteristics of banded iron formation and metavolcanics of Babina greenstone belt of the Bundelkhand craton, central India. J Econ Geol Georesour Manage 10:63–74

Singh VK, Slabunov A (2016) Two types of Archaean supracrustal belts in the Bundelkhand craton, India: geology, geochemistry, age and implication for craton crustal evolution. J Geol Soc India 88:539–548

Singh PK, Verma SK, Moreno JA, Singh VK et al (2019a) Geochemistry and Sm-Nd isotope systematics of metabasalts from the Babina and Mauranipur greenstone belts, Bundelkhand craton: implications for tectonic setting and Paleoarchean mantle evolution. Lithos 330–331:90–107

Singh PK, Verma SK, Singh VK et al (2019b) Geochemistry and petrogenesis of sanukitoids and high-K anatectic granites from the Bundelkhand craton: implications for the late-Archean crustal evolution. J Asian Earth Sci 174:263–282

Singh VK, Verma SK, Singh PK et al (2020) Archean crustal evolution of the Bundelkhand craton: evidence from granitoid magmatism. In: Archean Granitoids of India: Windows into Early Earth Tectonics. Geological Society, London, Special Publ 489:235–259. https://doi.org/10.1144/SP489-2018-72

Singh PK, Verma SK, Singh VK et al (2021a) Geochronology and petrogenesis of the TTG gneisses and granitoids from the Central Bundelkhand granite-greenstone terrane, Bundelkhand Craton, India: Implications for Archean crustal evolution and cratonization. Precambr Res 359:106210

Singh VK, Slabunov AI, Nesterova NS et al (2021b) Tectonostratigraphic terranes of the Bundelkhand craton (Indian Shield). In: Shandilya AK, Singh VK, Bhatt SC, Dubey CS (eds) Geological and Geo-environmental Processes on Earth, pp 155–164 (this volume)

Slabunov A, Singh VK (2018) Bundelkhand and Dharwar cratons (Indian shield): comparison of crustal evolution in Archean time. Arch Anthropol Open Acc. 3(suppl-2):327–333. https://doi.org/10.31031/AAOA.2018.03.000556

Slabunov AI, Singh VK (2019a) Meso-Neoarchaean crustal evolution of the Bundelkhand craton, Indian shield: new data from greenstone belts. Int Geol Rev 61:1409–1428. https://doi.org/10.1080/00206814.2018.1512906

Slabunov AI, Singh VK (2019b) The new tectonic division of the Bundelkhand craton Indian shield. Transactions of A. Fersman scientific session of Geological institute, Kola research centre, RAS. Apatity, Russia 16:521–524. https://doi.org/10.31241/FNS.2019.16.106

Slabunov AI, Singh VK (2020) Bundelkhand, Aravalli and Dharwar cratons Indian shield: comparison of Archean crustal evolution and location in the Kenorland supercontinent structure. Trans Karelian Res Centre Russ Acad Sci 2:5–17. https://doi.org/10.17076/geo1180

Slabunov A, Singh VK, Joshi KB, Li X (2017) Paleoarchean zircons from quartzite of South Bundelkhand Supracrustal complex: origin and implications for crustal evolution in Bundelkhand craton, central India. Curr Sci 112:794–801

Slabunov A, Egorova S, Singh VK et al (2018) Archean mafic-ultramafic Ikauna layered intrusion, Bundelkhand craton, India: petrography and geochemistry. Arch Anthropol Open Acc 3 (suppl-2):334–340. https://doi.org/10.31031/AAOA.2018.03.000557

Slabunov AI, Singh VK, Bayanova TB et al (2020) The felsic volcanics of the Central-Bundelkhand greenstone complex, Bundelkhand craton, India: new geochronological, Sm-Nd data and geodynamics setting. Abstracts of the recent trends in geoscientific research on Dharwar craton and other Indian Precambrian terrains. Hyderabad, India, pp 53–54

Sunder-Raju PV, Hanski E, Lahaye Y (2015) LA-MC-ICP-MS dating of zircon from chromitite of the Archean Bangur Gabbro complex, Orissa, India Ambiguities and constraints. Geol Acta 13:325–334. https://doi.org/10.1344/GeologicaActa2015.13.4.5

Upadhyay D, Chattopadhyay S, Kooijman E et al (2014) Magmatic and metamorphic history of Paleoarchean tonalite trondhjemite granodiorite (TTG) suite from the Singhbhum craton, eastern India. Precambr Res 252:180–190

Verma SK, Verma SP, Oliveira EP, Singh VK, More JA (2016) LA-SF-ICP-MS zircon U-Pb geochronology of granitic rocks from the central Bundelkhand greenstone complex, Bundelkhand craton, India. J Asian Earth Sci 118:125–137

Windley BF, Kusky T, Polat A (2021) Onset of plate tectonics by the Eoarchean. Precambr Res 35:1059–1080. https://doi.org/10.1016/j.precamres.2020.105980

Witze A (2006) The start of the world as we know it. Nature 442:128–131

Yang Q-Y, Santosh M (2015) Zircon U-Pb geochronology and Lu–Hf isotopes from the Kolar greenstone belt, Dharwar craton, India: implications for crustal evolution in an ocean-trench-continent transect. J Asian Earth Sci 113(2):797–811

Structure and Geological Processes of the Earth: Seismic Evidences from the Indian Shield

V. Vijaya Rao and Damodara Nara

Abstract

The Earth is a dynamic planet with a large number of geological processes taking place inside it. The internal processes lift and build the Earth's surface, whereas the external processes tend to destroy the shape of the Earth's relief by weathering, erosion and hydrological cycle. The paper places the crustal structure of the Indian shield in the context of global processes and evolution of the Earth. It summarizes internal structure of the Earth from core to crust and addresses numerous geological processes, like evolution of continental crust, operation of plate tectonics, convection currents, mantle plumes and supercontinental episodes using seismic and seismological images. Seismic images provide key evidences of the geological processes that are being taking place inside the Earth. Seismic images from the Archean and Proterozoic terrains of the Indian shield show evidence for operation of plate tectonics since Neoarchean. Further, they suggest differences in mode of subduction process during these periods. They provide evidences for the associated processes, like delamination, asthenospheric upwelling, and generation of younger Moho due to crust-mantle interaction. Evolution of sedimentary basins, passive continental margins, the Deccan Volcanic Province, Rajmahal traps and magmatic underplating are interpreted in terms of mantle plume activities using seismic images. Seismological data from Burmese arc region suggest an overturn of subducted Indian lithospheric slab at the transition zone (410–660 km depth) and its slow sinking leading to detachment. Further, remnants of detached lithospheric slabs from the Himalayan-Alps orogenic system are observed at various mantle depths ranging between 1000 and 2300 km beneath the Indian shield. Such displaced slab material pushes the other material to rise somewhere and thereby generate plumes/super plumes. These plumes break the supercontinent, which in turn cause plate movements on the surface. Tomographic studies suggest whole-mantle convection and provide key evidences for the mantle processes. The role of both the internal and external geological processes that shape the Earth are explained from the evolution of Himalayas and generation of huge sediments from it with the formation of world's largest Bengal fan.

Keywords

Geological processes • Crustal structure • Plate tectonics • Lithosphere • Convection currents • Mantle plumes • Large Igneous Provinces • Active–Passive seismics • Indian shield

1 Introduction

Geological processes contribute kinetic and dynamic forces that help to shape various zones including lithosphere, surface, and crusts of each planet. These processes may be linked to interaction with external environments like impact cratering, the atmosphere, the hydrosphere, and internal mechanisms like tectonism and volcanism. For example, impact cratering played a prominent role in forming and shaping the planetary crusts in the beginning of solar system history. The planet's geological history tells how these geological processes are involved in generating the products or deposits and their arrangements relative to one another. The geological processes are responsible for the Earth's landforms and surfaces. Some of the processes are destructive and others constructive. The geological processes are classified as internal and external.

Both internal and external processes need a source of energy to drive them. The Earth was very hot when it was formed at ~4.56 Ga (Ernst 2007). Slowly it has started cooling by releasing the internal heat at the mid-oceanic

V. V. Rao (✉) · D. Nara
CSIR-National Geophysical Research Institute, Uppal Road, Hyderabad-500007, India

A. K. Shandilya et al. (eds.), *Geological and Geo-Environmental Processes on Earth*, Springer Natural Hazards,
https://doi.org/10.1007/978-981-16-4122-0_5

ridges, volcanoes, mantle plumes, etc. in the form of volcanism. Further, disintegration of radioactive elements also generated heat inside the Earth. This heat of the Earth drives the internal geological processes, which usually occur far from the surface. They are responsible for movement of lithospheric plates, occurrence of earthquakes and volcanos, opening of new oceans and closing of old ones, formation of mountain ranges, etc. Rock deformation, crystallization and metamorphism are other important internal processes. The internal processes build the Earth's landforms and surfaces.

External processes shape the relief created by internal processes. Internal processes lift and build the Earth's surface, whereas external processes tend to destroy the shape of the Earth's relief. Weathering and erosion, hydrological cycle, deposition and soil formation, glaciers, rivers, streams, mass movement, land-sea interaction etc. are the major external processes. Solar energy is responsible for the external geological processes. It creates difference in atmospheric temperature, causes winds, responsible for water-cycle- evaporation and rain fall.

The main activities of the geological processes are erosion, weathering, operation of plate tectonics and associated processes. Gravity also plays a vital role in both internal and external geological processes. Gravity contributed in forming the convection currents in the Earth's interior. These convection currents will sink the dense (cold) material by replacing the lighter (hotter) materials that rise towards the surface. On the Earth's surface, gravity plays a role in moving the accumulated water and ice on the continents towards the sea by forming rivers and glaciers, which shape the rocks along their path due to erosion in due course.

In this contribution we attempt to evaluate the role of convection currents, plate tectonics, mantle plumes and other geological processes on the structure, composition and evolution of continental crust. Surface manifestations of some of these geological processes are discussed. The main objective is to show the important deep seismic images acquired along various transects within the Indian plate that constitute key evidences of the geological processes that are being taking place inside the Earth. A brief discussion of the role of external geological process on the evolution of plate tectonics and biosphere and the relief created by the internal processes is included.

2 Structure and Properties of the Earth

It is believed that the Earth along with the solar system was evolved from the dense cloud of interstellar gas by condensation at ~4.56 Ga (Ernst 2007). As Earth started cooling slowly chemically distinct layers, namely, the core, mantle and the crust were formed due to differentiation process, where high-density (Fe, Ni) melts settle at the bottom (core) and lighter ones to the top (crust) (Fig. 1a). Evidence for the various layers of the Earth is provided by the earthquake and active source seismic data. Radius of the earth is ~6400 km.

Core The core is the innermost layer, located at the centre of the Earth that is rich in iron and nickel. It is divided into two layers—the inner and outer cores. The inner core is solid with a density of 13 g/cm^3 and a radius of ~1220 km. The outer core with a density of 11 g/cm^3 surrounds the inner core and has an average thickness of ~2250 km. It is identified as a fluid layer by the earthquake data using the P and S-waves (Fig. 1c). The boundary between the mantle and core is marked by a substantial drop in P-wave velocity and S-waves do not propagate beyond this boundary as the outer core is in fluid form.

Mantle The mantle is located just above the core and surrounds it. It has a thickness of ~2900 km and comprises about 83% of the Earth's volume. It is divided into upper mantle, lower mantle and asthenosphere based on composition, rheology and P and T conditions. The upper mantle exists from the base of the crust downwards to a depth of ~670 km. This region is believed to be composed of ultramafic peridotite made up of the olivine and pyroxene minerals. The rocks in the upper portion of the mantle are more rigid and brittle because of cooler temperatures and low pressures. The lower mantle located below the upper mantle extends from a depth of 670 to 2900 km. This layer is hot and plastic. Minerals that are formed here are different from those of the upper mantle due to the presence of higher pressure and temperatures.

Crust The cool, rigid, and brittle part located on top of the mantle is called the crust. There are two types of crust—the oceanic crust and continental crust (Fig. 2). Both types of crust are less dense than the rock found in the underlying upper mantle. The Oceanic crust is thin and measures between 5 and 10 km thick. It is composed of basalt and has an average density of about 3.0 g/ cm^3. The continental crust is 20 to 80 km thick and composed mainly of lighter granitic rocks (Christensen and Mooney 1995). The density of continental crust is about 2.84 g/cm^3. The continental crust is a layer of igneous, sedimentary and metamorphic rocks. The continental shelves are part of it. The continental crust is also less dense than oceanic crust although it is considerably thicker. About 40% of the Earth's surface is now covered by continental crust, but continental crust makes up about 70% of the volume of Earth's crust. The continental crust is

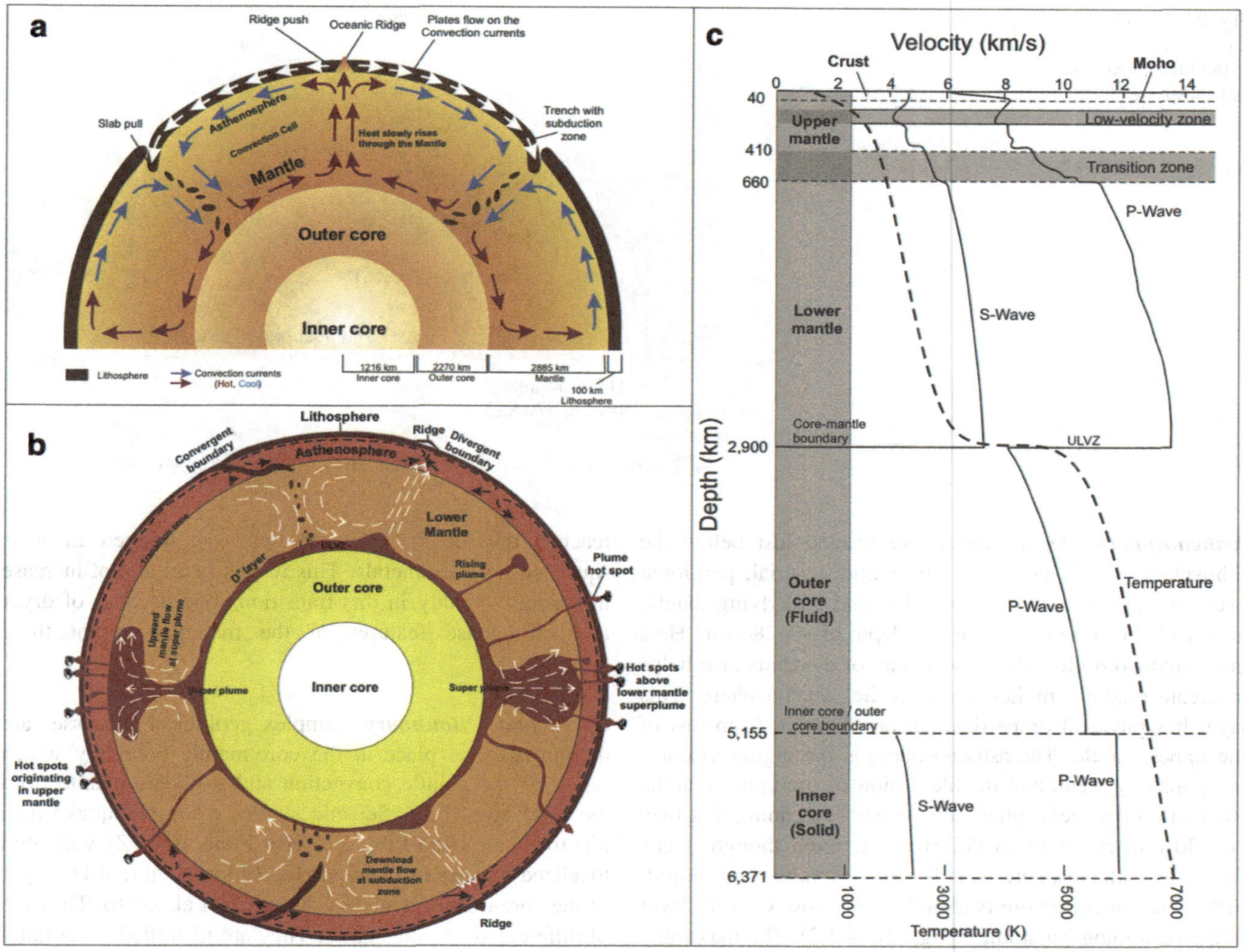

Fig. 1 **a** Structure of the Earth with convection currents and lithospheric plates, **b** Structure of the Earth along with mantle plumes and ultra-low-velocity-zones (ULVZs) located at the core-mantle boundary (D" layer). Partially molten rocks are identified as ULVZ by seismic tomography. Presence of ULVZ beneath the broad mantle plume indicate a close dynamical link between them, **c** P- and S-wave velocities-depth along with temperature profile of the Earth

thinnest in areas like the rift valleys of East Africa and the Basin and Range Province, USA and thickest beneath mountain ranges like the Himalayas. The Mohorovicic discontinuity/the Moho is a compositional (chemical) boundary that separates silica-rich rocks of the crust from olivine-rich rocks of the mantle.

Lithosphere The lithosphere is a solid layer that includes the crust and the upper most portion of the mantle, which constitute the hard and rigid outer layer of the Earth (Figs. 1a and 2). The thickness of the lithosphere is the depth to the ~1300 °C isotherm where olivine begins to deform viscously. It is associated with the transition between brittle and viscous behaviour. Lithospheric thickness varies depending on the tectonic and magmatic activity of the region. It is between 80 and 280 km for continental regions, 50 to 140 km for oceanic regions and very thin beneath the mid-ocean ridges. The lithosphere is buoyant because it is attached to the crust of lower density and glides over the rheologically weak asthenosphere. Heat transfer in the lithosphere is by conduction process, unlike convection in the mantle.

The Earth exhibits two types of lithospheres—the oceanic and continental. New oceanic lithosphere is constantly being produced at mid-ocean ridges and is recycled back to the mantle at subduction zones, so oceanic lithosphere is much younger than its continental counterpart. The oldest oceanic lithosphere is about 200 Ma compared to the continental lithosphere, which is ~4.00 Ga old. The oceanic lithosphere thickens as it ages and moves away from the mid-ocean ridge. It mainly consists of mafic crust and ultramafic mantle and is denser than continental lithosphere.

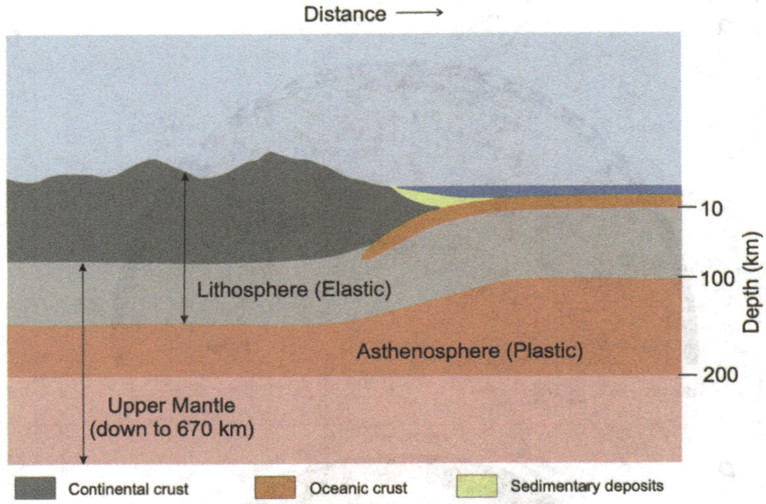

Fig. 2 Structure of the crust and topmost layer of the upper mantle—the lithosphere and asthenosphere

Asthenosphere The asthenosphere located just below the lithosphere is a layer where the mantle material, peridotite become weaker than both overlying and underlying mantle at ~1300 °C, corresponding to a depth of 80–280 km. Heat generated from the radioactive decay of elements also helps to create magma (molten rock) in the asthenosphere. This layer has physical properties that are different from rest of the upper mantle. The asthenosphere is the highly viscous, mechanically weak and ductile region of the upper mantle. Rocks in the asthenosphere are "plastic", meaning that they can flow in response to deformation. Even though it can flow, the asthenosphere is still made of solid (not liquid) rock. The asthenosphere is identified as a low-velocity layer in the seismological studies (Figs. 1c and 2). The maximum thickness of the asthenosphere can be around 350 km.

Lithosphere-Asthenosphere Boundary (LAB) LAB is defined as a rheological boundary (mechanical) boundary and its upper and lower units, namely, the lithosphere and asthenosphere respond differently to the applied stress. For very long periods of geological time, the lithosphere deforms elastically and remains rigid through brittle failure. In contrast, the asthenosphere deforms viscously and accommodates strain through plastic deformation. Although the thermal boundary differentiates the lithosphere and asthenosphere, they are chemically similar.

Transition Zone (Upper-Lower mantle boundary) At shallow depths (<410 km) mantle peridotite contains olivine. But at greater depths (410–660 km) the atoms in the olivine lattice rearrange to form more compact denser material called β-phase without affecting the overall chemical composition. This zone between 410 and 660 km is called the "Mantle Transition Zone". The bottom of this transition zone is called as 660 km discontinuity. When a subducting slab reaches this depth, seawater that was trapped in it is squeezed out of minerals. This results in an abrupt increase in seismic velocity in this transition zone because of dryer and ultra-dense features of the material present there (Fig. 1c).

Core-Mantle Boundary Complex geological processes are thought to take place at the core-mantle boundary which separates the mantle convection and a silicate mantle from the fluid outer core. Seismic studies from earthquake data identified an Ultra-Low Velocity Zone (ULVZ) with thin localized patches of material (15–30 km height) at D" layer of the core-mantle boundary (Garnero et al. 2016). They are of different sizes and shapes. They are identified as partially molten roots of plumes of hot rock (Fig. 1b) that slowly rise through the mantle to feed volcanoes (Yuan and Romanowicz 2017). These ULVZ are identified as the signature of the base of the world's plumes and upwellings (Fig. 1b and c). Swirling convection currents in the mantle drive the plumes. The seismic studies provided connections between the ULVZ and the plumes beneath the Hawaii, Samoa and Iceland.

3 Geology of India

Indian shield is a mosaic of cratons and mobile belts with 3.8 Ga geological history. It shows a variety of geological features of different periods manifested by different geotectonic processes. The Archean Dharwar, Bastar, Singhbhum, Bundelkhand and Mewar cratons were collided and sutured with the evolution of several orogens/mobile belts between them during different periods since Archean (Fig. 3). Some of these collision/suture zones are reactivated as rifts. The Dharwarian, Aravalli, Delhi, Satpura, Eastern Ghats are the important orogens evolved during the collisional process (Vijaya Rao, 2009). The Western Ghats, the Karnataka, Deccan, Chotanagpur and Shillong plateaus are the important epirogenic uplifts of the Indian shield. The Vindhyan (VB), Chattisgarh (Ch B), Bastar (Ba) and the Cuddapah (CB) basins are the undeformed and unmetamorphosed Proterozoic basins (Fig. 3). The Kutch (KB), Saurashtra (SB), Narmada (NB) Cambay (CB), and west Bengal (BB) are the sedimentary basins formed during the Mesozoic (Fig. 3). These petroliferous basins are located either between the cratons or continental margins. The granulite terrain located in the southernmost part of the Indian shield is evolved during the Late Neoarchean and also Late Neoproterozoic (during the East-African orogen at ~550 Ma). The Indian plate is unique as it was traversed by four major plume activities, namely, the Kergulean, Crozet, Marion and Reunion mantle plumes between ~130 and 65 Ma after separating from the Gondwanaland at ~200 Ma (Fig. 4). Separation of India from the Gondwana fragments, its rapid northward movement (15–18 cm/year) and final collision with the Asian plate at ~55 Ma is responsible for the evolution of Himalayas. A 2500 km long world's highest mountain chain and highest plateau Tibet makes the Himalaya a natural laboratory and a unique feature of the Indian shield.

Internal Geological Processes

Many geological processes operate inside the Earth. But, only a few relevant ones, like evolution of continental crust, operation of convection currents, plate tectonics, mantle plumes and supercontinental episodes are discussed here with reference to the Indian shield. Deep crustal seismic reflection images, velocity-depth models and the geological cross-sections derived from various seismic transects (vertical rectangles numbered 1–6 in Fig. 3) are interpreted in terms of the geological processes. They provide the knowledge of the Earth's working pattern.

4 Evolution of the Continental Crust

The continental crust is evolved from the mantle during the fractional differentiation process, which involves chemical separation by partial melting and outgassing of volatiles. Initially, due to very high temperatures the basaltic lava was erupted, and magma ocean was formed on the surface of the Earth. In an active planet the surface layers are returned back into the mantle and form a continuous distillation like process. Such a recycling process over a very long period produced highly differentiated magma that is distinctly different from basaltic composition and closer to the igneous rock granite. This process was primarily a result of volcanism and subduction. Evolution of the continental crust of granitic composition is a continuous process, and it has been taking place since 4.0 Ga, even after 4.56 Ga years of formation of the planet. The crust produced at the mid-oceanic ridges is cycled back into the mantle at the subduction zones and then upward again. The crustal material continues to undergo chemical processing from basalt to andesite and eventually to granite.

5 Convection Currents

Pressure and temperature increase with depth inside the Earth. The core being at higher temperature than mantle, the heat liberated by it is transferred to the mantle. Further, heat is also transferred from the mantle to the crust and finally to the surface of the Earth. Average temperature of the mantle is ~3700 °C, and the crust is around 500 °C. There exists a huge temperature difference between the mantle and crust. As temperatures rise with depth, rocks reach temperatures that would cause them to melt if they were at the surface. The rocks remain solid at depth despite their temperature because of the extreme pressures acting upon them. However, they do become plastic. Subjected to immense forces, and with vast amounts of time, such rocks will flow very slowly over a very long period.

The hot material being lighter (less-denser) raises and the heavier (more-dense) cold ones sinks. Similar process generates the convection currents in the mantle. Heat from the core and mantle itself causes convection currents. Mantle convection is the very slow creeping motion of Earth's solid silicate mantle caused by convection currents carrying heat from the interior to the planet's surface. Convection currents are movements of the mantle material that are less dense. The hot convection cells start from the core-mantle boundary and slowly rise up to the bottom of the lithospheric. They form mid-oceanic rides above it, dissipate heat and thereby cooldown. These cooled convection cells elongate in the asthenosphere and may be aligned in long cylinders that drive the overriding lithosphere plates along like a conveyor belt (Fig. 1a). A larger-scale circulation transports heat from the volcanic mid-ocean ridge to a deep-ocean subduction trench, where the convection cells travel their backward journey to the lower mantle. Large-scale convection currents may lift and/or lower entire continents. This results in earthquakes, volcanic activity, mountain building and ocean recycling. Thus, the Earth's convective system exert major influence on mantle as well as on continental dynamics. It is the engine that drives the plates to move across the surface.

There is a debate as to whether convection is "layered-mantle" or "whole-mantle" convection. Even though this debate continues, results from seismic tomography, numerical simulations of mantle convection and examination of Earth's gravitational field all suggest the

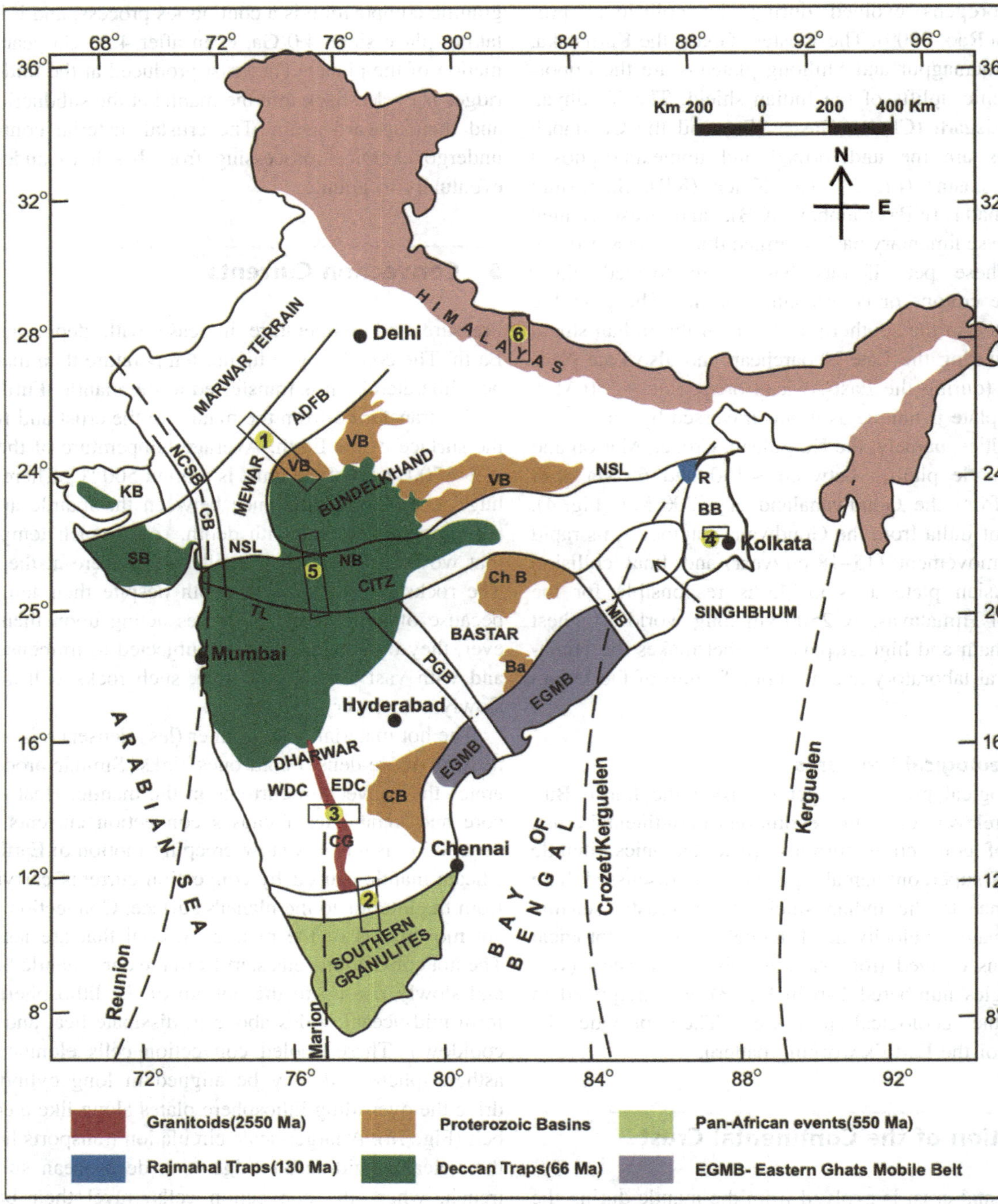

Fig. 3 Geological map of India along with various mantle plume paths superimposed over it. Numbers 1–5 indicate the areas discussed in the text. 1: Aravalli-Delhi fold belt; 2: Dharwar craton-southern granulite terrain boundary; 3: across western and eastern Dharwar cratons (WDC-EDC); 4: West Bengal sedimentary basin; 5: across the Narmada-Son and Tapi lineaments; 6: across the Himalayas

existence of 'whole' mantle convection, at least at the present time (Fig. 1a). In this model, cold, subducting oceanic lithosphere descends from the surface to the core–mantle boundary (CMB) and hot plumes rise from the CMB to the surface (van der Hilst et al. 1997). This picture is strongly based on the evidence from the global seismic tomography models, which identified remnants of subducted lithospheric slabs based on high seismic velocity anomalies in the mantle (van der Meer et al. 2018). An evidence from India further constrain the subducted slabs reaching the core-mantle boundary. Tomographic imaging of India and adjacent regions revealed high P-wave velocities at depths between 1000 and

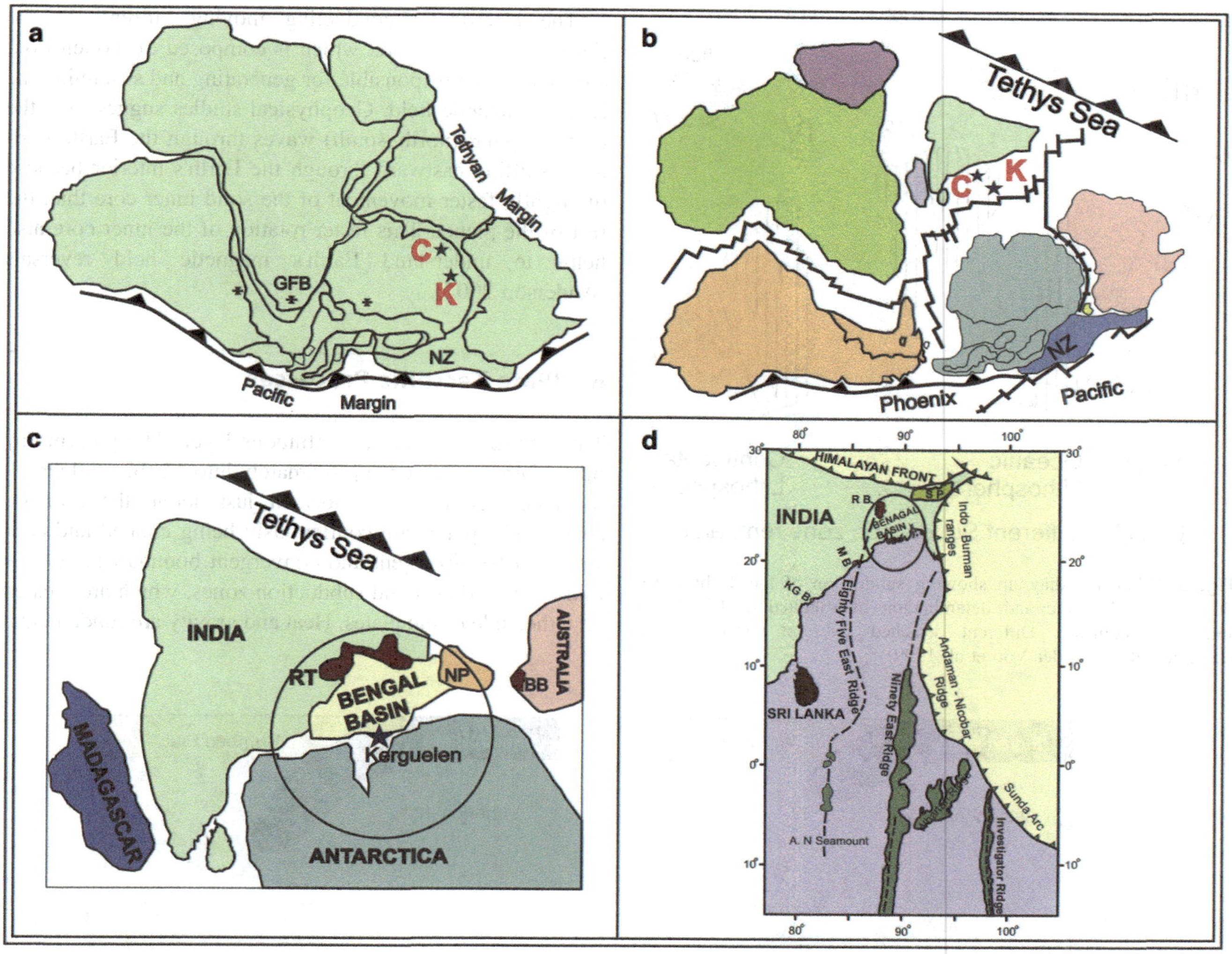

Fig. 4 Breakup of Gondwanaland due to plume activities. **a** Gondwanaland during ~200 Ma, **b** Separation of combined Antarctica and Australia from India due to the Kergulean-K and Crozet-C mantle plume activities during ~130 Ma, **c** Formation of the Bengal basin and evolution of the Rajmahal traps during ~117 Ma, **d** Formation of the Himalayan orogeny during the Eocene. (modified after Damodara et al. 2017)

2300 kms (Fig. 5). They are interpreted as remnants of detached oceanic lithospheric slabs from the Mediterranean-Himalayan-Indonesian orogenic system at various depths (Van der Voo et al. 1999). They were developed due to subduction of the Tethyan oceanic lithosphere that sink into the lower mantle. The Indian plate appears to have overridden its own detached and sinking lithospheric slabs. Similar conclusions are drawn from the paleomagnetic data (Torsvik et al. 2006). Tomographic images also show the origin location of mantle plumes from the ultra-low velocity-zone located at the D" of the core-mantle boundary, then rising and crossing the mantle transition zone (Fig. 1b).

Some recent studies indicate slab stagnation in the mantle transition zone (410–660 km depth) may be due to reduced viscosity (Fukao et al. 2009; Fukao and Obayashi 2013). The reduced viscosity provides some sort of lubrication on the slabs, which make the slab to slip and slide sideways instead of continuing their downward plunge. Seismological data using 60 earthquakes has suggested an overturn of the subducted Indian lithospheric slab at the transition zone in the Burmese arc region (Purnachandra Rao and Kalpana 2005). The data also suggests slow sinking of the overturned slab leading to detachment which is presented in Fig. 6. These deflected slabs move horizontally to a distance of ~1500 km in the transition zone. These overturned lithospheric slabs can be responsible for extension, tectonism and volcanism in overriding continental/intra-plate regions. The 3D convection model studies (Mao and Zhang 2018) suggest that the slabs are trapped only for 20 Ma. Subsequently, they penetrate the lower mantle indicating a transit nature of slab stagnation. It supports whole mantle convection models.

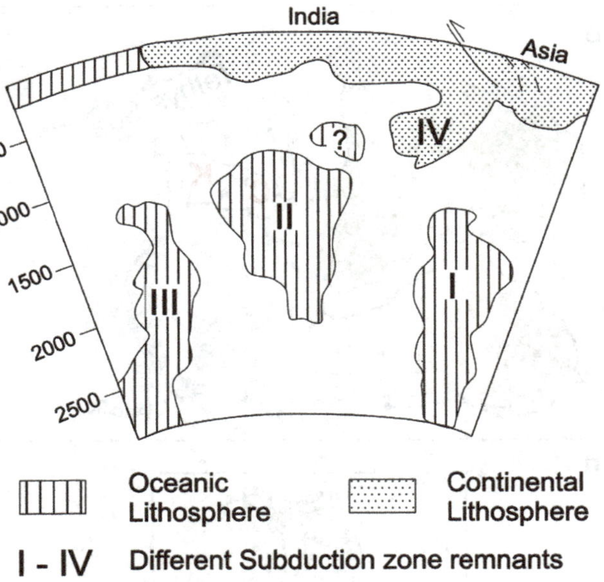

Fig. 5 Schematic diagram showing subduction of the Tethyan sea during the Mesozoic and delamination of subduction slab during India-Asia collision. Different detached slabs at various depths (modified after van der Voo et al. 1999)

The electrically conducting moving currents in the Earth's fluid outer core, which is composed of molten iron, are found to be responsible for generating and sustaining the Earth's magnetic field. Geophysical studies suggest that the route of polar (north–south) waves through the Earth gradually shifting eastward through the Earth's interior because of slightly faster movement of the solid inner core than the rest of the planet. This faster rotation of the inner core also helps to understand Earth's magnetic field reversals (Anderson 2007).

6 Plate Tectonic Processes

The lithosphere is not a continuous layer. The present-day lithosphere is divided approximately into 12 (big and small) movable tectonic plates located just above the asthenosphere. The plates are continuously being created and consumed at the divergent and convergent boundaries, i.e., the mid-oceanic ridges and subduction zones, which are located on either side of the plates. Heat and gravity are fundamental

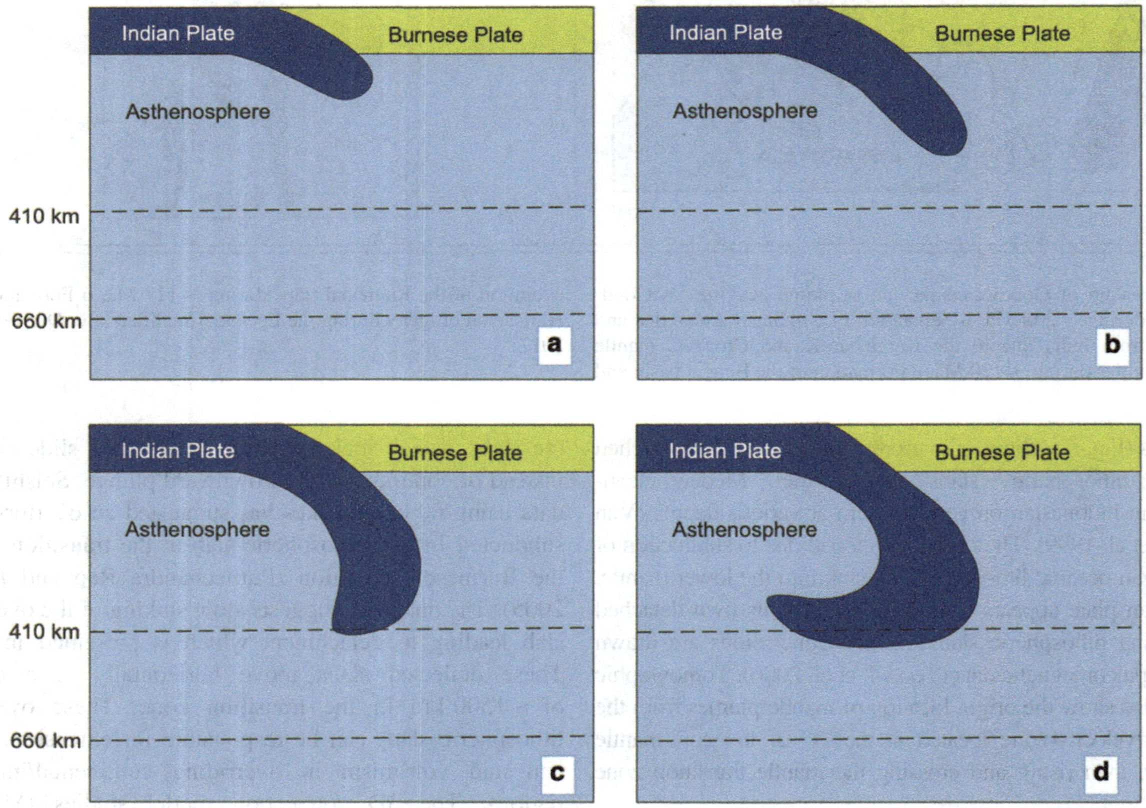

Fig. 6 Schematic model illustrating various stages of subduction of Indian plate beneath the Burmese arc. **a** Subduction of Indian plate, **b** slow subduction and sinking of the slab, **c** overturn of subduction slab at the transition zone located at 410 km depth discontinuity, **d** cessation of subduction and slow sinking of overturned slab (after Purnachandra Rao and Kalpana 2005)

to the process of plate tectonics. Lithospheric plates are part of a planetary scale gravity-driven thermal convection system. The energy source for plate tectonics is Earth's internal heat, whereas the gravity forces move the young hot plates away from spreading ridges (ridge push) and pulls old cold plates down into subduction zones (Slab pull) (Anderson 2007). The plate tectonics are responsible for opening of new oceans, closing of old ones, generation of volcanism and earthquakes, formation of mountain ranges, continental drifting, generation of continental crust and many more activities. The theory of plate tectonics is one of the great scientific advances of the present period that can be compared with the Darwin's theory of evolution and the Einstein's theory of relativity. Mantle convection is the main mechanism for operation of plate tectonics.

Convection currents in the asthenosphere drive the overriding lithospheric plates from the mid-oceanic ridges to the subduction zone. When tectonic plates come together, the Oceanic plate invariably sinks due to its own weight during the processes of subduction at an oceanic trench underneath the overriding continental plate. Then two plates collide with each other forming mountain ranges, like the present-day Himalayas, the Alps, the Andes etc.

Many large-scale surface features are manifestations of deep-seated geological processes. Thus, an understanding of crust may resolve some of the fundamental questions related to the mantle dynamics and evolution of the Earth. In order to understand the origin of crustal structure, it is essential to know the crust-mantle interaction and chemical exchange processes between them. In this context, seismic refraction/wide-angle reflections and high-resolution deep crustal normal-incidence seismic reflection images from several parts of the Indian shield provide necessary inputs. Few such studies, namely the DHARSEIS experiment of the Archean Dharwar craton and the Nagaur-Jhalawar transect across the Paleoproterozoic Aravalli Fold Belt (AFB) of the NW India (Fig. 3) are discussed to understand the evolution of continental crust and the geological processes responsible for the operation of plate tectonics since Neoarchean (Tewari et al. 1997; Vijaya Rao et al. 2000; Krishna and Rao 2011; Vijaya Rao and Krishna 2013; Mandal et al. 2018).

6.1 Paleoproterozoic Aravalli Fold Belt

NW part of the Indian shield exhibits 3.3 Ga geological history. It consists of Paleoproterozoic Aravalli and Late Mesoproterozoic Delhi orogens extending linearly to a distance of 700 km (rectangle-1, Fig. 3). The region consists of intensely folded, deformed and metamorphosed rocks deposited over the Archean basement. It contains ophiolites, high-grade granulites and granitic plutons of different periods starting from Mesoarchean. The region has unique distinction of having the highest reserves of Pb–Zn, phosphorites and evaporate minerals in the country.

A 400 km long multi-fold seismic reflection study across the Paleoproterozoic Aravalli fold belt imaged a crustal-scale dipping reflection fabric extending from the surface to a depth of 38 km (Fig. 7). It is interpreted as the signature of subduction of the Bundelkhand craton beneath the Mewar craton (Fig. 8). Subsequent collision of these two cratons resulted in the formation of Aravalli Fold belt (Vijaya Rao et al. 2000). During this period various crustal blocks in different parts of the world, including NW India, were involved in the process of accretion and formation of Paleoproterozoic Columbia supercontinent (Rogers and Santosh 2004). Generally, mountain belts represent regions where oceans might have opened and closed and they are the products of continental collision (Dewey and Bird 1970).

During the initial stage of collision process there was an increase in crustal thickness at the subduction zone (Fig. 8b). Once the oceanic crust has been consumed in the subduction zone, a continent-to-continent (Mewar-Bundelkhand) collision took place resulting in a further increase in crustal thickness (Fig. 8c). This large crustal thickness becomes gravitationally unstable. Considering the pressure and temperature conditions at depth, metamorphic process that took place within the crustal root as well as the subduction derived magmas resulted in the eclogitization of the crustal root. The high density of the eclogite initiate a delamination process, sink into the mantle and finally thinning out the crust (Fig. 8d). Finally, the orogen collapsed under its own weight due to gravitational instability (Dewey 1988). Then the rise of magma produced by decompression melting of upwelling asthenospheic creeped into shallow depth thereby post-collisional extension took place (Fig. 8d). During this crust-mantle interaction process basaltic magma accreted at the crust-mantle boundary, referred as magmatic underplating. The injected mantle material into the lower-crust results in a layered structure with alternating physical properties, consisting in high and low rock densities and seismic velocities (Mooney and Meissner 1992). Such geological structures with layered mafic intrusions (Deemer and Hurich 1994) resulted in a laminated sub-horizontal seismic signature in the lower crust (Fig. 7). The bottom of this reflection fabric represents the Mohorovicic (Moho) discontinuity. It is an important mechanism of the crustal growth in the vertical direction. This horizontal smoothening process is manifested by the presence of 12 km thick sub-horizontal reflection band at the crust-mantle boundary. Geodynamic processes responsible for the evolution of orogen and its subsequent stabilization are presented in Fig. 8. These processes are similar for most of the orogens observed in any part of the world.

During the collisional episode, crustal material is transferred to the mantle at the subduction zone due to

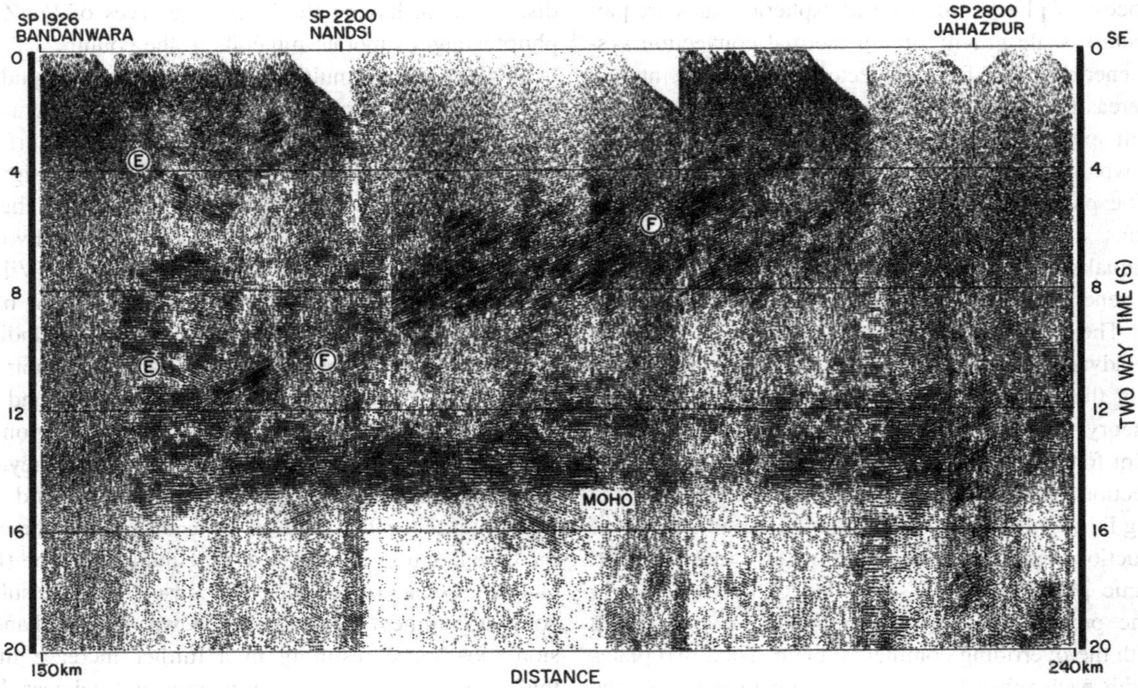

Fig. 7 Crustal-scale dipping reflection fabric identified from seismic reflection study along Nagaur-Jhalawar transect (vertical rectangle-1, Fig. 3). It represents a collisional suture in NW India and responsible for the evolution of Paleoproterozoic Aravalli Fold Belt. Sub-horizontal lower-crustal reflection band indicate post-collisional extensional processes in the region

compressional activity and due to delamination of thickened crust. The mantle material is transferred to the crust as magmatic underplating during the post-collisional extensional processes. Such a dynamic crust-mantle interaction process creates an equilibrated younger Moho that is represented by a sub-horizontal lower-crustal feature in a seismic reflection image (Cook 2002; Carbonell et al. 2013; Reddy and Vijaya Rao 2013). Thus, a two-way material exchange created a new crust-mantle boundary in the Aravalli fold belt region. The life cycle of an orogen, from its formation to destruction, provides the main mechanism for the evolution of continental crust at the convergent margins and its growth at the crust-mantle boundary. The present study suggests that the crustal structure as revealed by seismic data preserves the record of both the past tectonic processes responsible for its formation as revealed by the presence of deeply-penetrating crustal-scale reflection fabric and structural and chemical modification as exhibited by sub-horizontal lower crustal fabric. Vijaya Rao (2009) provides a detailed description of the geodynamics of the collision zones of the Indian shield.

6.2 Archean Dharwar Craton

Operation of plate tectonics during the Neoarchean has been suggested using the deep crustal wide-angle and normal-incidence seismic reflection images of the Indian continental plate (Reddy et al. 2003; Vijaya Rao et al. 2011). Results from two seismic profiles (rectangles-2 and 3, Fig. 3) across the Archean blocks are briefly discussed here.

6.2.1 Dharwar Craton—Southern Granulite Terrain

The seismic data across the Dharwar craton and the Southern Granulite Terrain reveal a gently dipping lower-crustal reflection band extending from 33 to 45 km depth across the Dharwar-Southern Granulite Terrain boundary in the southern part of India (rectangle 2, in Figs. 3 and 9). It is interpreted as a Neoarchean mantle subduction zone of the Dharwar craton (Vijaya Rao et al. 2006). High thermal regime of the Archean period might have decoupled the crust and mantle due to differences in the rheological properties, such as velocity, density and composition between

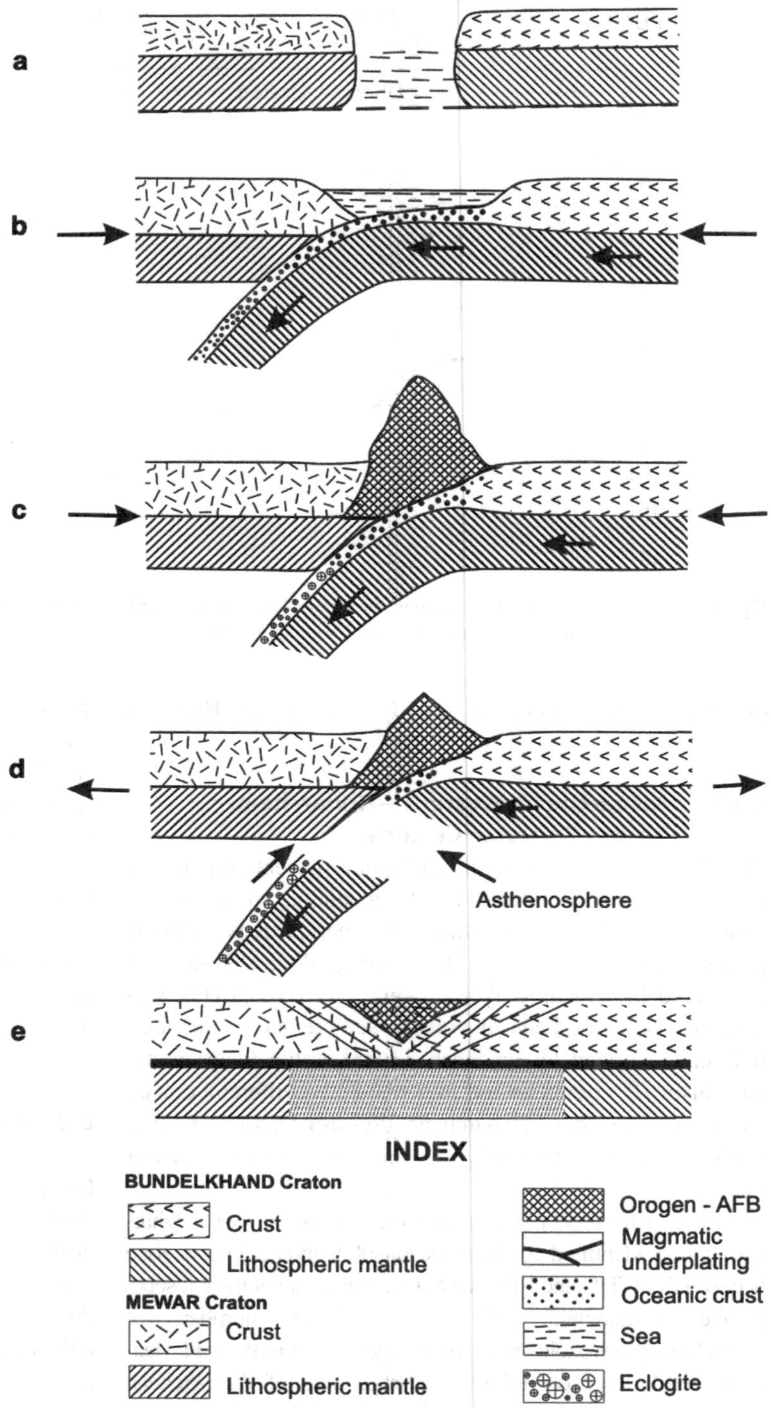

Fig. 8 A cartoon illustrating orogenic cycle involving the transition from compressional to extensional tectonic regimes. **a** Different cratonic blocks (Bundelkhand and Mewar) of NW India separated by an ocean between them. **b** Compression leads to subduction of the Bundelkhand craton beneath the Mewar craton. **c** Further convergence leads to collision and formation of an orogenic belt (Aravalli Fold Belt) along with thickening of crust and eclogatization of the lower crustal material. **d** Transfer of crustal material to mantle—delamination of thickened eclogitized lower crust and uppermost mantle. **e** Lithospheric extension, orogenic collapse and crustal thinning. Transfer of mafic/ultramafic mantle material to the crust and accretion at the lower-crust as magmatic underplating with the formation of sharp and flat Moho as depicted in Fig. 7

them. The Moho here acted as a detachment boundary and thereby only mantle subducted, unlike the whole crustal subduction observed at the Proterozoic Aravalli fold belt region as discussed above (Fig. 7). The lower-crustal/mantle subduction is found to be a common phenomenon during the Archean as revealed by several deep seismic reflection studies from different parts of the globe (BABEL Working Group 1990; Calvert et al.1995; van der Veldon and Cook 2005). Release of carbonic fluids from the supracrustal rocks and volcanic arc environment of the subduction zone are found responsible for the formation of granulites in the region. P–T-t estimates and seismic reflection images suggest that collision processes are responsible for the exhumation of granulites from the lower crustal depth

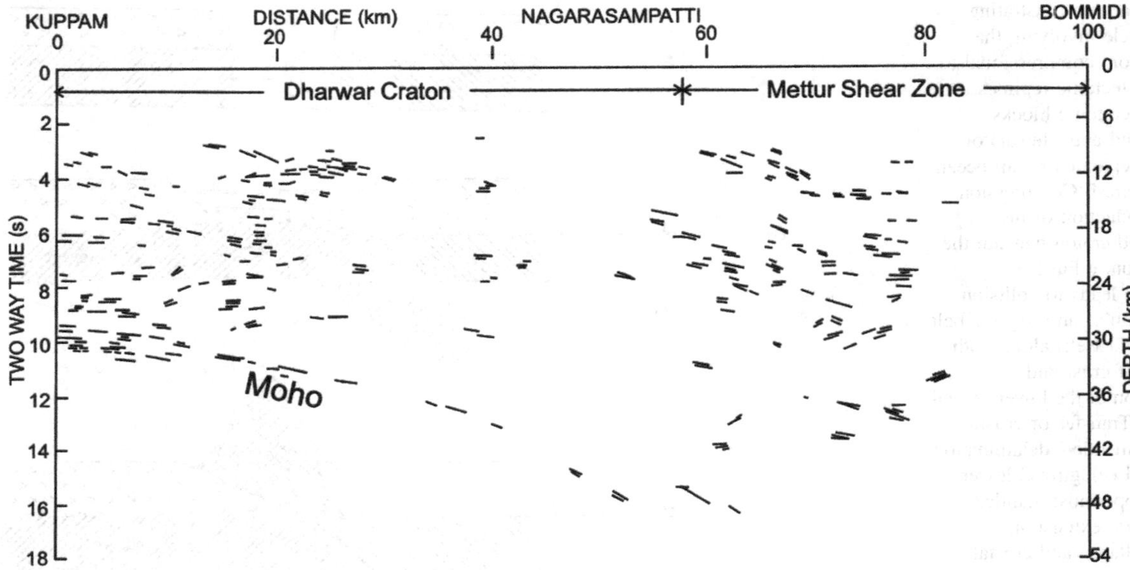

Fig. 9 Neoarchean subduction zone identified across the Dharwar craton-southern granulite terrain using high-resolution seismic reflection study along the Kuppam-Palani geotransect (vertical rectangle-2, Fig. 3)

(Vijaya Rao and Rajendra Prasad 2006; Vijaya Rao and Reddy 2009).

6.2.2 Mesoarchean Western—Neoarchean Eastern Dharwar Cratons

The Dharwar craton of southern India represents one of the largest and oldest cratonic blocks available on the Earth, similar to the Superior province of America. It is a classic granite-greenstone terrain with 3.5 Ga geological history. It is covered by Tonalitic Trondjhemite Gneisses (TTG) with several greenstone belts of volcano-sedimentary sequences. It is characterized by greenschist-amphibolite facies metamorphism. It comprises of two distinctly different crustal blocks, namely, the Mesoarchean Western Dharwar Craton (WDC) and predominantly Neoarchean Eastern Dharwar Craton (EDC).

The data from deep crustal seismic studies across the Dharwar craton identified a Neoarchean subduction-collision zone between the WDC and EDC with a mantle suture at the Closepet granite (rectangle-3, Fig. 3). A large number of Paleo-Mesoproterozoic mafic dikes and sills (French and Heaman 2010) are observed over a large area of the Archean Dharwar craton, southern India (Figs. 3 and 10). These dykes constitute large amounts of magma added to the crust. The relatively vertical dyke geometry is indicative of crustal/lithospheric extension processes which might be a result of post-collision (Vijaya Rao et al. 2015; Mandal et al. 2018). Much of the crustal growth in the craton is ascribed to the Neoarchean collision process.

Seismic studies from the Archean Dharwar craton described above suggest operation of plate tectonics at least from the Neoarchean. It is further constrained from similar seismic studies from different parts of the world (Calvert et al. 1995; van der Veldon and Cook 2005; Cook 2002). Plate tectonic operation is extended further back upto ~3.6 Ga by several researchers using the geological studies (de wit 1998; Smithies et al. 2005; Moyen et al. 2006). Operation of plate tectonics provide one of the important links to understand the mantle dynamics in addition to the knowledge from paleomagnetic data, tomographic models of the mantle, geodynamic modelling, volcanic eruptions etc. Further it links the deep earth processes with surface processes.

6.3 Cretaceous-Tertiary Himalayas

India after separating from the east Gondwana fragments, the Australia, Antarctica and Madagascar started moving in the northward direction during the cretaceous (Fig. 4). The Tethys ocean located north of the Indian plate started subducting and gradually disappeared resulting in continental collision between India and Asia in the early Tertiary at ~55 Ma with the evolution of Himalayan orogen. Surface area of the "Greater India" before collision was around 1000 km more than the present day. This along with the seafloor spreading likely occurred within the Neo-Tethys (during India-Australia separation) suggest at least 4000 km of Tethys oceanic lithosphere has gone down into the mantle during the subduction (van der Voo et al. 1999). Remnants of slab material derived from such a huge subduction are identified in the deep mantle between 1000 and 2300 km depth by tomographic studies (Fig. 5).

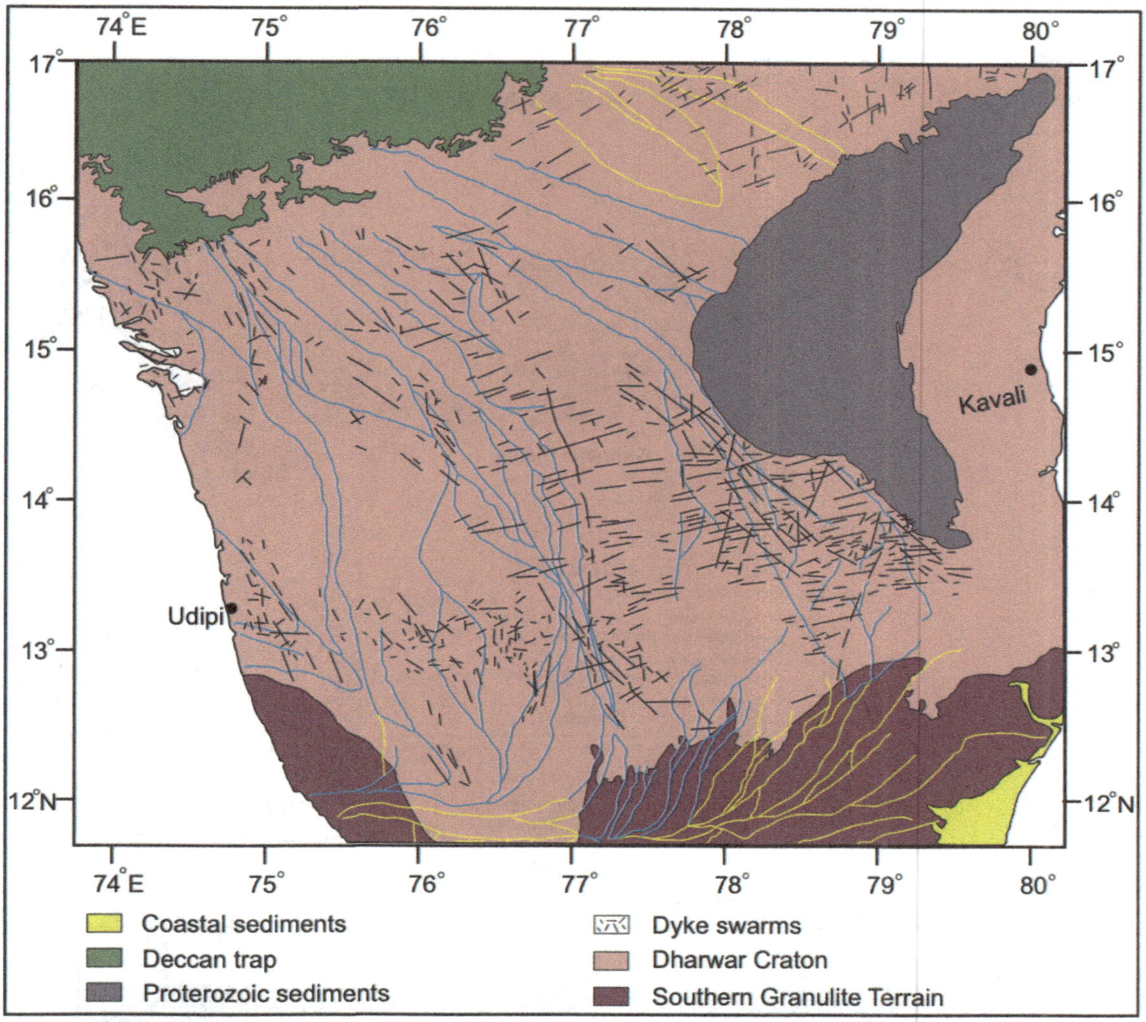

Fig. 10 Proterozoic Dike swarms of the Archean Dharwar craton. They are the manifestation of extensional activity in the region (modified after Vijaya Rao et al. 2015)

To the east of Eastern Himalayan syntaxis, India collided with the Burmese plate with the evolution of Indo-Burmese ranges in the north–south direction (Figs. 4d and 6). Further south of Burmese ranges continental collision transitions into oceanic–continental subduction along the Andaman-Sumatra subduction zone (Fig. 4d). During the continental convergence India rotated anticlockwise direction at an angle of ~20° and generated western and eastern syntaxes.

Structural architecture of the Himalayan fold-thrust belt is dictated by the underthrusting of Indian plate beneath the Tibetan plateau along the plane of detachment, known as the Main Himalayan Thrust (MHT) (Zhao et al. 1993). The Himalayan orogen joins different tectonostratigraphic domains, namely, the Sub-Himalaya, Lesser Himalaya, Greater Himalaya and Tethyan Himalaya (Fig. 11a). These domains are separated by the Main Boundary Thrust (MBT), Main Central Thrust (MCT) and South Tibetan Detachment (STD). They lie between Main Frontal Thrust (MFT) in the south and Indo-Tsango Suture (ITS) in the north (Fig. 11b). The Sub-Himalaya is the foreland basin consisting of clastic syn-orogenic sediments derived from the uplift, erosion and deposition by the Himalayan rivers and subsequently folded and faulted to form the Siwalik Hills at the foot of the mountain chain. The imbricated thrust stacks like MFT, MBT and MCT are terminated at the detachment layer, the MHT (Fig. 11b).

Several geophysical and geological studies were conducted to understand the structure, geometry and deformation of the Himalayan collision zone and its causative mechanisms (Zhao et al. 1993; Nelson et al. 1996; Kind et al. 2002; Rajendran and Rajendran 2005; Kumar et al. 2006, 2013; Kayal 2008; Gao et al. 2016 and the references there in). A high-resolution deep seismic reflection study

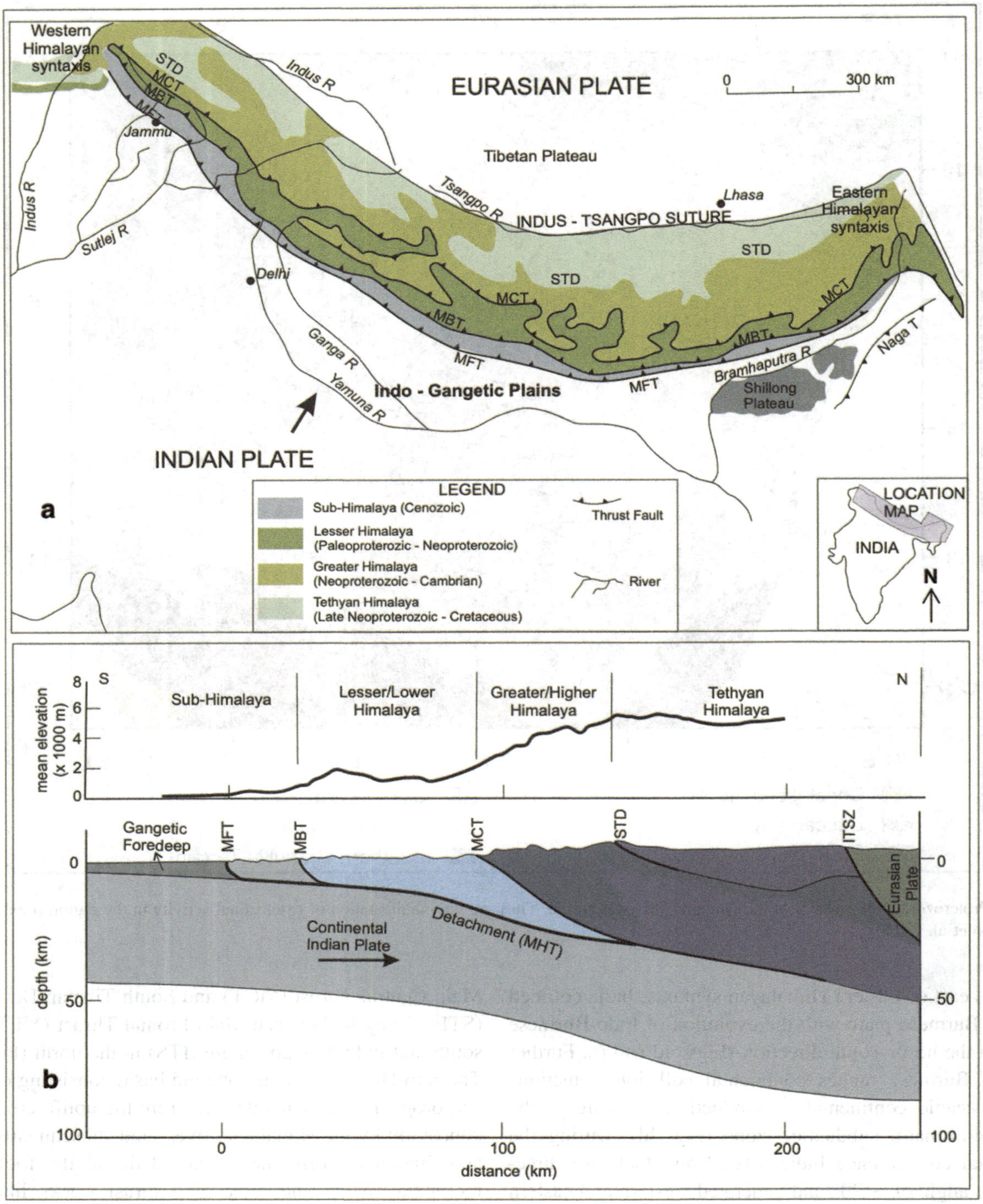

Fig. 11 a Generalized geological map of Himalaya along with the Indian-Eurasian plate boundary marked (vertical rectangle-6, Fig. 3). MFT-Main Frontal Thrust; MBT-Main Boundary Thrust; MCT- Main Central Thrust; STD-South Tibetan Detachment; ITS- Indus Tsangpo suture. **b** Generalised cross-section across the Himalaya in the general NE-SW direction. Sub-Himalaya, Lesser Himalaya, Greater Himalaya and Tethyan Himalaya along with their tectonic boundaries marked over it

conducted in the western part of Tethyan Himalaya of the southern Lhasa block near the Manasarovar has identified a crustal thickness of 70–75 km beneath the Indo-Tsangpo suture (Gao et al. 2016). Further, the above study imaged the MHT dipping with an angle of $\sim 20°$ to a depth of 60 km beneath the Indo-Tsangpo suture (ITS). The Indian crust has reduced from original ~ 40 km thickness to ~ 15 km beyond ITS with transfer of material from the lower plate to the upper plate by crustal duplexing. The lithospheric thickness in the foredeep region of the Himalaya is around 140 km (Kumar et al. 2006).

The underthrusted Indian plate generates a lot of stress, and the accumulated strain is released through structures evolved during the collision. Some of these structures generated most damaging earthquakes of Mw ~8.0. They have caused severe damage in the Indo-Gangetic alluvial plains. Great earthquakes generally initiate at brittle-ductile transition of the MHT and propagate along ramp-flat structures (MFT, MBT, MCT) of the brittle part of the crust. However, significant earthquakes have also been observed above and below the MHT. Seismological data indicate presence of two distinct earthquake zones (Chen and Yang 2004; Monsalve et al. 2006) one above the MHT in the upper crust (<20 km depth) and the other below the MHT in the crust and upper mantle (60–100 km depth). For example, the 2011 Sikkim earthquake of Mw 6.9 is located at the crust-mantle boundary. Similar intracontinental earthquakes ranging between 4.9 and 6.0 magnitude are observed in the mantle portion of the lithosphere around 100 km depth. They are observed beneath the western Himalayan syntaxies in the Kunlun Mountains and the southern Tibet at Xigaze. Arora et al. (2014) using the aftershocks of 2011 Sikkim earthquake and gravity models suggested a strong eclogite body in the lower crust which serves as a repository of high stress. Occurrence of earthquakes in the lower-crust and upper mantle indicates evidence for strong lithospheric lower-crust and upper mantle in the Himalaya to sustain brittle failure.

Some of the processes within the Himalaya are loss of Indian crust by subduction, transformation of mafic crust into eclogites, transfer of Indian crust into the Himalayan thrust sheets by crustal-scale thrust duplexing or partially molten channelized return flow and earthquake generation (Nelson et al. 1996; Sapin and Hirn 1997; Gao et al. 2016).

7 Mantle Plumes—Intraplate Volcanism

Most of the volcanoes occurred at the plate boundaries, e.g., the boundary of the Pacific Ocean, referred as "**the Ring of Fire**". However, some volcanoes occur in isolation, e.g., the big island of Hawaii erupted in the middle of the Pacific Ocean/plate, far distant from any volcanoes. Similarly, the Deccan and the Rajmahal volcanics of the Indian shield occurred on land in isolation (Fig. 3). The isolated volcanic sites are referred as the hot spots. There are about 100 such hot spots currently active on the surface of the Earth (Zhao 2007). It is believed that hot-spot volcanoes form above mantle plumes with columns of hot mantle rising from just above the core-mantle boundary (Coffin and Eldholm 1994; Duncan and Richards 1991).

The subducting oceanic lithosphere being denser than the mantle above, it tends to accumulate at the 660 km discontinuity. A large number of such subducting slabs reaching the mantle transition zone destabilize the 660 km discontinuity layer and sink into the lower mantle (van der Holst et al. 1997; Bijwaard et al. 1998; van der Meer et al. 2018). This process is known as a "slab avalanche" (Condie 1998). These displaced slab material pushes the other material to rise somewhere and thereby generate plume or super plumes (Fig. 1). Deep mantle tomography, numerical models and paleomagnetic study (Zhao 2004; Torsvik et al. 2006) envisaged the plumes as having narrow central conduit arising from D" thermal disturbance layer at the core-mantle boundary. Plume heads may reach diameters of the order of ~1000 km on reaching base of the lithosphere (Fig. 1b).

Based on the source of the plumes, three types of plumes are identified (Courtillot et al. 2003). (1) Primary-deep plumes rising from D" layer of the core-mantle boundary: (2) Secondary-plumes originating from the top of deep plumes; (3) Tertiary-plumes originating from 660 km discontinuity. Plumes have the capacity to generate large quantities of basaltic magma (e.g., continental flood basalts) before continental breakup and to produce chains of volcanic islands when oceanic plates move over fixed hot spots, e.g., Hawaiian Islands. They also cause plate movements on the surface (Gurnis 1988).

Insights brought out on the geological processes from seismic studies from plume affected areas of the Indian shield are presented in the following sections.

7.1 Mesozoic Mantle Plume Activity over the Indian Shield

India, Antarctica, Australia, Madagascar, Africa and South America were part of Gondwanaland, which was major constituent of the Pangea supercontinent during the Phanerozoic (~250 Ma). These continental blocks got separated due to several mantle plumes, namely, the Kerguleon, Crozet, Marion and Reunion at different times (Fig. 4). Breakup of the Gondwnaland into smaller continents during the Mesozoic (Storey 1995) is the best example to understand the role of mantle plume activity, mantle dynamics and crustal evolution.

Breaking of the Pangea supercontinent during the Mesozoic developed passive continental/volcanic rifted margins on both the eastern and western part of India. The margins are characterized by seaward dipping reflectors (SDRs), significant volcanism, high-velocity lower-crustal layers, formation of rift basins and presence of mafic dyke swarms.

7.1.1 East-Coast of India

The East Gondwana, one of the major constituents of the Pangea supercontinent started breaking up due to Kerguelen hotspot activity during the early Cretaceous (Storey 1995). India separated from the combined Antarctica-Australia with the evolution of Indian Ocean during ~117 Ma (Gaina et al. 2007). Breaking of the East Gondwana due to mantle plume activities is shown in Fig. 4. The east Gondwana breakup episode has developed the Bengal basin and influenced the architecture of the Mahanadi basin in the eastern margin of India (rectangle-4, Fig. 3). During this period widespread volcanism occurred with extrusion of continental flood basalts, the Rajmahal volcanics/traps. Seismic studies identified the Rajmahal traps in the Mahanadi and West Bengal basins. The basin architecture of the west Bengal sedimentary basin derived from the seismic tomographic study (Damodara et al. 2017) is presented in Fig. 12. Deep crustal refraction/wide-angle reflection study identified 10 km thick high-velocity layer at the base of the crust, usually referred as the magmatic underplating, which is related to the plume activity in the region (Mall et al. 1999).

7.1.2 West-Coast of India

The extensional activity during the supercontinent Pangea beakup event developed the Kutch, Saurashtra, Cambay and the Narmada peri-cratonic rift basins on the western continental margin of India (Biswas 1987). Subsequently, the Madagascar was separated from India due to the Marion plume and the Seychelles due to the Reunion mantle plume activity during 88 Ma and ~65 Ma respectively.

Continental rifting and associated basin formation due to the Reunion plume activity are related to lithospheric thinning with ascent of asthenospheric mantle. Mafic magma generated during such process extruded as the Continental Flood Basalts (CFB) resulting in the formation of Deccan Volcanic Province at the Cretaceous-Tertiary (K-T) boundary (Fig. 13a). Such flood basalt regions worldwide are referred as the Large Igneous Provinces (Coffin and Eldholm 1994). The Deccan trap of India is one of the largest known CFB provinces on the earth. The Karoo, Columbia River and Siberian traps are other important CFB regions of the world. The Deccan flood basalts are spread over 500,000 sq. km and covered central and western India. The lithospheric extension during this event emplaced 65 Ma old widespread mafic dikes and dike swarms (Fig. 13b) in the general N-S and E-W direction (Bhattacharji et al. 1996). The Narmada-Tapi (E-W), Cambay and North Cambay-Sanchor (N-S) and the western margin (N-S) rifts are evolved or reactivated during the formation of triple junction (Figs. 3 and 13a). The location where all the three rifts cross each

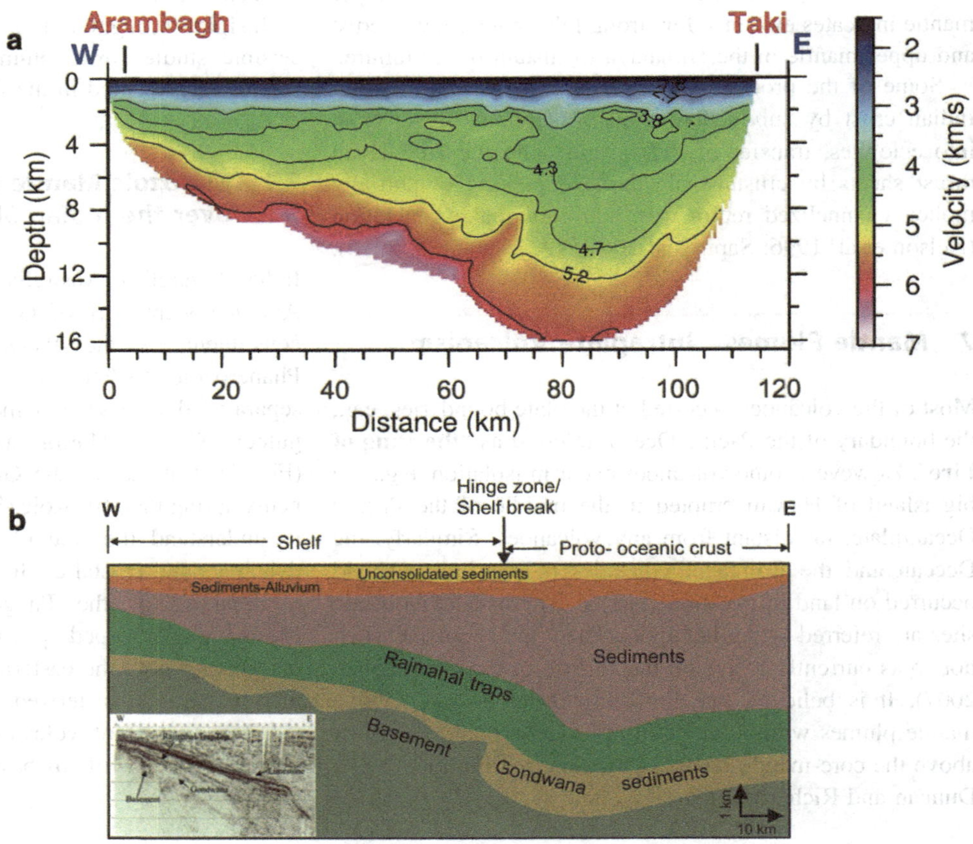

Fig. 12 **a** Basin configuration of the West Bengal sedimentary basin (vertical rectangle-4, Fig. 3) showing 15 km thick sediments in the deeper part of the basin. **b** A schematic diagram showing the basin architecture of the West Bengal sedimentary basin (modified after Damodara et al. 2017)

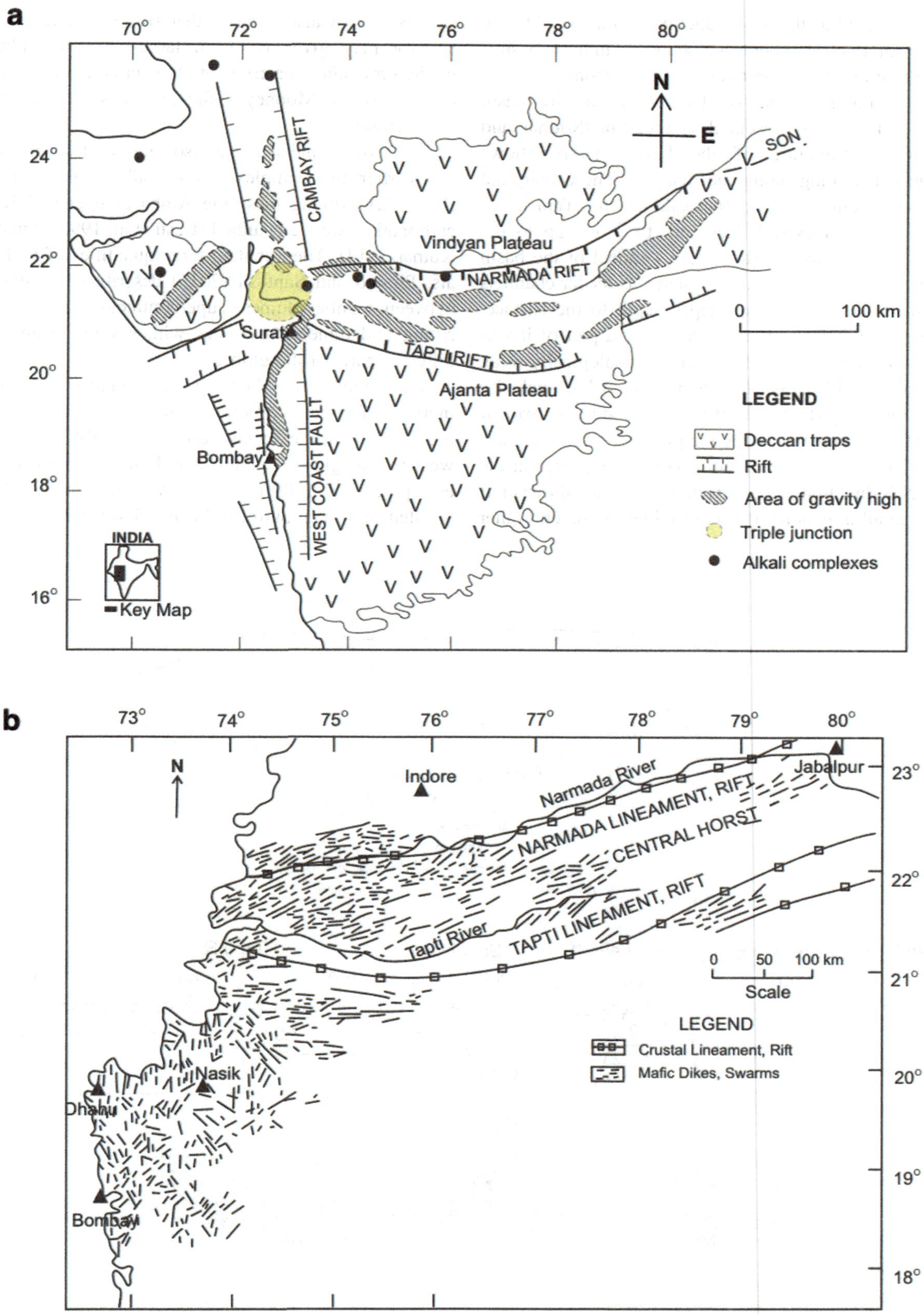

Fig. 13 **a** Deccan volcanic province located in the central and western India is shown along with gravity high locations marked over it. The Narmada-Tapi, Cambay, western margin rifts marked over it, **b** Mafic dikes in Deccan volcanic province. Note E-W and N-S trending dikes associated with the Deccan flows and the direction of rifting (modified after Bhattacharji et al. 1996)

other is considered as the triple-junction (Burke and Dewey 1973). 1–5 km thick sediments are deposited in these basins. Some of these basins are producing hydrocarbons.

Basement configuration of the Narmada rift basin (rectangle-5, Fig. 3) is presented in Fig. 14a (Sridhar and Tewari 2001). Structure of the basin clearly shows high-angle rift faulting along the margins. These marginal faults are coinciding with the Narmada and the Tapi lineaments. These faults extend deep into the crust up to the Moho as shown by the crustal velocity model of the basin (Fig. 14b). These faults might have acted as feeder channels to transport the magma from deeper depth to the surface. During the basaltic flow through the crust, a part of it was accreted as magmatic underplating at the rheologically weak lower-crust (Fig. 14b), which is identified as high-velocity layer (7.1 km/s) by deep crustal refraction studies (Sridhar et al. 2007). It is constrained by the presence of mafic dikes for a long distance, covering entire Deccan trap area in the E-W and N-S direction (Fig. 13b) indicating lateral transport of magma in all directions. It is constrained by several other studies, which also suggest that rifting and mantle plume activity may give rise to a large-scale underplating of mafic/ultramafic material at the crust-mantle boundary (Catchings and Mooney 1988; Fyfe 1993; Coffin and Eldholm 1994).

The extensional activity also emplaced several alkaline and carbonatite complexes along all the three rift basins (Fig. 13a). Among these the Amba Donger and Kamathai carbonatites are well studied (Basu et al. 1993; Bhushan and Kumar 2013). They contain large quantities of REE minerals. Pirajno and Santosh (2015) described the relationship between mantle plumes, supercontinents, intra-continental rifting and mineral resources with several examples from different parts of the globe.

When plume rises beneath a continent the dynamic and thermal effects may produce surface uplifts up to 2.0 km in the centre of the extensive dome (Cox 1989). The present western margin of peninsular India is marked by the escarpment of 1500 km long Western Ghats, which are uplifted to a maximum height of 1.5 km due to the Marion

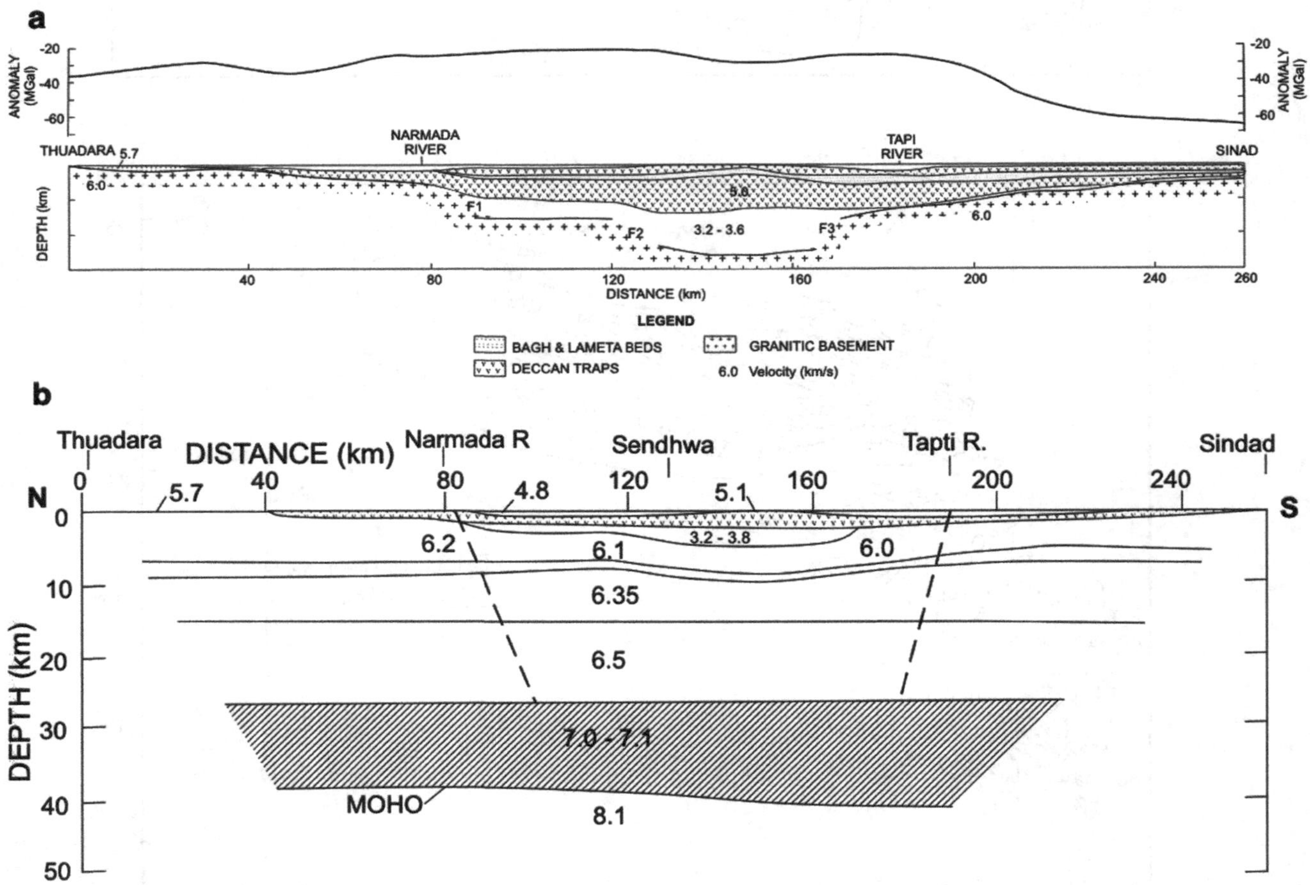

Fig. 14 **a** Configuration of the Narmada rift basin (vertical rectangle-5, Fig. 3). It is formed during the late Cretaceous period and related to the breakup of Pangea supercontinent (modified after Sridhar and Tewari 2001), **b** Crustal structure of the Narmada rift basin (Sridhar et al. 2007 with the permission from JGSI)

and Reunion mantle plume activities. Most of the rivers rise close to the west-coast, drain eastwards into the Bay of Bengal after travelling a distance of more than 1000 km. It suggests the role of mantle plumes in the development of geomorphology and drainage patterns of the continental crust (Raval and Veeraswamy 2000). A similar phenomenon is also observed in southern Brazil and southeastern Africa (Cox 1989). Period of the Reunion mantle plume activity at the KT boundary is contemporaneous with the global mass extinction (Dinosaurs) event.

8 Supercontinents

A supercontinent is the assembly of most or all of Earth's continental blocks/cratons to form a single large landmass. It is the most important development in Earth Science since the advent of plate tectonics. Supercontinents have assembled and dispersed multiple times in the Earth's geologic history (Rogers and Santosh 2004). They have profoundly influenced the evolution of the geosphere, hydrosphere, atmosphere and biosphere. The causes of supercontinent assembly and dispersal are thought to be driven by convection processes in the Earth's mantle. The subducting slab after reaching 660 km discontinuity plunges into the lower mantle due to slab avalanche process and generates plumes or super plumes. These rising plumes cause continental rifting and break the plates/supercontinent located above it. The plates will be moved towards a geoidal low or the down going limbs of convection cells, perhaps where the slab avalanche occurred and pushed away from the geoidal high that can be caused by the plumes or superplumes (Condie 1998). Pure oceanic lithospheric plates would probably consumed and disappear into the mantle. Then, more likely all the unsubductable continental blocks would be pushed together in one area to form supercontinents. The effect of mantle plumes on breakup and assembly of supercontinents is shown in Fig. 15. Orogenic belts are formed due to the accretion of continents during the supercontinental assembly and rift basins during the breakup.

At least four supercontinents Superia-Sclavia, Columbia, Rodinia, and Pangea were identified around 2.7 Ga, 1.8 Ga, 1.1 Ga and 250 Ma in the geological history of the Earth. A supercontinent cycle is the break-up of one supercontinent and the development of another, which takes place on a global scale (Fig. 15). Both these events overlap because the convection currents that cause breakup automatically begin the accretion process elsewhere. For example, separation of the Seychelles from India due to Reunion hot spot activity and formation of the Rocky Mountains due to subduction of Pacific Ocean beneath the western margin of North America were happed at the same time during the Cretaceous-Tertiary (KT boundary) period. The supercontinent cycle exhibits long-term episodicity in tectonic processes, orogeny and continental crust formation, and all of them are embodied in a cycle.

8.1 Supercontinental Evidences from the Indian Shield

India is one of the regions in the world for evaluating the assembly, evolution and dispersal of supercontinents because of its long 3.8 Ga geological history. Results from deep crustal near-vertical seismic reflection and refraction studies carried out in different geological provinces and constrained from geological data provide important details regarding the supercontinental episodes.

8.1.1 Neoarchean Superia-Sclavia Supercontinent

Extensive Proterozoic mafic dyke swarms that occurred in most of the Archean cratons may be genetically linked to major geodynamic processes, such as mantle plume activity, intracontinental rifting and supercontinental breakup (Davis 1997; LeCheminant and Heaman 1989). Paleoproterozoic radiating mafic dyke swarms of 2.37 to 2.18 Ga age identified in the Dharwar craton (Fig. 10) covers an area of $\sim$100,000 sq.km (Halls et al. 2007) and some of them extend over hundreds of kilometers. Occurrence of similar dyke swarms of 2.2 Ga worldwide (Antarctica, Australia, Finland, Greenland and North America) has been implicated in the breakup of an Archean supercontinent (French and Heaman 2010). Based on the similar Neoarchean greenstone belts, mafic dyke swarms and other geological evidences, French and Heaman (2010) suggested that the Dharwar and Slave cratons were part of Neoarchean Superia-Sclavia supercontinent. Rifting and breakup of this supercontinent at $\sim$2.2 Ga may be responsible for the observed dyke swarms in these regions.

8.1.2 Late Paleoproterozoic Columbia Supercontinent

Several geological and geophysical studies including deep crustal seismic studies have identified late Paleoproterozoic ($\sim$1.8 Ga) orogenic activities in the NW, central and eastern part of India with the evolution of Aravalli, Satpura and Eastern Ghat orogenies (Vijaya Rao and Reddy 2002; Mandal et al. 2013; Vijaya Kumar et al. 2010). During the same period a large number of orogenic episodes, like the Trans-Hudson, Wopmay, Sveco-Fennian, Capricorn etc., took place in different parts of the world (Hoffman 1988; BABEL Working Group 1990; Myers 1993). All these orogenies formed a network of late Paleoproterozoic orogenies, which resulted in the formation of the Columbia (Rogers and Santosh 2002; Zhao et al. 2004).

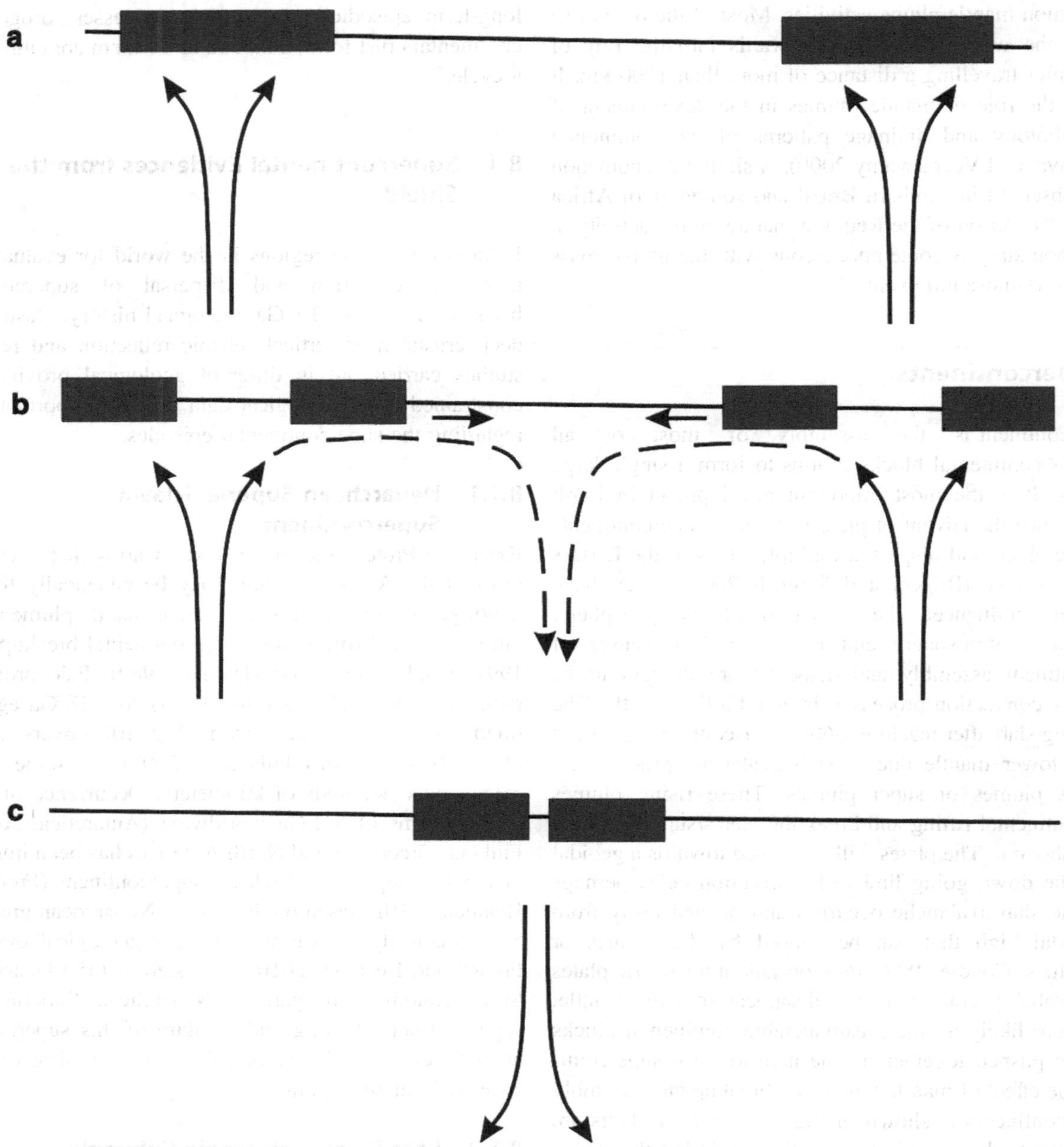

Fig. 15 Effect of mantle plumes on breakup and making of supercontinents. **a** Plumes and super plumes rise under continents, **b** Rising plumes cause continental rifting, **c** continents collide over regions of descending mantle

8.1.3 Late Mesoproterozoic Rodinia Supercontinent

Continental growth and global orogenic activities are found to be episodic and related to the supercontinental episodes (Condie 1998; Nance et al. 2014). The South Delhi fold belt of NW India, the Sausar orogeny of central India and the Eastern Ghat orogeny of eastern India were formed during the early Neoproterozoic (~ 1.1 Ga). They are contemporaneous with the Grenvillian orogeny, Pinjara Orogeny, the Circum-East Antarctic orogeny and many other orogenic belts in different parts of the world (Hoffman 1991; Mezger and Cosa 1999; Vijaya Rao and Krishna 2013). A network of all these early Neoproterozoic orogenic belts formed the Rodinia supercontinent at ~ 1.1 Ga (Rogers 1996; Meert and Torsvik 2003).

External Geological Processes

Erosion and weathering are two main external geological processes that modify the landscape of the Earth by wearing down mountains, filling in valleys and making rivers appear and disappear.

Erosion Erosion is a natural geological process usually made by the rock in which soil being loosened from the Earth's surface at one location and transported to another location by natural forces such as wind or water. This transported material is called as sediment. Erosion is opposite to deposition. It is usually a slow process that occurs over thousands or millions of years. Liquid water is the main agent of erosion on the Earth. Rain, rivers, floods, lakes and ocean carry away bits of soil and sand. Some of the natural factors impacting erosion in a landscape include climate, topography, vegetation and tectonic activity. Erosion can be speeded up by human activities such as farming and mining.

Weathering Weathering processes are similar to erosion, where they cause breakup of rock material in the Earth's uppermost layer but does not involve movement. Air and water are agents of weathering. Some of these processes are mechanical, e.g., the expansion and contraction caused by sudden large changes in temperatures, the expansive force of water freezing in cracks etc. Plants growth can also contribute to weathering by breaking the earthen material as they take root and can create cracks in rocks they encounter. Oxidation, hydration, carbonization and loss of chemical elements by solution in water are some of the chemical weathering processes.

9 Implication of Weathering and Erosion

A large quantity of sediments generated by erosion and weathering of the Himalayan mountain range are being carried away by the Ganges and Brahmaputra rivers during the last several million years to the Bay of Bengal. Thick column of Cretaceous, Tertiary and Quaternary sediments deposited in the Bengal basin covering the west Bengal, India and Bangladesh. A total of ~18 km thick sediments were deposited in the deeper part of the basin. The excess sediments of finer clastics were carried further deep in the Bay of Bengal as submarine fan deposits. The Bengal fan (Fig. 16) with a length of ~3000 km, a width of ~1000 km and a maximum thickness of 16.5 km is the largest submarine fan in the world (Curray et al. 2002). There are number of oils producing wells in the Bengal basin. Basin configuration of west Bengal sedimentary basin is shown in Fig. 12. Similarly, the Indus River originating from Himalaya and flowing through India and Pakistan is also responsible for erosion and weathering of Himalayan ranges. At 1 mm/ year erosion rate, the topography is removed at a rate of 1 km per 10 Ma. Thus, great mountain ranges will survive only for a few hundred million years and the high-topography is transient on water–rich planet like the earth.

Even though surface erosion seems to be a simple and insignificant process, it has great impact on the evolution of global plate tectonics. Recently, Sobolev and Brown (2019) proposed a new theory, based on the geological, geochemical and modelling studies, regarding the onset of present-day plate tectonics. The theory suggests that the surface erosion events control the evolution of plate tectonics and in turn the life on the Earth. Rise of the continents their erosion, generation of sediments and their accumulation at continental edges, subsequently in trenches has provided lubrication for stabilization of subduction. The above process is responsible for emergence and evolution of present-day large-scale plate tectonics on the Earth. Lubrication by sediments could enable both formation of new subduction zones by increasing number of passive margins and extension of existing subduction zones. Subduction should have less stable during the early part of the Earth due to high mantle temperatures. It suggests evidence for the absence of large-scale subduction events prior to ~3.0 Ga as suggested by Dhuime et al. (2012). The theory further suggests stable plate tectonics follow widespread surface erosion events (more sediments), whereas diminished surface erosions (less sediments) lead to reduced subduction and possibly intermittent plate tectonics. It is mainly because sediments reduce the friction between the plates and lack of sediments means higher friction. The new theory indicates two largest surface erosion and subduction lubrication events —one after global glaciation during 2.45–2.2 Ga, leading to the formation of the Columbia supercontinent and other after "Snowball Earth" glaciation during 0.75–0.65 Ga, forming the Gondwana supercontinent.

10 Geological Processes: Evolution of Atmosphere and Biosphere

Geological processes and geochemical environment have played a key role in the origin of life on earth. Though the details of the origin of life are still unresolved, serpentinization in hydrothermal vents, which are mainly distributed along the plate boundaries, is considered the most likely sites to harbour the chemistry necessary for life (Martin et al. 2008). These hydrothermal vents support two key processes necessary for life—(i) a consistent source of chemical energy and (ii) the formation organic molecules (such as methane and other small hydrocarbons) from the inorganic molecules (such as carbon dioxide and hydrogen). It has also been proposed that some of the amino acids could have formed deep inside the crust and have risen via the hydrothermal vents (Tunnicliffe 1991).

Fig. 16 The Bengal and Indus Fans located on the eastern and western continental margins of India are formed due to the sediments from the Himalayan ranges and carried away by the Ganges, Brahmaputra and Indus rivers

Plate tectonics seemed to have played a key role in the evolution of complex lifeforms by supporting the oxygenation of atmosphere similar to the hydrothermal vents in the origin of life (Blaustein 2016). Presence of granitic crust on the earth is the main source of nutrient supply to evolve and sustain life by stimulating photosynthesis activity and generation of oxygen (Lowe and Tice 2004). The continent–continent collisional episodes during the formation of supercontinents are believed to be responsible in increasing the Earth's atmospheric oxygen content. Due to the natural erosion process, these supercontinents would have eroded and the nutrients like phosphorus and iron would have washed into oceans (Santosh et al. 2014). Then the oceans rich in nutrients that are essential to photosynthetic organisms would have helped to respire large amounts of oxygen. It appears to exist a direct relationship between the orogeny and atmospheric oxygen content. The increased sedimentation concurrent with mass oxygenation events probably buried the organic carbon and pyrite beneath sediments. Thus, these are unable to react with the free oxygen leading to sustained increase of the atmospheric oxygen.

Beginning of the Earth to the end of Proterozoic, the lifeforms were basically small bacteria. The second important event in the birth of life on the earth was during the Ediacaran-Cambrian period. Beginning of the Phanerozoic witnessed the origin of multi-cellular organisms from the unicellular organisms and appearance of metazoans (Santosh et al. 2014). This period is termed as the "Cambrian Explosion". This period is coinciding with the assembly of the Gondwana supercontinent with the evolution of major ~3000 km long East-African orogen during ~550 Ma. Further, immediately prior to this period, the Earth was in a frozen state termed as "Snowball Earth". There is a considerable debate over the relation of these major events. The Cambrian Explosion is attributed to the geochemical, ecological and tectonic changes occurred during this period.

11 Summary

Active-source deep seismic studies from different tectonic domains of the Indian shield imaged several crustal structures, which are very important to understand the geological processes of the Earth. Some of these structures are due to horizontal plate tectonic processes, and others are due to vertical plume tectonics. These processes involve several intermediate processes, like the delamination, asthenospheric upwelling, generation of younger Moho etc. During these processes material can undergo physico-chemical changes, like phase change, transformed into different materials, change in velocity/density and viscosity, dehydration etc. due to different heat flow, pressure and temperature conditions. Seismic images from Neoarchean and Proterozoic terrains of the Indian shield show mantle and whole-crustal subduction respectively. It indicates differences in the mode of subduction during the Archean and Proterozoic.

Breaking of the Pangea supercontinent/Gondwanaland due to several mantle plume activities developed passive continental margins with the formation of Deccan volcanic province in the west and the Rajmahal traps in the eastern part of India. Further, in the western part rift-basins were formed in the E-W and N-S directions. Crustal velocity model of the Deccan trap covered E-W trending Narmada basin indicate high-velocity (7.1–7.4 km/s) lower-crustal layer formed due to the magmatic underplating in the region, which is attributed to the Reunion mantle activity at $\sim$65 Ma. The crust-mantle interaction played a vital role in forming rift-basins and post-collisional extension as well as generation of mafic dyke swarms and alkaline complexes.

Mantle convection process is found to be responsible for bringing most of the earth's terranes at one location, which will accrete to form supercontinents. Dispersion of supercontinent is caused by accumulation of heat under the supercontinent probably above the rising convection cells due to mantle plume activity. Evidences for the presence of Columbia, Rodina and Pangea are suggested from the Indian shield.

Erosion and weathering of the Himalayan mountain ranges are responsible for the generation of huge sediments and their deposition in the west Bengal basin and for the formation of world's largest Bengal fan.

Acknowledgements We thank V.M. Tiwari, Director, NGRI for providing the facilities, encouragement and permission to publish the work (vide reference no: NGRI/Lib/2021/Pub-25 and contribution towards MLP-6402-28(PK)) and Prof. Vinod K. Singh for his invitation to contribute the research article. We also thank W.D. Mooney and Ramon Carbonell for their useful suggestions.

References

Anderson DL (2007) New theory of the Earth (2nd edn). Cambridge University Press. doi https://doi.org/10.1017/CBO9781139167291

Arora BR, Prajapati SK, Reddy CD (2014) Geophysical constraints on seismotectonics of the Sikkim Himalaya. Bull Seis Soc Am 104 (5):2278–2287

BABEL Working Group (1990) Evidence for early Proterozoic plate tectonics from seismic reflection profiles in the Baltic Shield. Nature 348:34–38

Basu AR, Renne PR, Dasgupta DK et al (1993) Early and late igneous pluses and a high-3 He plume origin for the Deccan flood basalts. Science 261:902–906

Bhattacharji S, Chatterji N, Wampler JM et al (1996) Indian intraplate and continental margin rifting, lithospheric extension and mantle plume upwelling in Deccan flood basalt volcanism near the K/T boundary: evidence from mafic dike swarms. J Geology 104:379–398

Bhushan SK, Kumar A (2013) First carbonate hosted REE deposit from India. J Geol Soc India 81:41–60

Bijwaard HW, Spakman W, Engdahl ER (1998) Closing the gap between regional and global travel time tomography. J Geophy Res 103:30055–30078

Biswas SK (1987) Regional tectonic framework, structure and evolution of the western margin basins of India. Tectonophysics 135:307–327

Blaustein R (2016) The great oxidation event. Bioscience 66:189–195

Burke K, Dewey JF (1973) Plume-generated triple junction: key indicators in applying plate tectonics to old rocks. J Geology 81:406–433

Calvert AJ, Sawer EW, Davis WJ et al (1995) Archaean subduction inferred from seismic images of mantle suture in the superior provinces. Nature 375:670–674

Carbonell R, Levander A, Kind R (2013) The mohorovicic discontinuity beneath the continental crust: an overview of seismic constraints. Tectonophysics 609:353–376

Catchings RD, Mooney WD (1988) Crustal structure of the Columbia plateau: evidence for continental rifting. J Geophy Res 93:459–474

Chen WP, Yang ZH (2004) Earthquakes beneath the Himalayas and Tibet: evidence for strong lithosphere mantle. Science 304:1949–1952

Christensen NI, Mooney WD (1995) Seismic velocity structure and composition of the continental crust: a global view. J Geophy Res 100:9761–9788

Cook FA (2002) Fine structure of the continental reflection Moho. Geol Soc Am Bull 114:64–79

Coffin MF, Eldholm O (1994) Large igneous provinces: crustal structure, dimensions and external consequences. Rev Geophy 32:1–36

Condie KC (1998) Episodic continental growth and supercontinents: a mantle avalanche connection? Earth Plant Sci Lett 163:97–108

Courtillot V, Davaille A, Besse J et al (2003) Three distinct types hotspots in the Earth's mantle. Earth Plant Sci Lett 205:295–308

Cox KG (1989) The role of mantle plumes in the development of continental drainage patterns. Nature 342:873–876

Curray JR, Emmel FJ, Moore DG (2002) The Bengal fan: morphology, geometry, stratigraphy, history and processes. Mar Pet Geol 19:1191–1223

Damodara N, Vijaya Rao V, Sain K et al (2017) Basement configuration of the West Bengal sedimentary basin, India as revealed by seismic refraction tomography. Geophys J Int 208:1490–1507

Davis WJ (1997) U-Pb zircon and rutile ages from granulite xenoliths in the Slave province: evidence for mafic magmatism in the lower crust coincident with Proterozoic dike swarms. Geology 25:343–346

Deemer SJ, Hurich CA (1994) The reflectivity of magmatic underplating using the layered mafic intrusion analog. Tectonophysics 232:239–255

Dewey JF, Bird JM (1970) Mountain belts and new global tectonics. J Geophy Res 75:2625–2647

Dewey JF (1988) Extensional collapse of orogens. Tectonics 7:1123–1140

Duncan RA, Richards MA (1991) Hotspots, mantle plumes, flood basalts and true polar wander. Rev Geophy 29:31–50

de Wit MJ (1998) On Archean granites, greenstones, cratons and tectonics: does the evidence demand a verdict? Precam Res 91:181–226

Dhuime B, Hawkesworth CJ, Cawood PA et al (2012) A change in geodynamics of continental growth 3 billion years ago. Science 335:1334–1336

Ernst WG (2007) Speculations on evolution of the terrestrial lithosphere-asthenosphere system—plumes and plates. Gondwana Res 11:38–49

French JE, Heaman LM (2010) Precise U-Pb dating of Paleoproterozoic mafic dyke swarms of the Dharwar craton, India: implications for the existence of the Neoarchean supercraton Sclavia. Precam Res 183:416–441

Fukao Y, Obayashi M, Nakakuki T, Deep Slab Project Group (2009) Stagnant slab: a review. Annu Rev Earth Planet Inter 173:197–206

Fukao Y, Obayashi M (2013) Subducted slabs stagnant above, oenetrating through and trapped below the 660 km discontinuity. J Geophy Res 118:5920–5938

Fyfe WS (1993) Hot spots, magma underplating and modification of continental crust. Can J Earth Sci 30:908–912

Gaina C, Muller RD, Brown B et al (2007) Breakup and early seafloor spreading between India and Antarctica. Geophys J Int 170:151–169

Gao R, Lu Z, Klemperer SL et al (2016) Crustal-scale duplexing beneath the Yarlung Zangbo suture in the western Himalaya. Nat Geosci 9:555–560

Garnero EJ, McNamara AK, Sim SH (2016) Continental-sized anomalous zones with low seismic velocity at the base of Earth's mantle. Nat Geosci 9:481–489

Gurnis M (1988) Large-scale mantle convection and aggregation and dispersal of supercontinents. Nature 332:695–699

Halls HC, Kumar A, Srinivasan R et al (2007) Paleomagnetism and U-Pb geochronology of easterly trending Dykes in the Dharwar craton, India: feldspar clouding, radiating dyke swarms and the position of India at 2.37 Ga. Precam Res 155:47–68

Hoffman PF (1988) United plates of America, the birth of a craton: early Proterozoic assembly and growth of Laurentia. Ann Rev Earth Planet Sci 16:543–603

Hoffman PF (1991) Did the breakout of Laurentia turn Gondwanaland inside out? Science 252:1409–1412

Kayal JR (2008) Microearthquake seismology and seismotectonics of south Asia. Springer, Dordrecht. https://doi.org/10.1007/978-1-4020-8180-4

Kind R, Yuan X, Saul J et al (2002) Seismic images of crust and upper mantle beneath Tibet: evidences for Eurasian plate subduction. Science 298(5596):1219–1221

Krishna VG, Rao V (2011) Velocity modelling of a complex deep crustal structure across the Meso-Proterozoic south Delhi fold belt, NW India, from joint interpretation of coincident seismic wide-angle and near-offset reflection data–an approach by utilizing unusual reflections in wide-angle records. J Geophy Res 116: B01307. https://doi.org/10.1029/2009JB006660

Kumar P, Yuan X, Kind R et al (2006) Imaging the colliding Indian and Asian lithospheric plates beneath Tibet. J Geophy Res 111 (B6):10:1029/2005JB003930

Kumar P, Ravi Kumar M, Srijayanthi G et al (2013) Imaging the Lithosphere-asthenosphere boundary of the Indian plate using converted wave technique. J Geophy Res 118. https://doi.org/10.1002/jgrb.50366

LeCheminant AN, Heaman LM (1989) Mackenzie igneous events, Canada: middle Proteerozoic hotspot magmatism associated with ocean opening. Earth Planet Sci Lett 96:38–48

Lowe DR, Tice MM (2004) Geological evidence for Archean atmosphere and climate evolution: fluctuating levels of Co_2, CH_4, O_2 with an overriding tectonic control. Geology 32:493–496

Mall DM, Rao VK, Reddy PR (1999) Deep sub-crustal features in the Bengal basin: seismic signatures from plume activity. Geophys Res Lett 26:2545–2548

Mandal B, Sen MK, Vijaya Rao V (2013) New seismic images of the Central Indian Suture Zone and their tectonic implications. Tectonics 32:1–14. https://doi.org/10.1002/tect.20055

Mandal B, Vijaya Rao V, Sarkar D et al (2018) Deep crustal seismic reflection images from the Dharwar craton, Southern India-evidence for the neoarchean subduction. Geophy J Int 212:777–794. https://doi.org/10.1093/gji/ggx427

Mao W, Zhang S (2018) Slab stagnation due to reduced viscosity layer beneath the mantle transition zone. Nat Geosci 11:876–881

Martin W, Baross J, Kelley D et al (2008) Hydrothermal vents and the origin of life. Nat Rev Microbiol 6:805–814

Meert JG, Torsvik TH (2003) The making and unmaking of a supercontinent: Rodinia revisited. Tectonophysics 375:261–288

Mezger K, Cosa MA (1999) The thermal history of the Eastern Ghats Belt (India) as revealed by U-Pb and 40Ar/39Ar dating of metamorphic and magmatic minerals: implications for the SWEAT correlation. Precambr Res 94:251–271

Monsalve G, Sheehan A, Schulte-Pelkum V et al (2006) Seismicity and 1-dimentional velocity structure of the Himalayan collision zone: earthquake in the crust and upper mantle. J Geophy Res 111: B10301

Mooney WD, Meissner R (1992) Multi-genetic origin of crustal reflectivity: a review of seismic reflection profiling of the continental lower crust and Moho. In: Fountain DM, Arculus R, Kay RW (eds) Continental lower crust, Elsevier Publication Company

Moyen HF, Stevens G, Kisters A (2006) Record of mid-Archean subduction from metamorphism in the Barberton terrain, South Africa. Nature 442:559–562

Myers JS (1993) Precambrian history of west Australian craton and adjacent orogens. Ann Rev Earth Planet Sci 21:453–485

Nance RD, Murphy JB, Santosh M (2014) The supercontinent cycle: a retrospective essay. Gond Res 25:4–29

Nelson KD, Zhao W, Brown LD et al (1996) Partially molten middle crust beneath the southern Tibet: synthesis of project INDEPTH results. Science 274:1684–1688

Pirajno F, Santosh M (2015) Mantle plumes, supercontinents, intracontinental rifting and mineral systems. Precambr Res 259:243–261

Purnachandra Rao N, Kalpana (2005) Deformation of the subducted Indian lithospheric slab in the Burmese arc. Geph Res Lett 32: L05301. https://doi.org/10.1029/2004GL022034

Rajendran CP, Rajendran K (2005) The status of central seismic gap: perspective based on the spatial and temporal aspects of the large Himalayan earthquakes. Tectonophysics 365:19–39

Raval U, Veeraswamy K (2000) The radial and linear modes of interaction between mantle plume and continental lithosphere: a case study from western India. J Geol Soc India 56:525–536

Reddy PR, Vijaya Rao V (2013) Seismic images of the continental Moho of the Indian shield. In: Artemieva I, Brown L, Kennett BLN,

Thybo H (eds) Moho: 100 years after Andrija Mohorovicic, Tectonophysics 609:217–233

Reddy PR, Rajendra Prasad B, Vijaya Rao V et al (2003) Deep seismic reflection and refraction/Wide-angle reflection studies along Kuppam-Palani transect in southern granulite terrain of India. In: Ramakrishnan M (ed) Tectonics of southern granulite terrain, Kuppam-Palani geotransect, Geol Soc India Mem 50:79–106

Rogers JJW (1996) A history of continents in the past three billion years. J Geol 104:91–107

Rogers JJW, Santosh M (2002) Configuration of Columbia, a Mesoproterozoic supercontinent. Gondwana Res 5:5–22

Rogers JJW, Santosh M (2004) Continents and supercontinents. Oxford University Press, 289

Sapin M, Hirn A (1997) Seismic structure and evidence for eclogitization during the Himalayan convergence. Tectonophysics 273:1–16

Santosh M, Maruyama S, Sawaki Y et al (2014) The Cambrian explosion: plume driven birth of the second ecosystem on Earth. Gonndwana Res 25:945–965

Sridhar AR, Tewari HC (2001) Existence of a sedimentary graben in the western part of Narmada zone: seismic evidence. J Geodynamics 31:19–31

Sridhar AR, Tewari HC, Vijaya Rao V et al (2007) Crustal velocity structure of the Narmada-Son lineament along the Thuadara-Sendhwa-Sinddad profile in the NW part of central India and its geodynamic implicartions. J Geol Soc India 69:1147–1160

Smithies RH, Champion DC, Van kranendonk MJ, et al (2005) Modern style subduction processes in Mesoarchean: geochemical evidence from the 3.12 Ga Whundo intra-oceanic arc. Earth Planet Sci Lett 231:221–237

Sobolev S, Brown M (2019) Surface erosion events controlled the evolution of plate tectonics on Earth. Nature 570:52–59

Storey BC (1995) The role of mantle plumes in continental breakup: case histories from Gondwanaland. Nature 377:301–308

Tewari HC, Dixit MM, Rao NM et al (1997) Crustal thickening under the Paleo/Mesoproterozoic Delhi fold belt in NW India: evidence from deep reflection profiling. Geophys J Int 129:657–668

Torsvik TH, Smethurst MA, Burke K et al (2006) Large igneous provinces generated from the margins of the large low-velocity provinces in the deep mantle. Geophys J Int 167:1447–1460. https://doi.org/10.1111/j.1365-246X.2006

Tunnicliffe V (1991) The biology of hydrothermal vents: ecology and evolution. Oceanogr Mar Biol Annu Rev 29:319–407

Van der Hilst RD, Widiyantoro S, Engdahl ER (1997) Evidence for deep mantle circulation from global tomography. Nature 386:578–584

Van der Meer DG, van Hinsbergen DJJ, Spakman W (2018) Atlas of the underworld: slab remnants in the mantle, their sinking history, and a new outlook on lower mantle viscosity. Tectonophysics 723:309–448

van der Velden AJ, Cook FA (2005) Relict subduction zones in Canada. J Geophy Res 110. https://doi.org/10.1029/2004JB003333

Van der Voo R, Spakman W, Bijwaard H (1999) Tethyan subducted slabs under India. Earth Plant Sci Lett 171:7–20

Vijaya Kumar K, Ernst WG, Leelanandam C et al (2010) First Paleoproterozoic ophiolite from Gondwana: geochronologic-geochemical documentation of ancient oceanic crust from Kandra, SE India. Tectonophysics 487:22–32

Vijaya Rao V, Rajendra Prasad B, Reddy PR et al (2000) Evolution of proterozoic Aravalli Delhi fold belt in the northwestern Indian shield from seismic studies. Tectonophysics 327:109–130

Vijaya Rao V, Reddy PR (2002) A mesoproterozoic supercontinent: evidence from the Indian Shield. In: Rogers JJW, Santosh M (eds) Special volume on mesoproterozoic supercontinent, Gondwana Research 5:63–74

Vijaya Rao V, Sain K, Reddy PR et al (2006) Crustal structure and tectonics of the northern part of the Southern granulite Terrain, India. Earth Planet Sci Lett 251:90–103

Vijaya Rao V, Rajendra Prasad B (2006) Structure and evolution of the Cauvery shear zone system, southern granulite terrain, India: evidence from deep seismic and other geophysical studies. Gondwana Res 10:29–40

Vijaya Rao V (2009) Geodynamics of the collision zones: phases of lithospheric evolution from the Indian Shield. In: Arora BR, Sharma R (eds) Geodynamics of the collision zones. J Geol Soc India Memoir 72: pp165–194

Vijaya Rao V, Reddy PR (2009) Deep crustal seismic images of the southern granulite terrain, India. In: Reddy PR (ed) Seismic imaging of the Indian continental and oceanic crust. Professional Books Publishers, Chennai, pp 297–323

Vijaya Rao V, Khare P, Raju S et al (2011) Deep crustal seismic probing of the Dharwar craton by DHARSEIS Experiment. DST Newsletter 21(2):5–10

Vijaya Rao V, Krishna VG (2013) Evidence for the Neoproterozoic Phulad suture zone and genesis of Malani magmatism in the NW India from deep seismic images: implications for assembly and breakup of the Rodinia. Tectonophysics 589:172–185

Vijaya Rao V, Murty ASN, Sarkar D et al (2015) Crustal velocity structure of the neoarchean convergence zone between the eastern and western blocks of Dharwar Craton, India from seismic wide-angle studies. Precambr Res 266:282–295

Yuan K, Romanowicz B (2017) Seismic evidence for partial melting at the root of major hot spot plumes. Science 357:393–397

Zhao D (2004) Global tomographic images of mantle plumes and subducting slabs: insight into deep earth dynamics. Phy Earth Planet Inter 146:3–34

Zhao D (2007) Seismic images under 60 hotspots: search for mantle plumes. Gondwana Res 12:335–355

Zhao W, Nelson KD, Project INDEPTH Team (1993) Deep seismic reflection evidence for continental underthrusting beneath the Tibet. Nature 366:557–559

Zhao GC, Sun M, Wilde SA et al (2004) A paleo-Mesoproterozoic supercontinent: assembly, growth and breakup. Earth-Sci Rev 6:91–123

World Class Hutti Gold Deposit—An Archean Orogenic Gold Deposit in Hutti-Maski Greenstone Belt, Karnataka, India

Prabhakar Sangurmath

Abstract

Archean orogenic gold deposits are widespread in the Dharwar Craton and account for almost 80% of cumulative Indian gold production. The Dharwar Craton hosts significant Gold prospects and the Hutti Gold deposit stand as the only Gold producing Mine. A close examination of the geological characteristics of world class gold deposit reveals many common features: (i) strong structural control of orebodies, (ii) relative timing of gold mineralization with respect to peak metamorphism, (iii) consistent metal association and (iv) broadly uniform, low-moderate salinity, mixed aqueous-carbonic hydrothermal fluid. In the past 15 years, however, detailed investigations reveal that the world class deposits in the Dharwar Craton display a diverse range of depositional site characteristics (host lithologies, structural setting, mineralization styles), at both the camp- and deposit-scales, and a range in relative- and absolute-timings. This contribution explores the variability in deposit characteristics, and its implications for generic genetic models for this deposit-class. Gold mineralization is confined to sub-parallel reefs hosted within sheared metabasalts. Mining is in progress in six out of nine reefs with a metallurgical plant having capacity of processing 2500 tons of ore per day. The results of exploration, research and existing geological knowledge together indicate further persistence of gold mineralization to depth and strike.

Keywords

Gold deposit • Archean orogeny • Dharwar craton • Greenstone belt

1 Introduction

India's name is associated with gold from the ancient times. Gold occurs at many places in the country but South India, particularly Dharwar region is best endowed with the metal. Known Gold occurrences in various geological domains are compiled and tabulated at Table 1. Though there are occurrences/prospects reported from different metallogenic epoch, majority of Gold deposits/occurrences are reported from Dharwars (Fig. 1).

Hutti-Maski greenstone belt of Karnataka craton have retained a unique place in the history of gold mining in India. At present, Hutti, Uti and Hira-Buddinni Gold Mines of M/s. Hutti Gold Mines Co Ltd (HGML) are the only major gold producing regions. The Hutti gold mine is situated (Fig. 1) in the north-western fringes of Hutti-Maski greenstone belt. Earlier workers Robert Bruce Foote (1876, 1882, 1889), Munn (1934), Maclaren (1906, 1908), Smeeth and Sampart Iyenger (1916), Rama Rao (1962, 1963), Sangurmath (1991, 1996, 2003a, b, 2005, 2015), Raghunandan and Vidyadharan (1993), Vasudev et al. (2021) have studied the Hutti-Maski greenstone belt and published number of literatures on the subject but detailed geological attributes of the Hutti Gold Deposit have not been fully discussed.

The Hutti and adjacent gold deposits have to-date contributed 97.84 tonnes of gold to the Indian economy, and therefore the Late Archean Hutti-Maski Greenstone Belt (HMGB) of the Eastern Dharwar Craton (EDC) continues to be the focus of recent gold mining and exploration activity.

In this paper, attempt has been made to discuss the history of exploration/mining, gold mineralization, structure, pattern of ore shoots, wall rock alterations, mineragraphy, paragenesis, controls of minerlisation, etc., of Hutti Gold Mine.

P. Sangurmath (✉)
HGML, #51, Lakeshore Garden, Vidyaranyapura, Bangalore, Karnataka 560097, India

A. K. Shandilya et al. (eds.), *Geological and Geo-Environmental Processes on Earth*, Springer Natural Hazards,
https://doi.org/10.1007/978-981-16-4122-0_6

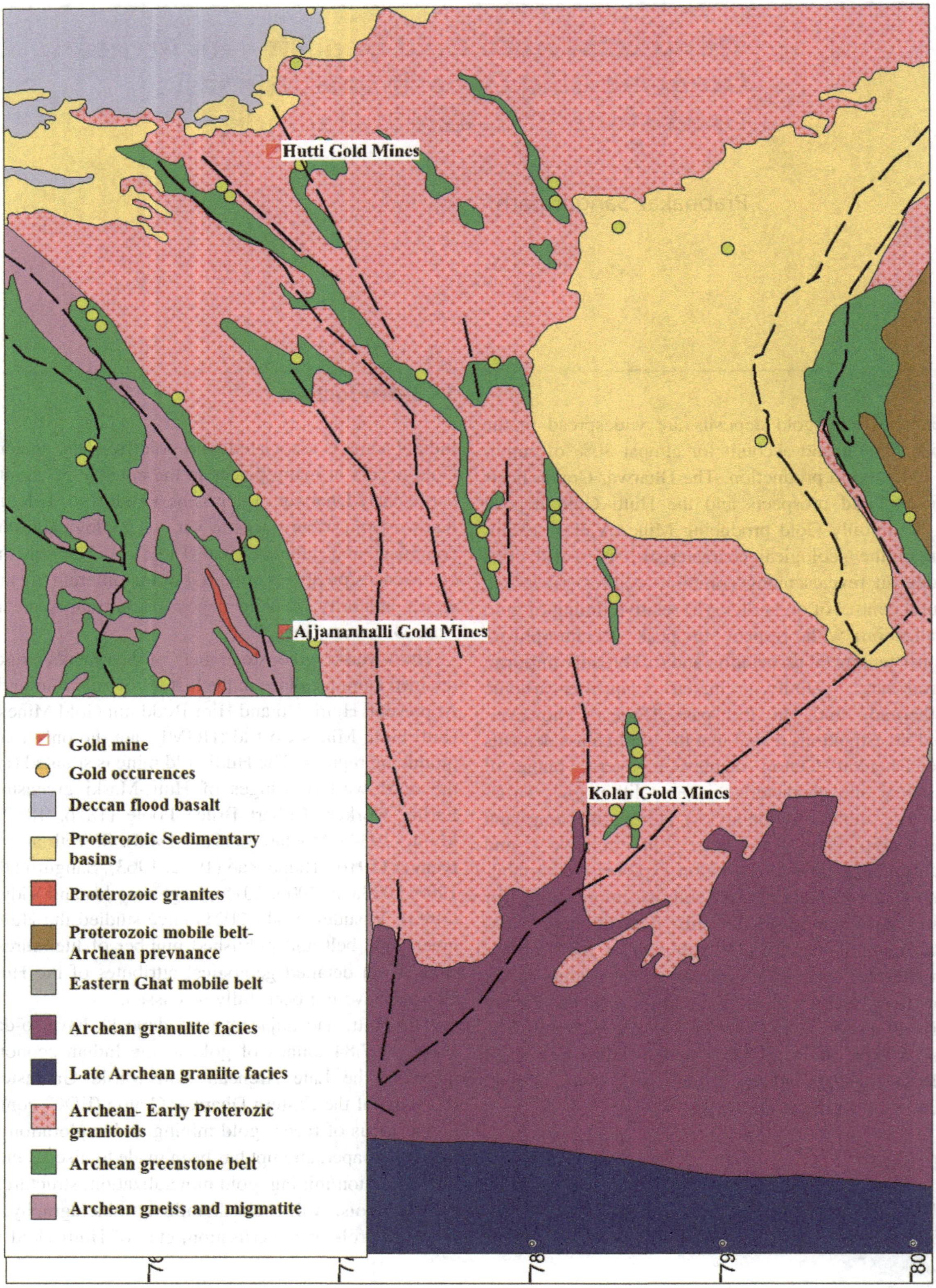

Fig. 1 Location of gold mines in Karnataka (after GSI)

Table 1 Indian Gold deposits by Metallogenic epoch

Deposit	Geo-tectonic province	Metallogenic epoch	Total resource reserves/	
			Million tonnes	Au (g/t)
Kolar(Champion)	Kolar Schist Belt	Dharwarian	47.66	5.00
Chigargunta	Kolar Schist Belt	Dharwarian	45.16	4.00
Bisnatham	Kolar Schist Belt	Dharwarian	0.03	4.00
Uti	Hutti Schist Belt	Dharwarian	0.88	3.50
Hutti	Hutti Schist Belt	Dharwarian	10.46	5.00
Wandali	Hutti Schist Belt	Dharwarian	0.65	3.50
Kadoni	Hutti Schist Belt	Dharwarian	0.07	4.00
Chennabhavi	Ramgiri Schist Belt	Dharwarian	0.04	4.00
Yeppamana	Ramgiri Schist Belt	Dharwarian	0.70	3.50
Jubitil	Ramgiri Schist Belt	Dharwarian	1.50	4.00
Mysore	Gadag Schist Belt	Dharwarian	1.00	3.50
Sangli	Gadag Schist Belt	Dharwarian	1.68	4.00
Hosur-Champion	Gadag Schist Belt	Dharwarian	1.09	3.00
Imalia	Mahakoshal	Mahakoshal	2.99	2.62
Pondi	Mahakoshal	Mahakoshal	2.99	2.62
Sunehra	Mahakoshal	Mahakoshal	2.99	2.62
Amgaon	Mahakoshal	Mahakoshal	2.99	2.62
Nawalia	Mahakoshal	Mahakoshal	2.99	2.62
Ajjanahalli	Chitradurga	Dharwarian	0.68	2.50
Bellara	Chitradurga	Dharwarian	0.05	3.00
Bhikia-Jagpura	Jhamarkotra-Delwara	Aravallies	38.50	1.40

2 Geology

2.1 General Geology of the Dharwar Craton

The Dharwar Craton is one of the major crustal segments that constitute the South Indian shield (Fig. 2), which records geological activity for more than 1.0 billion years between 3.5 Ga and 2.5 Ga (Radhakrishna and Naqvi 1986). The craton is primarily constituted by: (1) Early to Middle Archean TTG basement (3400–2900 Ma), regionally known as the Peninsular Gneissic Complex (PGC); (2) two generations of volcano-sedimentary greenstone belts, which include an older, high-grade and highly deformed "Sargur-type" sedimentary and mafic igneous rock sequence, and a younger, low-grade "Dharwar-type" sequence, made up of volcano-sedimentary rocks to the north (Viswanatha and Ramakrishnan 1976) and (3) Late Archean K-rich granitic intrusions. The Dharwar Craton is divided into the Western Dharwar Craton (WDC) and Eastern Dharwar Craton (EDC), which are separated by a thrust contact along the eastern margin of the Chitradurga greenstone belt (Swaminath and Ramakrishnan 1981; Naqvi and Rogers 1987). The WDC includes rocks of the older PGC and large greenstone belts that predominantly comprise metasedimentary rocks and komatiitic-tholeiitic volcanics, whereas the EDC is dominated by calc-alkaline granitoids and several linear greenstone belts. Importantly, most of the greenstone belts of the EDC host numerous gold prospects, the resource potential of which is not yet fully estimated. Many of the 2.7 to 2.5 Ga granitoids in the EDC have collectively been termed as the Dharwar batholith (Chadwick el al. 1996).

2.1.1 Geology of the Hutti-Maski Greenstone Belt

The Hutti-Maski Greenstone Belt (HMGB), located in the northern part of the EDC, is currently the sole lode gold producer in India, with an annual gold production of = 1.7 to 2.0 tonnes. Unlike other linear greenstone belts in the EDC, the HMGB is a horseshoe-shaped, elongate belt (Fig. 3). The major lithological units in the HMGB are pillowed metabasalts with lesser felsic to intermediate volcanic rocks and quartz rhyolite porphyry with minor intercalations of pelitic rocks. These rocks, which are metamorphosed from upper greenschist to amphibolile facies, are intruded by two generations of granitoids. The western boundary of the belt is tectonically juxtaposed against older TTG basement, whereas the northern, southern and eastern margins are cut by younger granitoids.

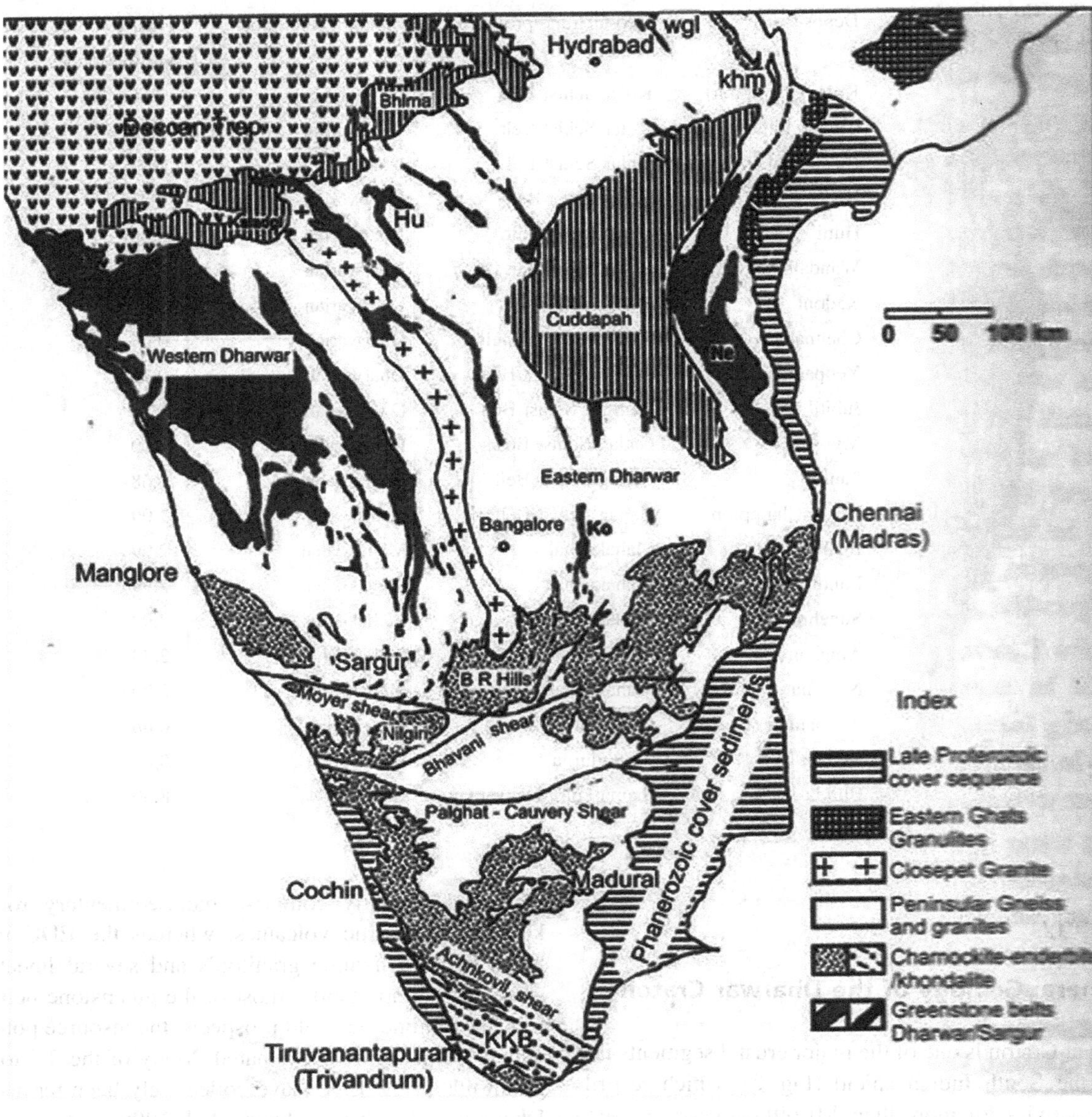

Fig. 2 Greenstone belts of South India (after GSI)

The tectonomagmatic evolution of the HMGB is a matter of debate, and several different models have been proposed (Vasudev et al. 2000; Rogers et al. 2007). Geochemical data for the volcanic rocks and surrounding TTG suggest an oceanic island arc setting for their evolution (Ram Mohan et al. 2008). However high La/Yb and Sr/Y ratios, and the Ni and Cr concentrations of the felsic volcanic rocks led Naqvi et al. (2008) to propose a slab melting origin in a subduction zone system. Identification of associated adakites, high-Mg basalts, and depleted and enriched basalts from the central and southern part of HMGB led Manikyamba et al. (2009) to propose an oceanic arc to back-arc setting, possibly during plume events.

The timing of volcanism, granitic magmatism, and gold deposition is relatively well defined (Sarma et al. 2008). The felsic volcanic rocks have yielded an U–Pb zircon crystallization age of 2587 ± 7 Ma., which is about 40 m.y. older than the age of gold mineralization at 2547 ± 10 Ma. as determined from hydrothermal monazite. The syntectonic Kavital granitoid has a U–Pb zircon age of 2545 ± 7 Ma. Whereas the Yeelagatti granitoid has yielded a titanite U–Pb age of 2532 ± 3 Ma (Anand et al. 2005). Based on the chronology of Hutti, it is evident that gold mineralization occurred throughout the Dharwar craton is 80 to 120 million years later than the world famous orogenic gold minerlisation in other late Archean cratons (Sarma et al. 2008).

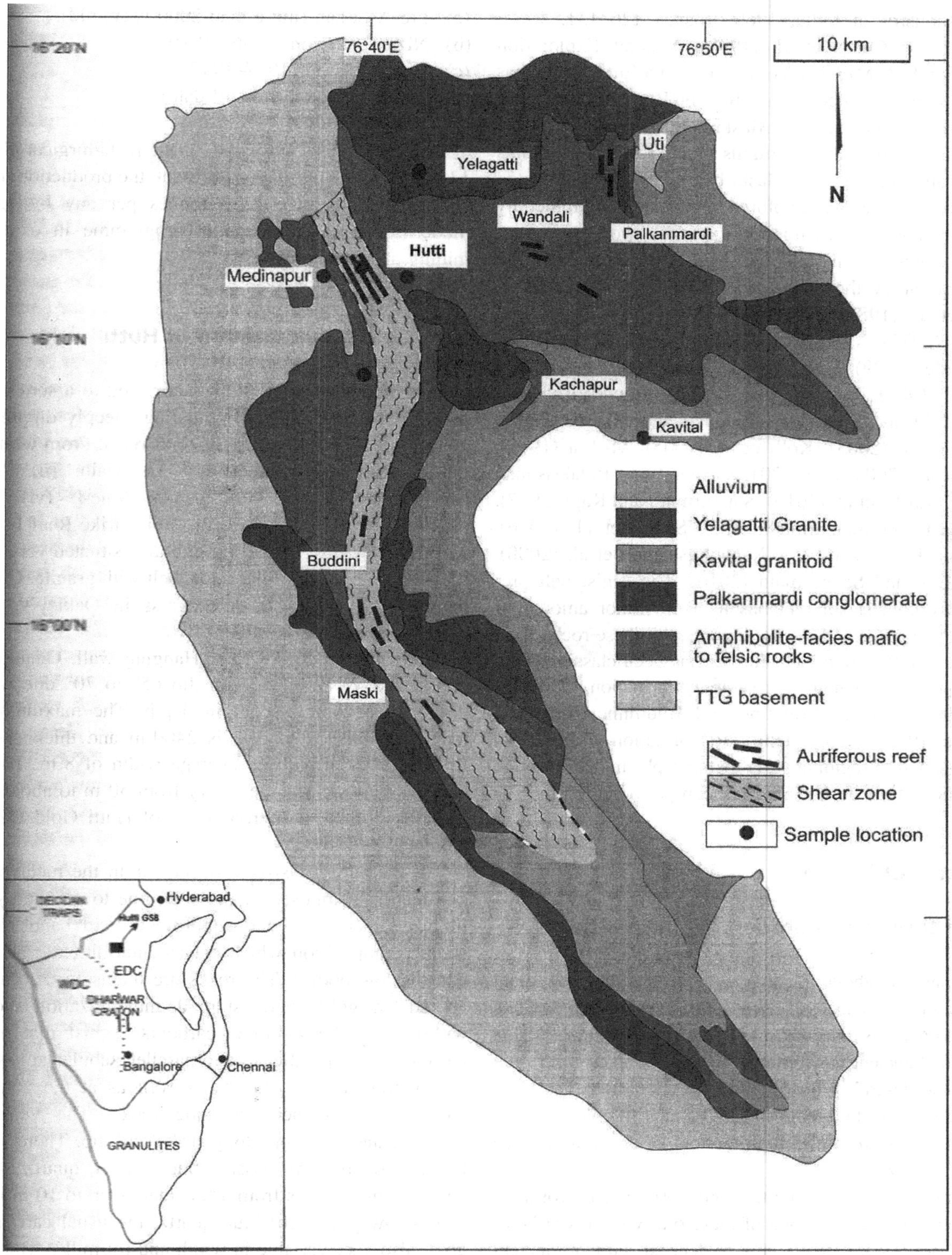

Fig. 3 Generalized geological map of hutti-maski greenstone belt (after Srikantia 1995)

Preliminary exploration activity of the Geological Survey of India (GSI) from the field season 1961–62 onwards has resulted in a volume of unpublished reports such as Krishnamurthy and Krishnamurthy (1962), Krishnamurthy (1964), Dixit (1971). A development report on the Hutti Gold Mine was published after the symposium on "Gold

Mining Industry in India, 1960" (pages 126–131, 1963) Curtis (1963); Curtis et al. (1990). Mineral Exploration Corporation Ltd (MECL). (A Govt of India Enterprises) has carried out the detailed exploration in Buddinni, Wondali and Uti Gold Mines in Hutti-Maski belt between July, 1985 and January, 1995. The enormous number of old workings scattered all over the Hutti-Maski belt area clearly indicated the fact that the area was explored, prospected and mined by the ancient and modern miners. Extensive work on exploration, metamorphism, magmatism, deformation and gold mineralization in the schist belt have been furnished in Biswas et al. (1988), Biswas (1990, 1991), Vasudev and Naganna (1973), Mahanti and Raju (1974), Anantha Iyer and Vasudev(1979), Roy(1979), Safonov et al. (1980), Roy and Raju (1980), Naganna (1987), Pathan and Virupakshi (1987), Raju and Sharma (1991), Pal and Mishra (2002), Pandalai et al. (2003), Kolb et al. (2005), Mishra (1990), Rogers et al. (2003, 2007, 2013), Anand and Balakrishnan (2010), Hazarika et al. (2016), Sangurmath and Raju (2017), Subba Rao and Sangurmath(2015), Sahoo et al. (2016), Badhe and Pandalai (2015), Nagabhushanam et al. (2008), Sundar Raju and Sangurmath (2016). The schist belt predominantly consists of metabasalts with minor amount of acid volcanic rocks and metasediments. All these rocks were termed as "Hutti group" and further has been classified into four formations such as Hussainpur Formation, 530 Hill Formation, Bullapur Formation and Buddinne Formation (Srikantia 1995). Subsequently to formation these rocks were subjected to regional scale metamorphism and intruded by granitic bodies during late-Archean period.

3 Hutti Gold Mine

3.1 History of Exploration/Mining at Hutti

Hutti is situated about 80 kms west of Raichur and about 500 km north of Bangalore, Karnataka State. The World Class Hutti Gold Mine is considered as one of the prestigious stable underground ancient metal mines in India and is an "Industrial Oasis in the desert" in the Hutti-Maski schist belt. The Hutti Gold Mines has so far produced 93.33 tonnes of gold metal (as on March, 2020) from 17.58 million tons of ore at an average grade of 5.30 g/t.

In Hutti-Maski Schist belt, about 300 ancient workings have been identified and one of these old workings at Hutti Main Mine (Main Reef) is the deepest known ancient workings in the world with a depth of almost 1.10 km from surface.

There are four periods of exploration and mining in Hutti with widely separated historical time spans as follows:

(a) Pre-Ashokan (more than 2000 years old).
(b) Nizam's Period (1886–1920).
(c) British Period (1937–1947).
(d) Fourth period (1947–till date).

The initial (1947) capacity of the metallurgical plant was 13 tonnes of ore per day. At present, the production has been enhanced to average 2500 tonnes per day by virtue of modernization and expansion programme in exploration, mining and metallurgy.

3.2 Gold Mineralization at Hutti

Gold mineralization at Hutti is confined to a series of nine parallel N-S to NNW-SSE trending, steeply dipping shear zones, that are locally referred to as reefs. From west to east the name of these reefs are (i) Main, (ii) Prospect, (iii) Oakley's, (iv) Middle, (v) Zone-I, (vi) Village, (vii) Strike Reef Hanging wall, (viii) Strike Reef Foot Wall and (ix) New East Reef (Fig. 4) and is situated very close to the western contact of the schist belt with granite. Currently, the mining activity is in progress in Oakley's, Middle, Zone-I, Village, Strike Reef.

Footwall and Strike Reef Hanging wall. Generally, the reefs strike NNW–SSE with dip 65° to 70° due west and gradually become steeper in depth. The maximum strike length established to date is 2400 m and thickness varies from 2 to 40 m with an average width of 8 m. The waste partings between the reefs vary from 50 m to about 100 m. Summary of the auriferous reefs of Hutti Gold Deposit is tabulated at Table 2.

All the auriferous reefs occur within the metavolcanics, dominantly composed of massive, fine to medium grained, pillowed metabasalt and subordinate acid bands of rhyodacitic composition which occurs intermittently as concordant narrow bodies. The rocks are intensely deformed and altered. The gold–quartz–sulphide mineralization is localised within the chlorite–biotite–carbonate schist that occur as narrow linear, parallel to sub-parallel, tabular zones occurring within the metabasalts as well as at the contacts of metabasalts and acid volcanic rocks. The veins show pinching and swelling along strike and dip. There are three types of quartz: (a) massive fine-grained, quartz of brown colour, which varies from a few meters up to 10 m in width and is low grade; (b) blue quartz, the usual carrier of the gold, also very variable in width and (c) milky white quartz occupying late tension fractures along and across the lodes and filling numerous joints and veinlets in the country rock. Besides these above mentioned reefs, several anomalous auriferous zones have been identified by the author by

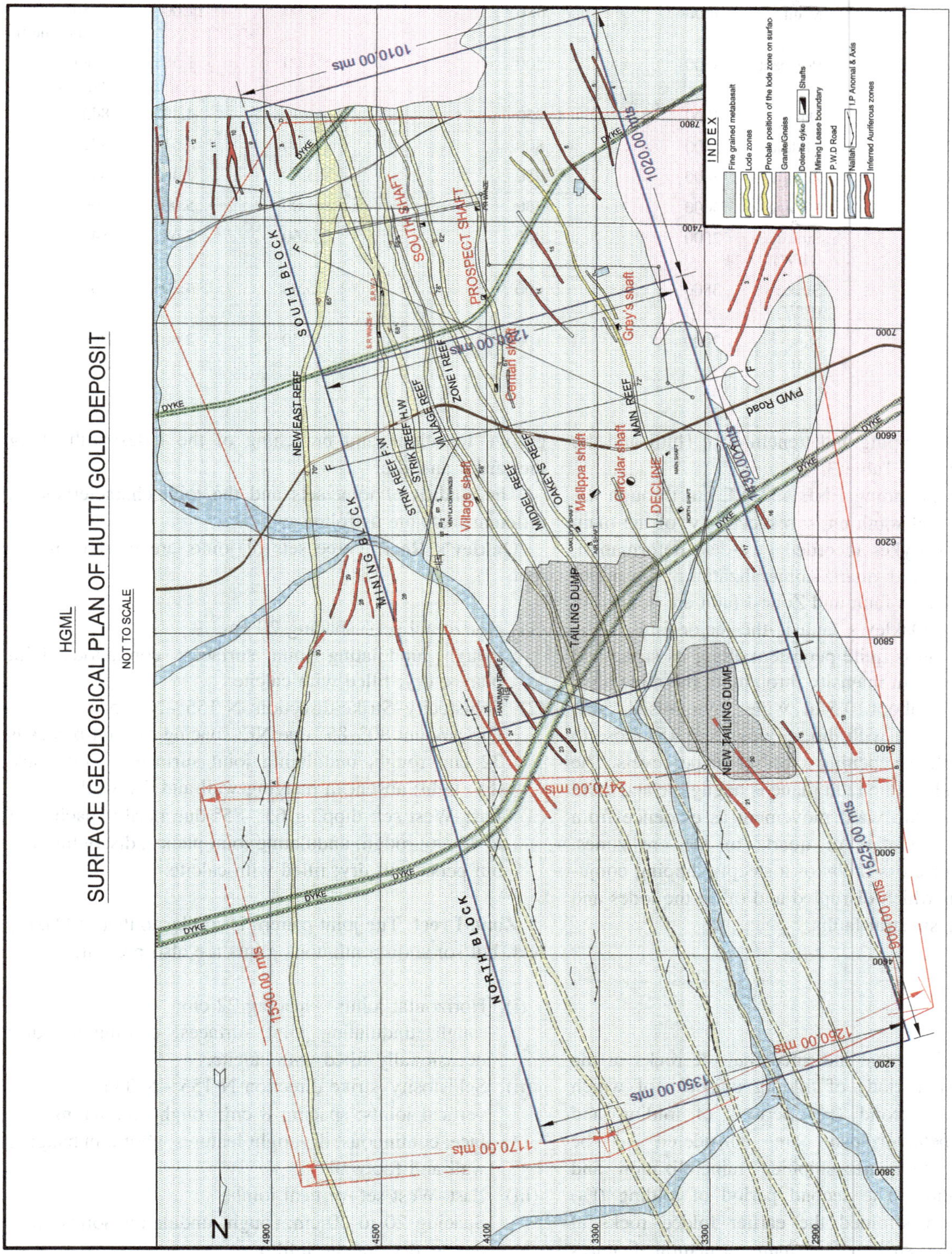

Fig. 4 Surface geological plan of hutti gold deposit

Table 2 Summary of the auriferous reefs

Reef name	Trace at Surface Strike length (m)	Underground Development. Strike length (m)	Average width (m)	Grade. Au (g/t)	Depth of working (m)
Main	4000	480	3.00	12.00	1056 (abandoned)
Prospect	3500	200	4.00	1.58	Under exploration
Oakley's	4000	950	4.40	4.49	842
Middle	3900	800	5.00	8.51	842
Zone-I	3800	950	5.80	4.46	842
Village	3000	350	4.00	5.00	300
Strike (H/W)	2500	900	6.60	7.68	842
Strike (F/W)	3800	600	5.00	4.20	450
New East	3000	100	2.50	2.00	30

surface geological mapping and trenching in Hutti and the details are furnished in Table 3.

Dolerite dykes extending ENE and NE cut through the lodes at the north and south ends, respectively, of the main mining area. These dykes recorded post-date deformation, metamorphism and gold–quartz–mineralization. Two major faults namely, Oakley's fault and Zone-I fault are noticed in the mine area. The Oakley's fault is the major fault in the Hutti gold mines and is quite persistent along its strike and dip almost with uniform intensity throughout the mine area. The fault zone is about 0.5 m wide mylonitised zone trending west-northwest with dip varying from 45° close to surface and gradually steepening to 85° at deeper levels. The lateral movement along the fault plane is negligible and there appears to be some dip ward movement as evidence from faulting of lensoid bodies of quartz in the ore zones. A number of minor cross-faults of a steeply dipping conjugate set also occur which disrupted and offset the lodes and are predominantly sinistral faults.

4 Structure

The structural studies have indicated that the rocks of the area suffered three periods of folding episodes of which the first (F_1) and second (F_2) periods of folding and accompanying metamorphism are considered to be responsible for gold–quartz–sulphide mineralization and ore–shoots formation. The second period of folding (F_2) has refolded or cross-folded the earlier folded rocks at intervals along the strike. The main structural features seen in the lodes are shearing, mylonitisation, en-echelon folds (generally dextral and occasionally sinistral), shear and tension fractures, inter-mineralization fracturing, groove lineations and branching of the lodes both along the strike and dip.

The studies of joint sets and the joint characteristic in different reefs are as follows:

Oakley's Reef: Three sets of joints are present in this reef.

(1) Horizontal set—spacing 20 cm;
Rough, undulating joint surfaces; continuous joint planes; dry; filled with calcite.
(2) Schistocity–Strike direction: N 155°–N 335°.
Dip amount 80°–85° due NE; spacing of joint planes is 32 cm; rough, undulating joint surfaces; continuous; dry; tight; absent in hanging wall and footwall.
(3) East–West set: dipping 65°–85° due north; spacing 15 to 45 cm; rough, undulating joint planes; discontinuous, but persistent; dry; filled with calcite.

Zone-I reef: The joint pattern is similar to that of Oakley reef, but some discontinuous joints are also present.

(i) Horizontal joints—spacing 22 cm;
rough, undulating joint surfaces; continuous; dry; occasionally filled with calcite.
(ii) Schistosity–strike direction N 155°–N 335°;
vertical joints; spacing 8 cm; rough, undulating surface; continuous; dry; tight features; absent in hanging wall and footwall.
(iii) East–West set–vertical joints;
Spacing 20 to 30 cm; rough, undulating joints; continues; dry; calcite infilling.
(iv) Minor joints–striking—approx: N 10°–N 190°;
Vertical; spacing 15 to 100 cm; rough, planar joints; discontinuous; dry; occasional filling.

Table 3 Summary of the anomalous zones identified by surface geological mapping at Hutti

Zone	Strike Length (m)	Area (Survey grid)	Blocks
SR-1	256	South West near magazine area (3300–7400)	South block
SR-2	896		
SR-3	400		
SR-4	224	South West near water reservoir (3700–7800)	
SR-5	352		
SR-6	160		
SR-7	576	South West near M.D. bungalow (3700–7000)	
SR-8	736		
SR-9	240	South East near ram rahim colony (4900–7800)	
SR-10	192		
SR-11	448	South East near wesly world school (5100–7400)	
SR-12	544		
SR-13	368		
SR-14	369		
SR-15	240	South West near cementry (3300–6900)	Mining block
SR-16	416	Western Area West of North shaft (3100–6200)	
SR-17	224		
SR-18	384	North West near new tailing dump (2900, 3800) and (3300–5400)	North block
SR-19	384		
SR-20	448		
SR-21	240		
SR-22	240	North Area hanuman temple (3700–4100-5400)	
SR-23	640		
SR-24	704		
SR-25	705		
SR-26	460	North East area (4200–5400)	
SR-27	800	North East area (4500, 4900–5400, 5800)	
SR-28	544		
SR-29	240		
SR-30	320		
SR-31	288	West of grey shaft (3400–7000)	
	Total = 12, 811		

Strike and Middle reefs: The structures in these two reefs are similar. There are two major sets and one random joint set.

(i) Horizontal joints—spacing 15 cm;
Rough, undulating joint planes; continuous; dry; occasionally filled with calcite.

(ii) Schistocity–strike: N 155°–N 335°;
Vertical joints; spacing 30 to 40 cm; rough, undulating joints; continuous; dry; tight; absent in hanging wall and footwall.

(iii) East–West set–Strike ranging from N 50°–N 230° to N 75°–N 255° for Middle reef and 65°–245° for Strike reef and 65°–85° due North; spacing 20 cm to wider than 1 m; rough, undulating joints; discontinuous; dry; filled with calcite.

The joint pattern is similar in the ore body, hanging wall and footwall regions, except that the schistosity is absent in the country rock. In Zone-I reef, the horizontal joint planes in the hanging wall region are attributed to slickensids, while in all other places majority of the joints are rough undulating.

The rocks are highly jointed with varying frequency due to the regional structure and stresses. There are three prominent sets of joints, i.e. the strike joint, dip joint and the sub horizontal joint. The sub-horizontal joints are most prominent and have mainly caused the ground control problems.

5 Pattern of Ore Shoots at Hutti

Localisation of gold–quartz mineralization in economic quantities occur in the form of ore—shoots and the ore boundaries generally are well defined in the "ore–shoots" portions. The lode zones are characterised by about 6–7 m thick alteration zone with consistent, well fractured and folded quartz veins showing mylonitisation, pinching and swelling, en-echelon distribution both along the strike and dip. In contrast to this, the sub-economic zones of lodes are characterized by narrow zones of very weak mineralization less than a metre width and continue to some extent. Generally, quartz is absent or insignificant in such zones.

The ore shoots are generally tongue-shaped bodies with longest axis having moderate to steep pitch. A definite geometric pattern of distribution for the ore shoots is noticed. In most of the ore shoots, the pitch of the shoots almost tally with the plunge of the corresponding dilatant structures in which they are localised. The important dilatant structural features reflected in the ore shoots regions are folds, en-echelon fractures, intersecting shears, groove lineations and shear zones.

Based on the geological information obtained so far by underground mine development in recent years on various reefs, it was observed that these ore shoots have got definite pitch (Fig. 5). The distribution of various ore shoots also indicate that there is a geometric pattern for their distribution. Further, it was observed that these are structurally controlled, have a definite geometric pattern of distribution along the strike and dip, viz. (1) Ore shoots on strike reef show clear northerly pitch of about 55°–60° and exhibit en-echelon dextral pattern of arrangement, (2) Ore shoots on Zone-I reef show steep northerly pitch of about 75–80° and exhibit en-echelon pattern of distribution and (3) Ore shoots on Middle Reef show steep southerly pitches of about 70° and exhibit typical en-echelon pattern of distribution.

Based on the genetic model and geometric pattern of distribution of the ore shoots evolved, viz. two ore shoots on Zone-I Reef and each on Strike Reef and Middle Reef were predicted, tested, developed, mining started and also in the other ore shoots.

The main observation on the ore shoots are (i) there is no consistent pitch in the shoots, (ii) while some pay shoots terminate in depth, others terminate upwards and appear to be improving with depth.

6 Wall Rock Alterations at Hutti

Chloritization, biotitization, sericitization, silicification and carbonatization constitute major mineralogical variations in these rocks. Two phases of alterations could be distinguished, (1) chlorite–biotite phase in the outer margins and (2) biotite–sericite phase towards the centre, where the quartz veins are localized. The physical changes are represented by changes in colour, compactness and structure (schistosity) of the altered wall rock. The mineral assemblage of these altered wall rocks along the shear zones indicate that they were subjected to retrograde metamorphism and alteration.

Within the altered wall rocks (chlorite, biotite, sericite schist) enveloping the quartz bodies and in the thin partings and horses of altered wall rocks occurring within the quartz bodies occupying in the central part of the auriferous zone. They are characterised by alteration zone of biotitization and sericitization with consistent well-fractured mylonitized quartz veins showing typical pinching and swelling and en-echelon distribution along strike and dip. In contrast with this the lean zones are characterised by narrow zones of weak alteration mostly chloritization and devoid of quartz veins in general.

7 Mineragraphy

The Hutti Gold Mine ore consists of (a) Metallic minerals, which are represented by native elements, sulphide and oxides, and (b) rock forming (gangue) minerals. The bulk of the metallic minerals consist of arsenopyrite, pyrrhotite, pyrite, sphalerite, chalcopyrite, native gold, scheelite, ilmenite and iron oxides. The sulphide minerals are such as arsenopyrite, pyrrhotite, pyrite, coarser grained constituents and little sphalerite and chalcopyrite (in the order of abundance). The total sulphides content of the lodes vary from 2 to 5%. The grains range from 0.1 mm to 2 mm, the average being about 0.4 mm. Arsenopyrite occurs as silver white, euhedral, irregular and altered crystals, pyrrhotite as massive anhedral and irregular grains and pyrite as perfect cubes and not generally well developed and extremely rugged in outline. These three are the principal sulphide minerals and there are commonly small amounts of sphalerite and chalcopyrite, ilmenite, iron oxide, etc. Reef-wise percentage of Sulphides are tabulated at Table 4.

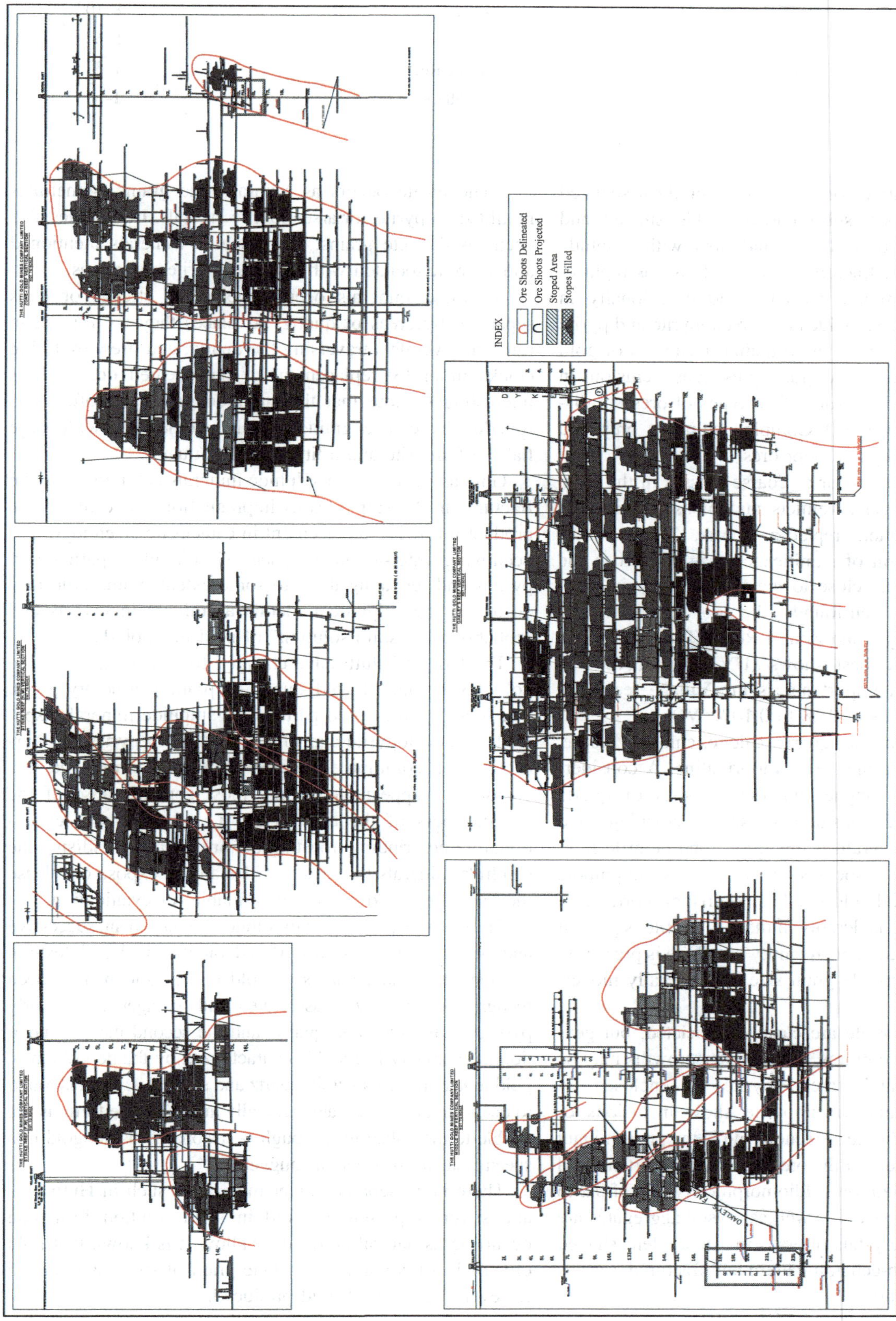

Fig. 5 Geometry of the ore shoots of the hutti gold mine

Table 4 Percentage of Sulphides in reefs

Sl.no	Reef	Sulphides (%)
1	Oakley's	10–15
2	Zone-I	8–10
3	Village	5–6
4	Strike Reef	1–3
5	Middle Reef	1–3

Gold occurs in native form and is generally associated with quartz as angular, sub-rounded, sickle shaped and wire-shaped (1 to 3 mm) forms and also with sulphide minerals as inclusions, fracture fillings and also as replacement. It occurs as minute particles and occasionally in spectular visible form. Sulphides like arsenopyrite and pyrite along with quartz constitute the principal localisers of gold in Hutti ores. The size of gold grains present as inclusions in sulphides is around 25 microns. Two types of ores are distinguished at Hutti, viz. (1) Complex sulphide ores and (2) Quartz rich and sulphide poor ores.

Scheelite is found to occur as coarse masses in the form of small lenses, stringers, veinlets and blebs and also as disseminations. It often replaces quartz, felsic bands. Scheelite mineralisation of significance was noticed in the strike and middle reefs, close to the portions rich in gold. Scheelite content has been found to be poor in the sulphide rich lodes, viz. Oakley's and Zone-I Reefs. The Middle and Strike reefs carry the best shows (UV lamp survey) of Scheelite and next in importance is the Village reef. The grade of Scheelite is about 0.08 to 0.10% WO_3.

The gangue minerals are quartz, calcite, chlorite, biotite, sericite, actinolite, hornblende and tourmaline. A considerable portion of the gangue minerals consists of quartz showing evidence of considerable strain, crushing, and re-crystallisation. Fine grained micaceous and chlorite minerals are intimately associated with fibrous amphibole (actinolite), sericite and chlorite. The relative proportions of these minerals vary considerably in the ore. Biotite is present in abundance. A considerable quantity of calcite is present in the ore and shows variable grain size but generally moderately fine.

Arsenopyrite and pyrite are closely associated, but generally occur as separate grains, though inter-growths may be seen occasionally and rarely pyrite is enclosed in arsenopyrite. Pyrite and arsenopyrite show inclusions of chalcopyrite and pyrrhotite. Arsenopyrite is the predominant sulphide mineral present in the ore and is the important gold bearing sulphide of Hutti ores. Idiomorphic crystals of 0.2 to 3 mm as individual crystals as well as crystal aggregates are common. Deformed crystals of arsenopyrite are lens shaped and are arranged in en-echelon pattern and fractured crystals are present within quartz.

Chalcopyrite occurs as separate fine grains, sometimes moulded on pyrites or arsenopyrite crystals. Iron oxides also occur as discrete grains. The metallic minerals mentioned above are associated with calcite, chlorite and micas.

All these minerals occur together in streaks or clots separated by areas of quartz. At places they are formed in ramifying veinlets traversing the quartz, and grew by filling cracks in or by replacement of the sheared quartz. It is interesting to note that the sulphide grains are often surrounded by coarse grained calcite, sometimes with large flakes of chlorite and a little lamellar quartz.

Gold associated with sulphide minerals is found along the fractures in the sulphides as impregnations, as cementation product following replacement in cataclastic arsenopyrite as occupying inter-granular space of sulphides, particularly pyrites and replacing them to some extent. Some gold also occurs in the sulphide minerals as minute inclusions and probably as a "solid solution" of gold in sulphides.

The lodes of Hutti have changed from gold–quartz (free gold) to the one of gold–quartz–sulphide (refractory gold). The locking-up of gold particles in sulphide minerals especially in arsenopyrite and to some extent in pyrite and pyrrhotite is a common feature in Hutti gold ores. And this encasement appears to be the most general cause of the refractoriness of the ores of Hutti. It necessitates exceptionally fine grinding, roasting the ore, thereby oxidising the sulphide minerals to porous oxides and to expose or release the locked-up gold making it available for cyanidation.

The ore minerals exhibit exsolution, replacement, successive and deformational textures. Based on the field and textural relations, two main phases of gold mineralization have been recognised. The first is associated with paragenetically early pyrite, arsenopyrite and quartz, and the second major generation is invariably late and is characterized by the deposition of gold in late fractures within quartz and sulphides. Gold is found to occur in the native state, generally in a very fine form, is not visible to the naked eye, though occasionally visible gold in a spectacular form occurs as nuggets.

There is no separate silver mineral as such in Hutti ores and silver is present in gold in solid solution form and occurring as natural gold–silver alloy. It is known that gold ores of Hutti contain silver. One gram of silver is obtained for every 12 gram of gold produced.

8 Paragenesis

Two main phases of gold mineralization have been recognised. The first is associated with paragenetically early pyrite, arsenopyrite and quartz and the second major generation is invariably late and is characterized by the deposition of gold in the late fractures within the quartz and sulphides. The locking of gold particles in sulphide minerals, especially in arsenopyrite and pyrite, is a common feature in Hutti ores. This encasement of gold appears to be the most general cause for the refractoriness of the Hutti ores. Gold is found to occur in a native state, generally in a very fine form. Gold ores of Hutti contain small quantities of silver and is occurring as a natural gold–silver alloy.

The ore processing plant integrates two circuits one for the recovery of free gold and the other for the gold contained in sulphides. This is done by comminution followed by gravity processing and finally by cyanidation in vats.

Gold is generally formed associated in quartz veins as free gold and with sulphide minerals mostly encased in sulphides. About 60% of the gold produced comes by Gravity separation as free gold and the remaining 40% of refractory type by leaching by cyanidation process. The encasement of gold particle size as small as 2 microns in the sulphides may be the main cause of the refractoriness of the Hutti gold. It necessitates exceptionally fine grinding to 300 mesh to release gold. However, 100% recovery is not possible because some particles which are much finer size and probably the gold present in the solid solution form in sulphides are not released from the sulphide matrices.

Very rarely gold occurs in the ore spectacularly as visible form. In general, most of gold present in the quartz veins along the parallel fractures is free milling and gold encased in the sulphide minerals is refractory and may be taken as 50% in each case.

9 Controls of Mineralization

The gold deposits of Hutti-Maski Schist belt have definite stratigraphy, lithological and structural control in their localization on regional scale. Locally, gold mineralisation and ore shoots formation is governed mainly by the small-scale structure. The gold deposits in this area are confined to the litho–stratigraphic units of Buddinni Formation of Hutti Group in the Hutti Greenstone Belt. Lithological control is expressed by the occurrence of gold quartz and gold quartz sulphide lodes in chlorite, biotite carbonate schists traversing the metabasalts. Mineralisation is exclusively confined to such schistose units in the Hutti mines area. The linear pattern of distribution of gold bearing veins confines to their intimate relationship with regional structural lineaments.

Though, stratigraphic, lithologic and broad structural features controlled gold mineralization on a regional scale, locally small-scale structures, such as en-echelon folds, shear and tension fractures, inter-mineralization fractures, shear zones and other dilatants structures consequential to the major folds and faults of the area, appear to have mainly controlled the gold mineralization and ore–shoots formation. Detailed study of the ore – shoots on different lodes has revealed that there is a uniform regular pattern of distribution for the ore shoots on each lode. The general patterns indicated are en-echelon dextral and en-echelon sinistral disposition of the ore shoots, along strike and dip. The ore shoots that are controlled by minor cross-folds and other dilatants zones have definite pitch and run almost parallel to the plunge of the folds. The Hutti Gold Mine area has indicated certain important features and conditions which serve as clues or guides to locate gold mineralization and ore shoots. The important guides noted are 1. Stratigraphic, 2. Lithologic, 3. Mineralogical, 4. Fracture patterns, 5. Vein patterns, 6. Folds, 7. Old workings and 8. Established pattern of ore shoots. Since, these features, though they are local, have regional bearing; they could be utilized for ore search not only in Hutti but also in its extensions and adjacent areas.

10 Discussions

There are nine auriferous reefs in the main block, and the mining activity is restricted to six reefs (viz. Oakley's, Middle, Zone-I, Village, Strike Reef Footwall and Strike Reef Hanging wall). All these reefs of Hutti mines persist at depth and there are no indications of any ore bottoming in any of the reefs to 27 level (933 MRL). Based on the detailed underground geological studies, it is clearly established that there is a regular distribution of the gold zone within the lode zones and definite pay shoots could be delineated. In Hutti deposit, the shear zones are the main structural controls for localisation of gold–quartz mineralization. Generally, the shear zones being very penetrative structural grain are expected to have greater depth persistence. The Kolar Gold Field, where the mine workings reached depths of about 3400 m, is a classic example for this view. The Main Reef of Hutti Gold Mine was mined to a depth of 1052 m during the period 1902–1908. So, it is inferred that all the parallel lodes to the east of Main Reef are expected to have such greater depth persistence.

The author with the Geological Survey of India (GSI) conducted gravity surveys during 1986–88, to determine the depth of rocks of Hutti-Maski Greenstone Belt.

These geophysical exploration studies indicated a depth of about 3.50 km for the rocks of Hutti-Maski Greenstone Belt. So, based on the above geological projection, it is expected that gold mineralisation at Hutti persist to depths at least about 3,000 m. As it clearly reveals from the surface geological plan that all the gold bearing reefs are known to extend further north beyond mining lease boundary for about 2 km. This has been indicated by Geophysical anomaly zones by GSI. However, this needs probing either by surface trenching/ diamond drilling or by underground mine development by extending the underground mine levels. Detailed underground geological studies on individual lodes, with the help of underground geological plans and assay plans reveals that ore shoots on different lodes are expected to continue along the strike as well as at depth, immediate north and south of present mining activity and therefore essential that one deep (+1.5 km) bore hole each in the immediate southern and northern extensions of mine to prove the depth and strike continuity of the reefs as suggested below as the Hutti Gold Mine is a suitable area to test the following exploration concepts/targets, i.e.

- Evaluation of behavioural pattern (geometry of the ore zones) for gold lodes at depth.
- Testing of conceptual models for mineralization at depth.
- Study of greenstone belts and mineralization at depth and for ore resources estimation.

Acknowledgements The author is very much thankful to Management of M/s. Hutti Gold Mines Co Ltd., for their continued inspiration and encouragement. The author wishes to place on record his sense of gratitude for the help and background information derived from the professional work done by his colleagues. The data/information gathered by other agencies were necessary to present a comprehensive review attempted in this note. The author wishes to thank the authors of such reprints/paper/notes and reports. The author thanks to both anonymous reviewers for their critical comments on the manuscript.

References

Anand R, Balakrishnan S (2010) Pb, Sr and Nd isotope systematic of metavolcanic rocks of the Hutti greenstone belt, Eastern Dharwar craton: constraints on age, duration of volcanism and evolution of mantle sources during Late Archean. J Asian Earth Sci 39:1–11

Anand R, Balakrishnan S, Mezger K (2005) Geochemical and geochronoligical studies on titanites from the granitoid rocks of the Eastern Dharwar craton, South India. Geophys Res Abstr 7, SRef ID: 1607–7962/gra/EGU05-A-01565

Badhe KV, Pandalai HS (2015) Investigation on the Possible Re—equilibration of aqueous fluid inclusions in barite: a study of barite and calcite from the Hutti gold deposit, Karnataka, India. J Geol Soc China 89(3):715–725

Biswas SK (1990) Gold mineralization in Hutti-Maski greenstone belt, Karnataka. India. Indian Miner 44(1):1–14

Biswas SK (1991) Geological characteristics and timing of gold mineralization in Hutti-Maski greenstone belt, India. BRAZIL Gold 91. In: Ladeira EA (ed) Balkema Publication, pp 187 – 189

Biswas SK, Shushidharan K, Shashimolian AL et al (1988) Regional exploration for gold in Hutti—Maski Schist belt. Raichur District. Karnataka. Rec Geol Sur India 122:117–118

Chadwick B, Vasudev VN, Ahmed N (1996) The Sandur schist belt and its adjacent plutonic rocks: implications for late Archaean crustal evolution in Karnataka. J Geol Soc India 47:37–57

Curtis LC (1963) The Hutti gold mines company limited—note on working results. Development and expansion. Mem Geol Sec India 1:126–131

Curtis LC, Radhakrishna BP, Naidu GK (1990) Hutti gold mine with a future mineral resources of India-2. Geological Society of India, Bangalore, p 123

Dixit RLN (1971) Present status of Knowledge on the old workings for gold in the Hutli-Maski Schist belt, Raichur district, Mysore state, Geol Surv India Rep

Foote RB (1876) The gological features of the southern Mahrastra country and adjacenet district. Memo Geol Surv India 7:259

Foote RB (1882) Notes on a traves across some goldfields of Mysore state. Rec Geo Surv India 15:4

Foote RB (1889) Dharwar System, the chief auriferous rocks series in South India. Rec Geo Surv India 22:1

Hutti Gold Mines Co. Ltd: Annual Reports, Brouchers, Website, and published press notes

Iyer GVA, Vasudev VN (1979) Geochemistry of the Archaean metavolcanic rocks of Kolar and Hutti gold fields, Karnataka. J Geol Soc Ind 20:419–432

lyer GVA, Vasudev VN, Jayaram S (1980) Rare earth element geochemistry of metabasalls from Kolar and Hutti gold bearing volcanic belts. Karnataka carton. J Geol Soc India I 21:603–608

John Taylor and Sons: The Old Gold Mines of India. Part-VIII

Kolb J, Rogers A, Michael F (2005) Relative timing of deformation and two stage gold mineralization at the Hutti Mine, Dharwar Craton, India. Miner Deposita 40:156–174

Krishnamurthy M (1964) The geology and gold-bearing lodes of South Hutti and Kodoni areas. Raichur district, Mysore state. Progress report of the G.S.I. for the field season 1963–64 (Unpublished)

Krishnamurthy M, Krishnamurthy AV (1962) The Geology and Gold bearing lodes of the Huai gold fields area. Raichur district. Mysore state. Progress report of the G.S.I. for the field season 1961–62 (Unpublished)

Maclaren JM (1906) Notes on some auriferous tracts in southern India, Rec Geol Surv India 34:2

Maclareen JM (1908) Gold: Its geological occurrence and geographical distribution. Min J London

Mahanti SC, Raju KK (1974) Geology and methods of working at Hutti gold mines. Mining and metallurgical society of India, Golden Jublee Volume. The Geological

Manikyamba C, Kerrich R, Khanna Tarun C, Satyanarayanan M, Keshav Krishna A (2009) Enriched and depleted arc basalts, with Mg-andesites and adakites: a potential paired arc—back –arc of the 2.6 Ga Hutti greenstone terrane, India. Geochim Cosmochim Acta 73(6):1711–1736

Mishra RN (1990) Banded iron formation of Karnataka. In: Studies in earth sciences. C Naganna felicitation Volume, Bangalore University

Munn L (1934) Section B Economics: dealing especially with the gold mining activity of the area with suggestions for further development. J Hyd Geol Surv 2(l):77–104

Nagabhushanam SP, Sangurmath P, Patil ML, Sukhija BS (2008) Ancient gold mining activity in the Hutti-Maski greenstone belt, Karnataka, India: Radiocarbon perspective. Curr Sci 95(3):389–391

Naganna C (1987) Gold mineralisation in the Hutti mining area, Karnataka, India. Econ Geol 82:2008–2016

Naqvi SM, Rogers JJW (1987) Precambrian geology of India. Oxford University Press

Naqvi SM, Srinivasa Sarma D, Ram Mohan M, Rana Prathap JG, (2008) Role of adakitic magmatism and subduction in gold endowment of Dharwar Neoarchaean greenstone belts, J Geol Soc India 71:875–888

Pal N, Mishra B (2002) Alteration geochemistry and fluid inclusion characteristics of the greenstone-hosted gold deposit of Hutti, Eastern Dharwar Craton, India. Miner Deposita 37:722–736

Pandalai HS, Jadhav GN, Mathew B, Panchapakeson RKK, Patil ML (2003) Dissolution channels in quartz and role of pressure changes in gold and sulphide deposition in the Archaean greenstone hosted Hutti gold deposit, Karnataka, India. Miner Deposita 38:597–624

Pathan AM, Virupakshi P (1987) Mineralisation of schealite from Hutti gold mines Raichur district. Karnataka. India. The Indian Mineral 28(1–2):79–85

Pranjit H, Hazarika, Mishra B, Pruseth KL (2016) Scheelite, apatite, calcite and tourmaline compositions from the late Archean Hutti orogenic gold deposit: implications for analogous two stage ore fluids. Ore Geol Rev 72:989–1003

Radhakrishna BP, Naqvi SM (1986) Precambrian continental crust of India and its evolution. J Geol 94:145–166

Raghunandan KR, Vidyadharan KT (1993) A status note on the geology along the Kolar-Raichur-Geotransect. Memoir Geol Soc India. 25:69–89

Raju KK, Sharma JP (1991) Geology and mineralisation of the Hutli gold deposit, Karnataka, India. In: Ladeira EA (ed) Brazil Gold 91

Ram Mohan M, Sarma DS (2016) Geochemistry and evolution of the Hutti-Maski greenstone belt: link between Archean crustal evolution and gold mineralization. In: Deb M, Goldfarb Richard J (eds) Gold metallogeny India and beyond

Ram Mohan M, Sarma DS, Charan SN, Balaram V, Rajasekhar VB, Ahmad T (2008) Geochemistry and petrogenesis of ambhibolites from the southern part of Gadag greenstone belt, Karnataka. J Geol Soc India 72:484–494

Rama Rao B (1962) An outlin survey of mineral resources of Mysore. Mys Geo Dept Bull 22

Rama Rao B (1963) Gold occurrences in Mysore and their prospects for large scale exploration. Geol Soc Ind Mem 1:21–30

Rogers AJ, Kolb J, Meyer FM (2003) The techno-magmatic evolution of the Hutti-Maski Greenstone belt and relative timing of gold mineralization. Appl Earth Sci (Trans Inst Min Metall B)

Rogers AJ, Kolb J, Meyer FM, Armstrong RA (2007) Tectonomagmatic evolution of the Hutti-Maski greenstone belt, India: constrained using geochemical and geochronological data. J Asian Earth Sci 31:55–70

Rogers AJ, Kolb J, Meyer FM, Vennemann T (2013) Two stages of gold mineralization at Hutti mine, India. Miner Deposita 48: 99–114

Roy A (1979) Polyphase folding deformation in the Hutti-Maski belt, Karnataka. Jour Geol Soc Ind 20:598–607

Roy A, Raju KK (1980) A note on the occurrence of olivine dolerite dyke at Hutti, Ratchur district, Karnataka. J Geol Soc India 21:151–154

Safonov YuG, Radhakrishna BP, Krishna Rao B et al (1980) Mineralogical and geochemical features of endogenous gold and copper deposits of south India. J Geol Soc India 21:365–378

Sahoo AK, Krishnamurthy R, Sangurmath P (2016) Auriferous Lode of Hira—Buddinni gold mine, Hutti-Maski Schist belt, Dharwar craton: mineralogy, alteration, types and mechanism of Vein emplacement. J Geol Soc India 88:675–684

Sangurmath P (1991) Hutti–Maski greenstone belt of Karnataka—a gold scene of Indian future. Ind J Geochem 6(1–2):39–54

Sangurmath P (1996) Status of gold exploration hot spots in Hutti–Maski greenstone belt, Karnataka, India. “Exploration and Exploitation of Gold Resources in India” 2 to 4 Dec 1996. NGRI, Hyderabad

Sangurmath P (2003a) Detailed exploration for epigenetic gold deposit: a case history: Hira–Buddinni gold deposit. Hutti–Maski Schist belt, India. Mining in the 21st Century Quo Vadis? In: Ghose AK, Bose LK (eds) 19th WMC, New Delhi, pp 315–329

Sangurmath P (2003b) The geology setting structure and gold mineralisation of Uti open-cast mine, Hutti–Maski greenstone belt, Karnataka. Min Eng J 5(1):19–21

Sangurmath P (2005) Geology and gold mineralization of the Buddinni gold deposit, Hutti-Maski greenstone belt. Karnataka. J Geol Soc India 66(5):552–560

Sangurmath P (2015) Geology, gold mineralization and exploration of mangulur gold mine, Yadgir Dist, Karnataka, India. J Econ Geol Geo Res Manag 10:103–110

Sangurmath P (2017) Harnessing gold resources for development in Karnataka: the sustainability dimension. J Ind Geol Cong 9(2):167–172

Sangurmath P, Raju PVS (2017) Scheelite bearing veins with enrichment of light rare earth element (LREE’s) from Hutti gold mines, Eastern Dharwar craton. India. J Appl Geochem 19(2):203–207

Sarma DS, McNaughton NJ, Fletcher IR et al (2008) Timing of gold mineralization in the Hutti gold deposit, Dharwar craton, South India. Econ Geol 103:1715–1727

Smeeth WF, Sampat Iyengar P (1916) Mineral resources of Mysore. Mys Geol Dept Bull 7

Srikantia SV (1995) Geology of the Hutti-Maski greenstone belt, In: Curtis LC, Radhakrishana BP (eds) Hutti gold mine-into the 21st Century. Geol Soc India, 8–27

Subba Rao DV, Prabhakar Sangurmath (2015) Major, trace and REE characteristic of the high-grade gold geochemical reference material from middle reef, Hutti gold mines, Karnataka. J Geol Soc India, 85(3)

Sundar Raju PV, Sangurmath P (2016) Scheelite geochemical signatures by LA-ICP-MS and potential for rare earth elements from Hutti gold mines and fingerprinting ore deposits. J Afr Earth Sci 114:220–227

Swaminath J, Ramakrishnan M (eds) (1981) Early precambrian supracrustals of southern Karnataka: memoir geological survey of India, 112:308

Vasudev VN, Chadwick B, Nutman AP et al (2000) Rapid development of the late Archean Hutti schist belt, Northern Karnataka: implications of new field data and SHRIMP U/pb zircon ages. J Geol Soc India 55:529–540

Vasudev VN, Naganna C (1973) Mineragraphy of gold-quartz-sulphide reefs of Hutti gold mines, Raichur District. Mysore State. J Geol Soc India 14:378–383

Vasudev VN, Ramachandra HM, Rajendran N (2021) Gold Mining in India: The Way Forward. Sp. Pub. 11: Geol. Soc. India,Bangalore, pp 170

Viswanatha MN, Ramakrishnan M (1976) The Pre Dharwar supracrustal rocks of Sargur schist complex in southern Karnataka and their tectono-metamorphic significance. Indian Mineral 16:48–65

Ziauddin M, Narayanswami S (1974) Gold resources of India. Geol Surv India Bull 38

Indian Rare-Earth Deposits: Related Industry, Balance Problem and *Atmnirbhar Bharat*

Yamuna Singh

Abstract

A synthesis of rare-earth deposits hosted in various geological domains of India is presented. Major rare-earth deposits in beach placers occur on eastern and southern coasts of India. The Chotanagpur Granite Gneiss Complex terrain constitutes as significant province for heavy rare-earth-bearing stream placers in parts of Jharkhand and Chhattisgarh. The two carbonatite-hosted rare-earth deposits in western India are Amba Dongar and Kamthai. The Samchampi carbonatite complex revealed low-grade Y ore. (Per)alkaline microgranite dykes and tuffs within Siwana Ring Complex in western Rajasthan host low-grade HREE mineralisation, whereas Kanigiri granite and associated soils have attractive REE, Nb, Ta, and Th concentrations. REE-bearing apatite-magnetite hydrothermal veins along the Singhbhum Shear Zone and South Purulia Shear Zone appear significant. Such veins in the Kasipatnam area also appear interesting for rare earths. Future augmentation of the Rare Earth Industry in India can be achieved by judicious use of different natural resources in a planned way. Large-scale mines may be preferred where vast stretches of deposits occur, whereas artisanal to small-scale mining would be appropriate in the areas where deposits are of small dimensions and scattered over a large area. Both downstream and upstream sectors should be strengthened to identify and quantify additional resources to sustain the RE Industry chain. This endeavour may also involve innovative groups at the interface of information technology and the minerals, metals, and materials engineering domains. The development of the RE Industry should be an integral part of the *Atmnirbhar Bharat* through the Mine in India and Make in India programme.

Y. Singh (✉)
Centre for Earth, Ocean and Atmospheric Sciences, Earth Science Building, University of Hyderabad, Dr. C.R. Rao Road, P.O. Central University, Gachibowli, Hyderabad, 500 046, India

Keywords

Indian rare-earth deposits • REE industry • Balance problem • *Atmnirbhar Bharat* • Mine in India and make in India

1 Introduction

According to the International Union of Pure and Applied Chemistry (IUPAC), the term "rare earths" represents the group of 15 geochemically coherent metallic elements. On account of similar geochemical properties and coherence, Scandium (Sc) and Yttrium (Y) are also grouped with rare earths (Spedding 1978; Connelly 2005). All the elements of the lanthanide series, from atomic numbers 57 (La) to 71 (Lu), occur in nature, except promethium (Pm). The rare-earth elements (REE) are divided into two groups: La to Sm as Light REE (LREE) and Eu to Lu as Heavy REE (HREE). Both Y and Sc are also clubbed with HREE (Connelly 2005; US Dept of Energy 2017). The discovery of all the rare earths has long-lasting history of about 160 years from 1788 to 1941 (Gupta and Krishnamurthy 2005; Fig. 1; Zepf 2013). From the 1960s onwards, much progress was made in the large-scale production of purer rare earths, in establishing unknown properties, and their uses in various significant commercial domains. The usable forms of rare earths are naturally occurring oxide mixtures, their synthesised products, high purity individual metals, alloys, and compounds.

After initial discovery and extraction, REE uses have substantially moved upwards to high-purity separated rare-earth metals finding applications in advanced electronics, lighting, power generation, and military applications (Fig. 2; Zepf 2013). Accordingly, the requirement for REE moved up from 75,500 tonnes (t) of rare-earth oxides (REO) in 2000 to 123,100 t REO in 2016 (Roskill 2016a). Due to the shift of end-users, the nature of REE products required has also varied. Applications of individual REE

A. K. Shandilya et al. (eds.), *Geological and Geo-Environmental Processes on Earth*, Springer Natural Hazards,
https://doi.org/10.1007/978-981-16-4122-0_7

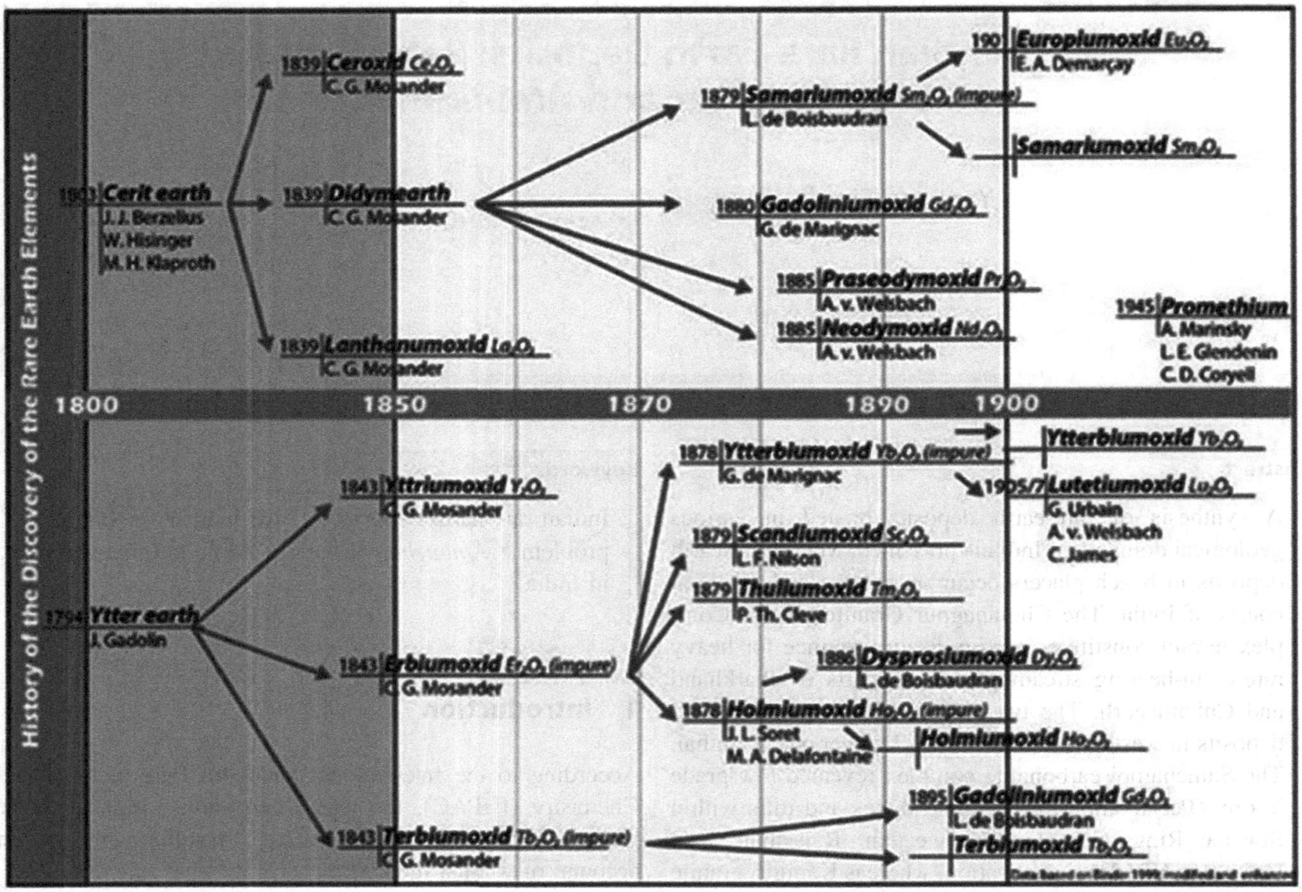

Fig. 1 Insights about long lasting history of REE discoveries (After Zepf 2013)

with respect to their functional usages are shown in Fig. 3 (Zepf 2013). In advanced countries, REE are essential to several industrial, commercial, and residential appliances and in the increasing electrification of vehicles. Although REE may only be required in very less amounts, they will impart life to certain products, rendering them difficult to replace (Smith Stegen 2015). From now to the coming years up to 2026, conventional usages will continue to lead demand growth for REE, though the availability of new products and technologies could change the requirement for REE, both in amount and the specific requirement of the individual elements used. In the 10 years to come, substantial demand for REE is expected for the manufacturing of hybrid electric vehicles (HEVs) and full electric vehicles (EVs) that will require a substantial quantity of REE (Weng et al. 2015; Goodenough et al. 2017). HEVs' and EVs' anticipated growth is from 2.3 million units in 2016 to over 10.1 million units in 2026 (Roskill 2016b). The accentuated growth is likely to increase the requirement of neodymium-iron-boron (NdFeB) magnets. Electric vehicles are believed to show substantial growth in demand. Also, another usage of NdFeB magnets is in renewable energy generation, which would become gradually imperative, as governments and industries aim to fulfil stringent climate change and emission standards (Weng et al. 2015; Goodenough et al. 2017). Today's high-tech world is governed by REE that are considered essential for achieving sustainable development targets (Bertinelli et al. 2019).

Natural REE ores do not contain all rare earths in equal amounts. For example, La, Ce, and Nd dominate the bulk of the ores, relative to much lesser amounts of HREE. Accordingly, due to requirements of different beneficiation routes, such ores pose many challenges in beneficiating them to recover the individual metals needed by the market (Jordens et al. 2013), besides employing extra steps to separate individual REE (Xie et al. 2014; Machacek and Fold 2014). Unequal amounts of individual REE in ore do not match with the market requirement, thereby leading to a balance problem (Binnemans et al. 2013; Binnemans and Jones 2015). In this article, an overview of prominent REE deposits hosted in India's diverse geological settings has been presented. Furthermore, an attempt has also been made to assess the relative commercial potential of different types of REE deposits, mining and rare-earth production,

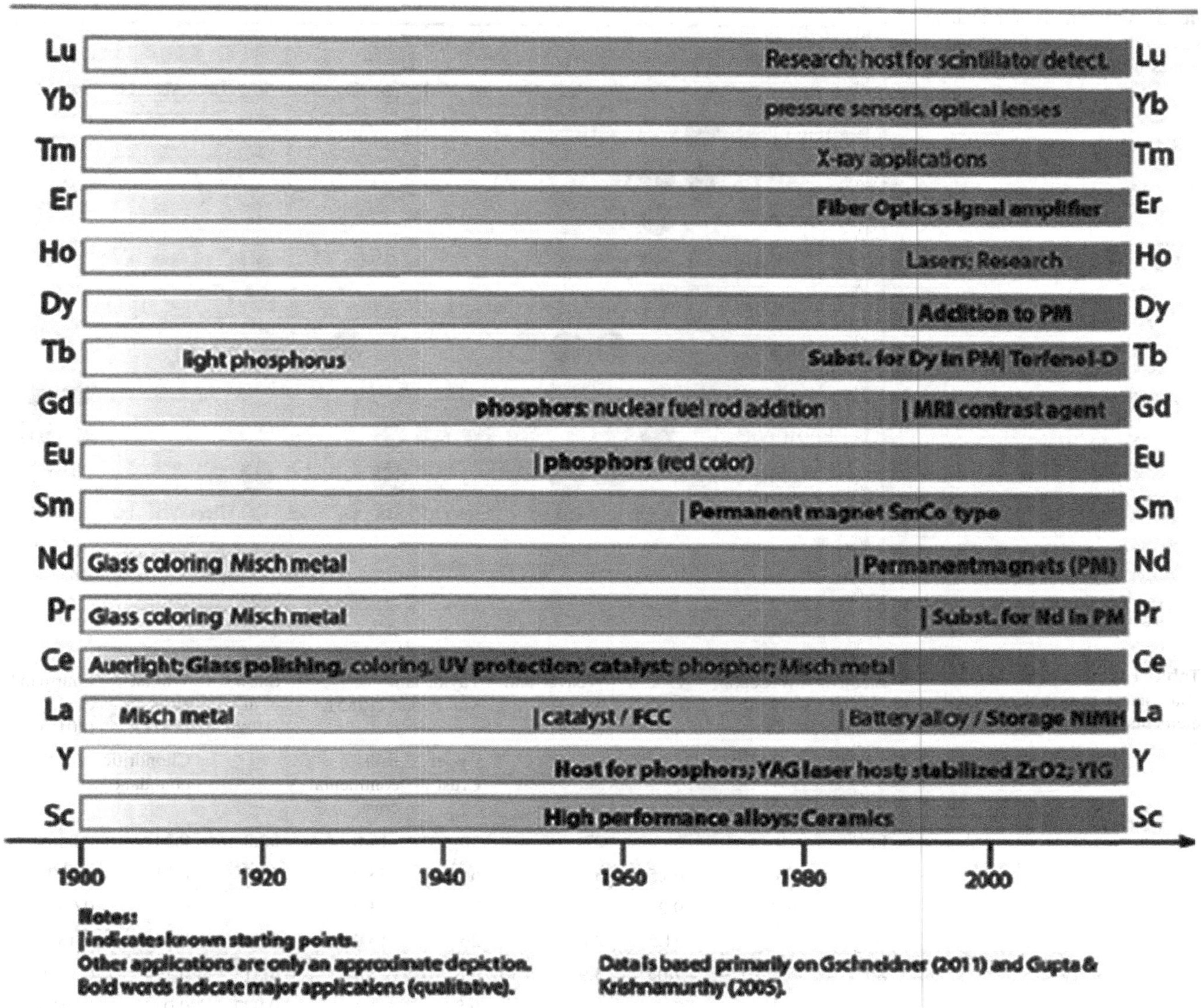

Fig. 2 Main applications of individual REE with tentative years of inventions for commercial applications (After Zepf 2013)

rare-earth industry and ways of addressing individual rare-earth element balance problems in the Indian context, with a focus on *Atmnirbhar Bharat*.

2 Crustal Abundances

In the Earth's crust, REE tend to decrease in abundances with increasing atomic numbers. Furthermore, the Oddo-Harkins effect of elements is responsible for the higher abundance of an even atomic number compared to those with odd atomic numbers. Therefore, the most abundant REE in the Earth's crust is Ce, whereas Lu is highly rare. Despite very low crustal abundances of REE, relative to rarest elements like gold and mercury, REE abundances are high. Additionally, abundances of seven REE are near to few other elements of economic importance, e.g., W, Sn, As and Br, still latter are not termed as rare (Henderson 1996). Also, four REE abundances are in the range of 15–100 ppm that is comparable with Cu, Co, Rb and Zn. Out of all the REE, the average 30 ppm content of Ce makes it most abundant in the crust (Henderson 1996). REE abundances as estimated by various authors are presented in Table 1.

3 Mineralogy

There are innumerable REE-bearing minerals. The rare earths occur as complex polymetallic oxides and hydroxides, carbonates, fluocarbonates and hydroxyl-carbonates and

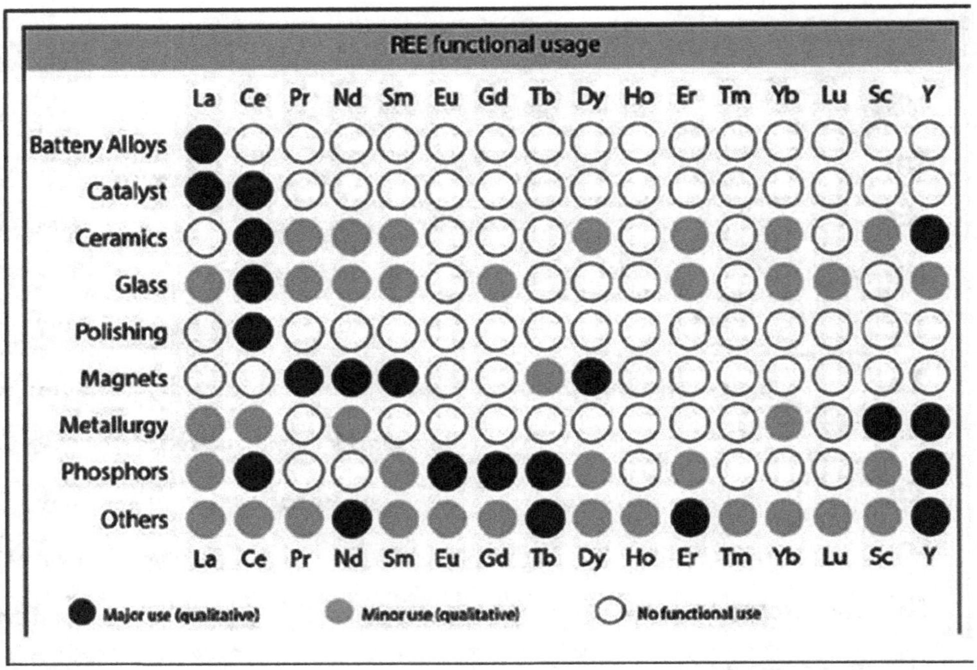

Fig. 3 Applications of individual REE with respect to functional usages (After Zepf 2013)

Table 1 Crustal and chondritic abundances (ppm) of rare-earth elements

Element	Wedepohl (1995)	Lide (1997)	Weaver and Tarney (1984)	Taylor and McLennan (1985)		Shaw et al. (1986)	Wakita et al. (1971)	Pourmand et al. (2012)
				Upper Crust	Bulk continental crust		Chondritic abundances	
La	30	39	27	30	16.0	32.3	0.34	0.2469
Ce	60	66.5	55	64	33.0	65.6	0.91	0.6321
Pr	6.7	9.2	–	7.1	3.9	–	0.121	0.0959
Nd	27	41.5	23	26	16.0	25.9	0.64	0.4854
Sm	5.3	7.05	3.9	4.5	3.5	4.51	0.195	0.1556
Eu	1.3	2	1.07	0.88	1.1	0.937	0.073	0.0599
Gd	4	6.2	–	3.8	3.3	2.79	0.26	0.2093
Tb	0.65	1.2	0.50	0.64	0.60	0.481	0.047	0.0378
Dy	3.8	5.2	–	3.5	3.7	–	0.30	0.2577
Ho	0.8	1.3	–	0.80	0.8	0.623	0.078	0.0554
Er	2.1	3.5	–	2.3	2.2	–	0.20	0.1667
Tm	0.3	0.52	0.23	0.33	0.3	–	0.032	0.0261
Yb	2	3.2	1.46	2.2	2.2	1.47	0.22	0.1694
Lu	0.35	0.8		0.32	0.3	0.233	0.034	0.0256
Y	24	33	13	22	20.0	21	–	1.395
Sc	16	22	–	–	30.0	–	–	5.493
Total	184.3	242.17	–	–	136.9	–	–	9.5118

Note: stands for not available.

Table 2 Name, chemical composition and approximate total REO content (in wt.%) in selected rare-earth minerals (After Binnemans and Jones 2015; Dostal 2017; Singh 2018, 2020a, b)

Mineral	Chemical formula	Total REO
Aeschynite	$(Ce,Ca,Fe,Th)(Ti,Nb)_2(O,OH)_5$	32
Allanite	$(Y,REE,Ca)_2(Al,Fe^{3+})_3(SiO_4)_3OH$	38
Apatite	$(Ca,REE)_5(PO_4)_3(F,Cl,OH)$	19
Bastnaesite	$(REE,Y)(CO_3)F$	53–79
Brannerite	$(U,Ca,Y,Ce)(Ti,Fe)_2O_5$	9
Britholite	$(Ce,Ca)_5(SiO_4.PO_4)_3(OH,F)$	32
Cebaite	$Ba_3Ce_2(CO_3)_5F_2$	32
Churchite	$REEPO_4.2H_2O$	43–56
Eudialyte	$Na_{15}Ca_6(Fe,Mn)_3Zr_3(Si,Nb,REE)Si_{25}O_{73}(OH,Cl,H_2O)_5$	< 10
Euxenite	$(REE,Ca,U,Th)(Nb,Ta,Ti)_2O_6$	< 30
Fergusonite	$(REE,Y)(Nb,Ta,Ti)O_4$	43–52
Florencite	$CeAl_3(PO_4)_2(OH)_6$	32
Gadolinite	$REEFeBe_2Si_2O_{10}$	< 52
Gittinsite	$CaZrSi_2O_7$	NA
Huanghoite	$BaCe(CO_3)_2F$	39
Iimoriite	$Y_2(SiO_4)(CO_3)$	NA
Kainosite	$Ca_2(REE,Y)_2Si_4O_{12}(CO_3).H_2O$	38
Knopite	$(Ca,Ti,Ce_2)O_3$	NA
Loparite	$(REE,Na,Ca)(Ti,Nb)O_3$	28–38
Monazite	$(REE,Th)PO_4$	38–71
Mosandrite	$(Na,Ca)_3Ca_3REE(Ti,Nb.Zr)(Si_2O_7)_2(O,OH,F)_4$	NA
Parisite	$CaREE_2(CO_3)_3F_2$	58–63
Pyrochlore	$(Ca,Na,REE)_2Nb_2O_6(OH,F)$	NA
Samarskite	$(Ce,Y,U,Fe^{3+})_3(Nb,Ta,Ti)_5O_{16}$	24
Steenstrupine	$Na_{14}REE_6Mn_2Fe_2(Zr,Th,U)(PO_4)_7Si_{12}O_{36}(OH)_2.3H_2O$	< 31
Synchisite	$CaREE(CO_3)_2F$	48–52
Xenotime	YPO_4	43–65
Yttrocerite	$(Ca,Ce,Y,La)F_3.nH_2O$	53
Zircon	$(Zr,REE)SiO_4$	NA

NA = Not available

silicates, phosphates, arsenates, sulphates and vanadates, uranyl-carbonates and uranyl-silicates and borates (Jones 1996; Singh 2020a). A list of very common REE minerals and their REO content are given for ready reference (Table 2). Nevertheless, out of these, bastnaesite and monazite for light REE and xenotime for heavy REE are commonly exploited for commercial production of REE. Monazite theoretically contains 70% combined REO (with ~2% yttrium oxide), but generally, monazite yields 55–65% REO. The theoretical content of bastnaesite is ~75% REO (with ~0.05% Y), but floatation concentrates generally yield 60% REO. While xenotime may theoretically contain ~67% Y_2O_3, xenotime concentrates on an average yield of 25% Y_2O_3 (Hedrick 1985). On account of similar geochemical behaviour, all the rare-earth elements, generally, occur together in complex minerals and behave like a single chemical entity.

4 Indian Rare-Earth Deposits

According to IBM (2018), the total estimated REO reserves of the world are 121 million tonnes, out of which, China accounts for over 36%, Brazil and Vietnam ~18% each and Russia ~15% (IBM 2018). India's share of REO reserve is ~5.7% (IBM 2018; Singh 2020a, b). World production of REE is dominated mainly by China (IBM 2018; Singh 2020a; Verma 2020). Adopting prevailing concepts in the classification, the Indian rare-earth deposits are briefly described below in decreasing order of their commercial potential. However, many comparatively minor natural rare-earth resources are not described. Furthermore, secondary sources of rare earths have also not been dealt with.

4.1 Coastal Beach Sand Deposits

The discovery of monazite-bearing black sand was made in 1909 by Schomberg, a German chemist along the southern and south-western coast of India (Overstreet 1967). Subsequently, several workers have contributed to the understanding and knowledge of beach sand deposits (Dhana Raju et al. 2001; Rajamanickam 2000, 2001; Rajamanickam et al. 2004; Loveson et al. 2005; Mallik 2007; Nageswara Rao et al. 2007; Nambiar et al. 2007; Roonwal 2009). Exploration efforts by employing both routine and innovative drilling and laboratory methods brought to light several heavy mineral placer deposits in different coastal areas. Some inland placer bodies in Kerala, Tamil Nadu, Odisha, Andhra Pradesh, Maharashtra, Gujarat and West Bengal states have also been recognised. Some of the coastline beach placer deposits of India are extensive and richest, comprising ilmenite, rutile, garnet, monazite, zircon and sillimanite heavy mineral assemblages. Ilmenite-rich major beach and dune sand deposits are located in Kerala (Chavara); Tamil Nadu (Manavalakurichi, Midalam, Vayakallur); Andhra Pradesh; Odisha; and Maharashtra. The ilmenite commonly contains 50–60% TiO_2 and is suitable for different process technologies. Ilmenites from the Kerala coast have up to 70% TiO_2 on account of the conversion of ilmenite to leucoxene (Ali et al. 2001a; IBM 2018). Zircon, monazite and sillimanite constitute potential co-products.

A few beach placer deposits are under exploitation over the last several decades. Mining and beneficiation of beach sand deposits are currently carried out by IREL (India) Limited (Formerly Indian Rare Earths Limited; IREL), a public sector undertaking of the Department of Atomic Energy, Government of India, and Kerala Mineral and Metals Limited (KMML), a Kerala State Government undertaking (IBM 2018). Additionally, two private producers, namely, M/s V.V. Mineral, Thoothukudi, and M/s Beach Minerals Co. Pvt. Ltd., Kuttam (both in Tamil Nadu), also played a key role in beach sand mining (IBM 2018). Abundances of REE-bearing minerals reveal considerable variation in different beach sand deposits.

The approximate resources in the Indian placer mineral deposits are 682 Million tonnes (Mt) of ilmenite (including rutile and leucoxene), 187 Mt of garnet, 36 Mt of zircon, 12.47 Mt of monazite and 255 Mt of sillimanite (IBM 2018; Singh 2019, 2020a, b). India's placer heavy mineral resources are nearly 35% of world resources of ilmenite, 10% of rutile, 14% of zircon and 71.4% of monazite. Furthermore, India makes up nearly 10% of the world garnet supply (Ali et al. 2001a). Different beach placer deposits of India are briefly described below.

4.1.1 Coastal Beach Placer Deposits of Odisha

The deposits have beach and dune sand origin, and show a nearly uniform mineralogical composition. Various studies have revealed the presence of ilmenite, garnet, sillimanite, rutile, zircon, monazite, magnetite and pyrobole (Rao 1989a; Rao et al. 2001; Behera 2003; Acharya et al. 2015; Das 2015; Ghosal et al. 2020). Based on coastal geomorphology, nature of source rocks, mineral distribution and grade, Odisha coast is divided into three sectors, southern, central and northern. In the southern segment, three important deposits, namely, Gopalpur, Chatrapur and Rushikulya-Bajrakot, have been established. These are located from south to north, and extend for over a coastal length of 43 km. The northern 18-km long tract between Rushikulya river and Gopalpur town is named Chatrapur Mineral Sands Deposit, [under mining by Orissa Sand Complex, OSCOM, of the IREL (India) Limited], whereas the southern 25-km contiguous area is called Gopalpur Mineral Sands Deposit. Furthermore, in the central sector, wider sand bodies with the richest grade of total heavy minerals (THM, 81.36% as at Baulapura; comprising 41.3% ilmenite, 20.76% garnet, 2.57% monazite, 7.77% rutile, 5.73% zircon and 0.56% kyanite) are also present along the Puri coast (Jagannadha Rao et al. 2008; Kathiravan et al. 2018). In the Ganjam coast, approximate THM in beach sand is 5.6%, as against 9.5% in dune sand (Satpathy et al. 2010). The distribution of particle size of beach sand is relatively more than the dune sand (Routray et al. 2009). Deltaic sediments brought by Mahanadi with the widest beach and dune complex comprise the central part of 230 km. Four palaeo-ridges, identified in the Mahanadi delta, contain widely varying (2–64%) heavy minerals; the inferred heavy mineral reserve is of the order of 12.89 Mt with an average grade of 10.48% heavies. The northern segment of the 132 km coastline comprises low-lying coast of tidal flats, beach and dune sands and palaeo-beach ridges parallel to the coast. The bulk of the sediments are supplied by, among others, Subarnarekha and Burhabalanga rivers. Heavy minerals show a fairly varying grade of 0.71–12.28% (Rao et al. 2001). The grade of monazite in the selected beach placer deposits of Odisha is highly variable (0.001–1.21%). Similarly, grades of zircon vary from 0.001% to 1.09% and of garnet from 0.02% to 13.5%.

One bulk sand sample of Jhatipodar from the southern segment reveals 0.08% CeO_2 and 0.09% ZrO_2 (Acharya et al. 2015), whereas from close to Brahmapur town analysed on an average 2091.98 ppm TREE (Khan et al. 2018). The LREE is dominated by Ce, La, Nd and Sm, whereas Dy and Yb dominate HREE (Khan et al. 2018). Ghosal et al. (2020) reported 104–3041 ppm TREE, with an average of 637 ppm TREE in beach sand samples. Monazite from Chatrapur deposit contains 59.76% TREO (Rao et al. 2001). In another study on monazite, Mohanty et al. (2003) reported 51.57–59.35% LREO, mainly comprising Ce, La and Nd in decreasing order. On the whole, content of HRE and Y together (1.43–2.06% HREY) is very low (Mohanty et al. 2003). Baddeleyite (80.84–82.72% Zr) and cheralite are also present in Chhatrapur beach sand deposit (Rao and Acharya 2015; Rao et al. 2016a). Apatite and garnet also occur as inclusion within ilmenite (Rao et al. 2016b).

4.1.2 Coastal Beach Sand Deposits of Andhra Pradesh

Innumerable mineral sands deposits of varied nature, size and grade occur along the coastline of Andhra Pradesh (Rao et al. 1989, 2001). Samples from badlands topography also contain heavy minerals (Mishra et al. 2019). This coast has also been divided into three sectors, southern, central and northern. In the northern segment, with sands of beach and dune origin, high-tonnage deposits are located in Bhavanapadu, Kalingapatnam and Srikurmam areas, whereas those of Donkuru-Barua, Koyyam, Bhimilipatnam and Pentakota belong to the category of medium-tonnage deposits. The average total heavy mineral (THM) in these deposits ranges from 5.7 wt.% to 34.36 wt.%. Ilmenite, garnet and sillimanite form 90 wt.% of THM. Out of the remaining 10%, gangue minerals and magnetite comprise 5.60% and 0.40% of THM, whereas the rest (4%) of the THM contains monazite, zircon and rutile (Dwivedy 1995; Ravi et al. 2001). Monazite (0.1–0.85%), zircon (0.68–1.50%) and garnet (19.33–37.10%) reveal large variations.

In the central sector, in Kakinada and Nizampatnam heavy mineral deposits are of medium-grade and large tonnage. Ilmenite, rutile, monazite, zircon, garnet, sillimanite and magnetite comprise heavy mineral assemblage. Abundances of REE-bearing minerals, e.g., monazite (0.25–1.14%), zircon (0.10–8.9%) and garnet (0.34–20.86%) are highly variable (Subrahmanyam et al. 2004; Jagannadha Rao et al. 2007). Xenotime is also present in beach placers of Narasapur coast, West Godavari district (Subrahmanyam et al. 2004). Here, THM content is 14.7% (av.), in which grade of xenotime is low (0.09–1.32 wt.%), with an overall grade of 0.61 wt.% (Subrahmanyam et al. 2004).

The southern sector falls in Prakasham and Nellore districts, and comprises dominantly salt marshes. Coastal sand contains 3–12.6% THM. Heavy minerals comprise pyrobole, ilmenite, magnetite, with notable amount of monazite, zircon garnet and rutile (Ravi et al. 2001). Like other sectors, here also monazite (0.02–0.54%), zircon (1.63–2.99%) and garnet (2.30–27.48%) show notable grade variations.

Samples from Bhimlipatnam and Ramakrishna beaches are enriched in LREE, especially La, Ce, Pr and Nd (Palaparthi et al. 2017). However, placers from Bhimlipatnam show more enrichment of REE (av. 2402 ppm), relative to placers (av. 1491 ppm) from Ramakrishna beach (Palaparthi et al. 2017). Although low, yet HREE, particularly Gd is also more in Bhimlipatnam beach (av. 48 ppm Gd) compared to Ramakrishna beach (av. 30 ppm Gd).

Monazite from coastal sands between Bhimunipatnam and Konada contains 49.5%–59% REO (av. 54.7%), dominated by LREO (47.7–57.8%, av. 53.3%) relative to HREO (0.77–3.42%, av. 1.38%) content (Naidu et al. 2016). Additionally, it also contains significantly higher amounts of Nd, e.g., average 10.53% Nd_2O_3 (Naidu et al. 2016). Reddy and Prasad (1997) reported cheralite from Kalingapatnam-Baruva coast with low REO (37.48%) content. Monazite from Visakhapatnam beach sand contains more rare earths (61.8% REO) relative to monazites from Kalingapatnam-Baruva coast (37.48–58.45% REO). Still low amount of rare-earth content is revealed by monazite from Visakhapatnam-Bhimunipatnam beach (36.56% REO).

Monazite grains from beach placers of Kalingapatnam coast (northern sector) contain 46.3–61.3% REO (av. 54.2%), with up to 0.4% Y_2O_3 (Panda et al. 2003). Significantly, Nd content (9.3–11.6% Nd_2O_3, av. 10.3% Nd_2O_3) is also more, with attractive amounts of Sm (1.4–1.8% Sm_2O_3, av. 1.6% Sm_2O_3).

Garnet from Kalingapatnam coast revealed 0.15–0.53% REO, bulk of which is contributed by heavy rare earths, e.g., 0.12–0.47% (Panda et al. 2017). A similar preponderance of HREO over LREO is also revealed by garnets present at various depths (Panda et al. 2017). This consistency is also maintained by the overall very low average LREO/HREYO ratio (0.16). Garnets of various grain sizes irrespective of their depth of occurrences also reveal similar behaviour of HREE concentration (Panda et al. 2017). In view of the vastness of its resources and considerable grade, garnets from beach and dune sands of Kalingapatnam coast may form low-grade REE resource (Panda et al. 2017).

4.1.3 Coastal Beach Sand Deposits of Tamil Nadu

Innumerable heavy mineral deposits occur in coastal stretches of Tamil Nadu and in the Teri sands of the southern coastal plains. Ilmenite, garnet, sillimanite, pyrobole, zircon, rutile, monazite and kyanite with minor amounts of spinel, tourmaline, epidote, apatite and staurolite constitute heavy mineral assemblage (Chandrasekharan and Murugan 2001). Anguswamy and Rajamanickam (2000) gave an account of heavy mineral distribution in the beaches of southern Tamil Nadu, whereas Mohan and Rajamanickam (2000) brought out buried placer mineral deposits along the east coast between Chennai and Pondicherry. Teri sands have been studied by Menon (1950), Gardner (1981) and others. One of the oldest known heavy mineral deposits, Manavalakuruci, is located in Tamil Nadu, and the same is under exploitation by IREL (India) Limited. Baddeleyite is also known from the sand of this deposit (Overstreet 1967). The deposit is spread over about 6 km length, with a width of 44 m, and THM of 39% (Ali et al. 2001a). Manavalakurichi is next to Chavara in terms of TiO_2 content that is more than 55% (IBM 2018). Monazite from this deposit contains 58% REO (Ali et al. 2001a). Two famous large deposits of Teri sand and soil (Red sediments) also occur along the southern coast in Sathankulam and Kuthiraimozhi in southern districts of Tamil Nadu. Ilmenites from Teri sand reveal intergrowth of ilmenite and hematite (Rao et al. 2005). In areas of Tirunelveli, Chidambaranar, Ramanathapuram and Kanyakumari districts, this type of four deposits have been reported with an average THM of 10% (Ali et al. 2001a). Grades of heavy minerals and associated monazite, zircon and garnet from heavy mineral deposits of Tamil Nadu reveal considerable variations (Singh 2020a).

Zircon from the southern coast of Tamil Nadu contains 163.1–317.2 ppm REE (Angusamy et al. 2004). Zircon is enriched in LREE (126.1–291.2 ppm) relative to HREY (51.5–75.7 ppm). Zircon may form low-grade REE resource-base (Singh 2020a).

4.1.4 Coastal Beach Sand Deposits of Kerala

In Kerala state, the largest and richest heavy mineral deposit is located along the Neendakara-Kayamkulam coastal stretch, namely, Chavara deposit. Concentrations of heavy minerals have also been noted in lake- and sea-bed sediments (Rao 1989b; Krishnan et al. 2001). The deposits/occurrences at Azhikode-Chavakkad, Chavakkad-Ponnani and Valarpattnam-Azhikode, falling in northern Kerala, are dominated by pyrobole with very low ilmenite content. Grade of THM and monazite, zircon and garnet of Chavara major deposit, and associated onshore land resources, and lake- and sea-bed resources and sand deposits in the northern continuity of Chavara deposit and those in southern Kerala and northern Kerala reveal significant variation (Krishnan et al. 2001; Singh 2020a).

On an average, commercial monazite from Kerala contains 59.5% REO (Overstreet 1967). REE contents of monazites from Chavara and Varakala (data in wt.%) are 10.3 and 10.5 La, 17.4 and 17.6 Ce, 9.4 and 10.2 Nd, 1.4 and 1.45 Sm, besides 5.8 and 5.9 Th and 4.5 ppm and 11.0 ppm Sc, respectively (Krishnan et al. 2001). Clearly, monazite also reveals notable amounts of Nd and some Sm. A similar geochemical nature is also revealed by monazite from the beach sand situated at the confluence of the river Achankovil, Alleppey, Kerala (Pruseth et al. 2017).

4.1.5 Coastal Beach Placers of Maharashtra and Gujarat

In Ratnagiri district (Maharashtra), several heavy mineral deposits/occurrences are known (Ali et al. 1989). Towards the north of Ratnagiri deposits is generally rich in ilmenite, whereas due south of it both ilmenite and magnetite occur together (Ali et al. 2001b). Ilmenite (and magnetite) is fine-grained and has smaller tonnage (Ali et al. 2001a). In Talashil beach placer of south Maharashtra coast, THM reaches up to 53.8% by weight (Rao et al. 2010). Further in the northwest extension of coastal deposits of Maharashtra, monazite occurs in the estuary of the Narbada River on the Gujarat coast. The monazite contains 5–11% ThO_2 (Overstreet 1967).

4.1.6 Coastline of West Bengal

The eastern portion of the coastal area in West Bengal contains monazite, zircon, apatite, garnet, sphene and ilmenite, besides hornblende, pyroxene, sillimanite, kyanite, apatite, tourmaline in various proportions. Abundances of ilmenite and magnetite are 4% and 5% in the Bakkhali area, with still more abundance in dunes. Ilmenite revealed up to 24.77% TiO_2 (Mallik and Sensarma 2009). More concentrations of heavies have been found around Digha, Shankarpur, and north of Rasulpur, especially in river mouths (Mallik and Sensarma 2009). In the mouth of Hooghly river, monazite, zircon and garnet, including sillimanite, kyanite, staurolite, hornblende and pyroxene occur in discontinuous patches of shelf sediments (Mallik 1976).

4.2 Inland Stream Placers

Inland stream placers are attractive especially for Y and HREE resources. According to potentiality, stream placers are grouped into major and minor resources. Major stream placer deposits occur in Chotanagpur granite gneiss complex (CGGC) terrain in Jharkhand and Chhattisgarh, whereas minor resources are scattered in other parts of India. About 3500 tonnes of polymineral concentrate, with xenotime grade of 3 wt.%–8 wt.%, was worked out in stream placers of India till 1995 (Banerjee et al. 1995).

4.2.1 Stream Placers (Major Sources) in CGGC Terrain

The streams containing REE-bearing placers flow through Precambrian CGGC terrain, being dominated by granitoid gneisses and migmatites (Mazumdar 1988; Mahadevan 2002; Ghose 1992; Singh 1992, 2012a; Singh and Reddy 2008; Singh and Krishna 2009; Singh et al. 2012). The nature of stream placers formed in different parts of CGGC terrain is briefly given below.

Ranchi-Purulia Area, Jharkhand and West Bengal

Highly radioactive inland sand deposits occur around Hururi, Purbatoli and Patratoli (Mohammad and Nandi 1971). The lower portion of the deposit is coarse-grained, greenish-coloured, arkosic sand with intermixed granitic rock fragments. It grades upward into medium to fine-grained, dark brown intermediate horizon, whereas the uppermost portion contains fine-grained, light green sand. Monazite, zircon, ilmenite, magnetite, tourmaline, rutile, sphene, apatite and sillimanite constitute heavy minerals (Mohammad and Nandi 1971). Around Kataholdih, monazite-rich sand has a thickness reaching 3 m, with an average thickness being less than 1 m, covering about 12 km^2 area (Overstreet 1967).

Gumla-Simdega District, Jharkhand

Sizeable REE-bearing placers occur in four streams, namely, Deo, Girma, Halwai and Pojenga. All these flow mainly through pink and grey felsic rocks (Singh and Rai 2008). Fluvial placer deposits are present in the easterly flowing upstream portion of Deo River extending for a length of 18 km (Rai et al. 1990, 1991). Towards downstream, placer bodies show changeable size, and persist up to a depth of 1 m. Potential placers are spread over about 135,000 m^2 area. Available quantity of ore gradually reduces from flood plain sediments to point bars through side bars to braid bars. However, weight percentage of total heavy minerals (THM) is appreciably high in various bars (1.5–2.5%) compared to flood plain sediments (0.46–1.0%). Clearly, in the top 50 cm (0.22–4.85 wt.% THM) higher amounts of heavies occur, compared to the lower 50 cm (0.08–3.45 wt. % THM). Approximate resource of THM is 1145 (av. grade 4.95 wt.%) tonnes (Bagora et al. 2006). Non-magnetic heavies analysed 1.44–1.94% Y_2O_3 (Rai et al. 1991). Heavy mineral concentrate contains ilmenite (44 wt.%), monazite (48 wt.%), xenotime (4 wt.%), zircon (0.7 wt.%), garnet (0.3 wt.%) and others (Bagora et al. 2006). Bagora et al. (2006) also reported $\sim$550 tonnes monazite and $\sim$46 tonnes xenotime. Magnetic fraction at 0.45–0.60 amperage revealed 33.75% Y_2O_3 (Rai et al. 1990, 1991). In Girma River, the significant placer occupies 174,000 m^2 areas (Rai et al. 1991). Significant xenotime concentration is noted in braid bars and flood plain sediments, whereas side bars are rather poorly developed. The weight percentage of THM, containing ilmenite, monazite, magnetite, zircon and xenotime, varies from 0.60 to 1.70% with 1.19–1.71% Y_2O_3. In Halwai River, placer deposit occurs intermittently over a length of about 9.5 km. Placer bodies show variable width, with a thickness of up to 1 m. Potential placer is present over 162,000 m^2 area (Rai et al. 1991). Different fluvial landforms, with heavy minerals in decreasing abundances, are colluvial deposits, side bars, braid bars, flood plain sediments, and minor amounts in point bars and river terrace sediments. THM varies from 0.50 to 3.65 wt.%, with 1.02–1.78% Y_2O_3. Ilmenite, magnetite, sphene, monazite, zircon and xenotime form heavy minerals. Observed fluvial landforms along the Pojenga river course include braid bars/island and side bars, with small areal extents. Approximate area of potential placers is 21000 m^2. THM varies from 0.50 to 3.65 wt.%, with 1.02–1.78% Y_2O_3. Heavy minerals comprise ilmenite, magnetite, sphene, monazite, zircon and xenotime (Rai et al. 1991).

Jashpur District, Chhattisgarh

Rare-earth and yttrium mineral resources are genetically connected to polyphase “S”-type granites of CGGC. Fine-grained grey granite (FGGG) and pink granite (PG) are identified to be the main sources of heavy REE and Y minerals (Singh 1990). On the other hand, apart from FGGG and PG, grey and porphyritic granites also contribute light REE mineral (Singh 1992). Riverine placers along streams draining through such granites contain economic concentrations of REE and Y minerals (Mahadevan 1988; Singh 1990; Singh and Rai 1992; Singh and Singh 1996; Singh et al. 1996, 2012, 2018). Workable concentrations of placers are noted in Siri-Chapajharia-Dhob and Baljora rivers, located in Chhattisgarh bordering Jharkhand (Fig. 4).

A major portion of the Siri-Champajharia-Dhob river basin is covered by alluvial plains, and occasionally by small granitic mounds and hillocks and schistose rocks. Placers of interest occur discontinuously over a total length of about 12 km along Siri river, and over about 2-km length in the downstream sections along its tributary, Champajhariya-Dhob. However, over about 31-km length along both the streams potential placer deposits are strewed in an area of about 403,300 m^2 (Singh 2020a). The THM content varies from 1.0 to 2.0 wt.%. About 300 tonnes of xenotime in preconcentrates, with 3–5% xenotime, has been inferred. Xenotime shows well-developed crystals of short by-pyramidal habit, without prism faces (Rajasekaran et al. 1994). Point chemistry of xenotime grains revealed 42.9–44.7% Y_2O_3 and 33.3–34.2% P_2O_5. Xenotime also contains high amounts (average) of 3.1% Gd_2O_3, 4.4% Dy_2O_3, 4.6% Er_2O_3 and 5.4% Yb_2O_3 (Sarbajna et al. 2007). Monazite grains analysed 12.2% La_2O_3, 26.6% Ce_2O_3, 2.7% Pr_2O_3, 8.8% Nd_2O_3 and 1.6% Sm_2O_3 (Sarbajana et al. 2007).

Several patches on either side of the Baljora river have been converted into cultivated fields leading to mutilation of original placer bodies. However, placer bodies of economic significance occur intermittently over a length of about 5.2 km in the middle portion of the river. Less-consequential, largely scattered, small-scale braid bar placers are also present towards downstream. Nevertheless, substantial deposition is restricted to point bars and flood plain sediments on

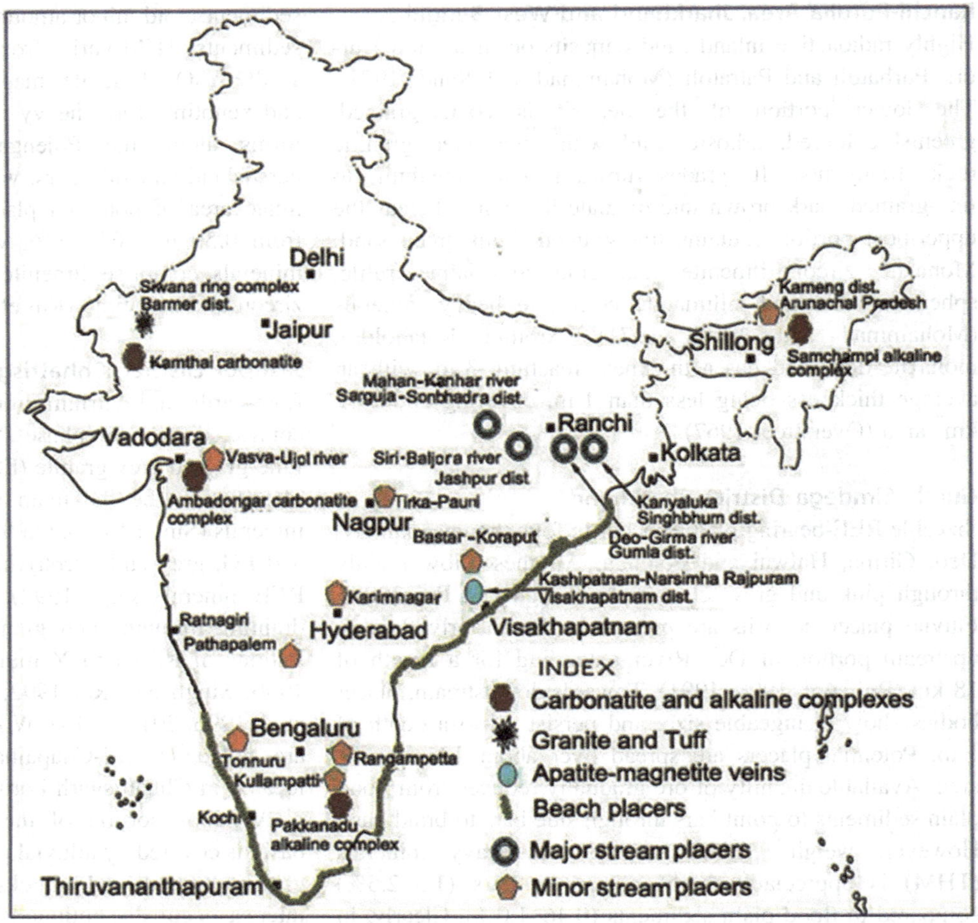

Fig. 4 Geographical locations of selected REE deposits/occurrences of India (After Singh 2020a).

either banks, but being intense along the right bank. Like elsewhere, in the case of Baljora River also placer bodies have variable dimensions, with thickness of 0.5–1.0 m. Economic placer occupies an area of about 171,000 m^2. Accumulation of heavy mineral is high in point bars, especially in top 0.5 m of the bars (Singh and Rai, 1992). In a study on polymineral concentrate, Rai et al. (1991) reported ilmenite 32%, monazite 38%, zircon 11.6%, xenotime 2.8%, magnetite 0.6%, rutile and hematite. Non-magnetic table concentrates revealed (average) 2.34% Y_2O_3, 0.238% Er_2O_3 and 0.02% Eu_2O_3, with mutual ratio of these being 10:1:0.1 that is similar to those shown by heavy mineral concentrates from the source granites. Xenotime contains up to 42.30% Y_2O_3 and is radioactive (Singh and Rai 1992).

Surguja-Balrampur Districts, Chhattisgarh

Xenotime-bearing placers occur along several streams flowing through the Precambrian crystalline rocks in Chhattisgarh and Uttar Pradesh, where REE and U-Th concentrations are present in various areas (Seth 1989; Saxena et al. 1990; Sinha et al. 1992; Singh et al. 1999), some qualifying low tonnage U deposits (Singh 2020a). These granitic rocks reveal anomalously higher concentrations (Mahadevan 1992; Saxena et al. 1992; Sinha et al. 1992; Singh and Singh 1996) of Y (33–356 ppm), Zr (114–533 ppm), Nb (16–161 ppm), U (10–141 ppm) and REE (666–891 ppm). In the catchment area of the Mahan river basin, grey and pink granitic rocks are present (Saxena et al. 1990, 1992), and have acted as provenance for rare-earth-bearing placers (Singh 1990; Rai and Banerjee 1995; Singh and Singh 1996; Ramesh Babu et al. 1998). Various streams of the basin contain placers over a length of more than a kilometer with THM of 0.50 wt.% analysing 1.0% Y_2O_3 (Ramesh Babu et al. 1998). Significant Y-rich placers are noted in tributaries of Kanhar river, namely, Devhar, Karniyan and Behri (Balrampur, Sonbhadra district, U.P.) and Sarsoti, Bairamua, Teenbhitiya and Righwa (Bhandaria, Garhawa district, Bihar). Heavy minerals from the Mahan river basin revealed 0.20–4.13% Y_2O_3 and 0.59–23.24% Ce_2O_3. Studies by Rai and Banerjee (1995) indicated the upgradation of yttrium (up to 20.92% Y_2O_3) and cerium (20.13% Ce_2O_3).

Surguja (Chhattiagarh) and Sonbhadra (Uttar Pradesh) Districts

Several rare-earth-bearing patches, measuring 10–50 m in length and 5–10 m in width, occur as side bars in the Kanhar river basin around Khoka, Sarpatwa, Amtiyatola, Bhagtaniya and Kota villages. The THM ranges from 1.2 to 5.3 wt.%. Heavy minerals are represented by ilmenite, zircon, rutile, xenotime, monazite, garnet and magnetite (Sharma et al. 2001). Non-magnetic heavy mineral concentrates revealed 0.20–0.41% Y_2O_3 and 0.26–1.75% Ce_2O_3 (Sharma et al. 2001). Approximately 250 tonnes of xenotime concentrates has been worked out in the Kanhar river basin (Sharma et al. 2001).

4.2.2 Stream Placers (Minor Sources) in Other Granitic Terrains

Bastar Craton, Chhattisgarh and Odisha

Placers in the environs of the Jhiram stream revealed 0.13–2.26% Y_2O_3 and 0.05–3.21% Ce_2O_3 (Singh 2007), whereas bulk samples contain 0.15–0.38% Y_2O_3 and 0.063–0.33% Ce_2O_3. Xenotime is fine-grained. Anomalous concentrations of Y and Ce are also known from Dumam streams around Kawadgaon-Muskel (Singh and Singh 2001). Notable thickness and preservation of soil found throughout the granitic terrain in the area (Singh 2012b), with various fluvial landforms along the Mari river course. Magnetite, hematite, ilmenite, goethite, epidote, zircon, monazite and occasionally xenotime form heavy minerals. Polymineral concentrates revealed 0.006–0.67% Y_2O_3 and 0.01–4.729% Ce_2O_3 (Singh and Shiv Kumar 2009). Placers of the Kawara and Nakti streams contain zircon, rutile, monazite, xenotime epidote, garnet, ilmenite, goethite, hematite and magnetite, with 0.02–0.99% Y_2O_3 and 0.01 2.03% Ce_2O_3 (Singh 2002). The Dussanada stream in Koraput district of Odisha is a tributary of Sabari (Kolab) river, and its placer contains xenotime, monazite, ilmenite, zircon, magnetite, hematite and epidote (Singh 2020a).

Bhandara Sector, Maharashtra

In the Tirka area of Gondia district, radioactive soils occur over 6 km^2 area, and contain 10–300 ppm Y and 5–418 ppm La. The THM is 0.31–34%. A bulk sample revealed 5.77% Y_2O_3 and 31.66% total REO. Monazite, xenotime, zircon, rutile, anatase, ilmenite and magnetite represent heavies. Monazite contains 52.57–63.65% REO and 1.11–3.35% Y_2O_3, whereas xenotime revealed 41.86% Y_2O_3 and 18.72% REO (Shobhita et al. 2007). Zircon analysed 0.03–0.93% Y_2O_3 and 0.02–0.97% REO. Sediments from various streams of the Pauni-Tangla area of Nagpur district contain monazite and xenotime, and analysed 3.59% Y_2O_3 and 13.71% Ce_2O_3 (Ramesh Babu 1999).

Godhra Granitic Terrain, Gujarat

Stream placers are known in parts of Panchmahal and Vadodara districts. The Godhra granitic rock contains 130 ppm Y_2O_3, and hosts xenotime, monazite, allanite, thorite and zircon (Maithani et al. 1993; Gupta et al. 2018), with several REE anomalies on its eastern part (Gupta et al. 2018). The several streams which flow through Godhra granite terrain contain anomalously radioactive placers along their courses (Muralidharan et al. 2001). Among them, Vasva and Ujol streams are prominent. Along the course of Vasva River in Vadodara district, placer bodies occur over 2.5 km length (Maithani et al. 1993). Heavy mineral concentrates, comprising xenotime, zircon, ilmenite, sheelite, rutile, hematite and fluorapatite, revealed 0.51–0.96% Y_2O_3. Significantly, the presence of radioactive placers over about 5 km length has also been noted in side bars, point bars and islands of the Ujol river in Panchmahal district, with variable THM (1.25–11.21wt.%). Heavy mineral assemblage of xenotime, monazite, ilmenite, rutile, zircon and garnet revealed 0.76–1.10% Y_2O_3 and 6.58–10.49% Ce_2O_3 (Goyal et al. 1997).

Dharwar Craton, Southern India

Dharmawaram Stream, Karimnagar District, Telangana

The pink granite, dated at 2237 Ma (Singh et al. 2004), with monazite, xenotime, thorite, allanite, zircon, epidote, ilmenite, magnetite, apatite and sphene, contains 296 ppm–1024 ppm REE and 51 ppm–85 ppm Y (Singh, 2004). The stream sediments derived from it, with similar heavy mineral assemblage, revealed 0.19%–0.40% Y_2O_3 (Singh et al. 1991).

Pathapalem River, Mehboobnagar District, Telangana

Pink granite forming provenance for REE placers revealed 203–345 ppm Y. Fluvial bedforms of various dimensions, with a THM of 0.61–1.46%, occur along the Pathapalem stream discontinuously over 10-km length (Rai et al. 1997). Polymineral concentrates, comprising ilmenite, zircon, monazite and xenotime, hematite and garnet, analysed 0.63%–5.26% Y_2O_3 and 5.29–13.40% Ce_2O_3. Bulk of such stream sediments fall in Telangana, whereas a few in the adjacent Raichur district, Karnataka.

Granitic Soils of Tonnur-Pandavpura area, Hassan and Mandya Districts, Karnataka Granitic bodies which form provenance for stream sediments contain 122–670 ppm Y (Rai et al. 1991), whereas heavy mineral concentrates of soils derived from them revealed 0.25–1.37% Y_2O_3. Polymineral concentrate comprises xenotime, monazite, zircon, rutile, ilmenite, hematite and magnetite, with THM of 1.43–3.34 wt.%.

Mincheri, Raichur District, Karnataka One brecciated pegmatite vein, of 1.9 km length and 30 m width, contains 0.135–0.605% REE, with one isolated value 3.202% REE. Another such body (1.3 km length and 1–8 m width) reveals variably high concentrations, e.g., 0.23–25.30% REE (Deori and Reghu 2014). In these pegmatites, REE is contributed by yttrofluorite, gadolinite, britholite, cerianite, allanite, monazite and bastnaesite. In adjacent area, Kesaratti aegirine-syenite plug analysed 0.01–0.03% REE (Deori and Reghu 2014).

Goddaru Vanka Stream, Rangampetta, North Arcot District, Tamil Nadu The biotite granite of the area is radioactive due to thorium and carries allanite, monazite, xenotime and zircon. Sediments derived mostly from granite, present along the Goddaru Vanka nala, show higher radioactivity over 1-km length (Rai et al. 1991; Kumar et al. 1995), and contain 0.003–0.034% Y_2O_3 and 0.034–0.35% Ce_2O_3. The THM is 3.68–12.43 wt.%, with xenotime (0.07–0.5%), allanite (0.3–2.02%), monazite (0.09–0.65%), zircon (0.01–0.2%), fergusonite, magnetite, ilmenite, rutile, garnet and hornblende (Rai et al. 1991).

Kullampatti, Salem District, Tamil Nadu The granites and pegmatites of the area revealed 0.05–1.52% HREE, and carry uraninite, uranothorite, brannerite, uranophane, allanite, columbite-tantalite, samarskite, fergusonite and zircon (Rai et al. 1991; Roy and Dhana Raju 1999; Singh 2020a). The THM (1.67–2.3 wt.%) from source rock analysed 0.81–1.26% Y_2O_3 (Rai et al. 1991). Granitic soils from the vicinity of source rocks revealed 1–4 wt.% THM and carry fergusonite (6–15%), samarskite (1–4%), zircon (2–9%), monazite (0.1–0.5%), xenotime, uraninite, columbite and garnet, with a separated fraction recording 10% Y_2O_3 (Rai et al. 1991).

4.3 Carbonatites

Assessment of economic potentiality of carbonatite bodies by various exploration agencies in India has revealed some carbonatite-hosted REE resources (Singh 2020c). Summaried account of the same is presented below.

4.3.1 Western India

Ambadongar Carbonatite, Gujarat The Ambadongar carbonatite in Chhota Udaipur district (Fig. 4) was identified in 1960s (Sukheswala and Udas 1963), and was believed to be a fluorite deposit (Subramaniam and Parimoo 1963). Early investigations on Ambadongar carbonatite brought to light 0.05–1.04% REO (Sukheswala and Udas 1963, 1964). Presently ongoing exploration (Nagabhushnam et al. 2018; Rai et al. 2018) in this complex revealed significant enrichment of rare earths (1.16% REO). More concentration of REE is revealed by ferro-carbonatites (ankerites) compared to calcio-carbonatites (sovites).

The ACC contains bastnaesite, synchysite, parisite, florencite, cerite, monazite and wakefleldite-(Ce, La) in association with barite, strontianite, pyrochlore, columbite and vanadinite (Doroshkevich et al. 2009; Nagabhushnam et al. 2018; Rai et al. 2018; Magna et al. 2020). The abundance of REE phases is variable in different carbonatite phases. Majority of the REE phases are fine to extremely fine-grained (Rai et al. 2018). Substantial REE is contributed by synchysite, parisite and monazite, followed by florencite and bastnaesite (Rai et al. 2018).

Initially, 1,40,000 tonne REO, contained in 12 million tonne of ore, was worked out (Nagabhushnam et al. 2018; Rai et al. 2018; Singh 2019). The continuity of REE mineralisation up to 120 m depth was traced (Rai et al. 2018). Afterwards, the resource has been revised upward to 3,46,460 lakh tonnes REO, with average grade of 1.34% REO (TOI 2019; Singh 2020a, c). The deposit is amenable for open-pit mining as mineralisation starts from the surface itself.

Kamthai Carbonatite, Barmer District, Rajasthan The Kamthai REE deposit is a part of the Tertiary Alkaline Complex (TAC) of Sarnu-Dandali, Barmer district, Rajasthan (Fig. 4). The anomalous concentration of REE (3–6%) due to carbocernaite, britholite, allanite and monazite in carbonatite dyke of Sarnu-Dandali was brought to light by Wall et al. (1993) and Srivastava et al. (1993). The REE potentiality of the TAC of Sarnu-Dandali came to light after the discovery of Kamthai REE deposit (Bhushan and Kumar 2013; Bhushan 2015). The ijolite plug intrudes Malani rhyolite and foidal syenite. The second phase of calciocarbonatite is soviet (Bhushan 2015), and comprises high-grade bastnaesite (>8% REE) with accessory amount of either synchysite or ancylite, relative to first phase calciocarbonatite (alvikite type), both phases intruding the ijolite plug. The dominant ore mineral is carbocernaite (low-grade REE up to 5%) in calciocarbonatite (Bhushan 2015). The calcio-carbonatite is essentially carbocernaite-rich fine-grained alvikite.

The dominant REE phases in the mineralised carbonatites present are bastnaesite-(La), bastnaesite-(Ce), synchysite-(Ce), carbocernaite-(Ce), cerianite-(Ce), with lesser daqingshanite-(Ce), britholite-(Ce), parisite and occasionally monazite and REE-bearing apatite (Bhushan 2015). All the REE minerals contain calcite and minor amounts of celestine, siderite, strontianite, fluorite and apatite. REE minerals are dominated by carbocernaite, with some ancylite, synchysite, bastnaesite and parisite.

The deposit notices the highest content of 17.31% LREO, and an average of 3.33% LREO (Bhushan and Kumar 2013). The carbonatite plug extends over an area of about 19,475

m^2. It revealed calculated resource of about 4.91 million tonne ore. The resources of individual rare earths are 52,196 tonne La, 66,026 tonne Ce, 13,663 tonne Nd, 5,415 tonne Pr, 920 tonne Sm and 207 tonne Eu (Bhushan and Kumar 2013). The current REE resource of Kamthai deposit is 7.36 million tonnes ore, at an average grade of 1.62% REO (Bhushan 2015).

4.3.2 Southern India

The Proterozoic alkaline province in northern Tamil Nadu, south India, occurs in the rift zone and is located along a major NE-SW lineament (Udas et al. 1974; Rao et al. 1978; Krishnamurthy 2019; Krishnamurthy et al. 2000). The carbonatite bodies include Samalpatti and Sevattur, besides other alkaline rocks. REE abundances in carbonatite of Tamil Nadu province vary from 0.01 to 3.0%. In general, the REO content is highest in benstonite carbonatites and apatite-magnetite bands in Jogipatti carbonatites, whereas the lowest REO is revealed by near-contact carbonate skarns.

Samalpatti, Dharampuri District, Tamil Nadu In the Samalpatti area, the outcrops of carbonatites occur in discontinuous bodies, which are either arcuate shape, or as dykes and veins within pyroxenite or syenite (Subramaniam et al. 1978). It is dolomite and calcite soviet. The carbonatite contains 8.2% REE (Schleicher et al. 1998). The REE mineral is dominantly monazite with some bastnaesite. Other associated REE-bearing phases are fergusonite, chevkinite and allanite (Rao et al. 1978). The contents of total REE oxides are 69.8% in monazite, 39.35% in chevkinite, 37.58% in fergusonite and 24.59% in allanite (Semenov et al. 1978).

Pakkanadu-Mulakkadu, Salem District, Tamil Nadu The Pakkanadu-Mulakkadu carbonatites, associated with pyroxenite, occur in the form of discontinuous lenticular bodies and intrude syenites. They are calcite-rich soviet type involving biotitic and ankeritic types. Apart from REE minerals like bastnaesite, monazite (bigger crystals), cerianite and allanite, other heavy minerals, viz., apatite, barite, magnetite and zircon are also present (Rao et al. 1978). Schleicher et al. (1998) reported 3.22% REE from Pakkanadu carbonatite, whereas recently Balasubramani et al. (2017) brought to light up to 5.83% TREE, with an average of 14,836 ppm TREE. A preponderance of monazite is known to the west of Pakkanadu in portions of carbonatites dominated by biotite, in contrast to dominance of allanites in pyroxenites west of Mulakkadu (Rao et al. 1978). In Pakkanadu, bigger crystals of monazite are commonly present. The bulk of the REE is contributed by monazite. Eschynite occurs in association with allanite and baryte in Pakkanadu carbonatites. Eschynite from Pakkanadu revealed 24.02% REO, whereas allanite contains 24.55% REO (Semenov et al. 1978).

Sevattur, North Arcot District, Tamil Nadu The Sevattur carbonatite complex is a crescent shaped body. Sovite occurs as thin fringes or bands within raughaugite. It is fine-grained with calcite containing pods of dolomite. The sovites are considered relatively older, compared to beforsites which are present in central portion of the complex. The REE abundances in carbonatites are 103–631 (av. 271) ppm La, 520–720 (av. 641) ppm Ce, 5–59 (av. 41) ppm Y and 7.9–28 (av. 16) ppm Sc (Krishnamurthy 1977). Significantly, uranopyrochlore from this carbonatite contains 0.35–1.78% REO (Borodin et al. 1971).

4.3.3 Eastern India

Beldih-Kutni Carbonatite, Purulia District, West Bengal Various rocks of an alkaline-carbonatite complex (ALCC) including alkali pyroxenite, nepheline syenite, phoscorite, carbonatite, syenitic fenite and glimmerite together with REE concentration are present at different locales along WNW-ESE trending South Purulia Shear Zone (SPSZ) in southern parts of Purulia district of West Bengal in the environs of Singhbhum crustal province (Basu and Ghosh 1999; Basu and Bhattacharyya 2014). Phoscorite comprises fluorapatite, dahllite, collophane, magnetite, hematite, goethite, phlogopite, calcite, sphene, monazite, pyrochlore, chlorite and quartz. The constituent minerals of carbonatite are Sr-calcite, including dolomite, tetraferriphlogopite, phlogopitic biotite, aegirine augite, richterite, fluorapatite, altered magnetite, sphene and monazite. With an increase in depth, REE concentrations in carbonatites show a slight increase. The source of REE is monazite in phoscorite and carbonatite. Phoscorite assayed 4181 ppm REE (Basu and Ghosh 1999). Fluorapatite contains a low amount of REE relative to dahllite. Pyrochlore analysed 0.14–0.26% Ce_2O_3 and 1.14–1.24% Gd_2O_3 (Basu and Bhattacharyya 2014). Monazite contains high amounts of LREE, e.g., 14.94–23.38% La_2O_3, 33.91–36.43% Ce_2O_3 and 7.61–13.36% Nd_2O_3 (Singh 2020a). Twelve (12) million tonnes of ore with 11% of P_2O_5 up to a depth of 30 m has been estimated in the Kutni-Beldih tract (Basu 1993). Apatite production from here averaged 12,750 tonnes during 1992–94 (IBM 1996). In view of considerable apatite resources, substantial quantity of REE is expected to come from the same.

4.3.4 North-East India

Some four carbonatite complexes have been recognised in the Shillong-Plateau-Mikir-Hills area (Samchampi, Sung Valley, Jasra, Barpung and Swangkare; Yusuf and Saraswat 1977; Nambiar and Golani 1985; Kumar et al. 1996). These complexes are located along major E-W, NE-SW and N-S trending faults and lineaments. In two complexes REE concentration is known.

Samchampi Carbonatite, Karbi Anglong District, Assam It covers an area of ~16.5 km^2, and is a fairly circular and stock-like body, emplaced along the NE-SW trending Kalyani Lineament (Acharya et al. 1986; Kumar et al. 1996). Archaean granite gneisses appear to have been intruded in a sequence of pyroxenite, titano-hematite rocks, syenite and carbonatite. Late stage phosphatic rock reprecipitated as secondary phosphate along fault zones in the eastern and southern periphery. Crust of weathered thick residual soil covers the complex; accordingly surface exposures of carbonatite are scanty. Sovite (REE 841 ppm) contains less REE than phosphatic breccia (REE 1062 ppm). REE content in fenitised granite gneiss (REE 8135 ppm) is substantially high relative to fenitised granite (REE 1407 ppm). Compared to potassic syenite (REE 106 ppm) and sodic-potassic syenite (REE 147 ppm), potassic-sodic syenite (REE 661 ppm) contains a noteworthy amount of REE.

Krishnamurthy et al. (2000) reported 1833 ppm REE in hematite ore, 1857 ppm REE in phosphatic rock and 1968 ppm REE in residual soil of Samchampi complex. Significantly, Nb-bearing residual soil of Samchampi alkaline complex (SAC) also contains 1867 ppm REE. The REE content varies from 2.4 to 5.38% REO in pyrochlore, 0.27–1.01% REO in crandallite and 0.18%–1.18% REO in apatite (Hoda and Krishnamurthy 2016a, b, 2020). Thus, it is apparent that the highest REE is contained in pyrochlore (av. 3.9% REO) followed by crandallite and apatite, all being dominated by LREE. Apart from monazite (61.96% REO), notable amounts of REE are also present in perovskite (2.37% REO and 0.04% Y_2O_3) and wavellite (5.99% REO and 0.14% Y_2O_3). Resource evaluation revealed 3644 tonnes of Y contained in 41.88 million tonnes of ore (Hoda and Krishnamurthy 2016b).

Sung Valley Carbonatite, Jaintia Hills District, Meghalaya In the Shillong Plateau, the first carbonatite discovered is Sung Valley carbonatite (Yusuf and Saraswat 1977). The carbonatite is located at the intersection of two major faults in the Wah-Sing Valley of Jaintia Hills district, Meghalaya (Chattopadhyay and Hashimi 1984; Krishnamurthy 1985). It is nearly oval-shaped stock and intrudes Shillong Group metasediments. Peridotite, ijolite, carbonatite, pyroxenite, syenite, magnetite-apatite rocks and melilite-bearing rocks are the important rock types of the Sung Valley complex (Kumar et al. 1996). Notable concentrations of cerium are reported from sovite (467 ppm Ce), apatite sovite (244–471 ppm Ce), dolomitic sovite (308 ppm Ce) and beforsite (149 ppm Ce), with 50–100 ppm Y (Krishnamurthy 1985).

4.4 (Per)alkaline Felsic Rocks

The geological environment for (per)alkaline suite of felsic rocks are generally a continental anorogenic tectonic setting linked to rifting (Bailey 1964). These are typically "A"-type granites of Collins et al. (1992) and Within Plate Granites (WPG) of Pearce et al. (1984) related to rifting process. High concentration of Zr, Nb, Y, Th, Zn, Ni and REE and depletion of Ba, Sr, Ca and Mg are generally strong indication of (per)alkaline granites (Bhushan and Somani 2019).

4.4.1 Siwana Granites, Barmer District, Rajasthan

The Siwana granites represent A-type magmatism of anorogenic nature (Kochhar 1984; Bhushan and Chittora 1999; Singh and Vallinayagam 2002; Vallinayagam 2004), and are peralkaline to mildly peraluminous, with locally anomalously high REE, Y, Nb, Ta, Zr, Zn, Pb, Hf, Ta, U and Th enrichments (Narayan Das et al. 1978; Jain et al. 1996; Mohanty and Bhushan 2004). Due to mineralogy, geochemistry and magmatic affiliation, the Siwana (per)alkaline granites are believed (Bhushan and Somani 2019) to be similar to Strange Lake Deposit Canada (Kerr and Rafuse 2012).

Several workers have reported high REE concentration (Bidwai et al. 2014, 2015; Das et al. 2015, 2016; Rastogi et al. 2015; Bhushan and Somani 2019). Preliminary exploration inputs have revealed a cumulative resource of more than one million tonne of REE in the alkaline Siwana Ring Complex (Fig. 4), hosted mainly in agpaitic granitoids and rhyolitic tuffs (Singh 2020a, b). REE distribution pattern in variants of the Siwana Ring Complex is shown (Fig. 5; Bidwai et al. 2015). Brief details of prominent occurrences are outlined below.

Gudanal The fine-grained granites are enriched in REE. Due west of Gudanal, fine-grained granites contain average 4652 ppm REE, with average HREE content of 1003 ppm. In addition, average Y content is 1562 ppm (Bhushan and Somani 2019). The critical HREE, Dy and Tb range 74–383 ppm and 13–55 ppm, respectively. Fine-grained granite from the eastern part of Gudanal revealed (average) 5997 ppm REE, with average 1284 ppm HREE, and 1283–2935 ppm Y, 246–498 ppm Dy and 45–80 ppm Tb (Bhushan and Somani 2019). REE is contributed by allanite, REE silicates, xenotime and monazite. The average modal abundance of REE minerals in Gudanal east and west is 2.23% and 2.04%, respectively. Gittinsite ($CaZrSi_2O_7$) shows notable abundance (3.07–4.13 vol%). Sodic amphibole (Fe-Na amphibole) appears to have formed by hydrothermal alteration of sodic pyroxene.

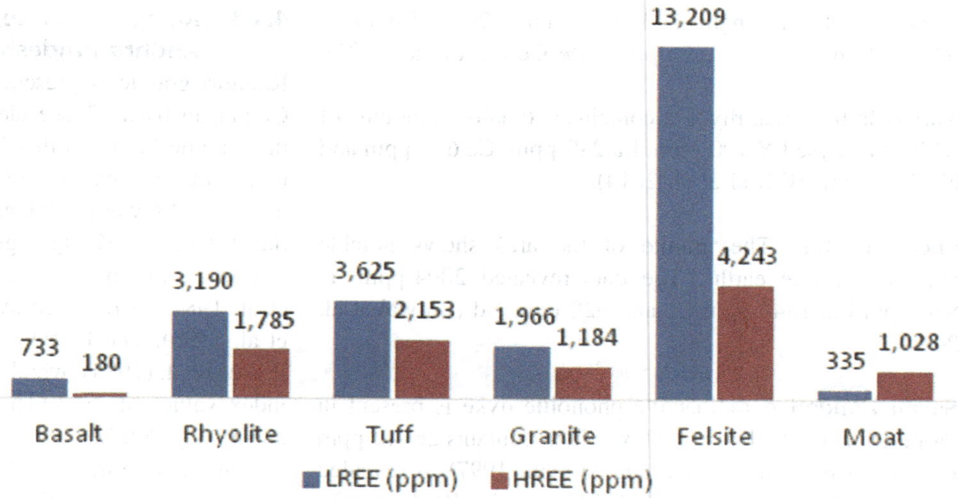

Fig. 5 Pattern of LREE and HREE distribution in various lithounits of the Siwana Ring Complex (Modified after Bidwai et al. 2015)

Bhatikhera In this area, REE concentration is revealed by microgranite dykes, which measure 10–20 m in width and 800–1100 m in length (Bhushan and Somani 2019). They revealed 4842–6042 ppm REE, with 829–995 ppm HREE, 1126–1512 ppm Y, 234–283 ppm Dy and 38–43 ppm Tb (Bhushan and Somani 2019). Bidwai et al. (2014) reported higher concentrations (average) of Y (1469 ppm), La (892 ppm), Ce (2097 ppm), Nd (995 ppm), Zr (12,252 ppm) and Nb (649 ppm) in microgranite. In composite samples of borehole LREE/HREE ratio is 60:40 (Singh 2020a). Interestingly, subsurface correlatable mineralisation is traced over a strike length of about 2 km. Allanite, REE clays and carbocernaite occur in decreasing abundance, and REE minerals account 2.50% of all the constituent minerals (Singh 2020a).

Dantala Tuffaceous altered rhyolite reveals REE concentrations in this part over a dimension of 500 m × 70 m. Samples revealed 0.594% REE, with 0.218% HREE. Exploratory drilling has revealed depth persistence beyond 100 m depth with notable thickness of mineralisation (Singh 2020a).

Meli The microgranite dykes have 10–80 cm width and 100–200 m length, and trend E-W. They revealed 5657 ppm REE, with 745 ppm HREE, 1021 ppm Y, 195 ppm Dy, 29 ppm Th, 4514 ppm Zr and 117 ppm Hf (Bhushan and Somani 2019).

Phulan In this area, three microgranite dykes are present with a length of 250 m, 500 m and 250 m and a respective width of 0.1–0.8 m, 0.3–2.0 m and 2–2.5 m. The microgranite dykes contain 3122–31,147 ppm REE, with 665–2721 ppm HREE, 194–882 ppm Dy and 32–152 ppm Tb (Bhushan and Somani 2019). Das et al. (2016) reported 1.3–1.76% REE. Bidwai et al. (2014) also noted higher abundances (average) of Y (5793 ppm), La (5039 ppm), Ce (12,463 ppm) and Nd (5620 ppm). Rare-earth minerals present include REE-silicates, bastnaesite, REE-bearing clay, monazite, thorianite and carbocernaite, making up 10.70 vol.%.

Nal Microgranite in the Nal area contains monazite, parisite, eudialyte, tritomite, britholite and niobo-aeschynite (Singh 2020a). Britholite analysed 62.05–62.92% REO, niobo-aeschynite 27.22% and 33.11% REO, whereas zircon revealed 4.56% HREO and 6.77% Y_2O_3, and a lesser amount of LREO (1.29%). Significantly, higher concentration of REE is also recorded in quartzo-feldspathic rock of the Nal area by Bidwai et al. (2014), that is, Y (195 ppm), La (1973 ppm), Ce (4564 ppm) and Nd (2012 ppm).

Ramaniya The fine-grained radioactive dykes contain 2529–2804 ppm Y, 1481–2589 ppm La and 4101 to > 1.0% Ce (Jain et al. 1996). Further studies (Bidwai et al. 2014) also revealed more abundance (average) of Y (3659 ppm), La (1138 ppm), Ce (2825 ppm) and Nd (1761 ppm) in granites.

Mokalsar Fine-grained radioactive granite dyke is traced up to 50 m depth on an escarpment face. It analysed (average) 1593 ppm Y, 679 ppm La and 1337 ppm Ce (Jain et al. 1996).

Mamaji-Ka-Wala About eight major fine-grained, radioactive dykes, besides many smaller ones, are present as fracture-fillings within the pink granites, having 30 cm–2 m width and 300–500 m length. These dykes analysed 3220 ppm Y, 2113 ppm La and > 1.0% Ce (Jain et al. 1996).

Ludhrara Out of several feebly radioactive granite dykes traversing the granite hillock, the main dyke has 2–50 cm width and 200 m length. Dykes contain 197–5377 ppm Y, 68–3922 ppm La and 249 ppm–1.0% Ce (Jain et al. 1996).

Mawri In this area, rhyolite contains anomalous amounts of REE. It analysed Y 700 ppm, La 249 ppm, Ce 646 ppm and Nd 304 ppm (Bidwai et al. 2014).

Sheetala Mata The granite of the area shows notable amounts of rare earths. The data revealed 2094 ppm Y, 846 ppm La, 1848 ppm Ce and 928 ppm Nd (Bidwai et al. 2014).

Sarnu Dandali A radioactive phonolite dyke is present in Sarnu Dandali alkaline complex, which contains 26–42 ppm U and 37–49 ppm Th (Jain and Miglani 1997). It revealed moderate abundances of La (43–143) and Y (10–41 ppm).

Nakora Ring Complex The granites from Nakora ring complex in Barmer district, Rajasthan (Kumar and Kumar 2020), are radioactive and show anomalously higher concentrations of Zr, Y, La and Ce (Vallinayagam and Kochhar 2010). They revealed 447 ppm–8213 (av. 1480.2) ppm Zr, 102 ppm–809 (av. 203.5) ppm Y, 13 ppm–293 ppm La and 65 ppm–253 ppm Ce.

Aforementioned (per)alkaline granite of Siwana Ring Complex assumes significance in terms of HREE enrichment. A few bodies especially Gudanal, Nal, Dantala, Bhatikhera, Meli and Phulan contain considerable HREE and Y resources. Since the area is well exposed with outcrops of raised relief, they are amenable for open cast mine. As HREE enriched dykes in Gudanal, Phulan and Bhatikhera are thin, the stripping ratio would be high and during mining dilution is expected (Bhushan and Somani 2019; Singh 2020a). Notwithstanding these, due to the ever-growing demand, uncertain availability in world market and high price of the HREE, the cost of the product can be economically realised (Bhushan and Somani 2019). To strike a balance between demand and supply linked with REE economics, various options have been outlined (Binnemans 2014; Binnemans et al. 2013).

4.4.2 Amet Granite

Amet granite is believed to be a result of partial melting of migmatite gneisses of the Mangalwar Complex (Guha et al. 2020). It contains allanite crystals. Within Amet granite, bands of variable dimensions of amphibolite are present. At Chatrawan mata temple close to Barona, sysnite-alkali feldspar syenite occurs in association with amphibolite. Increasing concentrations of REE (Guha et al. 2020) have been noted from amphibolite (41.30 ppm REE), to migmatite (156.12–715.57 ppm REE) to syenite (up to 2071 ppm REE).

4.4.3 Kanigiri Granite, Prakasham District, Andhra Pradesh

Kanigiri granite is present along the eastern margin of the Cuddapah basin. Trace element data along with fluorite in this granite indicates that it crystallised from a fluorine saturated magma that was derived from the partial melting of dehydrated lower continental crust (Sarvothaman and Sesha Sai 2010). The Kanigiri granite has an elliptical body, and covers an area of 6.2×2.1 km^2. It is 995 Ma old (Gupta et al. 1984). It revealed average 697 ppm REE (Thirupathi et al. 1996). The Kanigiri granite is of "A"-type granite of Pearce et al. (1984), which is further corroborated by agpatic index values of $\geq$ 4 (Thirupathi et al. 1996; Dharma Rao and Reddy 2007).

Various rare metal and rare-earth minerals present in the KG pluton are columbite-tantalite, fergusonite, samarskite, allanite, monazite and thorite (Ramachar et al. 1979; Banerjee et al. 1983). In addition, bastnaesite and hydroxyl-bastnaesite are also known (Nagendra Babu et al. 2013; Singh et al. 2014). Residual soils derived from Kanigiri granitic pluton on mineral beneficiation indicated the presence of non-magnetics from 5 to 11 wt.%, which also revealed bastnaesite and hydroxyl-bastnaesite, besides columbite-tantalite, fergusonite, monazite, thorite, ilmenite, rutile and zircon (Singh et al. 2014).

4.4.4 Other (Per)alkaline Felsic Bodies

In other parts of India, many (per)alkaline felsic bodies have revealed elevated REE concentrations. The Pala Lahara granite (Odisha) revealed 509 ppm REE (av.), with up to 103 ppm Y (Topno et al. 2018). In the Kumarkunti-Jharnomal area of Nuapada district, Odisha, 120–1231 ppm Y, 129–1757 ppm La and 453–1876 ppm Ce have been reported, with thorite, allanite, fersmite and zircon (Mishra et al. 2006). Also, Kuilapal Granite, West Bengal, revealed (average) 636 ppm REE and 83 ppm Y (Sen et al. 2008). The Nongpoh granite, Meghalaya, revealed 698–986 (av. 836.6) ppm REE, with sphene, apatite, allanite, monazite and zircon (Sadiq et al. 2014). Younger radioactive pink granite from Songsak, East Garo Hills, Meghalaya, apart from uraninite, also contains monazite, xenotime, allanite, zircon, apatite, chaistolite and fluorite (Sunil Kumar 1990). In Karbi Hills of Assam, pink granite revealed anomalously high REE (3643 ppm), along with allanite, xenotime, bastnaesite, apatite, zircon and monazite (Majumdar and Dutta 2014).

4.5 Hydrothermal Veins

Hydrothermal-type deposits/occurrences occur in certain parts of India, notably Singhbhum shear zone, south Purulia shear zone and Eastern Ghats Belt. In addition, a 320-km long albitite belt in Western India, Rajasthan, contains both LREE- and HREE-bearing minerals along with several refractory phases. Uranium phases from hydrothermal vein-type uranium deposits of Gogi, Karnataka, have revealed significant concentrations of LREE. Most of them show variable REE concentrations (Singh 2020a). Brief details about the first three (prominent) of them are given below.

4.5.1 Singhbhum Shear Zone

The more prominent Singhbhum Shear Zone (SSZ), initially named as the Copper Belt Thrust (Dunn and Dey 1942), is a NW–SE trending arcuate shape shear zone from Purandungri in the east to Porhat in the west. Rare-earth concentrations of varying levels are common in ore zones, both in uranium-rich and copper-rich zones. Exotic veins of apatite-magnetite occur in close association with uranium and sulphide ores (Rao and Rao 1980, 1983). Furthermore, all along the shear zone apatite and magnetite are also ubiquitously present as gangue minerals in all the ore zones (Rao and Rao 1980). Despite the widespread apatite-magnetite vein-type mineralisation in major parts of the shear zone, significant deposits are located in three sectors, namely, (i) Dudra-Nandup-Ramchandrapahar sector in the western part, (ii) Kasidih-Surda-Pathargarah sector in the eastern part and (iii) Gohala-Khejurdari sector in the easternmost part. Apatite and magnetite are the main minerals of the apatite-magnetite veins, with minor to trace amounts of xenotime, allanite, monazite, chlorite and quartz. All these minerals are also common accessories to minor minerals in the ore zone schists hosting other mineralisations (Rao and Rao 1980). Due to widespread occurrences of apatite-magnetite veins along the SSZ with notable rare-earth concentration, these veins assume significance for REE resources. About 600 tonnes of xenotime reserve has been estimated in 60,000 tonnes of ore in the eluvial and colluvial gravels at Kanyaluka (Ramachandran et al. 1989).

4.5.2 South Purulia Shear Zone

The South Purulia Shear Zone (SPSZ) is located along northern periphery of the Singhbhum mobile belt with the CGGC on the north. Spatial position in the periphery of CGGC, bordering the Singhbhum Group of rocks, is believed to have controlled emplacement of Alkaline Carbonatite Complex (ACC), at the intersection of the NW–SE lineament with the WNW-ESE trending close to this shear zone (Basu 1993). The ACC comprises alkali syenite, apatite-magnetite rock, carbonatite and phlogopite-amphibolite. Apatite-magnetite rock, forming apatite ore, occurs as lenses and pockets of varying dimensions within the schists and phyllites. From east to west, several linear apatite bodies, associated with ACC, are present at Chirugora, Kutni, Mednitar and Beldih along the SPSZ. The alkali syenite intrudes within phyllites and schists, whereas carbonatites occur as veins and veinlets, intruding ultramafic suite, phyllites and schists in the apatite fields at Kutni and Beldih. Apatite-magnetite rocks have analysed average 1296 ppm Ce, and comprise mainly fluorapatite, with minor allanite, xenotime and zircon, and are believed to be of hydrothermal origin (Singh 2020a).

4.5.3 Kasipatnam, Visakhapatnam District, Andhra Pradesh

The apatite mineralisation in the form of veins and veinlets show widespread development in the charnockites of the Kasipatnam area in Eastern Ghats Group (EGG). The apatite-magnetite veins of various sizes (few cm to 2.50 m) occur as fracture fillings within the charnockites. Six phases of mineral associations have been reported from these apatite-magnetite veins (Rao 1976). Economic significance of these veins is well known (Krishnan 1941; Rao 1976). The possible affinity of apatite-magnetite veins with carbonatitic complexes and alkali syenite has also been suggested. Fine-grained apatite samples from mines, and green and brown phosphatic rocks contain anomalous amounts of cerium (up to >3.0 wt.% Ce_2O_3) with a notable amount of yttrium, up to 0.082 wt.% Y_2O_3. Variable amounts of cerium and yttrium are also present in calcite associated with veins and stream sediment concentrates of the area (Singh et al. 1994). Allanite and fluorapatite also contain anomalously high REE (Rao 1976). Significantly, the phosphatic phase (apatite) from the Narsimharajapuram area, located due NNE of the Kasipatnam area, have yielded still higher amounts of rare earths (av. 17,279 ppm REE). Relative to apatite, calcite revealed still higher concentrations of REE (20,164–23,531 ppm). Apart from fluorapatite, the REE is contributed by ancylite, hydroxyl-bastnaesite and monazite (Panda et al. 2015). Pegmatite veins from the Ellaswaram area contain allanite, monazite and zircon. Allanite contains up to 32.08 wt.% REO, whereas monazite revealed 58.2 wt.% REO (Rao et al. 2018a). Zircon contains < 1.0% REO. An unidentified "Y-rich mineral" contains 9.34–10.7 wt.% REO. Monazite hosted in cordierite-gneisses near Vizianagaram in the Eastern Ghats (Rao et al. 2018b) also revealed notable amounts of Y with its REE content akin to detrital monazite (Panda et al. 2003) from Kalingapatnam coastal beach sand deposit.

5 Discussion

5.1 Promising Rare-Earth Deposits

The most promising deposits for light rare earths are hosted in the beach sands and two carbonatite complexes (one each in Gujarat and Rajasthan; Fig. 4). In beach sand deposits source of REE is monazite, whereas rare earth is contained in bastnaesites and associated minerals in the case of carbonatites. Riverine placers in central India, and hydrothermal deposits of Kanyaluka (Fig. 4), constitute heavy REE resources. (Per)alkaline microgranite dykes and tuffs of the Siwana Ring Complex in western India also appear significant as a source of REE due to their high basket price on account of HREE enrichment in them. Accordingly, they can be planned for commercial mining. Nevertheless, in carbonatite complexes and (per)alkaline felsic bodies, beneficiation studies are needed for efficient recovery of REE before taking up mining.

5.2 Mining and Rare-Earth Production

The IREL (India) Limited, an industrial arm of Department of Atomic Energy (DAE), Government of India, has mandate for mining and rare-earth production. Accordingly, REE production in India is done by the said industrial arm of DAE mainly by mining of beach sands (Fig. 6) from Kerala, Tamil Nadu and Odisha coasts. In addition, the Atomic Minerals Directorate for Exploration and Research (AMD), an exploration arm of DAE, is also involved in mining of stream placers (Fig. 7) in central India and producing 7–8 tonnes per year HREE-bearing polymineral concentrates (Verma 2019; Singh 2020a). Flowsheet adopted by IREL (India) Limited for separation of monazite and associated industrial minerals from beach sand deposits of India is shown (Fig. 8; Das et al. 2020). Processing of monazite by IREL (India) Limited, recovered from beach sand deposits, yields mostly light rare earths, dominated mainly by Ce, La and Nd in decreasing order. Additionally, it also produces value-added rare-earth compounds (viz., carbonates of cerium and lanthanum, oxalates of neodymium, praseodymium, yttrium, samarium and gadolinium) and other strategic materials (salts of U and Th) from monazite. Various commercial reagents and ionic liquids used by various workers in REE separation have been enlisted by Parhi et al. (2020). Schematic flowsheet for the separation of REE by solvent extraction (SX) and supported liquid membrane (SLM) process is also shown (Fig. 9; Parhi et al. 2020). Furthermore, various adsorbent materials for adsorptive removal of REE are also known (Koduru et al. 2020). Various products of IREL (India) Limited are used in different industries (IREL 2016, 2018). Additionally, IREL (India) Limited has entered into a memorandum of understanding with Bhabha Atomic Research Centre (BARC), Defence and Metallurgical Research Laboratoy (DMRL) and International Advanced Research Centre for powder metallurgy and new material (ARCI) for manufacturing of intermediate products (IREL 2016, 2018; Singh 2020a).

5.3 Rare-Earth Industry

Approximate requirement of REE in India may be visualised according to the pattern of global market. Based on about 1,19,650 tonnes of global REO consumption during 2015, distribution of rare earth consumption in various sectors are 24% catalysts, 23% magnets, 12% polishing, 9% other applications, 8% each in metallurgy and batteries, 7% glass, 6% ceramics and phosphors and 3% pigments (Zhou et al. 2017). Due to sustained focus for augmenting green energy, need for REE in India is bound to maintain progressive growth in coming decades. Therefore, challenges would be to sustain supply chain. Furthermore, during forthcoming 15–20 years, renewables will have a 40% share of the energy mix, and the goal to achieve near-complete electrification of surface transport would lead to additional REE demand (Kumar and Bharadwaj 2018).

5.4 Rare-Earth Balance

Approximate demand for Nd may be doubled to more than 40,000 tonnes of Nd oxide by 2030, followed by Dy and Ce oxides (Zhou et al. 2017). Although the demand for REE in lighting market is expected to decrease gradually, the REE demand for wind turbines, electric vehicles and NiMH batteries is likely to keep growing (Zhou et al. 2017). Accordingly, with the envisaged decline in the lighting industry in the forthcoming decade, the criticality of Eu, Tb and Y may not persist as far as supply is concerned (Binnemans et al. 2018), and Nd and Dy would become critical, particularly Dy, in terms of supply position (Balaram 2019). Therefore, the availability of Nd and Dy has to be maintained particularly for carbon neutral energy.

The need for Nd may be taken care from beach sand deposits of India (Fig. 4), by recovering monazite, which contains on an average of 10% Nd oxide (Singh 2020b). Some amount of Nd may also come from carbonatites of Amba Dongar and Kamthai, with 0.13–1.2 wt.% Nd oxide content, once mining starts. To this source, we may add some Nd associated with monazite (av. 8.8% Nd oxide content) from stream placers of central India (Fig. 4). Xenotime from central India and Kanyaluka (av. content 4.5% Dy oxide) and (per)alkaline granites of Rajasthan (0.025–0.05% Dy oxide) may supply Dy from them. The LREE and HREE required for various industries (Zhou et al. 2017) may be maintained from the existing and upcoming sources. Vast amounts of garnet and zircon from beach placers may not only add up additional REE resources but also help in addressing balance problem.

Fig. 6 Extensive beach sand deposits: **a** General view of beach sand deposit with prominent layers of heavy minerals (black); **b** Mining of a vast stretch of beach sand deposit by deploying JCB (*Credit* Deepak Rathod)

5.5 *Atmnirbhar Bharat*: Mine in India and Make in India

The growth of the REE industry in India needs to have linkage with systematic use of many natural resources in a coordinated way. The gradual exploitation of natural resources is generally governed by congenial geology, ore grade, tonnage, available metal extraction technology, market economics, environmental issues and cooperation from the local people, which need to be assessed scrupulously before commencing exploitation of identified deposits. Large-scale mine (LSM) may be preferred where vast stretches of deposits occur (Fig. 6), whereas artisanal to small-scale mining (ASM) would be appropriate in the case where deposits are of small dimensions and scattered over large area in a region (Fig. 7). In the latter cases, even micro,

Fig. 7 Small dimension REE-bearing stream placer deposits: **a** Meandering stream with selectively mined highly radioactive placers (point bar, island/braided bar); **b** Partially mined blocks of stream placer deposits

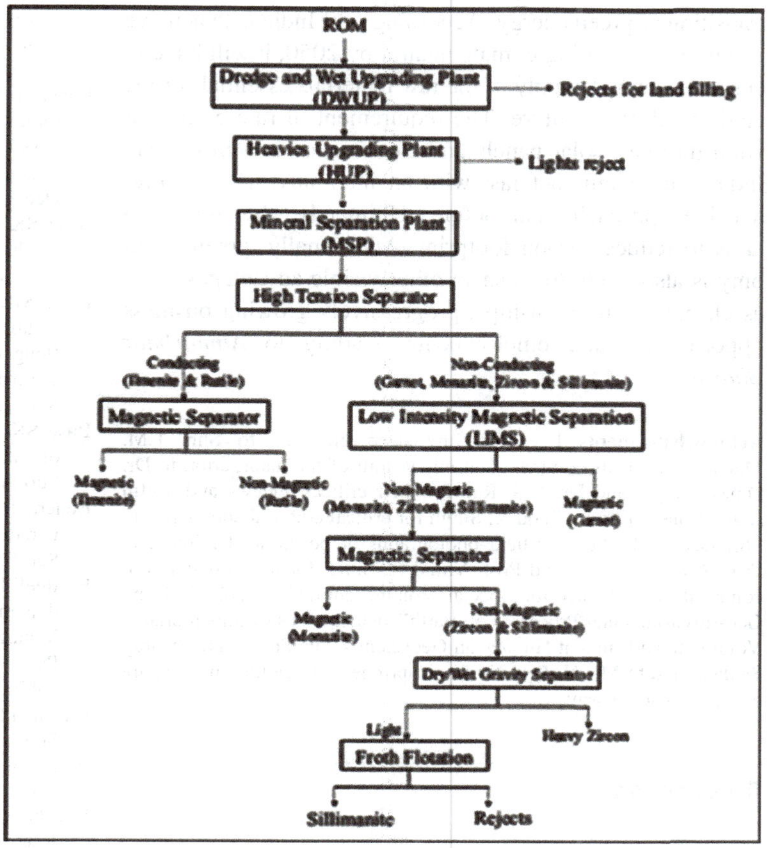

Fig. 8 Flowsheet adopted by IREL for separation of monazite from beach sand deposit of India (After Das et al. 2020)

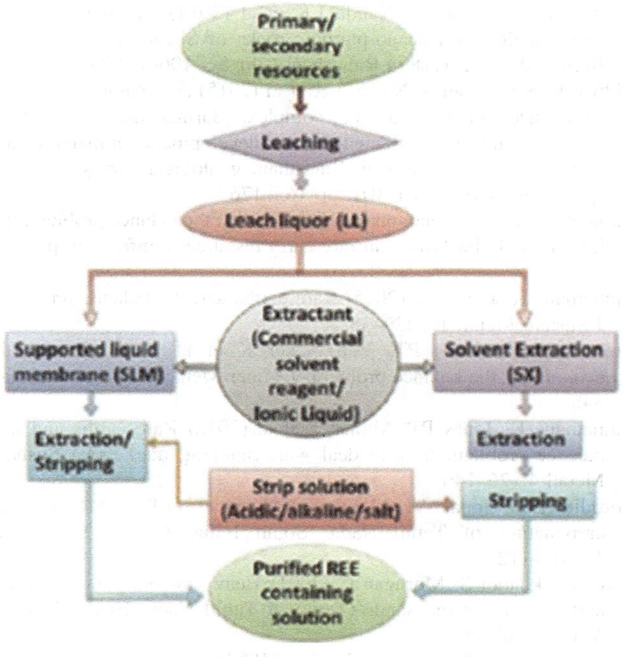

Fig. 9 Sematic flowsheet for separation of REE by solvent extraction (SX) and supported liquid membrane (SLM) process (After Parhi et al. 2020)

small and medium REE mining may be planned. Both downstream and upstream sectors should be developed to sustain the REE industry chain. This effort should integrate multidisciplinary groups (Pradip et al. 2019). The REE Industry should form an inseparable part of the *Atmnirbhar Bharat* through Mine to Metal in India including the Make in India programme, as envisaged by our Prime Minister of India. In this direction, GSI-AMD-IREL-BARC-NFC-DMRL-ARCI-CMAT together can make a significant contribution.

6 Concluding Remarks

It is a well-known fact that the global rare-earth market is dominated by China. According to REAI (2021), China is planning to impose serious restrictions on rare-earth export, especially to the USA to contain America's F-35 fighter jet manufacturing. Such action is considered as a "double-edged sword" by strategic experts, which although may contain the USA in short term on one side, but would be highly inspiring to the rest of the countries to develop efficient mining and extractive metallurgy of rare earths on the other side. Notwithstanding various strategic applications, the major build up for rare earths is linked with envisaged transition to green energy. Accordingly, if India is to achieve its target of attaining climate neutral by 2050, it will have to ensure sustained supply of the raw materials essential for the envisioned green future. The requirement of rare earths for wind turbines, solar panels and electric cars is well known. India should thus act fast with an integrated and comprehensive approach than before. Ultimately, this transition aims to reduce carbon footprint. Additionally, green economy is also likely to usher in other visible advantages, such as cleaner air in townships, progressively growing business opportunities and employments, leading to *Atmnirbhar Bharat*.

Acknowledgements I express my deep gratitude to Shri T.M. Mahadevan for his comments on initial draft of the manuscript; to Dr. Dinesh Gupta and Dr. D.S. Rao for their critical reviews and useful suggestions; to Prof. Vinod K. Singh for critical editorial comments; to Shri Deepak Rathod for field photographs of beach sand mining; to Prof. A.K. Shandilya and Prof. Vinod K. Singh for their invitation to contribute the chapter for Special Volume entitled, "Geological and Geo-Environmental Processes on Earth", dedicated as Commemorative Volume to an Eminent Himalayan Geoscientist late Prof. Prem Swarup Saklani; and to Ms. N. Jayanthi from Springer for handling manuscript in a professional way.

References

Acharya BC, Nayak BK, Das SK (2015) Mineralogy and mineral chemistry of placer deposit around Jhatipodar. Odisha J Geol Soc India 86(2):137–147

Acharyya SK, Mitra ND, Nandy DR (1986) Regional geology and tectonic setting of northeast India and adjoining region. Mem Geol Surv India 119:6–12

Ali MA, Naidu PS, Manjunath YS (1989) Studies on the beach placers of Ratnagiri district, Maharashtra, India. Explor Res At Miner 2:167–175

Ali MA, Krishnan S, Banerjee DC (2001a) Beach and inland heavy mineral sand investigations and deposits in India – an overview. Explor Res At Miner 13:1–21

Ali MA, Desapati T, Deshmukh RD (2001b) Heavy mineral resources in the beach placers of Maharashtra. Explor Res At Miner 13:147–155

Anguswamy N, Rajamanickam GV (2000) Distribution of heavy minerals along the beach from Mandapam to Kanyakumari, Tamil Nadu. J Geol Soc India 56:199–211

Bagora S, Mishra KS, Singh AK et al (2006) Xenotime placers as resources of REE in Deo river, Kolebira, Simdega district, Jharkhand. Explor Res At Miner 16:27–33

Bailey DK (1964) Crustal warping - a possible tectonic control of alkali magmatic. J Geophys Res 69:1103–1111

Balaram V (2019) Rare earth elements: a review of applications, occurrence, exploration, analysis, recycling, and environmental impact. Geosci Front 10:1285–1303

Balasubramani S, Sahoo P, Bhattacharya D et al (2017) A note on anomalous concentration of scandium in the Pakkanadu alkaline complex, Salem district, Tamil Nadu, India. In: Viladkar SG, Duraiswami RA and Krishnamurthy P (eds) Proceedings of International Seminar "Carbonatite-alkaline rocks and associated mineral deposits", Amba Dongar, India. December 8-11, 2017. Abstracts Volume, pp 46

Banerjee DC, Maithani PB, Ranganath N et al (1983) Rare metal mineralization in granite rocks of the Kanigiri area in the Prakasam district, Andhra Pradesh, India. Chem Geol 39:319–334

Banerjee DC, Sinha RP, Dwivedy KK (1995) Xenotime resource in India – an overview. In: Proceedings of seminar on recent developments in the science and technology of rare earths, organised by IREL, Udyogamandal, Cochin, Kerala, during December 14–16, 1995, p 1–9

Basu SK (1993) Alkaline-carbonatite complex in the Precambrian of south Purulia shear zone, eastern India: its characteristics and mineral potentialities. Indian Miner 47:179–194

Basu SK, Ghosh RN (1999) Rare earth elements and rare metal metallogeny in Precambrian terrains of Purulia district, West Bengal, and its status of exploration. In: Proceedings of national seminar on strategic rare and rare earth minerals and metals. Geol Surv India, Hyderabad, pp 129–140

Basu SK, Bhattacharyya T (2014) Petrography and mineral chemistry of alkaline-carbonatite complex in Singhbhum crustal province, Purulia region, eastern India. J Geol Soc India 83(1):54–70

Behera P (2003) Heavy minerals in beach sands of Gopalpur and Paradeep along Orissa coast line, east coast of India. Indian J Mar Sci 32(2):172–174

Bertinell L, Poncin S, Zou B (2019) The war of rare earth elements: a dynamic game approach. Centre for Research in Economic and Management (CREA), University of Luxembourg, Discussion Paper 2019–11: http://wwwfr.uni.lu/recherche/fdef/crea/publications/discussion_papers

Bhushan SK (2015) Geology of the Kamthai rare earth deposit. J Geol Soc India 85(5):537–546

Bhushan SK, Kumar A (2013) First carbonatite hosted REE deposit from India. J Geol Soc India 81(1):41–60

Bhushan SK, Chittora VK (1999) Late Proterozoic bimodal volcanic assemblage of Siwana subsidence structure, western Rajasthan, India. J Geol Soc India 53:433–452

Bhushan SK, Somani OP (2019) Rare earth elements and yttrium potentials of Neoproterozoic peralkaline Siwana granite of Malani igneous suite, Barmer district Rajasthan. J Geol Soc India 94(1):35–41

Bidwai R, Srinivasan S, Nanda LK et al (2014) Anomalous silver concentration in volcano-plutonic rocks of Siwana ring complex, Barmer district, western Rajasthan. Curr Sci 106(3):159–162

Bidwai R, Srinivasan S, Nanda LK et al (2015) Anomalous rare earth abundances in Siwana ring complex, Barmer district, western Rajasthan. In: Golani PR (ed) Recent developments in metallogeny and mineral exploration in Rajasthan, geological survey of India special publication, vol 101, pp 163–176

Binnemans K (2014) Economics of rare earths: the balance problem. In: ERES 2014: 1st European rare earth resources conference, pp 37–46

Binnemans K, Jones PT (2015) Rare earths and the balance problem. J Sustain Metall 1(1):29–38

Binnemans K, Jones PT, Van Acker K et al (2013) Rare-earth economics: the balance problem. J Miner Metals Mater 65(7):846–848

Binnemans K, Jones PT, Muller T et al (2018) Rare earths and the balance problem: how to deal with changing markets? J Sustain Metall 4:126–146

Borodin LS, Gopal V, Moralev VM et al (1971) Precambrian carbonatites of Tamil Nadu, South India. J Geol Soc India 12:101–112

Chandrasekharan S, Murugan C (2001) Heavy minerals in the beach and the coastal red sands (Teris) of Tamil Nadu. Explor Res At Miner 13:87–109

Chattopadhyay N, Hashimi S (1984) The Sung Valley alkaline-ultramafic-carbonatite complex, East Khasi and Jaintia Hills district Meghalaya. Rec Geol Surv India 113(4):24–33

Collins WJ, White AJR, Chappell BW (1992) Nature and origin of A-type granite with particular reference to south eastern Australia. Contrib Mineral Petrol 80:189–200

Connelly NC (2005) Nomenclature of inorganic chemistry: IUPAC recommendations. Third book, Royal Society of Chemistry, Cambridge

Das SK (2015) Characteristics of zircon of Chatrapur beach placer deposit, Ganjam district Odisha. J Geol Soc India 86(4):383–390

Das SK, Angadi SI, Kundu T et al (2020) Mineral processing of rare earth ores. In: Jyothi RK (ed) Rare-earth metal recovery for green technologies. https://doi.org/10.1007/978-3-030-38106-6_2

Das UK, Hussain S, Jena D et al (2015) Granite-hosted low grade REE prospect, Dhani area, Pali district, Rajasthan. In: Golani PR (ed) Recent developments in metallogeny and mineral exploration in Rajasthan, geological survey of India special publication, vol 101, pp 191–206

Das UK, Gantait A, Panda L et al (2016) Rare earth element potential of the felsite dykes of Phulan area, Siwana ring complex, Rajasthan. India Curr Sci 110(7):1157–1162

Dharma Rao CV, Reddy UVB (2007) Petrology and geochemistry of Paleoproterozoic A-type granite at Kanigiri in the Nellore-Khammam schist belt, Andhra Pradesh, India. J Asian Earth Sci 30:1–19

Deori LN, Reghu R (2014) Preliminary investigation for delineating the REE bearing zones around Mincheri, Inchanal and Hireupperi areas, Raichur district, Karnataka (Stage G-4). Unpubl. Progress Report for the Field Season 2013–14, Geol. Surv. India, SR, SU Karnataka & Goa, Project REE, Bengaluru (Code No. ME/REE/SR/KG/2013/052, July 2014, pp 1–69

Dhana Raju R, Ali MA, Krishnan S (eds) (2001) Special issue on beach and inland heavy mineral sand deposits of India. Explor Res At Miner 13:1–159

Dostal J (2017) Rare earth element deposits of alkaline igneous rocks. Resources 6(34):1–2

Doroshkevich AG, Viladkar SG, Ripp GS et al (2009) Hydrothermal REE mineralization in the Amba Dongar carbonatite complex, Gujarat, India. Can Mineral 47:1105–1116

Dunn JA, Dey AK (1942) The geology and petrology of Eastern Singhbhum and surrounding areas. Memoirs Geol Surv India 69 (2):281–456

Dwivedy KK (1995) Reserves, resources and geology of titanium deposits in India. Titanium 1:16–39

Gardner RAM (1981) Reddening of dune sands – evidence from southeast India. Earth Surf Proc Land 6:459–468

Ghosal S, Agrahari S, Banerjee S et al (2020) Geochemistry of the heavy mineral sands from the Garampeta to the Markandi beach, southern coast of Odisha, India: implications of high contents of REE and radioelements attributed to placer monazite. J Earth Syst Sci 129:152. https://doi.org/10.1007/s12040-020-01419-8

Ghose NC (1992) Chhotanagpur gneiss-granulite complex, eastern India. Present status and future prospect India. J Geology 64:100–121

Goodenough KM, Wall F, Merriman D (2017) The rare earth elements: demand, global resources and challenges for resourcing future generations. Nat Resour Res. https://doi.org/10.1007/s11053-017-9336-5

Goyal N, Varughese SK, Hansda PK et al (1997) Xenotime placers along Ujol river, Panchmahals district, Gujarat, India. J At Miner Sci 5:59–62

Guha DB, Raza A, Neogi S et al (2020) Allanite-britholite bearing migmatite gneiss-Amet granite and intrusive syenite-monzonite rocks of SW Rajasthan, India, and their REE potential. Episode 43 (2). https://doi.org/10.1881/epiiugs/2020/020044

Gupta CK, Krishnamurthy N (2005) Extractive metallurgy of rare earths, 2nd edn. CRC Press, London

Gupta JN, Pandey BK, Chabria T et al (1984) Rb-Sr geochronological studies on the granites of Vinukonda and Kanigiri, Prakasam district, Andhra Pradesh, India. Precamb Res 26:105–109

Gupta SK, Gandhi P, Bhatt NY (2018) Effective evaluation of geochemical data of stream sediments for targeting rare earth element potential area using multivariate analysis: a case study of Godhra granite in Panchmahal district Gujarat. J Appl Geochem 20 (2):170–180

Hedrick JB (1985) Rare earth elements and yttrium. Bureau of Mines Bulletin 675. Mineral Facts and Problems. United States Department of the Interior, Washington, DC, pp 647–662

Henderson P (1996) The rare earth elements: introduction and review. In: Jones AP, Wall F, Williams TC (eds) Rare earth minerals: chemistry, origin and ore deposits. Mineralogical society series, vol 7: Chapman & Hall Publishing Ltd., pp 1–17

Hoda SQ, Krishnamurthy P (2016a) Geochemistry, mineralogy and economic aspects of phosphatic rock of the Samchampi alkaline-carbonatite complex North East India. J Appl Geochem 19(1):21–27

Hoda SQ, Krishnamurthy P (2016b) Studies on residual soil geochemistry in the evaluation of niobium within Samchampi alkaline carbonatite complex, Karbi Anglong district, Assam India. J Appl Geochem 18(2):111–118

Hoda SQ, Krishnamurthy P (2020) Mineralogy, geochemistry and evolution of carbonatites of Samchampi alkaline complex, Assam India. J Appl Geochem 22(1):31–41

IBM (Indian Bureau of Mines) (1996) Indian Minerals Year Book 1995, I. Indian Bureau of Mines Press, Nagpur, India, General reviews

IBM (2018) Indian Minerals Year Book 2017 (Part-III: Mineral Reviews), 56th edn, Rare Earths. March 2018, Ministry of Mines, Government of India, Nagpur, 24–1 to 24–7

IREL (2016) Indian Rare Earths Limited, March 2016. http://www.irel.gov.in/scripts/unit.asp

IREL (2018) 68th annual report of the Indian rare earths limited for the year 2017–18. Indian Rare Earths Limited, pp 1–180. www.irel.co.in

Jagannadha Rao K, Muralidharan R, Srivastava PK et al (2007) On the effective use of integrated study of aerospace data for targeting heavy mineral placers – a case study from Krishna – Godavari coast, Andhra Pradesh India. J Geophys 28(3 & 4):97–104

Jagannadha Rao K, Subramanyam AV, Kumar A et al (2008) Discovery of heavy mineral-rich sand dunes along the Orissa-Bengal coast of India using remote sensing techniques. Curr Sci 94(8):983–985

Jain RB, Miglani TS, Kumar S et al (1996) Rare metal and rare earth rich peralkaline, agpatic granitoid dykes of Siwana ring complex, Barmer district Rajasthan. Current Sci 70(9):854–858

Jain RB, Miglani TS (1997) Niobium-, uranium-, and thorium-rich phonolite dyke from Sarnu Dandali alkaline complex, Barmer district, Rajasthan, India. J At Miner Sci 5:17–23

Jones AP (ed) (1996) Rare earth minerals. Mineralogical society series, vol 7. Chapman & Hall, pp 349–356

Jordens A, Cheng YP, Waters KE (2013) A review of the beneficiation of rare earth element bearing minerals. Miner Eng 41:97–114

Kathiravan R, Palanivel K, Rout SM et al (2018) Delineation of mineable heavy placer deposits using geospatial technology along Puri to Icchapuram coastal tract of Odisha State, India. J Appl Sci Comput 5(12):424–1438

Kerr A, Rafuse H (2012) Rare Earth element (REE) Geochemistry of the strange lake deposits: Implications for resource estimation and metallogenic models. Current Res. Nfld Labrador Dep Nat Resour Geol Surv Rep 12(1):39-60

Khan R, Ghoshal S, Sengupta D et al (2018) Studies on heavy mineral placers from eastern coast of Odisha, India. J Radioanal Nucl Chem. https://doi.org/10.1007/s10967-018-6250-1

Kochhar N (1984) The Malani Igneous Suite: Hot spot magmatism and cratonisation of the northern part of the Indian Shield. J Geol Soc India 25:155–161

Koduru JR, Lingamdinne LP, Kailasa SK et al (2020) Recent strategies on adsorptive removal of precious metals and rare earths using low-cost natural adsorbents. In: Jyothi RK (ed) Rare-earth metal recovery for green technologies. https://doi.org/10.1007/978-3-030-38106-6_5

Krishnamurthy P (1977) On some geochemical aspects of the Sevattur carbonatite complex, North Arcot district Tamil Nadu. J Geol Soc India 18(6):275–280

Krishnamurthy P (1985) Petrology of the carbonatites and associated rocks of Sung Valley, Jaintia Hills District, Meghalaya, India. J Geol Soc India 26:361–379

Krishnamurthy P (2019) Carbonatites of India. J Geol Soc India 94 (2):117–138

Krishnamurthy P, Hoda SQ, Sinha RP et al (2000) Economic aspects of carbonatites of India. J Asian Earth Sci 18:229–235

Krishnan MS (1941) Phosphates. Records geological survey of India. Bull 76:4

Krishnan S, Viswanathan G, Balachandran K (2001) Heavy mineral sand deposits of Kerala. Explor Res At Miner 13:111–146

Kumar P, Bharadwaj MD (2018) Indian rare-earth industry: need and opportunity for revival and growth. https://www.researchgate.net/publication/323656626

Kumar MK, Bhaskar DV, Laxminarayana Y et al (1995) Rare-earths and yttrium mineral stream sands at Rangampeta, North Arcot district, Tamil Nadu. J Geol Soc India 45:335–338

Kumar D, Mamallan R, Dwivedy KK (1996) Carbonatite magmatism in northeast India. J Southeast Asian Earth Sci 13(2):145–158

Lide DR (1997) Abundance of elements in the earth's crust and sea. In: CRC handbook of physics and chemistry, 78th edn. CRC Press, Boca Raton, p 14

Loveson VJ, Chandrasekar N, Sinha A (eds) (2005) Development planning of coastal placer minerals (Placer-2005). Allied Publishers Pvt. Ltd., New Delhi

Machacek E, Fold N (2014) Alternative value chains for rare earths: the Anglo-deposit developers. Resour Policy 42:53–64

Magna T, Viladkar S, Rapprich Vet al (2020) Nb-V-enriched sovites of the northeastern and eastern part of the Amba Dongar carbonatite ring dike, India – a reflection of post-emplacement hydrothermal overprint? Geochemistry 80:125534. https://doi.org/10.1016/j.chemer.2019.125534

Mahadevan TM (1988) Rare earth resources. Mater Sci Forum 30:13–32

Mahadevan TM (1992) Geological evolution of the Chhotanagpur Gneiss Complex in a part of Purulia district West Bengal. Indian J Geology 64(1):1–22

Mahadevan TM (2002) Geology of Bihar and Jharkhand. J Geol Soc India, Bangalore 563

Majumdar D, Dutta P (2014) Rare earth element abundances in some A-type Pan-African granitoids of Karbi Hills North East India. Current Sci 107(12):2023–2029

Maithani PB, Rathaiah YV, Varughese SK (1993) A note on the occurrence of xenotime-bearing placers of the Vasva river, Zoz area, Baroda district, Gujarat, India. J Indian Assoc Sedimentol 12:41–48

Mallik TK (1976) Self sediments of the Ganges delta with special emphasis on the mineralogy of the western part, Bay of Bengal, Indian ocean. Marine Geology 22(1):1–32. https://doi.org/10.1016/0025-3227(76)90007-4

Mallik TK (2007) Marine resources of the beaches and shallow sea around Indian coasts. Indian Geological Congress Bull 1(2):80–87

Mallik TK, Sensarma K (2009) Highlights of heavy mineral distribution patterns along the coastline of West Bengal. Indian Miner 63 (4):429–442

Mazumdar SK (1988) Crustal evolution of the Chhotanagpur gneissic complex and the mica belt of Bihar. Geol Soc India Memoir 8:49–84

Menon KK (1950) General features of the Teris of South Travancore. Indian Geophys J XXV(2): 1–8

Mishra KS, Rao MK, Hegde GN et al (2006) Geochemical petrographic studies on the granitoids of the Kumarkunti-Jharnomal area, Nuapara district, Orissa: a potential source for rare metal and rare earth minerals. Explor Res At Miner 16:61–72

Mishra B, Srikant SS, Routray S et al (2019) Preparation of nanomaterials from strategic placer heavy minerals recovered from red sediments of Badlands topography along the southeast coast of India. Curr Sci 116(8):1363–1372

Mohammad H, Nandi H (1971) A petrologic dimensional fabric and statistical study of Purbatili inland placers, district Ranchi. Bihar. Proc Indian Natl Sci Acad 37A(1):61–76

Mohan PM, Rajamanickam GV (2000) Buried placer mineral deposits along the east coast between Chennai and Pondicherry. J Geol Soc India 56:1–13

Mohanty AK, Das SK, Vijayan V et al (2003) Geochemical studies of monazite sands of Chhatrapur beach placer deposit of Orissa, India by PIXE and EDXRF method. Nucl Inst Methods Phys Res B 211:145–154. https://doi.org/10.1016/S0168-583X(03)01166-2

Mohanty M, Bhushan SK (2004) Neoproterozoic Peralkaline Siwana granite, Barmer district, Rajasthan and its potential for Rare earth element mineralization. Geol Surv India Spec Publ 72:363–376

Muralidharan R, Jagannadha Rao K, Dattanarayana TA (2001) Integration of airborne geophysical and remote sensing data for targeting the zones of rare earth minerals: a case study of Godhra area, Gujarat. J Geophys 22(2–4):83–89

Nagabhushnam B, Durai Raju S, Mundra KL et al (2018) LREE-Nb mineralisation in the southwestern part of Ambadongar carbonatite complex, Chhota Udaipur district, Gujarat. Curr Sci 114(8):1608–1610

Nagendra Babu G, Sai Baba M, Kumar MK et al (2013) Bastnaesite from Kanigiri granite, Prakasam district. Andhra Pradesh. J Geol Soc India 81(4):447–450

Nageswara Rao PV, Suryam RK, Ranga Rao V (2007) Heavy mineral potential and provenance of beach sand between False Devi point to Suryalanka, southeast coast of India. Indian Geol Congr Bull 1 (2):47–53

Naidu KB, Reddy KSN, Sekhar ChR et al (2016) REE geochemistry of monazites from coastal sands between Bhimunipatnam and Konada, Andhra Pradesh, east coast of India. Curr Sci 110(8):1550–1559

Nambiar AR, Golani PR (1985) A new find of carbonatite from Meghalaya. Curr Sci 54:281–282

Nambiar AR, Kumaran K, Dinesh AC et al (2007) Distribution and resources of offshore placer minerals in near shore area between Inayam and Midalam, north of Manavalakurichi, southwest coast, India. Indian Geol Congr Bull 1(2):54–58

Narayan Das GR, Bagchi AK, Chaube DN et al (1978) Rare metal contents, geology, and tectonic setting of the alkaline complexes across the Trans Aravalli Region, Rajasthan. Recent Res Geol Hindustan Publ Corp New Delhi 7:201–219

Kumar N, Kumar N (2020) Geochemistry of volcanic flows of Nakora area of Malani igneous suite, northwesten India: constraints on magmatic evolution and petrogenesis. Int J Eng Sci Technol 12 (1):66–82

Overstreet WC (1967) Geological occurrences of monazite. U.S.G.S. Prof Pap 530:1–327

Palaparthi J, Chakrabarti R, Banerjee S et al (2017) Economically viable rare earth element deposits along beach placers of Andhra Pradesh, eastern coast of India. Arab J Geosci 10:201. https://doi.org/10.1007/s12517-017-2973-5

Panda NK, Rajagopalan V, Ravi GS (2003) Rare earth element geochemistry of placer monazites from Kalingapatnam coast, Srikakulam district, Andhra Pradesh. J Geol Soc India 62:429–438

Panda NK, Rao AY, Ramesh Kumar K et al (2015) Anomalous REE concentration in carbonate-phosphate bearing phases from Narsimharajapuram area, Visakhapatnam district, Andhra Pradesh. Curr Sci 109(5):860–862

Panda NK, Sahoo P, Rao AY et al (2017) Concentration and distribution of rare earth elements in beach placer garnets of Kalingapatnam coast and their potential for heavy rare earths, Andhra Pradesh, India. J Geosci Res Spl 1:131–138

Parhi PK, Behera SS, Mandal D et al (2020) Fundamental principle and practices of solvent extraction (SX) and supported liquid membrane (SLM) process for extraction and separation of rare earth metal(s). In: Jyothi RK (ed) Rare-earth metal recovery for green technologies. https://doi.org/10.1007/978-3-030-38106-6_4

Pearce JA, Niger BWH, Tindle G (1984) Trace element discrimination diagrams for the tectonic interpretation of granitic rocks. J Petrol 25 (4):56–83

Pourmand A, Dauphas N, Ireland TJ (2012) A novel extraction chromatography and MC-ICP-MS technique for rapid analysis of REE, Sc and Y: revising CI-chondrite and Post-Archaean Australian Shale (PAAS) abundances. Chem Geol 291:38–54

Pradip GBP, Reddy S et al (2019) Future of mining, mineral processing and metal extraction. Trans Indian Inst Met 72(8):2159–2177. https://doi.org/10.1007/s12666-019-01790-1

Pruseth KL, Sahoo D, Kumar B et al (2017) Beach monazites from Alleppey: a window to the Trivandrum block, southern India. J Geol Soc India 89(3):240–246

Rai SD, Banerjee DC (1995) Xenotime placers in parts of Mahan river basin, Surguja district, Madhya Pradesh. J Geol Soc India 45:285–293

Rai SD, Hoda SQ, Parthasarathy TN (1990) Xenotime-bearing radioactive placers of Dev river, Gumla district, Bihar. In: Pai BC, Pillai RM, Damodaran AD (eds) Advances in Rare Earth Research, Proceedings of the Second India-USSR Symposium on Rare Earth Material Research, held during Nov. 5–7, 1990, at Trivandrum, India, 23–33

Rai SD, Shivananda SR, Tiwary KN et al (1991) Xenotime-bearing inland placers in India and their beneficiation. Explor Res At Miner 4:77–92

Rai SD, Desapati T, Sinha RP et al (1997) Occurrence of xenotime-bearing placers near Pathapalem, Krishna basin, in parts of Mahboobnagar district, Andhra Pradesh and Raichur district, Karnataka. Indian Miner 51:247–254

Rai SD, Mundra KL, Pandit PSC et al (2018) REE and Nb mineral potentialities of Ambadongar carbonatite complex and Kadipani fluorspar mine dumps. Explor Res At Miner 27:115–127

Ramachar TM, Krishnamurthy B, Jayaram KMV (1979) A note on the rare minerals bearing granite of Kanigiri in the Prakasam district, Andhra Pradesh. Curr Sci 48:860–861

Ramesh Babu PV (1999) Rare Metal and rare earth pegmatites of Central India. Explor Res At Miner 12:7–52

Ramesh Babu PV, Rajendran R, Vijaya Kumar K et al (1998) Exploration for xenotime placers and their source rocks in Balrampur-Bhandaria area, Surguja district, Madhya Pradesh and Garhwa district, Bihar, India. Explor Res At Miner 11:37–44

Rajamanickam GV (2000) Light-heavy minerals on the Indian continental shelf including beaches. In: Cronan DS (ed) Handbook of marine mineral deposits. CRC Press, London, pp 13–26

Rajamanickam GV (ed) (2001) Handbook of placer mineral deposits. New Academic Publishers, Delhi, p 327

Rajamanickam GV, Chandrasekar N, Angusamy N et al (2004) Status of beach placer mineral exploration in India. In: Loveson VJ, Misra DD (eds) National Workshop and Seminar on sustainable development of coastal placer minerals (PLACER-2002). Allied Publishers Pvt. Ltd., New Delhi, pp 9–21

Rajasekaran P, Senthilvel CNS, Sathyanarayana K et al (1994) Bi-pyramidal xenotime-(Y) from Siri river placers, Raigarh district, Madhya Pradesh, India. J At Miner Sci 2:1–17

Ramachandran S, Parthasarathy TN, Saraswat AC (1989) Xenotime from Kanyaluka, Singhbhum district, Bihar, India: rare-earth element abundances and economic potentiality. Explor Res At Miner 2:177–182

Rao AT (1976) Study of the apatite-magnetite veins near Kasipatnam, Visakhapatnam district, Andhra Pradesh, India. TMPS Min Pet 23:87–103

Rao AT, Dhanamjaya Rao EN, Jaydeep VA et al (2018) Mineral chemistry of a new occurrence of allanite in Eleswaram from Eastern Ghats Granulite Belt, India. J Appl Geochem 20(2):164–170

Rao AT, Dhana Raju R, Dhanamjaya Rao EN et al (2018) Electron micro-probe mineral chemistry and dating of monazite from the cordierite-gneisses near Vizianagaram in the Eastern Ghats Granulite Belt, India. J Appl Geochem 20(2):181–191

Rao KB (1989a) Origin and evolution of the sand dune deposits of the Ganjam coast, Orissa, India. Explor Res At Miner 2:133–146

Rao GD (1989b) Nature, distribution and evaluation of heavy minerals in the beach sand deposits between Mangalore and Cochin, West coast of India. Explor Res At Miner 2:157–166

Rao NK, Rao GVU (1980) Mineralogy of apatite-magnetite rocks in Singhbhum Shear Zone, Bihar. Proc 3rd Indian Geological Congress. Poona 1990:39–56

Rao NK, Rao GVU (1983) Uranium mineralization in Singhbhum shear zone, Bihar. III. Nature of occurrence of uranium in apatite-magnetite rocks. J Geol Soc India 24:555–561

Rao GD, Setty BK, Raminaidu Ch (1989) Heavy mineral content and textural characteristics of coastal sands in the Krishna-Godavari, Goshthani-Champavati, and Penna River deltas of Andhra Pradesh, India: a comparative study. Explor Res At Miner 2:147–155

Rao RG, Sahoo P, Panda NK (2001) Heavy mineral sand deposits of Orissa. Explor Res At Miner 13:23–52

Rao CS, Narayan Das GR, Setty BK et al (1978) Radioactive carbonatites of Pakkanadu and Mulakkadu, Salem district, Tamil Nadu. J Geol Soc India 19(2):53–63

Rao DS, Vijayakumar TV, Prabhakar S et al (2005) Alteration characteristics of ilmenites from south India. J Miner Mater Characterisation Eng 4(1):47–59

Rao DS, Mohapatra R, Vasumathi N et al (2010) Beneficiation studies of beach placer sample for steel making industries. J Min Metall 46A(1):11–21

Rao DS, Acharya BS (2015) Electron microscopic studies of monazite from Chhatrapur coast, Odisha, India. World Metall – ERZMETALL 68(2):102–106

Rao DS, Dash N, Mohanty S (2016a) Varieties and nature of zircnium bearing minerals from the Chhatrapur beach placer deposit, Odisha, India. World Min – Surf Undergr 68(6):390–393

Rao DS, Pani S, Dash N et al (2016b) Electron microscopic studies for the nature of mineral inclusions in the beach placer ilmenite sample. J Adv Microsc Reseah 11:1–4

REAI (2021) Beijing considers rare earth export controls targeted at U. S. defence. Rare Earths Association of India (REAI) News v. 21, No. 2, Pt. II, February 15–28, 2021, pp 1–221

Rastogi SK, Mukherjee T, Gupta GP (2015) Flow stratigraphy, geochemistry and REE mineralisation in northern part of the Siwana ring complex in Malani igneous suite (MIS), Barmer district, Rajasthan. In: Golani PR (ed) Recent developments in metallogeny and mineral exploration in Rajasthan, geological survey of india special publication, vol 101, pp 207–217

Ravi GS, Rao RG, Rao YA (2001) Coastal heavy mineral sand deposits of Andhra Pradesh. Explor Res At Miner 13:53–85

Reddy DR, Prasad VSS (1997) Thorium-rich monazites from the beach sands of Kalingapatnam-Baruva coast, Andhra Pradesh, east coast of India. Curr Sci 73(10):880–882

Roonwal GS (2009) Beach sand resources and industry in India. In: Shrivastava KL (ed) Economic mineralisation. Scientific Publishers (India) Jodhpur, pp 248–253

Roskill (2016a) Rare earths: global industry, markets and outlook, 16th edn. UK, Roskill, London

Roskill (2016b) Lithium: global industry, markets and outlook, 13th edn. Roskill, London, UK

Routray S, Rao DS, Rao RB (2009) Preliminary studies on recovery of total heavy minerals from Konark-Ramchandi beach. Int J Appl Eng Res 4(9):1715–1721

Roy M, Dhana Raju R (1999) Petrogenetic model of A-type granitoids of the Kullampatti area, Salem district, Tamil Nadu, India. Gondwana Res 2(1):127–135

Sadiq Md, Umrao RK, Dutta JC (2014) Occurrence of rare earth elements in parts of Nongpoh granite, Rio-Bhoi district, Meghalaya. Curr Sci 106(2):162–165

Sarbajna C, Paul AK, Rajagopalan V (2007) Mineral chemistry of columbite-tantalite, monazite and xenotime from Siri river placers, Jashpur district, Chhattiagarh. Gond Geol Mag Spl 10:161–171

Sarvothaman H, Sesha Sai VV (2010) Global correlation of A-type granites and related rocks, their mineralisation and significance in lithospheric evolution. IGCP-510, 2–14

Satpathy C, Routray S, Rao DS (2010) Heavy mineral recovery from beach and dune sands of Ganjam coast, Orissa, India. World Metall - ERZMETALL 63(1):5–13

Saxena VP, Bahukhandi NK, Bhatt AK et al (1990) Uranium metallogeny in central Surguja shear zone, Surguja district, Madhya Pradesh. Geol Surv India Spec Publ 28:618–638

Saxena VP, Krishnamurthy P, Murugan C et al (1992) Geochemistry of the granitoids from the central Surguja shear zone, India: geological evolution and implications on uranium mineralisation and exploration. Explor Res At Miner 5:27–40

Schleicher H, Kramm U, Pernicka E et al (1998) Enriched subcontinental upper mantle beneath south India: evidence from Pb, Nd, Sr and C-O isotopic studies on Tamil Nadu carbonatites. J Petrol 39 (10):1765–1785

Semenov EI, Upendran R, Subramanian V (1978) Rare earth minerals of carbonatites of Tamil Nadu. J Geol Soc India 19(12):550–557

Sen J, Murugan C, Dwivedi AK et al (2008) Geochemistry of A-type Kuilapal granite from north Singhbhum mobile belt, eastern India: implication for radioelemental concentration. Explor Res At Miner 18:179–199

Seth DK (1989) Structure and uranium mineralization in the Proterozoic rocks of Dumhath area, Surguja district, Madhya Pradesh, India. Explor Res At Miner 2:77–91

Sharma GS, Kothari PK, Parihar PS (2001) Xenotime-(Y) placers along Kanhar river, Sondhadra district, Uttar Pradesh, India. J At Miner Sci 7:35–38

Shaw DM, Cramer JJ, Higgines MD et al (1986) Composition of the Canadian Precambrian shield and the continental crust of the earth. Geol Soc Spec Publ 24, Blackwell, pp 275–282

Shobhita K, Rajagopalan V, Paul AK et al (2007) Occurrence of REE anomalous soil and compositional characterization of associated monazite, xenotime, zircon and rutile around Tirka, Gondia district, Maharashtra. Gond Geol Mag Spl 10:203–207

Singh Y (1990) Identification of xenotime-bearing Precambrian granites in Central India and its economic significance. In: Abstract of the Proc. National Seminar on Continental Margins of India, Geology Department, Andhra University, Visakhapatnam, March 25–26, 1990, pp 36–37

Singh Y (1992) Geochemistry of porphyritic granite from Kunkuri-Rajauti area, Raigarh district, Madhya Pradesh. Indian J Geol 64:23–36

Singh Y (2002) Fluvial landform, mineralogical and geochemical (Y, Ce, Nb, Ta, Sn) studies on grantic soil/alluvial deposits in the environs of Kawara and Nakti sub-basins around Amraoti, Bastar district, Chhattisgarh. J Indian Assoc Sedimentol 21(1–2):1–13

Singh Y (2004) Zr, Hf, U, Th and REE-fertile Lower Proterozoic potassic granite from parts of Andhra Pradesh, South India. Acta Geol Sin 78(4):921–930

Singh Y (2007) Granitic soil yttrium-cerium geochemistry with xenotime concentrate upgradation around Darba, Bastar district, Chhattisgarh. In: Pophare AM, Malpe DB (eds) Recent Developments in Geology and Mineral Resources of Central India. Gondwana Geol Mag Spec 10:173–178

Singh Y (2012a) Mineralogy and origin of the rare-earth-element-bearing granites from central India. In: Miroslava B, Bogdashka H (eds) Granites: occurrence, mineralogy and origin. Nova Science Publishers Inc., Hauppauge, New York, pp 1–26

Singh Y (2012b) Geochemical studies on granitoids of the Midtul area, Bastar craton, central India. J Appl Geochem 14(2):153–161

Singh Y (2018) Mineralogy, petrology and geochemistry of rare metal-bearing pegmatites. J Geosci Res 3(2):95–120

Singh Y (2019) Rare-earth resources of India. J Geol Soc India 94 (1):109. https://doi.org/10.1007/s12594-019-1276-8

Singh Y (2020a) Rare earth element resources: Indian context. Society of earth scientists series. Springer, pp 1–395

Singh Y (2020b) Rare earth element geochemistry of monazites from beach sand deposits of Indian coasts: implication for clean energy. J Appl Geochem 22(3):209–220

Singh Y (2020c) Strategic and critical element potential of Indian carbonatites. J Geosci Res 5(2):161–170

Singh Y, Krishna V (2009) Rb-Sr geochronology and petrogenesis of granitoids from the Chhotanagpur granite gneiss complex of Raikera-Kunkuri region, Central India. J Geol Soc India 74:200–208

Singh Y, Rai SD (1992) Yttrium-europium-erbium geochemistry of granitic soils of Kunkuri area, Raigarh district, M.P. India. J Geol Soc India 40:347–358

Singh Y, Rai SD (2008) Geochemistry of pink granite from a part of Gumla district, Bihar. Mem Geol Soc India 73:55–76

Singh Y, Reddy LSR (2008) Petrology of the REE- and Y-bearing granitoids from the Raikera – Kunkuri area, Central India. Explor Res At Miner 18:55–76

Singh Y, Shiv Kumar K (2009) A note on radioactive riverine placers around Midtul, Bastar Craton, Central India. J Geol Soc India 73:419–424

Singh Y, Singh KDP (1996) Mineralogical and geochemical studies on xenotime-bearing granites of Raikera area, Raigarh district, Madhya Pradesh, India. J At Miner Sci 4:37–47

Singh Y, Singh KDP (2001) X-ray mineralogy and cerium-yttrium-niobium-tantalum geochemistry of granitic soil from Muskel-Kawadgaon area, Bastar Craton. Central India. Indian Miner 55(1–2):47–54

Singh AK, Vallinayagam G (2002) Geochemistry and petrogenesis of granite in Kundal area, Malani igneous suite, western Rajasthan. J Geol Soc India 60(2):183–192

Singh Y, Rai SD, Lakshminarayana Y et al (1991) A note on the yttrium mineral potential of granitic soil in parts of Andhra Pradesh. J Geol Soc India 38:207–210

Singh Y, Krishnamurthy P, Balakrishnan SP (1994) Distribution of cerium and yttrium in apatite-magnetite deposits of the Kasipatnam area, Visakhapatnam district, Andhra Pradesh. Smarika 4:101–109

Singh Y, Rai SD, Dwivedy KK (1996) Environmental considerations of radioactivity in granitic rocks and mining of rare earth-bearing alluvial deposits near Kunkuri, Madhya Pradesh. Recent researches in geology, vol 15. Hindustan Publishing Corporation (India), New Delhi, pp 381–387

Singh Y, Bhatt AK, Thimmaiah M et al (1999) Uranium mineralization associated with Precambrian granite gneiss in Chowrdhawa and Sonarigadai areas, Bilaspur district. Madhya Pradesh. Indian Miner 53(3–4):175–186

Singh Y, Singh KDP, Prasad RN (2004) Rb-Sr whole-rock isochron age of Early Proterozoic potassic granite from Dharmawaram Karimnagar district, Andhra Pradesh. J Geol Soc India 64:93–96

Singh Y, Parihar PS, Maithani PB (2012) Rare-metal and rare-earth mineral resources of Chhattisgarh: a boon for development. In: Hari KR, Deshmukh SD, Shrivastava P (eds) Proc Natl Seminar on Emerging Trends in Scientific Research. Excellent Publishing House, New Delhi, pp 136–172

Singh Y, Babu GN, Viswanathan R et al (2014) X-ray crystallography and mineral chemistry of bastnaesite from Kanigiri granite, Prakasham district, Andhra Pradesh, India. Cent Eur J Geosci 6 (4):492–505

Singh Y, Mundra KL, Jagadeesan P et al (2018) Characteristics of source rocks of xenotime-bearing riverine placers in central India. Explor Res At Miner 27:145–162

Sinha KK, Upadhyay LD, Badri NSRK et al (1992) Uranium mineralization in granitic rocks of Binda-Nagnaha area, Palamau district, Bihar. Indian J Geol 64(1):61–70

Srivastava RK, Wall F, LeBas MJ (1993) Sr-REE rich carbonatite dyke from Sarnu-Dandali, Barmer, Rajasthan. Curr Sci 65(2):170–171

Subrahmanyam AV, Desapati T, Anil Kumar V et al (2004) Occurrence of xenotime in the Narasapur beach placers, West Godavari district, A.P. Curr Sci 87(10):1458–1461

Subramaniam NP, Parimoo ML (1963) Fluorspar mineralisation related to Deccan basalt volcanism at Ambadongar, Baroda district, India. Nature 198:563–564

Subramaniam V, Viladkar SG, Upendran R (1978) Carbonatite alkali complex of Samalpatti, Dharampuri, district, Tamil Nadu. J Geol Soc India 19(5):206–216

Sukheswala RN, Udas GR (1963) Note on the carbonatite of Ambadongar and its economic potentialities. Sci Cult 29:563–568

Sukheswala RN, Udas GR (1964) The carbonatite of Ambadongar, India – some structural considerations. 22nd Int Geol Cong Pt VII Sec 7:1–13

Kumar S (1990) Petrochemistry and geochronology of pink granite from Songsak, East Garo Hills, Meghalaya. J Geol Soc India 35 (1):39–45

Smith Stegen K (2015) Heavy rare earths, permanent magnets, and renewable energies: an imminent crisis. Energy Policy 79:1–8

Spedding FH (1978) Prologue. In: Gschneidner Gschneidner Jr, KA, Eyring L (eds) Handbook on the physics and chemistry of rare earths, vol 1. North Holland, Amsterdam, pp xv-xxv

Taylor SR, McLennan SM (1985) The continental crust: its composition and evolution. Blackwell, pp 1–312

Thirupathi PV, Sudhakar Ch, Krishna KVG et al (1996) Petrology and geochemistry of the Proterozoic A-type granite of Kanigiri, Prakasam district, Andhra Pradesh, India: implications for rare metal mineralisation. Explor Res At Miner 9:61–72

Times of India (TOI, July 2019) Rare earth metals used for atomic energy found in tribal village of Gujarat. TNN, July 30, 2019, 13.01 IST; timesofindia.com

Topno A, Dey S, Liu Y et al (2018) Early Neoarchaean A-type granitic magmatism by crustal reworking in Singhbhum craton: evidence from Pala Lahara area, Orissa. J Earth Syst Sci 127:43. https://doi.org/10.1007/s12040-018-0947-y

Udas GR, Narayan Das GR, Sharma CV (1974) Carbonatites of India in relation to structural setting. Proceedings international seminar on tectonics, metallogeny in South-East Asia and Far East; Geol Surv India Mis Publ 34(2):77–94

US Department of Energy (2017) Report on Rare Earth Elements from Coal and Coal Byproducts. Report to Congress, January 2017, 1–43

Vallinayagam G (2004) A report on rare metals and rare earths in the Siwana ring complex, Rajasthan. J Appl Geochem 2:387–391

Vallinayagam G, Kochhar N (2010) Mineralisation potential of the granites from Nakora ring complex, Barmer district, west Rajasthan, India: a geochemical perspective. J Appl Geochem 12(3):375–387

Verma MB (2019) Rare earth element resources of India in world's perspective. In: Proceedings of symposium on science, technology and applications of rare earths, Organised by Rare Earth Association of India, Dec 5–7, 2019, Mumbai. Abstracts Volume, pp 11–12

Verma SC (2020) Rare earths: a concise overview of their importance, availability and geological environments of occurrence. Indian J Geosci 74(3):275–279

Wall F, Le Bas MJ, Srivastava RK (1993) Calcite and carbonatite exsolution and cotectic texture in a Sr-REE rich carbonatite dyke from Rajasthan, India. Miner Mag 45:495–513

Weng ZH, Zowitt SM, Mudd GM et al (2015) Detailed assessment of global rare earth element resources: opportunities and challenges. Econ Geol 110:1925–1952

Xie F, Zhang TA, Dreisinger D et al (2014) A critical review on solvent extraction of rare earths from aqueous solutions. Miner Eng 56:10–28

Yusuf S, Saraswat AC (1977) A preliminary note on carbonatites in Wah-Sung Valley of Jaintia Hills district, Meghalaya. Curr Sci 46:703–704

Zepf V (2013) Rare Earth Eelements. Springer Theses. https://doi.org/10.1007/978-3-642-35458-8_2

Zhou B, Li Z, Chen, (2017) Global potential of rare earth resources and rare earth demand from clean technologies. Minerals 7:203. https://doi.org/103390/min7110203

Petrology and Geochemistry of Mafic–Ultramafic Rocks from Taka Area, Western Bastar Craton, Central India: An Implication for Their Genesis and Mineralization Potential

Vivek P. Malviya, Rajkumar R. Meshram, Sanjeet K. Verma, G. Gopalkrishna, M. Shareef, Tushar M. Meshram, M. L. Dora, Suresh A. Chore, and Pankaj Mehta

Abstract

The serpentinite and associated talc-tremolite-chlorite ± chromite occurs as enclaves and small lensoidal bodies within Amgaon granite gneiss around Taka village and adjoining areas of western Bastar craton, Central India. In the present study, we report petrological, mineralogical and geochemical data for these rocks. These data help to interpret the petrogenesis, composition of parental melt and their tectonic setting. These rocks are entirely serpentinized, showing well-developed mesh, hourglass and zoned textures containing bastite pseudomorphs. Serpentinite and talc-tremolite schist consist of chromite in the form of fine dissemination, cumulates and at places as bands. Cr-spinel is equigranular, euhedral to subhedral in shape and exhibits primary cumulus to intercumulus texture. Cr-spinel contains high Cr# and low TiO_2 content, which indicates arc tectonic setting. Two different normalized REE patterns, namely a flat pattern with low HREE and fractionated LREE depleted pattern are observed. The IPGE (Ir, Ru) enrichment relative to PPGE (Pd, Pt, Rh) and the relatively low Pd/Ir values suggest that their parental melts were produced by melting of a moderately depleted peridotite similar to the ones described for other IPGE rich chromitites originated in back arc basins in supra-subduction zones setting. Parental magma compositions of this high Cr-spinel suggest the reaction between boninitic melts and depleted peridotites for Taka serpentinite that got generated by partial melting of mantle in the supra-subduction zone setting.

Keywords

Ultramafic rocks • Western Bastar Craton • Central India • Boninite • Supra-subduction zone • PGEs

V. P. Malviya (✉)
Mineral Physics Division, Geological Survey of India, Northern Region, Lucknow, 226 004, India
e-mail: vivek.malviya@gsi.gov.in

R. R. Meshram · T. M. Meshram
Geological Survey of India, Central Region, Nagpur, 440 006, India

S. K. Verma
División de Geociencias Aplicadas, Instituto Potosino de Investigación Científicay Tecnológica (IPICYT), Camino a la Presa San José 2055, San Luis Potosí, 78216, México

G. Gopalkrishna · M. Shareef
Geological Survey of India, Southern Region, Bangalore, 500 078, India

M. L. Dora
Geological Survey of India, Southern Region, Hyderabad, 500 068, India

S. A. Chore
Trimurti Nagar, Khamla, Nagpur, 440 022, India

P. Mehta
Department of Environmental Sciences, Central University of Jammu Bagla (Rahya-Suchani), Samba, Jammu, J&K 181143, India

1 Introduction

Serpentinised mafic–ultramafic rocks are distributed worldwide and are commonly associated with subduction derived magma (Coleman and Jove 1992). Serpentinites can be formed in many ways in the subduction system, such as off-axis fractures and faults associated with seafloor spreading ridges, hydrated mantle wedge above a subducting slab (supra-subduction zone, SSZ), within the subduction channel, and along flexural faults associated with plate bending during subduction (Dilek 2003; Ranero et al. 2003). These rocks provide an invaluable insight into the nature of deeper undepleted portions of the sub-ocean lithosphere and asthenosphere and also comprise a significant portion of the upper ocean crust in slow spreading ridges.

A. K. Shandilya et al. (eds.), *Geological and Geo-Environmental Processes on Earth*, Springer Natural Hazards,
https://doi.org/10.1007/978-981-16-4122-0_8

In the Archean time, subduction drives plate tectonics and builds the lithosphere, and is one of the most important processes for shaping the present day Earth. Subduction zone magmatism in early to late Archean time is marked by a diverse compositional spectrum including island arc tholeiites (IAT), calc-alkaline basalts (CAB), basalt–andesite–dacite–rhyolite (BADR) associations, boninites, siliceous high-Mg basalts (SHMB) and high-Mg andesites (Polat and Kerrich 2006).

Serpentinized mafic–ultramafic rocks occurring as an enclave within basement Amgaon gneisses in Taka area from Western Bastar Craton (WBC) provides a window to look at the nature of subduction processes and its implication in late Archean crust formation in the Central part of India. Mafic–ultramafic rocks comprise of serpentinites, talc-tremolite schist, pyroxenites, gabbro and amphibolites. Serpentinite and talc-tremolite schist consist of primary Cr-spinel phase in the form of fine dissemination, bands and cumulates.

Serpentinites are commonly tricky to interpret in terms of their petrogenesis, because their primary mineralogy is rarely preserved, the hydration is assumed to have modified their bulk-rock compositions (Deschamps et al. 2013; Marchesi et al. 2013). Post magmatic processes such as alteration and metamorphism obliterate the nature of primary magma. Nevertheless, the ratio of some immobile trace elements and mineral chemistry of some of the primary phases such as Cr-spinel and relict olivine still preserve the nature of primary melt and provide information about the nature of the subduction process (Rollinson 1995).

In the present paper, we report petrography and mineral chemistry of serpentinites, talc-tremolite schist and chormitite occurring as an enclave within basement Amgaon gneisses in Taka area from Western Bastar Craton. We also present major, trace including rare earth elements (REE) and platinum-group element (PGE) compositions to evaluate their petrogenesis, geodynamic implications and mineralization potential.

2 Geological Setting and Field Observations

The Indian Peninsular shield is a mosaic of cratons and mobile belts. It consists of mainly four Archean cratons: Bundelkhand Craton in the north, Singhbhum craton in the east, Bastar craton in the centre and Dharwar craton in the south. These cratons are surrounded by Proterozoic mobile belts, viz. Eastern Ghat Mobile belt, Delhi-Aravalli Mobile belt, Singhbhum mobile belt and Kotri-Dongargarh mobile belt and Central Indian Tectonic Zone (CITZ) (Radhakrishna 1989; Yedekar et al. 1990; Roy and Prasad 2003). The Central Indian Tectonic Zone (CITZ) is a significant continental scale tectonic zone of Proterozoic age. It divides the Indian shield into two tectono-magmatic provinces, the northern crustal province comprising Bundelkhand craton and the southern peninsular crustal province comprising Bastar, Dharwar and eastern Singhbhum craton (Fig. 1a) (Naqvi and Rogers 1987).

The Bastar craton is limited by three prominent rifts, namely the Godavari rift in the SW, the Narmada rift in the NW and the Mahanadi rift in the NE. Its southeastern boundary is marked by the Eastern Ghats front. The Bastar craton consists of three major Archean- Paleoproterozoic and three Paleoproterozoic supracrustal groups. The Archean-Paleoproterozoic supracrustal groups are (i) the Bailadila Group, (ii) the Bengpal Group and (iii) the Sukma Group in the southern part of the craton, and Paleoproterozoic supracrustal groups are (i) the Sakoli Group, (ii) the Dongargarh Group and (iii) the Sausar Group in the northern part of the craton. The younger supracrustal rocks occur in two major Neoproterozoic basins, namely (1) the Chhattisgarh basin and (2) the Indravati basin within the Bastar craton.

The older gneiss supracrustals and schist belts are intruded by granitoid plutons of varying dimensions and also traversed by numerous quartz-chlorite veins (Ramchandra 2004) (Fig. 1b). The Proterozoic Pakhal Supergroup of rocks, Gondwana Supergroup of rocks and Deccan traps occur to the west of Bastar craton. Gneisses, granitoid, granulites, supracrustals (older and younger) and mafic igneous suites constitute the bulk of the geology of Bastar carton (Ramchandtra et al. 1995; Ramchandra 2004; Srivastava et al. 2010).

Taka area is located towards the western margin of Bastar craton (WBC) (Fig. 1b). The mafic–ultramafic rocks of Taka area belong to Sukma Group of Archaean age (Roy et al. 1996) and occur as enclaves within basement Amgaon gneiss (ca ~ 3.2 Ga, Ahmad et al. 2009). The other enclaves comprised of banded magnetite grunerite quartzite, amphibolites and pyroxenites with vanadiferous magnetite bands (Ganguly et al. 2013; Chandradas and Patel 2006). They are overlain in the western part, by the Phanerozoic Talchir and Barakar Formations of Gondwana Supergroup, cherty limestone of the Lameta Group and Deccan Trap (Fig. 2a). Gneisses and ultramafics are intruded by younger pegmatite and quartz veins. Evidence of brittle-ductile deformation is ubiquitous in all the above units (Meshram et al. 2014a, b, 2016).

In the field, serpentinite and talc-tremolite schist occur as small lensoidal bodies and enclaves within granite gneiss. The contact between enclaves and granite gneiss is sharp (Fig. 2b, c). They are brown to flesh coloured, medium to fine grained and mainly consist of serpentinized olivine. These are intensely traversed by siliceous, ferruginous and calcareous veins and show mesh texture. Siliceous, calcareous and ferruginous layers have been formed by secondary processes capping over serpentinite (birbirite) (Fig. 2d). The altered ultramafic bodies contain disseminated grains and also lenses of chromite. Chromite occurs in the

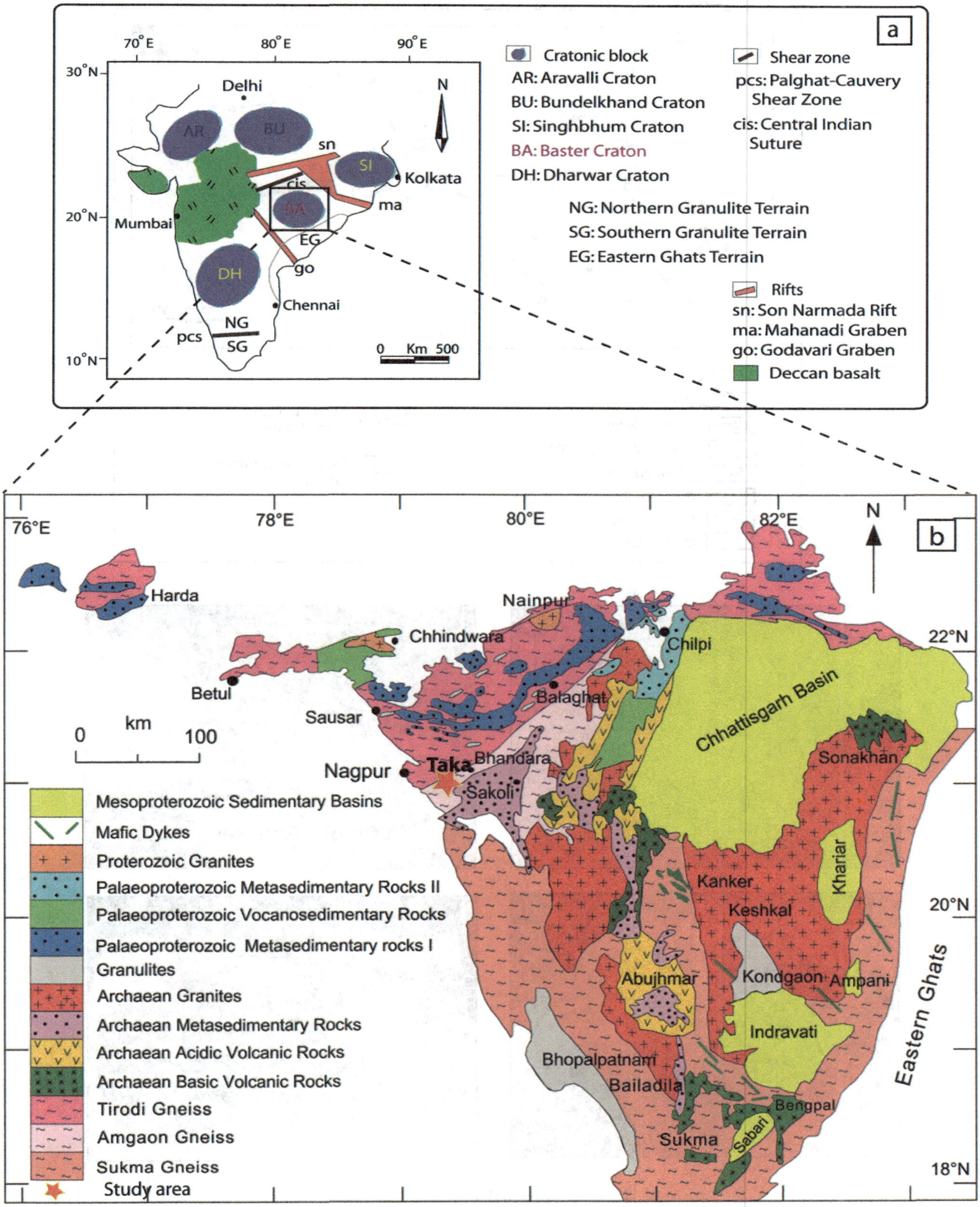

Fig. 1 **a** Regional geological map of India (After Naqvi and Rogers 1987). **b** Generalised geological map of Bastar Craton, Central India showing location of study area (After Mohanty 2015)

Figure-2

Fig. 2 **a** Geological map of Taka area (After Chandradas and Patel 2006; Meshram et al. 2014a); **b** field photograph showing enclave of talc tremolite schist within granite gneiss; **c** field photograph showing enclave serpentinite within granite gneiss; **d** Field photograph show mesh texture and siliceous veins in calcified serpentinite; **e** chrome-spinel cumulates in serpentinite. Note quartz- carbonate veins traversing the cumulate form of bands, pods, lenses, euhedral crystals and disseminations, besides minor magnetite (Fig. 2e). Large chromite grains show fracturing and are filled later by quartz-carbonate and serpentine veins. Talc-tremolite schist is composed of pyroxene with occasional amphibole and chlorite. Fine disseminated and black to brown coloured grains of chromite are also present in talc-tremolite schist.

3 Sampling and Analytical Methods

Fifty representative samples of both mafic and ultramafic bodies were collected and drill core samples were from the Regional Drill Core Repository, GSI, CR, Nagpur. Forty-five thin and polished sections were prepared and a petrographic study was carried out at Regional Petrology Laboratory at GSI, CR, Nagpur.

The mineral analysis has been carried out by powder XRD at Mineral Physics Division, GSI, NRO, Lucknow, on PANalytical X'pert Pro™ X-ray diffractometer (XRD) equipped with a copper target, operating at 45 kV and 40 mA. XRD patterns of random power have been recorded with a step size of 0.01° 2θ and a scan speed of 1° 2θ/min and scanned from 3°–70° 2θ. Mineral identification has been carried out using X'pert Highscore software using ICDD (PDF2 database). Relative semi-quantification of mineral phases has been carried out by Reference Intensity Ratio (RIR) method.

Mineral chemistry was determined by using *CAMECA SX 100* EPMA instrument at EPMA Lab, NCEGR, GSI, Bangalore and Kolkata. The analtical conditions used are accelerating current 15 kV, beam current 15 nA, beam size 1 µm and natural standards except for Ti, Cr and Al. The accuracy of each element is 0.05%. The raw electron microprobe data were processed by using probe software and ZAF corrections and are listed in Tables 1, 2 and 3.

The mafic–ultramafic rocks were analyzed for major oxide, trace element, rare earth elements (REE) and platinum-group elements (PGE) by XRF, ICP-MS and AAS. The major oxides were determined by PANalytical Magix 2424 XRF and trace element, REE and PGE were determined by ICPMS-ELAN-DRC methods at Chemical Division, GSI, Nagpur. PGE analysis for 5 nos. of samples was carried out using the nickel sulphide fire assay followed by ICPMS at Chemical Division, GSI, SR, Hyderabad. Details of chemical analysis are as per the Standard Operating Procedure (SOP) of GSI (source: www.portal.gsi.gov.in).

4 Results

4.1 Petrography

4.1.1 Serpentinite

Petrographic and x-ray diffraction studies indicate that the serpentinite is composed essentially of antigorite, lizardite and chrysotile with chromite, magnetite, secondary carbonate, chlorite and minor talc and sulphide minerals (Table 1). Serpentine is greyish and occurs in the form of fibrous, acicular, lamellar and platy crystals and as pseudomorphs with developments of talc in fine-grained serpentinite (Fig. 3a). Serpentinites are medium to coarse grained, and occurs in the form of mesh textured grains, which are pseudomorphs of olivine (Fig. 3b). Primary grain shapes and fractures are preserved in the same altered grains. However, no relict fresh olivine grains are found. Most are entirely serpentinized rocks showing well-developed mesh texture (Fig. 3c) and hourglass structures and bastite pseudomorphs. In the mesh textured serpentine, mesh rims are made up of fibrous serpentine (lizardite) and magnetite. Some of the mesh cores are made up of talc-chlorite or fine-grained silica-iron oxide matter.

Table 1 Mineralogical composition of Serpentinite, Talc-tremolite schist and Chromitite analysed by XRD

Sl. No	Sample Nos. and rock type	Constituents detected in semi-quantitative analysis of crystalline portion of samples by RIR values				
		Major	Minor	Good amount	Considerable amount	Trace amount
1	MNT-19-3 (Serpentinite)	Lizardite	Clinochrysotile	Calcite	Hematite	–
2	MNT-11-6 (Serpentinite)	Lizardite	Antigorite, Chrysotile, Calcite	Talc	Riebeckite, Hematite	Actinolite
3	MNT-11-3 (Talc-tremolite schist)	Edenite	Nimite	Cummingtonite, Hornblende	Talc	–
4	MNT-8-2 (Talc-tremolite schist)	Actinolite	Clinochlore, Nimite	Lizardite, Antigorite	Talc, Hematite, Calcite	Quartz
5	TK-4 (Chromitite)	Zinco-Chromite	Clinochlore, Edenite	Ni-chromite, Tremolite	Talc, Mg-Chromite, Riebeckite	Vermiculite

Table 2 Microprobe analysis of serpentine of representative samples (on the basis of 10 oxygen)

Sample nos	TK- 3 (5)				MNT-11/5							TK-7 (5)		TK-15	MNT-8/5					
Rock name	Serpentinite				Serpentinite							Chromitite		Chromitite	Serpentinite					
SiO_2	41.13	36.35	35.2	42.16	40.84	43.63	44.2	43.88	44.1	45.06	45	59.55	60.37	62.26	42.41	44	44.62	41.34	43.41	39.01
TiO_2	0.02	0.04	0.00	0.00	0.01	0.05	0.03	0.03	0.02	0.02	0.05	0.02	0.00	0.08	0.04	0.00	0.03	0.02	0.02	0.03
Al_2O_3	0.00	1.09	1.56	0.93	0.01	0.43	0.36	0.47	0.32	0.15	0.12	0.17	0.45	0.15	1.21	0.45	0.03	1.97	0.88	3.37
FeO	0.95	2.15	1.99	2.25	0.99	5.16	5.01	5.31	6.54	5.93	1.12	0.74	0.69	0.67	4.65	4.95	3.9	6.32	4.66	6.47
MnO	0.01	0.03	0.01	0.03	0.02	0.07	0.14	0.14	0.14	0.13	0.01	0.06	0.06	0.04	0.05	0.09	0.02	0.04	0.22	0.08
MgO	37.76	39.04	35.6	37.51	38.1	37.44	37.46	38.04	35.16	36.15	40.3	31.34	27.89	29.94	36.33	37.21	37.97	33.94	37.17	24
CaO	0.05	0.16	0.1	0.02	0.06	0.03	0.01	0.1	0.05	0.01	0.03	0.12	0.16	0.03	0.05	0.05	0.00	0.03	0.01	0.37
Na_2O	0.05	0.02	0.02	0.01	0.02	0.01	0.01	0.00	0.04	0.02	0.01	0.03	0.04	0.02	0.01	0.03	0.01	0.03	0.00	0.08
K_2O	0.00	0.02	0.03	0.02	0.05	0.00	0.02	0.01	0.01	0.01	0.01	0.01	0.07	0.02	0.01	0.01	0.00	0.02	0.01	0.07
Cr_2O_3	0.02	0.36	0.83	0.10	0.01	0.00	0.00	0.01	0.11	0.03	0.01	0.18	0.18	0.08	0.32	0.04	0.01	0.03	0.01	0.21
Total	79.97	79.21	75.3	83	80.08	86.82	87.2	87.95	86.45	87.47	86.65	92.2	89.92	93.3	85.08	86.8	86.51	83.73	86.37	73.69
Si	2.95	2.69	2.73	2.93	2.93	2.94	2.96	2.92	3.00	3.02	2.97	3.56	3.68	3.66	2.91	2.96	2.99	2.91	2.93	3.09
Al	0.00	0.09	0.14	0.08	0.00	0.03	0.03	0.04	0.03	0.01	0.01	0.01	0.03	0.01	0.10	0.04	0.00	0.16	0.07	0.32
Ti	0.00	0.00	0.00	0.00	0.00	0.00	0.00	0.00	0.00	0.00	0.00	0.00	0.00	0.00	0.00	0.00	0.00	0.00	0.00	0.00
Fe	0.06	0.13	0.13	0.13	0.06	0.29	0.28	0.30	0.37	0.33	0.06	0.04	0.04	0.03	0.27	0.28	0.22	0.37	0.26	0.43
Mn	0.00	0.00	0.00	0.00	0.00	0.00	0.01	0.01	0.01	0.01	0.00	0.00	0.00	0.00	0.00	0.01	0.00	0.00	0.01	0.01
Mg	4.03	4.30	4.11	3.88	4.07	3.76	3.74	3.78	3.56	3.61	3.97	2.79	2.53	2.62	3.72	3.73	3.79	3.56	3.75	2.84
Ca	0.00	0.01	0.01	0.00	0.00	0.00	0.00	0.01	0.00	0.00	0.00	0.01	0.01	0.00	0.00	0.00	0.00	0.00	0.00	0.03
Na	0.01	0.00	0.00	0.00	0.00	0.00	0.00	0.00	0.01	0.00	0.00	0.00	0.00	0.00	0.00	0.00	0.00	0.00	0.00	0.01
K	0.00	0.00	0.00	0.00	0.00	0.00	0.00	0.00	0.00	0.00	0.00	0.00	0.01	0.00	0.00	0.00	0.00	0.00	0.00	0.01
Cr	0.00	0.02	0.05	0.01	0.00	0.00	0.00	0.00	0.01	0.00	0.00	0.01	0.01	0.00	0.02	0.00	0.00	0.00	0.00	0.01
Total	7.05	7.26	7.18	7.03	7.07	7.04	7.02	7.06	6.99	6.98	7.02	6.43	6.31	6.34	7.03	7.02	7.01	7.01	7.03	6.75
Al +Cr	0.00	0.12	0.19	0.08	0.00	0.03	0.03	0.04	0.03	0.01	0.01	0.02	0.04	0.01	0.12	0.04	0.00	0.16	0.07	0.33
Fe(tot)+Mn+Zn +Mg +Ni	4.09	4.43	4.24	4.02	4.13	4.06	4.03	4.08	3.94	3.95	4.03	2.83	2.57	2.66	3.99	4.02	4.01	3.93	4.02	3.27

Table 3 Microprobe analysis of clinopyroxene of representative samples (on the basis of 6 oxygen)

Sample	MNT 8–1	MNT 8–1/1	TK 3 (5)	TK 3 (5)
Rock name	*Serpentinite*		*Serpentinite*	
SiO_2	57.57	55.67	57.82	57.6
TiO_2	0.00	0.17	0.01	0.05
Al_2O_3	0.01	1.62	0.09	0.23
FeO	3.22	5.22	6.01	6.03
Cr_2O_3	0.01	0.17	0.16	0.04
MnO	0.17	0.13	0.08	0.18
MgO	21.92	21.81	20.05	20.82
CaO	13.46	11.9	13.21	12.64
Na_2O	0.01	0.32	0.06	0.05
K_2O	0.00	0.05	0.02	0.02
Total	96.37	97.07	97.52	97.65
TSi	2.14	2.06	2.16	2.14
TAl	0.00	0.00	0.00	0.00
TFe^3	0.00	0.00	0.00	0.00
M1Al	0.00	0.07	0.004	0.01
M1Ti	0.00	0.01	0.00	0.00
$M1Fe^3$	0.00	0.00	0.00	0.00
$M1Fe^2$	0.00	0.00	0.00	0.00
M1Cr	0.00	0.01	0.01	0.00
M1Mg	1.00	0.92	0.991	0.987
M2Mg	0.22	0.28	0.12	0.16
$M2Fe^2$	0.10	0.16	0.19	0.19
M2Mn	0.005	0.00	0.00	0.01
M2Ca	0.536	0.47	0.53	0.50
M2Na	0.001	0.02	0.00	0.00
M2K	0.00	0.00	0.00	0.00
Sum Cation	4.00	4.00	4.00	4.00
Ca	28.88	25.63	28.80	27.21
Mg	65.44	65.37	60.83	62.36
Fe_2_Mn	5.68	9.00	10.37	10.44
WO	28.88	25.63	28.80	27.21
EN	65.44	65.37	60.83	62.36
FS	5.68	9.00	10.37	10.44
JD	0.078	2.434	0.471	0.388

The magnetite is released during serpentinization. Fine granular magnetite is concentrated in the cores and rims of mesh and hourglass structures. In the non-pseudomorphic variety, antigorite fibres and needles overgrow the relict olivine grains. Fractures are filled with cross-fibered chrysotile serpentine. Serpentine veins also cut across mesh rims and chromite grains.

Chlorite in the serpentinites is found associated with serpentine and in massive form. Tabular to platy crystals with green pleochroism and fine flaky aggregates are common varieties of chlorite. Fine aggregates are derived from tremolite due to alteration and further altered into fine dusty grains of talc. In serpentinite, it is formed from alteration of olivine and forms a rim around serpentine.

Chromite grains are highly fractured, and mesh rims are displaced. In coarse-grained serpentinite, olivine is entirely replaced by serpentine pseudomorphs. In case of complete replacement of olivine, serpentine crystal forms hourglass

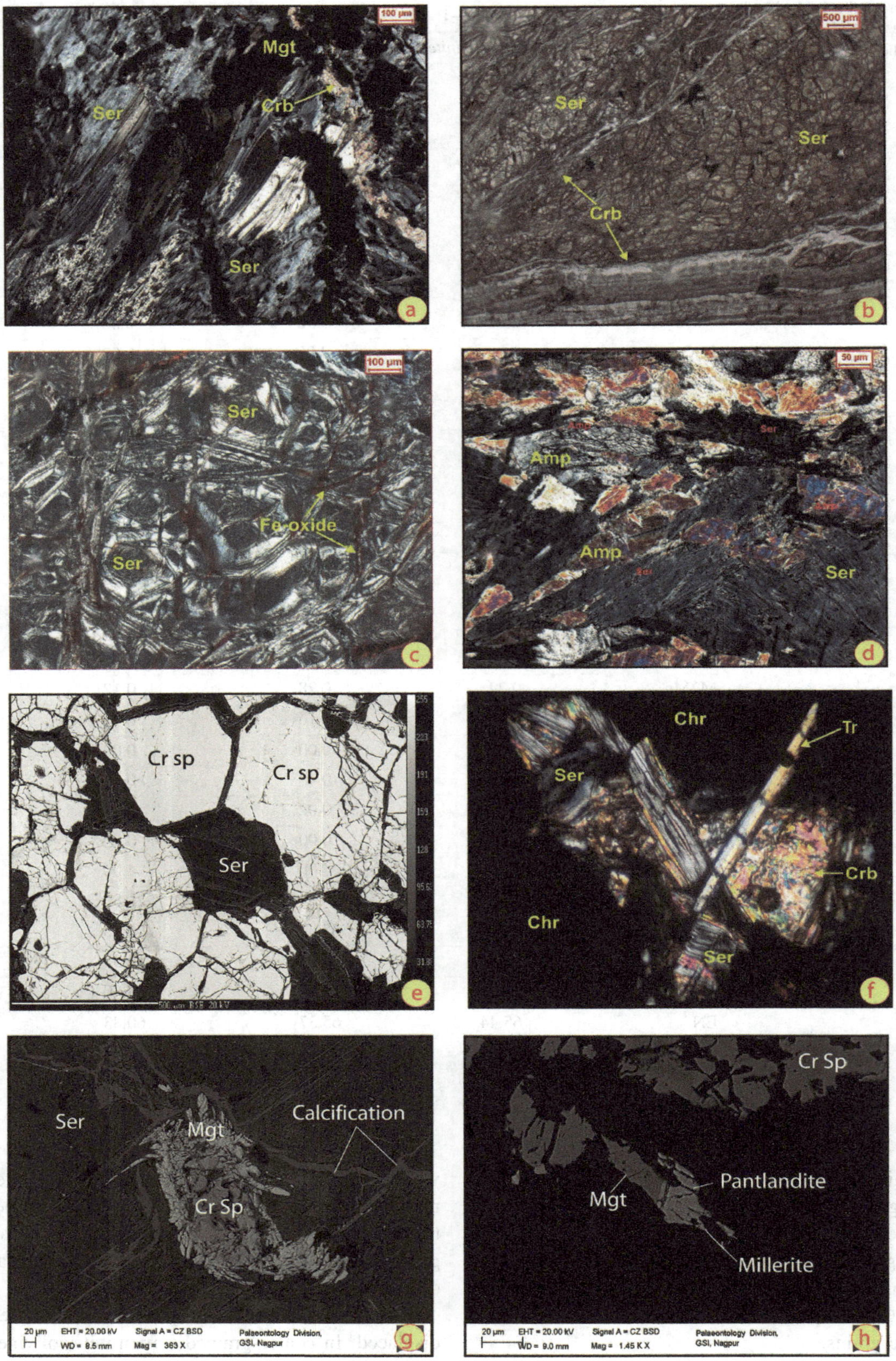

Fig. 3 **a** Lamellar sepentine showing foliation and crosscutting by magnetite and calcite veins; **b** Serpentine showing mesh texture, which is traversed by secondary carbonate veins and fine garined Chrysotile; **c** Serpintine showing mesh texture; **d** replacement of tremolite by serpentine; **e** Tremolite, serpentine and carbonate-talc matrix occurred as intercumulus mineral phases between chromite; **f** BSE image showing Cr-spinel cumulus texture; **g** BSE image showing Cr-spinel surrounded by magnetite, serpentine is traversed by secondary quartz and carbonate veins; **h** BSE image showing close association of magnetite and pentlandite. Abbreviations: Ser-Serpentine, Cpx-Clinopyroxene, Amp-Amphibole, Tr–Tremolite, Chl–Chlorite, Chr–Chromite, Crb–Carbonate, Cal–Calcite, Mgt–Magnetite, Qtz–Quartz

texture and zoning. Deformation and brecciation are noticed in the host rock. The occurrence of intense carbonate implies significant calcium mobility due to late-stage fluid activity. At places, serpentine occurs as platy and lamellar crystals indicating replacement of orthopyroxene (bastite) with magnetite trains along cleavage planes. It is also traversed by chrome spinel and sulphides bearing veins. Sulphide such as pentlandite and millerite occurs as disseminated grains and crystals which show mutual boundary texture with chromite crystals.

4.1.2 Talc-Tremolite Schist

Talc-tremolite schist is associated with serpentinite. It consists of talc, tremolite, chlorite, serpentine, carbonate, chromite and magnetite and is frequently traversed by quartz-carbonate veins. It shows prominent schistosity defined by layers of tremolite, chlorite and serpentine. Granular calcite aggregates and talc occur as fine dusty grains replacing serpentine minerals. Coarse grained and granular calcite shows pseudomorphism or replacement of serpentine, whereas fine-grained calcite aggregates show recrystallization. These features indicate the harzburgitic nature of protoliths. Platy chlorite shows schistosity with unaltered relict tremolite and fine dusty talc grains.

Tremolite occurs as colourless, elongated, prismatic and interlocking needles. Several discrete grains of magnetite are also observed and are possibly derived from tremolite during alteration. Quartz and calcite veins intruded possibly at the later stage. At places tremolite is replace by serpentine show ferruginous mass with mesh texture (Fig. 3d).

4.1.3 Chromitite

Chromitite is medium to coarse grained, entirely made up of chromite (Cr-spinel) crystals and layers along with tremolite, serpentine, chlorite and pyroxene in a minor amount as intercumulous phases. Cr-spinel is black to brown, equigranular, euhedral to subhedral and exhibits primary cumulus to intercumulus texture (Fig. 3e). These are formed due to early magmatic segregation during crystallization. The intercumulus spaces are occupied by platy serpentine (antigorite), amphibole (tremolite), chlorite, carbonate, cpx and talc matrix (Fig. 3f). Cr-spinel occurs as euhedral to hexagonal cumulate crystals (Fig. 3e) and defines layering in the groundmass of tremolite, talc and serpentine veins. Some of the grains are stretched and elongated, gives a planar fabric. At places, Cr-spinel is surrounded by magnetite and serpentine and is traversed by secondary quartz and carbonate veins (Fig. 3g).

Sulphide association is represented by pentlandite (Fig. 3h) millerite, chalcopyrite and pyrite. They occur as anhedral grains and are scattered. Meshram et al. (2014b) have reported Hollingworthite-Irarsite (IPGE mineral) in serpentinite of the Taka area.

4.2 Mineral Chemistry

The mineral chemistry of various mineral phases is carried out by EPMA analysis and mineral wise results are given below.

4.2.1 Serpentine

Serpentine minerals occur in three different forms namely pseudomorphs after olivine showing mesh texture, a platy habit with occasionally lamellar twinning as intercumulus mineral phase and fine fibrous clusters forming groundmass. The representative microprobe analysis of serpentine is given in Table 2. The analyzed samples show Si values in the moderate range from 2.69 to 3.09%; high Mg from 2.53 to 4.30%, Fe from 0.03 to 0.43%. The majority of studied samples plot in the Lizardite field and few plots in antigorite field in ternary atomic Mg-Si-Fe diagram after D'Antonio and Kristensen (2004) (Fig. 4a).

4.2.2 Clinopyroxene (Cpx)

In Talc-tremolite schist, a few serpentinites reveal the presence of Cpx as a relict phase. It occurs as euhedral to prismatic and subhedral crystals with prominent cleavage and twinning. However, some Cpx also occurs as intercumulus phase within chromitite. The microprobe analysis for representative samples is given in Table 3. The Si varies from 2.057 to 2.155%; Mg from 0.92 to 0.99% and Ca from 0.471 to 0.536%. The studied samples plot in the augite field in the ternary En-Wo-FS diagram (Morimoto et al. 1989) (Fig. 4b).

4.2.3 Amphibole

The representative microprobe analysis of amphiboles are given in Table 4. The amphibole composition shows high Si ranging from 6.203 to 8.078%, Mg ranging from 3.515 to 4.658% and Ca from 1.523 to 2.019%. Total Si versus Mg/(Mg+Fe_2) plot (Leake et al. 1997) indicates tremolite mineral chemistry while few samples plot transition between tremolite and actinolite field (Fig. 4c).

4.2.4 Chlorite

The microprobe analysis of representative chlorite is given in Table 5. The analyzed sample shows high Si ranging from 6.02 to 8.67%, Al from 0.88 to 3.96%, high Mg ranging from 6.22 to 9.75% and low Fe content varying from 0.14 to 2.83%. Atomic Si-(Al +Cr)–(Fe +Mn +Mg) diagram shows the composition of chlorite (Cressey et al. 2008), the studies samples reveals clinochlore type of chlorite (Fig. 4d).

4.2.5 Cr-Spinel

The representative microprobe analysis of Cr-spinel is given in Table 6. Some of the crystals of Cr-spinel are oxidized to magnetite and ferritchromite, but the cores of chromite

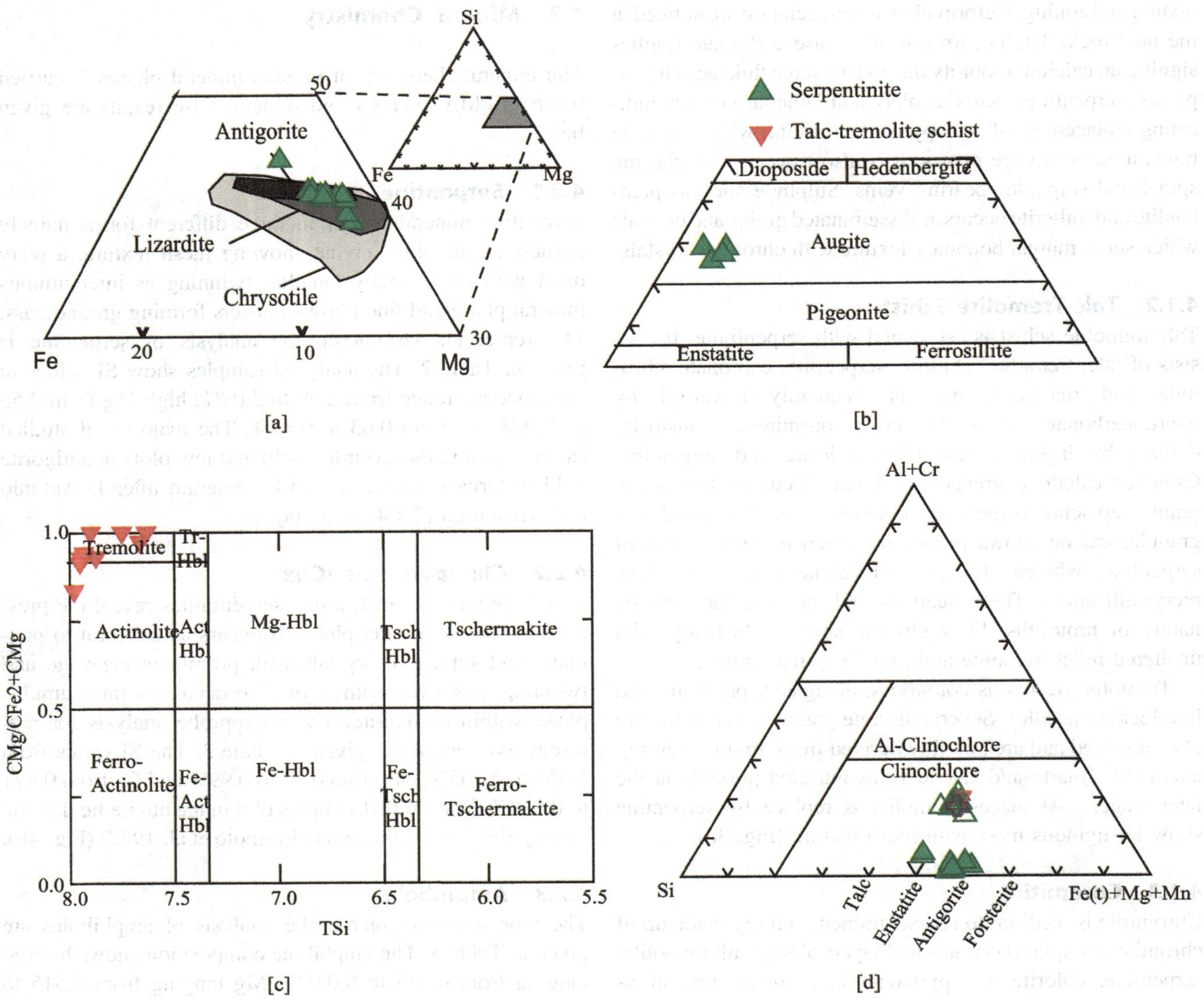

Fig. 4 **a** Fe–Si–Mg triplot for serpentine fields (after D'Antonio and Kristensen 2004); **b** En-Di-Hd-Fs Diagram for pyroxene (after Morimoto et al. 1989); **c** Si versus Mg/(Mg+Fe) diagram showing tremolite as major amphibole (after Leake et al. 1997); **d** Si–(Al +Cr)–(Fe +Mn +Mg) diagram showing composition of chlorite(unfilled symbol) and serpentine (filled symbol) (after Cressey et al. 2008)

grains retain their primary compositions. The spinel grains contain high Cr# from 0.70 to 0.94wt%, Mg# values range from 0.32 to 0.61wt %; TiO_2 from 0.04 to 0.23wt%; Fe^{3+} from 0.01 to 0.05wt%; high Al_2O_3 from 12.11 to 27.08wt% (Al from 0.46 to 1.06wt%).

Triangular Cr^{3+}–Fe^{3+}–Al^{3+} diagram represents the composition of spinel group minerals (Deer et al. 1992) and the spinel gap field is given by Barnes and Roeder (2001). The data of chromitite rich samples are plotted in the field of chromite to Al-rich chromite while Cr-spinel from serpentinite is plotted in picolite field (figure not shown). The composition of Cr-spinel is compared with that from the stratiform, ophiolitic and Ural-Alaskan chromitites in the Cr–Al–Fe^{3+} ternary diagram (Proenza et al. 2007). The composition of Taka Cr-spinel is plotted within ophiolitic chromite and slightly overlapping field of Ural-Alaskan type complexes (Fig.5a). In TiO_2 versus Cr# binary diagrams (Fig. 5b) data plots in boninites field and Island Arc tholeiitic (IAT) field (Arai 1992) and boninites and depleted peridotites field (Jan and Windley 1990). In Mg# versus Cr# binary diagram (After Stern et al. 2004) r-spinel data plotted in the overlapping field of podiform and stratiform spinel (Fig. 5c). The melt composition of Taka samples in Cr–Al–Fe^{3+} ternary diagram showing boninites and Mariana forearc peridities composition (Fig. 5d). Tectonic setting for chrome spinel is shown in Al_2O_3 versus TiO_2 diagram (Fig. 5e) and it clearly shows that the studied Cr-spinel plot is in the supra-subduction zone peridotites field (Kamenetsky et al. 2001).

Table 4 Microprobe analysis of amphibole of representative samples (on the basis of 23 oxygen)

Sample no	MNT 14-2	MNT 14-3	MNT 14-4	MNT 14-5	MNT 14-7	MNT 14-8	MNT 8-1	MNT 8-2	MNT 8-3	MNT 8-4	TK-13	TK-3	TK-4	TK-5
Rock type	TTS	SER	TTS	TTS	TTS	TTS	TTS	TTS	TTS	SER	SER	SER	SER	SER
SiO_2	56.49	57.68	50.57	55.71	56.99	56.95	57.57	56.37	56.42	55.67	55.8	55.19	57.82	57.6
TiO_2	0.01	0.01	0.01	0.04	0.05	0.05	0.00	0.01	0.01	0.17	0.01	0.07	0.01	0.05
Al_2O_3	1.09	0.22	0.21	0.4	0.22	0.13	0.01	0.19	0.04	1.62	0.24	1.70	0.09	0.23
FeO	4.47	4.68	3.59	5.43	4.43	4.15	3.22	6.9	7.7	5.22	0.78	7.21	6.01	6.03
Cr_2O_3	0.39	0.03	0.09	0.03	0.13	0.01	0.01	0.08	0.05	0.17	2.82	0.09	0.16	0.04
MnO	0.21	0.29	0.14	0.08	0.18	0.27	0.17	0.17	0.17	0.13	0.02	0.13	0.08	0.18
MgO	21.99	21.63	19.62	20.94	21.66	21.87	21.92	19.63	18.84	21.81	21.00	20.24	20.05	20.82
CaO	11.74	13.07	10.45	12.73	13.06	13.12	13.46	13.15	12.96	11.9	12.36	11.94	13.21	12.64
Na_2O	0.22	0.05	0.08	0.11	0.06	0.03	0.01	0.03	0.01	0.32	0.09	0.28	0.06	0.05
K_2O	0.07	0.01	0.00	0.00	0.02	0.03	0.00	0.01	0.01	0.05	0.01	0.10	0.02	0.00
TSi	7.75	7.95	7.90	7.87	7.93	7.93	8.04	7.98	8.05	7.63	8.03	7.66	8.08	7.95
Tal	0.18	0.04	0.04	0.07	0.04	0.02	0.00	0.02	0.00	0.26	0.00	0.28	0.00	0.04
TFe^3	0.08	0.02	0.07	0.06	0.04	0.05	0.00	0.00	0.00	0.11	0.00	0.06	0.00	0.01
Sum_T	8.00	8.00	8.00	8.00	8.00	8.00	8.04	8.00	8.05	8.00	8.03	8.00	8.08	8.00
Al	0.00	0.00	0.00	0.00	0.00	0.00	0.00	0.01	0.01	0.00	0.04	0.00	0.02	0.00
Cr	0.04	0.00	0.01	0.00	0.01	0.00	0.00	0.01	0.01	0.02	0.32	0.01	0.02	0.00
Fe^3	0.44	0.18	0.40	0.25	0.16	0.15	0.00	0.00	0.00	0.49	0.00	0.67	0.00	0.29
Ti	0.00	0.00	0.00	0.00	0.01	0.01	0.00	0.00	0.00	0.02	0.00	0.01	0.00	0.01
Mg	4.50	4.44	4.57	4.41	4.49	4.54	4.56	4.14	4.01	4.46	4.51	4.19	4.18	4.29
Fe^2	0.00	0.34	0.00	0.33	0.32	0.28	0.38	0.82	0.92	0.00	0.09	0.11	0.70	0.40
Mn	0.02	0.03	0.02	0.01	0.02	0.03	0.02	0.02	0.02	0.02	0.00	0.02	0.01	0.02
Ca	1.73	1.93	1.75	1.93	1.95	1.96	1.98	1.99	1.93	1.75	1.87	1.78	1.90	1.87
Na	0.06	0.01	0.02	0.03	0.02	0.01	0.00	0.01	0.00	0.09	0.03	0.08	0.02	0.01
Sum_cat	14.80	14.94	14.77	14.96	14.97	14.97	15.02	15.00	14.99	14.84	14.93	14.87	15.00	14.88

TTS–Talc-tremolite schist; SER–Serpentinite

4.3 Whole Rock Geochemistry

4.3.1 Major Oxides

The analyzed samples of serpentinite, and talc-tremolite schist show high MgO content varying from 37.17wt% to 7.41wt %; high FeO varying from 13.1wt% to 4.32wt% and high CaO ranging from 40.78 to 0.88wt%. SiO_2 ranges from 44.89 to 11.83wt% and Al_2O_3 vary from 6.75 to 4.72wt%. The analyzed drill core samples of serpentinite and talc-tremolite schist shows high MgO varying from 41.65 to 22.43wt%, high FeO varying from 17.01 to 6.45wt% and high CaO ranging from 14.27 to 1.96wt%. The SiO_2 is ranging from 41.62 to 25.27wt% and Al_2O_3 varying from 4.93 to 0.09wt% (Table 7).

The relationship between serpentinites and their protoliths in CaO–Al_2O_3–MgOwt% ternary diagram has been plotted. The studied samples show two different protoliths is indicated for dunite and hurzburgite, respectively, Li et al. (2004). In Al–(Fe_T +Ti)-Mg ternary diagram (Jensen 1976), the data of Taka serpentinite and talc-tremolite schist plotted in the komatiite field and chromitites in the basaltic komatiite field indicating komatitic nature of primary magma (Fig. 6).

Almost all samples have high loss-on ignition (LOI) values indicative of the extensive serpentinization of the samples. All samples show high # Mg values (0.723–0.947 wt. %) and low Al_2O_3 concentrations (0.09–2.71 wt. %) except sample no. TK-4, 5, 7, 8 and MNT-11-5, where very

Table 5 Microprobe analysis of chlorite of representative samples (on the basis of 28 oxygen)

Sample nos	TK-3 (5)									MNT-11/5	TK- 7(5)		TK-13	MNT-8/5	MNT-14/2				
Rock type	Serpentinite (SPT)												Chromitite	Serpentinite (SPT)					
SiO_2	29.7	31.5	31.1	32.14	30.88	31.82	32.15	30.11	28.43	32.08	30.72	30.89	33.71	39.01	32.23	32.94	32.07	32.03	31.89
TiO_2	0.08	0.22	0.97	0.20	0.18	0.13	0.11	0.08	0.12	0.15	0.05	0.05	0.04	0.03	0.19	0.13	0.23	0.09	0.17
Al_2O_3	14.87	14.91	14.4	15.14	14.33	14.59	13.38	14.7	15.89	12.39	14.65	14.84	12.59	3.37	14.6	13.74	14.68	14.87	14.11
FeO	15.27	16.68	15.75	15.34	15.72	15.63	14.95	2.05	1.49	5.85	1.75	1.53	0.9	6.47	10.14	11.42	11.49	10.78	10.64
MnO	0.13	0.23	0.14	0.05	0.07	0.06	0.14	-0.03	0.01	0.05	0.06	0.01	0.00	0.08	0.15	0.15	0.03	0.11	0.17
MgO	23.48	22.29	22.74	23.35	23.51	23.51	19.55	29.88	29.70	32.57	32.86	33.10	34.6	24.00	25.94	25.400	25.42	25.85	25.33
CaO	0.08	0.1	0.72	0.06	0.05	0.07	0.42	0.03	0.11	0.04	0.05	0.03	0.06	0.37	0.22	0.42	0.21	0.27	0.18
Na_2O	0.01	0.03	0.01	0.01	0.01	0.03	0.08	0.04	0.01	0.01	0.01	0.01	0.01	0.08	0.04	0.07	0.04	0.01	0.00
K_2O	0.53	0.3	0.21	0.65	0.68	1.03	0.93	0.02	0.02	0.34	0.01	0.01	0.01	0.07	0.12	0.17	0.12	0.06	0.16
Cr_2O_3	0.68	0.62	0.28	0.88	0.56	0.3	0.24	1.56	1.24	1.69	3.95	3.85	5.01	0.21	1.24	0.99	1.12	1.26	0.71
Total	84.83	86.88	86.32	87.82	85.99	87.17	81.95	78.44	77.02	85.17	84.11	84.32	86.93	73.69	84.87	85.43	85.41	85.33	83.36
Si	6.16	6.39	6.33	6.40	6.32	6.41	6.86	6.27	6.02	6.32	6.03	6.03	6.37	8.67	6.46	6.60	6.43	6.41	6.52
Al	3.64	3.56	3.45	3.55	3.46	3.47	3.36	3.61	3.96	2.88	3.39	3.42	2.81	0.88	3.45	3.25	3.47	3.51	3.40
Ti	0.01	0.03	0.15	0.03	0.03	0.02	0.02	0.01	0.02	0.02	0.01	0.01	0.01	0.01	0.03	0.02	0.03	0.01	0.03
Fe	2.65	2.83	2.68	2.56	2.69	2.63	2.67	0.36	0.26	0.96	0.29	0.25	0.14	1.20	1.70	1.91	1.93	1.80	1.82
Mn	0.02	0.04	0.02	0.01	0.01	0.01	0.03	0.00	0.00	0.01	0.01	0.00	0.00	0.02	0.03	0.03	0.01	0.02	0.03
Mg	7.26	6.74	6.90	6.93	7.17	7.06	6.22	9.27	9.37	9.57	9.61	9.64	9.75	7.95	7.75	7.59	7.60	7.71	7.72
Ca	0.02	0.02	0.16	0.01	0.01	0.02	0.10	0.01	0.02	0.01	0.01	0.01	0.01	0.09	0.05	0.09	0.05	0.06	0.04
Na	0.00	0.01	0.00	0.00	0.00	0.01	0.03	0.02	0.00	0.00	0.00	0.00	0.00	0.03	0.02	0.03	0.02	0.00	0.00
K	0.14	0.08	0.05	0.17	0.18	0.26	0.25	0.01	0.01	0.09	0.00	0.00	0.00	0.02	0.03	0.04	0.03	0.02	0.04
Cr	0.11	0.10	0.05	0.14	0.09	0.05	0.04	0.26	0.21	0.26	0.61	0.59	0.75	0.04	0.20	0.16	0.18	0.20	0.11
Total	20.02	19.79	19.80	19.81	19.97	19.95	19.57	19.80	19.88	20.13	19.97	19.96	19.85	18.90	19.71	19.71	19.73	19.74	19.72
Fe/(Fe+Mg)	0.27	0.30	0.28	0.27	0.27	0.27	0.30	0.04	0.03	0.09	0.03	0.03	0.01	0.13	0.18	0.20	0.20	0.19	0.19
Al+Cr	3.75	3.66	3.50	3.69	3.55	3.51	3.40	3.86	4.17	3.14	4.00	4.01	3.55	0.92	3.65	3.40	3.65	3.71	3.52
Fe(tot)+Mn+Zn+Mg+Ni	9.94	9.60	9.61	9.50	9.88	9.71	8.91	9.62	9.64	10.54	9.91	9.89	9.89	9.16	9.48	9.53	9.53	9.53	9.57

Table 6 Microprobe analysis of Cr-spinel of representative samples of serpentinite (on the basis of 4 oxygen)

Sample			TK-7							TK-15						TK-3				
Rock name			Serpentinite							Serpentinite						Serpentinite				
SiO_2	0.00	0.01	0.08	1.19	0.00	0.04	0.00	0.00	0.01	0.02	0.01	0.00	0.02	0.00	0.00	0.00	0.02	0.00	0.01	0.05
TiO_2	0.16	0.23	0.15	0.06	0.16	0.23	0.20	0.04	0.13	0.14	0.10	0.19	0.10	0.10	0.15	0.17	0.16	0.18	0.12	0.14
Al_2O_3	14.05	13.66	14.21	12.11	14.00	12.93	15.35	15.46	15.09	15.04	15.13	14.20	14.89	15.54	15.18	27.08	26.86	13.20	13.26	13.23
Cr_2O_3	57.70	57.48	58.06	59.34	57.47	55.81	55.10	55.18	55.33	55.03	55.42	55.37	53.36	53.72	53.56	34.90	35.30	57.41	57.55	57.54
FeO	21.40	21.45	22.65	16.04	21.42	23.28	24.44	24.04	24.00	23.45	22.87	22.85	21.78	20.68	21.00	16.63	15.24	15.82	16.03	16.55
MnO	0.03	0.11	0.21	0.92	0.12	0.24	0.10	0.12	0.10	0.20	0.11	0.34	0.19	0.06	0.50	0.09	0.11	0.08	0.19	0.11
MgO	8.68	8.48	7.52	10.94	8.61	6.77	6.42	6.68	6.57	6.61	7.30	6.95	8.22	9.18	8.65	11.05	11.99	12.11	12.11	11.48
CaO	0.00	0.00	0.00	0.00	0.00	0.01	0.01	0.00	0.00	0.00	0.01	0.00	0.00	0.02	0.01	0.04	0.01	0.01	0.00	0.00
Na_2O	0.01	0.00	0.00	0.03	0.00	0.02	0.03	0.01	0.00	0.01	0.05	0.00	0.03	0.02	0.04	0.02	0.02	0.01	0.00	0.00
K_2O	0.00	0.00	0.00	0.02	0.03	0.00	0.00	0.01	0.01	0.00	0.02	0.01	0.00	0.00	0.02	0.00	0.01	0.01	0.00	0.01
Total	102.03	101.42	102.88	100.66	101.81	99.32	101.65	101.54	101.24	100.50	101.01	99.91	98.58	99.32	99.10	89.98	89.72	98.83	99.27	99.11
O	4.00	4.00	4.00	4.00	4.00	4.00	4.00	4.00	4.00	4.00	4.00	4.00	4	4.00	4.00	4.00	4.00	4.00	4.00	4.00
Si	0.00	0.00	0.00	0.04	0.00	0.00	0.00	0.00	0.00	0.00	0.00	0.00	0.00	0.00	0.00	0.00	0.00	0.00	0.00	0.00
Ti	0.00	0.01	0.00	0.00	0.00	0.01	0.00	0.00	0.00	0.00	0.00	0.00	0.00	0.00	0.00	0.00	0.00	0.00	0.00	0.00
Al	0.53	0.52	0.54	0.46	0.53	0.51	0.59	0.59	0.58	0.58	0.58	0.55	0.58	0.60	0.59	1.06	1.05	0.51	0.51	0.51
Cr	1.47	1.47	1.47	1.50	1.47	1.48	1.42	1.42	1.43	1.43	1.43	1.45	1.40	1.39	1.39	0.92	0.93	1.48	1.47	1.48
Fe	0.58	0.58	0.61	0.43	0.58	0.65	0.67	0.65	0.66	0.64	0.62	0.63	0.60	0.57	0.58	0.46	0.42	0.43	0.43	0.45
Mn	0.00	0.00	0.01	0.03	0.00	0.01	0.00	0.00	0.00	0.01	0.00	0.01	0.01	0.00	0.01	0.00	0.00	0.00	0.01	0.00
Mg	0.42	0.41	0.36	0.52	0.41	0.34	0.31	0.32	0.32	0.32	0.35	0.34	0.41	0.45	0.42	0.55	0.59	0.59	0.58	0.56
Ca	0.00	0.00	0.00	0.00	0.00	0.00	0.00	0.00	0.00	0.00	0.00	0.00	0.00	0.00	0.00	0.00	0.00	0.00	0.00	0.00
Na	0.00	0.00	0.00	0.00	0.00	0.00	0.00	0.00	0.00	0.00	0.00	0.00	0.00	0.00	0.00	0.00	0.00	0.00	0.00	0.00
K	0.00	0.00	0.00	0.00	0.00	0.00	0.00	0.00	0.00	0.00	0.00	0.00	0.00	0.00	0.00	0.00	0.00	0.00	0.00	0.00
Total cation	3.00	3.00	2.99	2.98	3.00	3.00	2.99	2.99	2.99	2.99	3.00	2.99	3.01	3.00	3.01	3.00	3.01	3.01	3.01	3.00
Fe3+	0.00	0.00	0.00	0.00	0.00	0.00	0.00	0.00	0.00	0.00	0.00	0.00	0.01	0.01	0.01	0.01	0.02	0.01	0.02	0.01
Fe2+	0.58	0.58	0.61	0.43	0.58	0.65	0.67	0.65	0.66	0.64	0.62	0.63	0.59	0.55	0.56	0.45	0.41	0.42	0.42	0.44
XMg	0.42	0.41	0.37	0.55	0.42	0.34	0.32	0.33	0.33	0.33	0.36	0.35	0.40	0.44	0.42	0.54	0.58	0.58	0.57	0.55
XMg*	0.42	0.41	0.37	0.55	0.42	0.34	0.32	0.33	0.33	0.33	0.36	0.35	0.41	0.45	0.43	0.55	0.59	0.59	0.58	0.56
YCr	0.73	0.74	0.73	0.77	0.73	0.74	0.71	0.71	0.71	0.71	0.71	0.72	0.70	0.69	0.70	0.46	0.46	0.74	0.74	0.74
YAl	0.27	0.26	0.27	0.23	0.27	0.26	0.29	0.29	0.29	0.29	0.29	0.28	0.29	0.30	0.30	0.53	0.53	0.25	0.25	0.25
YFe3	0.00	0.00	0.00	0.00	0.00	0.00	0.00	0.00	0.00	0.00	0.00	0.00	0.01	0.01	0.01	0.01	0.01	0.01	0.01	0.00
Cr#	0.73	0.74	0.73	0.77	0.73	0.74	0.71	0.71	0.71	0.71	0.71	0.72	0.71	0.70	0.70	0.46	0.47	0.74	0.74	0.74

(continud)

Table 6 (continued)

Sample	TK-3																		
Rock name	Serpentinite																		
SiO_2	0.06	0.11	0.04	0.05	0.03	0.03	0.09	0.14	0.05	0.06	0.04	0.06	0.00	0.00	0.00	0.00	0.00	0.00	0.00
TiO_2	0.18	0.18	0.18	0.13	0.12	0.10	0.13	0.15	0.20	0.17	0.10	0.12	0.11	0.14	0.09	0.04	0.20	0.21	0.09
Al_2O_3	13.69	13.25	12.96	13.11	13.09	13.20	13.39	13.32	13.03	13.28	13.30	13.29	13.82	14.18	12.75	13.40	13.96	13.11	13.00
Cr_2O	56.44	56.84	57.19	58.56	57.63	57.57	56.30	57.07	58.18	57.04	56.94	57.47	56.44	55.52	56.26	56.81	56.27	58.06	56.81
FeO	17.78	18.12	17.81	18.04	18.14	17.67	18.16	17.99	18.53	18.06	18.52	18.74	17.09	17.26	18.20	18.07	17.11	18.77	18.15
MnO	0.36	0.44	0.34	0.10	0.40	0.46	0.29	0.05	0.30	0.13	0.27	0.17	0.22	0.26	0.36	0.23	0.33	0.34	0.27
MgO	12.68	12.58	12.31	12.53	12.09	12.33	12.44	12.35	12.00	12.34	12.32	12.16	12.56	13.25	11.62	12.28	12.99	12.31	12.44
CaO	0.00	0.00	0.00	0.01	0.00	0.00	0.00	0.04	0.03	0.00	0.00	0.00	0.00	0.00	0.00	0.00	0.00	0.00	0.00
Na_2O	0.03	0.05	0.10	0.00	0.04	0.04	0.02	0.00	0.05	0.00	0.00	0.00	0.02	0.00	0.00	0.00	0.00	0.00	0.00
K_2O	0.00	0.00	0.02	0.00	0.00	0.01	0.00	0.00	0.02	0.00	0.00	0.00	0.00	0.00	0.00	0.00	0.00	0.00	0.00
Total	101.22	101.57	100.97	102.57	101.54	101.41	100.85	101.11	102.39	101.27	101.49	102.01	100.26	100.61	99.52	100.83	100.86	102.80	100.89
O	4.00	4.00	4.00	4.00	4.00	4.00	4.00	4.00	4.00	4.00	4.00	4.00	4.00	4.00	4.00	4.00	4.00	4.00	4.00
Si	0.00	0.00	0.00	0.00	0.00	0.00	0.00	0.00	0.00	0.00	0.00	0.00	0.00	0.00	0.00	0.00	0.00	0.00	0.00
Ti	0.00	0.00	0.00	0.00	0.00	0.00	0.00	0.00	0.00	0.00	0.00	0.00	0.00	0.00	0.00	0.00	0.00	0.00	0.00
Al	0.51	0.50	0.49	0.49	0.49	0.50	0.51	0.50	0.49	0.50	0.50	0.50	0.52	0.53	0.49	0.51	0.52	0.49	0.49
Cr	1.42	1.43	1.45	1.46	1.45	1.45	1.43	1.44	1.46	1.44	1.44	1.44	1.43	1.40	1.45	1.44	1.42	1.45	1.44
Fe	0.47	0.48	0.48	0.48	0.48	0.47	0.49	0.48	0.49	0.48	0.49	0.50	0.46	0.46	0.50	0.48	0.46	0.50	0.49
Mn	0.01	0.01	0.01	0.00	0.01	0.01	0.01	0.00	0.01	0.00	0.01	0.00	0.01	0.01	0.01	0.01	0.01	0.01	0.01
Mg	0.60	0.60	0.59	0.59	0.58	0.59	0.59	0.59	0.57	0.59	0.59	0.58	0.60	0.63	0.57	0.59	0.62	0.58	0.60
Ca	0.00	0.00	0.00	0.00	0.00	0.00	0.00	0.00	0.00	0.00	0.00	0.00	0.00	0.00	0.00	0.00	0.00	0.00	0.00
Na	0.00	0.00	0.01	0.00	0.00	0.00	0.00	0.00	0.00	0.00	0.00	0.00	0.00	0.00	0.00	0.00	0.00	0.00	0.00
K	0.00	0.00	0.00	0.00	0.00	0.00	0.00	0.00	0.00	0.00	0.00	0.00	0.00	0.00	0.00	0.00	0.00	0.00	0.00
Total cation	3.03	3.03	3.03	3.02	3.02	3.02	3.03	3.02	3.02	3.02	3.03	3.02	3.02	3.03	3.02	3.03	3.03	3.03	3.03
Fe3+	0.06	0.06	0.05	0.05	0.05	0.05	0.06	0.05	0.05	0.05	0.06	0.05	0.04	0.07	0.05	0.05	0.06	0.06	0.06
Fe2+	0.41	0.42	0.42	0.43	0.44	0.42	0.42	0.43	0.44	0.43	0.43	0.44	0.41	0.39	0.44	0.43	0.40	0.44	0.42
XMg	0.56	0.55	0.55	0.55	0.54	0.55	0.55	0.55	0.54	0.55	0.54	0.54	0.57	0.58	0.53	0.55	0.57	0.54	0.55
XMg*	0.59	0.59	0.58	0.58	0.57	0.58	0.58	0.58	0.56	0.58	0.57	0.56	0.59	0.61	0.56	0.58	0.61	0.57	0.58
YCr	0.71	0.72	0.73	0.73	0.73	0.73	0.72	0.72	0.73	0.72	0.72	0.72	0.72	0.70	0.73	0.72	0.71	0.73	0.72
YAl	0.26	0.25	0.25	0.24	0.25	0.25	0.25	0.25	0.24	0.25	0.25	0.25	0.26	0.27	0.25	0.25	0.26	0.24	0.25
YFe3	0.03	0.03	0.03	0.02	0.02	0.02	0.03	0.03	0.02	0.03	0.03	0.03	0.02	0.03	0.03	0.03	0.03	0.03	0.03
Cr#	0.73	0.74	0.75	0.75	0.75	0.75	0.74	0.74	0.75	0.74	0.74	0.74	0.73	0.72	0.75	0.74	0.73	0.75	0.75

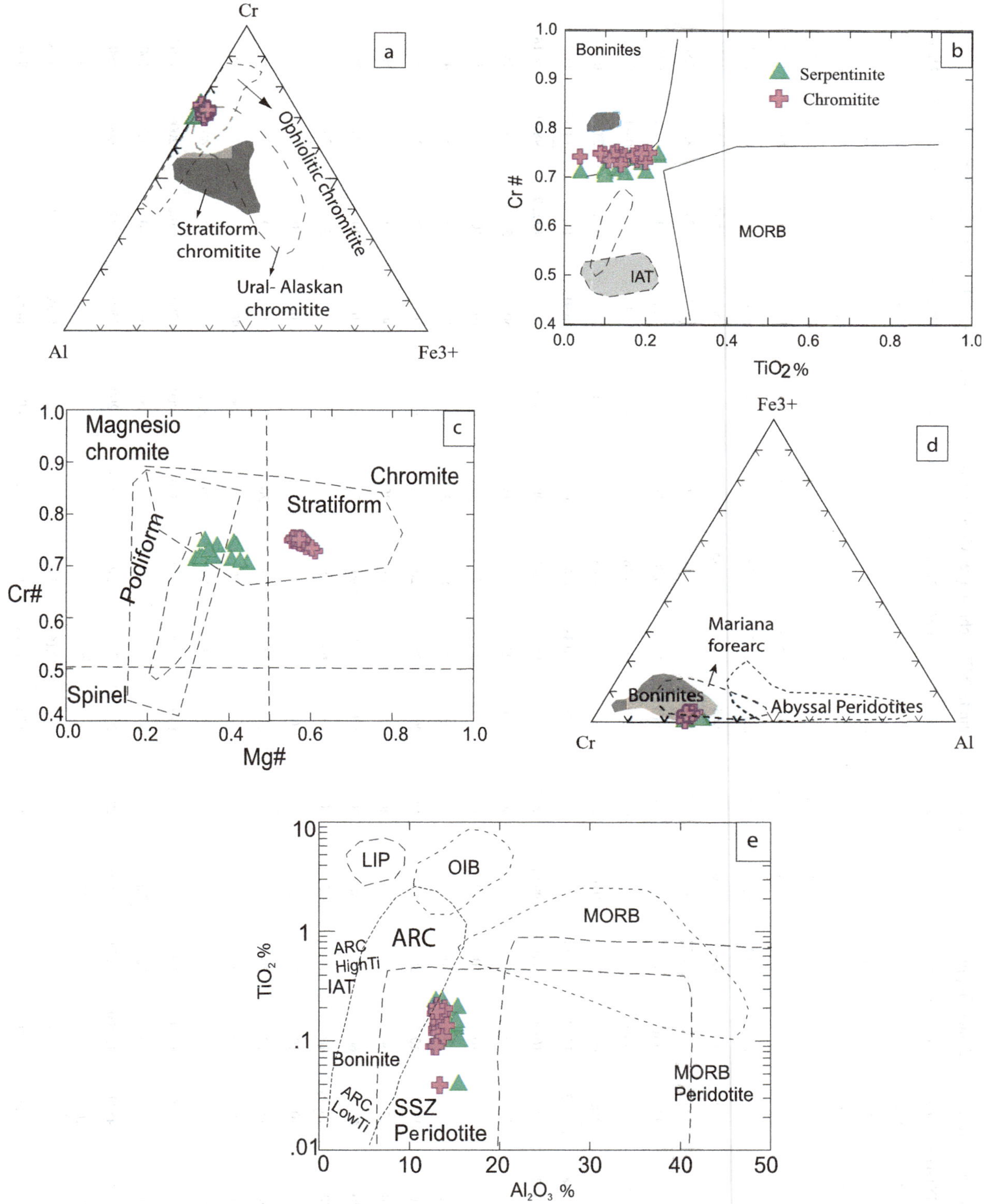

Figure-5

Fig. 5 **a** Cr–Al–Fe3+ ternary diagram, (Proenza et al. 2007) **b** TiO_2 versus #Cr diagram for Cr-spinel, (Arai 1992) showing boninitic composation of primary magma; **c** Mg# versus Cr# diagram for Cr-spinel (after Stern et al. 2004) showing podifrom-stratiform nature of Cr-Spinel; **d** Cr–Al–Fe3+ ternary diagram Cr–Al–$Fe3^{3+}$ ternary diagram, The field of Mariana forearc peridotites is from Ishii et al. (1992); Parkinson and Pearce (1998) Abyssal peridotites and bonninites are from Dick and Bullen (1984); Barnes and Roeder (2001) **e** Al_2O_3 versus TiO_2 diagram for Cr-spinel (after Kamenetsky et al. 2001) indicating supra-subduction zone setting for the formation of studies rocks

Table 7 Major oxide (in wt%), trace & REE data (in ppm) and PGE data(ppb) of serpentinite (SER), chromitite (CHRO) and talc tremolite schist (TTS) from Taka area

Sample no	MNT-11/5	MNT-11/6	TK-3 (5)	TK -6 (5)	TK-8 (5)	MNT 14/7	MNT 14/8	TK-4 (5)	MNT-8/2	MNT-11/3	MNT-14/3	MNT 14/10	MNT 14/11	MNT 14/12	MNT-8/7	MNT-8/4
Rock type	SER	SER	SER	SER	SER	SER	SER	CHRO	TTS	TTS	TTS	TTS	TTS	TTS	TTS	TTS
SiO_2	35.97	27.47	35.80	37.52	11.83	38.22	28.50	21.39	30.02	41.62	33.6	30.18	27.18	41.01	25.27	30.35
TiO_2	0.03	0.03	0.02	0.02	0.02	0.04	0.01	0.21	0.50	0.41	0.08	0.04	0.04	0.12	0.24	0.24
Al_2O_3	0.32	0.83	0.38	0.18	0.42	1.41	0.09	6.75	2.80	4.93	2.71	0.53	0.87	2.43	1.60	1.13
Fe_2O_3	9.41	10.65	7.79	3.80	4.32	7.25	7.43	4.91	12.35	17.01	6.45	6.97	7.08	7.10	13.01	13.2
MnO	0.12	0.15	0.04	0.09	0.53	0.11	0.07	0.41	0.22	0.30	0.13	0.07	0.11	0.11	0.17	0.14
MgO	36.21	38.43	37.17	34.31	7.41	34.09	39.46	9.68	27.58	22.43	27.26	41.65	35.94	33.78	35.70	36.96
CaO	2.70	1.96	4.81	9.17	40.78	7.16	8.25	2.09	14.27	6.81	13.44	5.19	12.49	5.39	8.00	4.86
Na_2O	0.00	0.10	0.10	0.10	0.10	0.10	0.10	0.10	0.00	0.00	0.00	0.10	0.10	0.10	0.00	0.00
K_2O	0.01	0.10	0.10	0.10	0.10	0.10	0.10	0.10	0.00	0.16	0.00	0.10	0.10	0.10	0.00	0.00
P_2O_5	0.01	0.01	0.03	0.02	0.03	0.01	0.02	0.02	0.04	0.03	0.01	0.01	0.02	0.01	0.03	0.03
LOI	11.84	18.48	15.43	15.81	35.14	11.12	15.61	1.48	10.33	4.61	14.42	15.21	16.91	10.14	14.09	11.27
Total	96.62	98.21	101.67	101.12	100.68	99.61	99.64	47.14	98.11	98.31	98.1	100.05	100.84	100.29	98.11	98.18
Ba	45	50	50	50	50	50	50	140	45	45	45	50	50	50	45	45
Nb	3	10	10	10	10	10	10	10	5	5	3	10	10	10	3	3
Rb	6	10	10	10	10	10	10	10	5	8	3	10	10	10	5	4
Sc	3	10	10	10	20	10	10	20	25	20	15	10	10	10	15	5
Sr	10	10	22	30	40	20	40	10	30	15	25	30	40	20	20	10
V	20	20	60	60	100	40	20	560	110	90	70	20	20	40	70	75
Zr	20	20	10	10	10	25	10	10	60	45	20	35	25	30	35	40
Y	5	10	10	10	10	10	10	10	15	15	10	10	10	10	10	10
Cu	30	50	15	10	20	15	15	20	55	15	20	10	20	20	20	35
Pb	40	60	50	30	50	40	80	40	25	20	40	100	90	40	25	25
Zn	60	150	70	110	60	125	150	60	30	70	20	225	210	60	40	35
Ni	50	1900	2200	1800	550	1100	1900	1000	1200	600	400	1400	1200	1100	2000	1800
Co	25	110	125	90	50	70	110	60	90	75	30	100	60	60	120	100
Cr	1130	3400	4500	15,000	3200	4500	2900	299,100	860	2400	2500	2200	3500	5000	600	590
Rb	6	20	10	9	9	25	9	9	5	8	3	35	25	30	5	4
La	3.38	1.80	1.02	1.00	2.61	1.87	1.55	1.56	9.16	5.67	4.48	3.09	11.06	3.79	7.48	4.74
Ce	3.27	2.10	1.29	2.00	4.00	2.40	1.17	3.75	17.18	12.62	6.28	4.04	14.57	7.29	12.45	9.48
Pr	0.32	0.23	0.15	0.10	0.24	0.30	0.10	0.56	2.24	1.99	0.69	0.52	1.90	0.86	1.41	1.25
Nd	1.28	0.97	0.55	0.30	1.15	1.28	0.30	2.40	9.60	9.31	2.50	1.90	7.06	3.29	5.59	5.06

(continued)

Table 7 (continued)

Sample no	MNT-11/5	MNT-11/6	TK-3 (5)	TK -6 (5)	TK-8 (5)	MNT 14/7	MNT 14/8	TK-4 (5)	MNT-8/2	MNT-11/3	MNT-14/3	MNT 14/10	MNT 14/11	MNT 14/12	MNT-8/7	MNT-8/4
Eu	0.03	0.02	0.03	0.02	0.07	0.12	0.05	0.10	0.57	0.49	0.03	0.06	0.21	0.12	0.15	0.11
Sm	0.17	0.11	0.10	0.10	0.17	0.21	0.10	0.59	2.12	2.15	0.36	0.33	1.35	0.57	1.07	0.97
Tb	0.02	0.02	0.02	0.01	0.03	0.04	0.01	0.10	0.37	0.37	0.06	0.06	0.19	0.09	0.19	0.16
Gd	0.17	0.11	0.11	0.05	0.21	0.22	0.05	0.58	2.31	2.27	0.37	0.38	1.21	0.56	1.16	1.04
Dy	0.08	0.09	0.09	0.05	0.17	0.27	0.03	0.71	2.21	2.24	0.38	0.36	0.96	0.62	1.13	0.96
Ho	0.02	0.02	0.02	0.01	0.04	0.06	0.01	0.15	0.41	0.43	0.08	0.07	0.18	0.13	0.22	0.18
Er	0.06	0.05	0.07	0.03	0.13	0.19	0.03	0.40	1.14	1.26	0.28	0.20	0.50	0.43	0.59	0.52
Tm	0.01	0.01	0.01	0.01	0.02	0.03	0.01	0.07	0.16	0.20	0.05	0.04	0.08	0.08	0.09	0.07
Yb	0.08	0.10	0.12	0.04	0.15	0.27	0.07	0.43	1.00	1.20	0.32	0.26	0.53	0.52	0.54	0.46
Lu	0.01	0.02	0.02	0.01	0.03	0.05	0.01	0.08	0.15	0.20	0.06	0.04	0.08	0.08	0.09	0.08
Pt	6.0	6.0	5.0	n.d	n.d	n.d	n.d	n.d	n.d	5.0	n.d	n.d	n.d	n.d	n.d	5.0
Pd	7.0	9.0	5.0	n.d	n.d	n.d	n.d	n.d	n.d	7.0	n.d	n.d	n.d	n.d	n.d	5.0
Ir	5.0	4.0	4.0	n.d	n.d	n.d	n.d	n.d	n.d	3.0	n.d	n.d	n.d	n.d	n.d	7.0
Ru	19.0	11.0	11.0	n.d	n.d	n.d	n.d	n.d	n.d	3.0	n.d	n.d	n.d	n.d	n.d	5.0
Rh	5.0	4.0	3.0	n.d	n.d	n.d	n.d	n.d	n.d	3.0	n.d	n.d	n.d	n.d	n.d	3.0

n.d.: not determined

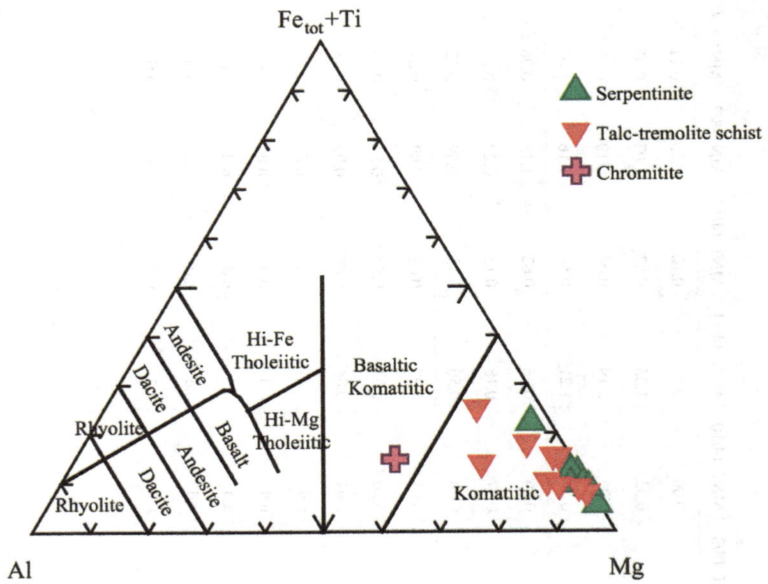

Fig. 6 Al–(Fe$_t$ +Ti)–Mg diagram showing komatiitic nature of magma (after Jensen 1976)

abnormal high Si or Ca value is observed because of profuse silicification and calcification in these samples.

4.3.2 Trace Elements

The trace element data are given in (Table 7). Serpentinites from Taka areas are enriched in Cr, Ni, V, Sc and Co and show a lower concentration of Zr, Y, Nb, Pb and Rb. These higher values are comparable to primitive mantle values, Sun and McDonough (1989). The Cr concentration of serpentinite samples is relatively higher (up to 29.91wt%), which could be due to the presence of Cr-spinel. The Ni concentration also showed a similar enrichment (up to 2000 ppm). Zr and Y also have moderate concentrations. Cr and Ni concentrations are high, ranging from 590 to 15,000 ppm and 600 to 2200 ppm, respectively, and three cumulate samples show very high Cr ranging from 24 to 48wt%.

It is well-established that serpentinization during seawater interaction results in enrichment in large ion lithophile elements (LILE) and LREE as well as carbonate hosted trace elements (Pb and Sr) when accompanied by carbonation (Kodolányi et al. 2012; Marchesi et al. 2013). We note that Sr is lost in these serpentinites, which could either be due to alteration or because plagioclase phenocrysts retained Sr in the magmas. Significant enrichment in mobile elements (viz. Pb, Sr, Ba) is observed due to the high concentration of these elements both in seawater and derived fluid the they are showing strong positive anomoly in the primitive mantle normalized diagram (Fig. 7a, b). Positive Y anomaly can be explained by its fractionation in the aqueous system due to chemical complexities and difference in electron configuration between Y and Ho, Sun and McDonough (1989).

4.3.3 Rare Earth Elements

The analyzed rocks display two different normalized REE patterns (Fig. 7c, d): a flat pattern with low HREE, and fractionated LREE depleted pattern $(La/Sm)_{cn}$ = 4.00–52.53 with mildly depleted HREE and Eu/Eu* = 0.54–1.13, Sun and McDonough (1989). The second group displays a relatively enriched fractionated LREE and depleted HREE pattern with negfv

ative Eu anomaly (Eu/Eu* = 0.22–0.78).

4.3.4 PGE Geochemistry

The PGE (Pt, Pd, Rh, Ru and Ir) analysis of 5 nos. of representative samples was carried out at Chemical Division, GSI, SR, Hyderabad and is given in Table 7. The studied samples have low PGE concentrations. The PGE normalized diagram (Fig. 8a) indicates that IPGE (Ir, Ru) is relatively higher than PPGE (Pd, Pt, Rh) in respect of chondrite value, Palme and Jones (2003). In Ir versus Ir/(Pt +Pd) diagram (Fig. 8b) serpentinite and talc tremolite schist samples are plotted in mantle peridotitic field. In Ni/Cu versus Pd/Rh diagram (Fig. 8c), samples plot in chromites field, while in Ni/Cu versus Pd/Ir diagram (Fig. 8d), samples plot in komatiite field. Moreover, the IPGE enrichment relative to PPGE and the relatively low Pd/Ir values of the serpentinite from Taka area suggests that their parental melts were produced by melting of a moderately depleted peridotite similar to described for other IPGE rich chromitite originated in back

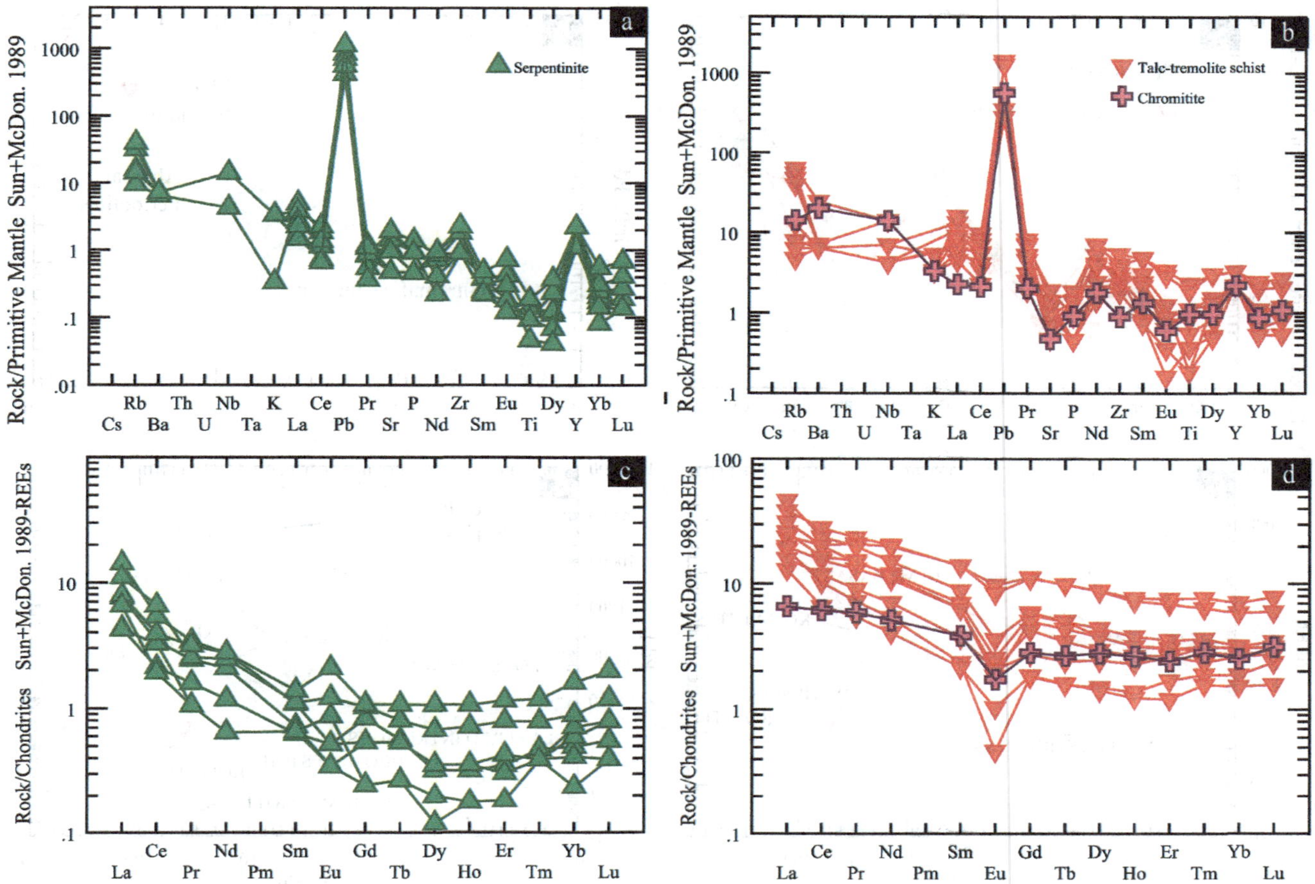

Fig. 7 **a** Trace element spider diagram for serpentinite; data normalised to primetive mantle (Sun and McDonough 1989); **b** Trace element spider diagram for talc-tremolite schist and chromitite, data normalised to primetive mantle (Sun and McDonough 1989); **c** REE spider diagram for serpentinite, data normalised to chondrite (Sun and McDonough 1989); **d** REE spider diagram for talc-tremolite schist and chromitite, data normalised to chondrite (Sun and McDonough 1989)

arc basins in supra-subduction zones (Gervilla et al. 2005; Kapsiotis et al. 2009).

5 Discussions

5.1 Implications of Serpentinite Textures

The serpentinite and talc-tremolite schist from Taka area are entirely altered, carbonated and silicified. The mineralogical assemblage suggests that the original protoliths were dunite and harzburgite. They have been serpentinized by talc-carbonate alteration. The serpentine was formed by the consumption of all the primary olivine/clinopyroxene (i.e. complete serpentinization) and, later the carbonate solution would react with the serpentine to produce talc ± carbonate.

The composition, mineralogy and textures of studied serpentinites indicate different processes of serpentinization. Serpentine formed by alteration of olivine showing mesh textures. The common presence of pseudomorphic pyroxene and olivine (hourglass texture and zoning), supports the harzburgites protolith for these rocks.

Talc-tremolite schist (tremolite rich rock) associated with the serpentinites indicates that the clinopyroxene may also have been altered from lherzolitic protolith, Frost and Beard (2007). Interpenetrative and interlocking textures, which are interpreted as having formed by retrograde metamorphism during exhumation of a previously existing non-pseudomorphic texture, are composed of prograde antigorite. Abundant magnetites were produced in a late-stage hydrothermal activity.

5.2 Parental Melts Composition Using Cr-Spinel Chemistry

The primary compositions of chrome spinels provide important information to constrain the petrogenesis of

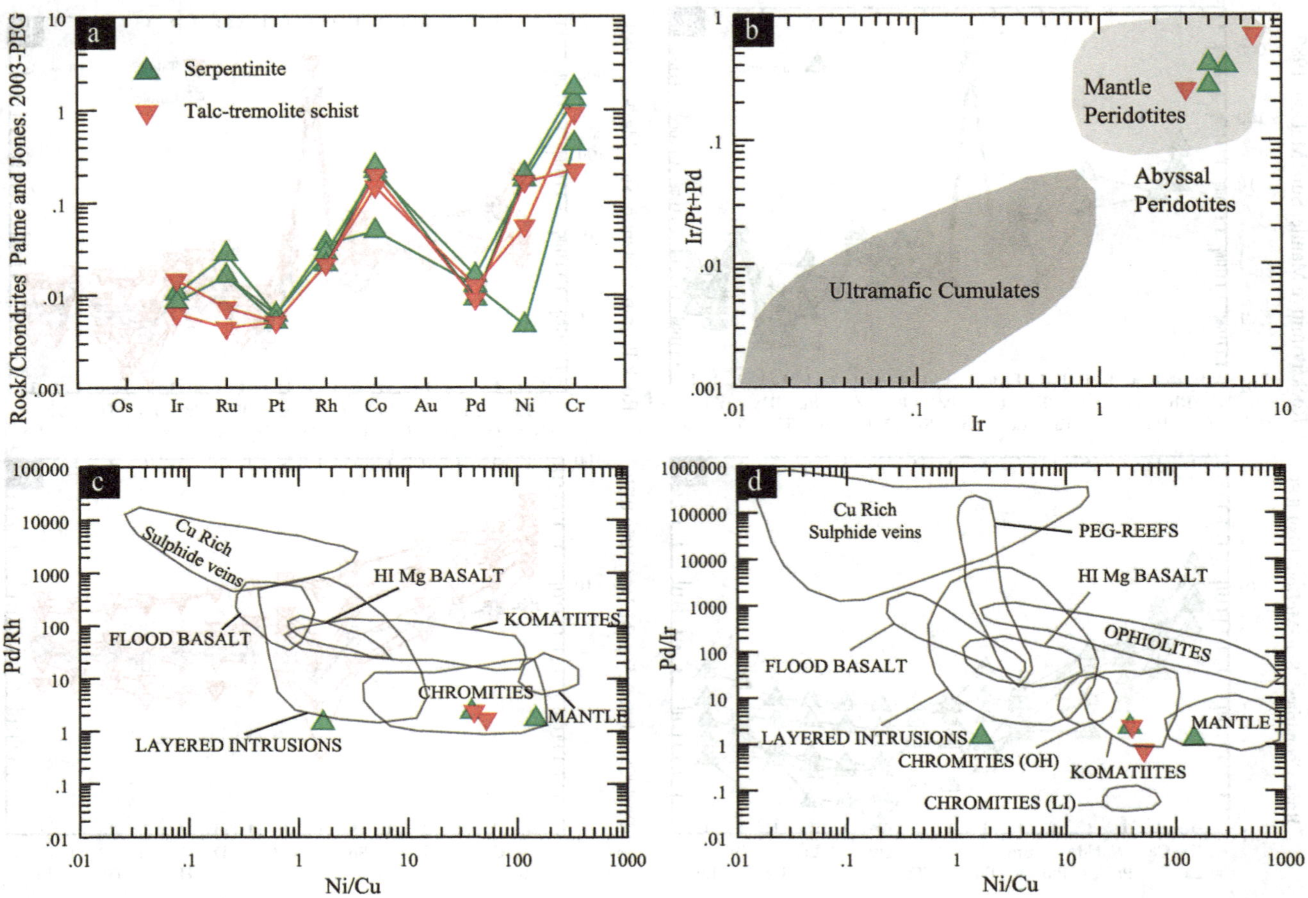

Fig. 8 **a** Chondrite normalised PGE diagram, normalization values are (after Palme and Jones 2003); **b** Ir versus Ir/(Pt +Pd) diagram for studied rocks; fields are (after Hattori and Guillot 2007); **c** Ni/Cu versus Pd/Rh and **d** Ni/Cu versus Pd/I diagram for studied rocks; fields are (after Barnes et al. 1985)

magmatic processes within the different geodynamic environments (Pal 2011; Gonzalez-Jimenez et al. 2009; Rollinson 2008; Arai et al. 2004; Kamenetsky et al. 2001). It is generally accepted that parental melts for chromitites are the result of the melt–rock interactions or melt–melt interactions previously depleted in mantle sources in supra-subduction zone settings (Uysal et al. 2005; Kamenetsky et al. 2001; Melcher et al.1999). Partial melting of source rock with high content of Cr-spinel suggest high-Mg tholeiitic or boninitic magma composition (Fisk 1986; Kelemen et al.1992; Arai 1994).

The chemistry of the parental melt for studied Cr-spinel/chromite can be gained using the equation of Maurel and Maurel (1982), namely $[(Al_2O_3wt\%)_{sp} = 0.035\ (Al_2O_3wt\%)_{Liquid}\ 2.42]$. The result shows that melt had an average Al_2O_3 content of 13.37% (Table 8). This value is similar to the mid ocean ridge and back arc basin basalts (Wilson 1989; Fryer et al. 1990).

Table 8 Average Al_2O_3 (wt %) contents and FeO/MgO ratio of melts in equilibrium with the Cr-spinel in serpentinite from Taka area (after Maurel and Maurel 1982)

	Al_2O_3(1)	FeO/MgO(2)	References
Taka Serpentinite	13.37	11.30	This Study
Tehuitzingo Serpentinite	13.74	0.93	Gonzalez-Mancera et al. (2009)
Tehuitzingo Chromitites	15.30	–	Proenza et al.(2004)
Boninite	10.64–14.40	0.7–1.4	Wilson (1989)
MOBR	16.00	1.2–1.6	Wilson (1989)
Back- arc basin basalts	>16.00	–	Fryer et al.(1990)

Table 9 Degree of partial melting modeled for peridotitic mantle, Taka area and other lithotectonic environments (after Hellebrand et al. 2001)

	Cr#	Partial melting	References
Taka Serpentinite	13.37	11.30	This Study
Tehuitzingo Serpentinite	13.74	0.93	Gonzalez-Mancera et al. (2009)
Supracrustal Zone	15.30	–	Pearce et al. (2000), Mellini et al. (2005)
Passive margin	10.64–14.40	0.7–1.4	Pearce et al. (2000);
Abyssal	16.00	1.2–1.6	Dick and Bullen (1984)

The degree of partial melting (F) based on the empirical equation proposed by Hellebrand et al. (2001) [i.e. $F = 10 \ln (Cr\#) + 24$], suggest partial melting up to 15–40%, which is within the range of peridotites from supra-subduction zones (Table 9).

High Cr# values confirm the refractory nature of the protoliths of serpentinites because Al is preferentially incorporated into the melt during partial melting. Parental magma compositions of this high chrome spinel specify the reaction between boninitic melts and depleted peridotites for Taka serpentinite that generated due to partial melting of the deep mantle (Rollinson 2008; Gonzalez-Jimenez et al. 2011). An increase in the TiO_2 contents indicates an interaction of parental melt and Ti–rich melts (Pearce et al. 2000).

5.3 Chemistry of Cr-Spinel and Tectonic Implication

The chemical composition of Cr-spinel can be used to deduce the origin of the parental magma and tectonic settings (Kamenetsky et al. 2001; Page and Barnes 2009). Cr-spinel with high Cr# (>0.6) and low TiO_2 wt. % (<0.3) contents are probably formed in arc-related tectonic settings (Arai et al. 2006), whereas chrome spinels formed in MORB settings generally have Cr# <0.6 (Arai et al. 2006, Arai and Ishimaru 2011). The Cr-spinels from Taka area contain high Cr# from 0.70 to 0.94wt% (i.e. >0.6) and TiO_2 from 0.04 to 0.23wt% (<0.3). On the TiO_2 versus Al_2O_3 plot (Fig. 5c) chrome spinel plot in arc-related tectonic field in supra-subduction zone, whereas few Al-rich data plot in MORB typesetting.

5.4 Mineralization Potential

The Cr concentrations in serpentinite samples are relatively higher, they range from 590 ppm to 29.91wt % and Ni concentration also showed a similar enrichment from 550 ppm to 0.22wt%. High Cr and Ni values are significant from a mineralization point of view, but the outcrops are scanty in the field. PGE and Au concentration is low and not encouraging. A low concentration of PGE value is observed (Pt is <10 ppb and Pd is <5 ppb; Ir varies from 3 to 7 ppb; Ru from 3 to 19 ppb and Rh from 3 to 5 ppb) (Table 7).

On the surface samples, there are no indications of sulphide mineralization, but in the borehole samples, occurrence of pyrite, pyrrhotite, pentlandite ($(Fe, Ni)_9S_8$) and Millerite(NiS) were observed along with Fe, Co, Ni and PGE bearing arsenide (i.e. holligworthite-Irasite).

The crystallization of Platinum Group Mineral (PGM) progresses along with the formation of different types of mineral assemblages depending upon the prevailing sulphur saturation conditions in time and space (Naldrett 2010). The serpentinisation process lowers both $fO2$ and $fS2$ giving place to partial desulfurization of primary magmatic sulphides and forming assemblages are characterized as S-poor sulphides, i.e. pyrrhotite and Fe, Co, Ni and PGE bearing arsenide, Meshram et al. (2014b).

6 Conclusions

On the basis of combined mineralogical, petrological and whole rock geochemical studies, it is suggested that serpentinite and talc-tremolite schist were formed from dunite, harzburgitic and lherzolitic protolith, respectively. Euhedral to cumulus Cr-spinel reveals early primary magmatic crystallization and later effect of deformations. High Cr# and low TiO_2 contents in Cr spinel composition indicate arc related tectonic setting for their genesis.

Parental magma compositions of this high Cr-spinel suggest the reaction between boninitic melts and depleted peridotites for Taka serpentinite that generated in partial melting in the mantle. High Cr # indicates, high degree of partial melting of mantle peridotite. Cr-spinel crystallized from boninitic melt at a high degree of partial melting generated in the supra-subduction zone setting.

The IPGE enrichment relative to PPGE and the relatively low Pd/Ir values of the Taka area suggests that their parental melts were produced by melting of a moderately depleted peridotite similar to described for other IPGE rich chromitites originated in back arc basins in supra-subduction zones tectonic setting.

Acknowledgements The authors are grateful to the corresponding editor Prof. Vinod K. Singh and both the reviewers Prof. B.C. Prabhakar, Bangalore University, Bangalore, India, and Dr. Alok Kumar, Banaras Hindu University, Varanasi, India, for their very critical but constructive comments that were most helpful during the revision of the paper. The present study is part of a research project of field season program (FS 2013-14) of Geological Survey of India, Central Region, an attached office to Ministry of Mines, Government of India. The authors are thankful to the DDG and HoD Central Region, Nagpur, for providing all necessary logistics and support for carrying out fieldwork and kind permission to collect drill core samples from the Regional Drill Core Repository, GSI, CR, Nagpur. VPM is thankful to Smt. Varsha A. Agalawe, DDG and HoD, GSI, Northern Region, Lucknow for providing infrastructural support for writing and giving permission to publish this work.

References

Ahmad T, Kaulina TV, Wanjari N et al (2009) U–Pb zircon chronology and Sm–Nd isotopic characteristics of the Amgaon and Tirodi Gneissic Complex, Central Indian shield: constraints on Precambrian crustal evolution. In: Precambrian continental growth and tectonism, Excel India Publishers New Delhi, pp 137–138

Arai S (1992) Chemistry of chromian spinel in volcanic rocks as a potential guide to magma chemistry. Mineral Mag 56:173–184

Arai S (1994) Compositional variation of olivine chromian spinel in Mg-rich magmas as a guide to their residual spinel peridotites. J Volcanol Geotherm Res 59:279–294

Arai S, Uesugi J, Ahmed AH (2004) Upper crustal podiform chromitite from the Northern Oman ophiolite as the stratigraphically shallowest chromitite in ophiolite and its implication for Cr concentration. Contrib Mineral Petrol 147:145–154

Arai S, Kadoshima K, Morishita T (2006) Widespread arc-related melting in the mantle section of the northern Oman ophiolite as inferred from detrital chromian spinels. J Geol Soc Lond 163:869–879

Arai S, Ishimaru S (2011) Zincian chromite inclusions in diamonds: possibility of deep recycling origin. J Mineral Petrol Sci 106:85–90

Barnes SJ, Naldrett AJ, Gorton MP (1985) The origin of the fractionation of platinum-group elements in terrestrial magmas. Chem Geol 53(3–4):303–323

Barnes SJ, Roeder PL (2001) The range of spinel compositions in terrestrial mafic and ultramafic rocks. J Petrol 42:2279–2302

Chandradas M, Patel K (2006) Report on exploration and assessment for chromite in Taka area, Chandrapur District, Maharashtra, GSI. Unpub. Rep. for FS 1998–2000 www.gsi.gov.in

Coleman RG, Jove C (1992) Geological origin of serpentinites. In: The vegetation of ultramafic (Serpentine) soils. In: Baker AJM, Proctor J, Reeves RD (eds) Proceedings of the first international conference on serpentine ecology. Intercept Hampshire, UK, pp 1–17

Cressey G, Cressey BA, Wicks FJ (2008) The significance of the aluminium content of a lizardite at the nanoscale: the role of clinochlore as an aluminium sink. Mineral Mag 72:817–825

D'Antonio MD, Kristensen MB (2004) Hydrothermal alteration of oceanic crust in the West Philippine Sea basin (ocean Drilling Programme Leg, 195, site 1201): inferences from a mineral chemistry investigation. Mineral Petrol 83(1–2):87–112

Deer WE, Howei RA, Zussman J (1992) The rock forming minerals, 2nd edn. Longmans, London

Deschamps F, Godard M, Guillot S et al (2013) Geochemistry of subduction zone serpentinites: a review. Lithos 178:96–127

Dick HJB, Bullen T (1984) Chromian spinel as a petrogenetic indicator in abyssal and alpine-type peridotites and spatially associated lavas. Contrib Mineral Petrol 86:54–76

Dilek Y (2003) Ophiolite concept and its evolution. In: Dilek Y, Newcomb S (eds) Ophiolite concept and the evolution of geological thought. Geological Society of America, Special Paper 373, pp 1–16

Fisk MR (1986) Basalt-magma interactions with harzburgite and the formation of high magnesium andesites. Geophy Res Lett 13:467–470

Frost BR, Beard JS (2007) On silica activity and serpentinisation. J Petrol 48:1351–1368

Fryer P, Taylor B, Langmuir CH et al (1990) Petrology and geochemistry of lavas from the Sumisu and Torishima backarc rifts. Earth Planet Sci Lett 100:161–178

Ganguly S, Paul P, Chowdhury AR et al (2013) Thematic mapping of Archean gneiss terrain, Western Bastar Craton in Bhisi and Nagbhid area, Chandrapur district, Maharashtra. GSI. Prog. Rep FS 2012–13. www.gsi.gov.in

Gervilla F, Proenza JA, Frei R et al (2005) Distribution of platinum-group elements and Os isotopes in chromite ores from Mayarí-Baracoa Ophiolilte Belt (eastern Cuba). Contrib Mineral Petrol 150:589–607

González-Jiménez JM, Kerestedjian T, Proenza JA et al (2009) Metamorphismon chromite ores from the Dobromirtsi ultramafic massif, Rhodope Mountains (SE Bulgaria). Geol Acta 7(4):413–429

González-Jiménez JM, Proenza JA, Gervilla F et al (2011) High-Cr and high-Al chromitites from the Sagua deTánamo district, Mayarí-Cristal ophiolitic massif (eastern Cuba): constraints on their origin from mineralogy and geochemistry of chromian spinel and platinum group elements. Lithos 125:101–121

Hattori KH, Guillot S (2007) Geochemical character of serpentinites associated with high- to ultrahigh-pressure metamorphic rocks in the Alps, Cuba, and the Himalayas: recycling of elements in subduction zones. Geochem Geophys Geosyst. https://doi.org/10.1029/2007gc001594

Hellebrand E, Snow JE, Dick HJB et al (2001) Coupled major and trace elements as indicators of the extent of melting in mid-ocean-ridge peridotites. Nature 410:677–681

Ishii T, Robinson PT, Maekawa H et al (1992) Petrological studies of peridotites from diapiric serpentinites seamounts in the Izu-Ogasawara-Mariana fore arc, Leg, 125. Proc Ocean Drill Program Sci Results 125:445–463

Jan MQ, Windley BF (1990) Chromian spinel silicate chemistry in ultramafic rocks of the Jijal complex, northwestern Pakistan. J Petro 131:667–715

Jensen LS (1976) A New cation plot for classifying subalkalic volcanic rocks. Ontario Geol Surv Misc Pap 66

Kamenetsky VS, Crawford AJ, Meffre S (2001) Factors controlling chemistry of magmatic spinel: an empirical study of associated olivine, Cr-spinel and melt inclusions from primitive rocks. J Petrol 42:655–671

Kapsiotis A, Grammatikopoulos TA, Tsikouras B et al (2009) Chromian spinel composition and platinum group element mineralogy of chromitites from the Milia area, Pindos ophiolite complex. Greece Can Miner 47:1037–1056

Kelemen PB, Dick HJB, Quick JE (1992) Formation of harzburgite by pervasive melt/rock interaction in the upper mantle. Nature 358:635–641

Kodolányi J, Pettke T, Spalnder C et al (2012) Geochemistry of ocean floor and fore-arc serpentinites: constraints on the ultramafic input to subduction zones. J Petro 153:235–270

Leake BE, Woolley AR, Arps CES et al (1997) Nomenclature of amphiboles: report of the subcommittee on Amphiboles of the international mineralogical association, commission on new mineral names. Am Miner 82:1019–1037

Li XP, Rahn M Bucher K (2004) Serpentinites of the Zermatt-saas ophiolite complex and their texture evolution. J Metamorp Geol 22:159–177

Marchesi C, Garrido CJ, Padron-Navarta JA et al (2013) Element mobility from seafloor serpentinisation to high-pressure dehydration of antigorite in subducted serpentinite: insights from the Cerro del Almirez iltramafic massif (southern Spain). Lithos 178:128–142

Maurel C, Maurel O (1982) Etude experimentale de la distribution de l'aluminium entre bain silicate basique et spinelle chromifere implications petrogenetiques: teneur en chrome des spineless. Bull Miner 105:197–202

Melcher F, Grum W, Thalhammer TV et al (1999) The giant chromite deposits at Kempirsai, Urals: constraints from trace element (PGE, REE) and isotope data. Miner Deposita 34:250–272

Mellini M, Rumori C, Viti, C (2005) Hydrothermally reset magmatic spinels in retrograde serpentinites: formation of "ferritchromit" rims and chlorite aureoles. Contrib Mineral Petrol 149: 266–275

Meshram R, Malviya VP, Chore SA (2016) Chemistry of Chrome-Spinel in serpentinite of the Taka, Central India: its genetic significance. In: 35th international geological congress, Cape Town, South Africa

Meshram RR, Malviya VP, Patil SB (2014a) A report on detailed petrography and geochemistry of mafic-ultramafic suite of rocks around Taka-Ballarpur-Puyardand area, Nagpur and Chandrapur district, Maharashtra with special emphasis on Cr, Ni and PGE mineralization. GSI Report (FS-2013-14). www.portal.gsi.gov.in

Meshram RR, Malviya VP, Chore SA (2014b) Report of IPGE mineral in serpentinite of Taka area, Nagpur and Chandrapur districts, Maharashtara, western Bastar Craton. Central India Indian J Geosci 68(2 & 3):237–244

Mohanty SP (2015) Palaeoproterozoic supracrustals of the Bastar Craton: Dongargarh Supergroup and Sausar Group. Geol Soc Lond Mem 43(1):151–164

Morimoto N, Fabries J, Ferguson AK et al (1989) Nomenclature of pyroxenes. Can Miner 27:143–156

Naldrett AJ (2010) Secular variation of magmatic sulfide deposits and their source magmas. Eco Geology 105(3):669–688; Naqvi SM and Rogers JJW (1987) Precambrian geology of India. Oxford University Press

Page P, Barnes SJ (2009) Using trace elements in chromites to constrain the origin of podiform chromitites in the thetford mines ophiolite, Quebec, Canada. Eco Geol 104(7):997–1018; Pal T (2011) Petrology and geochemistry of the Andaman ophiolite: melt-rock interaction in a suprasubduction-zone setting. J Geol Soc Lond 168:1031–1045

Palme H, Jones A (2003) Solar system abundances of the elements. In: Holland H, Turekian KK (eds) Treatise on geochemistry. Elsevier, New York 1:41–61

Parkinson IJ, Pearce JA (1998) Peridotites from the Izu-Bonin-Mariana forearc (ODP Leg 125): evidence for mantle melting and melt-mantle interaction in a supra subduction zone setting. J Petrol 139:1577–1618

Pearce JA, Barker PF, Edwards SJ et al (2000) Geochemistry and tectonic significance of peridotites fromthe South Sandwich arc-basin system, South Atlantic. Contrib Miner Petrol 139:36–53

Polat A, Kerrich R (2006) Reading the geochemical fingerprints of Archean Hot Subduction volcanic rocks: evidence for accretion and crustal recycling in a mobile tectonic regime. Am Geophys Union: Geophys Monogr Ser 164:189–213

Proenza JA, Zaccarini F, Lewis JF et al (2007) Chromian spinel composition and the platinum group minerals of the PGE-rich Loma Peguera chromitites, Loma Caribe peridotite, Dominican Republic. Can Miner 45:631–648

Radhakrishna BP (1989) Suspect tectonostratigraphic terrain elements in the Indian subcontinent. Geol Soc India 34:1–24

Ramchandra HM (2004) A review of terrane evolution in the Precambrian Dharwar and Bastar Craton. Geol Surv India Spec Publ 84:1–21

Ramchandra HM, Mishra VP, Deshmukh SS (1995) Dyke swarms of Peninsular India. Devraju TC (eds) Mem Geol Soc India 33:183

Ranero CR, Morgan JP, McIntosh et al (2003) Bending-related faulting and mantle serpentinization at the middle America trench. Nature 425:367–373

Rollinson H (1995) The relationship between chromite chemistry and the tectonic setting of Archean ultramafic rocks. In: Econ S-S (ed) TG Blenkinshop and P Tromps. Balkema, Geol Amsterdam, pp 7–23

Rollinson P (2008) The geochemistry of mantle chromitites from the northern part of the Oman ophiolite: inferred parental melt compositions. Contrib Miner Petrol 156:73–288

Roy A, Bandhyopadhyay BK, Huin AK et al (1996) A multidisciplinary studies in Sakoli Basin under Sakoli Basin project Unpublished GSI Report. www.gsi portal.gov.in

Roy A, Prasad MH (2003) Tectonothermal events in Central Indian Tectonic Zone (CITZ) had its implications in Rodinian crustal assembly. J Asian Earth Sci 22:115–129

Srivastava RK, Mondal S, Balram V et al (2010) PGE geochemistry of low-Ti high-Mg siliceous mafic rocks with the Archean Central Indian Bastar Craton: implications for magma fractionations. Miner Petrol 98:329–345

Stern RJ, Johnson RR, Kroner A et al (2004) Neoproterozoic ophiolites of the Arabian-Nubian Shield. In: Kusky TM (ed) Precambrian ophiolites and related rocks. Dev Precambrian Geo 113:95–128

Sun SS, McDonough WF (1989) Chemical and isotopic systematic of oceanic basalts: implications for mantle composition and processes. In: Saunders AD, Norry MJ (eds) Magmatism in the ocean basins London, U.K. Geological society special publication, vol 42, pp 315–345

Uysal I, Sadiklar B, Tarkian M et al (2005) Mineralogy and composition of the chromitites and their platinum-groupminerals from Ortaca (Muğla-SWTurkey): evidence for ophiolitic chromitite genesis. Miner Petrol 83:219–242

Wilson M (1989) Igneous petrogenesis. London U. K. Unwin hyman, 466p

Yedekar DB, Jain S, Nair KKK, Dutta KK (1990) The central Indian collision suture Precambrian of Central India, Geol. Surv. India Spec Pub 128:1–37

Metamorphism of the Central Bundelkhand Greenstone Complex of the Bundelkhand Craton, Indian Shield and Its Geodynamic Setting

Oleg S. Sibelev, Alexander I. Slabunov, Vinod K. Singh, and Sumit Mishra

Abstract

At least three stages of metamorphic processes were identified in the Central Bundelkhand terrane of the Bundelkhand Craton: (1) ca 2.8 Ga eclogite-facies; (2) ca. 2.7 Ga amphibolite-facies and metasomatism associated with its retrograde branch; (3) ca 1.9–1.8 Ga prehnite-pumpellyite-facies presumably associated with reef formation giant two-phase hydrothermal quartz veins. Data on the mineral associations, the chemical composition of minerals and the PT parameters of metamorphism are reported. The Mesoarchean (ca 2.8 Ga) eclogite-facies metamorphism is associated with subduction processes; Neoarchean (ca. 2.7 Ga) metamorphism is predetermined by accretion-collision events; metamorphic processes associated with 2.55–2.50 Ga subduction-accretion events are extremely poorly defined; the latest Paleoproterozoic (1.9–1.8 Ga) prehnite-pumpellyite-facies metamorphism is associated with rifting.

Keywords

Metamorphism • Metasomatism • Greenstone complex • Mineral associations • Garnet • PT conditions • Bundelkhand Craton • Indian Shield

O. S. Sibelev · A. I. Slabunov
Institute of Geology, Karelian Research Centre, RAS, Petrozavodsk, Russia
e-mail: Sibelev@krc.karelia.ru

A. I. Slabunov
Petrozavodsk State University, Petrozavodsk, Russia

V. K. Singh (✉)
Department of Geology, Institute of Earth Sciences, Bundelkhand University, Jhansi, India
e-mail: vinodksingh@bujhansi.ac.in

S. Mishra
Department of Geology, Hemvati Nandan Bahuguna Garhwal University, Srinagar, India

1 Introduction

The metamorphic evolution of geological complexes is critical for a better understanding of their geodynamic origin (Sklyarov 2001; Kornprobst 2003; Reverdatto et al. 2019). The Indian shield constitutes Bundelkhand, Aravalli, Dharwar, Bastar and Singhbhum Cratons (Fig. 1, Inset map) in which Bundelkhand Craton preserves mostly Archean TTGs, greenstone belts, sanukitoids and high-K anatectic granites with limited metamorphic studies (Malviya et al. 2006; Saha et al. 2011; Singh and Slabunov 2015; Kaur et al. 2016; Roy and Purohit 2018; Slabunov and Singh 2019a; Nasipuri et al. 2019; Jain et al. 2020; Pati and Singh 2020; Singh et al. 2019a, b; 2020, 2021a and references therein). It has generally been noted that the rocks of Bundelkhand Craton were metamorphosed under greenschist- to amphibolite-facies conditions (Basu 1986), but no systematic metamorphic studies of cratonic rocks have ever been attempted. Nevertheless, some areas were studied and were taken into account toward polymetamorphism (Saha et al. 2011; Sibelev et al. 2019, 2021). The innovative metamorphic evidence for the greenstone complex of the Central Bundelkhand Craton increases our knowledge of regional geology and makes it possible to test available geological models.

2 Geology of the Bundelkhand Craton

The Bundelkhand Craton consists dominantly of Neoarchean K-rich granite (Ramakrishnan and Vaidyanadhan 2010; Verma et al. 2016; Singh et al. 2020, 2021a; Pati and Singh 2020) with large fragments of 3.59–3.2 Ga TTG granitoids (Kaur et al. 2014) and greenstone belts (Malviya et al. 2006; Singh and Slabunov 2015; Slabunov and Singh 2019a). The Bundelkhand Craton is subdivided into three terranes on geophysical and geological characteristics (Fig. 1), i.e. the Central, Northern and Southern Bundelkhand (Slabunov and Singh 2019b; Singh et al. 2019c, 2021b). The Central

A. K. Shandilya et al. (eds.), *Geological and Geo-Environmental Processes on Earth*, Springer Natural Hazards,
https://doi.org/10.1007/978-981-16-4122-0_9

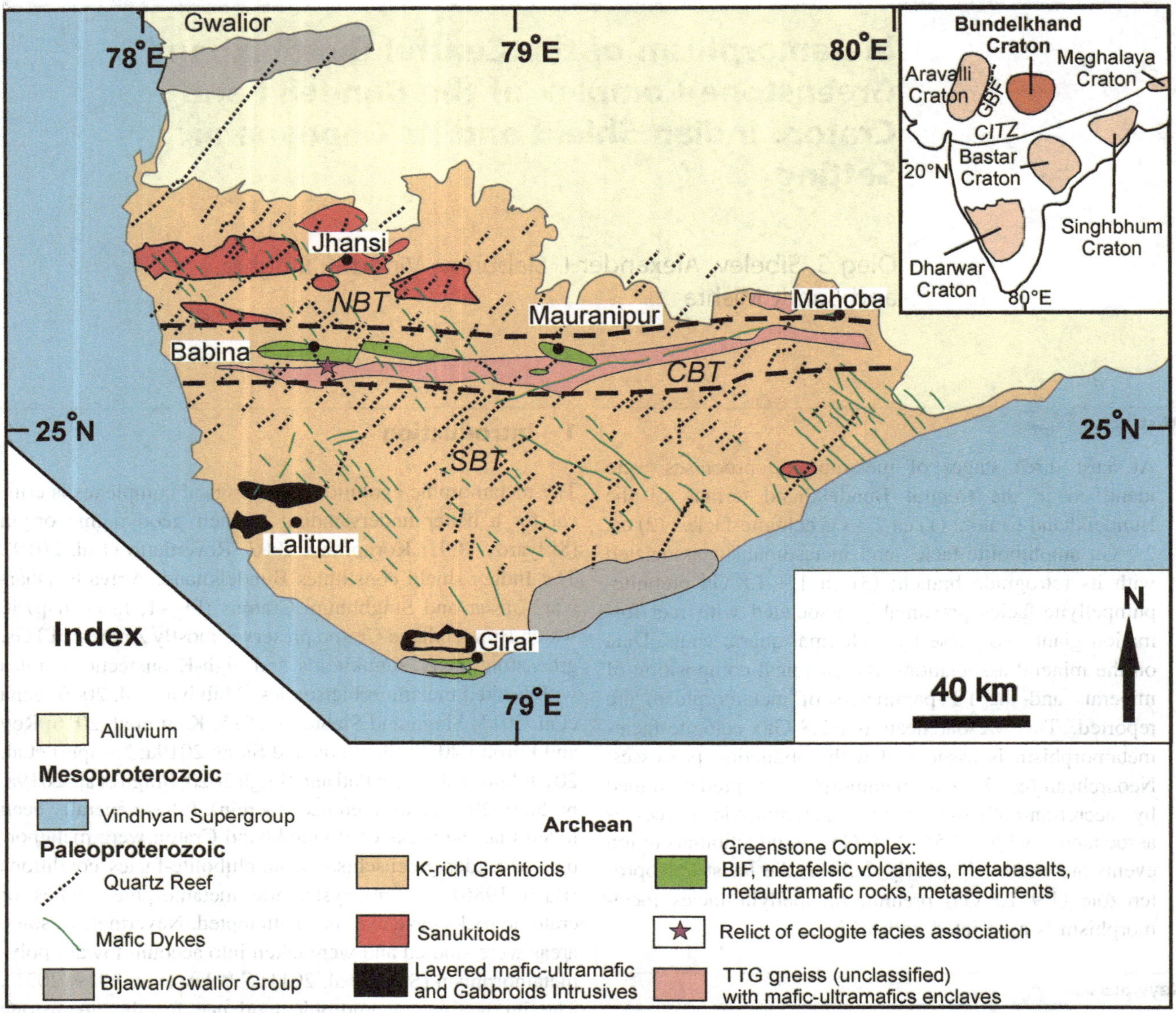

Fig. 1 General geological map of the Bundelkhand Craton (modified after Basu 1986; Saha et al. 2011; Joshi et al. 2017; Slabunov and Singh 2019a; Singh et al. 2020). NBT, CBT and SBT—Nothern-, Central- and Southern-Bundelkhand terranes (Slabunov and Singh 2019b). Inset shows the cratonic blocks of the Indian Shield (Ramakrishnan and Vaidyanadhan 2010; Jain et al. 2020), CITZ—Central Indian Tectonic Zone, GBF—Great Boundary Fault

Bundelkhand terrane consists of Paleo-Neoarchean TTG granitoids, a Meso-Neoarchean greenstone complex and lesser K-rich granites. Its earth crust has a relatively low resistivity and a thickness of about 60 km (Gokarn et al. 2013). The Northern Bundelkhand terrane is dominated by Neoarchean K-rich granites, sanukitoids and monzogranites; TTG and mafic rock fragments are scarce. The earth's crust of this terrane occurs as a homogeneous block with high resistivity and has a thickness of 65–70 km. The Southern Bundelkhand terrane not only is composed mainly of K-rich granites but also contains Paleoarchean TTG, Archean mafic–ultramafic intrusions, a schist belt and scarce Neoarchean sanukitoid massifs. The earth's crust displays a three-membered structure and is about 60 km thick.

Lenticular amphibolites, metaultramafic rock and harzburgite bodies, occurring among Paleoarchean TTG gneisses and Neoarchean granites in the Central Bundelkhand terrane, were described (Singh et al. 2019a; Sahu et al. 2020). They were dated (Sm/Nd whole-rock isochron) at 3.44 ± 0.16 Ga (Singh et al. 2019a). They are assumed to be early oceanic crust fragments (Singh et al. 2019a).

Greenstone belts, consisting of the Meso- to Neoarchean Central Bundelkhand and South Bundelkhand greenstone complexes, contribute markedly to the craton structure (Singh and Slabunov 2015, 2016; Slabunov et al. 2017a; Slabunov and Singh 2019a; Singh et al. 2020). The Central Bundelkhand greenstone complex (Fig. 1) makes up the Babina and Mauranipur belts that stretch from west to east over approximately 150 km (Malviya et al. 2006; Singh and Slabunov 2015; Slabunov and Singh 2019a; Singh et al. 2020, 2021a; Svetov et al. 2020).

The supracrustal rocks of the Central Bundelkhand greenstone complex consist of two stratotectonic associations (Fig. 2): an early sequence consisting of three units: (1) ultrabasic-metabasaltic, (2) metarhyolitic-dacitic and (3) iron formation (BIF), and a late sequence made up of felsic volcanics. The contacts between the units and associations are tectonic (Singh and Slabunov 2015; Slabunov and Singh 2019a). Neoarchean (2687 ± 17 Ma) epidote-quartz-plagioclase metasomatic rocks are formed locally in these contact zones of the Mauranipur greenstone belt. The early felsic volcanics are dated at 2810 ± 13 Ma from this belt (Slabunov and Singh 2019a). The high-Mg corundum-bearing phlogopite-chlorite schist lens at the southern flank of the Babina belt (Central Bundelkhand greenstone complex) has been subjected to eclogite-facies metamorphism (>18 kbar) dated at 2780 ± 64 Ma (Saha et al. 2011).

The late felsic volcanics are documented in the Central Bundelkhand greenstone complex and dated at 2542–2557 Ma (Singh and Slabunov 2015; Slabunov and Singh 2019a).

Paleoproterozoic hydrothermal processes, which gave rise to a unique giant (some of the veins, up to 700 m in

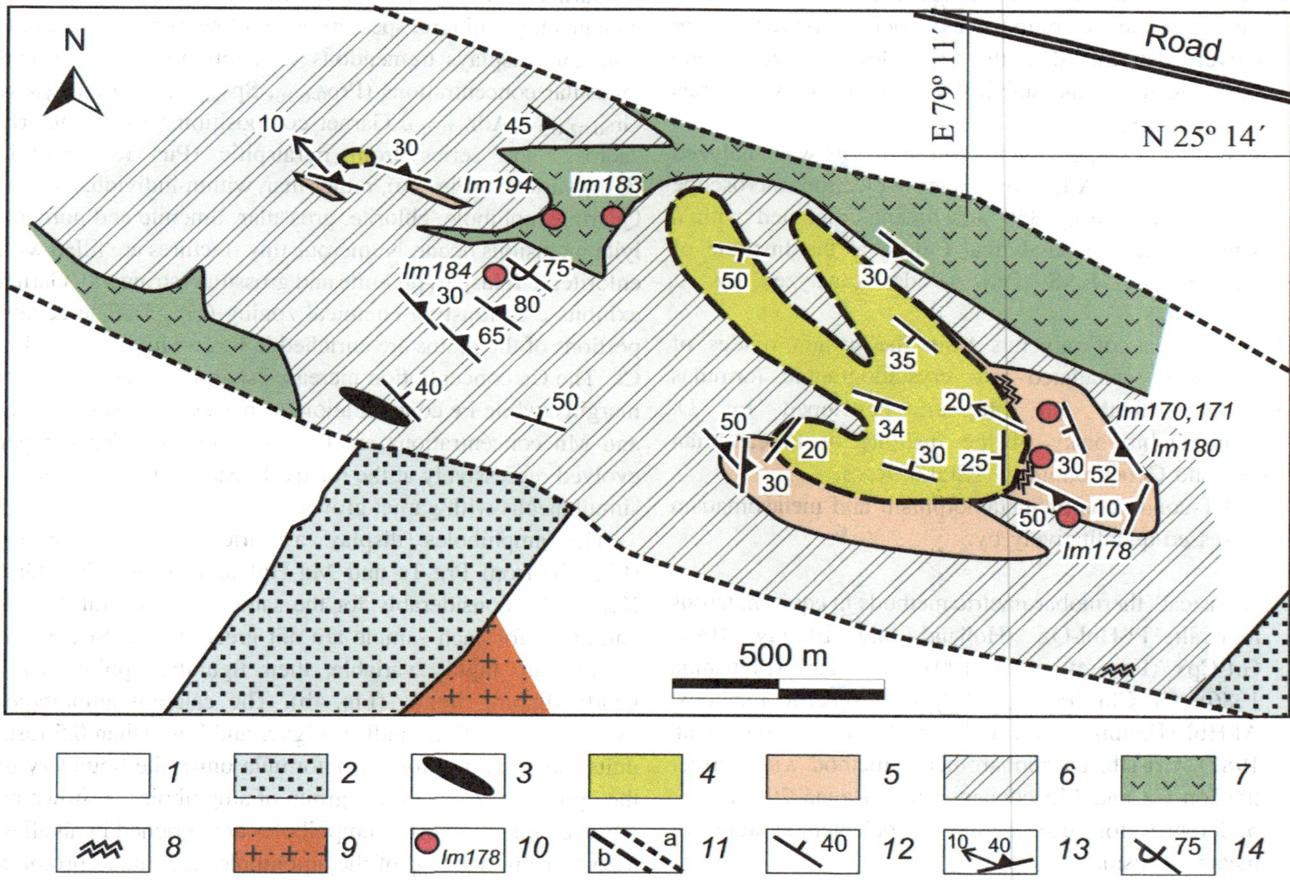

Fig. 2 Geological map of the Mauranipur belt, Central Bundelkhand Greenstone Complex, Bundelkhand Craton (simplified after Malviya et al. 2006; Slabunov and Singh 2019a) with sampling sites for petrological studies. Legend: 1: alluvial rocks; 2: 1.9–1.8 Ga quartz veins (reefs); 3–7 (greenstone complex); 3: 2.56 Ga metadacite dyke; 4: banded iron formation (BIF); 5: 2.81 Ga felsic volcanics; 6: metabasalt (locally with relics of pillow, amygdaloidal and variolitic texture); 7: metaultramafics and high-Mg mafic rocks; 8: ca.2.7 Ga metasomatic rocks; 9: 3.55–2.5 Ga granitoids; 10: sampling sites for metamorphic studies and their numbers; 11: assumed faults **a** and thrusts **b**; 12: bedding plane; 13: foliation and lineation; 14: overturned bedding

thickness, can be traced along the strike over about 100 km) quartz vein swarm (Ramakrishnan and Vaidyanadhan 2010; Pati et al. 2007), are also common (Fig. 1). The early stage of their formation is dated at 1866 ± 11 Ma and the late stage at 1779 ± 43 Ma (Slabunov et al. 2017b). The Archean granite-greenstone complexes are cut by three different-aged (ca. 1.98, 1.8 and 1.0 Ga) gabbroic dyke (Ramakrishnan and Vaidyanadhan 2010; Pradhan et al. 2012). Probably, 1.8 or 1.0 Ga dykes have cross-cut to quartz vein in some locations apparently.

3 Methods of Study

Over 60 samples of metavolcanic rocks were taken in the field for the study of metamorphic processes. Metamorphically and metasomatically altered rocks, containing superimposed minerals and mineral associations, were preferred. Mineral parageneses were separated by detailed petrographic studies and paragenetic analysis. The parameters of multiple alterations and their variations were qualitatively determined at the facies level, based on the mineral parageneses and stability regions of index minerals and/or their composition.

The chemical composition of the minerals was analyzed on a TESCAN VEGA II LSH scanning electron microscope with an INCA Emergy 350 attachment combined with a microanalyzer in the Analytical Centre at the Institute of Geology, KarRC, RAS, Petrozavodsk (analyzed by A. N. Ternovoy).

The formulas of minerals from the weight oxides of elements were calculated into crystallographic formulas using the Minal 3.0 program developed by D. V. Dolivo-Dobrovolsky of the Institute of Precambrian Geology and Geochronology (IGGD), RAS.

The PT-conditions of metamorphism and metasomatism were assessed quantitatively by.

(1) "classical" thermobarometric methods in equiponderous mineral: Pl-Hbl-Qz (Holland and Blundy 1994, Grt-Cpx (Fonarev et al. 1991), Cpx-Hbl (Plyusnina 1986), Grt-Chl (Perchuk 1989); barometric methods: Al-Hbl (Hammarstrom and Zen 1986; Hollister et al. 1987). Grt-Hbl thermobarometric method with correction on Ca and Mn concentration (Ravna 2000) gave preference for estimation of paleotemperature of metamorphism.
(2) multi-equiponderous thermobarometry (TWEEQU) methods (Berman 1991) using a combined approach. Calculations were made using TWQ 2.2 program with TWQ_Comb and TWQ_View supplements made by D. V. Dolivo-Dobrovolsky. JUN92 thermodynamic database was used in calculations (Berman 1988).

4 Mineral Associations in Metamorphic Rocks and Mineral Composition

To assess metamorphic parameters in the Central Bundelkhand greenstone complex, mineral associations were studied in the following metamorphic rocks: metadacites and rhyolites (Fig. 3a, b) of felsic volcanics, metaultramafics (Fig. 3c, d) and metabasalts (Fig. 3e) from the Mauranipur and Babina areas (Sibelev et al. 2019, 2021). These associations are multi-phase, often metastable and consist of relict, metamorphic and metasomatic minerals. Almost all of them persist as long as the last stages of recrystallization. As temperature decreases, retrograde minerals (Ms,[1] Ksp, Act, Ep, Chl, Cc, etc.) are added to relatively high-temperature parageneses that contain clinopyroxene, garnet, amphibole and plagioclase, and their crystallization becomes increasingly metasomatic. Parageneses with prehnite (Fig. 3f) and pumpellyite are formed at the final stages of metamorphic evolution.

Garnet preserves specific characteristics in these rocks. It contains high mineral spessartine and low pyrope concentrations and displays tremendous variations in almandine and grossular concentrations ($Prp_{0\text{-}0.16}$, $Sps_{0.17\text{-}0.46}$, $Alm_{0.06\text{-}0.68}$, $Grs_{0.02\text{-}0.53}$, $Adr_{0\text{-}0.25}$). Garnet composition varies regularly "across" the series from pyralspitic (Pir) to granditic (Grn) varieties (Fig. 4a), rather than within individual series. Quartz, amphibole, chlorite, grossular, ilmenite and pumpellyite inclusions occur. Numerous tiny fractures are filled with chlorite, K-feldspar, prehnite and grossular (stringers). Garnet exhibits a contrasting chemical zoning (Fig. 5). The central portions of the grains are enriched in Fe and impoverished in Ca. The Ca concentration increases sharply toward the grain margins, while Fe concentration decreases. Variations in Mg and Mn concentrations are less considerable. The stringer evolved regressively at the relatively early stages of blasts is simultaneous with garnet growth.

The amphiboles display a variety of compositions (Fig. 4b) from Prg-Ts and Mg-Hbl to Act and Cum-Gru. X_{Mg} varies considerably for the samples in general, but its variations for each sample are not great, and as Si concentrations are highly variable, their figurative points show gently dipping trends (Fig. 4b). The calcium amphiboles overlap the alkalinity indices higher and lower than 0.5 form units, i.e. a.p.f.u. (atoms per formula unit); the boundary of the figurative points of this group of amphiboles is shown by a broken line. Lowest Si amphiboles are enriched in alkalies. Variations in the X_{Mg} of the amphiboles are due to protolith composition, and substantial variations in their alumina content result from intense regressive-metamorphic (and metasomatic) reworking (Fig. 5).

[1] Mineral symbols are after Donna and Bernard (2010).

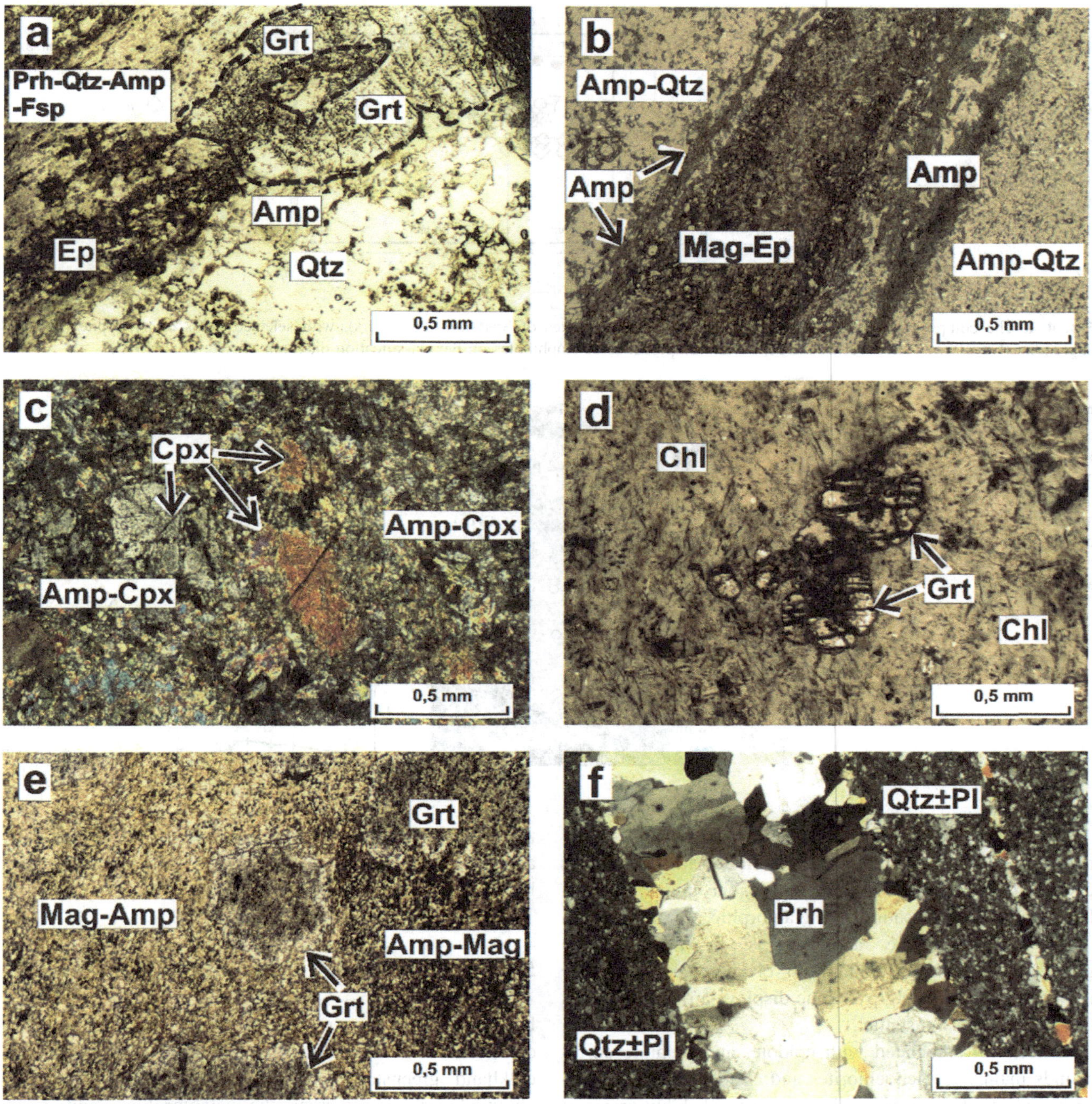

Fig. 3 Photomicrographs showing the evolution of metamorphic and metasomatic minerals after CBGC rocks. **a** Evolution of epidote and case-like garnet with rims of amphibole in prehnite-quartz-amphibole-K-feldspar matrix after felsic volcanics. The lower portion of the photo shows a relic of quartz porphyry phenocryst (in transmetted light). **b** "Lit-par-lit" zonal epidote and amphibole crystallization (+Mag) after felsic volcanics (in transmitted light). **c** Granulation (recrystallization) of relict (magmatic) clinopyroxene and evolution of amphibole after it in metaultramafic rocks (in crossed nicols). **d** Evolution of garnet-chlorite paragenesis (chloritite) after ultramafics (in transmitted light). **e** Crystallization of zonal garnet and magnetite-amphibole matrix after basalt (in transmitted light). **f** Prehnite veinlets cutting felsic metavolcanics; pumpellyite is crystallized in selvages (in crossed nicols)

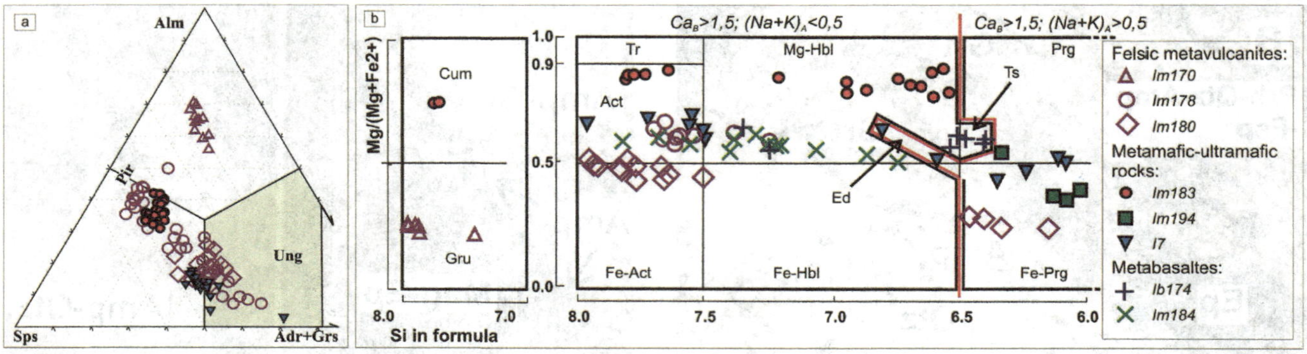

Fig. 4 **a** The composition of the garnets analyzed in the coordinates of Sps-Alm-(Adr + Grs), with selected fields of pyralspite (Pir) and ungrandite (Ung) series (Grew et al. 2013); **b** The compositions of amphiboles on the classification diagrams (Leake et al. 1997)

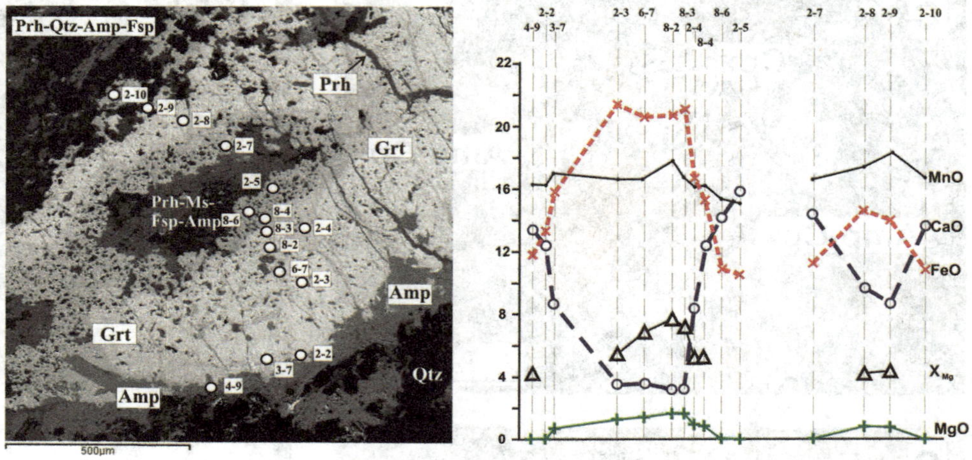

Fig. 5 BSE-images of a large case-like garnet grain (sample IM178; see Fig. 2) with microprobe analysis points and the profiles of changes in the main indices of chemical composition. Numbers above the profiles show the numbers of sites and analyses

Clinopyroxene as a typomorphic mineral of mafic–ultramafic is intensely replaced by amphibole and sometimes disappears completely. Fine clinopyroxene grains occur in felsic volcanics, indicating its metamorphic genesis (Fig. 6).

The clinopyroxene analyzed is in the diopside group. They actually contain no or any Na isomorphic impurities. This Fe–Mg pyroxene displays a rhombic and is thus identified as enstatite.

Chlorite is associated with epidote, and prehnite corresponds mainly to picnochlorites and clinochlores characteristic of propylites. No biotite has been revealed, but grains of micaceous habit, occurring as a mixture of minerals (Bt with Chl or with Ms?) not diagnosed by the microprobe method, are common. The few muscovites analyzed are similar in composition to a phengite subgroup. Epidotes (Czo-Zo) are common in the rocks discussed; they are involved in hydrothermal processes. Prehnite evolves mainly as cross-cutting veinlets, (M up to 1 cm) or fine grains in the matrix that replaces Pl. Prehnite is characterized by permanent chemical composition and the absence of impurities. It is paragenetic with calcium garnet and pumpellyite-(Mg). Accessory and ore minerals are varied and occasionally abundant. For example, Ttn, Mag or Mn-Ilm makes up at least 3% of some of the samples.

5 Metamorphic PT-Parameters

Paragenetic analysis was carried out and the chemical composition of minerals from volcanics of Central Bundelkhand greenstone complex was analyzed. The data obtained were used to assess the metamorphic conditions by classical barometric methods: maximum temperatures of 580–680 °C, pressures of 7.2–10 kbar and a wide range of PT values below this level, some of which are within the sillimanite stability field. Correct TWQ-diagrams (Berman 1991, TWQ_Comb TWQ_View D. V. Dolivo-Dobrovolsky) have only been obtained for retrograde stages at a moderate-gradient metamorphic trend (Fig. 7).

Early parageneses in the Mauranipur are Grt-Prg-Pl; Cpx-Prg-Pl; Cpx-Grt-Pl; Cpx-Prg (±Ksp, ±Qz, ±Mag). The presence of clinopyroxene together with high-alumina

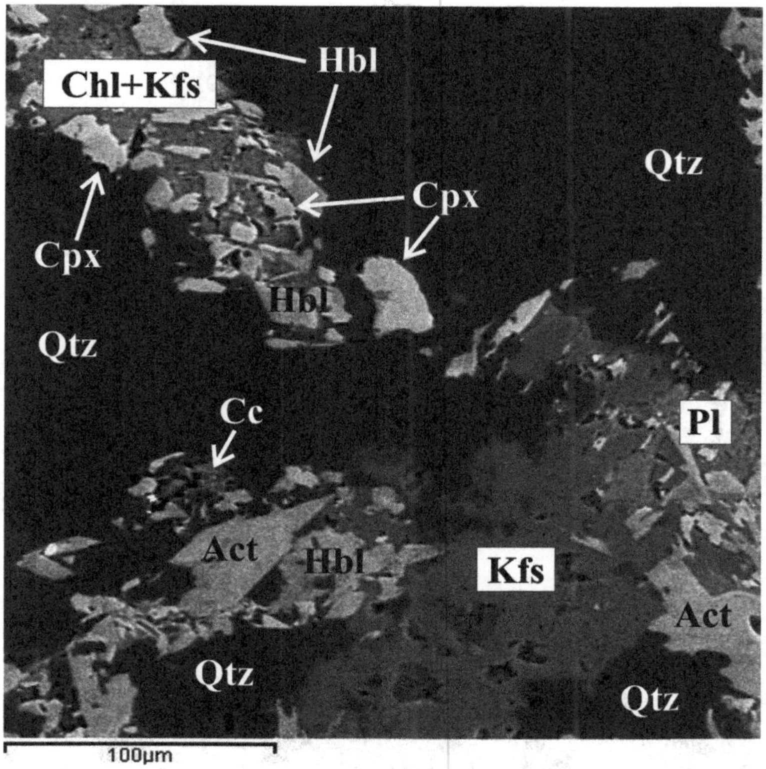

Fig. 6 Replacement of relatively high-temperature plagioclase-clinopyroxene-hornblende paragenesis (±Grt) by a mineral chlorite-actinolite-K-feldspar association (±Cc) in felsic metavolcanics (BSE-image of a thin section)

amphibolite suggests a metamorphic peak under amphibolites-facies conditions.

The retrograde pattern of this metamorphic stage has been traced by a sequence of parageneses (Fig. 8): Grt-Pl-Hbl; Grt-Chl-(Hbl-Act); Grt-(Zo-Czo)-(Hbl-Act); Grt-Chl-Pl (±Ksp, ±Qz, ±Mag, ±Ms). Garnet (Mg concentration decreases and Grs mineral concentration increases) and amphibole (silica concentration decreases to Act) compositions have changed.

Metasomatic processes are inseparably associated with the retrograde stage of this metamorphism. They are spatially confined to fault zones or tectonic rock contacts (Fig. 2), i.e. lithostatic pressure release zones and are consistent with the propylitie-facies metasomatism in the low-temperature range. The highest temperature parageneses of the metasomatic rocks Chl-(Gru-Cum)-Act, Grt-Chl-(Hbl-Act) are succeeded by relatively medium temperature Grt-Ep-Act, Ep-Pl-Act, Ep-Act-Alb, and low-temperature parageneses: Ep-Chl-Alb, Alb-Cc-Chl.

"Classical" propylites typically display T-200–350 °C at low pressures under hypabyssal conditions (Zharikov and Rusinov 1988; Rusinov 1972).

The late stage of metamorphism of the volcanics of Central Bundelkhand greenstone belt takes place under prehnite-pumpellyite-facies conditions. This process is separated in time from previous metamorphic and metasomatic stages. Prehnite-pumpellyite-facies parageneses display some characteristics of their own: the stability of grossular in the metavolcanics, together with barophility of pumpellyite and the absence of zeolites, are indicative of highest pressure-facies metamorphic conditions (T-150–250 °C, P-3–5 kbar). The parageneses of this metamorphism are as follows: Prh-Pmp-Alb; Prh-Pmp-Grs; Pmp-Grs-Alb; Prh-Cc-Act; Prh-Grs-Cc. The geothermal facies gradient is about 5–10°C/km.

6 Discussions

Paleoarchean (3.6–3.2 Ga) TTG gneisses in the Central Bundelkhand terrane (Kaur et al. 2016; Nasipuri et al. 2019) are fairly common. Furthermore, lenticular Paleoarchean amphibolite, harzburgite and metaultramafic bodies (Singh et al. 2019a; Sahu et al. 2020), tentatively identified as relics of early oceanic crust, were described from the above TTG gneisses and Neoarchean granitoids. The formation of TTG granitoids was most probably provoked by the subduction of such an oceanic crust (Singh et al. 2019a). This point of view is illustrated in (Fig. 9 (1, 2)), although there are other points of view in which plumes are assumed to have contributed greatly to the early crust formation here (Nasipuri et al. 2019; Sahu et al. 2020). However, we have failed to reveal metamorphic associations which formed in that period and thus to find independent criteria for the evaluation of Paleoarchean geodynamic processes.

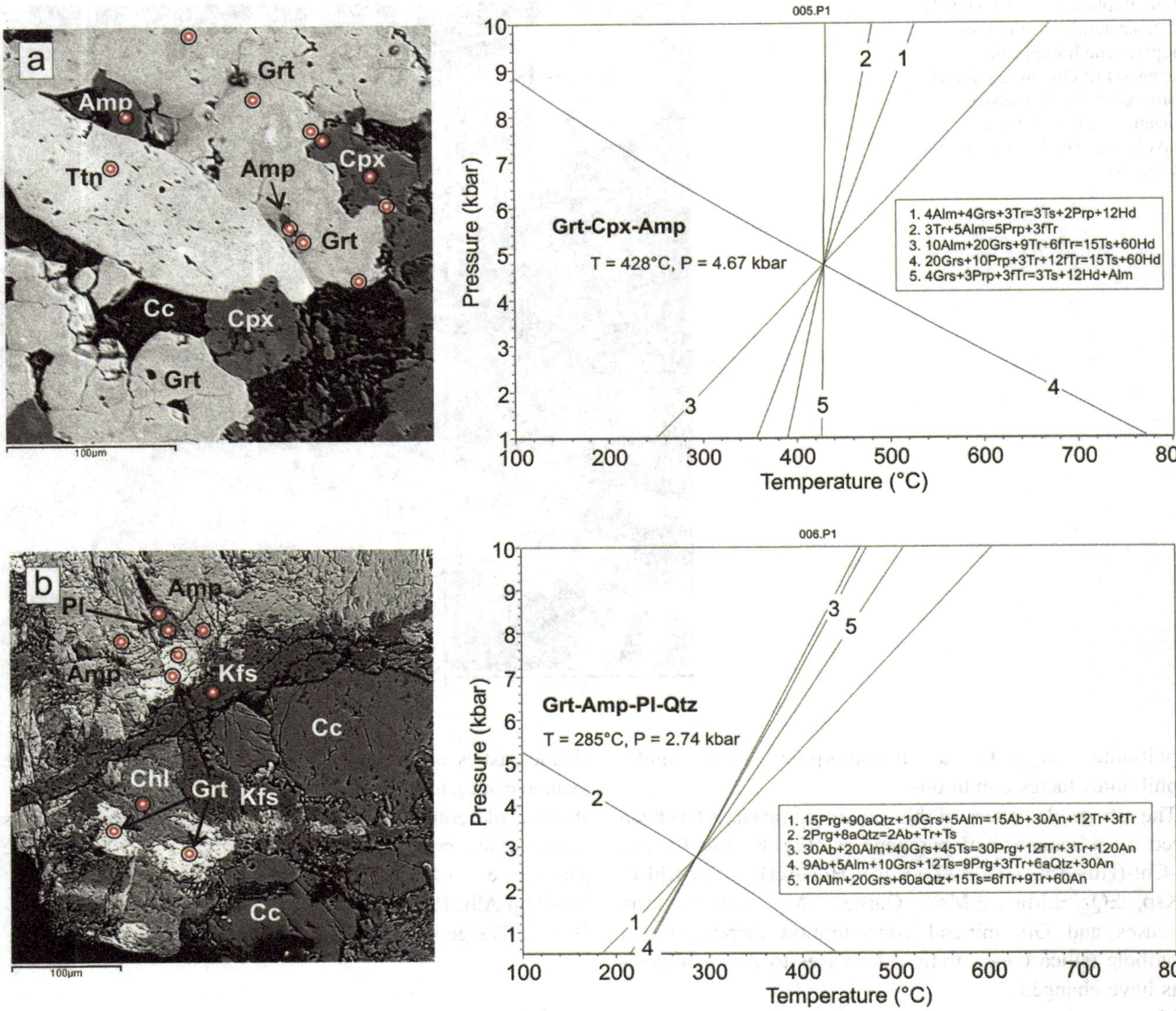

Fig. 7 BSE-images with analytical points and a PT-diagram obtained by multi-equiponderous thermobarometry for **a** metaultramafics and **b** metabasalt (calculations were made using a TWQ program (a JUN92 thermodynamic database was used in the calculations (Berman 1988)))

Mesoarchean geodynamic processes are better understood based on the study of metamorphic associations. First of all, it should be noted that in high-Mg corundum-phlogopite-chlorite schist, which seems to have formed after ultramafics, relics of metamorphic association, consistent with eclogite-facies conditions, are dated at 2780 ± 64 Ma (Saha et al. 2011). It takes place almost simultaneously (within the process age estimation error limits) with felsic volcanism (2810 ± 13 Ma, Slabunov and Singh 2019a). However, it should be noted that Mesoarchean metarhyolites discussed display geochemical characteristics typical of island-arc volcanics. These felsic volcanics in the Mauranipur belt (Fig. 2) are in direct contact with a boninite-bearing (Malviya et al. 2006) volcanic rock sequence. This contact is stratigraphic, suggesting that felsic and boninite volcanism associated with subduction are similar in age, i.e. they are Mesoarchean.

Thus, the association of Mesoarchean eclogites and island-arcs marks subduction processes in the Mesoarchean upon the formation of the continental crust of the Bundelkhand Craton (Fig. 9 (3)). A similar association has been reported from the Belomorian Province of the Fennoscandian Shield (Slabunov 2008; Slabunov et al. 2019).

Neoarchean (ca. 2.7 Ga) metamorphism is most common in the rocks of the Central Bundelkhand greenstone complex. It is consistent with amphibolite-facies moderate to slightly elevated pressure facies parameters (590–680 °C at 6.7–7.2 kbar), the metamorphic grade of belt rocks being

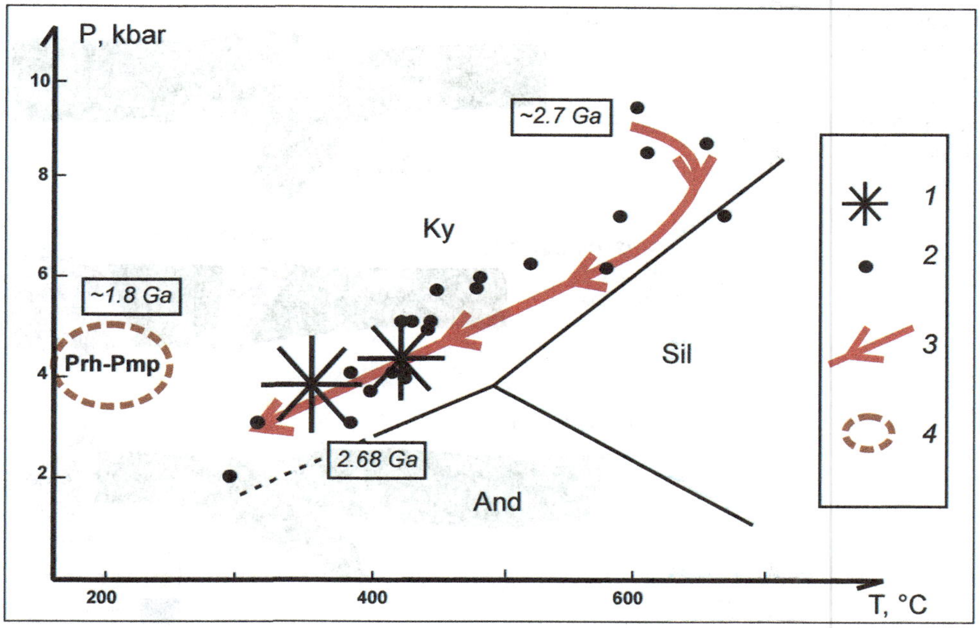

Fig. 8 P-T-t trends of isochemical metamorphism and metasomatism of metavolcanics studied in the Bundelkhand Craton. Legend: 1–2: figurative points of PT-parameters obtained using TWQ software (1) (Berman 1988; TWQ_Comb TWQ_View D.V. Dolivo-Dobrovolsky); and classical methods (2) (Ravna 2000; Holland and Blundy 1994; Blundy and Holland 1990; Plyusnina 1986; Fonarev et al. 1991; Hammarstrom and Zen 1986; Hollister et al. 1987); 3: PTt-trends of changes in metamorphic; 4: assumed PT-stability parameters of Prh-Pmp mineral parageneses

low. These processes to be correlated with 2.73 Ga metamorphic processes (Saha et al. 2011) reported also for the southern flank of the Babina greenstone belt (Fig. 1). Seemingly associated with this event are ca. 2.7 Ga metasomatic processes (Slabunov and Singh 2019a). These metasomatic rocks, interpreted as propylites, may have considerable metallogenic value.

The evolution of Neoarchean epigenetic processes in the complex studied is shown in Fig. 9 (4), which shows a general trend of retrograde metamorphism and associated metasomatism. The latter occurs at various stages when the pressure drops upon the accretion of tectonic slabs. This is probably responsible for a wide range of estimated PT-parameter values.

The final closure of the ocean, which separated the Southern and Northern Bundelkhand terranes (Fig. 9 (5)), was preceded by subduction processes which produced the volcanics of the late greenstone association, including a 2.56 Ga subvolcanic dacitic dyke in the Mauranipur belt (Slabunov and Singh 2019a), and later sanukitoids (Figs. 2, 9 (5)). Metamorphic processes, associated with Late Neoarchean accretion-collision events (Fig. 9 (6)), were revealed at the southern flank of the Babina belt, where metamorphogenetic monazite and zircons dated at 2.47 and 2.43 Ga, respectively, were revealed in high-Mg corundum-phlogopite-chlorite schist (Saha et al. 2011).

The latest metamorphism of the craton took place under prehnite-pumpellyite-facies conditions and seems to be associated with Paleoproterozoic rifting. In Paleoproterozoic time (1.9–1.7 Ga), the Bundelkhand Craton was rimmed by two rift-related structures (Fig. 9 (7)) and was extended and compressed by them and made magmatically and hydrothermally active (Pati et al, 2007; Slabunov and Singh, 2019c). In this case, near-east–west (WSW-ENE)-oriented rifting was probably discontinuous-impulse. NW–SE-oriented mafic dike swarms formed under compression conditions along the NNW-SSE-axis of the hard crust of the craton located between the rifts, and NE-SW-trending hydrothermal veins formed upon its extension along this axis (Figs. 1; 9 (7)). Conditions, conducive to the formation of these veins and associated metamorphism, seem to have arisen twice during the period 1870–1780 Ma (Slabunov et al. 2017b; Singh et al. 2019d; Slabunov and Singh 2019c).

7 Main Conclusions

1. The Central Bundelkhand greenstone belt displays a polymetamorphic evolution pattern. Mesoarchean eclogite-facies metamorphic events have been revealed locally (Saha et al. 2011), while Neoarchean amphibolite-facies metamorphism (at least 580–680 °C

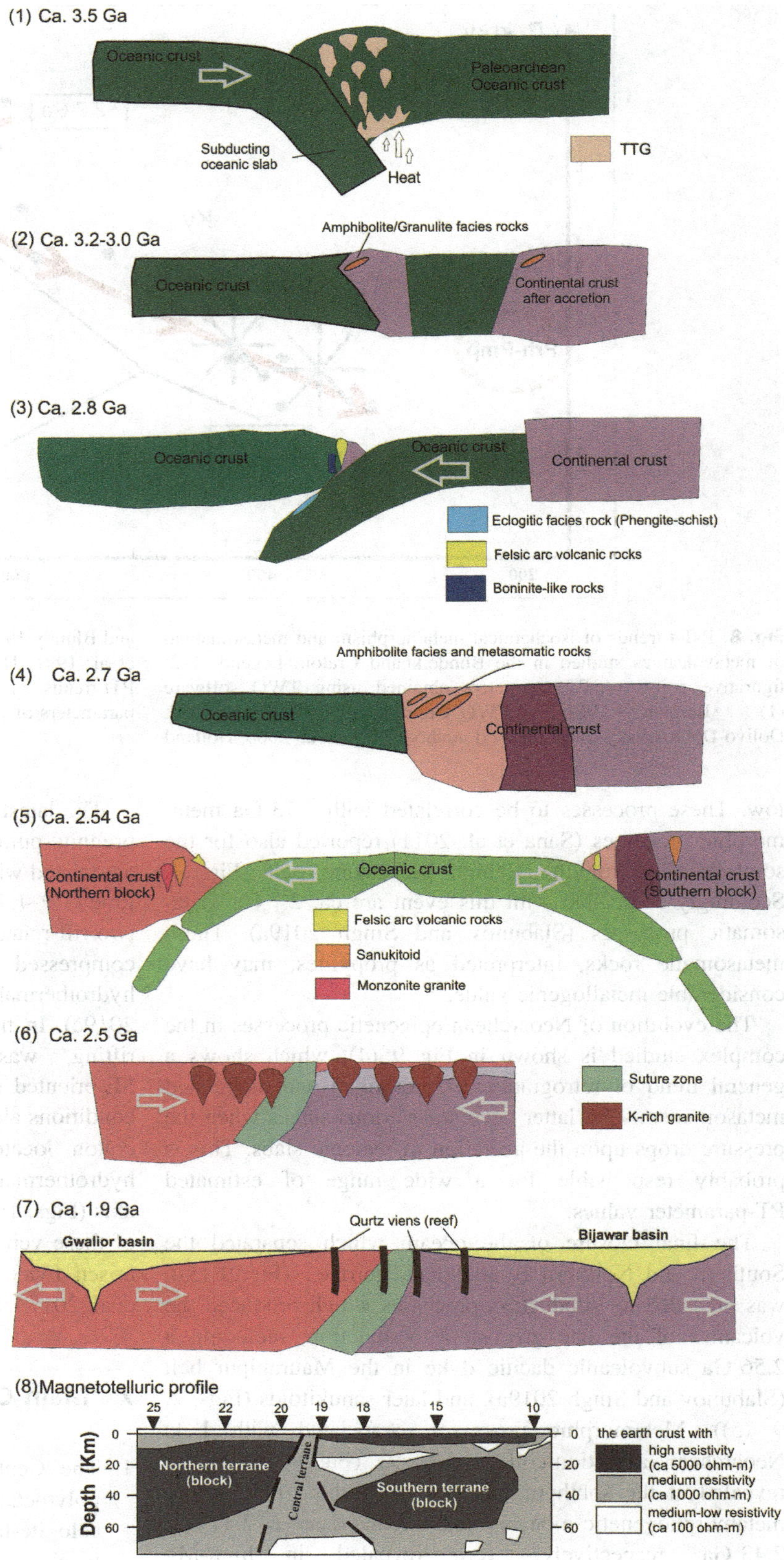

Fig. 9 Schematic tectonic model of 3.5–1.9 Ga crustal evolution of the Bundelkhand Craton (modified after Gokarn et al. 2013; Slabunov and Singh 2019c; Singh et al. 2019a, b, d, 2020, 2021a)

and 7.2–10 kbar) is more common. Also associated with this metamorphic stage are metasomatic events. The latest metamorphism of craton rocks took place locally under prehnite-pumpellyite-facies conditions.

2. The metamorphism of the rocks of the Central Bundelkhand greenstone complex under eclogite- to amphiboite-facies conditions is consistent with a Meso- and Neoarchean subduction-accretion model of the craton's continental earth crust formed. Metamorphic processes under prehnite-pumpellyite-facies conditions seem to be predetermined by Paleoproterozoic rifting.

Acknowledgements This work was carried out under a project funded by RFBR (grant 17-55-45005 IND-a). It is a contribution to a KarRC RAS State Assignment, Project A18-118020290085-4. VKS thanks Department of Science and Technology, Govt. of India, for grant (INT/RUS/RFBR/P-279). This work is carried out under MOU between Institute of Geology (thanks to Prof. Sergei A. Svetov and Prof. Vladimir V. Shchiptsov, Directors), Karelian Research Centre, Petrozavodsk, Russia, and Bundelkhand University, Jhansi, India. We also thank Divya Prahash and Vivek Malviya for their critical review comments and useful suggestions to improve the manuscript.

References

Basu AK (1986) Geology of parts of the Bundelkhand granite massif, Central India. In: Records of the geological survey of India (ed) pp 61–124

Berman RG (1988) Internally-consistent thermodynamic data for minerals in the system $Na_2O–K_2O–CaO–MgO–FeO–Fe_2O_3–Al_2O_3–SiO_2–TiO_2–H_2O–CO_2$. J Petrol 29:445–522

Berman RG (1991) Thermobarometry using multiequilibrium calculations: a new technique with petrological application. Can Mineral 29(4):833–855

Blundy JD, Holland TJB (1990) Calcic amphibole equilibria and a new amphibole-plagioclase geothermometer. Contrib Mineral Petrol 104 (2):208–224

Donna LW, Bernard WE (2010) Abbreviations for names of rock-forming minerals. Am Miner 95:185–187

Fonarev VI, Graphchikov AA, Konilov AN (1991) A consistent system of geotermometers for metamorphic complexes. Int Geol Rev 33 (8):743–783

Gokarn SG, Rao CK, Selvaraj C et al (2013) Crustal evolution and tectonics of the Archean Bundelkhand craton, Central India. J Geol Soci India 82:455–460

Grew ES, Locock AJ, Mills SJ et al (2013) Nomenclature of the garnet supergroup. Amer Mineral 98:785–811

Hammarstrom JM, Zen E (1986) An Aluminium in hornblende: an empirical igneous geobarometer. Amer Mineral 71(11/12):1297–1313

Holland T, Blundy J (1994) Non-ideal interactions in calcic amphiboles and their bearing on amphibole-plagioclase thermometry. Contrib Mineral Petrol 116:433–447

Hollister LS, Grissom GC, Peters EK et al (1987) Confirmation of the empirical correlation of Al in hornblende with pressure of solidification of calc-alkaline plutons. Am Miner 72:231–239

Jain AK, Banerjee DM, Kale VS (2020) Tectonics of the Indian Subcontinent. Springer, Cham, p 576

Joshi KB, Bhattacharjee J, Rai G et al (2017) The diversification of granitoids and plate tectonic implications at the Archaean–Proterozoic boundary in the Bundelkhand Craton, Central India. In: Halla J, Whitehouse MJ, Ahmad T, Bagai Z (eds) Crust–Mantle interactions and Granitoid diversification: insights from Archaean Cratons. Geological Society London, Special Publications, 449:123–157. https://doi.org/10.1144/SP449.8

Kaur P, Zeh A, Chaudhri N (2014) Characterisation and U–Pb–Hf record of the 3.55 Ga felsic crust from the Bundelkhand Craton, Northern India. Precambr Res 255:236–244

Kaur P, Zeh A, Chaudhri N et al (2016) Unravelling the record of Archaean crustal evolution of the Bundelkhand Craton, northern India using U-Pb zircon–monazite ages, Lu–Hf isotope systematics, and whole-rock geochemistry of granitoids. Precambr Res 281:384–413

Kornprobst J (2003) Metamorphic rocks and their geodynamic significance. A Petrological Handbook Kluwer Academic Publishers, p 208

Leake BE, Woolley AR, Arps et al (1997) Nomenclature of amphiboles; Report of the subcommittee on amphiboles of the international mineralogical association commission on new minerals and mineral names. Eur J Mineral 9:623–651

Malviya VP, Arima M, Pati JK et al (2006) Petrology and geochemistry of metamorphosed basaltic pillow lava and basaltic komatiite in the Mauranipur area: subduction related volcanism in the Archean Bundelkhand craton, Central India. J Mineral Petrol Science 101:199–217

Nasipuri P, Saha L, Hangqiang X et al (2019) Paleoarchean crustal evolution of the Bundelkhand Craton, north Central India, Earth's Oldest Rocks, 2nd ed. In: Van Kranendonk MJ, Bennett VC and Hoffmann JE (eds) Elsevier, Amsterdam pp 793–817

Nomenclature of amphiboles (1997) Report of the subcommittee on amphiboles of the international mineralogical association commission on new minerals and mineral names. Eur J Mineral 9:623–651

Pati JK, Singh AK (2020) Bundelkhand Craton. Proc Indian Natl Sci Acad 86(1):55–65. https://doi.org/10.16943/ptinsa/2020/49792

Pati JK, Patel SC, Pruseth KL et al (2007) Geology and geochemistry of giant quartz veins from the Bundelkhand Craton. Cent India Implic: J Earth Syst Sci 116:497–510

Perchuk LL (1989) P-T-fluid regimes of metamorphism and related magmatism with specific reference to the granulite-facies Sharyzhalgay complex of Lake Baikal. In: Daly JS, Cliff RA, Yardley BWD (eds) Evolution of metamorphic belts, vol 43. Special Publications, Geological Society London, pp 275–291

Plyusnina LP (1986) Experimental study on metabasite equilibria, geothermometry. Experiment in the solution of topical problems in geology. Nauka, Moscow, pp 174–183

Pradhan VR, Meert JG, Pandit MK et al (2012) Paleomagnetic and geochronological studies of the mafic dyke swarms of Bundelkhand craton, central India: implications for the tectonic evolution and paleogeographic reconstructions. Precambr Res 198–199:51–76

Ramakrishnan M, Vaidyanadhan R (2010) Geology of India, vol 1. Geological Society of India, Bangalore, p 556

Ravna EK (2000) Distribution of Fe2q and Mg between coexisting garnet and hornblende in synthetic and natural systems: an empirical calibration of the garnet–hornblende Fe–Mg geothermometer. Lithos 53:265–277

Reverdatto VV, Likhanov II, Polyansky OP et al (2019) The nature and models of metamorphism. Springer Geology, p 330. https://doi.org/10.1007/978-3-030-03029-2

Roy AB, Purohit R (2018) Indian shield: Precambrian evolution and Phanerozoic reconstitution. Elsevier, p 386

Rusinov VA (1972) Geological and physico-chemical propylitization pattern. M: Nauka, p 150 [in Russian]

Saha L, Pant NC, Pati JK et al (2011) Neoarchean high-pressure margarite–phengitic muscovite–chlorite corona mantled corundum in quartz-free high-Mg, Al phlogopite–chlorite schists from the Bundelkhand craton, north central India. Contrib Mineral Petrol 161:511–530

Sahu A, Vishwakarma N, Singh Y et al (2020) Mineral chemistry of high-Al chromian spinel from ultramafic rocks of the Babina-Prithvipur transect, Bundelkhand Craton, central India: Implication for petrogenesis and tectonic setting. J Earth Syst Sci 129:182–200. https://doi.org/10.1007/s12040-020-01448-3

Sibelev OS, Slabunov AI, Mishra S, Singh VK (2019) Metamorphism of the Central Bundelkhand greenstone complex of the Bundelkhand craton, Indian Shield. Transactions of A. (Fersman Scientific Session of Geological Institute. Kola Research Centre, RAS, Apatity, Russia) 16:512–516. https://doi.org/10.31241/FNS.2019.16.104

Sibelev OS, Slabunov AI, Mishra S, Singh VK (2021) Metamorphism of the Central Bundelkhand greenstone complex, Indian shield: mineral compositions, paragenesises, and P–T path. Petrology 29 (4):404–438. https://doi.org/10.1134/S0869591121040056

Singh VK, Slabunov A (2015) The Central Bundelkhand Archaean greenstone complex, Bundelkhand Craton, Central India: geology, composition, and geochronology of supracrustal rocks. Int Geol Rev 57(11–12):1349–1364

Singh VK, Slabunov A (2016) Two types of Archaean supracrustal belts in the Bundelkhand Craton, India: geology, geochemistry, age and implication for craton crustal evolution. J Geol Soc India 88:539–548

Singh PK, Verma SK, Moreno JA, Singh VK et al (2019a) Geochemistry and Sm-Nd isotope systematics of metabasalts from the Babina and Mauranipur greenstone belts, Bundelkhand craton: implications for tectonic setting and Paleoarchean mantle evolution. Lithos 330–331:90–107. https://doi.org/10.1016/j.lithos.2019.02.010

Singh PK, Verma SK, Singh VK et al (2019b) Geochemistry and petrogenesis of sanukitoids and high-K anatectic granites from the Bundelkhand craton: implications for the late-Archean crustal evolution. J Asian Earth Sci 174:263–282

Singh VK, Nesterova N, Slabunov A, Singh MM (2019c) Tectonic division of the Bundelkhand Craton Indian Shield: on the basis of geological, geophysical and Sm-Nd systematic of granitoids data. In: Abstract of the national seminar on geological aspects of Northwest Indian Shield. Department of Geology, M.S. University, Udaipur, p 32–33

Singh VK, Slabunov AI, Singh PK (2019d) Paleoproterozoic Giant quartz veins and rift-related basins of the Bundelkhand Craton, northern India: relationship and geodynamic consequences In: Abstract of the recent trends in earth science research. Department of Geology, Banaras Hindu University, Varanasi, p 81

Singh VK, Verma SK, Singh PK et al (2020) Archean crustal evolution of the Bundelkhand Craton: evidence from granitoid magmatism In: Archean Granitoids of India: windows into early Earth tectonics. Geological Society London Special Publ 489:pp 235–259. https://doi.org/10.1144/SP489-2018-72

Singh PK, Verma SK, Singh VK et al (2021a) Geochronology and petrogenesis of the TTG gneisses and granitoids from the Central Bundelkhand granite-greenstone terrane, Bundelkhand Craton, India: implications for Archean crustal evolution and cratonization. Precambr Res 359:106210

Singh VK, Slabunov AI, Nesterova NS et al (2021b) Tectonostratigraphic terranes of the Bundelkhand Craton (Indian Shield). In: Shandilya AK, Singh VK, Bhatt SC, Dubey CS (eds) Geological and Geo-environmental processes on earth, pp 155–164 (this volume)

Sklyarov EV (ed) (2001) Metamorphism and tectonics. Moscow: Intermet Engineering, p 216 [in Russian]

Slabunov AI (2008) Geology and geodynamics of Archean mobile belts (example from the Belomorian province of the Fennoscandian Shield. Petrozavodsk: KarRC RAS, p 281 [in Russian]

Slabunov AI, Singh VK (2019a) Meso-Neoarchaean crustal evolution of the Bundelkhand Craton, Indian Shield: new data from greenstone belts. Int Geol Rev 61:1409–1428. https://doi.org/10.1080/00206814.2018.1512906

Slabunov AI, Singh VK (2019b) The new tectonic division of the Bundelkhand Craton Indian Shield. Transactions of A. Fersman scientific session of Geological Institute, Kola Research Centre, RAS. Apatity, Russia 16:521–524. https://doi.org/10.31241/FNS.2019.16.106

Slabunov A, Singh VK (2019c) Giant quartz veins and rift-related basins as indicators of the Paleoproterozoic destruction of the earth crust of cratons in Northern India. In: Abstract of the formation and evolution stages of the proterozoic earth crust: stratigraphy, metamorphism and geodynamics, Proceedings of the 6th Russian conference on Precambrian geology and geodynamics. St. Petersburg, Russia, pp 213–215

Slabunov A, Singh VK, Joshi KB, Li X (2017a) Paleoarchean zircons from quartzite of South Bundelkhand supracrustal complex: origin and implications for crustal evolution in Bundelkhand Craton Central India. Curr Sci 112:794–801

Slabunov AI, Singh VK, Shchiptsov VV et al (2017b) A new Paleoproterozoic (1.9–1.8 Ga) event in the crustal evolution of the Bundelkhand Craton, India: the results of (SHRIMP) dating of zircons from giant quartz veins. In: Slabunov AI, Svetov SA, Baltybaev ShK (eds) Early precambrian versus modern geodynamics. extended abstracts and field trips guide, Petrozavodsk, KarRC RAS, pp 239–241

Slabunov AI, Balagansky VV, Shchipansky AA (2019) Archean-Paleoproterozoic crustal evolution of the Belomorian Province (Fennoscandian Shield) and the tectonic position of eclogites. In: Slabunov AI, Balagansky VV, Shchipansky AA (eds) Early Precambrian eclogites of the belomorian province, fennoscandian shield. Field Guidebook, Petrozavodsk: KarRC RAS, pp 5–10

Svetov SA, Singh VK, Chazhengina SY et al (2020) Variolithic lavas of basalts of the Mauranipur greenstone belt of the Bundelkhand craton, Indian Shield. Transactions of A. Fersman scientific session of Geological Institute, Kola Research Centre, RAS, Apatity, Russia, 17:482–486. https://doi.org/10.31241/FNS.2020.17.092

Verma SK, Verma SP, Oliveira EP, Singh VK, More JA (2016) LA-SF-ICP-MS zircon U-Pb geochronology of granitic rocks from the central Bundelkhand greenstone complex, Bundelkhand craton, India. J Asian Earth Sci 118:125–137

Zharikov VA, Rusinov VL (eds) (1988) Metasomatism and metasomatic rocks. Moscow: Nauchny Mir, p 492 [in Russian]

Tectonostratigraphic Terranes of the Bundelkhand Craton (Indian Shield)

Vinod K. Singh, A. I. Slabunov, N. S. Nesterova, M. M. Singh, and S. C. Bhatt

Abstract

The geological, geochronological, and isotopic geochemistry investigations complied with magnetotelluric sounding geophysical deep crustal structure compose three tectonic divisions, i.e., the Northern, Central, and Southern Bundelkhand terranes of the craton. The Central terrane comprises Paleo- and Neo-archean tonalite–trondhjemite–granodiorite (TTG), greenstone sequences (ultrabasic-basic, felsic volcanites, Banded Iron Formation (BIF)), and Neoarchean granodiorite-granite series considers it as a granite-greenstone terrane. The Northern terrane location north to the granite-greenstone terrane, mostly consists of Neoarchean potassic granitoids with the visible role of sanukitoids, monzogranites, and granodiorites (TTG), while the Southern terrane consists of mostly Neoarchean potassic granitoids with a minor amount of schist complex, sanukitoids, TTG, and mafic–ultramafic layered and mafic intrusion.

Keywords

Tectonostratigraphic terrane • Bundelkhand Craton • Central Bundelkhand terrane • Northern Bundelkhand terrane • Southern Bundelkhand terrane • Indian Shield

V. K. Singh (✉) · M. M. Singh · S. C. Bhatt
Department of Geology, Bundelkhand University, Jhansi, India
e-mail: vinodksingh@bujhansi.ac.in

M. M. Singh
e-mail: mmsgeol@bujhansi.ac.in

S. C. Bhatt
e-mail: scbhatt@bujhansi.ac.in

A. I. Slabunov · N. S. Nesterova
Institute of Geology, Karelian Research Centre, Russian Academy of Sciences, Petrozavodsk, Russia
e-mail: slabunov@krc.karelia.ru

A. I. Slabunov
Petrozavodsk State University, Petrozavodsk, Russia

1 Introduction

A terrane (or tectonostratigraphic terrane) is a block of Earth crust or fragment that preserves a distinctive geologic history that is different from the surrounding areas and that is usually bounded by faults/shears. Terrane analysis was developed for geodynamic reconstructions of Phanerozoic fold belts (Howell et al. 1985; Parfenov et al. 1993), but it also has been applied to the Precambrian (Slabunov et al. 2006). The algorithm of this study comprises: (1) identification of terranes, (2) the study of their formation, (3) comparison of the evolution of various terranes, and (4) the use of the data thus obtained for constructing a geodynamic model showing the evolution of the structure studied. Thus, identifying terranes is an important stage in studies aiming at geodynamic reconstructions.

The Bundelkhand Craton is a protocontinent situated in the center of the Indian shield, previously considered as a big granite massif consisting of 80–90% K-rich granites (Fig. 1; Basu 1986; Ramakrishnan and Vaidyanadhan 2010). Identified in the central portion of the craton is the Bundelkhand Tectonic Zone (BTZ), which cuts it into two blocks (Malviya et al. 2006; Jain et al. 2020). This craton has considered uniquely with similar structurally geological formations with other Indian cratons (Singh and Slabunov 2015a) than it was assumed before, where Neoarchean K-rich granitoids are dominantly occurring. Recently, new data suggest Paleo- and Neo-archean (3.6–3.2 Ga and 2.67 Ga) granitoids of tonalite-trondhjemite granodiorite (TTG) association (Mondal et al. 2002; Verma et al. 2016; Kaur et al. 2016; Singh et al. 2021), Paleoarchean (~3.44 Ga) mafic-ultramafic volcanic rocks (Singh et al. 2019a), Neoarchean (2.54–2.52 Ga) sanukitoids (Joshi et al. 2017; Singh et al. 2019b), Archean layered igneous intrusive (Slabunov et al. 2018) as well as greenstone complexes (Singh and Slabunov 2015a; Slabunov and Singh 2019a and references therein) exist in the Bundelkhand Craton. These Archean complexes are generally crosscut by several generations of Proterozoic

A. K. Shandilya et al. (eds.), *Geological and Geo-Environmental Processes on Earth*, Springer Natural Hazards,
https://doi.org/10.1007/978-981-16-4122-0_10

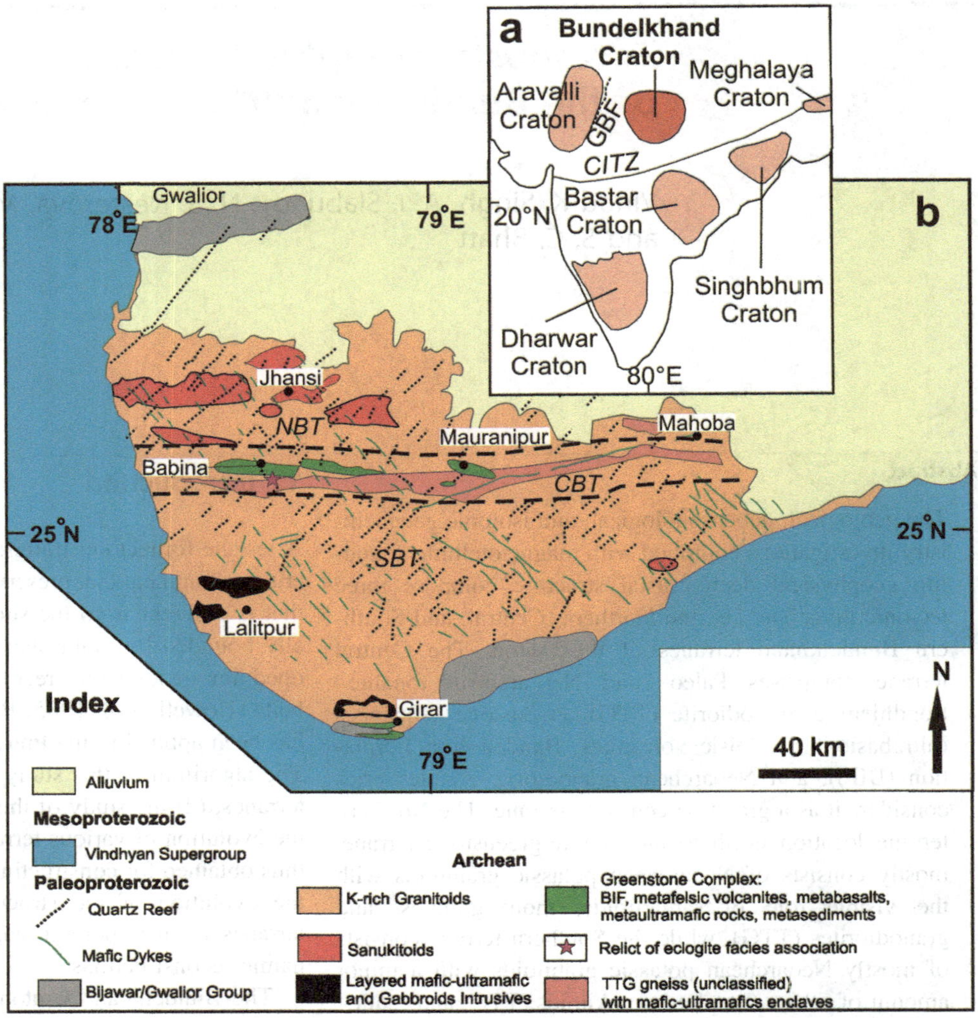

Fig. 1 **a** Shows the cratonic blocks of the Indian Shield (Ramakrishnan and Vaidyanadhan 2010), CITZ—Central Indian Tectonic Zone, GBF—Great Boundary Fault; **b** general geological map of the Bundelkhand Craton (modified after Basu 1986; Saha et al. 2011; Joshi et al. 2017; Slabunov and Singh 2019a; Singh et al. 2020, 2021). NBT, CBT and SBT—Nothern-, Central- and Southern-Bundelkhand terranes (Slabunov and Singh 2019b).

dykes (Pati et al. 2008; Pradhan et al. 2012) and giant quartz veins (Pati et al. 2007; Slabunov et al. 2017a). It forms a thick craton and the many multiple tectonic events that it contains are not well preserved, either because the evidence for them has been eroded or because the deformation is partitioned into narrow zones. Consequently, no attempts have been made previously to describe the structural feature and tectonic evolution of the ancient crust due to restrictions by Neoarchean granites.

The goal of the present paper is to discuss the available options of the tectonic division of the Bundelkhand Craton, to assess the validity of identification of terranes, and to test available geodynamic models.

2 Regional Geology

The Indian shield constitutes several segments of Archean cratons, i.e., Bundelkhand, Aravalli, Dharwar, Bastar, and Singhbhum Cratons (Fig. 1a). The ENE–WSW trending Central Indian Tectonic Zone (CITZ) or Narmada Son Lineament is a major lineament, which separates the Dharwar, Bastar, Singhbhum Cratons as a southern block (Basu 1986; Ramakrishnan and Vaidyanadhan 2010). The northern block constitutes Aravalli and Bundelkhand Cratons are also divided by NE–SW trending Great Boundary Fault into west as Aravalli and Bundelkhand Craton to the east respectively (Basu 1986; Mondal et al. 2002; Ramakrishnan and Vaidyanadhan 2010; Verma et al. 2016; Kaur et al. 2016).

The Bundelkhand Craton occurs in a semicircular shape of an area of about 29,000 km^2 mostly constitute acid granitoid complex, lesser amount of meta-sedimentary, meta-basics, with greenschist-amphibolite grade metamorphism of the older crust (Fig. 1b). However, craton also preserved the oldest Paleoarchean TTG crustal nuclei of the Indian shield (Kaur et al. 2014; Saha et al. 2016). There are two Archean supracrustal complexes in the Bundelkhand Craton, i.e., Central Bundelkhand greenstone and South Bundelkhand metasedimentary (schist) complexes (Malviya et al. 2006; Singh and Slabunov 2013, 2015a, b, 2016; Slabunov and

Fig. 2 Field occurrence of rock types in the Central Bundelkhand terrane. **a** Paleoarchean TTG (Babina); **b** Mesoarchean felsic amygdolitic volcanite (Mauranipur), **c** Neoarchean felsic volcanics (Babina); **d** Neoarchean post-tectonic porphyritic granite (Babina), (pen —14 cm and compass scale—10 cm)

Singh 2019a; Sibelev et al. 2021). Central Bundelkhand greenstone complex includes Paleo/Mesoarchean metabasites, BIF, and two generations of Meso- and Neo-archean felsic volcanites (Fig. 2). There are also 2.67 Ga TTG (consist of mostly granodiorites), 2.56–2.54 Ga sanukitoids and monzogranites, 2.53–2.52 Ga plagioclase–orthoclase granites, which are more widespread in the craton (Verma et al. 2016; Joshi et al. 2017; Singh et al. 2021).

The important post-Archean features of the craton are characterized by NW–SE trending mafic dyke swarms and NE–SW to NNE–SSW trending giant quartz reef. The NW–SE trending mafic dyke swarms occurs around >50 km long discontinuous exposures. Geochemical characteristics of these dykes indicating the involvement of mantle sources followed by fractional crystallization of magma with low Ti–tholeiite, while some studies suggested two phases of mafic magmatic events in the craton (Basu 1986; Mondal and Ahmad 2001; Rao et al. 2005; Pati et al. 2008; Pradhan et al. 2012; Radhakrishna et al. 2013). Samom et al. (2017) estimated a mineral–whole-rock Sm–Nd isochron age ~2.1 Ga for Gwalior basin mafic magmatism at the northern periphery of the Bundelkhand Craton which occurs in an extensional intra-continental rift tectonic setting. Similar types of rocks also occur in the Bijawar basin at the Southern fringe of the craton and these rocks were dominated by sedimentary deposits during Paleoproterozoic (Absar et al. 2009, Colleps et al. 2021).

The quartz veins (reefs) have Paleoproterozoic (1.9–1.8 Ga) age and indicate that the Bundelkhand Craton was affected by one of the orogenic belts in Late Paleoproterozoic (Pati et al. 2007; Slabunov et al. 2017a). Neoarchean granitic magmatism, Paleoproterozoic dolerite emplacement, and hydrothermal quartz fluid make complex geology and envelop the entire craton. The boundaries of the craton to the east, west, and south parts are covered by the Mesoproterozoic Vindhyan basin. The Paleoproterozoic Bijawar and Gwalior groups of rocks cover marginal parts toward the SE and NW of the craton, respectively. While the Deccan trap basalts occur in small outcrops at the southwestern peripheral of the Bundelkhand Craton, the Gangetic alluvium covers the northern part of it (Basu 1986).

The available geochronological data encourage that the Archean granitoid rocks of Bundelkhand Craton can be broadly grouped as: (i) 3.5–3.2 Ga TTG (consisting tonalites, trondhjemites and granodiorites), (ii) 2.65–2.67 Ga TTG (consist mostly granodiorites), (iii) 2.56–2.54 Ga sanukitoids and monzogranites, and (iv) ca. 2.54–2.49 Ga anatectic granites. Granitoids of group 1 are concentrated in the central portion of the craton, but they are occasionally encountered in other blocks, while Neoarchean TTG has been reported only from the central portion. Neoarchean sanukitoids and monzogranites are concentrated mainly to the north of BTZ, but they are scarce to the south of this zone. Anatectic K-rich granites are distributed throughout the craton in a different manner: they predominate in all of its portions, but their fraction in the central portion is slightly smaller.

Important data on the crustal structure of the craton were provided by the results of magnetotelluric studies (Gokarn et al. 2013). The earth crust in the central portion of the craton is identified by relatively low resistivity and a thickness of about 60 km. This narrow crustal block is separated from neighboring blocks by well-defined boundaries, which can be interpreted as faults. The earth crust in the northern portion of the craton is understood as a homogeneous block with high resistivity and a thickness of 65–70 km. The earth crust of the southern portion of the craton is different from the two other portions. It displays a three-membered structure and is about 60 km thick.

Thus, Bundelkhand Craton granites and supracrustal complexes are distributed throughout the craton in a non-uniform fashion. Furthermore, the earth's crust in its portions is structurally different. Therefore, the craton can be subdivided into three terranes: Central, Northern, and Southern Bundelkhand terrane (Slabunov and Singh 2019b).

3 The Geology of Terranes of Bundelkhand Craton

3.1 The Central Bundelkhand Terrane

The Central Bundelkhand terrane composed of mainly granite–greenstone complex, occur in Babina-Mauranipur-Mahoba area (Fig. 1b; Mondal et al. 2002; Saha et al. 2011, 2016; Singh and Singh 2011; Singh and Slabunov 2015a; Kaur et al. 2016; Verma et al. 2016; Slabunov and Singh 2019a; Singh et al. 2021). They include Paleo-Neoarchean TTGs, (Fig. 2a), greenstone rocks, and Neoarchean granodiorite-granite suite. The greenstone rocks occur mostly in the Babina and Mauranipur belts, with an E–W strike direction arise as Bundelkhand Suture/Tectonic Zone (Gokarn et al. 2013; Malviya et al. 2006). These belts composed of Paleoarchean sequence of mafic to ultramafic metavolcanics (Singh et al. 2019a), Mesoarchean felsic volcanics (Fig. 2b), which are found as tectonic slices in the Mauranipur belt (Slabunov and Singh 2019a), metasedimentary rocks (BIFs), and Neoarchean felsic volcanics (lava (Fig. 2c) and dykes) (Singh and Slabunov 2015a, 2016; Slabunov and Singh 2019a). Neoarchean (2542 ± 17 Ma) felsic volcanics (lava) in the Babina belt have Sm–Nd model age of 3.14 Ga, which means that these rocks are contaminated with pre-existing older crust (Singh and Slabunov 2015a). The 2557 ± 33 Ma felsic volcanics (dykes) in the Mauranipur belt (Slabunov and Singh 2017, 2019a) also show their mixing with Mesoarchean rocks as have inherent zircons there. Neoarchean porphyritic granites (Fig. 2d) and anatectic granites cut TTG and greenstone belts and common forms of big massifs (Singh and Slabunov 2015a; Slabunov et al. 2020). The Central terrane is separated from neighboring terranes through crustal faults/shear zones diagnosed on a magnetotelluric profile (Gokarn et al. 2013). The E-W trending Central-Bundelkhand terrane is traced through the entire craton at a width of up to 20 km (Fig. 1b). The Sm–Nd model ages of granitoids from the *Central Bundelkhand terrane* exhibit variations from ca. 2600 to 3600 Ma, and ε_{Nd} values range from −7 to +5 (Nesterova et al. 2019). This indicated that the terrane contains rocks varying in age from Paleoarchean to Neoarchean.

3.2 The Northen Bundelkhand Terrane

The Northern Bundelkhand terrane is located north of the Central Bundelkhand terrane, which mostly consists of granodiorite-granite series, with visible role of sanukitoids (Fig. 3a–c) (Joshi et al. 2017), monzogranites. There are mostly 2560–2540 Ma sanukitoids massifs situated in this part of the terrane (Joshi et al. 2017). The massifs of porphyritic trachydacite-rhyolites (micromonzonite- granite) occur as subvolcanic nature in the field (Fig. 3b). The andesitic nature of rocks is to be found in Sakrar-Nivari area in association with the porphyritic granite (Singh et al. 2020). There are some exposures of deformed amphibolites near Sakrar-Nivari area in granodioritic and sanukitoid rocks which are cut by K-rich granite confirm it as Archean mafic rocks (Fig. 3d). The vertical dipping fine-grained felsic rocks occur along with E-W shear zones. The granites were also emplaced along E-W to ENE-WSW trending strike-slip plane. The K-rich Neoarchean granitoids (2.6–2.49 Ga) were intruded as granitic complex and subvertical E-W strike-slip crustal shears, reveal sinistral top to SW (left lateral) shear movement occurring at intermediate to shallow depth within intracratonic domains in northern terrane of Bundelkhand Craton (Bhatt and Singh 2019). Singh and Patil (2011) establish westward dipping magnetic lineation in K-rich granite, and the anisotropy of the magnetic susceptibility show general strike of magnetic foliation along N120°E with local variation, confirms the sub-horizontal to inclined strike-slip tectonism in northern terranes of the Bundelkhand Craton. The Sm–Nd model age of granitoids from this terrane varies from ca. 3100 to 3300 Ma, and ε_{Nd} values range from −7 to −2 (Nesterova et al. 2019), suggesting that Paleo- and Meso-archean predominate here in the protolith of the continental crust.

3.3 The Southern Bundelkhand Terrane

The Southern Bundelkhand terrane consists mostly of Neoarchean anatectic granites with minor volume of schist complex (Fig. 4a, b), TTG, sanukitoids, mafic–ultramafic layered (Fig. 4c), and gabbroids (Fig. 4d) intrusions. The Girar schist (metasedimentary) belt consists of two groups of rocks (Berwar Formation of the Mehroni Group; Pascoe

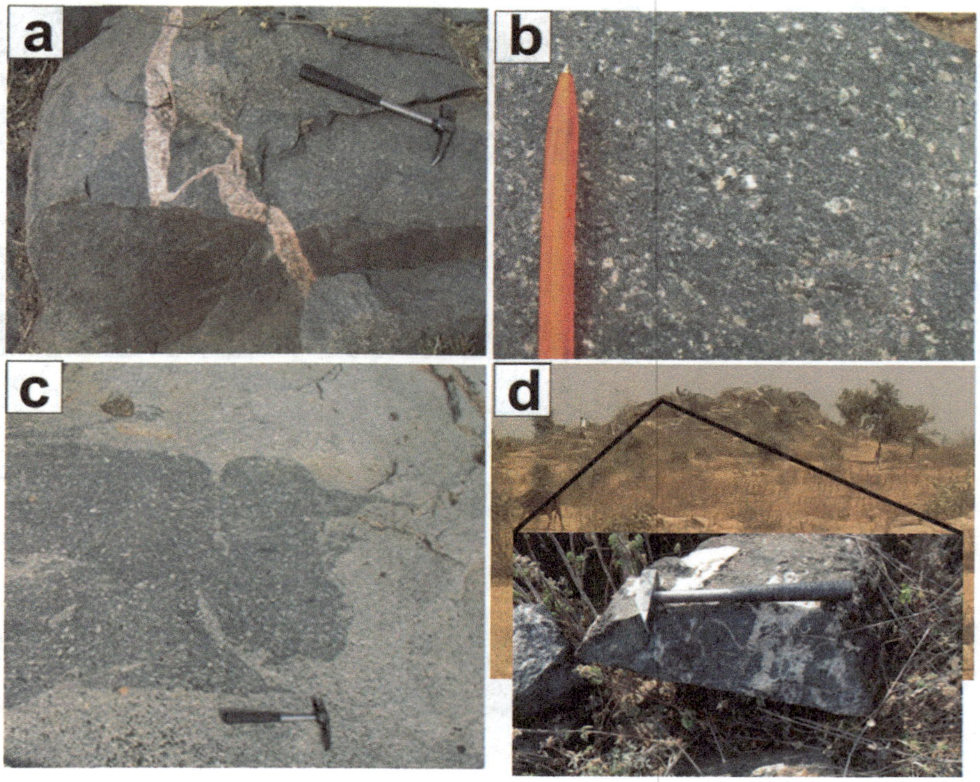

Fig. 3 Field occurrence of rock types in the Northern Bundelkhand terrane; **a** Sanukitoid granodiorite (Karera); **b** Subvolcanic sanukitoid (Nivari-Sakrar); **c** Mingling structure in granodiorite (Mahoba), (hammer 30 cm and pen—10 cm); **d** deformed amphibolites exposed near 5 km north of Nivari

1950): (i) quartzite (Fig. 4a), (ii) BIFs (Fig. 4b) and chlorite schist lenses near the quartzite/BIF boundary (Singh and Slabunov 2016; Slabunov et al. 2017b). Quartzites are represented by low-grade metamorphism of fuchsite- and hematite-bearing quartz arenite with thick meta-argillite (schist) laminae and lesser quartz pebble conglomerates (Singh and Slabunov 2016). BIF consists of thick-bedded quartz and hematite with magnetite. The quartzites of Girar schist (metasedimentary) belt consists of 3.43 and 3.25 Ga detrital zircons, indicating the existence of Paleoarchean granitoids (may be TTG) in the surrounding region (Slabunov et al. 2017b).

The Southern Bundelkhand terrane exposes a large intrusion, presumably a lopolith, known as the Ikauna complex (Farooqui and Singh 2010; Singh and Slabunov 2016; Slabunov et al. 2017b, 2018; Singh 2018). These mafic–ultramafic rocks lie among granitoids of Archean age, which apparently cut them, that determine the lower levels of their age as Archean. This intrusion indicates the influence of a mantle plume on the continental crust (Slabunov et al. 2018), but some scholars believe that it is an Alaska-type intrusion associated with subduction processes (Ramiz et al. 2019). Large-scale gabbroic rock and gabbro-diorite intrusions were identified near Lalitpur (Fig. 1b). The ~2.5 Ga granite has cross-cut to all rocks and indicates Archean age for other rocks. The Sm–Nd model age of granitoids varies from ca. 3100 to 4000 Ma, and ε_{Nd} values range from −15 to −2 in the Southern Bundelkhand terrane (Nesterova et al. 2019).

To identify the mineralogical composition, five samples of mafic–ultramafic rocks from Mauranipur greenstone belt (Central Bundelkhand terrane) and five samples of layered mafic–ultramafic and gabbroid rocks from Lalitpur region, Southern Bundelkhand terrane were collected and these samples were processed to X-ray diffraction analysis at the Innovative Centre of the Bundelkhand University, Jhansi, India. The phase formation of powder samples was confirmed by X-ray diffraction (XRD) technique using an X-ray powder diffractometer (Rigaku Corporation Japan, Smartlab 3 kW) with CuK α radiation (λ = 1.5405 Å) in slow scan in the 2θ range of 05–75°. Diffractograms were processed and mineral identification by using ICDD (pdf-2 release 2013) database. XRD analysis of selected rock samples is presented in Table 1 and diffractograms of two samples are shown in Figs. 5 and 6; signifying magmatic and post-magmatic minerals. It exhibits that the mafic–ultramafic rocks of Central terrane and layered mafic rocks of Southern terrane have different mineral phases.

Fig. 4 Field occurrence of rock types in the Southern Bundelkhand terrane; **a** Archean quartzite (Girar); **b** Archean BIF (Girar); **c** Mesoarchean pegmatitic gabbro (Ikauna intrusive); **d** Gabbro-diorite with fine gabbro xenolith (Lalitpur), (pencil—2 cm, hammer 30 cm and compass scale—10 cm)

Table 1 Mineral identified by X-ray diffraction (XRD) method to mafic–ultramafic rocks (Sl. No. 1–5) from Mauranipur greenstone belt, Central Bundelkhand terrane and layered mafic–ultramafic and gabbroid rocks (Sl. No. 6–10) from Lalitpur region, Southern Bundelkhand terrane

Sl. No	Sample no	Mineral identified by XRD in increasing order of abundance
1	MX20-1	Albite, Mg-Hornblende, Richterite
2	MX20-2	Edenite, Mg-Hornblende, Albite, Ferropargasite
3	MX20-3	Edenite, Albite, Quartz, Actinolite Mg-Hornblende
4	MX20-5	Edenite, Albite, Fe- Actinolite, Microcline
5	MX20-11	Fe- Actinolite, Albite, Mg Arfvedsonite, Quartz
6	19–21	Chlorite, Anorthite, Muscovite/Sericite, Augite, Fe-Actinolite
7	19–22	Chlorite, Quartz, Albite, Diopside, Muscovite/Sericite
8	19–25	Chlorite, Anorthite, Diopside, Quartz, Muscovite/Sericite, Mg-Hornblende
9	20–41	Chlorite, Anorthite, Augite, Muscovite/Sericite, Mg-Hornblende
10	20–46	Quartz, Albite, Riebeckite, Richterite, Diopside, Muscovite/Sericite

4 Discussions and Conclusions

The Babina and Mauranipur greenstone belts are composed of mafic to ultramafic lava, felsic volcanics, metasedimentary rocks (BIFs) of Mesoarchean (ca 2.81 Ga) sequence and Neoarchean felsic volcanics (lava and dykes) which combine the Central Bundelkhand greenstone complex. It recognize as: (a) the early (i) basic-ultrabasic, (ii) felsic volcanics, (iii) BIF, and (b) the late-felsic volcanics (2.54 Ga) assemblages. The fragments of Paleoarchean (3.44 Ga) oceanic mafic–ultramafic rocks are also present in these belts (Singh et al. 2019a). Slabunov and Singh (2019a) infer the SHRIMP dating of zircons of felsic volcanics are 3242 ± 65 Ma (xenocrystic zircons), 2810 ± 13, and 2542 ± 17 Ma (magmatic zircon). The Mauranipur felsic volcanics (dykes, 2557 ± 33 Ma) also show their mixing with Mesoarchean rocks as evidence of inherent zircons therein (Slabunov and Singh 2019a). The Sm–Nd model age of 2.54 Ga felsic volcanics indicates the existence of the old crust of 3.14 Ga, which verifies its crustal contamination (Singh and Slabunov 2015a). The felsic volcanics (calc-alkaline dacite-rhyolites composition) and basic-ultrabasic assemblage formed in subduction geodynamic settings. The ages of zircons from metasomatic rocks from tectonic contact are 2687 ± 11 Ma and this is the age of early accretion events. Thus, the Central Bundelkhand greenstone complex states the collage which forms in subduction setting at ∼2.83 and ∼2.54 Ga,

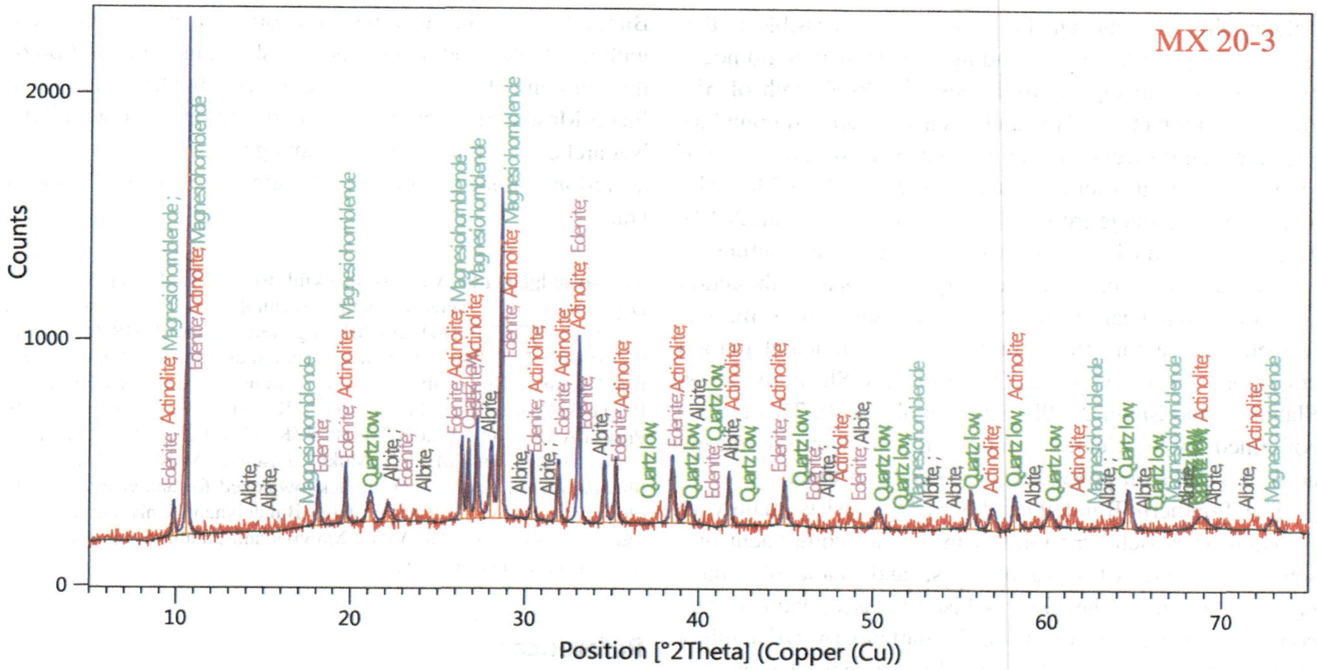

Fig. 5 X-RD analysis of deformed mafic–ultramafic sample (MX 20-3) near Kamla Sagar Dam, Kuraicha village, Mauranipur greenstone belt

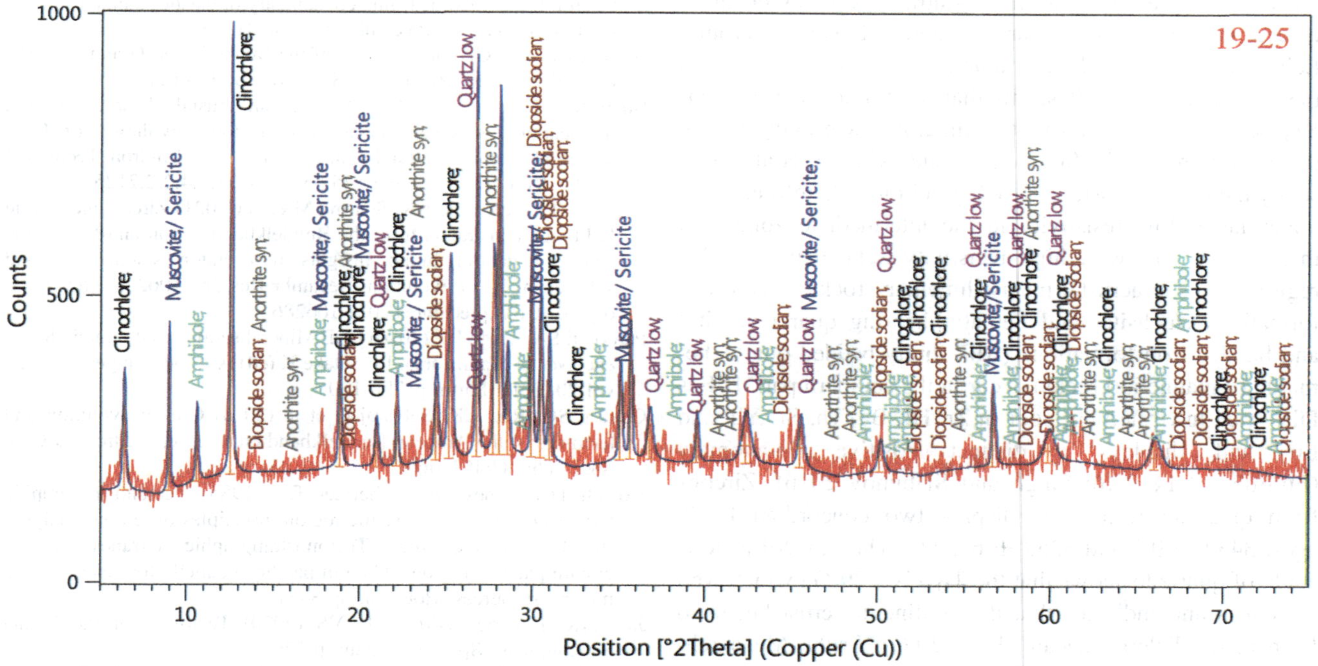

Fig. 6 X-RD analysis of layered mafic–ultramafic sample (19–25) near Jakhora, Lalitpur area in Southern Bundelkhand terrane

and accretion-collisional events during ~2.69 and ~2.51 Ga (Slabunov and Singh 2019b; Singh et al. 2020).

Granitoids (sanukitoids, granodiorites, and quartz monzonites) are more prominent in the Northern Bundelkhand terrane, which are exposed in north of the Central Bundelkhand granite–greenstone terrane. The sanukitoids massifs (2577–2560 Ma) are important components in northern terrane which occur as subvolcanic nature along with high-K granitic magmatism (Joshi et al. 2017; Singh et al. 2020). Around the Nivari Sakrar region, the deformed orthoamphibolite (metagabbro) bodies are accessible in this terrane. Magnetotelluric sounding has shown a homogeneous block with high crustal resistivity to a depth of 65–70 km (Gokarn et al. 2013) and has a well-defined boundary with the Central Bundelkhand terrane. Sm–Nd model age of granitoids from this terrane vary from ca. 3100 to 3300 Ma, and ε_{Nd} values range from −7 to −2 (Nesterova et al. 2019). This terrane consists mainly of Paleoarchean continental crustal rocks that made up the margin of a supersubduction slab in Neoarchean time. Numerous sanukitoid massifs formed in it at this stage under active continental margin conditions (Joshi et al. 2017; Joshi and Slabunov 2019; Slabunov and Singh 2019b; Singh et al. 2020). This terrane combined with Central and Southern terranes during accretion-collision processes in Neoarchean time.

The Southern Bundelkhand terrane consists mainly of Neoarchean K-rich granitoids with a small component of a schist complex, TTG, sanukitoids, and mafic–ultramafic layered intrusion. The Girar schist (metasedimentary) belt consists of two groups of rocks: (i) quartzite, (ii) BIFs, minor amount of dolomitic marble and chlorite schist lenses. The quartzite rocks consist of 3.43 and 3.25 Ga (detrital) zircons which are evident for the Paleoarchean granitoids (may be TTG) on this terrane. The mafic–ultramafic rocks occur as layered intrusive in this terrane too and well exposed around Lalitpur, Ikauna, and Madaura area due to mantle plume on the continental crust. It seems that the Girar metasedimentary belt was formed during this time. A low-gravity field in the central area of the Southern terrane (Gokarn et al. 2013) set by the thick deposit of K-rich granitoids; therefore, a thin upper layer, low resistivity in the intermediate zone, and thick lower unit with high resistivity zone depict in this region. The supracrustal metasedimentary rocks of quartzite formations (fuchsite- and hematite-bearing quartz arenite) and banded iron formations (BIF; thinly-bedded quartz–hematite (±magnetite) rocks) occur in the southern part. BIF is fairly rich in Cr and Ni, poor in Zr, Hf, Ba, Th, Sr, Yb, and Lu, and displays a distinct positive Eu-anomaly (Eu/Eu* = 1.14–2.46; Singh and Slabunov 2016). Zircons from quartzite formations display two concordian U–Pb ages: 3432 ± 9.7 and 3252 ± 6.4 Ma. The Sm–Nd isotope study of quartzite shows that the T_{DM} is 3.29 Ga which well correlates and indicates that the continental crust began to form in the Paleoarchaean (3.4–3.2 Ga). In the Mesoarchean, the continental crust of the South Bundelkhand Craton was affected by a mantle plume. As a result, numerous gabbroic rock intrusions and a layered mafic–ultramafic intrusion formed. Furthermore, the same geodynamic factor was responsible for the formation of a sedimentary basin, in which, quartzites and BIF were deposited (Singh and Slabunov 2016; Slabunov et al. 2017b).

Therefore, it seems that the Bundelkhand Craton consists of three terranes: Central, Northern, and Southern Bundelkhand. This tectonic division is in good agreement with available geodynamic models showing the evolution of the craton. Each of the cratons displays its own Paleo-Mesoarchean evolution pattern. And it was not until the Neoarchean that they amalgamated during an accretion-collision event and became one crustal block, a craton.

Acknowledgements VKS is thankful for financial support by the Department of Science and Technology, Govt. of India (INT/RUS/RFBR/P-279) and AIS is grateful to the RFBR for a grant (17-55-45005 IND-a). The work continues under MOU between Bundelkhand University and the Institute of Geology, Karelian Research Centre, RAS, Petrozavodsk, Russia, and is a contribution to Project AAAA-A18-118020290085-4 (Karelian Research Centre, RAS, Petrozavodsk, Russia). Dr. Mukul Pastore, Akash Raikwar, and Shubham Verma are thankfully acknowledged for analyzing the samples at the Innovation Centre of the Bundelkhand University, Jhansi, India. We also thank Dr. Vivek Malviya and Padum Singh (departed) for their critical comments.

References

Absar N, Raza M, Roy M et al (2009) Composition and weathering conditions of Paleoproterozoic upper crust of Bundelkhand craton, Central India: records from geochemistry of clastic sediments of 1.9 Ga Gwalior Group. Precambr Res 168:313–329

Basu AK (1986) Geology of parts of the Bundelkhand Granite Massif, Central India. In: Records, G.S.I. (Ed.), pp 61–124

Bhatt SC, Singh VK (2019) Neoarchean crustal shear zones and implications of shear indicators in tectonic evolution of Bundelkhand craton, central India. J Geosci Eng Environ Technol 4 (2):11–18. https://doi.org/10.25299/jgeet.2019.4.2-2.2125

Colleps CL, McKenzie NR, Sharma M et al (2021) Zircon and apatite U-Pb age constraints from the Bundelkhand craton andProterozoic strata of central India: Insights into craton stabilization and subsequent basin evolution Precambr Res 362:106286. https://doi.org/10.1016/j.precamres.2021.106286

Farooqui SA, Singh PK (2010) PGE Mineralisation in Ultramafic/Mafic enclaves of Ikauna area, In: Satake K (ed) Advances in geosciences, vol 20. Solid Earth, p 111–120

Gokarn SG, Rao CK, Selvaraj C et al (2013) Crustal evolution and tectonics of the Archean Bundelkhand craton, Central India. J Geol Soci India 82:455–460

Howell DG, Jones DL, Schermer ER (1985) Tectonostratigraphic terranes of the circum-pacific region: principles of terrane analysis. In: Howell DG (ed) Tectonostratigraphic terranes of the circum-pacific region. Circum-pacific council for energy and mineral resources, Houston, pp 3–31

Jain AK, Banerjee DM, Kale VS (2020) Tectonics of the Indian Subcontinent. Springer, Cham, p 576

Joshi KB, Slabunov A (2019) Neoarchean sanukitoids from the Karelian and Bundelkhand cratons: comparison of composition, regional distribution and geodynamic setting. Trans Karelian Res Centre Russian Acad Sci (2):5–25. https://doi.org/10.17076/geo841

Joshi KB, Bhattacharjee J, Rai G et al (2017) The diversification of granitoids and plate tectonic implications at the Archaean–Proterozoic boundary in the Bundelkhand Craton, Central India. In: Halla J, Whitehouse MJ, Ahmad T, Bagai Z (eds) Crust–Mantle interactions and Granitoid diversification: insights from Archaean Cratons. Geol Soc Lon (Special Publications) 449:123–157. https://doi.org/10.1144/SP449.8

Kaur P, Zeh A, Chaudhri N (2014) Characterisation and U-Pb–Hf record of the 3.55 Ga felsic crust from the Bundelkhand Craton, northern India. Precambr Res 255:236–244

Kaur P, Zeh A, Chaudhri N et al (2016) Unravelling the record of Archaean crustal evolution of the Bundelkhand Craton, northern India using U-Pb zircon–monazite ages, Lu–Hf isotope systematics, and whole-rock geochemistry of granitoids. Precambr Res 281:384–413

Malviya VP, Arima M, Pati JK et al (2006) Petrology and geochemistry of metamorphosed basaltic pillow lava and basaltic komatiite in the Mauranipur area: subduction related volcanism in the Archean Bundelkhand craton, Central India. J Mineral Petrol Science 101:199–217

Mondal MEA, Ahmad T (2001) Bundelkhand mafic dykes, central Indian shield: implication for the role of sediment subduction in Proterozoic crustal evolution. Island Arc 10:51–67

Mondal MEA, Goswami JN, Deomurari MP et al (2002) Ion microprobe 207Pb/206Pb ages of zircon from the Bundelkhand massif, northern India: implication for crustal evolution of Bundelkhand—Aravalli protocontinent. Precambr Res 117:85–100

Nesterova NS, Bayanova TB, Singh VK et al (2019) Sm–Nd mapping and tectonic division of the Bundelkhand Craton Indian Shield. Transactions of A. Fersman scientific session of Geological Institute, Kola Research Centre, RAS, Apatity, Russia (16):421–424. https://doi.org/10.31241/FNS.2019.16.085

Pascoe EH (1950) A manual of the geology of India and Burma, vol I, Chapter VII, Govt. of India Press, Calcutta, pp 246–293

Pati JK, Patel SC, Pruseth KL et al (2007) Geology and geochemistry of giant quartz veins from the Bundelkhand Craton. Cent India Implic: J Earth Syst Sci 116:497–510

Pati JK, Raju S, Malviya VP et al (2008) Mafic Dykes of Southern Uttar Pradesh, Bundelkhand Craton, Central India: field observations, mineralogy and petrochemical characterization In: Srivastava RK, Shivaji Ch (eds) Mafic dykes of India, Narosa publication, New Delhi, pp 547–569

Parfenov LM, Natapov LM, Sokolov SD et al (1993) Terrane analysis and accretion in northeast Asia. Island Arc 2:35–54

Pradhan VR, Meert JG, Pandit MK et al (2012) Paleomagnetic and geochronological studies of the mafic dyke swarms ofBundelkhand craton, central India: implications for the tectonic evolution and paleogeographic reconstructions. Precambr Res198–199:51–76

Radhakrishna T, Chandra R, Shrivastava AK et al (2013) Central/Eastern Indian Bundelkhand and Bastar cratons in thePalaeoproterozoic supercontinental reconstructions: a palaeomagnetic perspective. Precambr Res 226:91–104

Ramakrishnan M, Vaidyanadhan R (2010) Geology of India, vol. 1. In: Geological Society of India, Bangalore, p 556

Ramiz MM, Mondal MEA, Farooq SH (2019) Geochemistry of ultramafic–mafic rocks of the Madawara Ultramafic Complex in the southern part of the Bundelkhand Craton, Central Indian Shield: implications for mantle sources and geodynamic setting. Geol J 54:2185–2207. https://doi.org/10.1002/gj.3290

Rao JM, Rao GVSP, Widdowson M et al (2005) Evolution of Proterozoic mafic dyke swarms of the Bundelkhand Granite Massif, Central India. Curr Sci 88:502–506

Saha L, Pant NC, Pati JK et al (2011) Neoarchean high-pressure margarite-phengitic muscovite-chlorite corona mantled corundum in quartz-free high-Mg, Al phlogopite-chlorite schists from the Bundelkhand craton, north central India. Contrib Miner Petrol 161:511–530

Saha L, Frei D, Gerdes A et al (2016) Crustal geodynamics from the Archaean Bundelkhand Craton, India: constraints from zircon U-Pb–Hf isotope studies. Geol Mag 153:79–192

Samom JD, Ahmad T, Choudhary AK (2017) Geochemical and Sm–Nd isotopic constraints on the petrogenesis and tectonic setting of the Proterozoic mafic magmatism of the Gwalior Basin, central India: the influence of Large Igneous Provinces on Proterozoic crustal evolution. Geol Soc Lond (Special Publications) 463. https://doi.org/10.1144/SP463.10

Sibelev OS, Slabunov AI, Singh VK, Mishra S (2021) Metamorphism of the Central Bundelkhand greenstone complex of theBundelkhand Craton, Indian Shield and its geodynamic setting. In: Shandilya AK, Singh VK, Bhatt SC, Dubey CS (eds)Geological and Geo-environmental Processes on Earth, pp 143–154 (this volume)

Singh PK (2018) Platinum group of metals (PGM) occurrence and future prospective in ultramafic and associated rocks from Madaura-Ikauna-Pindar tract in Madaura igneous complex: signature of plume activity during evolution of Bundelkhand Craton, Central India. Arch Anthropol Open Acc 3(suppl-3):385–392. https://doi.org/10.31031/AAOA.2018.03.000563

Singh MM, Singh VK (2011) Geochemistry and tectonic setting of the TTG-gneiss and granite from central part of the Bundelkhand craton, India. In: Singh VK, Chandra R (eds) Proceeding of the 2nd international conference precambrian continental growth and tectonism. Angel Publication, New Delhi, pp 95–102

Singh VK, Patil SK (2011) Anisotropy of magnetic susceptibility studies of northern part of Bundelkhand craton: implication for tectonic evolution. In: Singh VK, Chandra R (eds), Proceeding of the 2nd international conference precambrian continental growth and tectonism. Angel Publication New Delhi, pp 59–71

Singh VK, Slabunov A (2013) The Greenstone belts of the Bundelkhand craton, Central India: new geochronological data and geodynamic setting. In: Singh VK, Chandra R (eds) International association for Gondwana research conference series No. 16. 3rd international conference precambrian continental growth and tectonism, Jhansi, India, pp 170–171

Singh VK, Slabunov A (2015a) The Central Bundelkhand Archaean greenstone complex, Bundelkhand Craton, Central India: geology, composition, and geochronology of supracrustal rocks. Int Geol Rev 57(11–12):1349–1364

Singh VK, Slabunov A (2015b) Geochemical characteristics of banded iron formation and Metavolcanics of Babina greenstone belt of the Bundelkhand Craton, Central India. J Econ Geol Geo Resour Manag 10:63–74

Singh VK, Slabunov A (2016) Two types of Archaean supracrustal belts in the Bundelkhand Craton, India: geology, geochemistry, age and implication for craton crustal evolution. J Geol Soc India 88:539–548

Singh PK, Verma SK, Moreno JA, Singh VK et al (2019a) Geochemistry and Sm-Nd isotope systematics of metabasalts from the Babina and Mauranipur greenstone belts, Bundelkhand craton: implications for tectonic setting and Paleoarchean mantle evolution. Lithos 330–331:90–107

Singh PK, Verma SK, Singh VK et al (2019b) Geochemistry and petrogenesis of sanukitoids and high-K anatectic granites from the Bundelkhand craton: implications for the late-Archean crustal evolution. J Asian Earth Sci 174:263–282

Singh VK, Verma SK, Singh PK et al (2020) Archean crustal evolution of the Bundelkhand Craton: evidence from granitoid magmatism. In: Archean Granitoids of India: windows into Early Earth tectonics. Geol Soc Lond (Special Publ 489):235–259. https://doi.org/10.1144/SP489-2018-72

Singh PK, Verma SK, Singh VK et al (2021) Geochronology and petrogenesis of the TTG gneisses and granitoids from theCentral Bundelkhand granite-greenstone terrane, Bundelkhand Craton, India: implications for Archean crustal evolution andcratonization. Precambr Res 359:106210. https://doi.org/10.1016/j.precamres.2021.106210

Slabunov AI, Singh VK (2017) Central Bundelkhand greenstone complex of the Bundelkhand Craton, India: new geochronological data, a geodynamic setting and the position of the craton in the Kenorland Supercontinent structure. In: Slabunov AI, Svetov SA, Baltibaev ShK (eds) Early precambrian versus modern geodynamics, extended abstracts and field trip guide, Petrozavodsk, Russia, KarRC RAS, pp 235–238

Slabunov AI, Singh VK (2019a) Meso-Neoarchaean crustal evolution of the Bundelkhand Craton, Indian Shield: new data from greenstone belts. Int Geol Rev 61:1409–1428. https://doi.org/10.1080/00206814.2018.1512906

Slabunov AI, Singh VK (2019b) The new tectonic division of the Bundelkhand Craton Indian Shield. Transactions of A. Fersman scientific session of Geological Institute, Kola Research Centre, RAS. Apatity, Russia 16:521–524. https://doi.org/10.31241/FNS.2019.16.106

Slabunov AI, Lobach-Zhuchenko SB, Bibikova EV et al (2006) The Archaean nucleus of the Fennoscandian (Baltic) Shield. Geol Soc Lond Mem 32:627–644

Slabunov AI, Singh VK, Shchiptsov VV et al (2017a) A new Paleoproterozoic (1.9–1.8 Ga) event in the crustal evolution of the Bundelkhand Craton, India: the results of (SHRIMP) dating of zircons from giant quartz veins. In: Slabunov AI, Svetov SA, Baltybaev ShK (eds) Early Precambrian versus modern geodynamics. Extended abstracts and field trips guide, Petrozavodsk, KarRC RAS, pp 239–241

Slabunov A, Singh VK, Joshi BK, Li X (2017b) Paleoarchean zircons from quartzite of South Bundelkhand Supracrustal complex: origin and implications for crustal evolution in Bundelkhand Craton, Central India. Curr Sci 112:794–801

Slabunov A, Egorova S, Singh VK et al (2018) Archean mafic-ultramafic Ikauna layered intrusion, Bundelkhand craton, India: petrography and geochemistry. Arch Anthropol Open Acc. 3(suppl-2):334–340. https://doi.org/10.31031/AAOA.2018.03.000557

Slabunov AI, Singh VK, Bayanova TB et al (2020) The felsic volcanics of the Central-Bundelkhand Greenstone complex, Bundelkhand Craton, India: new geochronological, Sm–Nd data and geodynamics setting. In: Abstracts of the recent trends in geoscientific research on dharwar craton and other Indian Precambrian terrains, Hyderabad, India, pp 53–54

Verma SK, Verma SP, Oliveira EP, Singh VK, Moreno JA (2016) LA-SF-ICP-MS zircon U-Pb geochronology of granitic rocks from the central Bundelkhand greenstone complex, Bundelkhand craton, India. J Asian Earth Sci 118:125–137

Tectonic Evolution of Babina–Prithvipur Crustal Shear Zones, Bundelkhand Craton, India: Implications of Shear Indicators

S. C. Bhatt and Vinod K. Singh

Abstract

The central Bundelkhand domains consisting of deformed gneisses and granitoids are delineated by E-W subvertical Neoarchean Babina–Prithvipur shear zone ($\sim$500 m wide). The tonalite-trondjhemite-granodiorite (TTG) gneisses (3.5–3.2 Ga), streaky to mafic gneisses, belonging to an oldest complex unit, and metabasic, banded iron formation (BIF), felsic volcanics, with metasedimentaries constituting greenstone complex were subjected to three sets of folding in D_1 and D_2 compressive phases. The S_1 foliation trending ESE-WNW to E-W directions were developed in tight to isoclinals F_1 folds during D_1 phase. The F_2 and F_3 open folds generated in the D_2 phase are characterized by S_2 and S_3 foliations. The NNE-SSW and NNW-SSE trends were observed in S_2 and S_3 foliations. The K-rich Neoarchean granitoids (2.6–2.49 Ga) intruded as large granitic complex were transected by E-W strike-slip Babina–Prithvipur shear zone (syn-to post-tectonic D_3 phase). These sheared rocks are characterized by the E-W to ESE-WNW trending mylonitic foliation and stretching lineation. The mylonitic foliation, asymmetrical rotated porhyroclasts (σa and σb mantled clasts), asymmetrical pressure shadows and deformation lamellae were identified as macroscopic and microscopic shear indicators in protomylonite and mylonites. The minor folds and crinkles developed due to back rotation of axial plane and reorientation of fold axis during progressive shearing were found important features. The scattered few porphyroclasts of quartz and feldspar were examined in a recrystallized quartz matrix of ultramylonite. The sinistral top–to–SW sense of shear movement was exhibited by most of the asymmetrical porphyroclasts. The presence of undulose extinction deformation lamellae in asymmetrical fabrics reveals that the crystal plastic process was dominant during the evolution of mylonite under low to medium temperature conditions.

Keywords

Shear sense indicators • Babina crustal shear zone • Bundelkhand Craton • Central India

S. C. Bhatt · V. K. Singh (✉)
Department of Geology, Institute of Earth Sciences, Bundelkhand University Jhansi, Jhansi, India
e-mail: vinodksingh@bujhansi.ac.in

S. C. Bhatt
e-mail: scbhatt@bujhansi.ac.in

1 Introduction

The micro tectonics dealing with the implications of macro and microstructures in crustal shear zones have become more significant in understanding the tectonic growth of continental crusts (Passchier and Trouw 2005). The existence of crustal shear zones on major and minor scales was well marked by the presence of scattered features preserved in cratons. Such crustal shear zones provide passage to emplace granitic magma into continental crust (Hutton 1988). The intense deformation with large shear strain is accumulated in these shear zones relative to surrounding rocks (Ramsay and Huber 1987). Small-scale structures that developed in response to progressive simple shear in a shear zone characterized by a protracted history of deformation are immensely useful in delineating the strain history and kinematics of a shear zone (Simpson and Schmid 1983; Passchier and Trouw 2005 and references therein).

The deformed rocks are considered as the direct source of preserving the imprints of deformation which depict the tectonic evolution of the rocks. The geometry, orientation and optical properties of fabrics of rocks are found important characters to reveal the structural history of the area (Passchier and Trouw 2005). The microstructural studies of such shear zones will provide important clues to understand the mechanism of deformation and metamorphic history of the rocks (Passchier and Trouw 2005). The Bundelkhand Craton located in the northern segment of the Indian shield is less

A. K. Shandilya et al. (eds.), *Geological and Geo-Environmental Processes on Earth*, Springer Natural Hazards,
https://doi.org/10.1007/978-981-16-4122-0_11

understood in terms of its crustal evolution and tectonics (Fig. 1; Basu 1986; Roday et al. 1993; Singh and Slabunov 2015a, 2016; Slabunov et al. 2017a; Singh et al. 2020, 2021a). The E-W crustal shear zones, characterized by intensive mylonitization in the Bundelkhand Craton, are considered potential zones to depict the tectonic evolution of Neoarchean crust (Senthiappan 1993; Roday et al. 1993, 1995; Bhatt and Hussain 2008; Bhatt and Mahmood 2008; Bhatt and Gupta 2009; Singh et al. 2020, 2021a, b). The earlier studies were based on structural and geological data collected from the field. Later on Bhatt and Mahmood (2012), Bhatt and Hussain (2012), Bhatt and Gupta (2014) and Bhatt (2014) pursued integrated field and microfabric studies along these shear zones and established new concepts for tectonic evolution.

In the present study, based on microstructural features and correlation of chronological status of each tectonic phase, the deformational history of the craton was reformulated. Therefore, to achieve this goal, attempts were made to explore the tectonic significance of shear indicators present in the Babina–Prithvipur crustal shear zone. The asymmetrical shear indicators (rotated porphyroclasts, mylonitic foliation, S-C planes, stretching lineation and minor faults/micro cracks) examined on mesoscopic and microscopic scales were used to decipher the sense of shear movement and a tectonic model is produced. Apart from this, the geological and tectonic events were correlated with reference to different geological age constraints.

2 Geological Setting

Bundelkhand Craton located in the north of Peninsular India spreads in a 29,000 square kilometer area in central India. The eastern and western margins are covered by Vindhyan and Bijawar basins and are separated from the southern protocontinental/Deccan protocontinental block by a Central Indian Tectonic Zone (Acharya and Roy 2000). The Aravalli and Bundelkhand Craton are tectonically separated by the Great Boundary fault. The northern fringes marked by deep Himalayan basins are tectonically delineated by the Yamuna fault. The oldest unit is considered as TTG-gneissic

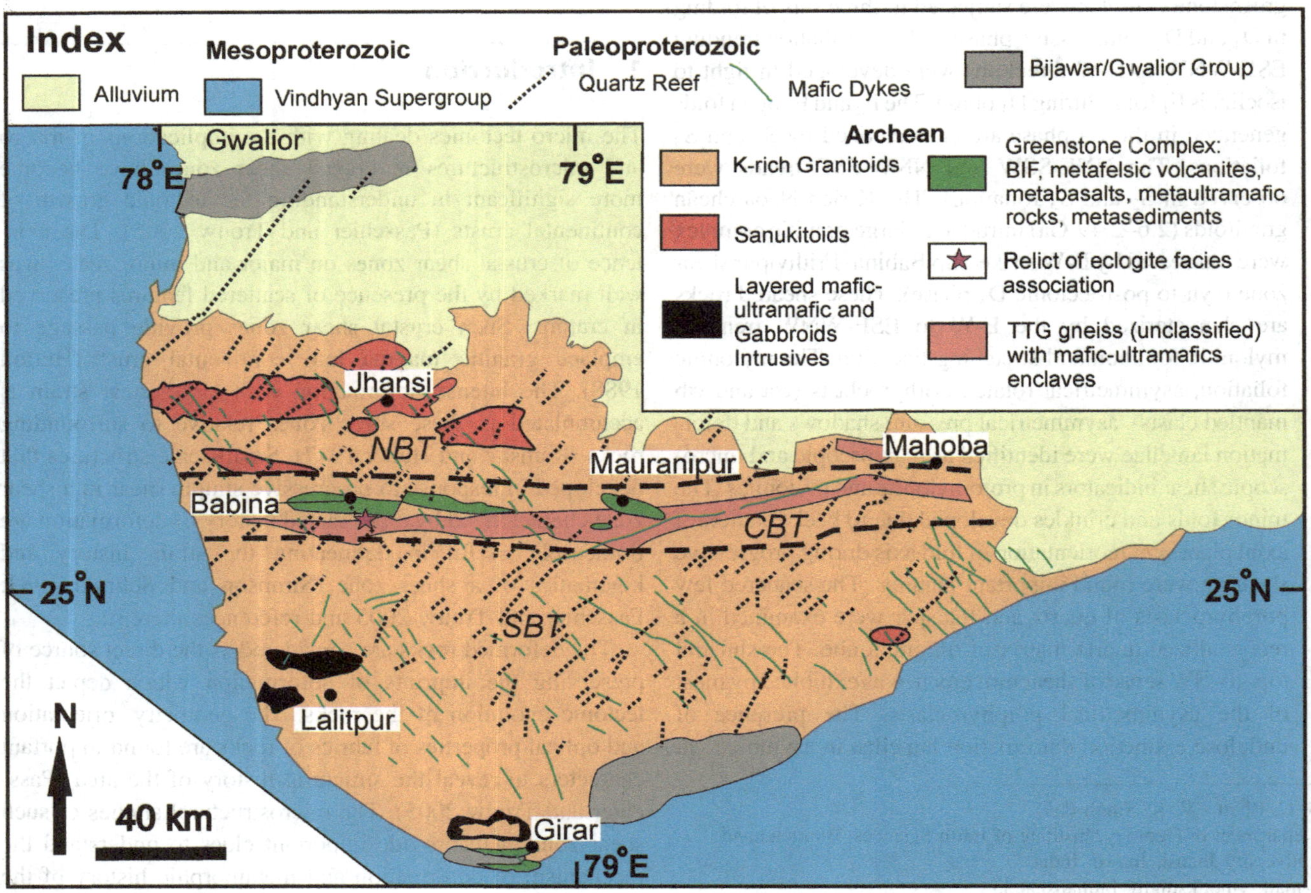

Fig. 1 General geological map of the Bundelkhand Craton. NBT, CBT and SBT—Nothern-, Central- and Southern-Bundelkhand terranes (modified after Singh et al. 2020, 2021a, b)

complex, constituting TTG and gneisses, associated with migmatites, schists and orthoamphibolites in the Bundelkhand Craton. The greenstone complex consists of basic, Banded Iron Formation (BIF) and less metamorphosed felsic volcanic-sedimentary rocks (Fig. 1). The granitic complex comprises of K-rich granitoids (various types of granites). Dolerite dyke, giant quartz veins (reefs), pegmatites, and aplite dykes are found important geological features in the Bundelkhand Craton (Jhingran 1958; Prakash et al. 1975; Mishra and Sharma 1975; Basu 1986, 2007; Roday et al. 1993, 1995; Pati et al. 2007; Bhatt and Hussain 2008, 2012; Bhatt and Mahmood 2008, 2012; Bhatt and Gupta 2009, 2014; Bhatt et al. 2011, 2017; Bhatt 2014; Singh and Slabunov 2015a, b, 2016; Slabunov et al. 2017a, b).

The oldest TTG rocks were reported in the central part of the Bundelkhand Craton (Sarkar et al. 1996; Kaur et al. 2014; Saha et al. 2016). Mondal et al. (2002) dated Mauranipur and Mahoba gneisses ca. 3.27 Ga by Pb–Pb model method and considered this event as the oldest deformation phase and one set of TTG magmatism. Singh et al. (2019a) suggest that the ca. 3.4 Ga mafic–ultramafic rocks from the central Bundelkhand greenstone complex have been interpreted to be derived from oceanic crust in a subduction-related setting with depleted mantle composition. The vast intrusive of sanukitoids, granodiorites, diorites, and high-K granites rocks were occurred during Neoarchean time (2.55–2.52 Ga; Mondal et al. 2002; Singh and Slabunov 2013, 2015a, 2016; Verma et al. 2016; Kaur et al. 2016; Joshi et al. 2017; Slabunov and Singh 2019; Singh et al., 2019b, 2020, 2021a). A NW–SE trending dolerite dyke swarm was intruded ca. 2000–1800 Ma (Rao et al. 2005). Patil et al. (2007) reveal that the dolerite dykes have a palaeomagnetic VGA position at about 2150 Ma. Keeping in view the complicated geological setup, detailed field investigations were carried out and tectonic set up has been reconstructed in Babina–Prithvipur sector.

3 Babina–Prithvipur Sector

The older basement consisting of light to dark grey gneisses belongs to TTG-gneissic complex. The light–dark grey TTG and banded mafic gneisses are exposed in and surroundings of Ghisauli, Badera and Jaunpur villages located in the south of the Babina town (Figs. 1 and 2; 3.5–3.3 Ga; Sarkar et al. 1996; Mondal et al. 2002). These multi folded (F_1-F_3) and foliated (S_1) gneisses exhibiting wider leucocratic (quartzo-feldspathic) and thinner melanocratic (mafic) bands show extensive shearing effects. Their foliation striking in E-W to WNW-ESE exhibits moderate to steep (45–60°) northerly and north-easterly dips. These gneisses transected by E-W brittle-ductile crustal shear zones, are represented by stretching lineation and mylonitic foliation (Fig. 3a). The shearing effects were also observed in pink granite exhibited by rotated fabrics of quartz and feldspar (Fig. 3b). The lensoidal bodies of dark green low-grade mafic and ultramafics associated with gneissic complex occur in small patches located in the north of Tangara, east of Badera village (Fig. 2). The outcrops of streaky and sheared gneisses are exposed in the west of Chamon village (Fig. 3c, d).

The large outcrops of sheared gneisses widely occur in the western part of Papawni and Dhaura villages and their small patches are also found in the south of Khiston and north of Gailwara villages. The superimposition of folding and thinner leucocratic and thicker melanocratic bands are found distinct features of this E-W to ESE-WNW striking rocks. The intensive shearing effects are noticed in the northwestern part of Papawni village (Fig. 2).

The linear hillocks ($\sim$300 m) of quartz magnetite rocks of Banded Iron Formation (BIF) sandwiched between the southern gneissic complex and northern granitic complex are exposed near Pura and Papawni villages (Fig. 2). The metabasic, ultrabasic, BIF and volcanic rocks are grouped under the Central Bundelkhand greenstone Complex (Singh and Slabunov 2015a). The BIF exhibiting irregular bands of quartz and magnetite are found in the west of Papawni village. The ESE-WNW trending foliated fuchsite quartzite consisting of alternating bands of white quartz (0.5–1 cm) and dark green fuchsite (1.5–3 cm), found as isolated patches within the older mafic gneissic terrain in the south of Babina, near Sukwan Dukwan reservoir (Fig. 2). Prasad et al. (1999) pointed out that the felsic volcano-plutonic complex was juxtaposed with southern gneisses and volcano-sedimentary sequence along a sinistral shear zone.

Near Papawni the tight and similar folds and sometimes the folded rafts of BIF are found to be associated with granite gneisses. At few places, these laminated bands of quartz (0.1–5 cm) and magnetite (0.4–1 cm) are displaced by the minor strike-slip faults. The mylonitised metavolcanics, mostly rhyolitic composition, occur as E-W trending linear belts in the north of Papawni village and exhibit extensive shearing effects (Fig. 2). The mylonite and ultramylonite zones are distinctly recognized in the south of Taparyan village. The ultramylonite zone is mainly characterized by a fine-grained recrystallized matrix of quartz with few protoliths of quartz and feldspar. The joints and fractures are commonly observed in these sheared rocks and at places the fractures are healed up by quartz veins. Sometimes these veins are displaced by the minor strike-slip faults. The mylonite zone exposed at Taparyan village and extending up to Dhaura exhibits well-developed mylonitic foliation and lineation (Fig. 3a). The mylonitic foliation is striking in E-W with steep (60–70°) northerly dips and stretching lineation represented by the elongated grains of quartz, feldspar lies subhorizontal to the main mylonitic foliation. The lensoidal

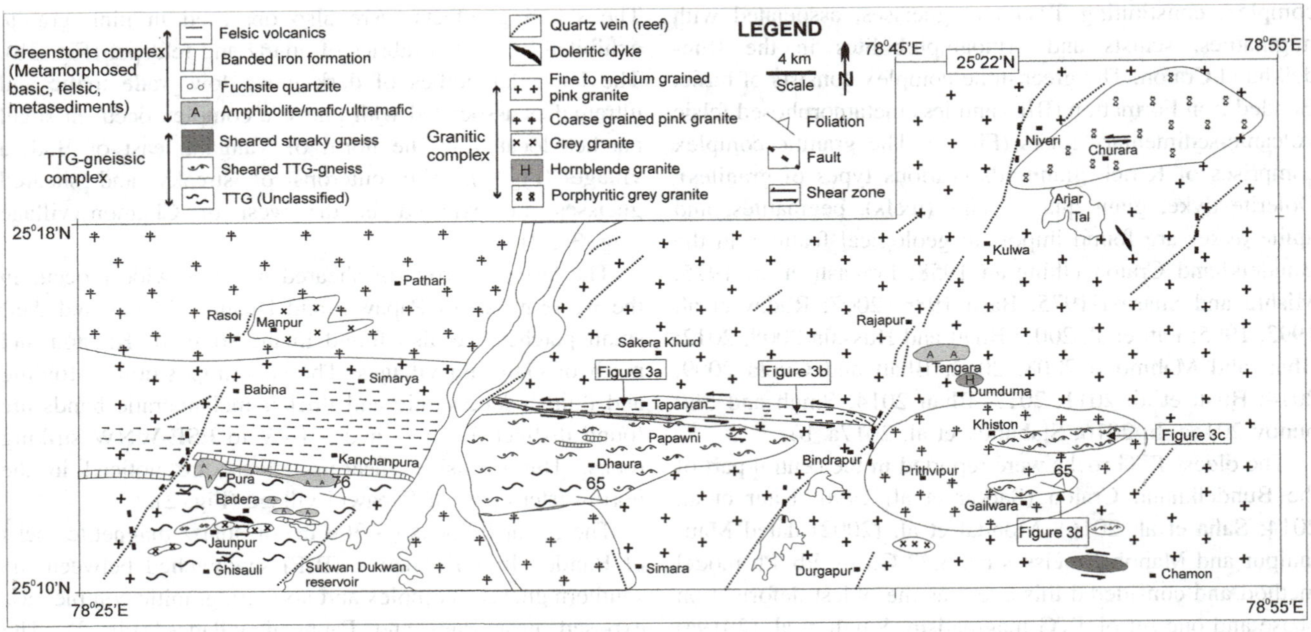

Fig. 2 Geological map of Babina–Prithvipur sector, Bundelkhand Craton, central India (After Bhatt and Mahmood 2012; Singh and Slabunov 2015a)

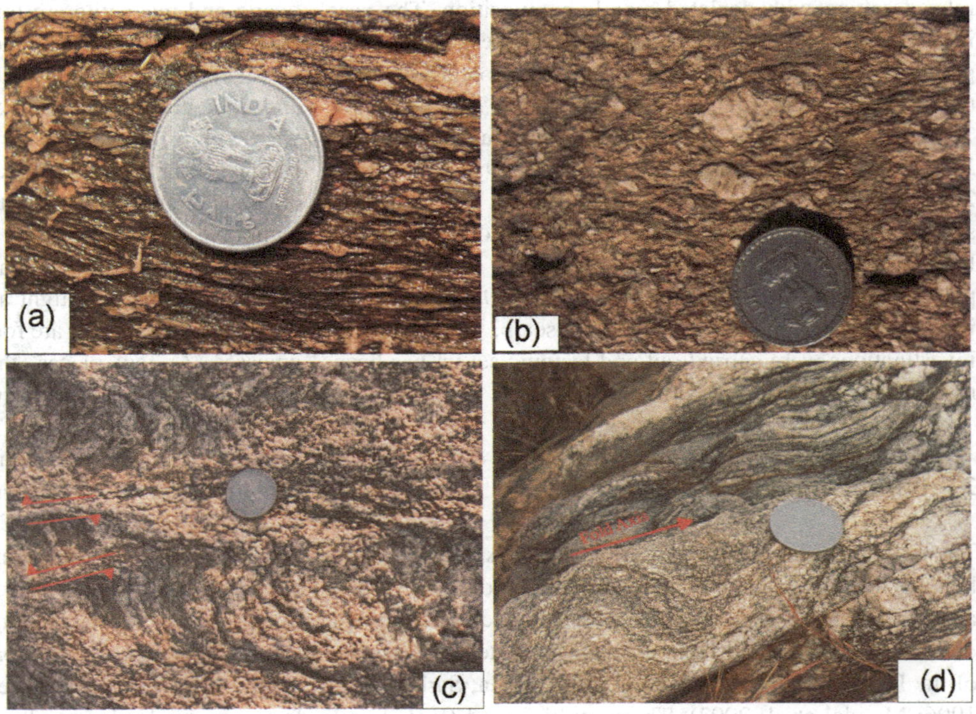

Fig. 3 **a** Well foliated sheared felsic volcanics showing mylonic foliation and lineation; south of Taparya village, diameter of coin is 2.4 cm; **b** Mylonitised grey granite showing mylonite zone represented by dextrally and sinistrally rotated quartz and feldspar grains, diameter of coin is 2.2 cm; **c** Sheared streaky gneisses exhibiting sinistral asymmetry in minor folds developed due to back rotation of closely spaced shear planes, west of Chamon village, diameter of coin is 2.2 cm; **d** The closely spaced shear planes developed parallel to fold axis in streaky gneisses, west of Chamon village, diameter of coin is 2.4 cm

bodies of amphibolite, metabasalt, magnetite quartz and schist were found at few places within the gneissic complex. These quartz magnetite rocks striking in E-W to WNW-ESE with steep northeast and north dips are characterized by alternate bands of sharp laminations and bedding. The schistosity is lacking in amphibolite and metabasalt.

Five types of granites: (i) porphyritic grey granite, (ii) hornblende granite, (iii) grey granite, (iv) coarse-grained pink granite and (v) fine to medium grained pink granite are identified on the basis of field relationship and petrographic characters (Fig. 2). The lensoidal bodies (2 km) of fine greyish green hornblende granite exposed on both sides of

Nivari–Prithvipur road near Dumduma are containing hornblende and quartz with rare phenocrysts of feldspar. The porphyritic dark grey granite consisting of large porphyroclasts (0.3–2 cm) of quartz and feldspar (mainly orthoclase) are found in the eastern flanks of Nivari and in the surroundings of the Churara village (Fig. 2). Few outcrops of massive and medium-grained granites, occasionally intruded by quartz veins are exposed in a dome-shaped hillock in the north of Durgapur village and southwest of Papawni village (Fig. 2). The small enclaves of fine to medium-grained grey granite, varying a few centimeters in size and occurred within the pink granite, reveals that the grey granites were formed earlier than pink granites. The outcrops of medium to coarse-grained grey granite are extensively exposed in low-lying hills located near Manpur village (Fig. 2). The medium to coarse-grained pink granite constituting the megacrysts of white quartz, K-feldspar are occupying the northern part. The large tracts of massive, pinkish-red coarse-grained pink granites widely occur around Prithvipur and Chandawni, Pathari and Rasoi villages (Fig. 2), and at places, these rocks are marked by one to two sets of tensional fractures and three sets of joints. These rocks show sharp contact with TTG-gneisses in the south of Khiston. The NE-SW to E-W trending quartz veins and the enclaves of mafic rocks (7–15 cm) are noticed in these granites. The huge outcrops of fine to medium grained, dark pink granites occurred in the surroundings of Nivari, Rajapur, Durgapur and Babina localities and are recorded as the youngest granitic activity in the Bundelkhand Craton (Fig. 2). The joints are seldom persistent in these rocks, but at some places, two to three sets of joints are noticed.

A medium-grained NW–SE dolerite dyke is exposed in the southeast of Arjar Tal (Fig. 2). The spectacular NE-SW linear of quartz reef passes from Babina and is truncated by oblique faults. Another NW–SE trending quartz reef (200–300 m) encompassing from west of Simara is intruded by several sigmoidal quartz veins. A quartz reef appeared as an elevated ridge and emerging from north of Nivari is extending to the southwest of Durgapur is a major reef in the area.

4 Shear Zones and Shear Indicators

The shear zones can be recognized from hand specimen to plate boundary scales and can be reactivated after a very long geological time span. The ductile and brittle-ductile shear zones were the product of high shear strain domains and are classified as important mechanical heterogeneities (Butler et al. 1997). The geometry of mylonitic foliation, asymmetrical fabrics and stretching lineation (Berthe et al. 1979; Matlauer et al. 1981) and implications of shear criteria (Simpson and Schmid 1983) strongly support the criteria for the determination of sense of shear movement in most of the crustal scale shear zones. The deformation of quartzo-feldspathic granitic rocks is controlled by crystal plasticity and pressure solution mechanism (Berthe et al. 1979; Watts and Williams 1979).

The protomylonite was initially produced under low to moderate strain conditions whereas the S-C mylonite evolved under moderate to high shear strain conditions (Passchier and Trouw 2005). Three types of mylonites: (1) Protomylonites, (2) S-C mylonites and (3) ultramylonites were noticed along Babina–Prithvipur shear zone of the Bundelkhand Craton.

5 Results and Discussions

5.1 Deformation Pattern

The geometrical and kinematic analysis of folds and study of asymmetrical structures (on meso and microscopic scale) in Babina–Prithvipur and Mauranipur area infer that the Archean gneissic complex was deformed and exhibits three sets of folding (F_1-F_3) under the influence of compressive tectonic episodes (D_1-D_2; Bhatt et al. 2011; Bhatt 2014). The F_2 and F_3 (open to reclined) folds were displaced orthogonally by sheared planes in the later episode of shearing (Bhatt et al. 2011; Bhatt and Mahmood 2008, 2012). The various types of granitoids were emplaced along with fractures and shear zones in the D_3 tectonic phase. The E-W crustal shear zones transecting the gneissic and granitic rocks of the area exhumed during syn-to post-tectonic D_3 phase under brittle-ductile environment (Bhatt and Mahmood 2012; Bhatt 2014; Singh and Slabunov 2015a, b). The dolerite dykes were emplaced along with NW–SE trending fractures in an extensional tectonic setting during the D_4 (magmatic) phase. The linear quartz reefs exhumed along NE-SW shears (riedel shears?) were evolved in the last endogenic activity (D_5) (Slabunov et al. 2017b; Slabunov and Singh 2018).

The TTG gneisses of this craton associated with metabasic underwent earlier magmatism and metamorphic event at ~3.3–3.1 Ga (Mondal et al. 2002). This phase may represented by F_1 folding and associated with the D_1 phase of deformation. The major compressive tectonic phase (D_2) responsible to generate three sets of folding (F_1-F_3) took place between 3.2 and 2.7 Ga. The Mesoarchean to Neoarchean TTG-gneisses along with metasedimentary units and felsic volcanics were folded and refolded during this compression phase (D_2). Saha et al. (2011) suggested that the outer collar of chlorite-phengitic muscovite of Chlorite schist located south of Babina was formed at 18–20 Kbar and ca. 630 °C and the inner collar yield lower metamorphic P–T conditions of 11 ± 3 Kbar ca. 630 °C. They further analyzed the U–Th–Pb chemical dating of metamorphic monozite and LA-ICPMS U–Pb dating of requilibrated zircon and recorded ca. 2.78 Ga ages. It implies that the Neoarchean high-pressure metamorphism in Bundelkhand Craton may be associated with the D_2 phase of tectonic compression. The multiple phased granitic rocks occurring between 2580 and 2530 Ma show extensive granitic magmatic phase (D_3) of Bundelkhand Craton (Verma et al. 2016; Joshi et al. 2017; Singh et al. 2019b). The field setting and other signatures reveal that the quartz reefs were exhumed from the intermediate to shallow depth at crustal levels along NE-SW trending shear zones. The mylonitised granitic rocks sinisterly displaced by NE-SW oblique faults may be formed due to the emplacement of huge quartz reefs (endogenetic movement) in ductile to brittle regime.

5.2 Babina–Prithvipur Shear Zone

The older rocks of TTG gneissic and greenstone complexes of Bundelkhand Craton were subjected to three sets of folding (F_1-F_3) under the influence of compressive tectonic episodes (D_1-D_2; Bhatt et al. 2011; Bhatt 2014). The various types of granitoids were emplaced along with fractures and shear zones in the D_3 tectonic phase. The E-W crustal shear zones transecting the gneissic and granitic rocks of the area emplaced during syn-to post-tectonic D_3 phase under brittle-ductile environment (Bhatt and Mahmood 2012; Bhatt 2014; Singh and Slabunov 2015a, b). Bhatt and Mahmood (2012) and Bhatt et al. (2011) pointed out that the NNW to NW plunging (40–50°) tight and isoclinal F_1 folds and open to reclined F_2 folds plunging 30° NNE are commonly noticed in BIF, mafic and biotite gneisses, migmatites and quartz-sericite schist. Their axial planes trend in N-S to NNW-SSE directions. The open to tight F_3 folds are plunging 45° to 50° in different directions. The elliptical folds observed in banded (mafic) gneisses occur as isoclinal plane pattern, and provide strong evidence of intensive ductile and brittle-ductile shearing. Due to intense shearing effects, the fold axes in tight-close folds were rotated and reoriented and at some places, extensional crenulation cleavage and shear bands were formed (Bhatt and Mahmood 2012).

An E-W trending Babina–Prithvipur crustal shear zone (500–1500 m wide) extending up to 30 km was evolved in syn-to post-tectonic D_3 phase of deformation (Fig. 2). It passes from south of Babina and Taparyan villages and extends up to the northwest of Prithvipur. The gneisses and metavolcanics were subjected to intensive mylonitisation and emplaced along this shear zone. The preferred orientation of quartz and feldspar and minute flaky minerals formed a mylonitic foliation. It strikes in E-W to ESE-WNW and shows steep northerly dips (50–70°). The mylonitic lineation lying oblique (10–15°) to mylonitic foliation is represented by parallel alignment of quartz, feldspar and flaky minerals (Bhatt and Mahmood 2012). The small-scale shear zones varying in width from few centimeters to several meters and extending up to a few hundred meters are also traced in grey granites near Churara village (Fig. 2). The mesoscopic and microfabric analysis implies that the crustal shear zones emplaced along central domains of Bundelkhand Craton were syntectonically deformed during progressive shearing under low temperature–pressure conditions in a brittle–ductile regime.

5.3 Mesoscopic Shear Indicators

The protomylonite zone in streaky gneisses, metavolcanics and granites containing phenocrysts (0.5–3.0 cm) of quartz and feldspar (>50%) with lesser amount of matrix, is exposed in the west of Chamon, south of Taparyan and in the north of Churara villages. An intensively deformed mylonite zone occurring in felsic metavolcanics along the Taparyan-Papawni shear zone. The mylonitised metavolcanic and grey granite zones are exposed in the south of Taparyan and in the north of Churara villages (Fig. 2). The ultramylonite zone (100–150 m) consisting of matrix and few protoliths of quartz and feldspar is traced in the south of Taparyan village and extends up to 1.5 km in its strike direction (Fig. 2). The phenocrysts of quartz and feldspar (0.3–2 cm) in mylonitised gneisses and granitoids display dominant sinistral top to SW sense of shear movement (Fig. 3b, c). The minor strike-slip faults disrupting quartz veins and displaying the dextral sense of shear movement were observed in pink granitoids.

A minor shear zone (20–30 m) is noticed in granites and is marked by S-C planes (Fig. 3b). The progressive shearing effects are represented by the development of S and C planes. The S-planes are mainly striking in ESE-WNW directions whereas C planes (shear planes) strike in NW–SE directions. The C planes showing obliquity (30° to 40°) to main mylonitic foliation are resembling with C–C' bands of Berthe et al. (1979) and constituting identical mineralogical composition may be produced in the same shearing event (Passchier 1991, Passchier and Trouw 2005). These shear bands (C–C' structures) are not equivalent to C' type equal extensional crenulation cleavage of Platt and Vissers (1980) but are corresponding with type I, S-C mylonites of Lister and Snoke (1984).

5.4 Microscopic Shear Sense Indicators

The microstructural studies confined to oriented thin sections are nominated as "Vortocity Profile Plane" by Passchier and Coelho (2006). The protomylonite constituting large porphyroclasts of quartz and feldspar (50–90%) and the smaller amount of matrix are characterized by two generations of quartz (porphyroclasts). The phenocrysts of tabular and polycrystalline quartz of the first generation, exhibiting irregular grain boundaries, undulose extinction and deformation lamellae (Fig. 4a, b, c), were evolved excessively due to grain boundary migration (Passchier and Trouw 2005). The secondary quartz widely occurring in margins of host quartz were produced due to excessive dynamic recrystallization. The K-feldspar (mainly orthoclase) showing deformational lamellae (Fig. 4d) constitute more than 10% of the total composition. The undulose extinction (Fig. 4b, c) and low-temperature grain boundary migration displayed by quartz were formed between 300 and 400 °C (Stipp et al. 2002), whereas the bending undulose extinction (Fig. 4a) and subgrain formation exhibited by feldspar were evolved under low to moderate temperature conditions (400–500 °C; Pryer 1993). Two types of mantled σ porphyroclasts (σ_a and σ_b) are recognized in protomylonite (Passchier and Simpson 1986). The σ_a mantled porphyroclast of quartz and feldspar showing undulose extinction and pressure shadows (Fig. 4a). The σ_b type of porphyroclast associated with C planes and narrow zones of high shear strain is rarely observed. The σ_a mantled porphyroclasts exhibit sinistral sense top –to- SW shear movement (Fig. 4 a). These σ (σ_a and σ_b) types of mantled clasts mainly occur in low to medium shear strain domains (Whitmeyer and Simpson 2003; Pennacchioni et al., 2001). The alteration effects and kink bands examined in a few phenocrysts of orthoclase and plagioclase (Fig. 4d) were evolved under medium shear strain conditions (Pryer 1993).

The micro cracks and extensional fractures examined in few ribbons of quartz grains are indicative of overprinting of brittle deformation on ductile deformation (Fig. 4d) and were developed under low-grade temperature conditions (below 300 °C). Such feldspar grains were possibly deformed by internal microfracturing process with minor dislocation glide under low grade but sustained metamorphic conditions (400–500 °C; Pryer 1993; Ji 1998a, b). The intensive granulation effects and dynamic recrystallization were attributed to reduce the percentage of relicts of quartz and feldspar (<10) and as result, the transition zone switches to superplastic ultramylonite zone.

The protomylonite containing a higher percentage (50–90%) of bigger porphyroclasts of quartz and feldspar were evolved under low to moderate shear/ strain conditions (Passchier and Trouw 2005). The asymmetrical porphyroclasts and pressure shadows and angular relationship between S-C fabrics indicate that the mylonites were produced under the influence of non-coaxial (simple shear) deformation in a ductile regime. The variations in the geometry of porphyroclast and pressure shadow (quartz, feldspar) and obliquity between S-C fabrics are suggestive of low to medium temperature conditions. The elongated and ribbon-shaped quartz and feldspar grains produced by

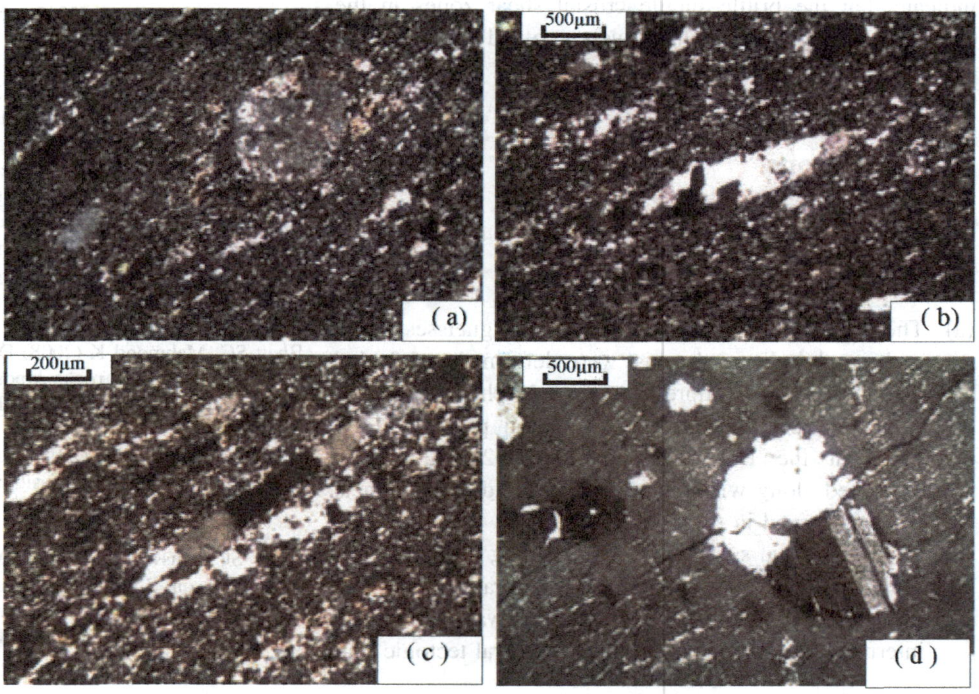

Fig. 4 **a** Sinistrally rotated σa mantled rotated porphyroclast of K- feldspar forming pressure shadow in stretched matrix; **b** Elongated fabric of quartz showing undulose extinction aligned parallel to mylonitic foliation; **c** Ribbons of quartz grains showing undulose extinction oriented parallel to mylonic foliation and surrounded by recrystallised matrix; **d** Orthoclase showing micro cracks and twinned plagioclase

pressure solution and dynamic recrystallization were developed due to intensive deformation under high strain conditions. It indicates that the degree and intensity of deformation was consistently increased from margin to central part of the shear zone. The microfaults/micrcracks (Fig. 4d) noticed in a few porphyroclasts of quartz and feldspar show sinistral movement and formed by brittle deformation in a cataclasite regime under low-temperature conditions.

The undulose extinction, deformation lamellae and dynamic recrystallization in quartz grains are characteristics of strain softening processes and were evolved in plastic deformation under ductile and low to medium temperature conditions (Pryer 1993; Ji 1998a, b; Hippertt and Hongen 1998; Passchier and Trouw 2005 and references therein). The crystal plastic deformation became dominant under moderate to high temperatures and led to initiate the dynamic recrystallization and stretching of quartz and feldspar (Pryer 1993; Stipp et al. 2002; Passchier and Trouw 2005; and many others). The mylonites evolved in non-coaxial flow under low to medium shear strain and temperature conditions show asymmetrical and S-C fabrics exhibiting sinistral top to SW shear movement. The progressive shearing took place in the initial phase of deformation and was subsequently followed by brittle shearing in the later phase. The σa mantled porphyroclasts produced in the plastic regime are indicative of moderate to high shear strain conditions and the kinking and twining examined in plagioclase are suggestive of low temperature and pressure conditions (Passchier and Trouw 2005). The presence of mylonitic foliation, stretching lineation and asymmetrical and S-C fabrics and sheath folds indicate that the brittle-ductile crustal shear zones in the central domain of Bundelkhand were evolved under moderate to high shear strain conditions.

6 Conclusions

Based on the above observations, the following conclusions can be drawn:

(a) The TTG gneisses, mafic and streaky gneisses along with amphibolite and schist were deformed and folded in two regional compressive tectonic phases (D_1-D_2). The main phase of granitic intrusion probably took place in the D_3 magmatic phase (2.56–2.44 Ga), emplaced along with major fractures and shear planes, were preserved in the craton.

(b) The subvertical strike-slip E-W trending Babina–Prithvipur crustal shear zone exhibits sinistral top- to- SW (left lateral) shear movement, and was formed at intermediate to shallow depth as a central tectonic unit within interacratonic domains of Bundelkhand Craton in the late D_3 phase.

(c) These studies infer that the mylonites were evolved under crystal plastic and strain softening processes in brittle-ductile (non-coaxial) environment.

Acknowledgements We are extremely thankful and highly indebted to Prof. Manoj Limaye and Ritesh Purohit for reviewing this paper and giving fruitful suggestions. The Department of Science and Technology (DST), Govt. of India, is thankfully acknowledged for providing financial support to SCB under DCS (ESS/16/255/2005) and to VKS under DST-RFBR (INT/RUS/RFBR/P-279), DST-ILTP (INT-ILTP/B-2.72) and DST (SR/S4/ES-399/2008) sponsored projects.

References

Acharya SK, Roy A (2000) Tectonothermal history of the Central Indian Tectonic Zone and reactivation of major faults/shear zones. J Geol Soc India 55:239–256

Basu AK (2007) Role of the Bundelkhand Granite Massif and the Son Narmada megafault in precambrian crustal evolution and tectonism in Central and Western India. J Geol Soc India 70:745–770

Basu AK (1986) Geology of parts of the Bundelkhand Granite massif Central India: Record Geological Survey of India, 117:61–124

Berthe D, Choukroune P, Jegouzo P (1979) Orthogenesis mylonite and non-coaxial deformation of granites: the example of the South Armorican shear zone. J Struct Geol 1:31–42

Bhatt SC, Gupta MK (2014) Microstructural analysis and strain pattern in mylonites and implications of shear sense indicators in evolution of Dinara- Garhmau Shear Zone, Bundelkhand Massif, Central India. Indian Mineralogist (J Mineral Soc India) 48(2):186–206

Bhatt SC, Hussain A (2008) Structural history and fold analysis of basement rocks around Kuraicha and adjoining area Bundelkhand massif. Central India: J Geol Soc India 72:331–347

Bhatt SC, Hussain A (2012) shear indicators and strain pattern in Quartz Mylonites in Chituad- Deori shear zone, northeastern part of Bundelkhand massif. Central India: Earth Sci India 5:60–78

Bhatt SC, Singh VK, Hussain A (2011) Implications of shear indicators for tectonic evolution of Mauranipur shear zone, Bundelkhand Craton, Central India. In: Singh VK, Chandra R (eds) 2nd Proceeding of precambrian continental growth and tectonism. Angel Publication, New Delhi, pp 36–49

Bhatt SC, Gupta MK (2009) Tectonic significance of shear Indicators in the evolution of Dinara-Garhmau shear zone, Bundelkhand Massif, Central India.In: Kumar S (ed) Magmatism tectonism and mineralization. Macmillan Publishers India Ltd., New Delhi, India, pp 122–132

Bhatt SC, Mahmood K (2008) Deformation pattern and kinematics of folds in basement rocks exposed around Babina and Ghisauli area, Bundelkhand Craton, Central India. Bull Indian Geol Assoc 41: 1–16

Bhatt SC, Mahmood K (2012) Deformation pattern and microstructural analysis of sheared gneissic complex and mylonitic metavolcanics of Babina-Prithvipur sector, Bundelkhand Massif, Central India. Indian J Geosci 66:79–90

Bhatt SC, Hussain A, Suresh M et al (2017) Geological structure control on Sukhnai Basin and land use/land cover pattern in Mauranipur and adjoining areas, Bundelkhand Craton, Central India. J Geol Soc Sri Lanka 18(2):53–61

Bhatt SC (2014) Geological and tectonic aspects of Bundelkhand Craton, Central India. Angel Publication, Bhagwati Publishers and Distributors C-8/77-B, Keshav Puram, New Delhi, p 190

Butler RWH, Holdsworth RE, Llyod GE (1997) The role of basement reactivation in continental deformation. J Struct Geol 154:69–72

Hippertt JE, Hongen FD (1998) Deformation mechanisms in the mylonite/ ultramylonite transition. J Struct Geol 20:1435–1448

Hutton DHW (1988) Granite emplacement mechanism and tectonic controls: inferences from deformation studies. R Soc Edinburgh Trans Earth Sci 79:245–255

Jhingran AG (1958) The problem of Bundelkhand granites and gneisses, Presidential Address. In: Proceedings of the Indian science congress 45th session, Madras, pp 48–120

Ji S (1998a) Deformation microstructure of natural plagioclase. In: Snoke A, Tullis J, Todd VR (eds) Fault related rocks- a photographic atlas. Princeton University Press, New Jersey, pp 276–277

Ji S (1998b) Kink bands and recrystallisation in plagioclase. In: Snoke A, Tullis J, Todd VR (eds) Fault related rocks-a photographic atlas. Princeton University Press, New Jersey, pp 278–279

Joshi KB, Bhattacharjee J, Rai G et al (2017) The diversification of granitoids and plate tectonic implications at the Archean–Proterozoic boundary in the Bundelkhand craton, Central India. In: Halla J, Whitehouse MJ, Ahmad T et al (eds) Crust–mantle interactions and granitoid diversification: insights from Archean Cratons. Geological Society, London, Special Publications, vol 449, pp 123–157

Kaur P, Zeh A, Chaudhri N (2014) Characterisation and U-Pb-Hf isotope record of the 3.55 Ga felsic crust from the Bundelkhand Craton, Northern India. Precambr Res 255:236–244

Kaur P, Zeh A, Chaudhri N et al (2016) Unravelling the record of Archaean crustal evolution of the Bundelkhand Craton, northern India using U-Pb zircon–monazite ages, Lu–Hf isotope systematics, and whole-rock geochemistry of granitoids. Precambr Res 281: 384–413

Lister GS, Snoke AW (1984) S-C Mylonites. J Struct Geol 6(6):617–638. https://doi.org/10.1016/0191-8141(84)90001-4

Matlauer M, Faure M, Malavieille J (1981) Transverse lineation and large scale structures related to Alpine obduction in Corsica. J Struct Geol 3:401–409

Mishra RC, Sharma RP (1975) New data on the geology of the Bundelkhand complex of the Central India. Recent Researches in Geology 2, Hindustan Publishing, Corporation, New Delhi, pp 311–346

Mondal MEA, Goswami JN, Deomurari MP et al (2002) Ion microprobe 207Pb/206/Pb ages of zircon from the Bundelkhand massif, northern India: implication for crustal evolution of Bundelkhand – Aravalli protocontinent. Precambr Res 117:85–100

Passchier CW (1991) Geometric constraints on the development of shear bands in rocks. Geol Mijnbouw 70:203–211

Passchier CW, Coelho S (2006) An outline of shear sense analysis in high grade rocks. Gondwana Res 10:66–76

Passchier CW, Simpson C (1986) Porphyroclast systems as kinematic indicators. J Struct Geol 8:831–843

Passchier CW, Trouw RAJ (2005) Microtectonics. Springer Berlin, Heidelberg, New York

Pati JK, Patel SC, Pruseth KL et al (2007) Geology and geochemistry of giant quartz veins from the Bundelkhand craton, central India and their implications. J Earth Syst Sci 116:497–510

Patil S, Nishad RK, Seena MS, Singh VK (2007) Palaeomagnetic, rock magnetic and low field AMS investigations on the Proterozoic dykes of Bundelkhand Craton: constraints on their age and mode of emplacement mechanism. In: International association of Geodesy, IUGG general assembly, July 2–13, 2007, Perugia, Itlay, ASI003, 3053

Pennacchioni G, Toro GD, Mancktelow NS (2001) Strain-insensitive preferred orientation of porphyroclasts in Mont Mary mylonites. J Struct Geol 23:1281–1298. https://doi.org/10.1016/S0191-8141(00)00189-9

Platt JP, Vissers RLM (1980) Extensional structures in anisotropic rocks. J Struct Geol 2:397–410. https://doi.org/10.1016/0191-8141(80)90002-4

Prakash R, Singh JN, Saxena PN (1975) Geology and mineralization in the southern parts of Bundelkhand in Lalitpur dist., U.P. J Geol Soc India 16:143–156

Prasad MH, Hakim A, Rao BK (1999) Metavolcanic and Metasedimentary inclusions in the Bundelkhand Granitic Complex in Tikamgarh District, Madhya Pradesh. J Geol Soc India 54:359–368

Pryer LL (1993) Microstructures in feldspar from a major crustal thrust zone: the Grenville Front Ontario, Canada. J Struct Geol 15:21–36

Ramsay JG, Huber MI (1987) The techniques of modern structural Geology, 2: folds and Fractures. Academic Press, London, pp 309–700

Rao JM, Rao GVSP, Widdowson M et al (2005) Evolution of Protoerozoic mafic dyke swarms of the Bundelkhand Granite Massif, Central India. Curr Sci 88:502–506

Roday PP, Diwan P, Pal A (1993) A two stage model for the development of Karitoran/shear zones, Lalitpur district, Uttar Pradesh, India. J Geol Soc India 42:481–492

Roday PP, Diwan P, Singh S (1995) A kinematic model of emplacement of quartz reef and subsequent deformation pattern in the Central Indian Bundelkhand batholiths. Proc Indian Acad Sci (earth Planet Sci) 104:465–488

>Saha L, Pant NC, Pati JK et al (2011) Neoarchean high-pressure margarite-phengitic muscovite-chlorite corona mantled corundum in quartz-free high-Mg, Al phlogopite–chlorite schists from the Bundelkhand craton, north central India. Contrib Miner Petrol 161:511–530

Saha L, Frei D, Gerdes A et al (2016) Crustal geodynamics from the Archaean Bundelkhand Craton, India: constraints from zircon U-Pb–Hf isotope studies. Geol Mag 153:79–192

Sarkar A, Paul DK, Potts PJ (1996) Geochronology and geochemistry of the Mid–Archean, Trondhjemitic gneisses from the Bundelkhand craton, Central India. In: Saha AK (ed) Recent researches in geology, 16:76–92

Senthiappan M (1993) Geology of the area along the Raksa Shear Zone, Jhansi district, UP. Rec Geol Surv India 73:73–76

Simpson C, Schmid SM (1983) An evaluation of criteria to deduce sense of movement in sheared rocks. Bull Geol Soc Am 94:1281–1288

Singh VK, Slabunov A (2013) The Greenstone belts of the Bundelkhand craton, Central India: new geochronological data and geodynamic setting. In: Singh VK, Chandra R (eds) International association for Gondwana research conference series No. 16. 3rd international conference precambrian continental growth and tectonism, Jhansi, India, pp 170–171

Singh VK, Slabunov A (2015a) The Central Bundelkhand Archaean greenstone complex, Bundelkhand Craton, Central India: geology, composition, and geochronology of supracrustal rocks. Int Geol Rev 57(11–12):1349–1364

Singh VK, Slabunov A (2015b) Geochemical characteristics of banded iron formation and metavolcanics of Babina greenstone belt of the Bundelkhand Craton, Central India. J Econ Geol Geo Resour Manage 10:63–74

Singh PK, Verma SK, Moreno JA, Singh VK et al (2019a) Geochemistry and Sm-Nd isotope systematics of metabasalts from the Babina and Mauranipur greenstone belts, Bundelkhand craton: implications for tectonic setting and Paleoarchean mantle evolution. Lithos 330–331:90–107

Singh PK, Verma SK, Singh VK et al (2019b) Geochemistry and petrogenesis of sanukitoids and high-K anatectic granites from the Bundelkhand craton: implications for the late-Archean crustal evolution. J Asian Earth Sci 174:263–282

Singh VK, Verma SK, Singh PK et al (2020) Archean crustal evolution of the Bundelkhand Craton: evidence from granitoid magmatism In: Archean Granitoids of India: windows into early earth tectonics. Geological Society, London, Special Publ 489, pp 235–259. https://doi.org/10.1144/SP489-2018-72

Singh PK, Verma SK, Singh VK et al (2021a) Geochronology and petrogenesis of the TTG gneisses and granitoids from the Central Bundelkhand granite-greenstone terrane, Bundelkhand Craton, India: Implications for Archean crustal evolution and cratonization. Precambr Res 359:106210

Singh VK, Slabunov AI, Nesterova NS et al (2021b) Tectonostratigraphic terranes of the Bundelkhand Craton (Indian Shield). In: Shandilya AK, Singh VK, Bhatt SC, Dubey CS (eds) Geological and Geo-environmental Processes on Earth, pp 155–164 (this volume)

Singh VK, Slabunov A (2016) Two types of Archaean supracrustal belts in the Bundelkhand Craton, India: geology, geochemistry, age and implication for craton crustal evolution. J Geol Soc India 88:539–548

Slabunov A, Singh VK (2018) Bundelkhand and Dharwar Cratons (Indian Shield): comparison of crustal evolution in Archean time. Arch Anthropol Open Acc. 3(suppl-2):327–333. https://doi.org/10.31031/AAOA.2018.03.000556

Slabunov A, Singh VK, Joshi KB, Li X (2017a) Paleoarchean zircons from quartzite of South Bundelkhand Supracrustal Complex: origin and implications for crustal evolution in Bundelkhand Craton, Central India. Curr Sci 112:794–801

Slabunov AI, Singh VK, Shchiptsov VV et al (2017b) A new Paleoproterozoic (1.9–1.8 Ga) event in the crustal evolution of the Bundelkhand Craton, India: the results of (SHRIMP) dating of zircons from giant quartz veins. In: Slabunov AI, Svetov SA, Baltybaev ShK (eds) Early precambrian vs modern geodynamics. Extended abstracts and field trips guide, Petrozavodsk, KarRC RAS, pp 239–241

Slabunov AI, Singh VK (2019) Meso-Neoarchaean crustal evolution of the Bundelkhand Craton, Indian Shield: new data from greenstone belts. Int Geol Rev 61:1409–1428. https://doi.org/10.1080/00206814.2018.1512906

Stipp M, Stünitz H, Heilbronner R et al (2002) The eastern Tonale fault zone a "natural laboratory" for crystal plastic deformation of quartz over a temperature range from 250 to 700°C. J Struct Geol 24:1861–1884

Verma SK, Verma SP, Oliveira EP, Singh VK, More JA (2016) LA-SF-ICP-MS zircon U-Pb geochronology of granitic rocks from the central Bundelkhand greenstone complex, Bundelkhand craton, India. J Asian Earth Sci 118:125–137

Watts MJ, Williams GD (1979) Fault rocks as indicators of progressive shear deformation in the Guingamp region Brittany. J Struct Geol 1:323–332. https://doi.org/10.1016/0191-8141(79)90007-5

Whitmeyer SJ, Simpson C (2003) High strain-rate deformation fabrics characterize a kilometers-thick Paleozoic fault zone in the Eastern Sierras Pampeanas central Argentina. J Struct Geol 25:909–922. https://doi.org/10.1016/S0191-8141(02)00118-9

‘Dudhmania Shear Zone’—As Interpreted from the Satellite Image and Its Ground Realities

D. C. Banerji

Abstract

A very prominent sinuous belt, having an average width of about 25 km and extending for more than 105 km in an overall E–W trend, covering parts of Sidhi and Singrauli districts of Madhya Pradesh and Sonbhadra district of Uttar Pradesh, attracts attention of the satellite image viewers. The belt exhibits a characteristic broom stick fabric. This uncommon, very close spaced fabric has a cross cutting relationship with the ENE-WSW trending fabric of the main Mahakoshal and appears to be superposed over it. Also glaring is a sinistral slip, manifested across its length and width. The sinuous belt, on ground, is occupied by a phyllite/phyllonite—meta-greywacke—chert association, exhibiting various minor structures which are common of a ductile regime. The ubiquitous presence of anticlockwise rotation of fabric elements, producing S-C structure in phyllite/phyllonite and in meta-greywacke, along with frequent presence of ‘alpha’ and ‘delta’ type structures, strongly asymmetric ‘S’ and at places ‘Z’ shaped folds and the sheath folds suggest that the ductile shearing, with an overwhelming sinistral shear sense, is spread over the entire sinuous belt. Continuity of this non-coaxial sinistral movement, even after hardening of the deforming mylonites, is evident by the presence of numerous local as well as regional scale sinistral slips/faults, spread over the entire belt. All the available evidences prove that the sinuous belt, having strong manifestation on the imagery, represents a wide and extensive shear zone, named as ‘Dudhmania Shear Zone’, and is unique in its characteristics.

Keywords

Mahakoshal Group • Sinuous belt • Anticlockwise rotation • Sinistral shearing • Failed ‘Z’ symmetry • Dudhmania Shear Zone

1 Introduction

The Paleoproterozoic rocks of Mahakoshal Group are exposed as a narrow belt, extending for about 500 km, along the Narmada and Son rivers, between Barmanghat (Narsinghpur) in the west to Palamau (Jharkhand) in the east. These rocks are bounded to the north by Son Narmada North Fault (SNNF) and to the south by Son Narmada South Fault (SNSF) and represent the Son Narmada Lineament (SONA) zone (Sharma 2009). Based on their highly deformed nature, lithology and tectonic setting, they are differentiated from the earlier considered Bijawar Series. The volcano sedimentary rocks of Mahakoshal Group have a rift basin type of tectonic setting (Roy 2008). Major part of the belt, extending between Narsinghpur and Sidhi, has a dominant ENE–WSW trend. The axial traces of D_1 and D_2 structures define the ENE extension of the belt (Roy and Bandopadhyay, 1990). The southeastern extension of the belt, restricted between Obra-Amsi-Jiawan fault (OAJ) and the Songarh-Kasar-Dudhi (SKD) fault, however, shows a WNW–ESE to NW–SE trend (Fig. 1). This part of Mahakoshals also hosts a number of low tenor gold occurrences, across its length and width, which eventually disappear to the west, where the belt hits against the ENE-WSW trend of main Mahakoshals. Gold mineralization in this eastern part of Mahakoshal sequence occurs in two modes: (i) the disseminated sulphide type and (ii) the quartz vein type (Sharma et al. 2000; Soni and Jha, 2001). The fluid inclusion studies carried out by Mangla Prasad et al. (2000), however, clearly indicate that

D. C. Banerji (✉)
21F, SSEA-2, Renaissance Township, Burdwan, 713102, India

A. K. Shandilya et al. (eds.), *Geological and Geo-Environmental Processes on Earth*, Springer Natural Hazards,
https://doi.org/10.1007/978-981-16-4122-0_12

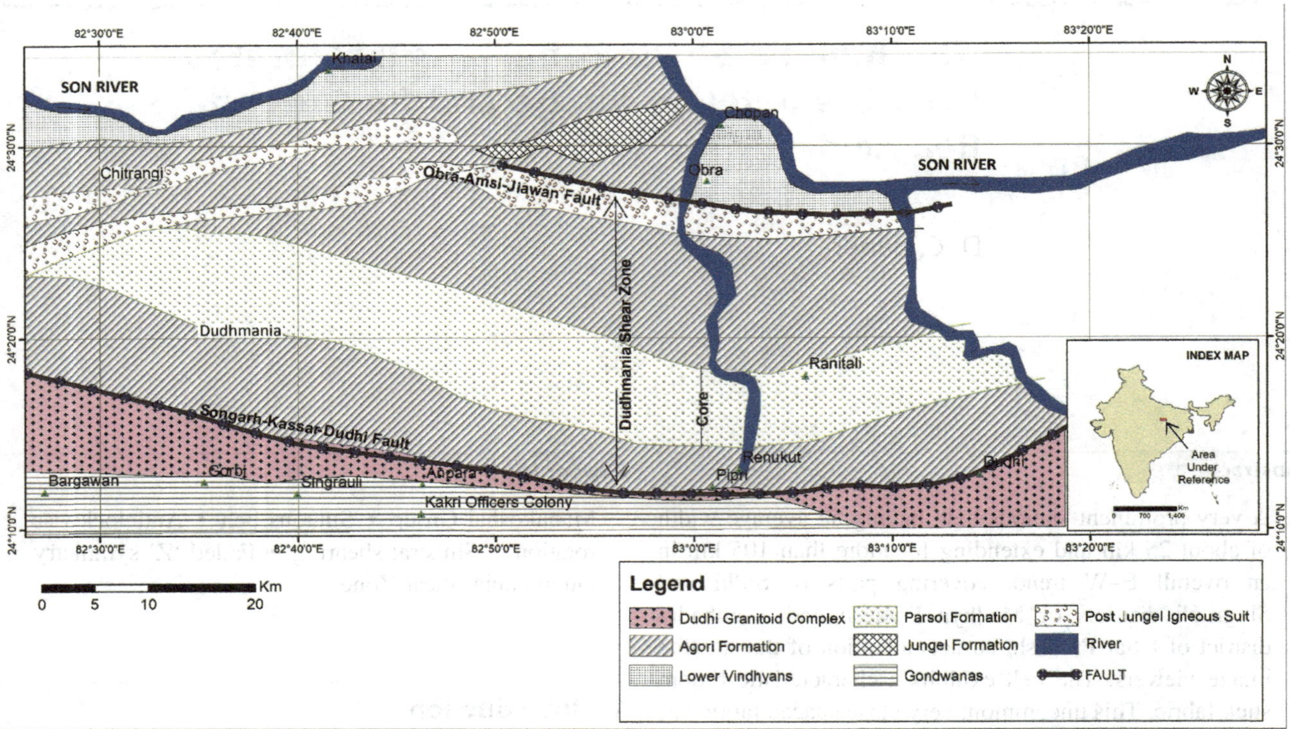

Fig. 1 Geological sketch map of parts of Sidhi and Singrauli districts of M.P. and Sonbhadra district of U.P

the gold mineralization of different types discussed above has resulted from compositionally similar fluids, genetically related to the same source.

Banerji (2011, 2019, and in Banerji and Prasad, 1997), however, considered the WNW-ESE trending Mahakoshals of southeastern part, involving the tract of Agori and Parsoi Formations, as mylonites representing his 'Dudhmania Shear Zone'. The WNW-ESE trend is proposed to have produced by mylonitization of the Mahakoshal rocks, affected by a non-coaxial sinistral movement between the northern 'Obra-Amsi-Jiawan fault' and the southern 'Songarh-Kasar-Dudhi fault', and the mylonitic foliation (trending WNW-ESE) thus produced is superposed over the ENE-WSW trend of the main Mahakoshals. Reporting of this non-coaxial sinistral shear zone is also recorded in the Annual General Report (1997) of Geological Survey of India. Passing reference of ductile shearing along the southern boundary of the Mahakoshal belt of this area is also available in a paper by Roy and Devrajan (2000). Roy et al. (2002) later documented shear zone structures along the southern boundary of the supracrustal rocks of Mahakoshals of Singrauli area and deduced a sinistral shear movement.

This paper analyses the images of the concerned area and presents corresponding ground truths, to further strengthen the ideas expressed by the author in his earlier work. Google Earth, an easily available alternative for developing regional bias, proved to be of immense help in this study and in reproducing parts of the images as supportive figures. It is necessary to mention that some of the field photographs are already published in the earlier papers of the author, but felt necessary to reproduce them for providing a better perception to the reader.

2 Analysis of Imagery

The main Mahakoshal belt, present in the north-western part of the image (Fig. 2a) as a continuous tract, made up mainly of coarse ridges of medium height, disappears against the Son River in the east. To the south, a very conspicuous sinuous belt, extending for more than 105 km in an overall E–W trend, covering parts of Sidhi and Singrauli districts of Madhya Pradesh and Sonbhadra district of Uttar Pradesh, attracts attention of the viewer. This sinuous belt is representative of the gold bearing Mahakoshals restricted between the northern Obra-Amsi-Jiawan fault and the southern Songarh-Kasar-Dudhi fault. The maximum width of the belt is about 27 km.

The sinuous belt hits against the ENE-WSW trending coarse ridges of the main Mahakoshals, present to the west. The belt has a characteristic broom stick fabric, especially glaring in aerial photographs of the area. This uncommon, very close spaced, fabric is different from rest of the Mahakoshals. The close spaced fabric has a cross cutting

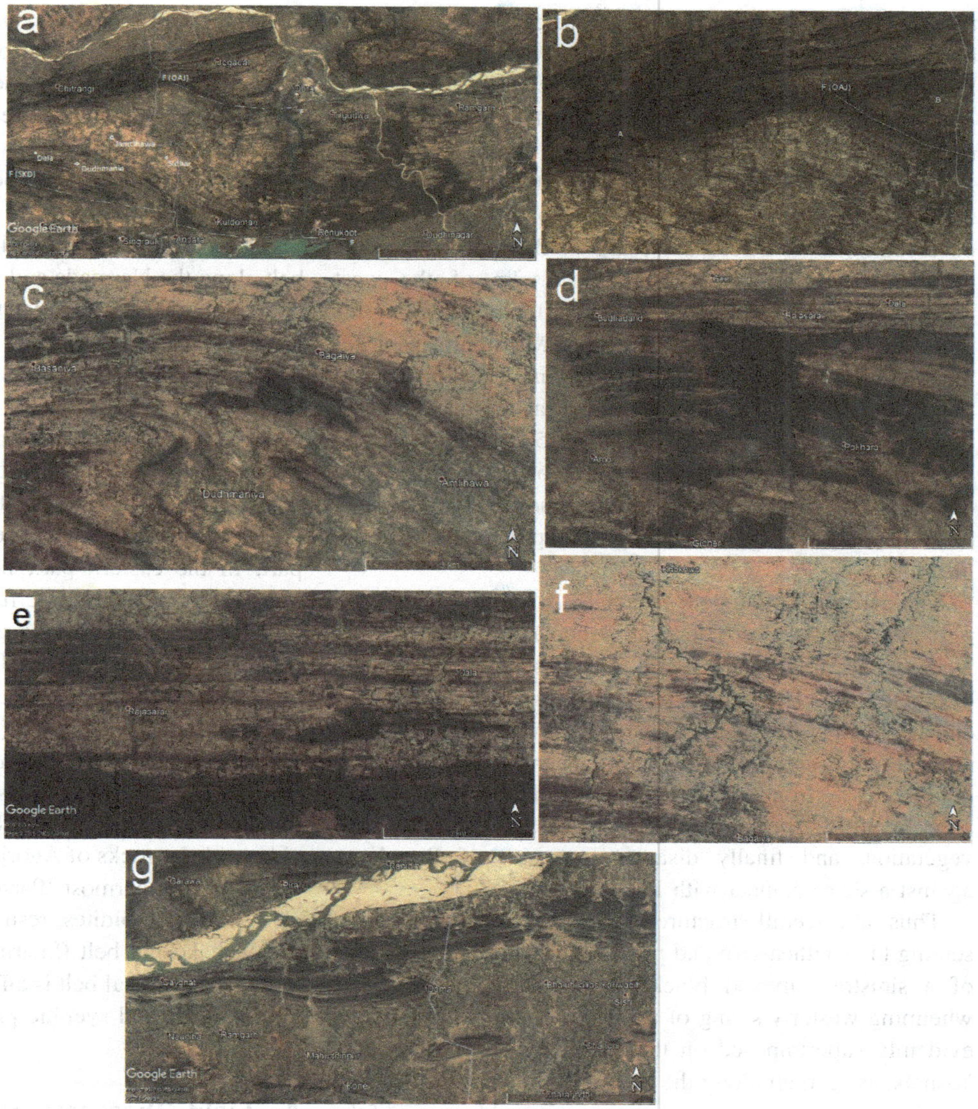

Fig. 2 **a** Satellite image showing the sinuous belt of the Sidhi and Singrauli districts of M.P. and Sonbhadra district of U.P. The northern Obra-Amsi-Jiawan (OAJ) fault (F) and the southern Songarh-kasar-Dudhi (SKD) fault is marked with broken white lines; **b** Cross cutting relationship of the fabric of the sinuous belt with the fabric of Main Mahakoshals, in the blown up part of northern margin of the belt (A and B sectors); **c** Failed 'Z' symmetry near Dudhmania. Note the superposed swerving around fabric, with a westerly swing, north of Dudhmania and Amlihawa; **d** A second set of superposed fabric over the initial fabric of the sinuous belt. Note the westerly swing of the later fabric and reduction in obliquity between the two sets, towards north; **e** Anticlockwise slip of the fabric elements, west of Dala. Note the curvilinear trend of the slip plane and dragging effect along it (enlarged image, west of Dudhmania); **f** Anticlockwise slip of the fabric elements, north-west of Dudhmania; **g** Anticlockwise slip of the fabric elements, north-east of Ramgarh (far east of the sinuous belt)

relationship with the ENE-WSW trending fabric of the main Mahakoshals, distinguishable at many places, especially west of Obra (Fig. 2b). In the strike continuity, south of Chitrangi, the WNW-ESE fabric of the sinuous belt abuts / merges into the main Mahakoshal belt. This fabric appears to be superposed over the fabric present in the northern main Mahakoshals. The central (core) part of the sinuous belt is characterized by a light toned smooth zone, having faded impression of the laminations. The southern edge of the belt is sharp and largely sub-parallel to the trend of the belt. The sharp nature of the edge is probably indicative of displacement along this margin.

Though, in most part, the sinuous belt has a sub-parallel fabric to its margin, a major kink in the trend is noticeable near Dudhmania. On closer examination, it is perceived that the fabric at this place is arranged in a disrupted 'Z' symmetry (Fig. 2c). There is no connectivity between the arms as they appear to be detached because of another set of fabric, swerving around the 'Z' symmetry. This superposed set of swerving around fabric, with a westerly swing, is very conspicuous north of Dudhmania and Amlihawa and appears to have rotated the 'Z' fold, sinistrally, causing detachment of the limbs.

This second set of superimposed fabric, present at an angle of 10°–15°, is also very clearly distinguishable in the southwestern corner (Fig. 2d) of the sinuous belt. It appears from the strong westerly swing of this fabric that it is imprinted later, by a very strong westerly (anticlockwise) movement and has rotated the earlier set of fabric to make it near parallel to the long axis of the direction of movement. Such a movement must have been in a ductile state, making the rocks to flow along. This superposed second set of fabric, thus, is representative of the long axis of the anticlockwise flow, across the entire sinuous belt. The presence of these two sets of fabric, at an acute angle to each other, is ubiquitous in the entire belt, as seen between Tilgudwa and Renukoot. In most cases, however, the earlier set is feeble.

A very prominent feature of this belt is a sinistral slip, represented at many places across its length and width. Amongst them, mention can be made of the one observed in the western part, west of Dala; a second sinistral slip is observed north of Dudhmania, and another, in the far east, north-east of Ramgarh. The sinistral slip of the west (Fig. 2e) is observed along a curvilinear trend, extending for more than 2 km in NW–SE direction. Maximum displacement along this plane is more than 500 m. The sinistral slip observed north-west of Dudhmania is also along NW–SE trend and has a displacement of about 500 m (Fig. 2f). To the far east, in the fringe area of this belt, sinistral slip is observed along WNW-ESE trend, extending for more than 3 km (Fig. 2g). Dragging effect is very prominent in all the locations.

The core part of the belt, especially in the western half, is light toned and smooth, largely peneplaned, and has only feeble traces of the linear fabric. This has a maximum width of about 5 km. The light toned central part is devoid of any elevated feature. However, towards the fringes of this central zone, a few low lying ridges start appearing and increase in number, away from core. Continuity of this light toned core zone is not that prominent further to the east, due to thick vegetation, and finally disappearing, east of Renukoot, against a sharp contact with Dudhi Granitoid.

Thus, the overall structure of the 'S' shaped belt, representing the southeastern end part of the Mahakoshals, is that of a sinistraly moved block, as indicated by the overwhelming westerly swing of the linear fabric. This fabric is evidently superimposed on the earlier fabric of the Mahakoshals, as seen all along the northern border between Obra and Chitrangi, and appears to have generated because of the strong sinistral movement affecting the pre-existing rocks. This sinistral movement, continuing till much late, is also evident by the sinistral slips, observed at many places within the belt.

However, presence of the structure resembling a failed 'Z' symmetry, west of Dudhmania, probably indicate that earlier their existed a strong dextral movement resulting in formation of 'Z' symmetry in the associated fabric. This earlier fabric appears to have been affected by a later, very strong, anticlockwise movement, which rotated the pre-existing fabric/structures sinistrally.

3 Geologic Setting

The study area, as represented in satellite image of Fig. 2a, is dominantly occupied by rocks of Mahakoshal Group, bordered to the north by Lower Vindhyan rocks and to the south by Dudhi granitoids (Fig. 1). Coal bearing Permian rocks of Gondwana Super Group also has its representation along the southern margin. Since the purpose of the work is to establish a correlation between the structural features observed in satellite imagery with that of the ground realities of the Mahakoshal tract, other geological formations are not discussed here.

The rocks of Mahakoshal Group are exposed as a narrow belt along the Narmada and Son rivers between Barmanghat (Narsinghpur) in the west to Palamau (Jharkhand) in the east. The rocks comprise quartzite, phyllite, dolomite, metabasics and banded ferruginous chert. In the western and central part of the belt, the Mahakoshals comprise a sequence of basal quartzite interbedded with metabasics and overlained by dolomites with chert interbands belonging to Agori Formation, representing its lower division. Dominance of dolomite characterizes the western and central part. In the eastern part, lithological assemblage includes metabasics, banded ferruginous chert, phyllite and meta-greywacke with near absence of carbonates. Nair et al. (1995) have considered the metabasics/ultramafic flows occurring near Chitrangi [marked as post Jungel Igneous suit in the map of Sharma et al. (2000)], to be the lowermost part of Mahakoshals and named them 'Chitrangi Formation'. The argillite meta-greywacke association occupying core part of the southeastern (WNW-ESE trending) belt, flanked on either side by rocks of Agori Formation, has been considered by them as uppermost 'Parsoi Formation'. The Parsoi rocks also include turbidites, restricted to the southeastern part of the Mahakoshal belt (Sharma, 2009). The southern margin of the Mahakoshal belt is affected by intrusions of granitic to granodioritic and syenitic plutons (Soni and Jha 2001).

4 Field Observations

Following the broad tonal differences in satellite image and for the sake of convenience, the sinuous belt has been divided into a 'Core' (Parsoi Formation) or zone of 'Ultramylonite', representing the central light toned smooth belt of the image; an 'inner margin' or 'Mylonite', present on either side of the core, and an 'outer margin' or zone of 'Protomylonite', represented on outer flanks of the 'mylonites', bordering the sinuous belt. The rock types present in the sinuous belt include phyllite/phyllonite, banded ferruginous and non-ferrous cherts, meta-greywacke, and the vein quartz. The southern edge of the belt has a contact with granitoids of intrusive origin. The entire belt is traversed by very characteristic discrete, thin quartz venations. All such quartz veins are, undoubtedly, quartz mylonites, conforming that the entire southeastern belt has undergone mylonitization. All along the southern margin, in the western half of sinuous belt, well cleaved, sheared, and at places brecciated, strike parallel, thick qurtz reef marks the contact between granitoids and the Mahakoshals. Continuity of the reef, east of Singrauli, is rare.

Evidences of ductile shearing are recorded all along the belt, affecting all the litho types. A very pervasive mylonitic foliation is conspicuous in all the rock types. The trend of the foliation is N80°W-S80°E at the western end, swerving to N60°W-S60°E in the middle part and again swinging to N80°W-S80°E in the eastern end of the belt. The dip of the foliation is sub-vertical. A second set of foliation, as observed in south western corner of the imagery, is also feebly distinguishable at an acute angle to the other.

West of Dudhmania, trend of the foliation is discontinuous and ever changing in phyllite/phyllonite, swinging between N80W-S80E and N80E-S80W in the flanks of a ridge, top of which is occupied by a few, thin, NE-SW trending bands of ferruginous chert. This frequent change in trend is attributed to the superimposition of one set of foliation over the other, in a regime where both the sets are fairly strong. The abnormally oriented (NE-SW) ferruginous cherts hit against the WNW-ESE trending phyllite/phyllonite, at either end. The chert bands present in vicinity, to the north and south, are extensively stretched and thin, and are oriented parallel to the regional trend. A classic example of anticlockwise rotation of 'Z' fold to 'S' symmetry (Fig. 3a), is observed in a chert band in northern continuity. This location falls near the northern hinge of the failed 'Z' symmetry reflected in satellite image.

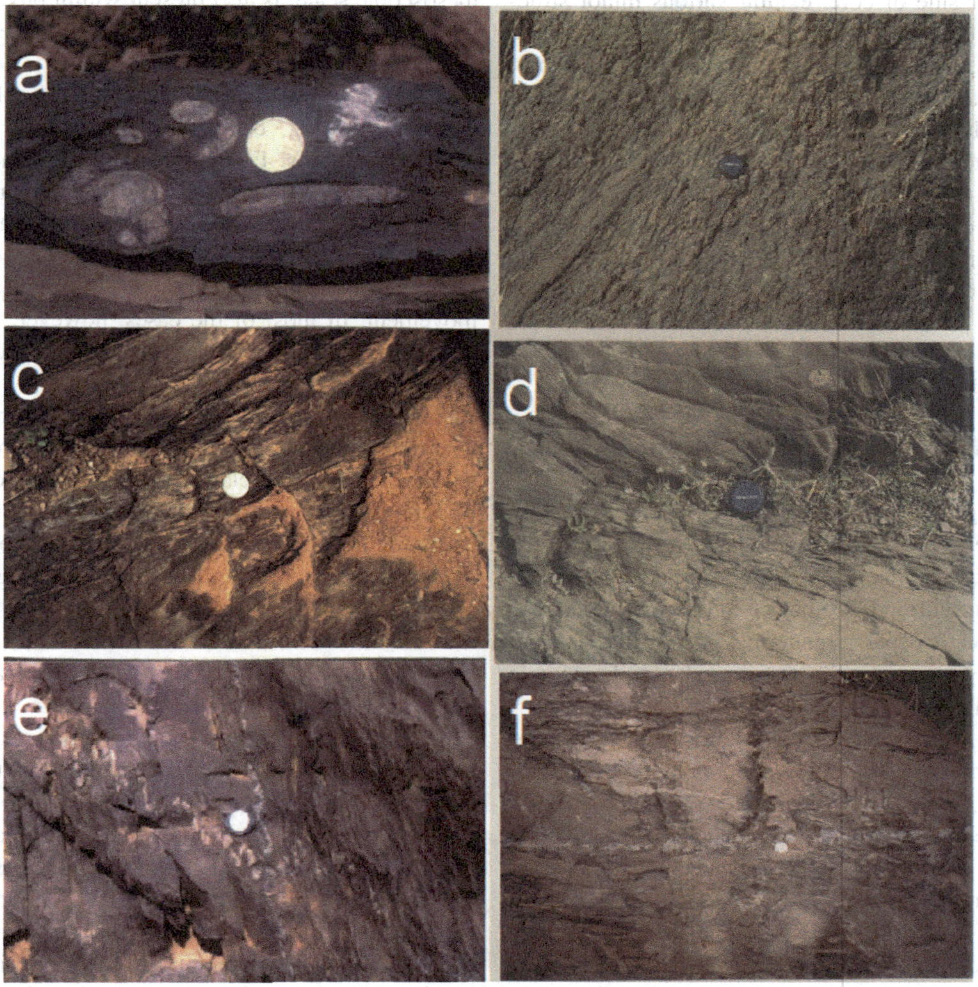

Fig. 3 **a** Reverse rotated 'Z' fold in the chert band. Because of reverse rotation, the lower hinge of 'Z' fold is rotated upward and the upper hinge has come closer to the lower hinge. The extended/boudinaged fragments of the upper and lower limbs have moved inward (towards center); **b** Sheared granitoid (pink gneiss) from Kanhana, north of Singrauli; **c** Strain sensitive fabric developed in phyllite/phyllonite. The foliation has rotated clockwise during deformation. The coin is placed on 'C' plane dominated area, while above and below the coin 'S' planes show rotation. Sense of shear is dextral; **d** Strain sensitive fabric developed in greywacke. The foliation has rotated anticlockwise during deformation. Sense of shear is sinistral; **e** and **f** Shortened and boudinaged (extended) quartz veins oriented oblique and parallel to mylonitic foliation respectively. Also, note the anticlockwise rotation of the boudins, and extended limbs and short hinges in obliquely oriented quartz veins. Sense of shear is sinistral

The core part of the sinuous belt is occupied by phyllite/phyllonite with near absence of meta-greywacke. The banded ferruginous chert which is otherwise ubiquitous in Mahakoshals, is nearly absent. The foliation in this part of the belt is very closely spaced and continuous. The quartz venations are regular, thin, continuous, and rendered to quartz mylonites. This zone is fairly consistent from the western end till the middle of the belt, where it abruptly ends against Dudhi granitoid, obviously juxtaposed by a fault representing SNSF. Away from this core part, on either side, frequency of the bands of ferruginous chert and meta-greywacke increases gradually to become abundant.

The rocks are distinctly banded and hard nearer to core, representing banded mylonite. In general, the phyllite/phyllonite, meta-greywacke and the chert bands, present on either side of core, exhibit various minor structures those are common of a ductile regime. Thus, these minor structures, overwhelmingly preserved in the rocks of the area, indicate existence of ductile shearing across the length and width of the sinuous belt. This ductile shearing has even affected the northern margin of the southern Granitoids present south of the Songarh-Kasar-Dudhi (SKF) fault. A detailed discussion about the minor structures preserved in different parts of the belt is included in the following section.

5 Shear Generated Structures

5.1 Outer Margin or Zone of Protomylonites

Ductile shearing of the granitoids, along the southern boundary of the shear zone, has rendered a distinct gneissosity to the rock, producing protomylonite and indicating that the shearing and emplacement of granitoids is coeval. Stretching of the feldspars, rendering gneissosity, is conspicuous along two obliquely running planes, having an angle of about 10°–12° between them (Fig. 3b).

Development of a strain sensitive fabric with 'S' planes rotated both clockwise (Fig. 3c) and anticlockwise (Fig. 3d) directions, in mesoscopic scale, is common within meta-greywacke and phyllite/phyllonites. This phenomenon is predominant along the southern flank of the belt and observed at many places north of SKD fault. The 'C' planes define mylonitic foliation, running parallel to the boundary of the sinuous belt. Maximum angle between the two planes, in such segments, is about 15°. It is notable that the clockwise rotated segments are comparatively lesser in number and mostly represented towards south-western margin of the belt. Such rotational features, though, persist across the width of the belt, are less conspicuous towards the core, with the angle between the two planes reducing to less than 10°. Also, the ubiquitously present Quartz veins and the chert bands of the peripheral area exhibit variety of shortening and extensional features. The bands which are oriented parallel to the pervasive foliation commonly show stretching and boudinaging, while those with oblique orientation display shortening by en-echelon type of folding (Fig. 3e). More towards inner margin, obliquely arranged bands show progressive anticlockwise shifts and parallelism of the most parts of the oblique bands, by rotation and extensive stretching, along the foliation (Fig. 3f). Such features, along with the presence of short hinges between the two stretched arms, clearly represent a strong flow regime. Also, the phenomenon is quite corroborative with the instantaneous extensional and shortening regimes, conforming flow along mylonitic foliation. Feeble traces of en-echelon folding, in mesoscopic scale, is also present within phyllite / phyllonite and meta-greywacke (Fig. 4a), having lamellar flow along axial traces.

The chert and quartz veins, lying parallel to foliation, exhibit stretching, boudinaging and, also, rotation of the stretched boudins. Formation of 'alpha' and 'delta' structures (Fig. 4b, c), within the stretched bands, is again a common phenomenon throughout the belt. The boudins formed by stretching are rotated either clockwise or anticlockwise, depending upon the prevailing dextral or sinistral movement. In most of the cases, however, the shear sense deduced from such rotated boudins indicate a sinistral movement.

The commonest micro shape fabric in meta-graywacke, near the margin of the belt, is a porphyroclastic texture where the 'porphyroclasts' are wrapped around by quartz ribbons and muscovite flakes (Fig. 4d). Quartz veins present in the entire belt are rendered quartz mylonite showing a micro fabric predominantly of S-C type (Fig. 4e).

5.2 Inner Margin or Zone of Mylonites

With the appearance of interlayered chert and haematitite/magnetite bands, ranging in thickness from 0.3 m to 2.0 m, the terrain passes into a zone of inner margin or zone of mylonite, where effect of shearing is further strong. These iron-oxide bearing chert bands are also referred as 'banded iron formation' (BIF) and are noticeable on either side of the central belt of Parsoi Formation (zone of ultramylonite), as discontinuous low lying ridges within the terrain of phyllite and meta-greywacke of Agori Formation, for few hundreds of meter along the WNW-ESE strike. They mostly occur as dislodged and tilted blocks, and at many a places thickness is increased because of tight en-echelon folding of the bands. Dislodged exposures of fold closures (Fig. 4f), and deviation in alignment of the bands, confirm

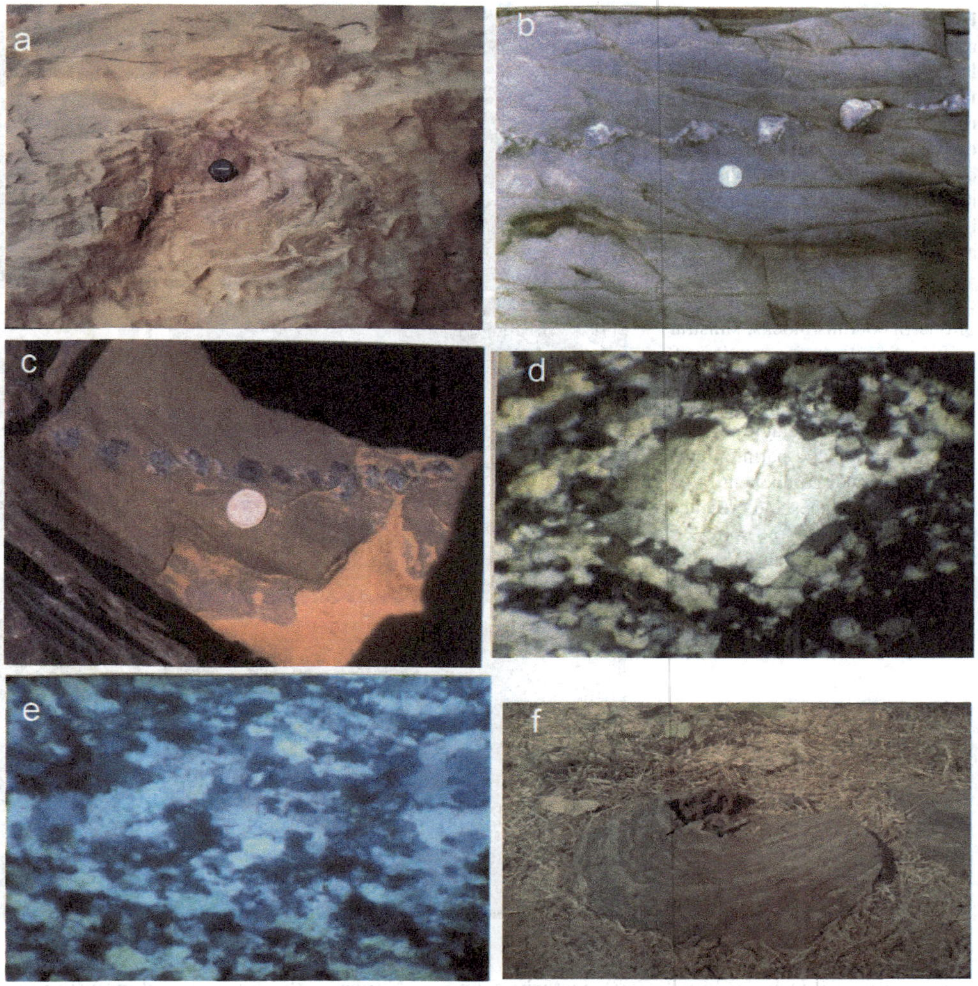

Fig. 4 **a** Traces of en-echelon folding within meta-greywacke, showing lamellar flow along long axis ('C' plane) of the fold, representing mylonitic foliation. Sense of shear is sinistral; **b** Quartz vein showing 'alpha' structure; **c** Quartz vein showing 'delta' structure. Sense of shear is sinistral; **d** Quartz porphyroclast wrapped by quartz ribbon and muscovite flakes. Crossed nicols, 32X; **e** S-C fabric shown by quartz mylonite. Crossed nicols, 3X. Shear sense is sinistral; **f** Folding in BIF band. Please note the thinning, thickening, boudinaging and en-echelon extension of the chert bands, resembling vein quartz

repetition because of such type of folding. However, repetition of thin bands, not caused by folding, and separated by greenish grey phyllite is also not uncommon.

Depositional structures like load casts, banding and colour banding are reported from within the 'banded iron formation'. The rocks are known to be thinly bedded and maintain a gradational relationship with other members (phyllite and meta-greywacke) of the Agori Formation. They are also known for along the strike gradation to chert, cherty quartzite, and brecciated quartzite/jasperite (Nair et al. 1995).

Closer examination of the chert bands of ferruginous as well as non-ferrous type, in the WNW-ESE trending arm of the Mahakoshals reveal that though the colour banding exhibited by the alternate arrangement of cherty and ferruginous layers are very prominent in ferruginous chert bands, at many places, however, two sets of cherty layers with cross-cutting relationship are prominent (Fig. 5a). One such set of discrete chert layers has a parallelism with the regional strike (mylonitic foliation or 'C' plane), whereas the other set makes an acute angle to it (representing 'S' planes). The strike parallel chert bands exhibit features like thinning and boudinaging, while the other set shows rotation at either end, in an en-echelon pattern, making the ends parallel to strike, exhibiting a sinistral movement. Many of the strike parallel chert layers, as represented in ferruginous as well as non-ferruginous type, also show en-echelon pattern of extension clearly caused by sinistral movement. The non-coaxial movement is also responsible for boudinaging and anticlockwise rotation of the chert bands (Fig. 5b) associated with BIFs. It is, however, a matter of further research, if the above features of the BIFs are supportive of a sedimentary parentage. Alignment of the BIFs along the superposed fabric (mylonitic foliation) is equally enigmatic from the point of their being sedimentary.

Chert bands also preserve a variety of en-echelon folds, tight to isoclinal and asymmetric in nature. In many cases, long axis of such folds trend oblique to the foliation. Both 'S' and 'Z' shaped folds are represented; however, 'S' shaped folds predominate over the other type. These

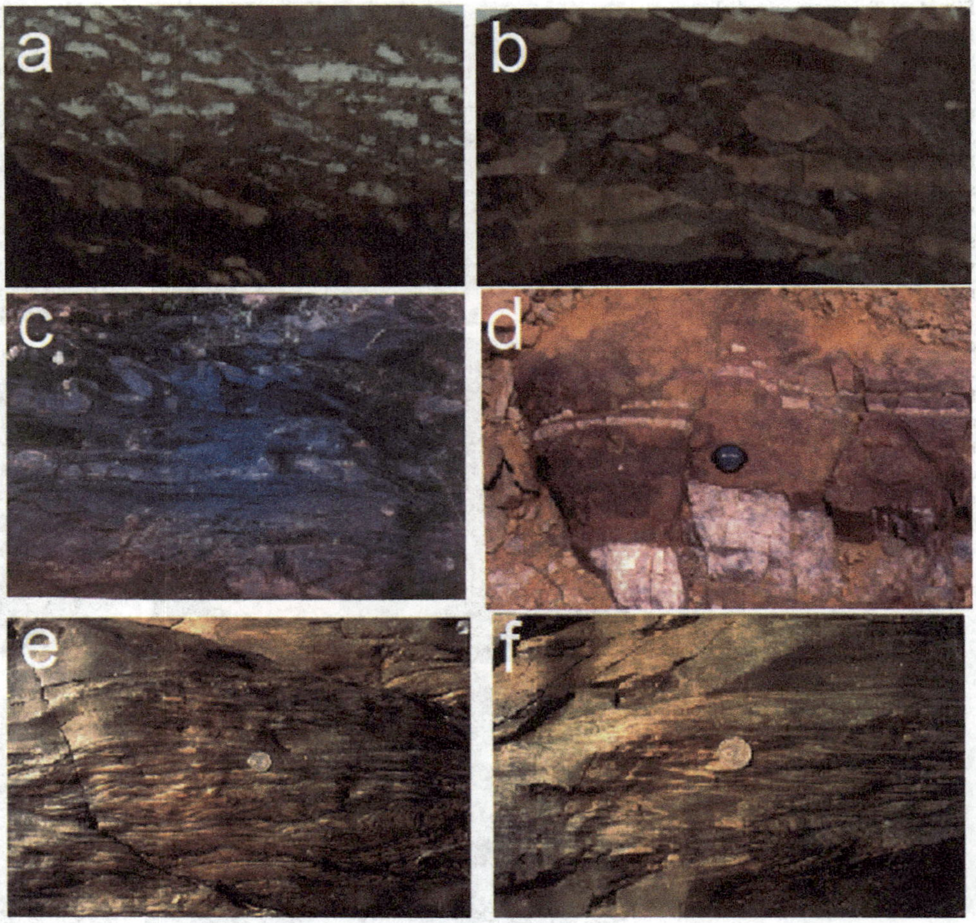

Fig. 5 **a** Cross cutting relationship amongst the cherty layers of BIF; **b** Banded iron formation showing boudinaging, rotation and en-echelon extension of chert bands, similar to those of quartz veins; **c** Anticlockwise rotated axial planes of near isoclinal folds and their amplification in chert band. Shear sense is sinistral; **d** Fault affecting quartz vein and resulting in trail of quartz boudins/blocks, sinistral shift; **e** Anticlockwise rotation of siliceous lamellae having obliquity to the flow plane (bottom right) in metagreywacke. Note the systematic partitioning of the rotational component by discrete shear planes and presence of rootless folds (below coin). Shear sense is sinistral; **f** Pseudo cross bedding associated with rootless folds (above and below the coin). Shear sense is sinistral

mesoscopic folds have low (<30°) easterly plunges. Nevertheless, steep (50°–70°) easterly plunges are also observed in the obliquely trending (NNE-SSW to NNW-SSE) shortened chert bands. Anticlockwise rotation of the axial planes of near isoclinal folds, making them parallel to mylonitic foliation, and their amplification (Fig. 5c) is a regular phenomenon of the area. Sheath folds with an eyed outcrop pattern and also the textbook type plane non-cylindrical sheaths are common. Incidentally, continuity of the non-coaxial sinistral movement till much late, even after hardening of the deforming mylonite, is evidenced by the minor faults with sinistral shift of the trail of vein quartz boudins/blocks (Fig. 5d) present near the southern margin.

Further towards the core, on either side, the meta-greywacke exhibit extensive presence of siliceous lamellae, concentrated as discontinuous bands. Extension of such bands is restricted to a few tens of meters, tapering on either ends before disappearance. These siliceous lamellae, in fact, get dissipated within the greywacke, leaving behind still traceable faint markings. Within the bands, the lamellae are stretched, rotated and shortened and also displaced along the long axis, depending upon the orientation of the lamellae with respect to the direction of flow. The rotational component of deformation is marked by rotation of the siliceous lamellae having obliquity to the flow plane (Fig. 5e). The rotation of these quartzose lamellae very often produce a pseudo cross bedding. Nevertheless, these pseudo cross bedded structures occur in close association with tight rootless miniature folds (Fig. 5f), indicating that such structures are the products of intense deformation (Passchier et al. 1991). These structures are likely to be mistaken for turbidites. The rocks of the inner margin thus represent mylonites.

Nearer to the core, the siliceous lamellae present within meta-greywacke are present only as small discontinuous patches, mostly as hinges of small folds with markedly obliqe relationship with the flow plane. Continuity of these obliquely oriented and highly folded lamellae are conspicuously broken along the discrete planes representing mylonitic foliation. The rock itself acquires a crudely banded nature with alternate light and darker bands (Fig. 6a).

Fig. 6 **a** Crudely banded greywacke with traces of obliquely running siliceous lamellae; **b** Banded Ferruginous Chert showing en-echelon folding; **c** Banded mylonite. Alternate light and dark bands are arranged largely parallel to the foliation and show en-echelon shortening/rotational features, wherever there is an obliquity; **d** Layered ultramylonite with rotated pressure shadow (near left margin); **e** Near complete transposition and rotation of planes/siliceous lamellae in phyllite/phyllonite. Note the presence of pseudo cross bedding (below lens cap); **f** Small, rounded, relatively stiff mass, present as rotated segment, wrapped within the intensely foliated ultramylonite. The rounded segment shows a crude sigmoidal trail of the internal foliation, oblique to the external foliation. Shear sense is sinistral; **g** Synthetic fracture/slip in chloritoid porphyroclast within phyllonite/ultra-mylonite. Crossed nicols, 3X. Shear sense is sinistral

Adjacent to core, the chert bands, both ferruginous and non ferrous types, reduce in number. They, however, continue displaying en-echelon folding (Fig. 6b). Rocks of the innermost margin, on either side of the core, attain a well-defined banded nature, very likely to be confused with turbidites. The alternate light and dark bands are largely arranged parallel to the foliation and show shortening/rotational features in an en-echelon pattern, wherever there is an obliquity to the flow plane, differentiating them from the turbidites. Thus, the rocks become a classic banded mylonite (Fig. 6c), with a marked sinistral shear sense.

5.3 Core Zone or Zone of Ultramylonites

In the core part of the belt, the banded mylonite passes into ultra mylonite with still retaining the banded nature. The rocks have more similarity to phyllite/phyllonite and turbidites. Rocks are friable and show faster denudation. The alternate dark and light bands become less prominent and show minimum obliquity to the flow plane. Rotational features still continue as pressure shadows (Fig. 6d), very helpful in differentiating them from a turbidite. A very strong and close spaced flow plane representing mylonitic foliation is preserved all through the area. Near-complete transposition and rotation of oblique planes produce pseudo cross bedding (Fig. 6e). Without a holistic approach, it is very difficult to differentiate them from a normal turbidite. Rarely, small, rounded, relatively stiff mass of rock appears as rotated segment wrapped within the intensely foliated mass. These rounded segments show a crude sigmoidal trail of the internal foliation, oblique to the external foliation, i.e., the foliation present in the wrapping mass (Fig. 6f). This represents the presence of a component of simple shear flow along the foliation present in the wrapping mass, causing rotation to the relatively stiff and isotropic material. Presence of synthetic fractures/slips in chloritoid porphyroclast of phyllonite/ultramylonite (Fig. 6g) also represents a component of simple shear (sinistral) affecting the rocks.

The innermost part of the belt is remarkably devoid of chert bands. Innumerable thin quartz veins, sub-parallel to the mylonitic foliation, however, traverse the core part of the belt. These quartz veins are also affected by shearing and rendered as quartz mylonites.

6 Discussions and Conclusions

The main belt of Mahakoshal, extending between Narsinghpur in the west to Sidhi in east, has a dominant ENE–WSW trend, representing the axial traces of D_1 and D_2 structures. However, the southeastern extension of the belt, lying between Obra-Amsi-Jiawan fault (OAJ) and the Songarh-Kasar-Dudhi (SKD) fault, shows a WNW–ESE to NW–SE trend. This trend is clearly superimposed on the main Mahakoshal trend, well depicted in imagery, and does not appear to have resulted by folding. Contribution of aerial photo-geology and remote sensing, especially in the field of regional structural interpretation, is well established and provide vital clue in unraveling geological complications.

The overwhelming presence of rotation of fabric elements all along its length, frequent presence of 'alpha' and 'delta' type structures, strongly asymmetric 'S' and at places 'Z' shaped en-echelon folds and the sheath folds suggest that the belt has undergone strong ductile deformation. It is also evident that the form surfaces for the lineation and asymmetric minor folds, present in this part of Mahakoshals, are the dominant WNW-ESE to NW–SE trending foliation. Transposition of fabric along this foliation is also amply clear all through the belt. This foliation, therefore, represents a 'mylonitic foliation'. The stretching and boudinaging of chert bands and quartz veins along this foliation, and shortening of the obliquely oriented features, is suggestive of bulk flow along the foliation. Anticlockwise rotation of the stretched boudins and of the axial planes of near isoclinal folds, however, indicate that the deformation was noncoaxial and progressive in nature. The ductile flow of the chert bands associated with BIFs and their alignment along 'S' and 'C' planes, especially in the context of the considered opinion of their being sedimentary, is to be resolved.

The dominance of anticlockwise rotated fabric and shape symmetry indicate existence of a sinistral sense of movement, though, at places, clockwise rotated fabric along with 'Z' folds are also observed. The mapable fold of dextral vergence, as reported by Roy and Devrajan (op. cit.), from near Dudhmania, appears to be a failed 'Z' structure, refolded to 'S' symmetry, because of reversal of shearing from dextral to sinistral, during the non-coaxial progressive deformation. It appears more likely that with the onset of granitic activity to the south, initial compressional force came from south-west, squeezing the southern part of Mahakoshals to north-east. This was followed by greater granitic activity to south-east and resultant direction of compression reversing to north-west. Presence of both 'S' and 'Z' shaped folds indicate, at least, the role of reverse shearing (Hippert and Tohar 1999). Gneissosity acquired by the granitoids, along northern margin, is supportive of shearing and granitic activity being coeval.

Fluid inclusion studies indicate that a fluid, carrying gold, has moved along shear planes and the associated transient tension fractures to give rise to low tenor gold occurrences of this belt. The structurally controlled gold mineralization of the area, therefore, is not expected to continue to further west, beyond the limit of this shear zone.

Finally, ignoring the structure depicted in satellite imagery and aerial photographs, and continuing with allotting formational status to the rocks produced under different intensities of shear deformation, is misrepresentation of the facts. It is obvious that the shear generated structures present in the sinuous belt were mistaken for turbidites, in earlier literature. It is worthwhile to mention that it has been advocated that the turbidites selectively occur only in the southeastern sector of Mahakoshals. Also, the pseudo current bedding, as reported in the present paper is never associated with the main Mahakoshals. Close association of these structures with shear sense indicators is of great help in differentiating them from normal syn-sedimentary structures (turbidites) formed in tectonically active basins. Overwhelming and ubiquitous presence of rotational features in chert bands and in quartz veins of the southeastern sector of the Mahakoshals, further ascertains that the terrain is representative of a flow regime. Also, the splendid representation of the belt in satellite imagery, clearly showing a superposed trend, leaves no ambiguity in mind in differentiating the sinuous belt from the ENE-WSW trending main Mahakoshals. The name 'Dudhmania Shear Zone' was proposed for this remarkable belt (Banerji and Prasad 1997). Minor structures recorded herein, nevertheless, represent only the tip of the iceberg. A detailed study of the minor structures will further unravel the secrets of this large shear zone.

Acknowledgements The author thankfully acknowledges the help rendered by Mr Abhijeet Chatterjee, an alumnus of Department of Geology, Kazi Nazrul University (previously part of University of Burdwan) in digitizing geological maps used in his papers in recent years. Also, the help rendered by two anonymous reviewers of the paper, in improving its quality, is gratefully acknowledged.

References

Banerji DC (2011) A discussion on structural aspects of gold bearing belt of Singrauli and Sidhi districts of Madhya Pradesh and Sonbhadra district, Uttar Pradesh. J Econ Geol Geo Resour Manage 8(1–2):85–96

Banerji DC, Prasad AVK (1997) A report on specialized thematic mapping of the Mahakoshal Group of rocks around Chakaria Kalan, Distt. Sidhi, M.P., Part-I, Unpub. Prog. Rep., Geol Surv India

Banerji DC (2019) Evidences of large-scale shearing from Southeastern extension of the Mahakoshal Belt, Covering Parts of Sidhi & Singrauli Districts of Madhya Pradesh and Sonbhadra District, Uttar Pradesh, India. Int J Conf Proc 2(1), ICP.000526

Hippert J, Tohar E (1999) On the development of zones of reverse shearing in mylonitic rocks. J Struct Geol 21:1603–1614

Mangla Prasad, Dhir NK, Sharma DP et al (2000) Fluid inclusion studies on Proterozoic Gold prospect of Gurar Pahar and Gulaldih, Sidhi and Sonbhadra districts, M.P. and U.P. Geol Surv Ind, Spl. Pub. No. 57:250–259

Nair KKK, Jain SC, Yedekar DB (1995) Stratigraphy, structure and geochemistry of the mahakoshal greenstone belt. GSI Mem 31:403–432

Passchier CW, Myors JS, Kroner A (1991) Field geology of high grade gneiss terrains. Narora Publishing House, New Delhi and Springer Verlag, New York

Roy A, Bandopadhyay BK (1990) Tectonic and structural pattern of Mahakoshal belt of Central India. In: Precambrian of Central India: Spl Pub Geol Surv Ind. 28:226–240

Roy A, Devrajan MK (2000) A reappraisal of the stratigraphy and tectonics of the Palaeo-proterozoic Mahakoshal supracrustal belt, Central India; Geol Surv Ind, Spl. Pub. No. 57:79–97

Roy A, Devrajan MK, Hanuma Prasad M (2002) Ductile shearing and syntectonic granite emplacement along the southern margin of the Palaeoproterozoic Mahakoshal supracrustal belt: Evidence from Singrauli Area, Madhya Pradesh. J Geol Soc Ind 59(1):9–21

Roy A (2008) Magma emplacement in Central Indian Tectonic Zone – an evidence for large scale crustal growth and recycling during Proterozoic: a review. In: National symposium on geodynamics and evolution of Indian Shield – through time and space (Golden Jubilee of the Geol. Soc. India at Centre for Earth Science Studies, Thiruvananthapuram) pp 1–2

Sharma DP, Sinha VP, Kannadasan T et al (2000) Gold mineralization in eastern part of Son valley greenstone belt, Sidhi and Sonbhadra districts, M.P. and U.P. Geol Surv Ind, Spl. Pub. No. 57:271–278

Sharma RS (2009) Cratons and Fold Belts of India, Lecture Notes in Earth Sciences 127, Springer, Verlag, Berlin Heidelberg. https://doi.org/10.1007/978-3-642-01459-8_5

Soni MK, Jha DK (2001) Mahakoshal greenstone belt and associated gold mineralization. Geol Surv Ind, Spl Pub No. 64:317–326

Geo-Environmental Processes

Shallow Structure and Seismic Hazard in the Coastal Odisha, India

A. K. Rai and Rama K. Nayak

Abstract

The coastal India, particularly the east coast, is more vulnerable to natural hazards such as earthquakes, tsunami, and cyclones. The east coast is dominated by NW–SE trending faults and lineaments and is classified under seismic zone-II and III. Neo-tectonic earthquake activities have reportedly caused damage to infrastructure and monuments. Study of the geomorphic features have identified areas of thick accumulation of unconsolidated soil in the study region. Analysis of spectral amplitude ratio of horizontal and vertical component (HVSR) of the ambient noise data and multi-channel surface wave phase velocity (MASW) data collected at various locations of coastal Odisha indicate that the dominant frequency varies between ~0.16 and ~3.0 Hz in the coastal and nearby region. Inversion of dispersion curves computed from multi-channel surface wave data indicates that the shear-wave speed within the top 30 m of soil, i.e., Vs_{30} are ~400 m/s in the coastal region and ~700 m/s in the interiors. These values classify the coastal region as class 'C' sites (360–760 m/s), i.e., very dense soil and soft rocks as per the NEHRP recommendations, which could be a valuable input for further development of the region. Teleseismic receiver function also indicate a region of lower velocity in the shallow crust. We conclude that shallow seismic parameters are critical to the assessment of impact of natural hazards in general and particularly the seismic hazard in any region.

A. K. Rai (✉) · R. K. Nayak
Centre for Oceans, Rivers, Atmosphere and Land Sciences, Indian Institute of Technology Kharagpur, Kharagpur, 721302, West Bengal, India
e-mail: abhishek@coral.iitkgp.ac.in

Keywords

Horizontal to vertical spectral ratio (HVSR) • Multichannel surface wave analysis (MASW) • Receiver functions • NEHRP site classification

1 Introduction

The coastal regions are more vulnerable to natural hazards such as earthquakes, tsunamis, and cyclones because of their proximity to oceans. About 40% of the world's population resides in less than 100 km from the coast line. The eastern part of India holds a number of major monuments of historical and archaeological significance. The region is affected by both in-land seismicity and earthquakes in the Bay of Bengal. Although prediction of earthquakes is still not possible with any practical accuracy, the possibility of seismic hazard mitigation in case of a future earthquake does exits by proper study of site-effects and shallow subsurface characteristics. The eastern coastal region of India is dominated by NE-SW trending tectonic features such as lineaments and faults of Eastern Ghats and classified as zone II and III in the seismic zonation map of India. Minor earthquakes are frequently reported by local news agencies and national earthquake monitoring agencies. Tectonic activity along Gondwana Mahanadi rift basin and associated features is a major source of small magnitude earthquakes. Reactivation of older faults and lineaments in Mahanadi graben area is suggested as cause low to moderate earthquakes (Gupta et al. 2014).

The east coast region is drained by a radial network of streams such as Kuakhai, Bharagavi and Daya rivers which are the main tributaries of Mahanadi river system. Due to repeated flooding, thick unconsolidated soil and sediment are deposited in many parts particularly along the river banks. An earthquake of magnitude ~6.0 occurred in Bay of Bengal (Fig. 1), about 300 km away from Odisha coast (Rai et al. 2015), and reportedly caused minor damage to

A. K. Shandilya et al. (eds.), *Geological and Geo-Environmental Processes on Earth*, Springer Natural Hazards,
https://doi.org/10.1007/978-981-16-4122-0_13

buildings and monuments situated on unconsolidated sediments. The region is dominated by mining activity which fuels the economy of the region. A few of the major cities in the region such as Bhubaneswar have been notified as future smart city. This is likely to attract a number of development activities and further accelerate the development of the region. It is an established fact that shaking due to earthquakes are more vigorous in regions situated on thick riverine sediments. Therefore, it is essential to conduct a study of site-effects and microzonation for proper and planned development of the coastal region. It is also important for minimizing secondary seismic hazards such as damage caused by collapse of building and infrastructure.

Study of the effects of geologic and shallow structure on intensity of ground motion due to earthquakes are some of the most important aspects of geotechnical and earthquake engineering. In the recent past, site characterization studies are considered important inputs for infrastructure development projects and have become very popular among the city planners and civil engineers. The subsurface material properties are often defined in terms of shear modulus, density, shear-wave speed etc. Several techniques have been developed to study these characteristics of the subsurface soil. Among those, the non-invasive geophysical techniques such as horizontal to vertical spectral amplitude ratio (HVSR) derived from ambient noise data, and inversion of dispersion curve derived from multichannel surface wave analysis (MASW) are the most popular among geotechnical engineers for estimating site characteristics and shallow subsurface shear-wave speed at a faster and practical way even in the populated city centres and urban areas (Mahajan 2009; Mahajan et al. 2011). Furthermore, the so-called "receiver functions" (Rai et al. 2009) derived from seismic converted phases also provide information about shallow and deep seismic structure that may be included in obtaining further insights about the subsurface lithology.

In a recent study of variations in topographic elevations, soil types, distribution of geological features such as faults and lineaments, using GIS techniques Nayak et al. (2017) and Dhar et al. (2017) identified regions that have relatively low groundwater level and thick sediments deposit. The authors suggested that since low elevation areas are expected to have higher ground water level, therefore these areas are prone to a greater degree of seismic hazard such as liquefaction. Similarly low slope areas are may have greater accumulation of unconsolidated sediment deposits. It was further suggested that structural features such as ridge, valley, faults, fractures, and lineaments may redirect the seismic

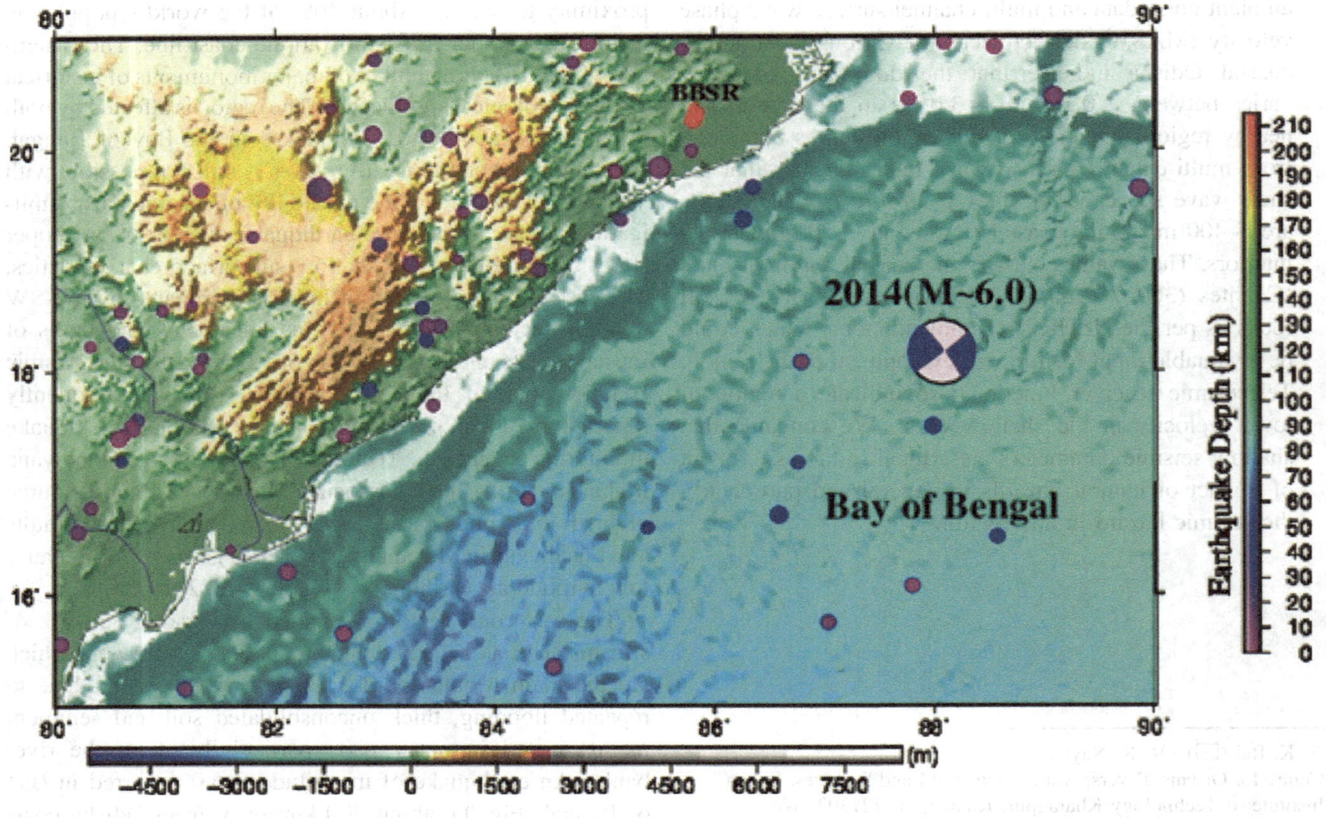

Fig. 1 Map of the study region. Location of Bhubaneswar (BBSR) is shown by red polygon. Earthquakes in Bay of Bengal and neighbouring region are also shown scaled by their respective magnitude, and colour corresponding to focal depths

energy. Using weighted overlay method and GIS technique (Dhar et al. 2017) tried to identify regions susceptible to unconsolidated sediment accumulation and presented a map for high susceptibility for seismic amplification which appears to be mostly confined in the eastern coastal region (Fig. 2). Low to moderate susceptibility areas are in the middle, north and western part of Mahanadi basin.

In the present work, we use ambient noise microtremor records to compute Horizontal to vertical amplitude ratio (HVSR) to estimate the peak frequency at the sites. We also use multi-channel surface wave (MASW) dispersion data to estimate shallow subsurface characteristics. In the following sections, we discuss details of the data collected, methodology, results and their

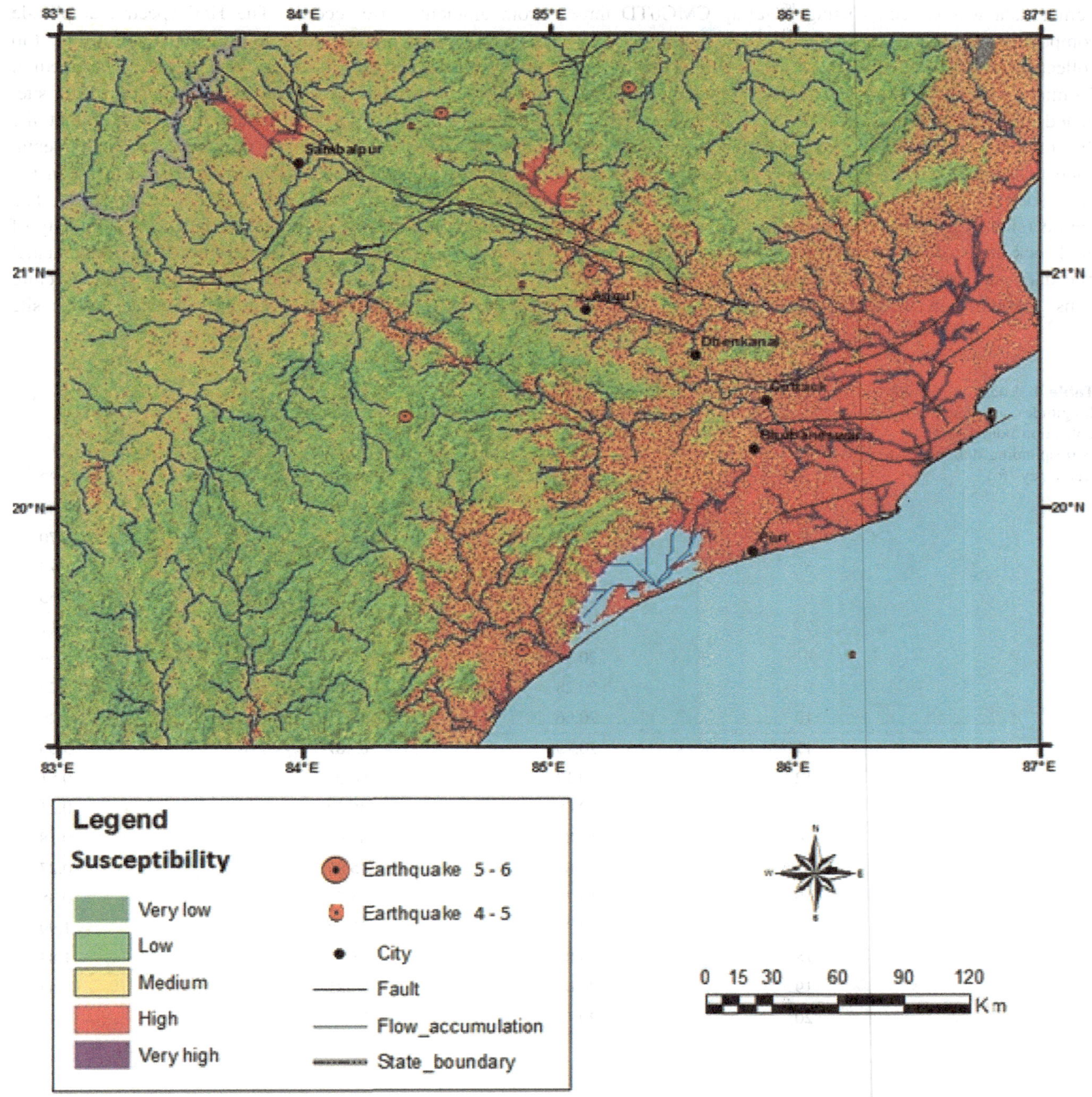

Fig. 2 Soft sediment deposits and ground motion susceptibility in the study region derived from GIS weighted overlay analysis (modified from Dhar et al. 2017)

implications for shallow seismic characteristics and seismic hazard.

2 Data and Methodology

The field works to collect the required data of ambient noise microtremors and multichannel surface wave were conducted between 2016 and 2019. The ambient noise microtremor data was collected using @Guralp CMG6TD three component broadband seismometers. The noise data was collected for 20 sites (Table 1) for a period of ~20 to 30 min for each of the sites. The data collection was completed in several visits of the field party in various parts of the study region. The multi-channel surface wave data were collected along three profiles located in various geological units such as in the coastal region (Puri), Mahanadi delta region (smart city Bhubaneswar) and Jatni area which are hard-rock areas of Khorda district. Thus, the collected seismic data and profiles sample the three main geologic terrains prevailing in the study region.

2.1 Horizontal to Vertical Spectral Amplitude Ratio (HVSR)

One of the most widely used methods to estimate the site effects is based on the Nakamura technique (Nakamura 1989). The method is relatively simple, faster and allows collecting the ambient noise data even in urbanized localities. In this technique, the spectral ratio of the vertical and horizontal components of ground motions, i.e., H/V is estimated from ambient noise records. The H/V spectral amplitude ratios (HVSR) are sensitive to the thickness of the top unconsolidated soil and shear-wave speed. The method provides an estimate of fundamental frequency of the site. Studies using H/V spectral method have shown that the ambient noise HVSR shows a peak near the fundamental S-wave frequency if there is an impedance contrast in the subsurface (Biswas et al. 2015; Mahajan et al. 2011). The phenomenon is related to frequency dependent ellipticity of Rayleigh waves (Lachet and Bard 1994). If the fundamental frequency of a site and that of an existing man-made structure are nearly similar, it may result in resonance at the site

Table 1 Location (latitude, longitude) of ambient noise data collection points, and corresponding dominant site frequency (fo)

Sl. No	Latitude (Deg.)	Longitude (Deg.)	fo (Hz)
1	20.24	85.86	0.17
2	20.21	85.85	3.82
3	20.24	85.83	3.82
4	20.25	85.82	0.70
5	20.24	85.82	1.73
6	20.26	85.78	0.53
7	20.29	85.79	1.11
8	20.31	85.71	0.64
9	20.28	85.85	5.29
10	20.06	86.35	2.20
11	20.15	85.67	3.03
12	20.17	86.05	0.51
13	20.00	86.20	0.21
14	19.85	85.96	0.18
15	19.81	85.85	0.27
16	19.79	85.79	0.35
17	20.27	85.88	1.99
18	19.87	85.82	1.83
19	20.00	85.82	0.64
20	20.13	85.84	1.31

causing severe damage to the buildings and infrastructure. Therefore, estimation of site frequency is an important step towards earthquake hazard evaluation. Empirical formulas relating soil thickness and resonance frequency allow us to delineate zones of vulnerability. It is seen that resonance frequency of soil (fo) is closely related to its thickness (h) by $H = a/f^b$, where a and b are empirical constants, and their values vary from one region to the other (Seht and Wohlenberg 1999; Parolai et al. 2002). Similarly, fundamental periods of building frames are also highly scattered; however, it is possible to get an estimate of natural frequency of tall buildings as a function of sediment thickness (H). Empirically, it is observed that the fundamental frequencies of buildings are approximately inversely proportional to the sediment thickness beneath it (Goel and Chopra 1997).

Synthetic tests indicate that as the thickness of the top layer increases, the peak of the H/V shifts toward the lower frequency side and vice-versa. On the other hand, the increase in shear-wave speed of the top layer shifts the peak towards the higher side of the frequency (Fig. 3). Furthermore, it is evident that the shape of the H/V curve is controlled by the impedance contrast between layers. As the contrast decreases, the amplitude of the peak decreases, and the shape of the H/V curve also changes accordingly. Therefore, horizontal to vertical spectral amplitude ratio (HVSR) is sensitive to unconsolidated soil thickness and shear-wave speed of the top layer and is widely used to estimate shear-wave speed profile of the subsurface.

The data was collected during a field work that was conducted in March–May of 2016 which marks the beginning of the summer in the region. To record the horizontal and vertical components of the ambient noise, the three component Guralp CMG-6TD (30 s–50 Hz) were used. The data was recorded at 20 sites at 100 samples/s for a continuous period of about ~30 min. The locations of the sites were determined using Global Positioning System (GPS) (Table 1).

The horizontal to vertical component spectral amplitude ratio were computed for variable window lengths of 15 s to 40 s. The amplitude spectra were smoothed (Konno and Ohmachi 1998). The final H/V ratio was obtained from averaging the H/V spectral amplitude ratio for all the windows. The horizontal to vertical spectral amplitude ratio (HVSR) for locations 16 and 20 are shown in Fig. 4 and are compiled in Table 1 for all the sites. The analysis of horizontal to vertical spectral ratios indicates that the dominant frequency at various sites vary between 0.16 and 3.0 Hz indicating wide variation in soil thickness in the coastal regions and away from the coast (Fig. 5).

2.2 Multi-Channel Surface Wave Analysis (MASW)

Dispersion of low frequency surface waves, particularly the Rayleigh wave phase velocity, is attributed to variations of shallow shear wave speed in the medium, and thus related to the stiffness property of the soil. The multi-channel analysis of surface wave (MASW) techniques estimate the phase velocity of Rayleigh waves as a function of frequency (Park et al. 1999). The dispersion curve can be forward and inverse modelled to obtain the shallow subsurface shear wave speed profile and classify the site on the International Building Code (IBC) standards.

The active source MASW data collection was conducted with a 24 channel vertical component Geometrics Geode system, with geophones having a natural frequency of 4.5 Hz. The geophones were deployed at a regular interval of 1.5 m for the three profiles that were acquired in different geological conditions. The length of the array is approximately equivalent to the maximum wavelength that can be retrieved, which also determines the maximum depth of investigation (Liu et al. 2000). The data was collected at a sampling interval of 0.5 ms for 1.024 s for each geophone. Multichannel surface wave analysis requires an active source such as a weight drop or a sledge hammer to generate low frequency surface waves. We used sledge hammer (~10 kg) as a source. The source was offset at a distance of −10 m from the first geophone. Similar configurations were used to acquire the profiles at each location. For the multichannel surface wave analysis (MASW), the shot gathers were processed to generate the dispersion curve (Fig. 6). The depth of penetration of ground roll is considered to be approximately equal to its wavelength (λ), whereas the maximum depth to which the dispersion curve constrains the model can be estimated as approximately half or one-third of the largest wavelength analysed (Richart et al. 1970; Rix and Leipski, 1991). The dispersion data collected during this experiments is sufficient to sample depths up to ~ 30 m of the shallow subsurface. Inverse modelling of dispersion curves indicates that the shear-wave speed varies between ~400 m/s and ~700 m/s in the study region (Figs. 6, 7). The profiles cover various geological terrains of the region, providing a clear picture of variation of shallow seismic shear-wave speed. On the basis of National Earthquake Hazard Reduction Program (NEHRP) classifications, the sites where data were collected may be classified as class ‘C’ sites.

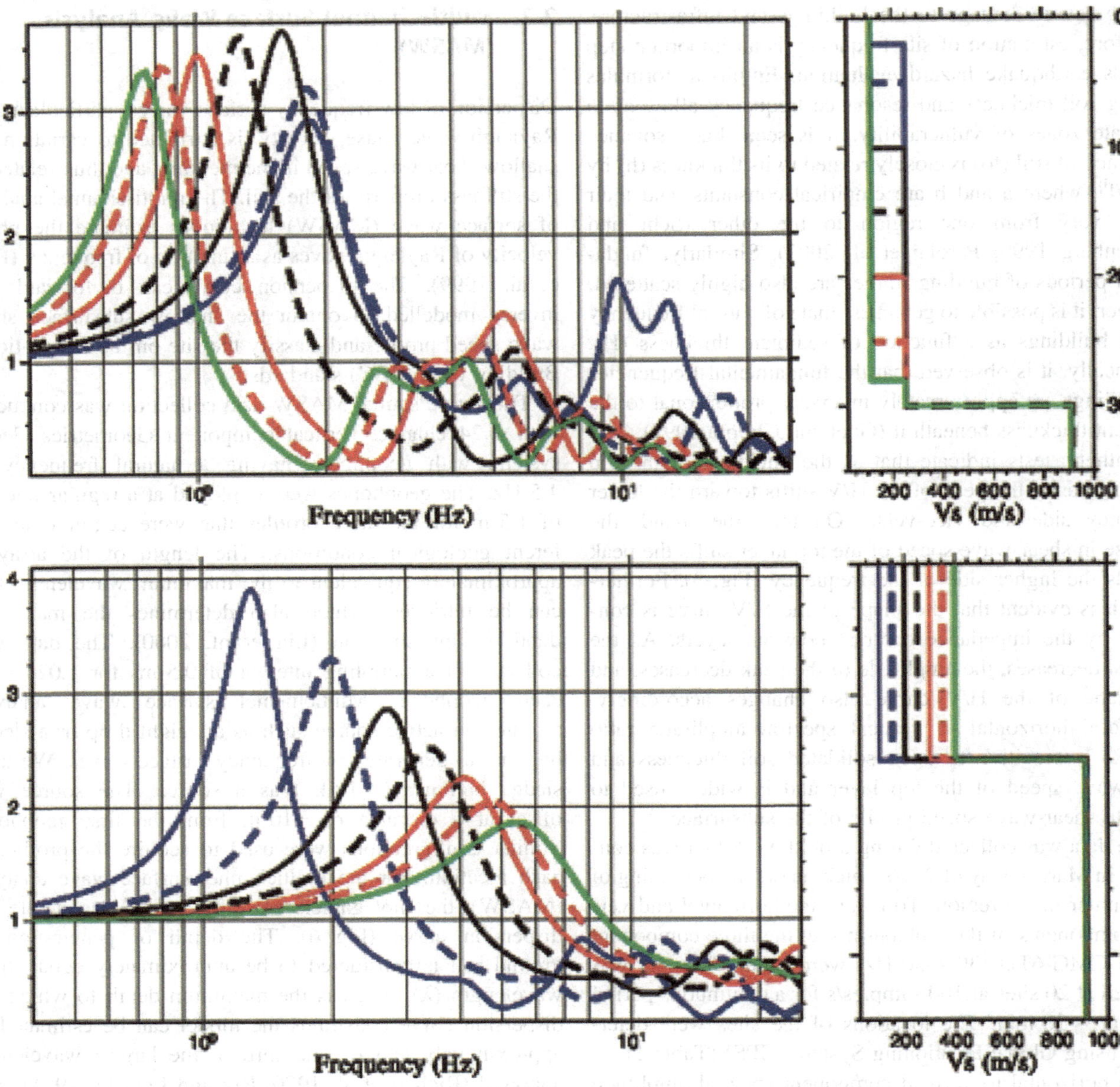

Fig. 3 Variation of H/V spectral ratio (top panel), as a function of soil thickness, and shear wave speed of the top soil (bottom panel). The H/V peak shift to lower value as the thickness of the top soil increases, and shear-wave speed of top layer decreases

3 Receiver Functions

Shallow structure of the subsurface affects the propagation of seismic waves (Rai 2017). Study of converted seismic phases also called receiver functions using P_s and S_p phases is a well-established seismological technique to study the shallow structure (Rai et al 2009). Receiver functions provided better insights into various layers and shear-wave speed of the shallow to the deeper crust. Seismological data recorded by observatory located at Jagatshinghpur district of Odisha which is about 40 km away from the Bay of Bengal coast was used to compute receiver functions from epicentral distance of 30° to 90°, which are free from interference from other arrivals. An example of the computed receiver function along with the synthetic modelling is shown in Fig. 8. Time-delay in P-wave arrival (at 0s) indicates significantly lower shear-wave speed in the top few kilometres of the crust.

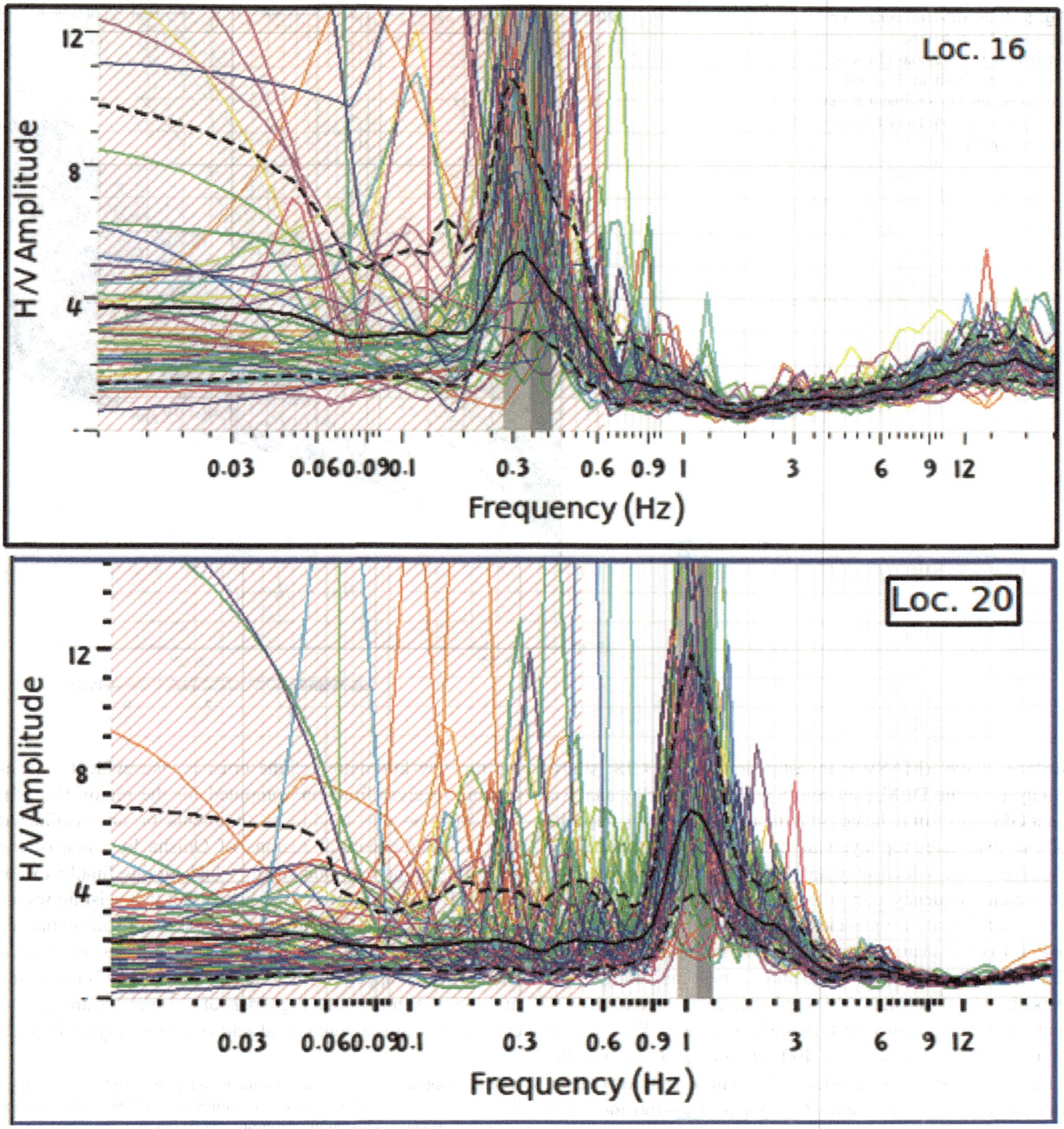

Fig. 4 The H/V spectral amplitudes (solid black lines) at two sites location 16 and 20 (top and bottom panels) (Table 1). The Variation in dominant frequency indicates change in thickness of the top unconsolidated soil

4 Conclusions

The site effect is a complex representation of several parameters such as shear wave speed, stiffness of the soil, soil thickness, and moisture content. The shallow subsurface characteristics such as unconsolidated soil thickness, characteristic frequency, and shear-wave speed are important for evaluating the geotechnical effects. We conducted study of the coastal Odisha and surroundings including the city of Bhubaneswar to determine the dominant frequency at various sites and complimented the results using multichannel

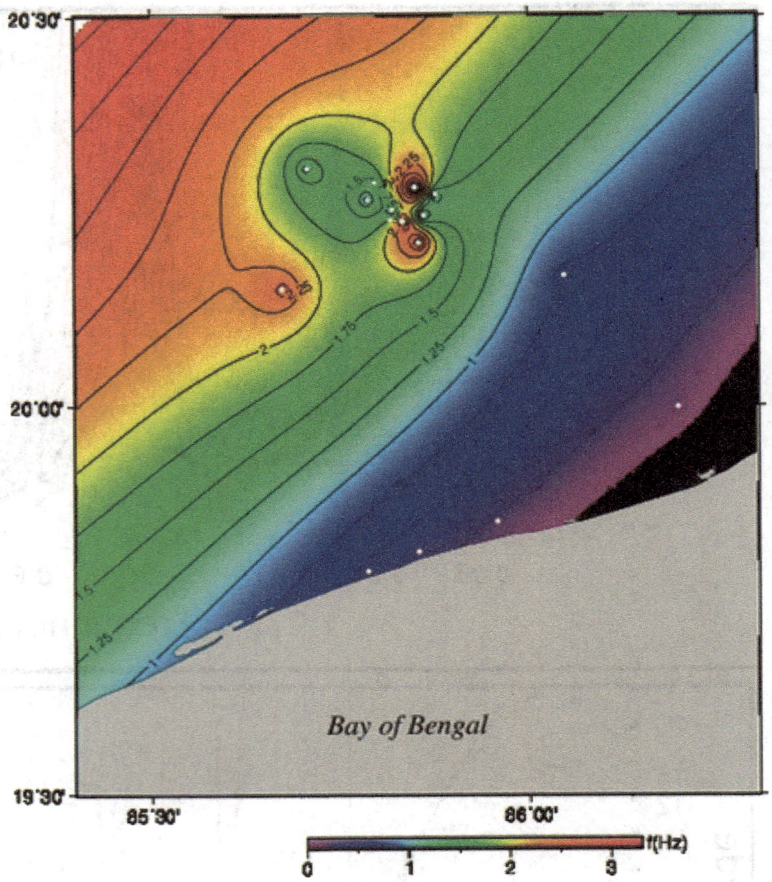

Fig. 5 Peak H/V frequency for the coastal Odisha and surroundings. The coastal areas show lower dominant frequency indicating thicker sediment cover that appear to vary away from the coastal region

surface wave (MASW) investigations, and GIS study. Analysis of the DEM data, and a weighted overlay analysis on a GIS platform indicates that the region has thick deposits of unconsolidated top layers in the coastal regions and along the river delta, which may amplify the seismic energy. The dominant frequency computed from HVSR analysis indicates that the site frequencies vary between a range of 0.17 and 3.0 Hz, indicating considerable variation in thickness of the top soil. The low frequency zones indicating thick deposits of unconsolidated layer is confined in coastal areas and are in agreement with GIS analysis results. The average of the Vs_{30} varies between 400 m/s and 700 m/s. On the basis of NEHRP classifications, the central zone of Odisha can be classified as class 'C' (Vs ~ 360–760 m/s) site which indicates hard or very hard soil, gravels and soft rocks. A major part of coastal Odisha also shows shear-wave speed of ~ 400 m/s and can be classified as class 'C' sites. However, lower shear wave speed is possible within the range of 180-360 m/s that are indicative of sands, clays and gravels and hard soil. Significantly lower seismic velocity in the top few kilometres of the crust is also evident in teleseismic receiver function computed for the region (Fig. 8). This may be due to thicker sediments or unconsolidated deposits in the coastal regions of Odisha by some of the large rivers such as Mahanadi, which drains finally to the Bay of Bengal. It is well established fact that seismic waves trapped in unconsolidated sediments or softer soils enhances the shaking several times. Therefore, regions of coastal Odisha which appear to have unconsolidated softer deposits may be at significantly high risk of seismic hazard due to earthquakes in the continental part or in the Bay of Bengal.

Acknowledgements The authors thank all participants who helped during the field work. This work was completed by financial assistance of the National Geospatial Program (NGP-DST), Govt. of India through its grant (NRDMS/CHRA/Rai/e-02/2019), Ministry of Education, Govt. of India, and SRIC-ISIRD grant from the Indian Institute of Technology Kharagpur. We sincerely thank and acknowledge the constructive comments by the two reviewers which significantly improved the original manuscript.

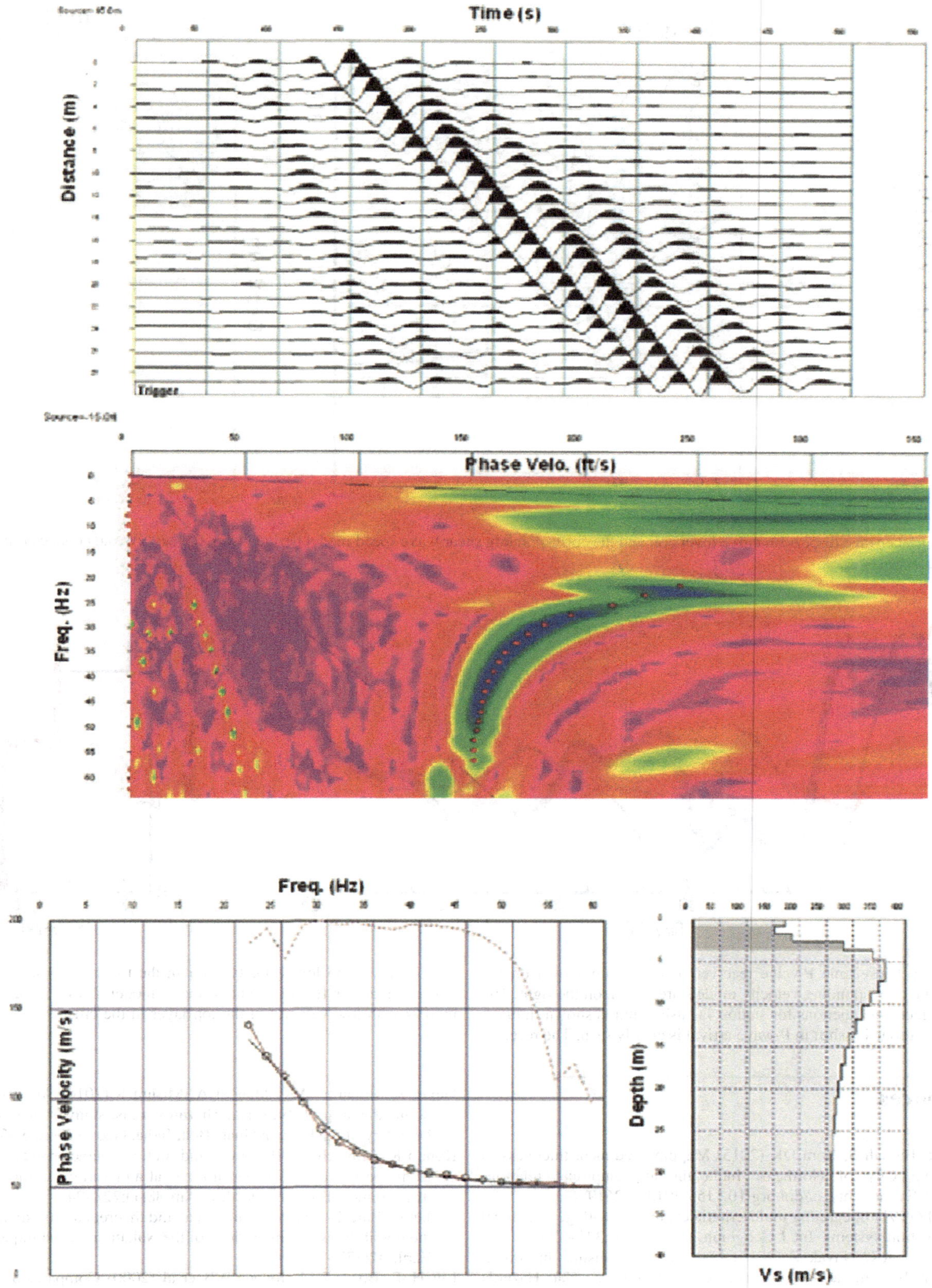

Fig. 6 Multi-channel surface wave records for coastal region (Puri). The multi-channel surface wave dispersion data (top), and the corresponding phase velocity dispersion curve (middle). The inversion of the surface wave phase velocity (bottom-left) and the corresponding inverted velocity model (bottom-right)

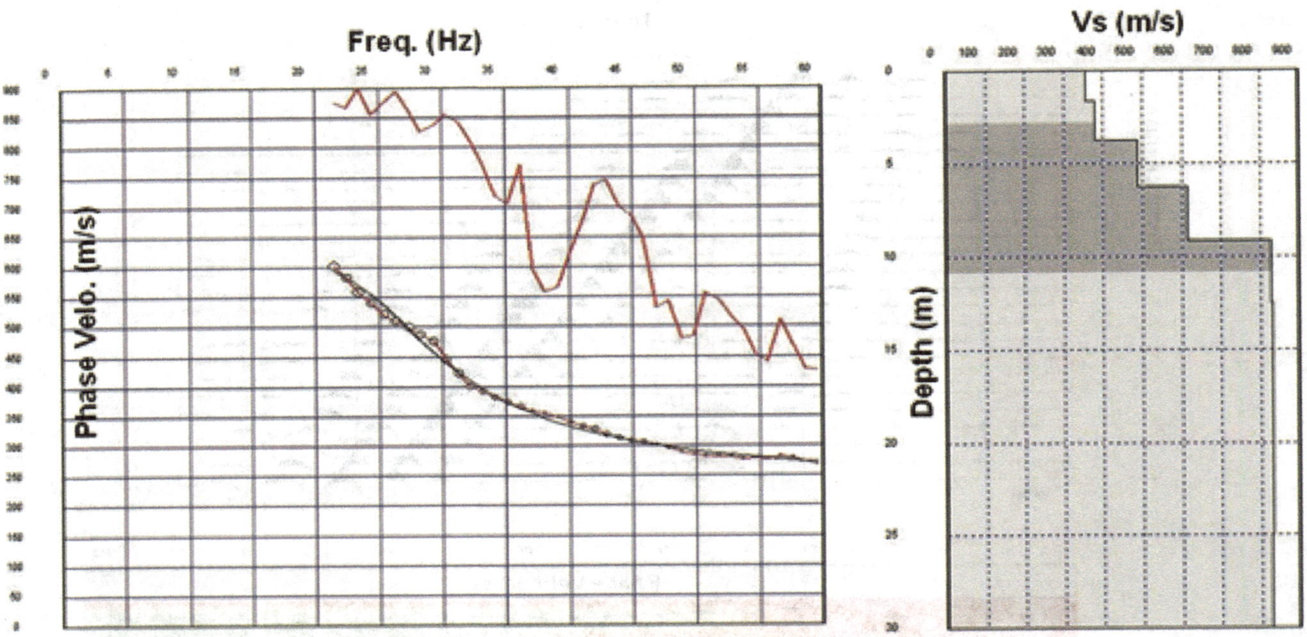

Fig. 7 Surface wave dispersion curve inversion and the corresponding shear wave speed model derived from the MASW data collected in the city of Bhubaneswar

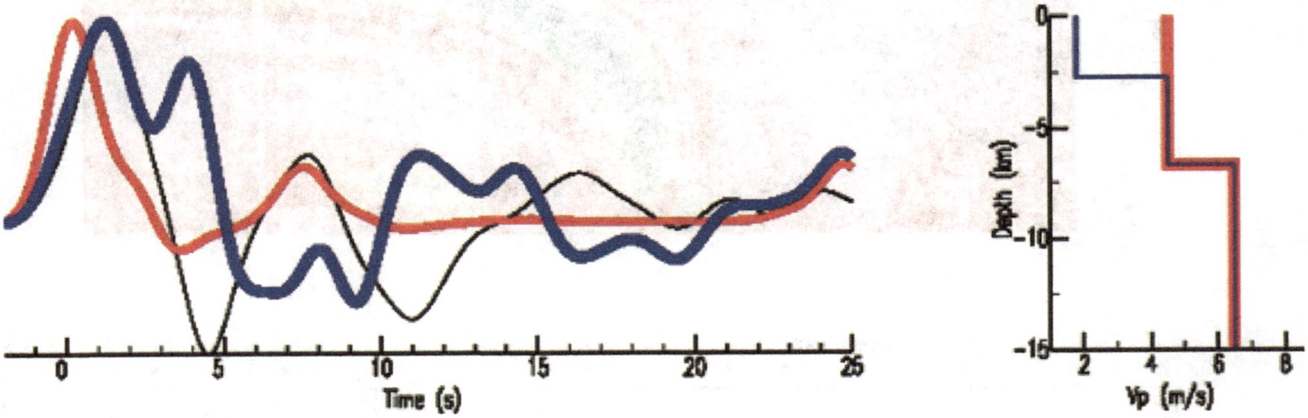

Fig. 8 The teleseismic P-wave receiver functions are shown on the left, and the corresponding velocity models are shown on the right. The observed receiver functions for station Jagatsinghpur is shown in black line (left). About 1 s shift in P-wave arrival is clearly seen. The receiver functions with no low velocity layer in the top curst is shown in red, whereas receiver functions and velocity model shown in blue are for models with low velocity in the top layer in the crust

References

Biswas R, Baruah S, Bora DK (2015) Mapping sediment thickness in Shillong City of Northeast India through empirical relationship. J Earthqs. https://doi.org/10.1155/2015/372619

Carver SJ (1991) Integrating multi-criteria evaluation with geographical information systems. Int J Geograph Inf Sys 5(3):321–339

Dhar S, Rai AK, Nayak P (2017) Estimation of seismic hazard in Odisha by remote sensing and GIS techniques. Nat Hazards 86(2):695–709. https://doi.org/10.1007/s11069-016-2712-3

Goel RK, Chopra AK (1997) Period formulas for moment resisting frame buildings. J Struct Eng 123(11):1454–1461

Gupta S, Mohanty WK, Mandal A, Mishra S (2014) Ancient terrene boundaries as probable seismic hazards: a case study from the northern boundary of the Eastern Ghats Belt, India. Geosci Front 5:17–24

Konno K, Ohmachi T (1998) Ground motion characteristics estimated from spectral ratio between horizontal and vertical components of microtremor. Bull Seismol Soc Am 88(1):228–241

Lachet C, Bard P-Y (1994) Numerical and theoretical investigations on the possibilities and limitations of the Nakamura's technique. J Phys Earth 42:377–397

Liu H.-P, Boore DM, Joyner WB et al (2000) Comparison of phase velocities from array measurements of rayleigh waves associated with microtremor and results calculated from borehole shear-wave velocity profiles. Bull Seism Soc Am 90(3):666–678

Lermo J, Chavez-Garcia FJ (1993) Site effect evaluation using spectral ratios with only one station. Bull Seis Soc Am 83:1574–1594

Mahajan AK, Galiana-Merino JJ, Lindholm C et al (2011) Characterization of the sedimentary cover at the Himalayan foothills using active and passive seismic techniques. J Appl Geophys 73:196–206

Mahajan AK (2009) NEHRP Soil classification and estimation of 1-D site effect of Dehradun fan deposits using shear wave velocity. Eng Geol 104:232–240

Nakamura Y (1989) A method for dynamic characteristics estimation of sub surface using-microtremor on the surface. Railw Tech Res Inst Rep 30:25–33

Nayak P, Rai AK, Tripathy S (2017) Evaluating groundwater prospects using GIS techniques. Sustain Water Res Manage 3(2):129–139. https://doi.org/10.1007/s40899-017-0082-y

Park CB, Miller RD, Xia J (1999) Multichannel analysis of surface waves. Geophysics 64(3):800–808

Parolai S, Bormann P, Milkereit C (2002) New relationships between Vs, thickness of sediments, and resonance frequency calculated by the H/V ratio of seismic noise for the Cologne area (Germany). Bull Seism Soc Am 92(6):2521–2527

Rai AK (2017) Frequency dependent scattering observed in P- and Surface wave arrivals from south India. Pure Appl Geophys 174 (3):895–905

Rai AK, Tripathy S, Sahu SC (2015) The May 21st, 2014 Bay of Bengal Earthquake: Implications for intraplate stress regime. Curr Sci 108(9):1706–1712

Rai AK, Gaur VK, Rai SS, Priestley K (2009) Seismic signatures of the Pan-African orogeny: implications for southern Indian high-grade terranes. Geophys J Int 176(2):518–528

Richart Jr FE, Hall JR, Woods RD (1970) Vibrations of soils and foundations. Prentice-Hall, Inc

Rix GJ, Leipski EA (1991) Accuracy and resolution of surface wave inversion. In: Bhatia Shobha K, Blaney Geoffrey W (eds) Recent advances in instrumentation, data acquisition and testing in soil dynamics. American Society of Civil Engineers, pp 17–32

Routray JK, Rath KC, Sahoo NN (1996) Growth, development and planning of Bhubaneswar. Cities 13(2):79–96

Seht Malte Ibs-von, Wohlenberg J (1999) Microtremor measurements used to map thickness of soft sediments. Bull Seismol Soc Am 89 (1):250–259

Theilen-Willige B (2010) Detection of local site conditions influencing earthquake shaking and secondary effects in Southwest-Haiti using remote sensing and GIS-methods. Nat Hazards Earth Syst Sci 10:1183–1196

A Critical Evaluation of the Role of Geotectonics in Groundwater Arsenic Contamination

Mohammad Ayaz Alam, Abhijit Mukherjee, and Prosun Bhattacharya

Abstract

It has been observed globally that primary groundwater arsenic (As) is derived from modern and ancient magmatic arcs at continental convergent margins of some of the most prominent orogenic systems worldwide (e.g., Andes, Himalaya). It ends up in arc-derived sediments in the adjacent foreland basin through rapid and continuous erosion of exhuming arc sequences, where groundwater could be contaminated through secondary As in the aquifer matrix, mostly allochthonous arc-derived sediments. Apart from the origin and development of magmatic arc, geotectonic control for other primary As sources is also evident in the form of (i) global distribution of geothermal and volcanic centers, where As-enriched hydrothermal and magmatic fluids contaminate surface and groundwater, (ii) ore (especially metal sulfide) deposits through natural leaching that releases As to surrounding environments, which is accelerated by metal mining and processing activities, and (iii) coal deposits and hydrocarbon reservoirs, together with the exploitation processes of both these fuel resources, are other geogenic sources of As contamination that is accelerated by the extraction and refining processes and subsequent usage. Whereas the aforementioned solid and liquid phase As inputs to groundwater are attributed to the lithology and geothermal and volcanic fluids, respectively, As speciation can also be associated with the basin's structural settings driven by geotectonics. Basin morphology, prolongation of faults, especially division of the underlying basement into blocks can produce discrete compartmentalized hydrodynamic environments, resulting in spatial variations in hydrochemical composition, including As speciation.

M. A. Alam (✉)
Departamento de Geología, Facultad de Ingeniería, Universidad de Atacama, Avenida Copayapu 485, Copiapó, Región de Atacama, Chile
e-mail: ayaz.alam@uda.cl

A. Mukherjee
Department of Geology and Geophysics, Indian Institute of Technology (IIT), Kharagpur, West Bengal, India
e-mail: abhijit@gg.iitkgp.ac.in

P. Bhattacharya (✉)
KTH-International Groundwater Arsenic Research Group, Department of Sustainable Development, Environmental Science and Engineering, KTH Royal Institute of Technology, Stockholm, Sweden
e-mail: prosun@kth.se

Keywords

Arsenic • Environment • Contamination • Groundwater • Geotectonics • Speciation • Arc magmatism • Geothermal • Volcanism • Metallogeny • Orogenesis • Foreland basin • Basement structure

1 Introduction

Although arsenic (As) content of crustal rocks varies widely, for most known and quite extensive instances of groundwater As contamination, its sources are geogenic (Smedley and Kinniburgh 2002; Bhattacharya et al. 2002, 2017; Nriagu et al. 2007; Coomar et al. 2019). Arsenic is the most ubiquitous toxic element present in all sorts of geological environments that gets transported by all kinds of media. Tracing all its origin and pathways for its human exposure has been a challenging task. However, this challenge has been met through multidisciplinary global collaborative and team efforts by deciphering As sources and pathways. Arsenic can get disseminated in air and carried by wind (e.g., volcanic ash in Argentina and Chile, Bhattacharya et al. 2006a, b, Quintanilla et al. 2009; Bundschuh et al. 2004, 2010, 2012; Nicolli et al. 2012; Morales-Simfors et al. 2020; Aullón Alcaine et al. 2020; Mariño et al. 2020), dissolved

A. K. Shandilya et al. (eds.), *Geological and Geo-Environmental Processes on Earth*, Springer Natural Hazards,
https://doi.org/10.1007/978-981-16-4122-0_14

and transported by water (e.g., groundwater in Argentina, Bangladesh, Bolivia, Chile, India, Peru, Tanzania, Bhattacharya et al. 1997, 2001; Biswas et al. 2014a, b, c; Chakraborti et al. 2015; Bhowmick et al. 2018; Tapia et al. 2019a, b, Chakraborty et al. 2015, 2020; Mukherjee et al. 2021; Quino Lima et al. 2020, Quino Lima 2021a, b, Ijumulana et al. 2020), or stay in solid form in As-bearing minerals (e.g., mineralized zones, coupled with the residual products of mining and metallurgical processes, e.g., Latin America and the Caribbean, Quintanilla et al. 2009; Ramos et al. 2012, 2014; Bundschuh et al. 2012, 2020).

The ubiquitous nature of As in the environment can be understood simply by considering that hypothetical leaching of its average crustal concentration (1.8 mg/kg, Henke 2009) into groundwater is sufficient to cause toxic levels of As concentration (0.01 mg/L, WHO 2017). Moreover, surface sediments and soils in various parts of the world contain higher and more variable concentrations (ranging from 2 to 20 mg/kg on average) than the average crustal concentration of 1.8 mg/kg (Taylor and McLennan 1985; Wedepohl 1995; Mandal and Suzuki 2002; Bhattacharya et al. 2002; Henke 2009; Mukherjee et al. 2009a, b, and references therein). The average and maximum As concentration in 3,024 marine sediment samples from coastal Japan was determined to be 14 and ~580 mg/kg, respectively, of which As concentration in 65% samples was <10 mg/kg, while in 87.5% it was <20 mg/kg (Masuda 2018). On the other hand, Ohta et al. (2010) attributed the sediments with high As concentrations in Japan's inland areas to coal and sulfide mines. Consequently, even a trace amount of As in solid-phase aquifer matrices is sufficient for groundwater under conducive environments (Korte and Fernando 1991; Stüben et al. 2003; Sracek et al. 2004; Guilliot and Charlet 2007; Henke 2009). All these imply that any lithological unit in the crust can be considered a potential As source (Matschullat 2000; Bhattacharya et al. 2002, 2007; Nordstrom 2002; Smedley and Kinniburgh 2002; Saunders et al. 2005). Nonetheless, considering "average value" being a theoretical concept and also, as a matter of fact, not all aquifers around the world are As-contaminated, as only certain segments of the lithosphere are enriched in As. Delineation and segmentation of such enriched portions are the basis for discussing the role of geotectonics in As contamination of aquifers in this chapter, based on some insightful works and representative examples from across the world (e.g., Mukherjee et al. 2007, 2008, 2009a, b; Mukherjee et al. 2014, 2019; Barringer and Reilly 2013; Giménez-Forcada and Smedley 2014; Masuda 2018; Morales-Simfors et al. 2020; Bundschuh et al. 2020).

2 Some Considerations and Constraints

Arsenic, a metalloid, being siderophile and chalcophile element, behaves as a metal and combines with sulfur; however, it shows volatile behavior in molten rock, similar to the light rare earth elements (Sims et al. 1990; Bhattacharya et al. 2002; Nriagu et al. 2007). The partition coefficient of As between sulfide and silicate melts varies as a function of oxygen fugacity, f_{O_2} (Li and Audétat 2012, 2015). Accordingly, at a specific temperature and FeO content, As fraction in silicate and monosulfide melts increases with f_{O_2} (Li and Audétat 2012, 2015). Although silicate and carbonate mineral phases do not host As as a major element, i.e., it does not appear in their formula (Majzlan et al. 2014), at low temperature, high f_{O_2} and low sulfur fugacity f_{S_2}, silicates and carbonates can incorporate As, e.g., filatovite $K(Al,Zn)_2(As,Si)_2O_8)$, a variety of feldspar first reported at Tolbachik volcano, Kamchatka, Russia (Vergasova et al. 2004). Charnock et al. (2007) found As incorporated into garnet in the overgrowth of a hydrothermal vein. Pascua et al. (2005) reported As-enriched smectite in hydrothermal precipitates at a geothermal plant in Akita, Japan. Moreover, Ryan et al. (2011) found fractured aquifers in ultramafic rocks contaminated with As derived from serpentine and magnesite through chemical weathering.

Arsenic sources that can potentially contaminate the aquifers, i.e., leading the As concentration to toxic levels (0.01 mg/kg, WHO 2017), may develop through complex interactions between tectonic, geochemical, and biological processes (e.g., Saunders et al. 2005; Zheng 2006; Nordstrom 2009; Barringer and Reilly 2013; Mukherjee et al. 2014, 2019; Masoodi and Rahimzadeh 2018; Masuda 2018; Polya et al. 2019). Although there is no controversy about As mobilization from the aquifer matrix leading to As enrichment of groundwater and there are plenty of works on it (e.g., Smedley and Kinniburgh 2002; Saunders et al. 2005 and references therein), the primary source of As is often controversial and studied by few researchers in detail (e.g., Mukherjee et al. 2011; Raychowdhury et al. 2014). Bundschuh et al. (2021) have identified different sources and pathways for As exposure for Latin America, namely, (i) volcanism and geothermalism, taking into account (a) volcanic rocks, fluids (e.g., gases), and ash, including large-scale transport of the latter through different mechanisms (b) geothermal fluids and their exploitation, (ii) natural lixiviation and accelerated mobilization from (mostly sulfidic) metal ore deposits by mining and related activities, (iii) coal deposits and their exploitation, (iv) hydrocarbon reservoirs and co-produced water during exploitation, (v) solute and sediment transport through rivers to the sea,

(vi) atmospheric As (dust and aerosol), and (vii) As exposure through geophagy and involuntary ingestion. Of these, authors identified the two most important and well-recognized sources and mechanisms for As release into environments affecting the Latin American population: (i) volcanism and geothermalism, and (ii) the strongly accelerated release from geogenic sources by mining and related activities. This is in line with the findings of Smedley and Kinniburgh (2002) that typical environments containing highly As-contaminated groundwater include (i) low temperature (including mining and non-mining) and (ii) high temperature (hydrothermal) settings.

Since As-contaminated groundwater is often distal from the (primary) As source(s), it has been argued continuously that understanding As mobilization is more critical than deciphering the (primary) source (e.g., Nordstrom 2002; Smedley and Kinniburgh 2002; Mukherjee and Fryar 2008; Scanlon et al. 2009; Ravenscroft et al. 2009). Consequently, many published peer-reviewed works deal with the hydrogeochemical processes controlling As mobilization processes in the contaminated areas, which hugely outnumber the studies on primary or ultimate As source. Saunders et al. (2005) emphasized the need for the latter, putting forward a case for a better understanding of the ultimate As source, considering that there is a consensus about mobilization processes leading to As enrichment within the aquifers (Saunders et al. 2005). van Geen (2011), on the other hand, put this view in another term. He emphasized the importance of understanding the etiological aspects of As enrichment in groundwater and the mobilization in aquifers and its control through in-situ reactive mobilization within the aquifer or transported advectively along flow paths from upstream enriched sources, i.e., ex situ, to areas with low As initially. van Geen (2011) proposed that mobilization processes, viz. water chemistry, with possible influence of local geology, lithology in particular, control the spatial distribution of As in aquifers. The duration of such hydrochemical processes leading to As enrichment of groundwater could be from few decades to thousands of years (van Geen 2011). Several workers have recognized the importance of ex situ sources, in addition to the in situ mobilization processes (viz. leaching) for As enrichment of groundwater, e.g., Smedley and Kinniburgh (2002), Nordstrom (2002), Mukherjee et al. (2009a, b), Ravenscroft et al. (2009). Accordingly, high As in young sedimentary aquifers should not be attributed only to local geology, as the time required for in situ hydrochemical processes might be more than the time elapsed since the deposition of those sediments and diagenesis, instead to larger, regional-scale geodynamic processes (Bhattacharya et al. 2004; Scanlon et al. 2009; Mukherjee et al. 2014, 2019). In this regard, although Ravenscroft et al. (2009) did highlight the proximity of young fold mountain belts to many As-enriched areas, they did not elaborate further on the relationship between the two. Saunders et al. (2005), on the other hand, proposed that As-contaminated groundwater is a product of interactions between tectonic, geochemical, and biological processes. Zheng (2006) and Nordstrom (2009) suggested a link between As enrichment and crustal evolution. Moreover, regional-scale geochemical processes lead to chemical weathering of a variety of As-bearing rocks, as noted by analyses of surficial material (viz. rocks, sediments, etc.) in different parts of the continental crust (e.g., Ahmed et al. 2004; Grosz et al. 2004; Saunders et al. 2005; Mukherjee et al. 2014, 2019). For example, As source in the contaminated aquifers of the major fluvial systems of the Indian subcontinent (e.g., Ganges–Brahmaputra, Indus, etc.) have been traced to the Himalayas, which acts as the sediment provenance, as well as the headwater for the rivers draining the basin (Stanger 2005; Hattori and Guillot 2003; Guilliot and Charlet 2007; Fendorf et al. 2010; Mukherjee et al. 2014, 2019; Chakraborty et al. 2015, 2020).

Arc magmatism and related hydrothermal activity, together with high rates of erosion due to rapid uplift of the arc zones and transport of the eroded sediments to the downstream basins, contribute to As contamination of the aquifers through in situ leaching of As-bearing minerals present in arc derived sediments (Mukherjee et al. 2019). However, geotectonics play a significant role in arc magmatism and erosion rates of the arc zone lithologies (Hecht and Oguchi 2017; Masuda 2018), which implies the tectonic control on geodynamic processes responsible for the geologic distribution and dissemination of As.

Another vital consideration is As speciation, since not all its species are equally toxic to humans, animals, and plants. Moreover, its toxicity varies widely with its oxidation state. WHO (1981, 2001) has documented that inorganic species are generally more toxic than organic ones, and arsenite [As(III)] is 60 times more toxic than arsenate [As(V)], which is 70 times more toxic than methylated species, monomethylarsonic acid (MMA) and dimethylarsinic acid (DMA), which are consequently considered to be moderately toxic, whereas arsenobetaine (AsB) and arsenocholine (AsC) are considered to be nontoxic (Kumaresan and Riyazuddin 2001). Therefore, for an accurate environmental and human health risk assessment, identification and quantification of the As species are essential, especially the more toxic ones, rather than simply reporting total As concentrations, which has been the norm for many years. In this regard, Eh–pH diagrams (also known as Pourbaix diagrams, Pourbaix 1974) are conventionally used to represent and describe the speciation of As and identify significant geochemical processes (Welch and Stollenwerk 2002; Nicolli et al. 2012). Figure 1 shows the Eh–pH diagram of aqueous As species in water at 25 °C.

Arsenic may exist with oxidation states of −3, 0, + 3, and +5; however, the oxidation states −3 and 0 (native As) are

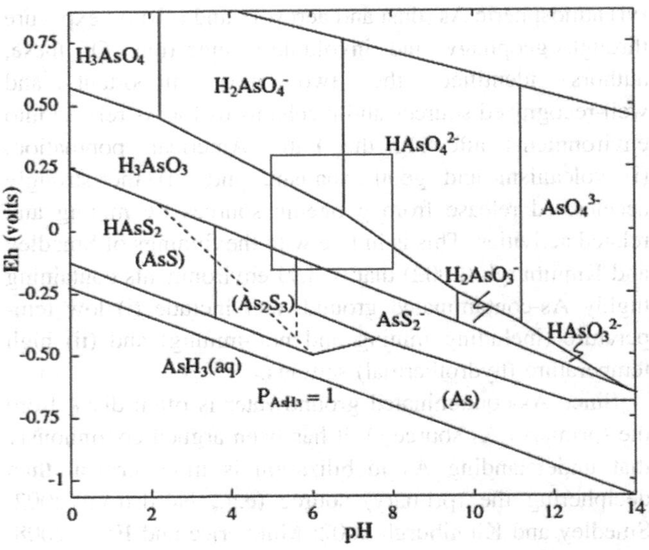

Fig. 1 Eh–pH diagram illustrating the stability fields of the oxidized and reduced species of As in water at 25 °C. The blue rectangle represents the general pH and redox condition of groundwater systems. Modified after Herath et al. (2016)

rare in nature (Panagiotaras et al. 2012). Consequently, As chemistry primarily focuses on As(V) or As(III), depending on redox potential and pH of groundwater (Sracek et al. 2004). Both As(III) and As(V) may form protonated oxyanions in aqueous solutions (Panagiotaras et al. 2012). However, the extent of protonation depends mainly on the pH of the medium. Under oxidizing conditions, As(V) hydrolyzes to four possible species, depending on the pH range encountered in surface and groundwater systems. In oxygenated water systems, arsenious acid (H_3AsO_4) is the predominant species only at extremely low pH (<2), whereas, As exists in the form of $H_2AsO_4^-$ and $HAsO_4^{2-}$ in a pH range of 2–11. In reducing environments and at low pH, arsenious acid converts into $H_2AsO_3^-$, and at pH > 12, it occurs in the form of $HAsO_3^{2-}$ (Welch and Stollenwerk 2002; Panagiotaras et al. 2012; Gong et al. 2006). In the pH range of 4–10, As(V) species is negatively charged in water, whereas the predominant As(III) species is neutral because of redox potential and the presence of complexing ions such as sulfur, iron, and calcium and microbial activity (Katsoyiannis and Zouboulis 2015). However, complete dissociation of As(V) is rare, because very few water systems reach a pH > 11.5 (Brookins 2012). In reducing environments, completely protonated arsenite species predominate over a wide range of pH without being ionized; hence, arsenite cannot adsorb As strongly as arsenate species, thus making the removal of the As species more difficult during remediation of As-contaminated water (Maity et al. 2011a). These authors found high As concentration in the geothermal fluids in southern Taiwan to be associated with low redox (Eh) conditions (Maity et al. 2011b).

Sulfide and arsenide minerals have low solubility in water and are insoluble in anoxic environments, e.g., the solubility coefficient (log K) of realgar (AsS) and orpiment (As_2S_2) is −83.13 and −180.43, respectively, under standard temperature and pressure (STP) conditions, however, these minerals readily oxidize and then dissolve in oxic waters at STP (Boyle and Jonasson 1973). During chemical weathering of ore deposits and subsequent soil formation, As-bearing mineral phases change as follows: arsenopyrite and löllinite in the ore body become scorodite, pharmacosiderite, and arsenosiderite in the weathered rock, and finally goethite in the soil, latter As-bearing phases being more susceptible to getting dissolved and may contaminate surface and groundwater (Majzlan et al. 2014). Under neutral to alkaline conditions, arsenate ion substitutes for sulfate ion in the alteration minerals (Fernández-Martínez et al. 2008; Raychowdhury et al. 2014). Arsenite and arsenate minerals (e.g., scorodite) are the most likely sources of As contamination for their high solubility in aqueous solution, except under strongly acidic conditions (Masuda 2018). Due to secondary enrichment and their higher solubility, weathering products of the primary As-bearing rocks and minerals are more critical (and direct) sources of As contamination (Masuda 2018). For example, Savage et al. (2000) found that As concentration in soils, sediments, and weathered crust (>2000 mg/kg) from the fluvial basins in the Mother Lode District, California, have higher than nearby source rocks (<600 mg/kg).

In contaminated Cenozoic aquifers, As is primarily present in iron oxyhydroxides/oxides (usually goethite) through chemical weathering of iron-bearing minerals (Masuda 2018). Some studies examining iron oxyhydroxides/oxide precipitation and As incorporation into the aquifer sediments are as follows. According to Berg et al. (2007), in the Red River delta, the principal mechanism of As contamination of groundwater is the reduction of As-bearing iron oxyhydroxide minerals in the delta sediments, which are likely chemical weathering products transported from an upstream source. Other mechanisms include in situ decomposition of As-bearing pyrite with seasonal changes of the groundwater level (e.g., Munshuganj, Bangladesh, Pollizzotto et al. 2005) and in situ chemical weathering of chlorite (e.g., Sonragaon, Bangladesh, Masuda et al. 2012, 2013). Both the Ganges–Brahmaputra delta examples imply that the source of As contamination in the aquifers is in situ dissolution of As from primary minerals rather than the dissolution of or desorption from As-bearing iron oxyhydroxides. Such dissolutions could be accelerated further through enhanced inductive recharge of surface water if there is overexploitation of the aquifers (Masuda 2018).

3 Geotectonic Arc Distribution and Recycling

The As concentration of Earth's primitive mantle is estimated to be 0.05 mg/kg (McDonough and Sun 1995). Hattori et al. (2002) found As highly concentrated in microscale sulfide minerals in mantle xenoliths from Lesotho, moreover, relatively higher As concentration in lherzolite (>200 mg/kg As) than harzburgite (<14 mg/kg As). Besides, comparative analyses of mantle xenoliths of Ichinomegata, Japan by Hattori et al. (2002) showed that As is depleted in wedge mantle, where As concentrations of <50 mg/kg were found in microscale sulfide minerals.

Noll et al. (1996) showed enriching arc magmas in As ($\sim$10 mg/kg) by hydrothermal fluids derived from dehydrating subducting slab during metamorphism and confirmed the As release from subducting slab through laboratory experiments as well as analyses of the field samples. Through a similar experiment, You et al. (1996) demonstrated As enrichment of a hydrothermal solution at temperatures $\sim$300 °C by leaching the sediments from the Nankai Trough (where the Philippine Sea plate is subducting beneath the Eurasian plate) with a synthetic NaCl–$CaCl_2$ solution. With the example of the relatively cool subduction zone associated with Catalina Schist on Santa Catalina Island of the Channel Islands, California, Bebout et al. (1999) proposed that devolatilization of As in fore-arc is low, so As is subducted at depth, resulting in As-enriched arc lavas in such subduction zones. On the other hand, Hattori et al. (2005) attributed the formation of As-bearing serpentine to alteration of the fore-arc mantle in the Himalayas, wherein As was derived from hydrothermal fluids releasing from the subducting slab, as observed in the aquifers of Ladakh Himalayas Indus-Tsangpo Suture Zone (Lone et al. 2020). This was corroborated by the documentation of an 85% (in some cases as much as 95%) decrease in the As concentration of arc lavas from fore-arc to back-arc along the convergent margins (Noll et al. 1996). Fujiwara et al. (2011) reported As release from deep continental and island-arc crust during the low-temperature and high-pressure metamorphism (forming pelitic schists) in the rocks from the Sambagawa metamorphic belt, Japan. These authors also reported a decrease in As concentration from $\sim$10 mg/kg to 0 (below detection limit) with the increasing temperature (from 300 to 450 °C). In striking contrast, Breuer and Pichler (2013) reported that As concentration in submarine hydrothermal fluids from the East Pacific Rise (up to 80.5 μg/L) and the Mid-Atlantic Ridge (up to 24 μg/L) one degree lesser than those from the back-arc basin (1386 μg/L) and island-arc (5850 μg/L). Sigufusson et al. (2011) found that As in hydrothermal solutions in basalts of Iceland are complexed with sulfur, i.e., $As(OH)S_2^{2-}$, AsS_3H^{2-}, AsS_3^{3-}, $As(SH)_4^{-}$. The chemical species formed by covalent bonds between As and sulfur (S) precipitate from the hydrothermal solution as arsenide and sulfide minerals; consequently, many sulfide minerals, viz. pyrite, contain As (Masuda 2018). With decreasing temperature and increasing f_{O_2} of the hydrothermal fluids, As–S complexes are decomposed and oxidized to arsenites, commonly found near the hot-spring water, where these fluids discharge on the surface (Masuda 2018).

High As concentration in hydrothermal systems than its average concentration in groundwater is reported globally (Maity et al. 2017). Arsenic-enriched hydrothermal discharges impact downstream surface and groundwater mixing, including related hydro(bio)geochemical processes (Maity et al. 2017). These processes are further influenced by respective changes in pressure, temperature, redox, and pH conditions (Mukherjee and Fryar 2008; Shimada et al. 2004). Such impacts of geothermal water on freshwater resources (surface and groundwater) lead to significant degradation, making affected resources unsuitable for drinking or irrigation without treatment (Maity et al. 2017). Hydrothermal discharges, even with similar geological/tectonic settings, may have very different As concentrations (Bundschuh and Maity 2015; López et al. 2012; Maity et al. 2011a, b; Maity et al. 2016, 2017). For example, significantly higher As concentrations (up to 75 mg/L) have been recorded for the hydrothermal discharges at the western margin of the Pacific Plate, e.g., El Tatio in northern Chile, Copper River and Yellowstone National Park in the USA, Los Humeros in Mexico (Ellis and Mahon 1967; Ball et al. 1998; Motyka et al. 1998; Gonzalez-Partida et al. 2001; Romero et al. 2003), as compared to those (up to 6.2 mg/L) at its eastern margin, e.g., Wairakei, Waiotapu, Ohaaki and Broadlands in New Zealand, Mt Apo in the Philippines (Ritchie 1961; Ellis and Mahon 1977; Webster 1990, 1999; Guo et al. 2007). A better understanding of As cycling in hydrothermal discharges

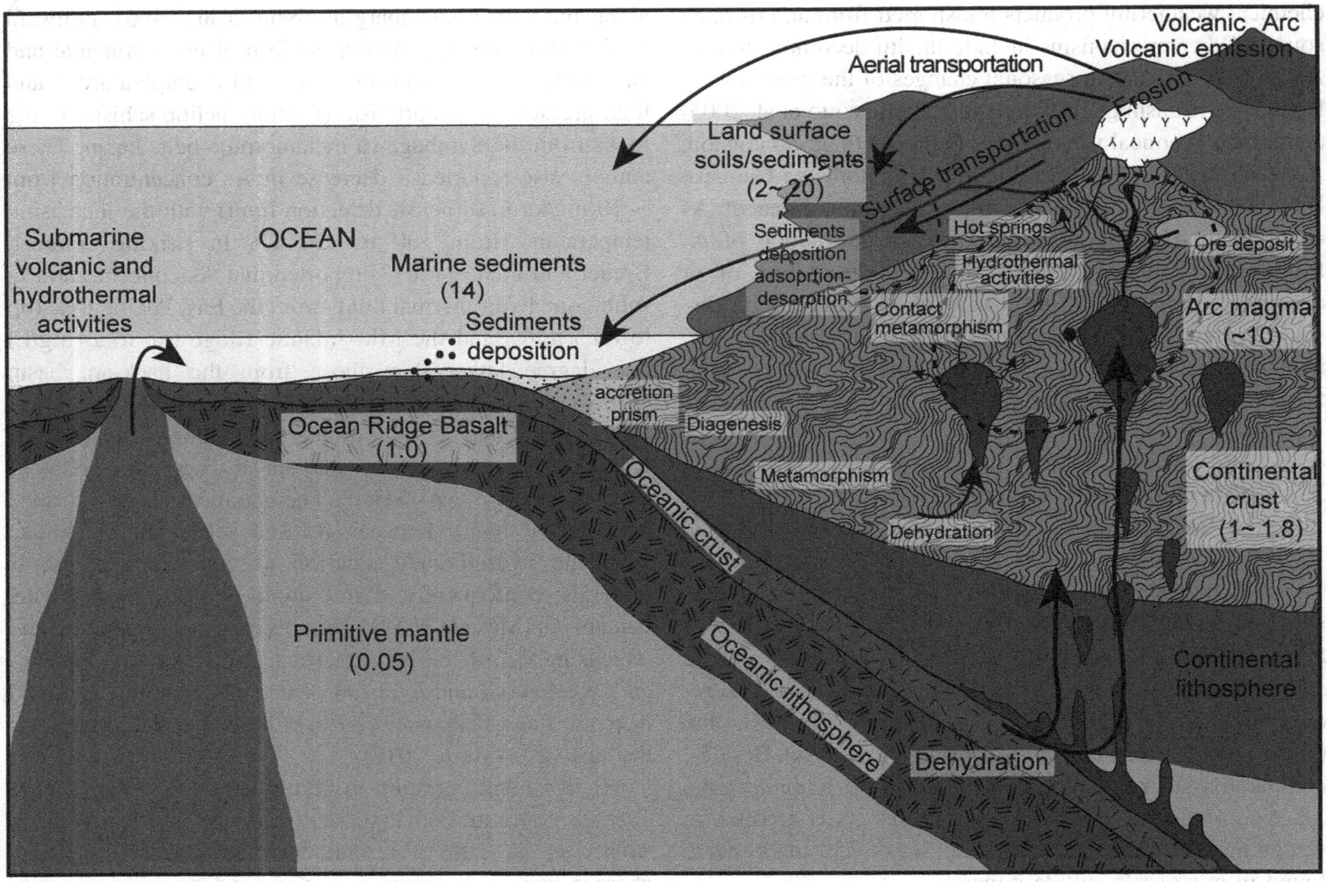

Fig. 2 Redistribution of As during endogenic and exogenic geologic processes at the convergent plate boundary. Numbers in parenthesis indicate average As concentrations in the representative geologic bodies (references within this section). Purple and blue colored letters indicate the releasing and fixing mechanism, respectively, of As with source to sink redistribution cycle. Modified after Masuda (2018)

will help better understand hydrothermal fluids' role in the geochemical evolution of global hydrological systems and the related environmental risk (Bundschuh et al. 2013; Maity et al. 2017) (Fig. 2).

4 Geotectonic Distribution and Disposition of As-Contaminated Aquifers

Despite intrinsic heterogeneity of As concentrations (primarily due to local geology) in more than 70 countries on six continents, worldwide distribution of naturally As-contaminated aquifers show their existence within foreland basins adjacent to the orogenic belts on the global tectonic map (Fig. 3; Smedley and Kinniburgh 2002; Mukherjee et al. 2014, 2019; Ravenscroft et al. 2009).

A large-scale source-sink relationship between orogenesis and As concentration in groundwater has motivated workers like Mukherjee et al. (2007, 2008, 2009a, b, 2011, 2012, 2014, 2019, 2021) to identify universal primary As source-aggradation mechanism(s) for contaminated aquifers across the globe, instead of explaining individual cases of As contamination. With three examples, these authors corroborated that the primary As source in groundwater is ex situ. Also, As in affected foreland basin aquifers is derived through a series of geodynamic processes, viz. tectonic cycles, sedimentation, and regional-scale water flow. This As mobilizes with the availability of conducive surface geochemical and biogeochemical environments (Mukherjee et al. 2014, 2019). In this context, it is essential to consider modern and ancient orogenic belts and corresponding foreland basins. Present-day examples include (i) the Alpine–Himalayan orogen that stretches from Western Europe (the Alps), through the Indian subcontinent, to eastern China (the Himalayas), (ii) the Cordilleran orogens in western North America, (iii) the Andean orogen of western South America, and the (iv) Lachlan orogen of southeastern Australia, etc. Ancient equivalents are the Appalachian of the eastern USA and the Caledonian orogen of northern Europe and East Greenland (Mukherjee et al. 2014, 2019).

The aforementioned orogenic belts have associated foreland basins, which are extensive depressions developed

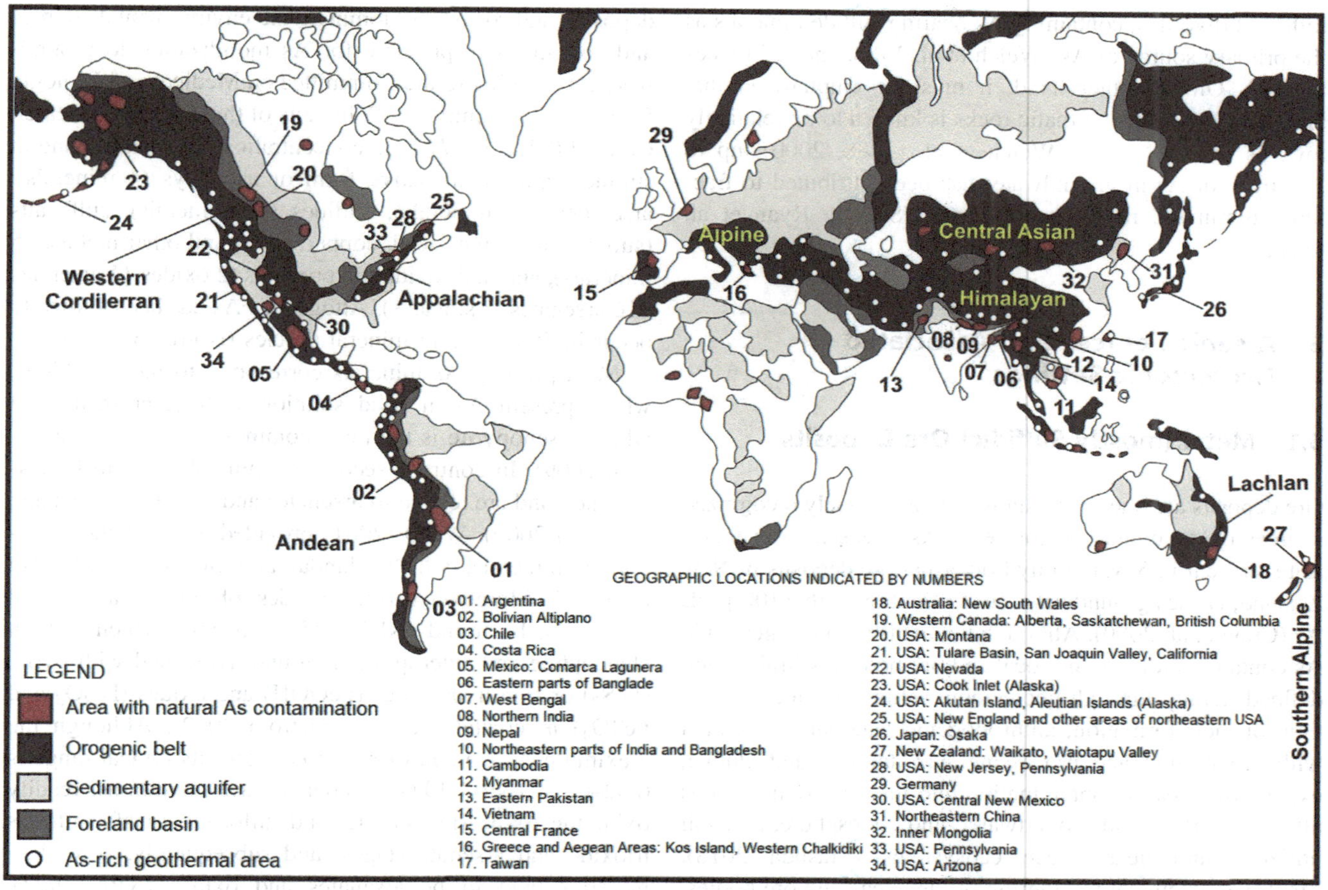

Fig. 3 Global distribution of major As contaminated aquifers (groundwater resources data from BGR & UNESCO under World-wide Hydrogeological Mapping and Assessment Programme (WHYMAP) available at http://www.whymap.org, Margat 2008), orogenic belts (based on the global tectonic map available on the website of the USGS Earthquake Hazard Program, http://earthquake.usgs.gov), in addition to naturally As-enriched aquifers and geothermal fluids (based on Smedley and Kinniburgh 2002; Nriagu et al. 2007; Mukherjee et al. 2007, 2008, 2009a, b, 2011, 2012, 2014, 2019; Ravenscroft et al. 2009; Morales-Simfors et al. 2020). Modified after Mukherjee et al. (2019)

during the mountain building process through lithospheric flexure (DeCelles and Giles 1996). These foreland basins act as sinks for the sediments eroded from the rapidly exhuming adjacent orogens and typically have thick sedimentary successions, thinning away from the orogenic belts. Examples of such basins include the Indus–Ganges–Brahmaputra, Northern Tarim, Southern Junggar basins of Asia, the Northern Alpine, Po, Ebro, Guadalquivir, Aquitaine basins of Europe, the Appalachian and Western Canada Sedimentary basins of North America. Based on their structures, DeCelles and Giles (1996) classified foreland basins as (i) peripheral, where the basins occur on the subducted plate, i.e., the outer arc of the orogen, e.g., the Indus–Ganges–Brahmaputra basin and (ii) retroarc, where the basins occur on the overriding plate during the convergence and are situated behind the magmatic arcs, e.g., the basins of the Andean or Western Cordilleran orogens.

Arsenic, being an incompatible element, is mainly concentrated in differentiated magmas; and as a result, volcanic rocks of intermediate (andesitic) to felsic (rhyolitic) composition have relatively higher As concentrations when compared to mafic (basaltic) and ultramafic rocks (Mukherjee et al. 2019), as shown by Welch et al. (1988, 2000) through a dataset of lava samples obtained from the western USA. This association of As with felsic rocks has been corroborated by Peters et al. (1999), Peters and Blum (2003), and Peters (2008), who attributed the As content in groundwater of Maine and New Hampshire, USA, to pegmatites (up to 60 mg/kg As), granites (46 mg/kg As) and granofels (39 mg/kg As), and also by Brown and Chute (2002), who attributed groundwater As to the weathering of pegmatite veins in Connecticut, USA. Moreover, in California, USA, As concentrations up to 48 mg/L have been reported in groundwater associated with acid volcanics, viz.

tuffs and rhyolites, containing As-bearing sulfide minerals as the primary source of As (Welch et al. 1988, and references therein). On the other hand, it must be emphasized that although As content of mafic rocks is known to be relatively low, as pointed out by Welch et al. (1988, 2000), up to 327 µg/L of As in groundwater has been attributed to fractured ultramafic rocks in Vermont, USA, by Ryan et al. (2011).

5 Arsenic Sources and Associated Geotectonic Settings

5.1 Metal (mostly Sulfidic) Ore Deposits

Ore deposits are one of the significant and widely recognized sources of As contamination, e.g., As released from marcasite, present in 5 Ma old hydrothermal ore deposits in New Zealand, causes groundwater contamination with >100 µg/L As (Craw et al. 2000). Although the distribution of geogenic As contamination around localized ore bodies is limited and defined mainly through geotectonic setting prevalent at the time of their formation, anthropogenic materials associated with ore mining and processing, viz. tailings and sludge, expand the area of contamination into the hydrosphere, thus affecting surface and groundwater, and pedosphere, i,e., soil horizons and near-surface ecosystems (Masuda 2018). Moreover, chemical weathering of the excavated ore bodies, exposed to exogenic forces of weathering and erosion, produce As-contaminated sediments, which might be the potential sources of As contamination of the environment and consequent human As exposure (Masuda 2018). Mining and metallurgical activities cause As contamination to spread to air, sediments, and streams, which might remain contaminated long after the cessation of these activities. For example, long after mining was ceased at Toroku mine, Miyazaki (Kyushu, Japan), 146 people were found to be suffering from arsenicosis due to inhalation of refining soot and drinking contaminated water at this site in 1990, where arsenite was mined from the 1820s to 1962 (Shimada 2009). On the other hand, in the absence of such activities, the contamination is restricted to a limited area (locality), close to the exposure or contact of ore bodies and alteration zones to water (surface or groundwater) or air.

Metal mining (mainly gold, copper, silver) in Latin America has been the key for As mobilization from the geologic materials and aggravating contamination of groundwater resources (McClintock et al. 2012; Tapia and Audry 2013; Tapia et al. 2012, 2018, 2019a, b, c, 2021). A summary table of the worldwide occurrence of As in groundwater by Nordstrom (2002) lists As bearing ores and other geologic sources of As, which include 24 As-bearing minerals that commonly occur in hydrothermal veins, ore deposits, and rocks and dominate the environmental sources and conditions responsible for As mobilization to groundwater. According to the Committee on Medical and Biologic Effects of Environmental Pollutants of the National Research Council (NRC 1977), high concentrations of As are found in sulfide deposits, in native form or as alloys (4 minerals), arsenides (27 minerals), sulfides (13 minerals), sulfosalts (sulfides of As with lead, copper, silver and other metals, 65 minerals), and their oxidation products (2 oxides 11 arsenites 116 arsenates, 7 silicates). Altogether, As has been shown to occur in 568 different mineral species (Nordstrom 2002).

Most primary As minerals correspond to metal sulfides, which present As in solid solution with other metals, of which arsenopyrite is the most common (Ehrlich and Newman 2009). In contrast, secondary minerals tend to be less common and are related to arsenates and oxides (Ehrlich and Newman 2009). WHO (2001) provided a list of these secondary minerals, which Mandal and Suzuki (2002) also tabulated. Moreover, arsenosulfides of almost any metal cation can be found (NRC 1977). Arsenic, which is both chalcophile and siderophile, is found associated with pyrite (FeS_2) or iron hydroxides (FeOOH) and oxides (Fe_2O_3 and $Fe3O_4$) in crustal rocks (Nordstrom 2002). Although the maximum As content of pyrite is ~5%, its typical range is 0.02–0.5% (NRC 1977). Under surface conditions, readily oxidizing As-bearing sulfides and sulfosalts transform to As trioxide and arsenate stages, and subsequently, secondary minerals tend to be arsenates and oxides (NRC 1977). Although igneous rocks show a relatively uniform distribution of As among the major constituent minerals, their sulfide mineral constituents show slight As enrichment (NRC 1977).

Through weathering, As present in sulfides ores are converted into a highly mobile [As(V)], which can enter the As biogeochemical cycle directly in either solid-state as dust particles or through dissolution in the rain, surface, and groundwater (Maity et al. 2011a). Consequently, soil overlying sulfide ore deposits commonly contain several hundred ppm of As, up to 8000 ppm (NRC 1977). In this way, As may be present in unweathered sulfide minerals or in an inorganic anion state. Inorganic arsenate is usually bound to iron and aluminum but could form bonds with any other cation present, viz. calcium, magnesium, lead, zinc (NRC 1977). The composition of the recharging water, water-aquifer matrix interactions (e.g., cation exchange, mineral dissolution), availability of reactive organic matter, and groundwater residence time within the aquifer primarily control groundwater chemistry (e.g., Edmunds et al. 1981; Stallard and Edmond 1983; Chapelle 1993; Drever 1997; Faure 1998; Mukherjee et al. 2009, 2012). Accordingly, elevated concentrations of chalcophile and siderophile As in contaminated aquifers may result from the dissolution of primary or secondary sulfidic minerals (viz. pyrite), or desorption from secondary concentrations in hydrous metal oxides (Nordstrom 2002), which is common in the aquifer matrix in many large unconsolidated sedimentary provinces (Mukherjee et al. 2014, 2019).

A simple analysis of the basic types of sulfide ore deposits based on associated geotectonic regimes reveals a systematic pattern. For example, calc-alkaline magmatism at convergent plate boundaries gives rise to (i) Kuroko-type (massive sulfide) deposits in submarine volcanic environments and (ii) Cordilleran-type (postmagmatic) ore deposits in epizonal plutonic settings (Sawkins 1990). Porphyry copper deposits, an important subtype of Cordilleran-type deposits, show a spatial relationship to present or ancient convergent plate boundary regimes (Sawkins 1990). Copper-nickel deposits occurring in ultramafic extrusive or penecontemporaneous intrusive rocks are also related to plate convergence (Sawkins 1990). Sulfide ore deposits generated at divergent plate boundaries (spreading-center systems) are rare as a function of their submarine habitat and are unknown at transform plate boundaries (Sawkins 1990).

On the other hand, stratiform copper deposits and most of the Mississippi Valley-type lead–zinc deposits are formed in continental intraplate environments. In contrast, magmatic deposits in layered mafic complexes also form in intraplate environments, but with spreading-center activity limited in time and space (Masuda 2018). Moreover, gold deposits in greenstone-type volcanic arc sequences can be attributed to a two-stage concentration process at inferred (ancient) convergent plate boundaries (Masuda 2018). Arsenic, being an integral part of all these deposits, implies that primary As distribution is tectonically controlled, transported by secondary processes in solution through leaching or solid-state through physical erosion to the sedimentary basins (see Sect. 5.2).

It is important to emphasize that As contamination, even in a mining district, can sometimes be associated with other lithologies instead of the ore deposits themselves. A study from Osaka Prefecture in western Honshu (Japan) pointed out that although many sulfide mines are present in the Hokusetsu Mountains, they are not essentially sources of As contamination as much as the arsenide disseminated in hornfels and shales (Masuda 2018). On the eastern edge of the Cretaceous–Neogene magmatic bodies in the area, Even et al. (2017) reported high As content in hornfels near the granitic complex that effectively led to As contamination of surface and groundwater. Similarly, As leaching and oxidation in disseminated Miocene sediment-hosted hydrothermal gold (Au) ore deposits led to As contamination of shallow groundwater in Sulawesi, Indonesia (Iskandar et al. 2012). In both cases, the As contamination mechanism was oxidation and decomposition reactions of oxic water infiltrating through a fault system with diffusively distributed sulfides and arsenides (Masuda 2018). Section 6 deals with the basement structure's role in controlling the hydrochemical environment, thus As speciation.

In the mining area of Wassa West district in Ghana, Bhattacharya et al. (2012) found As (besides manganese and iron) contamination in some wells potentially affected by acid mine drainage (AMD) likely to be controlled by the adsorption processes, evident by supersaturation of groundwater with respect to several mineral phases including iron-hydroxides/oxides in the hydrogeochemical models.

5.2 Arc Magmas

Reported concentrations of As in various lithologies worldwide suggest a close association with magmatic arcs at convergent continental margins (Fig. 2) and collision belts that inevitably contain arc-derived components (Mukherjee et al. 2014, 2019). Accordingly, high As content in arc magmas can be derived from (a) the subducting oceanic crust with its sediment cover, (b) the mantle component of the subducting slab, (c) the mantle wedge component overlying the slab, (d) the overlying continental crust through which the arc magma ascends to the surface (Hattori and Gulliot 2003; Hattori et al. 2005; Gulliot and Charlet 2007). Ocean-floor basalts, comprising the oceanic crust are formed by partial melting of the depleted mantle at the mid-oceanic ridge, are not enriched in As (<0.01–1 mg/kg, Deschamp et al. 2013). This implies that As in the oceanic crust is derived from the sediments derived from the continents (terrigenous detritus).

The As concentration in the continental crust is at least two orders of magnitude higher than in the depleted mantle, the source region for mid-oceanic ridge basalts (Matschullat 2000). It is worth mentioning that the arc magma of some volcanoes in the southern part of the Japan-Kurile volcanic chain has low As content, comparable to normal depleted mantle-derived magmas. In contrast, several volcanoes in the northern part of the same region, e.g., Esan (Japan) and Onekotan (Kuriles), which are underlain by thick arc basement, have very high As concentration in their magmas (Noll et al. 1996; Henke 2009). Similarly, medium to high As concentration have been reported from the arc extrusives at continent to ocean convergence-related volcanics of Cordilleran orogenic belts of British Columbia (Boyle et al. 1998), Alaska (Crock et al. 1999), Mexico (Armienta et al. 2001), Chile (Noll et al. 1996) and Perú (Noble et al. 1998). This tells that the As content is almost always high in the settings where the arc magma ascends through an overlying continental crust, implying that interactions of volatile-enriched arc magma with the continental crust is the most important source of primary As enrichment compared to other tectonic settings. This further suggests that (i) in any tectonic regime, the continental crust forms the ultimate source of groundwater As, (ii) As in continental crust is mobilized hydrothermally by young magmatic hydrous fluids of metasomatic origin, and (iii) As mobilization is enhanced through prolonged interaction between continental crust and hot hydrous fluids or magmas.

Arsenic mobilization from the sediment cover of the subducting plate to the fluid is most effective in continental domains with elevated geothermal gradients, such as near active, e.g., Andes (e.g., Raychowdhury et al. 2014) or ancient magmatic activity centers, e.g., Indus-Tsangpo Suture Zone lithologies of the Himalayas (e.g., Hattori et al. 2005; with up to As 257 mg/kg, Lone et al. 2020). Meteoric water in such domains percolate to depth, becomes heated, and forms hydrothermal systems, e.g., geothermal wells in Costa Rica (As ~25,900 μg/L, Hammarlund et al. 2009) that can efficiently remove As and other mobile elements from the crust. The limits of such As mobilization by thermal fluids are controlled and/or governed by the extent of water–rock interaction, which in turn is constrained by the maximum penetration depth of water can (Gleeson and Ingebritsen 2016). Arsenic can be mobilized from deeper parts of the continental crust only if it comes in contact with hot hydrous fluids and gets dissolved. This is evident at active continental margins (mostly hot orogens, Mukherjee et al. 2014, 2019), where hot hydrous fluids mobilize As derived from the subducting slab (Tatsumi 1989). These fluids may or may not be As enriched at the time of slab dehydration. The As content of these fluids depends on the proportions of continent-derived sediments (terrigenous detritus) and hydrated oceanic crust associated with the subducted slab (Mukherjee et al. 2014, 2019). The fluids, which may also be associated with slab-derived melts, subsequently metasomatize the depleted continental lithospheric wedge, lower the mantle solidus and initiate the production of hydrous arc magmas (Mukherjee et al. 2014, 2019). Fractionation of such magmas leads to the generation of high-Al basalts/gabbros, andesites/diorites and dacites/granodiorites (Sisson and Grove 1993) in active or ancient magmatic arcs, which are known to have high to very high (up to 1500 mg/kg) solid-phase As (Henke 2009). The chemistry of the arc magmas may thus be influenced by sedimentary input from the continent into the trench (Plank and Langmuir 1993). At active continental margins, the ascending magma invariably interacts with the (thick) continental crust as it rises, unlike at an ocean-ocean setting; thus, continental arc volcanics tend to have the highest primary As enrichment (Mukherjee et al. 2014, 2019).

As in the Andes, Arc volcanism leads to the deposition of As-rich volcanic ash and other continent-derived sediments (Fig. 2) that can be carried to downstream basins. Arsenic enriched rhyolitic glass, volcanic ash, and associated silicates in sediments and metamorphic deposits, with As up to 10 mg/kg (Raychowdhury et al. 2014), undergo hydrolytic dissolution and produce an influx of major (e.g., Na^+, K^+, Si^{4+}, and HCO_3^-) and minor (e.g., oxyanions of As, V, and Mo) solutes to groundwater, which leads to precipitation of clays (Mukherjee et al. 2007, 2008; Mukherjee and Fryar 2008; Raychowdhury et al. 2014), under favorable climatic conditions. Himalayan arc-derived lithologies also comprise significant proportions of mafic and intermediate rocks (predominantly andesites), enriched in As compared to their non-arc equivalents, owing to interaction with subduction-zone fluids (Hattori et al. 2005; Guillot and Charlet 2007).

Chemical weathering and leaching of more soluble and labile, solid-phase As from the metastable, high-temperature arc-derived mafic minerals (e.g., olivine, pyroxenes, serpentine) in these rocks (As ~19 mg/kg, Verma et al. 2016) facilitates the release of this As (up to 4100 μg/L, Mukherjee et al. 2009a, b) into the hydrological system. Some As-carrying (up to 25,900 μg/L (Hammarlund et al. 2009) hydrothermal fluids may also discharge on the surface as hot springs and fumaroles (For more details, see Sect. 5.4 Geothermal and volcanic fluids). Subsequently, with orogeny initiation at active continental margins, the As-bearing arc volcanics coalesce with sediments and rocks of the exhuming orogen. With continuing convergence, slices of serpentinized and, therefore, As-enriched mantle rock (As ~275 mg/kg, Hattori et al. 2005) are also incorporated into the orogen through thrusting (Hattori and Guillot 2003). Sediments derived from the rapidly exhuming orogenic belt by weathering of the As-bearing arc volcanics and obducted, serpentinized mantle rocks are subsequently deposited in adjacent foreland and even back-arc basins (Verma et al. 2016). During protracted convergence over geologic time, multiple generations of foreland basins can form and coalesce with the main orogenic belt, e.g., Siwaliks in the Himalayas (As ~10 mg/kg, Guillot et al. 2015).

Rivers or glaciers cutting through the orogenic belt erode the foreland sediments and exhumed arc derived magmatic rocks (Guillot and Charlet 2007; Verma et al. 2016), and deposit weathered As-rich sediments in the most recent foreland basins, e.g., Terai, Nepal (As up to 27 mg/kg, Guillot et al. 2015). Proximity to these orogenic As sources enhances the possibility of increased chemical weathering and leaching of solid-phase primary As, from meta-stable, high-temperature arc-derived rocks, which constitute primary As source (Verma et al. 2016). With increasing distance from these As sources, weathering and transport lead primary As to remain chemically sorbed mainly in the residual state, more recalcitrant or more strongly bonded, more crystalline or mineralogically mature phases, viz. low-temperature/late-phase silicates (Onishi and Sandell 1955), which are less prone to chemical weathering (Henke 2009). Conversely, with distance, the released As gets physisorbed in more labile secondary solid phases, which act as the immediate (and secondary) source of groundwater As enrichment in adjoining foreland basin aquifers by redox-dominated reactions (Mukherjee et al. 2014, 2019). This liberated As can co-precipitate with/or get adsorbed onto secondary Fe(III) (oxyhydr)oxides (Smedley and Kinniburgh 2002) or can be re-sequestered by re-sorption or co-precipitation with authigenic Fe(II) sulfides under sulfate-reducing conditions, i.e., aquifer sediment sink (Mukherjee and Fryar 2008; Mukherjee et al. 2008).

To integrate the knowledge of orogen-sourced sediments and groundwater chemistry of ancient (e.g., Himalayan system) and modern (e.g., Andean system) orogens, Mukherjee et al. (2014, 2019) correlated the tectonic settings and groundwater As-enrichment data from 63 major aquifers across the world, to show that a distinctly high proportion of As-affected aquifers are located in convergent ($\sim$90%), particularly magmatic arc settings. In stark contrast, only 5% of the basins in other tectonic settings contain groundwater As. However, rift-related basin aquifers are an exception to this generalization (Mukherjee et al. 2014, 2019), although all such rift basins receive magmatic arc detritus, commonly andesite (Peters 2008). Similarly, As-contaminated aquifers in the Red River and the Mekong basins have a mixed convergent/non-convergent setting, with sediment sources from ancient Qamdo-Simao arc rocks (Stanger 2005). Dissolution of volcanic glasses in ash layers and leaching of loess-type deposits in the Chaco-Pampean plain of Argentina have led to groundwater As concentrations from <10 to 5300 µg/L (Nicolli et al. 2012). These examples corroborate that magmatic arcs are the predominant source of primary As in groundwater across the globe.

5.3 Allochthonous As-bearing Sediments

The third primary source, preferably a secondary source (see the previous section), of contamination comprises the Cenozoic (mostly Holocene) aquifer sediments. It is worth highlighting that the total As concentration in contaminated groundwater is not as high when compared to the average concentration in soils and sediments, for example, despite total As concentration in Holocene bulk aquifer sediments in Bangladesh ranges from 5 to 16 mg/kg, groundwater has only little over 1 mg/L As (Seddique et al. 2011). This is because groundwater As contamination depends mostly on the chemical form of the As bearing solid phase (mineral species) and changes in the aquatic environment leading to the release of As from aquifer sediments to water (Seddique et al. 2011). The mechanisms for As release from Cenozoic, especially Holocene, aquifer sediments have been described in this section.

As mentioned in the previous section, As-contaminated aquifers are often hosted within extensive clastic sedimentary basins associated with young mountains (Ravenscroft et al. 2009) within or around foreland basins at both active and ancient convergent continental margins (Mukherjee et al. 2014, 2019, 2021), e.g., Indus–Ganges–Brahmaputra basin, the Mekong and Red River deltas in southeast Asia (Alpine–Himalayan Orogenic System), the North American foreland of the western USA (American and Pacific Cordillera) and Canada (Laramide-Nevadan-Sonoma Orogenic System), the Appalachian foreland in New England and adjacent areas of the eastern USA (Appalachian-Ouachita Orogenic System), parts of Argentina, Bolivia, Chile and Peru in the Andean forelands of South America (Andean-Pampean-Toco Orogenic System). Arsenic found in such aquifers can be attributed to the solid phase concentration of As-bearing clasts in their provenance located in major magmatic arcs and orogenic belts at convergent boundaries (Mukherjee et al. 2014, 2019). Thus, As sourced from magmatic arcs may be referred to as primary for both Andean (Ramos Ramos et al. 2012; Raychowdhury et al. 2014; Ormachea Muñoz et al. 2015) and Himalayan (Hattori and Guillot 2003; Hattori et al. 2005; Guillot and Charlet 2007; Mukherjee et al. 2014, 2019; Verma et al. 2016) type orogenic systems, at continental convergent margins across the globe, which after exogenic processes of erosion and transport, ends up in the sediments as a (secondary) source of As, in the aquifers hosted in adjoining foreland basins. Tectonic setting interpretations based on major, trace, and rare earth elements, in conjunction with field measurements of primary sedimentary structures, can help in discriminating provenance and tectonic setting (including ancient, e.g., Proterozoic sedimentary basin in Rajasthan, India, Banerjee and Bhattacharya 1994) of the As-bearing sediments.

5.4 Geothermal and Volcanic Fluids

Arsenic contamination of the aquifers—clastic or fractured—might not involve leaching of As from allochthonous clastics or any other host formation lithology. Instead, it might be from As rich geothermal and volcanic fluids that interact with groundwater (e.g., Latin America, Morales-Simfors et al. 2020; Western Iran, Keshavarzi et al. 2011), which are the sources as well as pathways of As. In a recent study covering 15 countries in Latin America (Morales-Simfors et al. 2020), 423 sites were characterized with As originating from geothermal sources, mostly related to present volcanic activity (0.001 < As < 73 mg/L, mean: 36.5 mg/L) and the transboundary Guarani Aquifer System (0.001 < As < 0.114 mg/L, mean: 0.06 mg/L).

High As concentrations in geothermal fluids indicate that As is a volatile magmatic component, which is highly soluble in water., e.g., hot springs in Yellowstone National Park, USA have up to 10 mg/L As (Stauffer and Thompson 1984; Webster and Nordstrom 2003), which is released to groundwater and surface water from geothermal fluids within and near the park (Ball et al. 1998, 2002; Nimick et al. 1998). Geothermal fluids in Dominica, the Lesser Antilles, have high As concentrations (>50 μg/L, McCarthy et al. 2005). Moreover, As from hot springs can be mobilized to surface waters and also enter riverbed sediments, e.g., arsenite released from hot springs along the Owens River (USA) gets oxidized to arsenate by microbial activity, and subsequently precipitates and settles to the bottom, thus becoming part of the riverbed sediments with high As content (up to 60 mg/kg, Nimick et al. 1998). It has been observed that geothermal well fluids occasionally contain higher As concentrations (up to 50 mg/L) than associated surface discharges through host springs (Webster and Nordstrom 2003 and references therein). Arsenic content in the outflow (drainage water) from Waikato Power Station, New Zealand was 145 t/year till 2000; however, it has been lowered to ~70 t/year (Webster-Brown et al. 2000 and references therein). This discharge has led to elevated As concentrations (61–1790 mg/kg) in the Waikato River sediments (Webster-Brown et al. 2000).

Geothermal areas occur along plate boundaries, e.g., the Pacific Ring of Fire, in tectonic rift areas, e.g., East Africa, Central India and at seafloor spreading centers, e.g., Iceland, and at hot spots, where mantle-derived plumes ascend, e.g., Hawaii and Yellowstone National Park, USA. Arsenic is one of the incompatible elements, together with antimony (Sb), boron (B), fluoride (F), lithium (Li), mercury (Hg), selenium (Se), and thallium (Tl), and hydrogen sulfide, in high-temperature geothermal settings (Webster and Nordstrom 2003). Moreover, As concentrations are high mainly in thermal waters that leach continental rocks, whereas thermal waters in basaltic rocks, e.g., in Iceland, contain lesser amounts of As. Stauffer and Thompson (1984) demonstrated that As in thermal fluids at Yellowstone National Park in Wyoming, USA, it is derived mainly from leaching host rocks rather than magmas. Arsenic in thermal water is present as [As(III)] in arsenious acid (H_3AsO_3). In contrast, in low-sulfide fluids, the [As(III)] in the arsenious acid is oxidized to [As(V)]. The ascending fluid interacts with cold oxygenated groundwater or comes in contact with the atmosphere. On the other hand, in high sulfide solutions, As could be present as thioarsenate complexes (Webster and Nordstrom 2003; Planer-Friedrich et al. 2007).

Since the 1950s, As contamination has been attributed to hydrothermal activity in many geothermal areas of Japan, as reflected by some old works, e.g., Beppu, Kyushu (Kawakami et al. 1956), Tamagawa, Akita, Northeast Japan (Minami et al. 1958), Hakone, Yugawara, Kanagawa (Awaya et al. 2002), and Toyohira, Hokkaido (Jin et al. 2012). In Kusatsu-Shirane (Gunma), thermal water discharges with up to 7 mg/L of As concentration was reported flowing into the Yukawa River for a decade until the 1990s (Kikawada et al. 2006, 2009). Up to 3.4 mg/L As was reported in thermal water at Kakkonda Geothermal Plant, Akita, Japan (Okada et al. 2002). Acidic (pH 1.2) thermal water with about 2600 μg/L As was reported by Noguchi and Nakagawa (1969) to precipitate out as As sulfides and lead (Pb) As sulfides in surface-water sediments containing ~5–56 wt.% of As. Up to 4.8 mg/L As in thermal waters from geothermal fields in New Zealand are attributed to the leaching of variable composition host rocks. This was considered the mechanism for As enrichment of thermal waters from various geothermal areas in New Zealand.

In Latin America, volcanic and geothermal activity along the convergent plate boundary discharges As-rich waters and gases on the surface in the form of hot springs and fumaroles (Raychowdhury et al. 2014; Mukherjee et al. 2014, 2019, 2021; Coomar et al. 2019). The As concentration of ~50,000 μ/L has been reported by López et al. (2012) in Na-Cl type thermal waters at El Tatio geothermal field in Chile (Landrum et al. 2009). Up to 73,600 μg/L of As concentration has been reported in thermal waters ascending through a wide range of lithologies—sandstones and shales, lava flows and pyroclastics, and metamorphosed carbonate rocks, basalts, and hornblende andesites—in Mexican geothermal fields (López et al. 2012). In the coastal volcanic zones of Central and South America, where the predominant lithological composition is andesitic or rhyolitic, As concentration in thermal waters varies widely, and although As concentrations of several thousand μg/L are frequently reported, none of them is as high as the highest reported As concentration in Mexican geothermal fields (i.e., 73.6 mg/L, López et al. 2012). Reported As concentrations, apparently is contributed by oxidation of sulfide deposits, are as high as 4600 mg/L (?) in the Bolivian Altiplano, where hot springs and fumaroles discharge water and gases in Poopó Lake (López et al. 2012).

In Europe, As concentrations >50 μg/L in groundwater are reported from the Massif Central's geothermal fields in France and in Greece (Brunt et al. 2004; Karydakis et al. 2005). Arsenic concentration in groundwater in Iceland often exceeds 10 μg/L (Arnorssón 2003), although basaltic lava typically contains less As than silicic lavas (Onishi and Sandell 1955; Baur and Onishi 1969; Ure and Berrow 1982). Geothermal systems in northern and northeastern Spain are characterized by high As concentrations and As-rich mineral deposits around surface geothermal manifestations (Navarro et al. 2011), e.g., hot springs with 50–80 μg/L As in the Caldes de Malavella geothermal field in northern Spain contributes a substantial amount of As (from <1 to

200 μg/L) to groundwater (Piqué et al. 2010). The As concentrations in groundwater (including brines) range from 1.6 to 6900 μg/L in the Phlegraean geothermal fields in southern Italy (Aiuppa et al. 2006). Quaternary volcanic rocks with hydrothermal activity on Ischia (a volcanic island offshore from Naples, Italy, in the Tyrrhenian Sea) contribute As to groundwater at concentrations ranging up to 3800 μg/L (Aiuppa, et al. 2006; Daniele 2004).

In the Middle East, reported As concentrations in thermal waters are as high as 3500 μg/L near Mt. Sabalan, a stratovolcano in northwestern Iran (Haeri et al. 2011). In western Anatolia, Turkey, natural leaching, aided by pumping and discharge of waste geothermal fluids from an active geothermal system, has mobilized As from metamorphic, igneous, and sedimentary rocks to groundwater in a shallow alluvial aquifer, where highest As concentrations in groundwater and geothermal fluids were 561 and 594 μg/L, respectively (Gunduz et al. 2010). Baba et al. (2021) reported that geothermal fluids circulating through volcano-sedimentary units in the Anatolian plate caused a more effective alteration, increasing As concentrations in geothermal systems. Moreover, they determined that high As concentrations within the convective-non-magmatic extensional geothermal play types, viz. Western Anatolian Extensional System and the North Anatolian Fault. Bundschuh et al. (2013) found the mixing of geothermal fluids primarily responsible for contamination of freshwater resources, making them unsuitable for drinking or irrigation, based on hydrogeochemical investigations of geothermal fluids from deep wells and hot springs in western Anatolia, Turkey. Baba and Sözbilir (2012) deciphered that the enhanced dilation ~ E–W-strikes of the graben faults in west Turkey, accommodating the deep circulation of hydrothermal fluids of meteoric origin, provide a suitable environment for the presence of high As levels in geothermal fluids.

Groundwater in early Proterozoic silicic volcanics and granites of the Chhattisgarh Basin (India) contains >10 μg/L of As, emplaced there by hydrothermal fluids (Acharyya 2005).

Geothermal systems are controlled by global tectonic and local structural settings, which ultimately define heat and fluid flow regimes. See the chapter Tectonic and Structural Control on Geothermal Systems in the same book for more details. This association of geothermal systems with geotectonic settings advocates for geotectonic control in distributing and disseminating As.

5.5 Coal and Hydrocarbon Resources

Together with both of these fuel resources' exploitation processes, coal deposits and hydrocarbon reservoirs are other geogenic sources of As contamination that is accelerated by the extraction and refining processes and subsequent usage. Geotectonics also controls their formation in ancient and modern sedimentary basins. The average world concentration of As in coal is 9.0 ± 0.8 and 7.4 ± 1.4 mg/kg for bituminous coals and lignites, while in their ash, it is 50 ± 5 and 49 ± 8 mg/kg, respectively (Yudovich and Ketris 2005).

Petroleum reservoir waters from sedimentary basins can contain As as a result of the evaporation of seawater that concentrates As in oil reservoirs (Birkle et al. 2010). For example, Birkle et al. (2010) reported As concentration up to 2000 μg/L at depths of 2900–6100 m below sea level in oil fields of southeastern Mexico. Moreover, formation water (fresh or saline water trapped in geological formations for thousands to millions of years) forms the most significant part of the volume of co-produced water in hydrocarbon production (Neff et al. 2011). Stromgren et al. (1995) found that the formation water has particularities of the geological formation and is polluted by metals, inorganic compounds, and minerals, including As. Also, their concentrations can differ considerably from seawater.

6 Structural Controls on As Concentration and Distribution in Aquifers

Apart from the aforementioned solid and liquid phase inputs of As, attributed to the lithology and geothermal and volcanic fluids, respectively, there is yet another essential aspect of As enrichment of aquifers, which has to do with the structural settings of the aquifers in a basin, e.g., Duero Basin, Spain (Giménez-Forcada and Smedley 2014), wherein basin morphology, prolongation of faults, especially division of the underlying basement into blocks, produces discrete hydrodynamic environments, resulting in spatial variations in hydrochemical composition, including As speciation (Fig. 4, Giménez-Forcada and Smedley 2014).

Although pH and redox conditions play essential roles in controlling the As occurrence, distribution, and mobility in groundwater, the basement structure plays a significant control, e.g., faults behave as effective conduits for groundwater flow and influence the recharge of the basin (Giménez-Forcada and Smedley 2014). The groundwater flow could be controlled by the faults (or fault zones), which act as preferential flow conduits and zones in tectonic depressions (Vilanova et al. 2008), particularly those faults oriented parallel to the general flow path in the basin (Giménez-Forcada and Smedley 2014). Moreover, horst and graben structures can generate discrete hydrogeological cells (Giménez-Forcada and Smedley 2014).

Groundwater chemistry varies laterally and with depth, depending upon climatic and hydrological factors, viz.

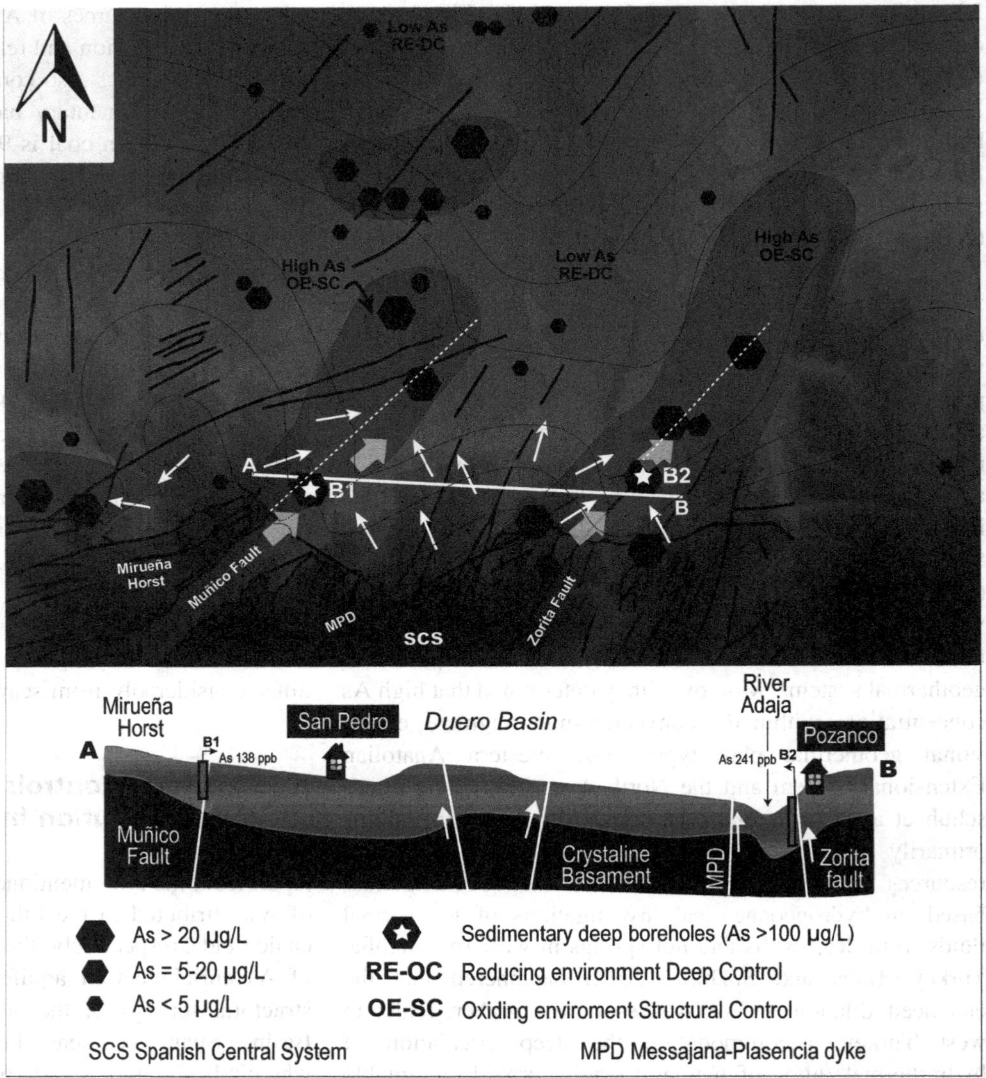

Fig. 4 Groundwater (springs and boreholes) Arsenic distribution, along with the prominent structural fractures mapped by Simón Gómez (1996), in Duero Basin. Deep boreholes B1 and B2 with the highest As concentrations are represented in the cross-section A–B. Blue contours indicate the piezometric surface. Symbol size reflects arsenic concentration. Modified after Giménez-Forcada and Smedley (2014)

evaporation (Giménez-Forcada and Smedley 2014). For example, in the shallow sedimentary aquifer (<40 m), evaporation can increase the concentrations of dissolved elements, including trace elements like As (e.g., Fujii and Swain 1995; Welch and Lico 1998; Smedley and Kinniburgh 2002). For shallow wells (<40 m) in Duero Basin, Spain, oxidizing conditions and evaporation control groundwater composition. At 40–160 m depth, reducing conditions affect the concentrations of As and other redox-sensitive species, viz. SO_4, NO_3, Fe, and Mn (Giménez-Forcada and Smedley 2014). The low As concentration at depth is consistent with As substitution into sulfide minerals under reducing conditions (Giménez-Forcada and Smedley 2014).

However, in deeper boreholes within the sedimentary aquifer (200–450 m) in Duero Basin, the dominant waters are Na-HCO_3 type with alkaline pH (>9) and high As (up to 241 µg/L, borehole B2), which implies the presence of an oxidizing environment (Fig. 4, Giménez-Forcada and Smedley 2014). These authors explained it in terms of the contribution of oxidizing flows following the main direction of faults in Duero Basin. These deep flow paths (coming directly from the basement and also through the discontinuities within the sedimentary basin) can also transport dissolved As from other As-enriched zones under oxic and alkaline groundwater conditions (Giménez-Forcada and Smedley 2014).

The basement structures consisting of an alternation of horsts and grabens, as in the case of Duero Basin, which is associated with late or post-Variscan tectonism (Ubanell 1985; Ares Yañez et al. 1995; Gómez Ortiz and Babín Vich 1996; Simón Gómez 1996; Herrero 1999; Giménez-Forcada and Smedley 2014), have their manifestation on the surface as large lineaments or areas with a high density of smaller lineaments (Fig. 4) and define subsequent sedimentation and lithological discontinuities (Colmenero et al. 2001; Armenteros et al. 2002; Alonso Gavilán et al. 2004), which in turn control the hydrogeological and hydrochemical features in the basin (Alberruche Del Campo 2010).

The major fault structures thus control the occurrence of groundwater As, as in Duero Basin (Giménez-Forcada and Smedley 2014), where groundwater extracted from the deep boreholes along the main fault lines, taking advantage of the direction of preferential recharge flows, are oxidizing, alkaline, Na-HCO_3 type with relatively high concentrations of As, up to 241 μg/L As, albeit with limited data, appear to have high percentages of [As(V)], along with U, V, and F (Giménez-Forcada and Smedley 2014).

Another structural control could be the hydraulic gradient, the lower one providing more residence time for groundwater (Bhattacharya et al. 2002; Nriagu et al. 2007). Smedley and Kinniburgh (2002) pointed out that whether released As remains at problematic levels in groundwater depends not only on whether there are biogeochemical reactions that retard the transport of As but also upon the hydrologic and hydrogeologic properties of the aquifer, such as flow velocity and dispersion. If the kinetics of As release is slow, and groundwater residence time is short, then As concentrations may not increase to the point where groundwater would be considered contaminated. Conversely, with rapid As mobilizing reactions and prolonged residence, As can accumulate in groundwater such that concentrations become hazardous, e.g., Bangladesh (Zheng et al. 2004; Bhattacharya et al. 2009). Eventually, if the biogeochemical conditions that lead to As release and mobilization continue to be present (within a geologic timeframe), the source could become exhausted.

Bhattacharya et al. (2012) highlighted the importance of groundwater divides, which leads to many local flows, thus forming different groundwater systems controlling As and other toxic element concentrations in Wassa West district's mining area in Ghana. The aquifers in this hilly area are not affected severely by the leaching of minerals because of short groundwater residence time and intense flushing because of the high hydraulic gradient. Thus, groundwater flow systems' local character prevents AMD's significant impact on groundwater systems on a regional scale.

7 Conclusions

Based on the average crustal concentration and presence in different geological materials, any aquifer is potentially susceptible to contamination. Instead, there is heterogeneity in the level of As contamination of groundwater across the globe, albeit with a pattern defined mainly by geotectonics. For example, major occurrences of contaminated aquifers within sedimentary basins located in the vicinity of active or ancient magmatic arcs at continental convergent margins are controlled by geodynamic processes operating at a regional-scale over millions of years. They are not mere manifestations of heterogeneous As-mobilizing (bio) geochemical conditions.

As enrichment of aquifer matrix in the forelands basins and subsequently the residing groundwater therein is also related to exhumation and erosion of arc-related lithologies that supply As-enriched sediments to these basins. Although rift basins seem to be apparent exceptions to the aforementioned conclusion of associating As contamination to convergent boundaries; however, those basins also derive sediments, at least substantially if not alone, from nearby magmatic arcs. So this observation regarding As contamination of aquifers in the rift basin does not contradict the previous conclusion. Arsenic contamination is widespread globally; however, contaminated areas tend to concentrate in and around active tectonic zones (modern/active and ancient), manifested along with previously mentioned arc magmatism, through volcanics center, geothermal fields, mineralized zones (especially metal sulfides), etc.

Among geogenic sources of As contamination, the most common occurrences are metal sulfides and sorbed species on Fe hydroxides, although As occurs within some silicate and carbonate rocks as well. Although geotectonics control the primary geologic sources of As through regional scale geodynamic processes, the basement structure of the basin, also driven by (paleo)geotectonics, may control As speciation locally through the development of compartmentalized hydrodynamic environments, resulting in spatial variations in hydrochemical composition, including As speciation.

Acknowledgements The authors would like to thank the two anonymous reviewers for their careful reading of the original manuscript and its revised versions. Their insightful comments and suggestions helped us tremendously in improving the quality of our work. Efficient editorial handling is much appreciated. Mohammad Ayaz Alam would like to thank the Universidad de Atacama for providing the necessary facilities and support to carry out this work, also Adolfo Muñoz for his help with the two figures. Prosun Bhattacharya gratefully acknowledges the financial support on arsenic research from the Swedish International Development Cooperation Agency (Sida) Contributions 75000553, 700 (Bolivia), 75000854 (Bangladesh), and 51170072 (Tanzania).

References

Acharyya SK, Shah BA, Ashyiya ID et al (2005) Arsenic contamination in groundwater from parts of Ambagarh-Chowki block, Chhattisgarh, India: source and release mechanism. Environ Geol 49:148–158. https://doi.org/10.1007/s00254-005-0074-3

Ahmed KM, Bhattacharya P, Hasan MA et al (2004) Arsenic enrichment in groundwater of the alluvial aquifers in Bangladesh: an overview. Appl Geochem 19:181–200. https://doi.org/10.1016/j.apgeochem.2003.09.006

Aiuppa A, Avino R, Caliro S et al (2006) Mineral control of arsenic content in thermal waters from volcanic-hosted hydrothermal systems: insights from the island of Ischia and Phlegrean fields (Campanian Volcanic Province, Italy). Chem Geol 229:313–330. https://doi.org/10.1016/j.chemgeo.2005.11.004

Alberruche Del Campo ME (2010) Atlas del medio natural y de los recursos hídricos de la provincia de Ávila. Instituto Geológico y Minero de España-Diputación de Ávila, Madrid

Alonso Gavilán G, Armenteros I, Carballeira J et al (2004) Cuenca del Duero. In: Vera JA (ed) Geología de España. SGE-IGME, Madrid, pp 550–556

Ares Yañez M, Gutiérrez Alonso G, Díez Balda MA et al (1995) La prolongación del Despegue de Salamanca (segunda fase de deformación varisca) en el Horst de Mirueña (Zona Centro Ibérica). Rev Soc Geol Esp 8:175–191

Armenteros I, Corrochano A, Alonso Gavilán G et al (2002) Duero basin (northern Spain). In: Gibbons W, Moreno T (eds) The geology of Spain. Geological Society, London, pp 304–315

Armienta MA, Villaseñor G, Rodríguez R et al (2001) The role of arsenic-bearing rocks in groundwater pollution at Zimapan Valle, Mexico. Environ Geol 40(4–5):571–581. https://doi.org/10.1007/s002540000220

Arnorssón S (2003) Arsenic in surface- and up to 90°C ground waters in a basalt area, N-Iceland: processes controlling its mobility. Appl Geochem 18:1297–1312. https://doi.org/10.1016/S0883-2927(03)00052-0

Aullón Alcaine A, Schulz C, Bundschuh J et al (2020) Hydrogeochemical controls on the mobility of arsenic, fluoride and other geogenic co-contaminants in the shallow aquifers of northeastern La Pampa Province in Argentina. Sci Total Environ 715:136671. https://doi.org/10.1016/j.scitotenv.2020.136671

Awaya T, Oyama M, Ishizaka N et al (2002) The amount of arsenic loads of river waters and hot springs in the Hakone-Yugawara area. Ann Rep Kanagawa Onsen Chigaku Kenkyujo 33:49–70. Japanese

Baba A, Sözbilir, H (2012) Source of arsenic based on geological and hydrogeochemical properties of geothermal systems in Western Turkey. Chem Geol 334:364–377. https://doi.org/10.1016/j.chemgeo.2012.06.006

Baba A, Uzelli T, Sözbilir H (2021) Distribution of geothermal arsenic in relation to geothermal play types: a global review and case study from the Anatolian Plate (Turkey). J Hazard Mat 414:125510. https://doi.org/10.1016/j.jhazmat.2021.125510

Ball JW, Nordstrom DK, Jenne EA et al (1998) Chemical analyses of hot springs, pools, geysers, and surface waters of Yellowstone National Park, Wyoming and vicinity 1972–1975. U.S. Geological Survey Open-File Report 98–182, U.S. Geological Survey, Reston, VA. https://doi.org/10.3133/ofr98182

Ball JW, McCleskey RB, Nordstrom, DK et al (2002) Water chemistry data for selected springs, geysers, and streams in Yellowstone National Park, Wyoming 1999–2000. U.S. Geological Survey Open-File Report 2002–382, U.S. Geological Survey, Reston, VA. https://doi.org/10.3133/ofr02382

Banerjee DM, Bhattacharya P (1994) Petrology and geochemistry of greywackes from the Aravalli Supergroup, Rajasthan, India and the tectonic evolution of a Proterozoic sedimentary basin. Precambrian Res 67(1–2):1–35. https://doi.org/10.1016/0301-9268(94)90003-5

Barringer JL, Reilly PA (2013) Arsenic in groundwater: a summary of sources and the biogeochemical and hydrogeologic factors affecting Arsenic occurrence and mobility. In: Bradley PM (ed) Current perspectives in contaminant hydrology and water resources sustainability. IntechOpen, Rijeka, Croatia. https://doi.org/10.5772/55354

Baur WH, Onishi BMH (1969) Arsenic. In: Wedepohl KH (ed), Handbook of geochemistry. Springer-Verlag, Berlin, pp 33-A-1-33-0-5

Bebout GE, Ryan JG, Leeman WP et al (1999) Fractionation of trace elements by subduction-zone metamorphism – effect of convergent-margin thermal evolution. Earth Planet Sci Lett 171:63–81. https://doi.org/10.1016/S0012-821X(99)00135-1

Berg M, Stengel C, Trang PTK et al (2007) Magnitude of arsenic pollution in the Mekong and Red River deltas – Cambodia and Vietnam. Sci Total Environ 372:413–425. https://doi.org/10.1016/j.scitotenv.2006.09.010

Bhattacharya P, Chatterjee D, Jacks G (1997) Occurrence of Arsenic-contaminated groundwater in alluvial aquifers from delta plains, Eastern India: options for safe drinking water supply. Int J Water Resour Dev 13(1):79–92. https://doi.org/10.1080/07900629749944

Bhattacharya P, Jacks G, Sracek A, et al (2001) Geochemistry of the holocene alluvial sediments in the Bengaldelta plains: implication on arsenic contamination in the groundwater. In: Jacks G, Bhattacharya P, Khan AA (eds) Proceedings of the KTH - Dhaka University Seminar on Groundwater Arsenic Contamination in the Bengal Delta Plains of Bangladesh, Department of Geology, Dhaka, Bangladesh, 7–8 February, 1999, KTH Special Publication TRITA-AMI Report 3084, Stockholm, Sweden. Available at: https://www.divaportal.org/smash/get/diva2:500376/FULLTEXT01.pdf

Bhattacharya P, Frisbie SH, Smith E et al (2002) Arsenic in the environment: a global perspective. In: Sarkar B (ed) Handbook of heavy metals in the environment. Marcell Dekker, New York, pp 147–215

Bhattacharya P, Welch AH, Ahmed KM et al (2004) Arsenic in groundwater of sedimentary aquifers. Appl Geochem 19(2):163–167. https://doi.org/10.1016/j.apgeochem.2003.09.004

Bhattacharya P, Welch AH, Stollenwerk KG et al (2007) Arsenic in the environment: biology and chemistry. Sci Total Environ 379:109–120. https://doi.org/10.1016/j.scitotenv.2007.02.037

Bhattacharya P, Hasan MA, Sracek O et al (2009) Groundwater chemistry and arsenic mobilization in the Holocene flood plains in south-central Bangladesh. Environ Geochem Health 31:23–43. https://doi.org/10.1007/s10653-008-9230-5

Bhattacharya P, Sracek O, Eldvall B et al (2012) Hydrogeochemical study on the contamination of water resources in a part of Tarkwa mining area, Western Ghana. J African Earth Sci 66–67:72–84. https://doi.org/10.1016/j.jafrearsci.2012.03.005

Bhattacharya P, Classon M, Bundschuh J et al (2006b) Distribution and mobility of arsenic in the Río Dulce alluvial aquifers in Santiago del Estero Province, Argentina. Sci Total Environ 358:97–120. https://doi.org/10.1016/j.scitotenv.2005.04.048

Bhattacharya P, Ahmed KM, Hasan MA et al (2006a) Mobility of arsenic in groundwater in a part of Brahmanbaria district, NE Bangladesh. In: Naidu R, Smith E, Owens G, Bhattacharya P, Nadebaum P (eds) Managing Arsenic in the environment: from soil to human health. CSIRO Publishing, Melbourne, Australia, pp 95–115

Bhattacharya, P, Polya DA, Jovanovic D (eds) (2017) Best practice guide for the control of arsenic in drinking water. International Water Association, London. https://doi.org/10.2166/9781780404929

Bhowmick S, Pramanik S, Singh P et al (2018) Arsenic in groundwater of West Bengal, India: a review of human health risks and assessment of possible intervention options. Sci Total Environ 612:148–169. https://doi.org/10.1016/j.scitotenv.2017.08.216

Birkle P, Bundschuh J, Sracek O (2010) Mechanisms of arsenic enrichment in geothermal and petroleum reservoirs fluids in Mexico. Water Res 44(19):5605–5617. https://doi.org/10.1016/j.watres.2010.05.046

Biswas A, Gustafsson JP, Neidhardt H, et al (2014a) Role of competing ions in the mobilization of arsenic in groundwater of Bengal Basin:

insight from surface complexation modeling. Water Res 55:30–39. https://doi.org/10.1016/j.watres.2014.02.002

Biswas A, Bhattacharya P, Mukherjee A, et al (2014b) Shallow hydrostratigraphy in an arsenic affected region of Bengal Basin: implication for targeting safe aquifers for drinking water supply. Sci Total Environ 485–486: 12–22

Biswas A, Neidhardt H, Kundu AK, et al (2014c) Spatial, vertical and temporal variation of arsenic in shallowaquifers of the Bengal Basin: controlling geochemical processes. Chem Geol 387:157–169. https://doi.org/10.1016/j.chemgeo.2014.08.022

Biswas A, Nath B, Bhattacharya P, et al (2012) Hydrogeochemical contrast between brown and grey sand aquifers in shallow depth of Bengal Basin: consequences for sustainable drinking water supply. Sci Total Environ 431, 402–412. https://doi.org/10.1016/j.scitotenv.2012.05.031

Biswas A, Majumder S, Neidhardt H, et al (2011) Groundwater chemistry and redox processes: depth dependent arsenic release mechanism. Appl Geochem 26:516–525. https://doi.org/10.1016/j.apgeochem.2011.01.010

Boyle RW, Jonasson IR (1973) The geochemistry of arsenic and its use as an indicator element in geochemical prospecting. J Geochem Explor 2:251–296. https://doi.org/10.1016/0375-6742(73)90003-4

Boyle DR, Turner RJW, Hall GEM (1998) Anomalous arsenic concentrations in groundwaters of an island community, Bowen Island, British Columbia. Environ Geochem Health 20:199–212. https://doi.org/10.1023/A:1006597311909

Breuer C, Pichler T (2013) Arsenic in marine hydrothermal fluids. Chem Geol 348:2–14. https://doi.org/10.1016/j.chemgeo.2012.10.044

Brookins DG (2012) Eh-pH diagrams for geochemistry. Springer, Berlin Heidelberg

Brown, CJ, Chute SK (2002) Arsenic in bedrock wells in Connecticut (Abstract). Arsenic in New England: a multidisciplinary scientific conference, national institute of environmental health sciences, superfund basic research program, Manchester, New Hampshire, 2002

Brunt R, Vasak L, Griffioen J (2004) Arsenic in groundwater: probability of occurrence of excessive concentration on a global scale. International Groundwater Resources Assessment Centre: Report no. SP 2004-1. https://www.un-igrac.org/resource/arsenic-groundwater-probability-occurrence-excessive-concentration-global-scale. Accessed 31 January 2021

Bundschuh J, Maity JP (2015) Geothermal arsenic: occurrence, mobility and environmental implications. Renew Sustain Energy Rev 42:1214–1222. https://doi.org/10.1016/j.rser.2014.10.092

Bundschuh J, Bhattacharya P, Hoinkis J et al (2010) Groundwater arsenic: from genesis to sustainable remediation. Water Res 44 (19):5511. https://doi.org/10.1016/j.watres.2010.10.028

Bundschuh J, Litter MI, Parvez F et al (2012) One century of arsenic exposure in Latin America: a review of history and occurrence from 14 countries. Sci Total Environ 429:2–35. https://doi.org/10.1016/j.scitotenv.2011.06.024

Bundschuh J, Maity JP, Nath B et al (2013) Naturally occurring arsenic in terrestrial geothermal systems of western Anatolia, Turkey: potential role in contamination of freshwater resources. J Hazard Mat 262:951–959. https://doi.org/10.1016/j.jhazmat.2013.01.039

Bundschuh J, Armienta MA, Morales-Simfors N et al (2020) Arsenic in Latin America: new findings on source, mobilization and mobility in human environments in 20 countries based on decadal research 2010–2020. Crit Rev Env Sci Tec. https://doi.org/10.1080/10643389.2020.1770527

Bundschuh J, Schneider J, Alam MA et al (2021) Seven potential sources of arsenic pollution in Latin America and their environmental and health impacts. Sci Total Environ. https://doi.org/10.1016/j.scitotenv.2021.146274

Bundschuh J, Farias B, Martin R et al (2004) Groundwater arsenic in the Chaco-Pampean Plain, Argentina: case study from Robles County, Santiago del Estero Province. Appl Geochem 19(2):231–243. https://doi.org/10.1016/j.apgeochem.2003.09.009

Chakraborti D, Rahman MM, Mukherjee A et al (2015) Groundwater arsenic contamination in Bangladesh—21 Years of research. J Trace Elem Med Biol 31:237–248. https://doi.org/10.1016/j.jtemb.2015.01.003

Chakraborty M, Mukherjee A, Ahmed KM (2015) A review of groundwater arsenic in the Bengal Basin, Bangladesh and India: from source to sink. Curr Pollution Rep 1:220–247. https://doi.org/10.1007/s40726-015-0022-0

Chakraborty M, Sarkar S, Mukherjee A et al (2020) Modeling regional-scale groundwater arsenic hazard in the transboundary Ganges River Delta, India and Bangladesh: infusing physically-based model with machine learning. Sci Total Environ 748:141107. https://doi.org/10.1016/j.scitotenv.2020.141107

Chapelle FH (1993) Ground-water microbiology and geochemistry. John Wiley and Sons, New York

Charnock JM, Polya DA, Gault AG et al (2007) Direct EXAFS evidence for incorporation of As5+ in the tetrahedral site of natural andraditic garnet. Am Min 92:1856–1861. https://doi.org/10.2138/am.2007.2541

Colmenero JR, Rodríguez JM, Gómez JJ et al (2001) Estratigrafía del subsuelo y evolución sedimentaria del sector sur de la cuenca terciaria del Duero. Geotemas 3:129–132

Coomar P, Mukherjee A, Bhattacharya P et al (2019) Contrasting controls on hydrogeochemistry of arsenic-enriched groundwater in the homologous tectonic settings of Andean and Himalayan basin aquifers, Latin America and South Asia. Sci Total Environ 689:1370–1387. https://doi.org/10.1016/j.scitotenv.2019.05.444

Craw D, Chappell D, Reay A (2000) Environmental mercury and arsenic sources in fossil hydrothermal systems, Northland, New Zealand. Environ Geol 38:875–887. https://doi.org/10.1007/s002549900068

Crock JG, Gough LP, Wanty RB et al (1999) Regional geochemical results from the analyses of rock, water, soil, stream sediment, and vegetation samples — Fortymile River Watershed, East-Central, Alaska 1998 Sampling. U.S. Department of the Interior, U.S. Geological Survey, Open-File Report 00-511. https://doi.org/10.3133/ofr00511

Daniele L (2004) Distribution of arsenic and other minor trace elements in the groundwater of Ischia Island (southern Italy). Environ Geol 46:96–103. https://doi.org/10.1007/s00254-004-1018-z

DeCelles PG, Giles KA (1996) Foreland basin systems. Basin Res 8 (2):105–123. https://doi.org/10.1046/j.1365-2117.1996.01491.x

Deschamp F, Godard M, Guillot S et al (2013) Geochemistry of subduction zone serpentinites: a review. Lithos 178:96–127. https://doi.org/10.1016/j.lithos.2013.05.019

Drever JI (1997) The geochemistry of natural waters, 3rd edn. Prentice-Hall, Upper Saddle River, NJ

Edmunds, WM., Walton, NRG, Howard MP J et al (1981). Geochemical estimation of aquifer recharge. British Geological Survey, Wallingford, Oxfordshire, Report WD/OS/80/17

Ehrlich HL, Newman DK (eds) (2009) Geomicrobiology. CRC Press, Boca Raton, FL

Ellis AJ, Mahon WAJ (1967) Natural hydrothermal systems and experimental hot water/rock interactions (Part II). Geochim Cosmochim Acta 31(4):519–538. https://doi.org/10.1016/0016-7037(67)90032-4

Ellis AJ, Mahon WAJ (1977) Chemistry and geothermal systems. Academic Press, New York

Even E, Masuda H, Shibata T et al (2017) Geochemical distribution and fate of arsenic in water and sediments of rivers from the Hokusetsu area, Japan. J Hydrol Reg Stud 9:34–47. https://doi.org/10.1016/j.ejrh.2016.09.008

Faure G (1998) Principles and applications of geochemistry, 2nd edn. Prentice Hall, Upper Saddle River, NJ

Fendorf S, Holly HA, van Geen A (2010) Spatial and temporal variations of groundwater arsenic in South and Southeast Asia. Science 328:1123–1127. https://doi.org/10.1126/science.1172974

Fernández-Martínez A, Cuello GJ, Johnson MR et al (2008) Arsenate incorporation in gypsum probed by neutron, x-ray scattering and density functional theory modeling. J Phys Chem A 112:5159–5166. https://doi.org/10.1021/jp076067r

Fujii R, Swain WC (1995) Areal distribution of trace elements, salinity, major ions in shallow ground water, Tulare basin, Southern San Joaquin Valley, California. U.S. Geological Survey Water Resources Investigations Report 95-4048, Sacramento, California. https://doi.org/10.3133/wri954048

Fujiwara S, Yamamoto K, Mimura K (2011) Dissolution processes of elements from subducting sediments into fluids: evidence from the chemical composition of the Sanbagawa pelitic schists. Geochem J 45:221–234. https://doi.org/10.2343/geochemj.1.0117

Giménez-Forcada E, Smedley PL (2014) Geological factors controlling occurrence and distribution of arsenic in groundwaters from the southern margin of the Duero Basin, Spain. Environ Geochem Health 36(6):1029–1047. https://doi.org/10.1007/s10653-014-9599-2

Gleeson T, Ingebritsen SE (eds) (2016) Crustal permeability. Wiley-Blackwell, Chichester

Gómez Ortiz D, Babín Vich RB (1996) La tectónica alpina en el sector centro-oriental del borde norte del Sistema Central, Provincia de Segovia, España. Geogaceta 19:19–22

Gong ZL, Lu XF, Watt C et al (2006) Speciation analysis of arsenic in groundwater from Inner Mongolia with an emphasis on acid-leachable particulate arsenic. Anal Chim Acta 555:181–187. https://doi.org/10.1016/j.aca.2005.08.062

Gonzalez-Partida E, Hinojosa ET, Verma MP (2001) Interraccion agua geothermica manantiales en el campo geotermico de Los Humeros, México. Ing Hidraul Mex XVI:185–194

Grosz AE, Grossman JN, Garrett R et al (2004) A preliminary geochemical map for arsenic in surficial materials of Canada and the United States. Appl Geochem 19:257–260. https://doi.org/10.1016/j.apgeochem.2003.09.012

Guilliot S, Charlet L (2007) Bengal arsenic, an archive of Himalaya orogeny and paleohydrology. J Environ Sci Health A 42:1785–1794. https://doi.org/10.1080/10934520701566702

Guillot S, Garcon M, Weinman B et al (2015) Origin of arsenic in Late Pleistocene to Holocene sediments in the Nawalparasi district (Terai, Nepal). Environ Earth Sci 74:2571–2593. https://doi.org/10.1007/s12665-015-4277-y

Gunduz O, Simsek C, Hasozbek A (2010) Arsenic pollution in the groundwater of Simav Plain, Turkey: its impact on water quality and human health. Water Air Soil Pollut 205:43–62. https://doi.org/10.1007/s11270-009-0055-3

Guo H-M, Yang S-Z, Li X-H et al (2007) Groundwater geochemistry and its implications for arsenic mobilization in shallow aquifers of the Hetao basin, Inner Mongolia. Sci Total Environ 393(1):131–144. https://doi.org/10.1016/j.scitotenv.2007.12.025

Haeri A, Strelbitskaya S, Porkhial S et al (2011) Distribution of arsenic in geothermal fluids from Sabalan geothermal field, N-W Iran. In: Proceedings 36th workshop on geothermal reservoir engineering, Stanford University, Stanford, California, USA, 31 January–2 February 2011, SGP-TR-191. https://pangea.stanford.edu/ERE/pdf/IGAstandard/SGW/2011/haeri2.pdf. Accessed 31 January 2021

Hammarlund L, Pionens J, Bhattacharya P et al (2009) Study of geothermal fluid-groundwater interaction and evolution in thermal fields of Costa Rica. Geological Society of America, Abstracts with Programs 41(7):219

Hattori KH, Guillot S (2003) Volcanic fronts form as a consequence of serpentinite dehydration in the forearc mantle wedge. Geology 31 (6):525–528. https://doi.org/10.1130/0091-7613(2003)031%3c0525:VFFAAC%3e2.0.CO,2

Hattori KH, Arai S, Clarke DB (2002) Selenium, tellurium, arsenic and antimony contents of primary mantle sulfides. Can Mineral 40:637–650. https://doi.org/10.2113/gscanmin.40.2.637

Hattori K, Takahashi Y, Guillot S et al (2005) Occurrence of arsenic (V) in forearc mantle serpentinites based on X-ray absorption spectroscopy study. Geochim Cosmochim Acta 69(23):5585–5596. https://doi.org/10.1016/j.gca.2005.07.009

Hecht H, Oguchi T (2017) Global evaluation of erosion rates in relation to tectonics. Prog Earth Planet Sci 4:40. https://doi.org/10.1186/s40645-017-0156-3

Henke KR (2009) Arsenic in natural environments. In: Henke KR (ed) Arsenic—environmental chemistry, health threats and waste treatment. John Wiley & Sons, Chichester, pp 69–236

Herath I, Vithanage M, Bundschuh J et al (2016) Natural arsenic in global groundwaters: distribution and geochemical triggers for mobilization. Curr Pollut Rep 2:68–89. https://doi.org/10.1007/s40726-016-0028-2

Herrero M (1999) El papel explicativo de las rocas filonianas en la evolución morfoestructural de áreas de zócalo cristalino: La Sierra de Ávila. Cuaternario Geomorfol 13:51–60

Ijumulana J, Ligate F, Bhattacharya P et al (2020) Spatial analysis and GIS mapping of regional hotspots and potential health risk of fluoride concentrations in groundwater of northern Tanzania. Sci Total Environ 735:139584. https://doi.org/10.1007/s40726-016-0028-2

Iskandar I, Koike K, Sendjaja P (2012) Identifying groundwater arsenic contamination mechanisms in relation to arsenic concentrations in water and host rocks. Environ Earth Sci 65:2015–2026. https://doi.org/10.1007/s12665-011-1182-x

Jin X, She Q, Ang X et al (2012) Removal of boron and arsenic by forward osmosis membrane: influence of membrane orientation and organic fouling. J Membr Sci 389:182–187. http://dx.doi.org/10.1016/j.memsci.2011.10.028

Karydakis G, Arvanitis A, Andritsos N et al (2005). Low enthalpy geothermal fields in the Strymon Basin (Northern Greece). In: Proceedings world geothermal congress 2005, Antalya, 24–29 Turkey April 2005, Paper Number: 0258. https://pangea.stanford.edu/ERE/pdf/IGAstandard/EGC/2007/258.pdf. Accessed 31 January 2021

Katsoyiannis IA, Mitrakas M, Zouboulis AI (2015) Arsenic occurrence in Europe: emphasis in Greece and description of the applied full-scale treatment plants. Desalin Water Treat 54(8):2100–2107. https://doi.org/10.1080/19443994.2014.933630

Kawakami H, Nozaki H, Koga A (1956) Chemical study on Beppu Hoto spring (II) – trace elements of Peppu hot spring (II) distribution of arsenic. Nihon Ishigaku Zasshi 77:1785–1789. Japanese

Keshavarzi B, Moore F, Mosaferi M et al (2011) The source of natural arsenic contamination in groundwater, west of Iran. Water Qual Expo Health 3:135–147. https://doi.org/10.1007/s12403-011-0051-x

Kikawada K, Kawai S, Oi T (2006) Long term changes in the concentration of dissolved arsenic and its present supply in the Kusatsu hot springs Gunma Japan. Chikyukagaku (Geochemistry) 40:125–136. Japanese with English abstract

Kikawada K, Kyomen K, Oi T (2009) Behavior of arsenic in Yukawa River of the Kusatsu hot spring resource area, Gunma prefecture, Japan. J Hot Spring Sci 59:81–87. Japanese with English abstract

Korte NE, Quintus F (1991) A review of arsenic (III) in groundwater. Crit Rev Environ Sci Technol 21(1):1–39. https://doi.org/10.1080/10643389109388408

Kumaresan M, Riyazuddin P (2001) Overview of speciation chemistry of arsenic. Curr Sci 80(7):837–846. https://www.jstor.org/stable/24105734. Accessed 31 January 2021

Landrum JT, Bennet PC, Engel AS et al (2009) Partitioning geochemistry of arsenic and antimony, El Tatio Geyser Field, Chile. Appl Geochem 24:664–676. https://doi.org/10.1016/j.apgeochem.2008.12.024

Li Y, Audétat A (2012) Partitioning of V, Mn Co, Ni, Cu, Zn, As, Mo, Ag, Sn, Sb, W, Au, Pb, and Bi between sulfide phases and hydrous basanite melt at upper mantle conditions. Earth Planet Sci Lett 355–356:327–340. https://doi.org/10.1016/j.epsl.2012.08.008

Li Y, Audétat A (2015) Effects of temperature, silicate melt composition, and oxygen fugacity on the partitioning of V, Mn Co, Ni, Cu, Zn, As, Mo, Ag, Sn, Sb, W, Au, Pb and Bi between sulfide phases and silicate melts. Geochim Cosmochim Acta 162:25–45. https://doi.org/10.1016/j.gca.2015.04.036

Lone SA, Jeelani G, Mukherjee A et al (2020). Geogenic groundwater arsenic in high altitude bedrock aquifers of upper Indus river basin (UIRB), Ladakh. Appl Geochem 113:104497. https://doi.org/10.1016/j.apgeochem.2019.104497

López DL, Bundschuh J, Birkle P et al (2012) Arsenic in volcanic geothermal fluids of Latin America. Sci Total Environ 429:57–75. https://doi.org/10.1016/j.scitotenv.2011.08.043

Maity JP, Chen C-Y, Bundschuh J et al (2017) Hydrogeochemical reconnaissance of arsenic cycling and possible environmental risk in hydrothermal systems of Taiwan. Groundwater Sustainable Dev 5:1–13. https://doi.org/10.1016/j.gsd.2017.03.001

Maity JP, Kar S, Liu J-H et al (2011a) The potential for reductive mobilization of arsenic [As(V) to As(III)] by OSBH2 (Pseudomonas stutzeri) and OSBH5 (Bacillus cereus) in an oil-contaminated site. J Environ Sci Health 46:1239–1246. https://doi.org/10.1080/10934529.2011.598802

Maity JP, Liu C-C, Nath B et al (2011b) Biogeochemical characteristics of Kuan-Tzu-Ling, Chung-Lun and Bao-Lai hot springs in southern Taiwan. J Environ Sci Health, Part A 46(11):1–11. https://doi.org/10.1080/10934529.2011.598788

Maity JP, Chen C-Y, Bundschuh J et al (2016) Investigation of arsenic contamination from geothermal water in different geological settings of Taiwan: hydrogeochemical and microbial signatures. In: Bhattacharya P, Vahter M, Jarsjö, J et al (eds) Arsenic research and global sustainability as 2016. Interdisciplinary book series: "Arsenic in the Environment—Proceedings". Series Editors: Bundschuh J, Bhattacharya P, CRC Press/Taylor and Francis, London, pp 84–85

Majzlan J, Drahota P, Filippi M (2014) Parageneses and crystal chemistry of arsenic minerals. Rev Mineral Geochem 79:17–184. https://doi.org/10.2138/rmg.2014.79.2

Mandal BK, Suzuki KT (2002) Arsenic round the world: a review. Talanta 58:201–235. https://doi.org/10.1016/S0039-9140(02)00268-0

Margat J (2008) Les eaux souterraines dans le monde [Groundwater around the world]. BRGM, Orleans and UNESCO, Paris

Mariño EE, Teijó Ávila G et al (2020) The occurrence of arsenic and other trace elements in groundwaters of the southwestern Chaco-Pampean plain, Argentina. J S Am Earth Sci 100:102547. https://doi.org/10.1016/j.jsames.2020.102547

Masoodi M, Rahimzadeh M (2018) Tectonic controls on groundwater geochemistry in Hormozgan Province, Southern Iran. Arab J Geosci 11:141. https://doi.org/10.1007/s12517-018-3478-6

Masuda H (2018) Arsenic cycling in the Earth's crust and hydrosphere: interaction between naturally occurring arsenic and human activities. Prog Earth Planet Sci 5:68. https://doi.org/10.1186/s40645-018-0224-3

Masuda H, Shinoda K, Okudaira T et al (2012) Chlorite – source of arsenic groundwater pollution in the Holocene aquifer of Bangladesh. Geochem J 46:381–391. https://doi.org/10.2343/geochemj.2.0208

Masuda H, Okabayashi K, Maeda S et al (2013) Sequential chemical extraction of arsenic and related elements from the Holocene sediments of Sonargaon, Bangladesh, in relation to formation of arsenic-contaminated groundwater. Geochem J 47:651–666. https://doi.org/10.2343/geochemj.2.0269

Matschullat J (2000) Arsenic in the geosphere – a review. Sci Total Environ 249(1–3):297–312. https://doi.org/10.1016/S0048-9697(99)00524-0

McCarthy KT, Pichler T, Price RE (2005) Geochemistry of Champagne Hot Springs shallow hydrothermal vent field and associated sediments, Dominica, Less Antilles. Chem Geol 224:55–68. https://doi.org/10.1016/j.chemgeo.2005.07.014

McClintock TR, Chen Y, Bundschuh J et al (2012) Arsenic exposure in Latin America: Biomarkers, risk assessments and related health effects. Sci Total Environ 429:76–91. https://doi.org/10.1016/j.scitotenv.2011.08.051

McDonough WF, Sun S-s (1995) The composition of the Earth. Chem Geol 12:223–253. https://doi.org/10.1016/0009-2541(94)00140-4

Minami H, Sato G, Watanuki K (1958). Concentrations of arsenic and lead of Tamagawa hot spring waters, Akita prefecture. Nippon Kagaku Zasshi 79:860–865 Japanese

Morales-Simfors N, Bundschuh J, Herath I et al (2020) Arsenic in Latin America: a critical overview on the geochemistry of arsenic originating from geothermal features and volcanic emissions for solving its environmental consequences. Sci Total Environ 716:135564. https://doi.org/10.1016/j.scitotenv.2019.135564

Motyka RJ, Poreda RJ, Jeffrey AWA (1998) Geochemistry, isotopic composition, and origin of fluids emanating from mud volcanoes in the Copper River basin, Alaska. Geochim Cosmochim Acta 53(1):3302–3309. https://doi.org/10.1016/0016-7037(89)90270-6

Mukherjee A, Fryar AE (2008) Deeper groundwater chemistry and geochemical modeling of the arsenic affected western Bengal basin, West Bengal, India. Appl Geochem 23(4):863–892. https://doi.org/10.1016/j.apgeochem.2007.07.011

Mukherjee A, Fryar AE, Howell PD (2007) Regional hydrostratigraphy and groundwater flow modeling in the arsenic affected areas of the western Bengal Basin, West Bengal, India. Hydrogeol J 15(7):1397–1418. https://doi.org/10.1007/s10040-007-0208-7

Mukherjee A, von Brömssen M, Scanlon BR et al (2008) Hydrogeochemical comparison and effects of overlapping redox zones on groundwater arsenic near the Western (Bhagirathi sub-basin, India) and Eastern (Meghna sub-basin, Bangladesh) margins of the Bengal Basin. J Contam Hydrol 99:31–48. https://doi.org/10.1016/j.jconhyd.2007.10.005

Mukherjee A, Fryar AE, Scanlona BR et al (2011) Elevated arsenic in deeper groundwater of the western Bengal basin, India: extent controls from regional to local scale. Appl Geochem 26:600–613. https://doi.org/10.1016/j.apgeochem.2011.01.017

Mukherjee A, Scanlon BR, Fryar AE et al (2012) Solute chemistry and arsenic fate in aquifers between the Himalayan foothills and Indian craton (including central Gangetic plain): influence of geology and geomorphology. Geochim Cosmochim Acta 90:283–302. https://doi.org/10.1016/j.gca.2012.05.015

Mukherjee A, Verma S, Gupta S et al (2014) Influence of tectonics, sedimentation and aqueous flow cycles on the origin of global groundwater arsenic: paradigms from three continents. J Hydrol 518:284–299. https://doi.org/10.1016/j.jhydrol.2013.10.044

Mukherjee A, Gupta S, Coomar P et al (2019) Plate tectonics influence on geogenic arsenic cycling: From primary sources to global groundwater enrichment. Geochim Cosmochim Acta 683:793–807. https://doi.org/10.1016/j.scitotenv.2019.04.255

Mukherjee A, Fryar AE, O'Shea BM (2009a). Major occurrences of elevated arsenic in groundwater and other natural waters. In:

Henke KR (ed) Arsenic—Environmental Chemistry, Health Threats and Waste Treatment. John Wiley & Sons, Chichester, pp 303–350

Mukherjee A, Fryar AE, Thomas WA (2009b). Geologic, geomorphic and hydrologic framework and evolution of the Bengal basin, India and Bangladesh. J Asian Earth Sci 34(3):227–244. https://doi.org/10.1016/j.jseaes.2008.05.011

Mukherjee A, Sarkar S, Chakraborty M et al (2021) Occurrence, predictors and hazards of elevated groundwater arsenic across India through field observations and regional-scale AI-based modeling. Sci Total Environ 759:143511. https://doi.org/10.1016/j.scitotenv.2020.143511

Navarro A, Font X, Viladevall M (2011) Geochemistry and groundwater contamination in the La Selva geothermal system (Girona, Northeast Spain). Geothermics 40:275–285. https://doi.org/10.1016/j.geothermics.2011.07.005

Neff J, Lee K, Deblois EM (2011) Produced water: overview of composition, fates and effects. In: Lee K, Neff J (eds) Produced water. Springer, New York

Nicolli HB, Bundschuh J, Blanco MdC et al (2012) Arsenic and associated trace-elements in groundwater from the Chaco-Pampean plain, Argentina: results from 100 years of research. Sci Total Environ 429:36–56. https://doi.org/10.1016/j.scitotenv.2012.04.048

Nimick DA, Moore JN, Dalby CE, Savka MW (1998) The fate of geothermal arsenic in the Madison and Missouri Rivers, Montana and Wyoming. Water Resour Res 34:3051–3067. https://doi.org/10.1029/98WR01704

Noble DC, Ressel MW, Connors KA (1998) Magmatic As, Sb, Cs and other volatile elements in glassy silicic rocks. Geological Society of America, Abstracts with Programs 30(7):377

Noguchi K, Nakagawa R (1969) Arsenic and arsenic-lead sulfides in sediments from Tamagawa hot springs, Akita Prefecture. Proc Jpn Acad 45:45–50. https://doi.org/10.2183/pjab1945.45.45

Noll PDJr, Newsom HE, Leeman WP et al (1996) The role of hydrothermal fluids in the production of subduction zone magmas: evidence from siderophile and chalcophile trace elements and boron. Geochim Cosmochim Acta 60, 587–611. https://doi.org/10.1016/0016-7037(95)00405-X

Nordstrom DK (2002) Worldwide occurrences of arsenic in ground water. Science 296:2143–2145. https://doi.org/10.1126/science.1072375

Nordstrom DK (2009) Natural arsenic enrichment: effects of diagenetic-tectonic hydrothermal cycle. Geological Society of America, Abstracts with Program 41(7):217

NRC (1977) 3. Distribution of arsenic in the environment. In: Arsenic: medical and biologic effects of environmental pollutants. National Research Council (NRC) Committee on medical and biological effects of environmental pollutants. National Academies Press, Washington, DC. https://www.ncbi.nlm.nih.gov/books/NBK231016/. Accessed 31 January 2021

Nriagu JO, Bhattacharya P, Mukherjee AB et al (2007) Arsenic in soil and groundwater: an overview. In: Bhattacharya P, Mukherjee AB, Bundschuh, J et al (eds) Trace metals and other contaminants in the environment, vol 9. Elsevier, Amsterdam, pp 3–60. https://doi.org/10.1016/S1875-1121(06)09001-8

Ohta A, Imai N, Terashima S et al (2010) Factors controlling regional spatial distribution of 53 elements in coastal sea sediments in northern Japan: comparison of geochemical data derived from stream and marine sediments. Appl Geochem 25:357–376. https://doi.org/10.1016/j.apgeochem.2009.12.003

Okada H, Tada T, Chiba et al (2002) Decontamination of geothermal water – removal of arsenic. Low Temp Eng 37:331–337. Japanese with English abstract

Onishi H, Sandell EB (1955) Geochemistry of arsenic. Geochim Cosmochim Acta 7:1–33. https://doi.org/10.1016/0016-7037(55)90042-9

Ormachea Muñoz M, Bhattacharya P, Sracek O, Ramos Ramos O, Quintanilla Agurre J, Bundschuh J, Maity JP (2015) Arsenic and other trace elements in thermal springs and in cold waters from drinking water wells on the Bolivian Altiplano. J S Am Earth Sci 60:10–20. https://doi.org/10.1016/j.jsames.2015.02.006

Panagiotaras D, Papoulis D, Panagopoulos G et al (2012) Arsenic geochemistry in groundwater system. In: Panagiotaras D (ed) Earth's system processes. IntechOpen, Rijeka, Croatia. https://doi.org/10.5772/39384

Pascua C, Charnock J, Polya DA et al (2005) Arsenic-bearing smectite from the geothermal environment. Mineral Mag 69:897–906. https://doi.org/10.1180/0026461056950297

Peters SC (2008) Arsenic in groundwaters in the Northern Appalachian Mountain belt: a review of patterns and processes. J Contam Hydrol 99:8–21. https://doi.org/10.1016/j.jconhyd.2008.04.001

Peters SC, Blum JD (2003) The source and transport of arsenic in a bedrock aquifer, New Hampshire, USA. Appl Geochem 18:1773–1787. https://doi.org/10.1016/S0883-2927(03)00109-4

Peters SC, Blum JD, Klaue B, Karagas MR (1999) Arsenic occurrence in New Hampshire drinking water. Environ Sci Technol 33:1328–1333. https://doi.org/10.1021/es980999e

Piqué A, Grandia F, Canals A (2010) Processes releasing arsenic to groundwater in the Caldes de Malavella geothermal area, NE Spain. Water Res 44:5618–5630. https://doi.org/10.1016/j.watres.2010.07.012

Planer-Friedrich B, London J, McCleskey RB et al (2007) Thioarsenates in geothermal waters of Yellowstone National Park: determination, preservation, and geochemical importance. Environ Sci Technol 41:5245–5251. https://doi.org/10.1021/es070273v

Plank T, Langmuir CH (1993) Tracing trace-elements from sediment input to volcanic output at subduction zones. Nature 362 (6422):739–743

Pollizzotto ML, Harvey CF, Sutton S et al (2005) Processes conductive to the release and transport of arsenic into aquifers of Bangladesh. Proc Natl Acad Sci USA 102:18819–18823. https://doi.org/10.1073/pnas.0509539103

Polya DA, Sparrenbom C, Datta S, Guo H (2019) Groundwater arsenic biogeochemistry – key questions and use of tracers to understand arsenic-prone groundwater systems. Geosci Front 10(5):1635–1641. https://doi.org/10.1016/j.gsf.2019.05.004

Pourbaix M (1974) Atlas of electrochemical equilibria in aqueous solutions 2nd edn. National Association of Corrosion Engineers, Houston, TX

Quino Lima I, Ramos Ramos OE, Ormachea Muñoz M et al (2020) Spatial dependency of arsenic, antimony, boron and other trace elements in the shallow groundwater systems of the Lower Katari Basin, Bolivian Altiplano. Sci Total Environ 719:137505. https://doi.org/10.1016/j.scitotenv.2020.137505

Quino Lima, I., Ramos Ramos, O.E., Ormachea Muñoz, M et al (2021b) Geochemical mechanisms of natural arsenic mobility in the hydrogeologic system of Lower Katari Basin, Bolivian Altiplano. J Hydrol 594:125778. https://doi.org/10.1016/j.jhydrol.2020.125778

Quino Lima I, Ormachea Muñoz M, Ramos Ramos OE et al (2021a) Hydrogeochemical contrasts in the shallow aquifer systems of the Lower Katari Basin and Southern Poopó Basin, Bolivian Altiplano. J S Am Earth Sci 105:102914. https://doi.org/10.1016/j.jsames.2020.102914

Quintanilla J, Ramos OE, Ormachea M et al (2009) Arsenic contamination, speciation and environmental consequences in the Bolivian Plateau. In: Bundschuh J, Armienta MA, Birkle P et al (eds) Natural arsenic in groundwaters of Latin America. Taylor & Francis, Boca Raton, NJ, pp 91–99

Ramos Ramos OE, Rötting T, French M et al (2014) Geochemical processes controlling mobilization of arsenic and trace elements in shallow aquifers and surface waters in the Antequera and Poopó mining regions, Bolivian Altiplano. J Hydrol 518:421–433. https://doi.org/10.1016/j.jhydrol.2014.08.019

Ramos Ramos OE, Cáceres LF, Ormachea Muñoz MR et al (2012) Sources and behavior of arsenic and trace elements in groundwater and surface water in the Poopó Lake Basin, Bolivian Altiplano. Environ Earth Sci 66(3):793–807. https://doi.org/10.1007/s12665-011-1288-1

Ravenscroft P, Brammer H, Richards KS (2009) Arsenic pollution: a global synthesis. Wiley-Blackwell Publication, Chichester

Raychowdhury N, Mukherjee A, Bhattacharya P et al (2014) Provenance and fate of arsenic and other solutes in the Chaco-Pampean Plain of the Andean foreland, Argentina: from perspectives of hydrogeochemical modeling and regional tectonic setup. J Hydrol 518:300–316. https://doi.org/10.1016/j.jhydrol.2013.07.003

Ritchie JA (1961) Arsenic and antimony in some New Zealand thermal waters. NZ J Sci 4:218–229

Romero L, Alonso H, Campano P et al (2003) Arsenic enrichment in waters and sediments of the Rio Loa (Second region, Chile). Appl Geochem 18(9):1399–1416. https://doi.org/10.1016/S0883-2927(03)00059-3

Ryan PC, Kim J, Wall AJ et al (2011) Ultramafic-derived arsenic in a fractured rock aquifer. Appl Geochem 26:444–457. https://doi.org/10.1016/j.apgeochem.2011.01.004

Saunders JA, Lee MK, Mohammad S (2005) Natural arsenic contamination of Holocene alluvial aquifers by linked tectonic, weathering, and microbial processes. Geochem Geophys Geosyst 6(4):Q04006. https://doi.org/10.1029/2004GC000803

Savage KS, Tingle TN, O'Day PA et al (2000) Arsenic speciation in pyrite and secondary weathering phases, Mother Lode Gold District, Tuolumne County, California. Appl Geochem 15:1219–1244. https://doi.org/10.1016/S0883-2927(99)00115-8

Sawkins FJ (1990) Metal deposits in relation to plate tectonics, 2nd edn. Minerals and Rocks Series 17, Springer-Verlag, New York

Scanlon BR, Nicot JP, Reedy R et al (2009) Elevated naturally occurring arsenic in a semiarid oxidizing system, Southern High Plains aquifer, Texas, USA. Appl Geochem 24:2061–2071. https://doi.org/10.1016/j.apgeochem.2009.08.004

Seddique AA, Masuda H, Mitamura M et al (2011) Mineralogy and geochemistry of shallow sediments of Sonargaon, Bangladesh and implications for arsenic dynamics: focusing on the role of organic matter. App Geochem 2:587–599. https://doi.org/10.1016/j.apgeochem.2011.01.016

Shimada Y, Stute M, van Geen A et al (2004) Redox control of arsenic mobilization in Bangladesh groundwater. Appl Geochem 19(2):201–214. https://doi.org/10.1016/j.apgeochem.2003.09.007

Shimada N (2009) The essence of problems on groundwater and soil pollutions caused by naturally occurring heavy metals and harmful elements: arsenic. Oyo Technical Report 29, pp 31–59. Japanese with English abstract

Sigufusson B, Gislason SR, Meharg AA (2011) A field and reactive transport model study of arsenic in a basaltic rock aquifer. Appl Geochem 26:553–564. https://doi.org/10.1016/j.apgeochem.2011.01.013

Simón Gómez JL (1996) Estudio estructural de la Comarca de La Moraña (provincia de Ávila). Universidad de Zaragoza, Zaragoza, Spain, Departamento de Geología

Sims KWW, Newsom HE, Gladney ES (1990) Chemical fractionation during formation of the Earth's core and continental crust: clues from As, Sb, W, and Mo. In: Newson HE, Jones JH (eds) Origin of the earth. Oxford University Press, New York, pp 291–317

Sisson TW, Grove TL (1993) Experimental investigations of the role of H2O in calc-alkaline differentiation and subduction zone magmatism. Contrib Mineral Petrol 113:143–166. https://doi.org/10.1007/BF00283225

Smedley PL, Kinniburgh DG (2002) A review of the source, behavior and distribution of arsenic in natural waters. Appl Geochem 17(5):517–568. https://doi.org/10.1016/S0883-2927(02)00018-5

Sracek O, Bhattacharya P, Jacks G et al (2004) Behavior of arsenic and geochemical modeling of arsenic contamination. Appl Geochem 19(2):169–180. https://doi.org/10.1016/j.apgeochem.2003.09.005

Stallard RF, Edmond JM (1983) Geochemistry of the Amazon: 2. The influence of geology and weathering environment on the dissolved load. J Geophys Res Oceans 88:9671–9688. https://doi.org/10.1029/JC088iC14p09671

Stanger G (2005) A palaeo-hydrogeological model for arsenic contamination in southern and south-east Asia. Environ Geochem Health 27:359–367. https://doi.org/10.1007/s10653-005-7102-9

Stauffer RE, Thompson JM (1984) Arsenic and antimony in geothermal waters of Yellow Stone National Park, Wyoming, USA. Geochim Cosmochim Acta 48:2547–2561. https://doi.org/10.1016/0016-7037(84)90305-3

Stromgren T, Sørstrøm SE, Schou L et al (1995) Acute toxic effects of produced water in relation to chemical composition and dispersion. Mar Environ Res 40:147–169. https://doi.org/10.1016/0141-1136(94)00143-D

Stüben D, Berner Z, Chandrasekharam D et al (2003) Arsenic enrichment in groundwater of West Bengal, India: geochemical evidence for mobilization of As under reducing conditions. Appl Geochem 18:1417–1434. https://doi.org/10.1016/S0883-2927(03)00060-X

Tapia J, Davenport J, Townley B et al (2018) Sources, enrichment, and redistribution of As, Cd, Cu, Li, Mo, and Sb in the Northern Atacama Region, Chile: implications for arid watersheds affected by mining. J Geochem Explor 185:33–51. https://doi.org/10.1016/j.gexplo.2017.10.021

Tapia JS, Audry (2013) Control of early diagenesis processes on trace metal (Cu, Zn, Cd, Pb and U) and metalloid (As, Sb) behaviors in mining- and smelting-impacted lacustrine environments of the Bolivian Altiplano. Appl Geochem 31:60–78. https://doi.org/10.1016/j.apgeochem.2012.12.006

Tapia J, Audry S, Townley B et al (2012) Geochemical background, baseline and origin of contaminants from sediments in themining-impacted Altiplano and Eastern Cordillera of Oruro, Bolivia. Geochem-Explor Eenv A 12(1):3–20. https://doi.org/10.1144/1467-7873/10-RA-049

Tapia J, Audry S, van Beek P (2019a) Natural and anthropogenic controls on particulate metal(loid) deposition in Bolivian highland sediments, Lake Uru Uru (Bolivia). Holocene 30(3):428–440. 10.1177%2F0959683619887425

Tapia J, Murray J, Ormachea M et al (2019b) Origin, distribution, and geochemistry of arsenic in the Altiplano-Puna plateau of Argentina, Bolivia, Chile, and Perú. Sci Total Environ 678, 309–325. https://doi.org/10.1016/j.scitotenv.2019.04.084

Tapia J, Rodríguez MP, Castillo P et al (2019c) Arsenic and copper in Chile and the development of environmental standards. In: Alaniz AJ (ed) Chile: environmental history, perspectives and challenges. Nova Publishers, Hauppauge, NY

Tapia J, Schneider B, Inostroza M et al (2021) Naturally elevated arsenic in the Altiplano-Puna, Chile and the link to recent (Mio-Pliocene to Quaternary) volcanic activity, high crustal thicknesses, and geological structures. J S Am Earth Sci 105:102905. https://doi.org/10.1016/j.jsames.2020.102905

Tatsumi Y (1989) Migration of fluid phases and genesis of basalt magmas in subduction zones. J Geophys Res Solid Earth 94(B4):4697–4707. https://doi.org/10.1029/JB094iB04p04697

Taylor SR, McLennan SM (1985) The continental crust: its composition and evolution. Blackwell, Oxford

Ubanell AG (1985) Características principales de la fracturación tardihercinica en un segmento del Sistema Central Español. Cuad Geog 7:591–605

Ure A, Berrow M (1982) Chapter 3. The elemental constituents of soils. In: Bowen HJM (ed) Environmental chemistry, vol 2. Royal Society of Chemistry, London, pp 94–204

van Geen A (2011) International drilling to recover aquifer sands (IDRAs) and arsenic contaminated groundwater in Asia. Sci Drill 12:49–52. https://doi.org/10.2204/iodp.sd.12.06.20112011

Vergasova LP, Krivovichev SV, Britvin SN et al (2004) Filatovite, K [(Al, Zn)2(As, Si)2O8], a new mineral species from the Tolbachik volcano, Kamchatka peninsula, Russia. Eur J Mineral 16:533–536. https://doi.org/10.1127/0935-1221/2004/0016-0533

Verma S, Mukherjee A, Mahanta C et al (2016) Influence of geology on groundwater–sediment interactions in arsenic enriched tectono-morphic aquifers of the Himalayan Brahmaputra river basin. J Hydrol 540:176–195. https://doi.org/10.1016/j.jhydrol.2016.05.041

Vilanova E, Mas-Pla J, Menció A (2008) Determinación de sistemas de flujo regionales y locales en las depresiones tectónicas del Baix Empordà y La Selva (NE de España) en base a datos hidroquímicos e isotópicos. Bol Geol Min 119(1):51–62

Webster JG (1990) The solubility of As2S3 and speciation of As in dilute and sulphide bearing fluids at 25 and 90°C. Geochim Cosmochim Acta 54:1009–1017. https://doi.org/10.1016/0016-7037(90)90434-M

Webster JG, Nordstrom K (2003) Chapter 4. Geothermal arsenic the source, transport and fate of arsenic in geothermal systems. In: Welch AH, Stollenwerk, KG (eds) Arsenic in groundwater. Kluwer Academic Publishers, Boston, MA

Webster, JG (1999) The source of arsenic (and other elements) in the Marbel – Matingo river catchment, Mindanao, Philippines. Geothermics 28(1):95–111. https://doi.org/10.1016/S0375-6505(98)00046-7

Webster-Brown JG, Lane V (2005) The environmental fate of geothermal arsenic in a major river system, New Zealand. In: Proceedings world geothermal congress 2005, Antalya, Turkey, 24–29 April 2005, Paper Number: 0263. https://www.geothermal-energy.org/pdf/IGAstandard/WGC/2005/0263.pdf. Accessed 31 January 2021

Webster-Brown JG, Lane V, Webster KS (2000) Arsenic in the Waikato River – an update. In: Proceedings of the 22nd New Zealand Geothermal Workshop, pp 63–68

Wedepohl KH (1995) The composition of the continental-crust. Geochim Cosmochim Acta 59:1217–1232. https://doi.org/10.1016/0016-7037(95)00038-2

Welch AH, Lico MS (1998) Factors controlling As and U in shallow ground water, southern Carson Desert, Nevada. Appl Geochem 13:521–539. https://doi.org/10.1016/S0883-2927(97)00083-8

Welch AH, Stollenwerk KG (2002) Arsenic in ground water: geochemistry and occurrence. Springer Science & Business Media, New York

Welch AH, Lico MS, Hughes JL (1988) Arsenic in groundwater of the western United States. Ground Water 26:333–347. https://doi.org/10.1111/j.1745-6584.1988.tb00397.x

Welch AH, Westjohn DB, Helsel DR et al (2000) Arsenic in groundwater of the United States: occurrence and geochemistry. Ground Water 38:589–604. https://doi.org/10.1111/j.1745-6584.2000.tb00251.x

WHO (1981) Arsenic: environmental health criteria 18. World Health Organization, Geneva, Switzerland

WHO (2001) Arsenic and arsenic compounds (Environmental Health Criteria 224), 2nd edn. World Health Organization, Geneva, Switzerland

WHO (2017) Guidelines for drinking-water quality, 4th edn, incorporating the 1st addendum. World Health Organization, Geneva, Switzerland

Yudovich Y, Ketris MP (2005) Arsenic in coal: a review. Int J Coal Geol 61:141–196. https://doi.org/10.1016/j.coal.2004.09.003

You C-F, Castillo PR, Gieskes JM et al (1996) Trace element behavior in hydrothermal experiments: implications for fluid processes at shallow depths in subduction zones. Earth Planet Sci Lett 140:41–52

Zheng Y (2006) The heterogeneity of arsenic in the crust: a linkage to occurrence in groundwater. Geological Society of America, Abstracts with Programs 38(7):179

Zheng Y, Stute M, van Geen A, et al (2004) Redox control of arsenic mobilization in Bangladesh groundwater. Appl Geochem 19 (2):201–214. https://doi.org/10.1016/j.apgeochem.2003.09.007

Tectonic and Structural Controls on Geothermal Systems

Mohammad Ayaz Alam

Abstract

Although heat source, in terms of enthalpy and/or temperature, is usually considered the prime factor driving geothermal systems and has traditionally been the basis of their classification, this approach undermines the importance of the tectonic settings they exist in. The tectonic setting defines the regional stress regime that controls the permeability structure and determines the nature of heat source—magmatic or non-magmatic—and thermal regime—convective or conductive or a combination of the two, and also prevailing geothermal gradient and heat flow. Moreover, despite the tectonic setting being favorable, the local stress regime may make a geothermal system either highly productive or uneconomic, depending upon whether it aids the fluid circulation or not. Understanding the tectonic setting and local structural conditions may help enhance a low-performing exiting geothermal system's performance or even create a new one by developing artificial fractures to facilitate fluid circulation if a heat source—magmatic or non-magmatic (viz. radioactive)—is available.

Keywords

Geothermal fluid • Tectonics • Structure • Faults • Basement • Play type • Heat source • Magmatic • Non-magmatic • Convection-dominated • Conduction-dominated • Stress • Strain

1 Introduction

A geothermal system has been defined quite precisely by Boden (2016) as a combination of processes involving heat transfer to the Earth's surface through convection or conduction from source to sink, which envisages its present-day understanding and application. This can be elaborated further as follows, encompassing the processes and practices involved therein. A geothermal system is a combination of favorable physicochemical conditions within a confined volume of the Earth's crust, viz. heat and mass transfer, temperature and pressure gradients, permeability (natural, artificial, or enhanced), hydrology and fluid flow, fluid composition, and mineral equilibria that together transfer heat through convection and/or conduction) from a heat source (magmatic or non-magmatic) to a heat sink, usually in the form of surface manifestations (viz. hot springs, fumaroles, geysers, mud pools) or wells for extracting hot fluids—vapor or liquid—for direct (i.e., using heat itself) or indirect (i.e., using heat for some process, viz. electricity generation) use.

Furthermore, a geothermal system is referred to as (i) blind or hidden in the absence of any surface manifestation and (ii) enhanced or engineered, when the flow rate of the hot fluids (which is naturally very low or virtually absent due to low permeability) is increased to a rate sufficient for economical use (viz. district heating, electricity generation) by augmenting the natural permeability through technological solutions, viz. hydraulic fracturing, stimulation or fracking. In enhanced geothermal systems (EGS), the circulating fluid can be natural hot fluid if a low permeability formation hosts a hydrothermal system; or it can be an artificially injected fluid if the formation of the geothermal system does not contain enough fluid volume for heat extraction, and these are referred as Hot Dry Rock (HDR) systems. Aforesaid physicochemical conditions and processes are controlled by prevailing regional and local stress regimes, which will be discussed in this chapter.

M. A. Alam (✉)
Departamento de Geología, Facultad de Ingeniería, Universidad de Atacama, Avenida Copayapu 485, Copiapó, Región de Atacama, Chile
e-mail: ayaz.alam@uda.cl

A. K. Shandilya et al. (eds.), *Geological and Geo-Environmental Processes on Earth*, Springer Natural Hazards,
https://doi.org/10.1007/978-981-16-4122-0_15

There are efforts to denominate geothermal systems based on fundamental geological parameters that dictate geothermal resources' characteristics, the tectonic environment in which they were formed, and how those resources might best be explored and developed (e.g., Walker et al. 2005; Erdlac et al. 2008; Moeck 2014; Boden 2016). This approach is significantly different from the traditional classification of geothermal systems based on temperature and enthalpy—low, medium, and high (e.g., Muffler 1979; Muffler and Cataldi 1977; Hochstein 1988; Benderitter and Cormy 1990; Haenel et al. 1988; Sanyal 2005). Several workers considered these classifications inconsistent and insufficient to categorize a geothermal system from the development point of view (e.g., Lee 2011; Moeck 2014).

2 Importance of Structural and Tectonic Controls

The tectonic setting and geologic structure of geothermal systems, which are the results of the prevailing regional and local stress regimes, control the physicochemical characteristics of geothermal resources. An understanding of these controls not only provides valuable input for geothermal resource development but has also been the basis of a relatively new classification of geothermal systems as a catalog of "play types" similar to the oil industry (Moeck 2014). This catalog has evolved from the early ideas of Muffler (1973, 1976) to their refinement by Walker et al. (2005), Erdlac et al. (2008), and Boden (2016). This classification or cataloging is from a geothermal developer's perspective, based either on the geographical extent of favorable settings (Philips et al. 2013) or repeating sets of prospects with common characteristics defining a "play type" (King et al. 2013).

On the other hand, this classification is governed by the plate tectonic setting on a large scale and structural elements (local stress fields, rock mechanics, fracture systems) on a smaller scale. For example, whether a geothermal system or "play type" is related to convection or conduction-dominated heat transfer and/or is magmatic or non-magmatic depends on its tectonic setting. On the other hand, the local stress field's orientation controls fluid flow along the faults, and rock mechanics defines permeability anisotropy of the fractured reservoirs. Moeck's (2014) cataloging of geothermal "play types" has been used here as the basis for discussing the tectonic controls on the geothermal systems.

Since EGS and HDR development involves the creation of new fractures to increase permeability and the orientation as well as the growth of these fractures are strongly controlled by the stress field and rock mechanics, comprehending the stress fields and rock mechanics is particularly important for (i) designing the stimulation process and define injection rates for creating these artificial fractures (Moeck et al. 2009), (ii) keeping the induced fractures open during production and subsequent formation pressure drop (Moeck 2014) and (iii) risk assessment during injection through fault reactivation potential analysis using the slip and dilation tendency technique (Moeck et al. 2009), which also includes reinjection (Moeck and Backers 2011). This is achieved through 3D structural geological modeling, stress field analysis, and fault stress modeling during all the three stages of geothermal field development—exploration, drilling, and reservoir engineering (Moeck et al. 2009; Moeck 2014).

Understanding and characterizing the tectonic and structural controls on geothermal systems has been an ongoing focus on different scales, from plate tectonics (e.g., Muffler 1973, 1976; Heicken 1982) to the local structural regime (e.g., Rowland and Sibson 2004; Faulds et al. 2010a 2010b 2010c, Rowland and Simmons 2012). In short, it is the geologic setting that constrains the temperature, fluid composition, and reservoir characteristics of a geothermal system and establishes whether it is a convective or conductive system (Moeck 2014). The locations of geothermal fields worldwide are invariably tectonically controlled. They are often associated with block faulting, grabens or rifting, and collapsed caldera structures, with reservoir depths of around 1–3 km (Nicholson 1993). Typical settings are around active plate margins such as subduction zones (e.g., Pacific Rim), spreading ridges (e.g., Mid-Atlantic), rift zones (e.g., East Africa, Central India), and orogenic belts (e.g., Himalayas, Mediterranean). The characteristics of a geothermal system (viz. its thermal and hydrological regimes, fluid chemistry and dynamics, faults and fractures, stress regime, regional heat flow, lithology, rheology) are controlled by its tectonic framework.

Moreover, locally, faults can act not just as fluid conduits for a geothermal system but quite often form barriers for fluid circulation. In some cases, such barriers may lead to a segmentation of the system, some parts being more productive than others and some even unproductive. Mariposa geothermal field in south-central Chile has two lobes like wings of a butterfly (*mariposa* in Spanish), having different characteristics. This difference is evident from the magnetotelluric survey and structural studies for an ongoing geothermal development project (Hickson et al. 2011, 2012; Fox Hodgson 2012). Moreover, cap rocks that contain the heat of the geothermal system and overlie the reservoir may be breached due to high rates of uplift and ensuing erosion, e.g., in the Andes (Coolbaugh et al. 2015).

3 Convective and Conductive Heat Transfer

The heat transfer mechanism of a geothermal system is seldom either convective or conductive. It is instead a combination of both, dominated by one of them. Moreover, convection-dominated systems could be (i) magmatic, induced by a magmatic heat source (Fig. 1), or (ii) non-magmatic, in which the geothermal fluid originates as meteoric water that gains heat through the circulation to depth within a fault zone (Fig. 2). Consequently, structural controls have a significant effect on fluid flow pathways in such systems. Thus, besides a high-temperature gradient, high permeability is necessary to facilitate convection. The high geothermal gradient, natural fluid flow, and fluid dynamics are the characteristics of convection-dominated geothermal systems (Moeck 2014). So they host high enthalpy resources at shallow depth, making them most attractive for geothermal developers and investors.

On the other hand, conduction-dominated geothermal systems are marked by the absence of or limited convective flow of fluids and form low to medium enthalpy resources. Such systems are usually located in passive tectonic settings, with the geothermal gradient being average. They are less attractive for the developers and even more so for the investors, as the exploitable reservoirs are deeper (often more than 2 km) than convection-dominated geothermal systems. Moreover, conduction-dominated geothermal systems are hosted in low permeability rocks, viz. compacted sandstones, carbonates, massive granites. They need to create fractures to develop or enhance permeability to facilitate fluid circulation using EGS technology to make such "deep geothermal" resources economical (Moeck 2014). The conduction-dominated geothermal systems could be of magmatic or non-magmatic origin.

Magmatic activities can give rise to both conduction and convection-dominated geothermal systems, depending primarily upon the prevailing permeability conditions and the heat source. It has been reported that conduction-dominated systems hosted in or above igneous bodies are related to high radiogenic heat production, e.g., high heat-producing radioactive elements rich granites (e.g., Singh et al. 2018). Such systems occur in areas with no active volcanism and without or absence of present day tectonic activities. Moreover, the lack of large volumes of natural fluids marks the conduction-dominated magmatic geothermal systems due to low permeability conditions. As a result, such fluid-less systems require EGS technology, i.e., hydraulic fracturing and injection-induced circulation of fluids to transfer heat from depth to surface, for their development. On the other hand, convection-dominated magmatic geothermal systems require a magma chamber as the heat source in tectonically active areas.

In convection-dominated geothermal systems, whether magmatic or non-magmatic, fluids transport heat from the reservoir to the surface. As a result, temperature and volume of the fluids—vapor, brine, or both—that can be extracted from the reservoir, and the depth of the latter determines whether a geothermal resource is economical, as drilling of geothermal wells is expensive and risky, and even more so if the reservoirs are deeper. Moreover, to make a geothermal system sustainable, optimization between production and injection of the remnant thermal fluid is vital, extending a geothermal development project's life. Nonetheless, for the latter, origin of the reservoir fluid and their chemistry, recharge characteristics are also important but are beyond this chapter's scope.

Another aspect related to structural controls is the effect of topography on geothermal systems, the influence of steep topography, particularly highlighted by Hochstein (1988), which causes large volumes of meteoric water recharge convective geothermal systems through high infiltration rates. However, the effect of steep terrain on the hydraulic head is not limited to the convection-dominated geothermal systems in volcanic areas, as envisaged by Hochstein (1988). It applies to the sedimentary basins as well, e.g., Alberta Basin in Canada with a low enthalpy conduction-dominated systems hosted in carbonate and compacted sandstone reservoirs (Bachu 1995; Weides et al. 2012). Thus, steep

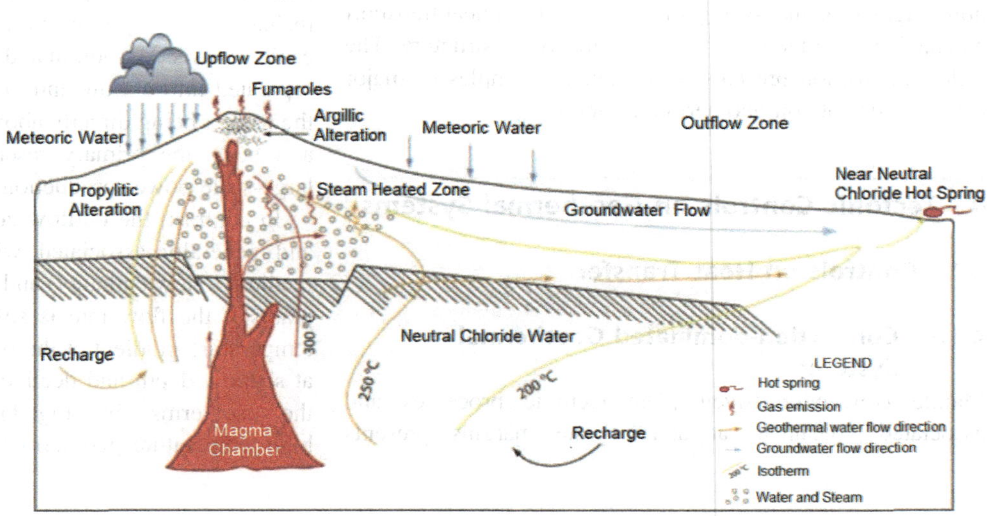

Fig. 1 Magmatic convection-dominated geothermal system (modified from Henley and Ellis 1983)

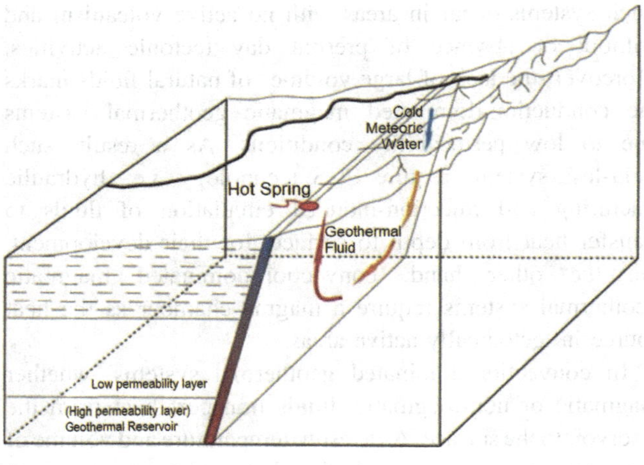

Fig. 2 Non-magmatic convection-dominated geothermal system (modified from Reed 1983)

terrain can affect infiltration in both high and low enthalpy systems.

Most of the geothermal power plant complexes across the world have convection-dominated high enthalpy geothermal systems with shallow magma chambers as the heat source. They are transacted by faults facilitating the rapid circulation of geothermal fluids and recharging the reservoir (Boden 2016). Considering that deep faults and active volcanism are associated with active plate tectonic margins, understanding dynamic tectonic processes at different scales is crucial to characterize convection-dominated high enthalpy geothermal resources. On the other hand, for conduction-dominated low enthalpy geothermal systems, understanding the present-day stress field particularly and the geodynamic evolution process is essential to develop or enhance permeability by creating new fractures. For the reasons and observations described above, modern cataloging of geothermal systems (e.g., Moeck 2014) is based on the plate tectonic setting (Fig. 3), a heat source (magmatic or non-magmatic), and local geologic controls on heat transport mechanism, storage system, and permeability structure. The following section presents some salient examples of major tectonic controls on geothermal systems.

4 Tectonic Controls on Geothermal Systems

4.1 Controls on Heat Transfer

4.1.1 Convection-dominated Geothermal Systems

Mantle convection-driven plate tectonic processes and associated volcanism at active plate margins presents favorable conditions for high enthalpy, convection-dominated geothermal system (Moeck 2014), viz. (i) magmatic arcs above subduction zones at convergent plate margins (e.g., the Andean Volcanic Arc), (ii) divergent margins—intra-oceanic (e.g., Mid-Atlantic Ridge) and intra-continental (e.g., East African Rift, Central Indian Rift), (iii) transform plate margins with strike-slip faults (e.g., the San Andreas) and (iv) intraplate ocean islands formed by hot spot magmatism (e.g., Hawaii). Major fault zones (e.g., Liquiñe Ofqui Fault Zone in Chile, Alam et al. 2010) can act as principal fluid conduits that lead to crustal regions of elevated heat flow. Upwelling asthenosphere and asthenospheric bulge may cause such high heat flow at subduction zones and beneath the rifts, respectively. Rapid exhumation and tectonic denudation in extensional terrains may also lead to increased heat flow, e.g., Rechnitz Window, Eastern Alps (Dunkl et al. 1998).

In convection-dominated geothermal systems, the circulating fluids transport heat from depth to shallow reservoirs or discharge to the surface. The igneous activity (e.g., magma chamber) in volcanic areas, faults in extensional terrains, or both (e.g., intrusive bodies at fault zones) control such systems. The circulating geothermal fluids, originating from infiltration of meteoric water from a high elevation, may receive contributions from the magmatic fluids in such systems (Moeck et al. 2014).

Magmatic Convection-dominated Geothermal Systems
Such systems can be associated with volcanism—(i) basaltic (e.g., divergent plate margins, as in Iceland, Arnórsson et al. 2008), (ii) basaltic to andesitic (e.g., island arcs, as in Java, Pambudi 2018), and (iii) andesitic to dacitic (e.g., convergent margin, as in Chile, Alam et al. 2010)—or plutonism (e.g., continent–continent convergent margins, as in Tuscany, Italy, Bertini et al. 2006). Magma chambers, the heat source for such systems in volcanic areas, control their fluid chemistry, which depends on the parental melt, magmatic recharge, and crystallization in the magma chamber. Volcanic convection-dominated geothermal systems can be separated into upflow and outflow zones (Hochstein 1988), the former being directly above the magma chamber (Fig. 2) and hosts the primary reservoir, which is the target for large-scale power production (Moeck 2014).

In contrast, the outflow zone is distal to the heat source and generally associated with a secondary reservoir (of medium to low enthalpy) and can be utilized for small power plants if the flow rate is sufficient (Hochstein 1988). The temperature gradient at the outflow zone typically increases at shallow depth and declines below the outflow layer (see the isotherms in Fig. 1), thus do not reflect a high-temperature geothermal reservoir directly beneath the

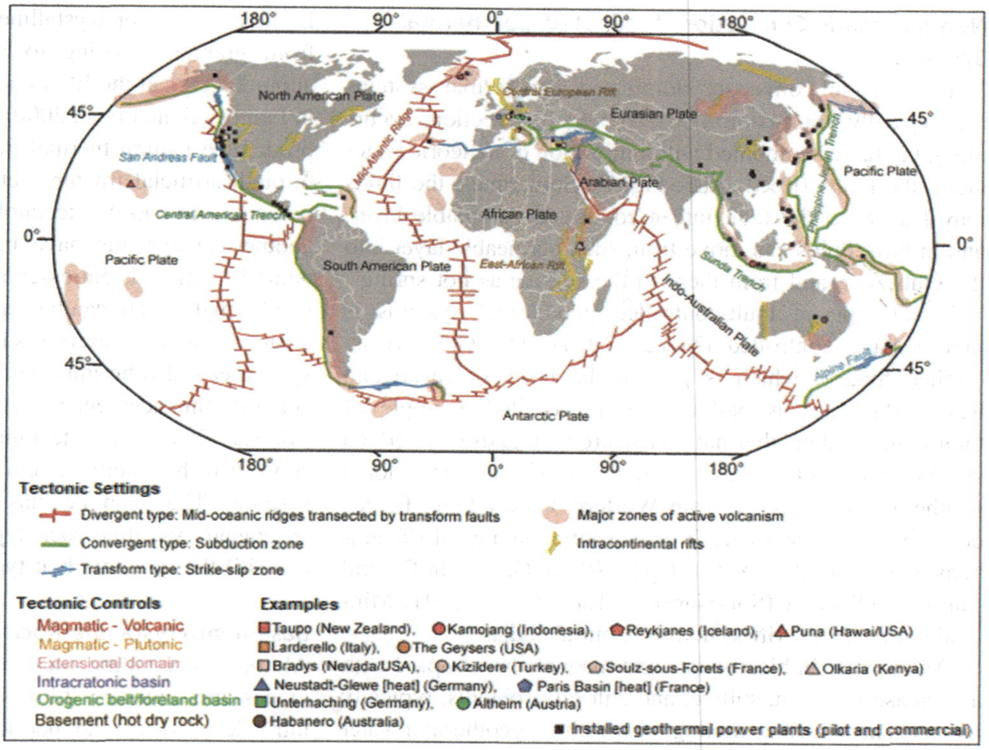

Fig. 3 Tectonic controls on geothermal systems with examples of major fields across the world and location of installed power plants (modified from Moeck 2014)

surface discharges, which are typically springs with or without travertine deposits. On the other hand, geothermal manifestations in the upflow zone are acidic springs with varying degrees of altered rock-forming alteration clays, indicating high-temperature reservoirs immediately underlying them. The upflow zone often has a vapor-dominated part above a liquid-dominated part.

In volcanic areas, the condensate layers in steep terrains can conceal high-temperature reservoirs formed by upwelling acidic fluids that get condensed and neutralized at a shallow depth above the heat source (Fig. 1). The surface discharges in the outflow zone from such condensate layers acquire the condensate layer's cation content. Thus the geochemistry of such discharges are different from that of the original vapor (Schubert et al. 1980; Hochstein 1988). For the formation of condensate layers, a low permeability domain at a depth of the steam-water boundary in vapor-dominated systems is necessary (Schubert et al. 1980).

A geothermal system with a cooling pluton or even an extinct one with radioactive elements as a heat source, on the other hand, can have extensions varying from a few hundred square meters to several square kilometers, depending upon the size of the batholith and other associated intrusive bodies, viz. stocks, dikes, sills, laccoliths, and lopoliths. Such systems are highly dependent on the age and size of emplacement of the intrusive bodies and/or the presence of radioactive elements, e.g., large-scale granitic bodies supplying remnant and radioactive heat to an overlying geothermal system. Geothermal systems associated with plutonism could be related to recent plutonism and extension, as in Laderello, Italy (Minissale 1991), or active volcanism, typically at magmatic arcs along convergent margins, as in Java (Indonesia) or mid-oceanic ridge settings at divergent plate margins, as in Iceland. At Laderello, young (0.3–0.2 Ma) magmatism related to granite intrusions generates a fluid-dominated layer above the granite and a vapor-dominated layer above the former. Moreover, Pliocene extension associated with magmatic rocks' emplacement generates low-angle normal faults that control meteoric water's recharge (Bertini et al. 2006).

On the other hand, at the Geysers, a large felsic pluton is the heat source for a vapor-dominated fluid in a porous meta-sedimentary reservoir capped by a low permeability serpentinite mélange and meta-greywacke (Ingebritsen and Sorey 1988). These low permeability lithologies impede the meteoric recharge of the geothermal system. They have necessitated the injection of treated sewage water to sustain the system and keep the heat recovery at optimum level (Majer and Peterson 2007).

Non-magmatic Convection-dominated Geothermal Systems

Non-magmatic convection-dominated geothermal systems are typically fault-controlled, wherein convection occurs along the faults, combined with infiltration of meteoric water along the faults (Reed 1983; Fig. 1). Sometimes, the fluids may leak from the fault into a concealed permeable layer, and in turn, fluids can move from such permeable layer into the fault zone and from there to the surface as hot springs. This sub-type of fault-controlled geothermal systems is fault-leakage controlled (Moeck 2014). The Great Basin (United States), which is part of the northern Basin and Range Province, is perhaps the most cited example of fault-controlled geothermal systems (e.g., Faulds et al. 2010a 2010b 2010c). Other salient examples of such extensional geothermal systems are from Western Turkey (e.g., Faulds et al. 2009) and tectonically active intracontinental rift grabens, viz. East African Rift, Upper Rhine Graben in Central Europe, SONATA (Son-Narmada-Tapi; Verma 1991; Minissale et al. 2000) rift system in Central India.

An increase in bicarbonate and magnesium, coupled with a decrease in boron, sulfate, and chloride contents, typically indicates near-surface mixing of upwelling geothermal water with groundwater or meteoric water (Flynn and Ghusn 1983; Nicholson 1993). In such systems, the stress regime controls the fluid circulation along the faults. Consequently, stress modeling helps identify the faults that favor the circulation of the geothermal systems in a complex fault system (Faulds et al. 2010b, Moeck et al. 2010), e.g., Bruhn et al. (2010) and Jolie et al. (2010) found the dilational or shear dilation faults the most favorable. Moreover, proper reinjection and maintenance of reservoir pressures are crucial to the management of geothermal field with the presence of fossil geothermal fluids, e.g., at Great Basin (Faulds et al. 2010b), for which a comprehensive study of the fault systems is vital. The reinjection well-sites for a fault-controlled geothermal system must avoid thermal breakthrough of the injected cooled water along permeable faults to the production wells. To ensure this, injection and production wells should not be along the same or interconnected fault(s) in the same fault block.

4.1.2 Conduction-dominated Geothermal Systems

Geothermal systems located at passive plate tectonic settings (e.g., passive continental margins and intracontinental tectonically inactive areas) are mostly conduction-dominated in the absence of asthenospheric anomalies. The conductive settings of sedimentary basins exemplify them. In such systems, a near-normal heat flow heats the deep reservoirs. In the basement or crystalline igneous rocks, heat originates from granites, leading to a significant positive thermal anomaly, e.g., at the EGS reservoir in granitic rock at Soultz in France (Genter et al. 2000). HDR type EGS resources lack producible natural thermal fluids and require fluid injection through artificial fracture networks. Due to the new developments in EGS technologies, conduction-dominated geothermal systems have become quite significant. Thus naturally non-existent essential conditions associated with a geothermal system can be generated (as in HDR systems) by creating fractured reservoirs in crystalline rocks. An existing system can also be improved through reservoir and fracture network enhancement in tight, i.e., low permeability rocks (Moeck 2014). Permeability anisotropy, predominantly controlled by faults and/or lithology, characterizes such systems. They can be classified into three types: (i) the basement/crystalline rock type, (ii) the intracratonic basin, and (iii) the orogenic belt type (Moeck 2014).

Basement/crystalline Rock Type Systems

Although crystalline rocks (e.g., granites) in igneous provinces are potential heat sources themselves, these low intrinsic porosity and permeability rocks require reservoir development, i.e., enhancing permeability through stimulation techniques. This measure facilitates circulation between the injector and producer wells in HDR systems, wherein the rock mass acts as the heat exchanger (Moeck 2014). The main challenge of EGS development in crystalline rocks is creating an augmented permeability structure between the two wells, which is overcome by a thorough understanding of the stress field, e.g., the magnitude of the intermediate principal stress, as it controls the in situ stress regime. Further considerations include geomechanical parameters and failure models of the reservoir rocks under stimulation conditions (discussed in Sect. 5).

Intracratonic Basins and Orogenic Belts Type Systems

The conduction-dominated geothermal systems may occur in different geologic settings where there is no active igneous activity. In such cases, the tectonic activity within the geothermal system is low, feeble or absent. Such systems could be (i) within intracratonic basins, (ii) orogenic belts, and associated foreland basins. In the first setting, a near-normal heat flow heats deep aquifers in the sedimentary basins at great depth (>3 km, e.g., Québec, Eastern Canada, Majorowicz and Minea 2015). In the second setting, advective heat transport plays a key role. High permeability domains and deep-rooted faults allow deep circulation of meteoric water, often associated with the subsequent formation of hot springs (e.g., Manikaran in the Himalayas,

Alam et al. 2004; Chandrasekharam et al. 2005). Two basin types hosting geothermal systems (Moeck 2014) are (i) extensional or lithospheric subsidence basins, viz. Central European Basin (e.g., Scheck-Wenderoth et al. 2014), and (ii) foreland basins within orogenic belts, viz. the Molasse Basin of the Alps (e.g., Chelle-Michou et al. 2017), Western Canada Sedimentary Basin associated with the Rocky Mountains (Higley et al. 2005).

4.2 Major Tectonic Settings for Geothermal Systems

The tectonic setting of geothermal systems firmly controls their thermal regime and chemistry. Primary tectonic settings and associated geothermal systems are described here.

4.2.1 Divergent Boundaries

Midoceanic Ridges

Such boundaries have exceptionally high volcanism rates along the ridge or plate boundary because seafloor-spreading concurs with a geologic hot spot or mantle plume in this case. The examples include Iceland, which lies across the Mid-Atlantic Ridge (MAR), a divergent boundary marking the eastern edge of the North American tectonic plate and the western edge of the Eurasian plate (Arnórsson 1995a, b, Arnórsson et al. 2008; Boden 2016).

Continental Rifts

Continental rifting marks significant crustal extension, which in due course of geologic time gives rise to a new ocean basin, e.g., the formation of the Red Sea about 25 million years since the rifting started about 30 million years ago (Boden 2016). Two main types of continental rifts can be associated with the geothermal systems, with (e.g., Olkaria, Menengai, Longonot, and Eburru in the East African Rift) or without (e.g., Dixie Valley in the Basin and Range Province of the USA, Tattapani, Bakreswar, Tantloi, Surajkund, Rajgir, Munger in SONATA geothermal province in Central India) magmatic heat source.

4.2.2 Convergent Boundaries—Continental and Island Volcanic Arcs

The subduction of the oceanic lithosphere leads to partial melting of the overlying mantle rocks. The volatiles released from the former aid in lowering the melting point of the latter. The generated magma rises to the upper crust and produces volcanoes. The subduction of the oceanic lithosphere beneath the continental lithosphere gives rise to continental volcanic arcs (e.g., Andes, Ramos 2009). Relatively older and colder oceanic lithosphere subducting beneath younger and warmer oceanic lithosphere gives rise to the island arcs with volcanic island chains (e.g., Japan, Philippines, Marianas, Condie 2011, 2016). Shallow magma chambers in continental and island volcanic arcs serve as heat sources for the overlying geothermal systems; e.g., producing geothermal fields at Los Azufres and Los Humeros in Mexico (Martinez 2013; Elders et al. 2014), Miravalles in Costa Rica (Ruiz 2013), San Jacinto Tizate in Nicaragua (Chin et al. 2013), and Berlin and Ahuachapan in El Salvador (Herrera et al. 2010). The strain in volcanic arcs may vary from compressional to extensional in a direct (head-on) convergence; e.g., subduction zone rollback at Taupo Volcanic Zone (Seebeck et al. 2014; Villamor et al. 2017). However, it may be transtensional or transpressional with oblique convergence; e.g., Andean Volcanic Zone (Dewey and Lamb 1992; Dewey et al. 1998; Cembrano and Lara 2009; Sielfeld et al. 2019).

Although producing geothermal systems occur in both compressional and extensional strain conditions in the volcanic arcs, Wilmarth and Stimac (2014 2015) found that those associated with the arcs having complex structural settings induced by either oblique convergence, involving transtension in particular (e.g., Salak, Indonesia; Aprilina et al. 2015) or intra-arc rift-related extension (e.g., Wairakei, New Zealand; Villamor et al. 2017) are more favorable from a geothermal development point of view (Boden 2016). Where convergence is oblique, strike-slip faults can form in the overlying plate; e.g., Liquiñe Ofqui Fault Zone in south-central Chile (Alam et al. 2010). In areas of fault step-overs, zones of transtension can occur, forming possible pull-apart basins that can foster crustal dilation (improved permeability for convection of geothermal fluids) and the rise of magma into the upper crust forming heat source (Boden 2016). Such high-level intrusions of magma can also thermally weaken overlying rocks, leading to gravitational collapse and dilation, generally orthogonal to the direction of plate convergence (Downs et al. 2014; Holden et al. 2015; Boden 2016). As a result, a series of extensional basins or grabens can form that, in association with heat from volcanism, can help create highly productive geothermal systems, viz. Los Humeros and Los Azufres in the Trans-Mexican Volcanic Belt, Miravalles in Costa Rica (e.g., González-Partida 2005; Bernard-Romero et al. 2010; Bernard et al. 2011).

The producing geothermal field at Miravalles (DiPippo 2012), an excellent example of a geothermal system developed in a local zone of extension (transtension) in a continental volcanic arc, is located within an N-NW trending graben on the southwest flank of the Quaternary Miravalles stratovolcano (Chavarría-Rojas 2003; Chavarría-Rojas 2003; Boden 2016). This graben may be related to possible left steps in W-NW striking left-lateral faults (transtension) arising from the oblique (left-lateral) convergence between the downgoing Cocos plate and overriding Caribbean plate (e.g., DeMets 2001; Symithe et al. 2015; Rosas et al. 2016; Boden 2016). The graben's bounding and internal faults have produced secondary fracture permeability of the volcanic rocks, facilitating convection of geothermal fluids and developing this field as a significant power producer (Boden 2016).

A highly productive Hatchobaru–Otake geothermal field (e.g., Taguchi and Nakamura 1991; Taguchi 2001), located on the flanks of an active Mt. Kuju volcano, lies within the arc-parallel, E-NE trending Beppu–Shimbara graben (Ehara 1989). This graben transects the island, and its N-NW-directed extension reflects slab rollback of the subducting Philippine oceanic plate (Ehara 1989). Moreover, both northeast-striking graben-parallel faults and northwest-striking normal faults control the flow of geothermal fluids at Hatchobaru (Momita et al. 2000). The northwest-striking normal faults may be a consequence of strike-slip motion on the northeast-striking graben-parallel faults, resulting in local NE-directed extension in areas where northeast faults stepover (Momita et al. 2000).

4.2.3 Convergent Boundaries—Back-Arc or Intra-Arc Extension

In this setting, the extension is oriented perpendicular to the arc, resulting in elongated grabens that run parallel for a significant part of the arc (Feuillet et al. 2002; Sdrolias and Müller 2006; Arai et al. 2018; Magni 2019). Such dilation is more extensive than the localized extension related to transtension in the case of oblique convergence, as discussed in the previous section. The development of back-arc or intra-arc spreading is more common in island arcs than continental arcs (Sdrolias and Müller 2006). Back-arc or intra-arc spreading (Feuillet et al. 2002; Sdrolias and Müller 2006; Condie 2011, 2016; Boutelier and Cruden 2013; Nakakuki and Mura 2013; Arai et al. 2018; Magni 2019) are significant because the extensional forces promote secondary rock permeability and crustal dilation (Boden 2016). This further aids intrusion of magma to high crustal levels that can serve as a heat source and development of overlying convecting and potentially exploitable geothermal systems (Boden 2016); viz. rift-related systems at Wairakei and Rotokawa in the Taupo Volcanic Zone, New Zealand (Seebeck et al. 2014; Villamor et al. 2017).

4.2.4 Convergent Boundaries—Continental Convergence

Due to the continental convergence, viz. at the Himalayas (Kious and Tilling 1996), the crust is over-thickened, viz. Tibetan plateau, which lowers the geothermal gradient (Vanderhaeghe et al. 2003; Beaumont et al. 2006; Shi et al. 2017). Moreover, the resultant compressional stress decreases the fractures and faults' dilation, reducing permeability (Rogers 2003). Furthermore, in the absence of subduction, as the continental plates merely colloid, there is no magma generation as in the case of convergent continental and island arcs, so a magmatic heat source is absent (Fucheng et al. 2018). However, high temperature and pressure metamorphism due to continental collision can cause partial melting of rocks, resulting in magma forming a heat source to drive a geothermal system (Bea 2012).

Moreover, buried radiogenic granitic and high-grade metamorphic rocks (Bea 2012) can also be the source of heating up of deeply circulating meteoric water. Geothermal systems in such a setting could be associated with localized extension zones within an otherwise compressional stress regime due to fault-valve action (Sibson 2020). In the absence of subduction, the crust is thickened during the continental collision, and extensional forces are orthogonal to the main compression direction (Mo et al. 2007). This occurs because the uplifted and thickened crust commences to collapse or buckle under its own weight. Depending on the rock type and strength, normal faults can develop in localized extensional zones and bind the horsts and grabens perpendicular to the central mountain range (Brun 2002). Such extensional normal faults and associated dilation are responsible for permeability in this otherwise tight zone. This helps the circulation of meteoric water circulate at great depths. This water is heated and then returns to the surface due to buoyancy through other normal faults to form hot springs and shallow geothermal reservoirs, e.g., Yangbajing and Yangyi geothermal fields, both in Tibet (Boden 2016).

4.2.5 Transform Boundaries

Transform boundaries are typically found in the ocean floor associated with mid-ocean ridge spreading zones. However, some well-known ones on the continent, e.g., San Andreas Fault extending from northwestern Baja Mexico to northwestern, California Anatolian Fault of northern Turkey on land, the Alpine Fault on the South Island of New Zealand (Boden 2016). Geothermal systems associated with transform boundaries are generally limited. However, some critical exceptions occur when extension with or without magmatism occurs together with the transform motion (Boden 2016).

For example, there is a zone of discontinuous north to northwest-striking right-lateral faults constituting the Eastern California Shear Zone and Walker Lane of western Nevada within the San Andreas Fault (Faulds et al. 2005a, b). Thus, the southern part of the San Andreas Fault (Fig. 4) has two favorable conditions for hosting geothermal systems, evident by geothermal fields of Cerro Prieto (Mexico), the Salton Sea (USA), and the Imperial Valley (USA). The northeast-striking normal faults, developed due to the local transtensional regime, provide conduits for geothermal fluids

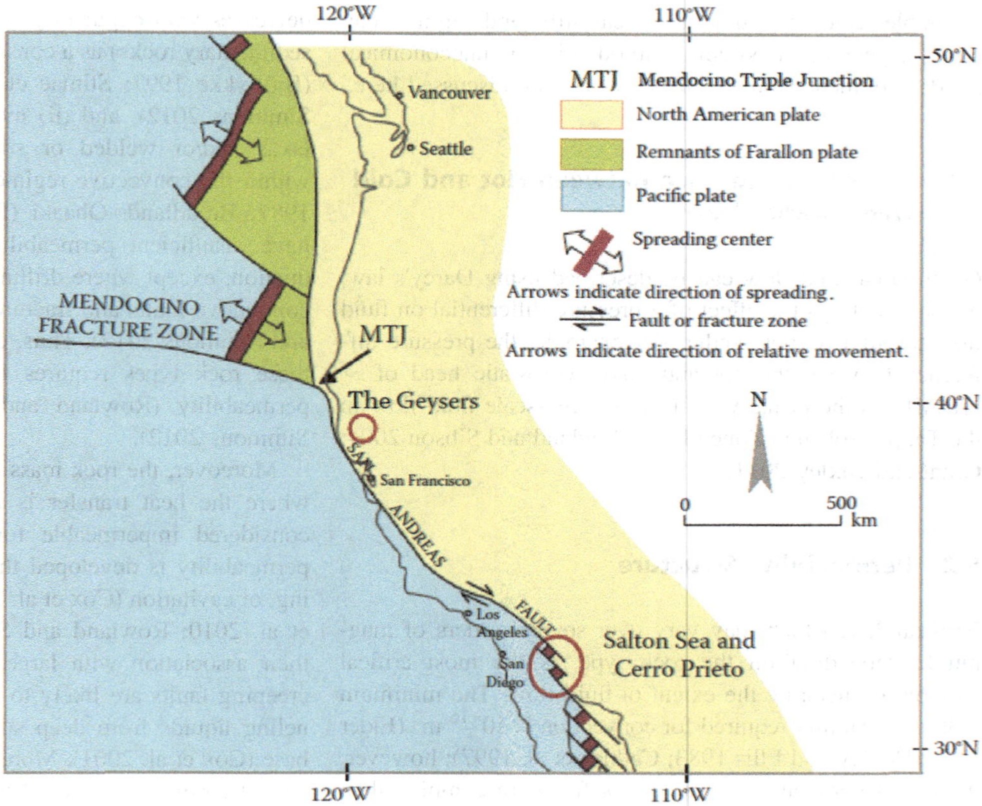

Fig. 4 Tectonic setting of the San Andreas Fault with the location of important geothermal systems located on its either end (modified from Boden 2016)

circulation (Bennett 2011). Another favorable condition is the presence of shallow magmatic heat sources (Bennett 2011). This is because of the buried spreading ridge segments and thinning of the crust due to the extension leading to the lowering of pressure, which induces the melting of the heated rocks (Bennett 2011).

Although the source of heat at The Geysers, located near the north end of the San Andreas Fault (Fig. 4), is magmatic, unrelated to spreading or extension, as is the case at the southern end of the San Andreas Fault. Instead, it is related to the transition from a previous subduction margin of the now-extinct Farallon plate to the current transform margin (Boden 2016).

4.2.6 Hot Spots

Upwelling of relatively stationary mantle plumes constitutes hot spots, where the ascent of mantle material and lowering of pressure together induce partial melting and subsequent rise of magma to form volcanic centers. Then, the tectonic plate's movement over a mantle plume gives rise to a series of volcanic centers that may lead to a series of volcanic islands, e.g., the Hawaiian Islands within the Pacific tectonic plate (e.g., White 2016). Since the Big Island of Hawaii currently lies over the plume, it hosts five active volcanoes. Kilauea is the most active one in Hawaii for being directly above the plume, close to which Puna geothermal field is located (Boden 2016).

4.2.7 Stable Cratons

Located away from plate boundaries within the continents, they are geologically stable, as evident from the absence or rare occurrence of seismic events or volcanism. They include: (i) deep sedimentary basins often with producing oil and gas fields (e.g., Teapot Dome oil field in Wyoming, Curry 1977; Williston Basin in North Dakota, Drake II et al. 2017), (ii) deep (3 to 5 km), old but still hot, granites due to radiogenic decay of U and Th (e.g., Cooper Basin in Australia, Holl 2015; Peninsular India, Singh et al. 2014). Although there is a general lack of interconnected fracture systems in such a setting, they are potential areas for developing HDR systems due to enough heat (usually > 200 °C).

5 Structural and Other Local Factors Controlling Geothermal Systems

Apart from the tectonic factors described in previous sections, local factors also play an essential role in defining geothermal systems. Sometimes, despite being located in a favorable tectonic setting, the local structural regime may make a geothermal system unproductive or uneconomical for development. Some of these factors are discussed here.

5.1 Pressure Difference between Hot and Cold Hydrostatic Head

Geothermal fluid flow can be described using Darcy's law, which quantifies the effect of a pressure differential on fluid flow through porous media. For example, the pressure difference between the hot and cold hydrostatic head of ~ 10 MPa is the primary control on large-scale fluid flow in the Taupo Volcanic Zone (TVZ, Rowland and Sibson 2004; Grant and Bixley 2011).

5.2 Permeability Structure

Permeability, which may vary over several orders of magnitude depending on the rock type, is the most critical variable in dictating the extent of fluid flow. The minimum bulk permeability required for convection is 10^{-16} m^2 (Elder 1981; Henley and Ellis 1983; Cathles et al. 1997); however, the actual permeability structure is far more complex than estimated bulk permeability (Rowland and Simmons 2012). Moreover, the total rock mass directly involved with the transmission of fluid is minimal compared to the bulk volume of rock which hosts the flow network, wherein high-flux conduits occur under exceptional situations (Donaldson and Grant 1981; Elder 1981; Donaldson 1982; Grant et al. 1982; Hanano 2004).

5.3 Crustal Heterogeneity and Anisotropy

The heterogeneous assemblage of lithologic units results in significant lateral and vertical variations in permeability. For example, the stratified Quaternary volcanic sequence approximating a layered medium induces a strong contrast between layer parallel and layer perpendicular permeability at TVZ (Manning and Ingebritsen 1999; Rowland and Simmons 2012). The latter is controlled by the low-permeability layers (Manning and Ingebritsen 1999; Rowland and Simmons 2012). The vertical to horizontal permeability ratio is estimated as 1:40 for the volcanic stratigraphy at 3 km depth based on numerical reservoir modeling (Mannington et al. 2004; Rowland and Simmons 2012). Despite the abundance of granular layers conducive to diffuse flow within this Quaternary volcanic sequence at TVZ, macroscopic faults and fractures must comprise an essential component of crustal-scale permeability for two reasons: (i) intergranular porosity, and thus permeability, decreases with depth in granular materials (pyroclastic and sedimentary rocks) as a consequence of diagenetic processes (Björlykke 1997; Stimac et al. 2004, 2008; Rowland and Simmons 2012), and (ii) metasedimentary rocks, andesitic lavas and/or welded or silicified ignimbrites, if present within the convective regime (e.g., TVZ, Stern and Davey 1987; Broadlands-Ohaaki (Rowland and Simmons 2012), have insufficient permeability to sustain geothermal production, except where drilling has intercepted hydraulically conductive faults and fractures (Wood et al. 2001; Rowland and Simmons 2012). Thus, convective flow through any of these rock types requires fault and/or fracture-controlled permeability (Rowland and Sibson 2004; Rowland and Simmons 2012).

Moreover, the rock mass near the magmatic heat source, where the heat transfer is mostly through conduction, is considered impermeable for advective fluid flow unless permeability is developed through fracturing, microfracturing, or cavitation (Cox et al. 2001; Cox 2005; Micklethwaite et al. 2010; Rowland and Simmons 2012). In addition to their association with large earthquakes, shear zones and creeping faults are likely to be an essential means of channeling liquids from deep sources to the seismogenic zone base (Cox et al. 2001). Moreover, seismic events may cause episodic rupturing of the brittle-ductile transition processes near the magmatic source at the bottom of high-temperature convection cells, allowing magmatic fluids' entrainment into the meteoric convection regime (Rowland and Simmons 2012).

5.4 Hydrothermal Alteration and Mineral Deposition

Permeability, porosity, and rock strength that control mineral dissolution, transformation, and precipitation are continually modified as a function of time and space (e.g., Browne and Ellis 1970; Hedenquist and Browne 1989; Simmons and Browne 2000; Rowland and Simmons 2012). In a particularly evolved (also referred as long-lived, > 50Ky) geothermal systems, three types of alteration effects have been be recognized: (i) clay-rich alteration that forms in shallow steam-heated aquifers and on the periphery of the upflow zone reduces permeability as well as rock strength by increasing the proportion of clay minerals that replace volcanic glass and feldspars (e.g., Hedenquist and Browne 1989; Hedenquist 1990; Simmons and Browne 2000; Rowland and Simmons 2012), (ii) silicification and K-metasomatism due to deposition of quartz and adularia from rising and cooling chloride waters reduce porosity and permeability but increase rock strength (e.g., Ohakuri; Henneberger and Browne 1988), which may in turn enhance the development of fault-fracture–related permeability (e.g.,

Broadlands-Ohaaki, Simmons and Browne 2000; Rowland and Simmons 2012), and (iii) mineral deposition can line and therefore isolate high-permeability pathways from incursion of fluid from the surrounding country rock (Rowland and Simmons 2012). Thus, in the upflow zone, fracture permeability becomes increasingly important with the evolution of a geothermal system. Intergranular permeability is gradually decreased by pervasive silicification and mineral deposition, reflected in pore fluid pressure fluctuations in producing geothermal fields (Rowland and Simmons 2012).

5.5 Brittle Deformation and Conditions for the Development of High-Flux Fluid Conduits

The fluid flow in a geothermal system depends on fractures' distribution—fracture network—within the upper crust. However, characterizing fracture networks' three-dimensional geometry and defining a quantifiable connection between geometrical and hydraulic connectivity to understand fractured media flow is quite complex (Berkowitz 2002; Rowland and Simmons 2012). To deal with this complexity, Rowland and Simmons (2012) suggested a qualitative approach for understanding the role of various factors essential for developing high-flux conduits. This approach considers the macroscopic mode of brittle failure, lithology, stratigraphy, the seismic cycle, and the degree of sealing through hydrothermal cementation (Sibson 2000). Brittle structures rarely form in isolation, and their cumulative hydrologic effect additionally exerts an essential control on permeability heterogeneity and anisotropy (Rowland and Sibson 2004). Three macroscopic modes of brittle failure (Fig. 5, Table 1; Sibson 1998, 2004, 2020) are possible: shear failure (faulting, i.e., displacement parallel to the fracture surface), extensional failure (generation of dilational fractures perpendicular to the least principal stress, $\sigma3$), and hybrid extensional-shear failure (involving components of shear and dilation). The mode of brittle failure depends upon pore fluid pressure, P_f, differential stress defined by the difference between the greatest and least principal stresses ($\sigma_1 - \sigma_3$), and tensile strength, T, of the deforming rock volume, modulated by hydrothermal alteration and mineral deposition (Rowland and Simmons 2012). In tectonically active regimes (Fig. 6) that promote fluid flow, P_f, ($\sigma_1 - \sigma_3$) and T vary temporally and spatially. The effect on a failure mode is illustrated in the pore fluid factor (Fig. 7) and differential stress space (Fig. 8), where the pore fluid factor, λ_v, is the ratio between fluid pressure and overburden σ_v (Cox 2010).

5.6 Permeability in Fault Zones

Normal faulting with a minor strike-slip component is a favored brittle failure mode that exerts the principal structural influence on fluid redistribution, as in TVZ, where faulting and subsidence of the graywacke basement played a prominent role in controlling the structural development in the cover sequence (Rowland and Sibson 2001; Acocella et al. 2003; Rowland and Simmons 2012).

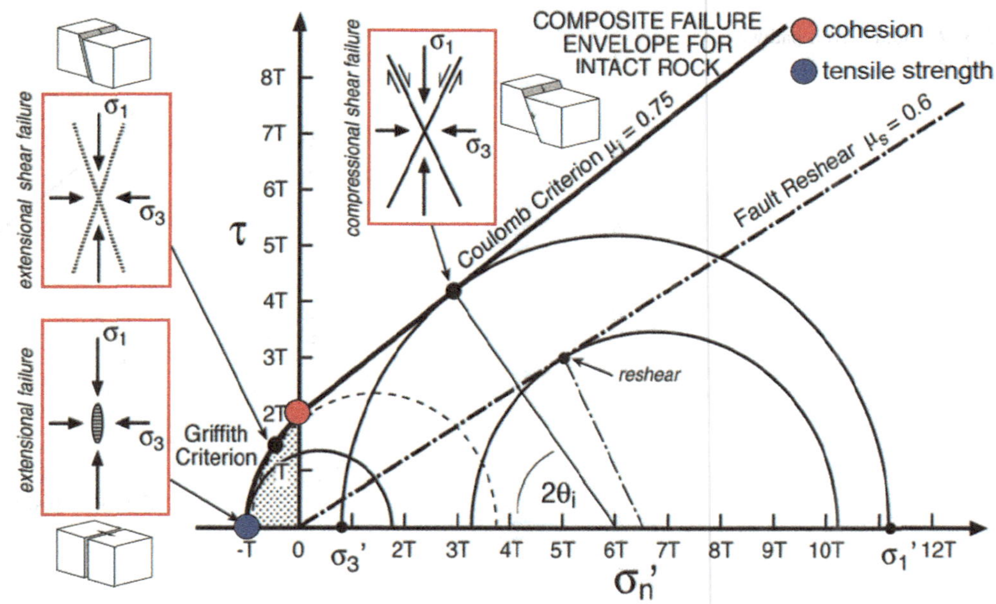

Fig. 5 Three macroscopic modes of brittle failure (modified from Sibson 2004)

Table 1 Relevant failure criteria, assumptions, and limitations in pore fluid factor-differential stress space (Sibson 1998; Cox 2010; Rowland and Simmons 2012; Ferrill et al. 2020; also see Figs. 4, 5, 6, 7)

Macroscopic mode of brittle failure	Failure criteria	Assumptions	Limitations
Brittle shear failure on optimally oriented faults in intact rock	Coulomb criterion $\tau = C + \mu(\sigma_n - P_f)$, where τ = shear stress, C = cohesive strength, μ = coefficient of friction in an isotropic rock mass, σ_n = normal stress, P_f = pore fluid pressure; for a fault inclined θ_{opt} to σ_1 $\lambda_v = [4C - \sigma_1 + 4\sigma_3]/3\sigma_v$, where σ_v = overburden, pore fluid factor $\lambda_v = P_f/\sigma_v$ and $\theta_{opt} = ½ \tan^{-1}\mu^{-1}$	Plane strain (σ_2 lies in the fault plane) $\sigma_v = \sigma_1$ assuming an Andersonian extensional stress regime $\mu = 0.75$ C ~ 2 T	$(\sigma_1 - \sigma_3) \geq 5.66\,T$
Extension failure	$\lambda_v = (\sigma_3 + T)/\sigma_v$		$(\sigma_1 - \sigma_3) < 4\,T$, where T = tensile strength $P_f = \sigma_3 + T$
Hybrid extensional-shear failure	Griffith criterion $(\sigma_1 - \sigma_3)2 = 8\,T(\sigma'_1 + \sigma'_3)$, where $\sigma'_1 = (\sigma_1 - P_f)$ and $\sigma'_3 = (\sigma_3 - P_f)$ $\lambda_v = [8\,T(\sigma_1 + \sigma_3) - (\sigma_1 - \sigma_3)^2]/16T\sigma_v$		$4\,T < (\sigma_1 - \sigma_3) \leq 5.66\,T$

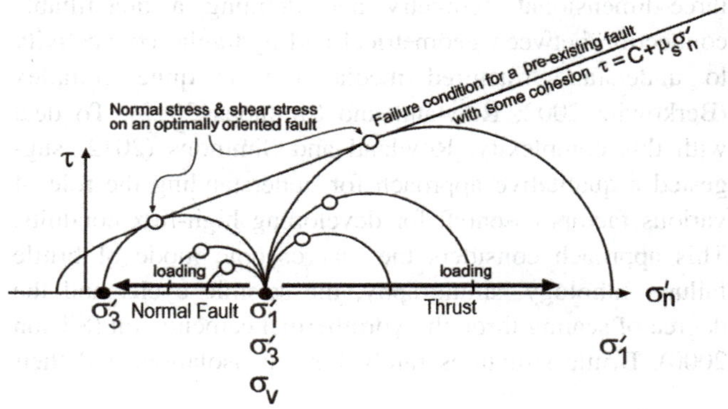

Fig. 6 Tectonically driven pathways to failure

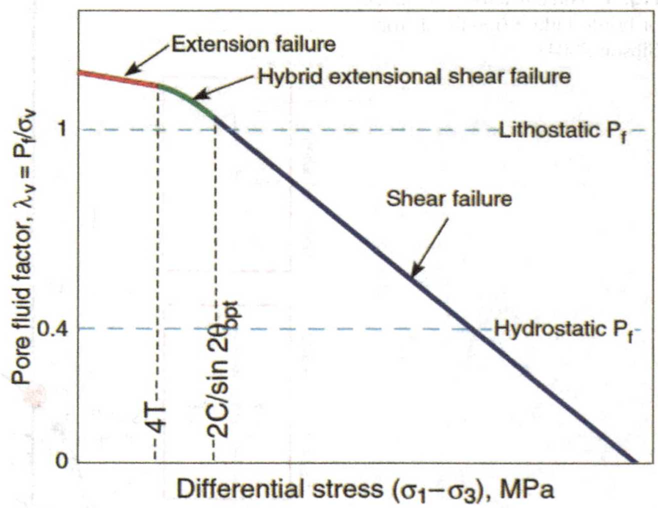

Fig. 7 Fluid driven pathways to failure

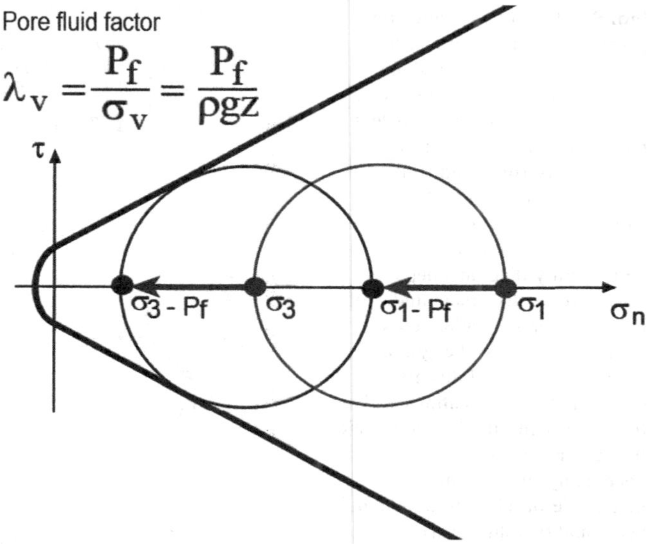

Fig. 8 Generic failure curve at some depth in the crust (modified from Cox 2010)

5.7 Fault Zone Complexity, Rift Architecture, and Directional Permeability

The permeability distribution within fault zones is affected by fault growth and interaction (Curewitz and Karson 1997) and structural overprinting (e.g., Berger et al. 2003). Geothermal fluid flows through interconnected fault-fracture networks within spatially more extensive fault and/or fracture systems (Rowland and Simmons 2012). Fault irregularities in the direction of slip can cause the development of highly permeable zones due to dilation within extensional stepovers and jogs (Sibson 2001). In extensional settings, jogs direct the fluid flow along strike (parallel to σ_2), while in strike-slip settings, jogs form subvertical pipes through which fluid flows (Sibson 2000). Moreover, such piped zones of enhanced permeability can also be developed due to normal fault growth and linkage, leading to dilated subvertical zones in some extensional settings (e.g., Nortje et al. 2006). Thus fault growth and linkage generate subvertical zones of enhanced permeability between normal fault segments, transfer fault intersections with rift faults, and lateral fault tips (Curewitz and Karson 1997; Micklethwaite 2009; Rowland and Simmons 2012; Fig. 9).

Additionally, rift architecture may lead to directional permeability. Moreover, extension in an area may cause the formation of segmented blocks with subparallel arrays of normal faults, which might host dyke swarms in the presence of magma (e.g., Dabbahu Rift in Afar, Rowland et al. 2007). In general, segmentation scales with the thickness of the mechanical layer that is breaking (Ebinger et al. 1999). The displacement between resulting segments must be accommodated, i.e., it should be coupled or linked with another movement. Accordingly, these segments are either "hard-linked" (Gibbs 1984) with transfer faults oriented at a high angle to the axis of rifting or "soft-linkage" (Rosendahl et al. 1986; Morley et al. 1990) with distributed deformation and small-scale faulting within the blocks between adjacent segments. The bulk permeability is a function of the cumulative effect of subparallel faults and fractures in the former and structurally favorable sites for enhanced vertical permeability in the latter (Rowland and Simmons 2012). Its value in rift segments contrasts with that in accommodation zones (Rowland and Sibson 2004). An array of faults and fractures has the same effect on bulk permeability as stratigraphic layering: permeability across the strike of the array is lower relative to other directions, regardless of whether faults behave as conduits or baffles to flow (Rowland and Simmons 2012). When superimposed upon a layered sequence, the combined effect reduces vertical and across-strike permeability relative to along-strike permeability (Fig. 9). In contrast, all favorable structural sites for focused vertical flow occur within accommodation zones (e.g., rift fault-transfer fault intersections, lateral fault tips on first-order structures, and linkage zones between first-order structures, Fig. 9). Thus, rift architecture may modulate fluid-flow paths such that upflow zones are favored in accommodation zones, and recharge and axial flow are selected in rift segments (Rowland and Sibson 2004).

The tendency of geothermal fields to occur around the basin's margin (e.g., Broadlands-Ohaaki, Rotokawa, Waiotapu, Te Kopia, Orakeikorako) in accommodation zones may produce basement highs (Wairakei-Tauhara, Ngatamariki), which suggests deep-seated control by inherited basement faults. These structures are of particular importance, because they are likely rooted in potentially permeable shear zones within the ductile lower crust, thus channeling geothermal fluids across the brittle-ductile transition zone (Cox et al. 2001).

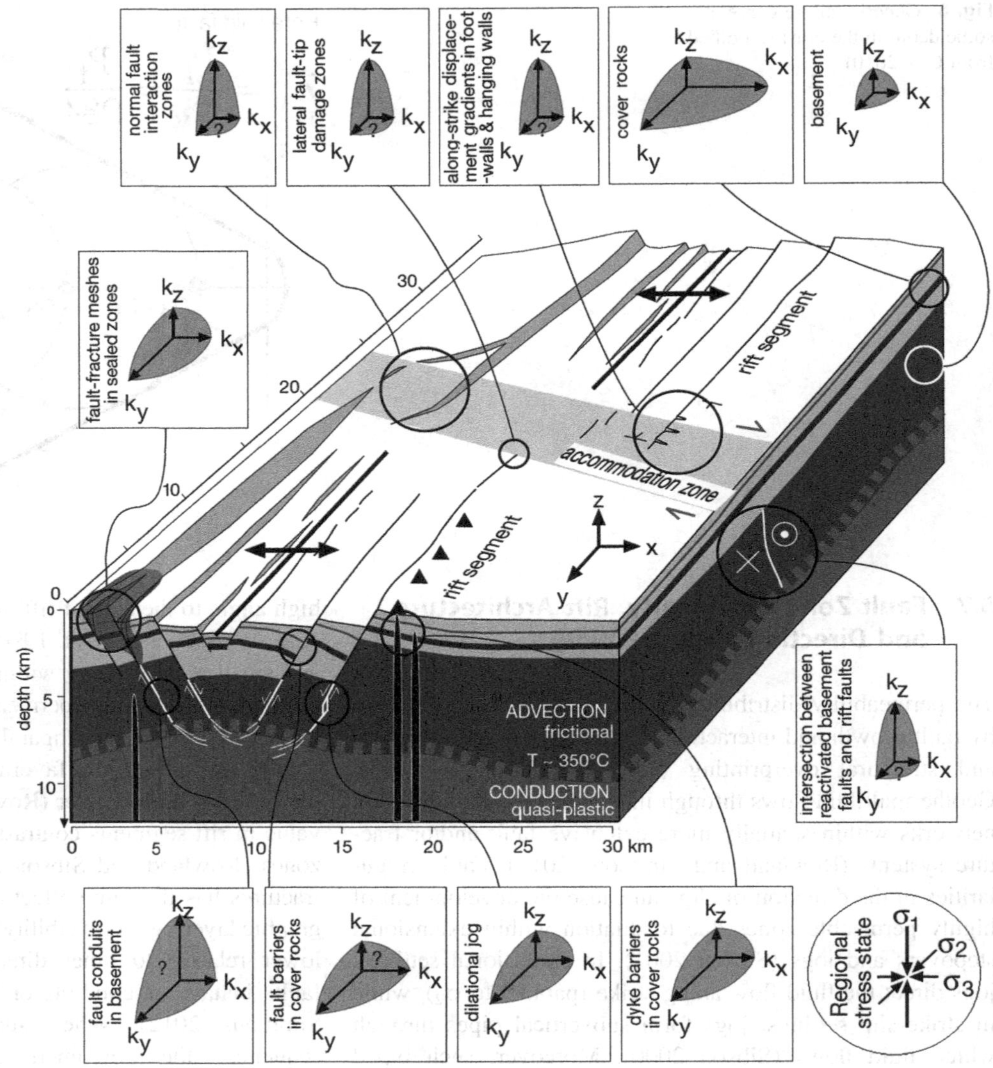

Fig. 9 Effect of structures on permeability anisotropy in the central Taupo Volcanic Zone, assuming a simple layered cover sequence overlying a competent basement with low intrinsic permeability (modified from Rowland and Sibson 2004, Rowland and Simmons 2012) [Notes: (i) Three-dimensional permeability diagrams depict indicative relative magnitudes of mutually perpendicular across strike (x), along strike (y), and vertical (z) permeability for different structural settings within the rift system. (ii) Question mark between arrows indicates uncertainty in the relative magnitude of adjacent directional permeability values. (iii) Settings with enhanced and localized vertical permeability are favored in accommodation zones.]

Thus, the main factors constraining fluid flow are (i) magmatic intrusion, which mainly supplies heat and energy, but which can also facilitate fracture extension at the tip of a dike; (ii) proximity to the brittle-ductile transition, which limits the downward flow of water; (iii) the tensile strength of host rocks which is generally high; and (iv) hydrostatic fluid pressure, which is maintained because brittle failure relieves the build-up of fluid overpressures.

Lastly, it is essential to distinguish between paleostress and recent stress since an area's overall structuration is primarily controlled by paleostresses. The overprint of local and regional current stresses can modify the permeability of the existing fracture system. For example, fractures aligned along the SHmax are more likely to keep themselves "open," and the ones perpendicular to it would most likely be "closed." This underscores the importance of dealing with tectonic and structural controls together for the geothermal systems, not just tectonic or structural controls alone or separately.

6 Conclusions

Tectonic and structural settings control the inherent variability in the nature and evolution of a geothermal system, both of them being dynamic. A combination of these in time and space being unique makes each system unique. Apart from controlling hydrological conditions, viz. permeability structure governs the fluid—both recharging cold fluid and hot geothermal fluid—flow, pressure head difference between cold and hot water flow that makes a system sustainable establish whether the thermal regime is conductive or convective. Despite being complex and difficult to predict, the fluid-flow paths in geothermal systems are quite self-regulating, as evidenced by long-lived (>50,000 years) systems worldwide. This is particularly true for the systems under natural conditions with recharge and discharges in harmony and the natural stimulation of fluid conduits through seismic events, creating new fractures and reopening fractures clogged by hydrothermal deposits.

In the case of geothermal fields under production, besides optimizing the geothermal discharge from wells, proper care should be taken for ensuring the continuance of the permeability structure through geophysical studies, viz. microseismic, to make the geothermal systems sustainable.

The conditions for the development of permeability structure that facilitates geothermal fluid flow and circulation, viz. dilation of the conduits in the fault zones, accommodation zones that transfer extension between rift segments, can be summarized as follows. Geothermal fluid flow is primarily controlled by (i) heat source—magmatic or non-magmatic—that drives convective circulation, (ii) intergranular porosity and permeability, (iii) permeability due to fault-fracture network produced by tectonism or magmatism (volcanism and/or plutonism), (iv) conduits of volcanic and hydrothermal eruptions, and (v) hydrothermal alteration and mineral deposition causing porosity and permeability heterogeneity in the geothermal reservoir and conduits.

Acknowledgements The author would like to express his gratitude to the two anonymous reviewers for their insightful observations and suggestions reflecting a careful and meticulous reading of the manuscript that has helped tremendously improve the manuscript's original version. The author appreciates the efficient editorial handling of the reviewing process. Finally, the author would like to thank the Universidad de Atacama for the necessary facilities to carry out this work and funding through DIUDA Regular Project No. 22401.

References

Acocella V, Spinks K, Cole J, Nicol A (2003) Oblique back arc rifting of Taupo Volcanic Zone. New Zealand. Tectonics 22(4):1045. https://doi.org/10.1029/2002TC001447

Alam MA, Chandrasekharam D, Minissale A (2004) Geothermal potential of thermal waters of Manikaran, Himachal Pradesh, India. In: Wanty RB, Seal II RR Jr (eds) Water-Rock Interaction. Proceedings of the 11th International Symposium on Water-Rock Interaction WRI-11, Saratoga Springs, NY, USA, 27 June – 2 July 2004. AA Balkema Publishers, Taylor & Francis Group, London, p 327–331

Alam MA, Sánchez P, Parada MA (2010) Interplay of volcanism and structural control in defining the geothermal system(s) along the Liquiñe-Ofqui Fault Zone, in the south-central Chile. GRC Transactions, vol 34. Geothermal Resources Council, Davis, California, p 747–750. https://publications.mygeoenergynow.org/grc/1028732.pdf. Accessed 17 December 2020

Aprilina NV, Satya DY, Rejeki S et al (2015) Geologic modeling workflow for volcanic hosted geothermal reservoirs: Case study from Salak Geothermal Field. Proceedings World Geothermal Congress 2015, Melbourne, Australia, 19–25 April 2015, Paper Number: 12029. https://pangea.stanford.edu/ERE/db/WGC/papers/WGC/2015/12029.pdf. Accessed 17 December 2020

Arai R, Kodaira S, Takahashi T et al (2018) Seismic evidence for arc segmentation, active magmatic intrusions and syn-rift fault system in the northern Ryukyu volcanic arc. Earth Planets Space 70:61. https://doi.org/10.1186/s40623-018-0830-8

Arnórsson S (1995) Geothermal systems in Iceland: Structure and conceptual models—I. High-Temperature Areas. Geothermics 24 (5–6):561–602. https://doi.org/10.1016/0375-6505(95)00025-9

Arnórsson S (1995) Geothermal systems in Iceland: Structure and conceptual models—II. Low-Temperature Areas. Geothermics 24 (5–6):603–629. https://doi.org/10.1016/0375-6505(95)00026-7

Arnórsson S, Axelsson G, Samundsson K (2008) Geothermal systems in Iceland. Jökull 58:269–302

Bachu S (1995) Synthesis and model of formation-water flow, Alberta Basin. Canada. AAPG Bull 79(8):1159–1178. https://doi.org/10.1306/8D2B2209-171E-11D7-8645000102C1865D

Bea F (2012) The sources of energy for crustal melting and the geochemistry of heat-producing elements. Lithos 153:278–291. https://doi.org/10.1016/j.lithos.2012.01.017

Beaumont C, Nguyen MH, Jamieson RA et al (2006) Crustal flow modes in large hot orogens. Special Publications, 268, Geological Society, London, p 91–145. https://doi.org/10.1144/GSL.SP.2006.268.01.05

Benderitter Y, Cormy G (1990) Possible approach to the geothermal research and relative cost estimate. In: Dickson MH, Fanelli M (eds) Small Geothermal Resources. UNITAR/UNDP Centre for Small Energy Resources, Rome, pp 61–71

Bennett S (2011) Geothermal potential of transtensional plate boundaries. Geothermal Resource Council Transactions, vol 35. Geothermal Resources Council, Davis, California, p 703–707. https://publications.mygeoenergynow.org/grc/1029317.pdf Accessed 17 December 2020

Berger BR, Tingley JV, Drew LJ (2003) Structural localization and origin of compartmentalized fluid flow, Comstock Lode, Virginia City, Nevada. Econ Geol 98:387–408. https://doi.org/10.2113/gsecongeo.98.2.387

Berkowitz B (2002) Characterizing flow and transport in fractured geological media: A review. Adv Water Resour 25:861–884. https://doi.org/10.1016/S0309-1708(02)00042-8

Bernard-Romero RA, Taran YA, Pennisi M (2010) Geochemistry of Boron in Fluids of Los Humeros and Los Azufres Hydrothermal Systems, Mexico. In: Birkle P, Torres Alvarado, IS (eds) Water-Rock Interaction. Proceedings of the 13th International Symposium on Water-Rock Interaction WRI-13, Guanajuato, Mexico, 16–20 August 2010. CRC Press, Taylor & Francis Group, London, p 145–148

Bernard R, Taran Y, Pennisi M et al (2011) Chloride and boron behavior in fluid of Los Humeros geothermal field (México): a model based on the existence of deep acid brine. Appl Geochem 26:2064–2073. https://doi.org/10.1016/j.apgeochem.2011.07.004

Bertini G, Casini M, Gianelli G et al (2006) Geological structure of a long-living geothermal system, Larderello, Italy. Terra Nova 18:163–169. https://doi.org/10.1111/j.1365-3121.2006.00676.x

Björlykke K (1997) Lithological control on fluid flow in sedimentary basins. In: Jamtveit B, Yardley BWD (eds) Fluid Flow and Transport in Rocks: Mechanisms and Effects. Chapman and Hall, London, p 15–137. https://doi.org/10.1007/978-94-009-1533-6_2

Boden DR (2016) Geologic Fundamentals of Geothermal Energy. CRC Press, Boca Raton, Florida

Boutelier D, Cruden A (2013) Slab rollback rate and trench curvature controlled by arc deformation. Geology 41(8):911–914. https://doi.org/10.1130/G34338.1

Browne PRL, Ellis AJ (1970) The Ohaaki-Broadlands geothermal area, New Zealand: Mineralogy and related geochemistry. Am J Sci 269:97–131. https://doi.org/10.2475/ajs.269.2.97

Bruhn D, Manzella A, Vuataz F et al (2010) Chapter 2: Exploration methods. In: Huenges E (ed) Geothermal Energy Systems, Wiley-VCH publisher, Weinheim, p 37–111. https://doi.org/10.1002/9783527630479.ch2

Brun JP (2002) Deformation of the continental lithosphere: Insights from brittle-ductile models. In: De Meer S, Drury MR, De Bresser JHP et al (eds) Deformation Mechanisms, Rheology and Tectonics: Current Status and Future Perspectives. Geological Society, London, p 355–370. https://doi.org/10.1144/GSL.SP.2001.200.01.20

Cathles LM, Erendi AHJ, Barrie T (1997) How long can a hydrothermal system be sustained by a single intrusive event? Econ Geol 92:766–771. https://doi.org/10.2113/gsecongeo.92.7-8.766

Cembrano J, Lara L (2009) The link between volcanism and tectonics in the Southern Volcanic Zone of the Chilean Andes: A review. Tectonophysics 471(1–2):96–113. https://doi.org/10.1016/j.tecto.2009.02.038

Chandrasekharam D, Alam MA, Minissale A (2005) Thermal discharges at Manikaran, Himachal Pradesh, India. In: Proceedings World Geothermal Congress 2005 Antalya, Turkey, 24–29 April 2005, Paper Number: 0881. https://www.geothermal-energy.org/pdf/IGAstandard/WGC/2005/0881.pdf. Accessed 17 December 2020

Chavarría-Rojas L (2003) Miravalles geothermal field, Costa Rica: Evidence of thermal evolution and a comparison of the mineralogy of an acid well and a neutral well. The United Nations University Geothermal Training Programme, Orkustofnun, Reykjavík, Iceland, Reports 2003, UNU-GTP-2003–06. https://orkustofnun.is/gogn/unu-gtp-report/UNU-GTP-2003-06.pdf. Accessed 17 December 2020

Chelle-Michou C, Do Couto D, Moscariello A et al (2017) Geothermal state of the deep Western Alpine Molasse Basin, France-Switzerland. Geothermics 67:48–65. https://doi.org/10.1016/j.geothermics.2017.01.004

Chin C, Wallace K, Harvey W et al (2013) Big Iron in Nicaragua: a muscular new geothermal plant! The San Jacinto–Tizate geothermal project. GRC Transactions, vol 37. Geothermal Resources Council, Davis, California, p 687–691. https://publications.mygeoenergynow.org/grc/1030642.pdf. Accessed 17 December 2020

Coolbaugh M, Shevenell L, Hinz NH et al (2015) Preliminary ranking of geothermal potential in the Cascade and Aleutian volcanic arcs. Part III. Regional data review and modeling. GRC Transactions, vol 39. Geothermal Resources Council, Davis, California, p 677–690. https://publications.mygeoenergynow.org/grc/1032207.pdf. Accessed 17 December 2020

Condie KC (2011) Earth as an Evolving Planetary System, 2nd edn. Elsevier, Amsterdam

Condie KC (2016) Earth as an Evolving Planetary System, 3rd edn. Elsevier, Amsterdam

Cox SF, Knackstedt MA, Braun J (2001) Principles of structural control on permeability and fluid flow in hydrothermal systems. Reviews in Economic Geology, vol 14. Society of Economic Geologists, Boulder, Colombia, p 1–24. https://doi.org/10.5382/Rev.14.01

Cox SF (2005) Coupling between deformation, fluid pressures and fluid flow in ore-producing hydrothermal environments. In: Hedenquist JW, Thompson JFH, Goldfarb RJ, Richards JP (eds) Economic Geology, 100th Anniversary Volume. Society of Economic Geologists Economic Geology, Littleton, pp 39–75

Cox SF (2010) The application of failure mode diagrams for exploring the roles of fluid pressure and stress states in controlling styles of fracture-controlled permeability enhancement in faults and shear zones. Geofluids 10:217–233. https://doi.org/10.1111/j.1468-8123.2010.00281.x

Curewitz D, Karson JA (1997) Structural settings of hydrothermal outflow: Fracture permeability maintained by fault propagation and interaction. J Volcanol Geotherm Res 79:149–168. https://doi.org/10.1016/S0377-0273(97)00027-9

Curry WH Jr (1977) Teapot dome — Past, present, and future. AAPG Bull 61(5):671–697. https://doi.org/10.1306/C1EA3DC0-16C9-11D7-8645000102C1865D

DeMets C (2001) A new estimate for present-day Cocos Caribbean plate motion: Implications for slip along the Central American volcanic arc. Geophys Res Lett 28:4043–4046. https://doi.org/10.1029/2001GL013518

Dewey JF, Lamb SH (1992) Active tectonics of the Andes. Tectonophysics 20(1–3):79–95. https://doi.org/10.1016/0040-1951(92)90419-7

Dewey JF, Holdsworth RE, Strachan RA (1998) Transpression and transtension zones. In: Holdsworth RE, Strachan RA, Dewey JE (eds) Continental Transpressional and Transtensional Tectonics. Special Publications, 135, Geological Society, London, p 1–14. https://doi.org/10.1144/GSL.SP.1998.135.01.01

DiPippo R (2012) Geothermal Power Plants: Principles, Applications, Case Studies, and Environmental Impacts, 3rd edn. Butterworth-Heinemann, Waltham, Massachusetts

Donaldson IG, Grant MA (1981) Heat extraction from geothermal reservoirs. In: Rybach L, Muffler LJP (eds) Geothermal Systems: Principles and Case Histories. John Wiley and Sons, Chichester, pp 145–179

Donaldson IG (1982) Heat and mass circulation in geothermal systems. Annu Rev Earth Planet Sci 10:377–395. https://doi.org/10.1146/annurev.ea.10.050182.002113

Downs DT, Rowland JV, Wilson CJN et al (2014) Evolution of the intra-arc Taupo-Reporoa Basin within the Taupo Volcanic Zone of New Zealand. Geosphere 10(1):185–206. https://doi.org/10.1130/GES00965.1

Drake II RM, Schenk CJ, Klett TR et al (2017) Assessment of undiscovered continuous oil and gas resources in the Heath Formation, central Montana and western North Dakota, 2016. U. S. Geological Survey Fact Sheet 2017–3032, U.S. Geological Survey, Reston, Virginia. https://doi.org/10.3133/fs20173032

Dunkl I, Grasemann B, Frisch W (1998) Thermal effects of exhumation of a metamorphic core complex on hanging wall syn-rift sediments: an example from the Rechnitz Window, Eastern Alps. Tectonophysics 297:31–50. https://doi.org/10.1016/S0040-1951(98)00162-0

Ebinger CJ, Jackson JA, Foster AN et al (1999) Extensional basin geometry and the elastic lithosphere. Phil Trans R Soc A 357:741–765. https://doi.org/10.1098/rsta.1999.0351

Ehara S (1989) Thermal structure and seismic activity in central Kyushu. Japan. Tectonophysics 159(3–4):269–278. https://doi.org/10.1016/0040-1951(89)90133-9

Elder JW (1981) Geothermal Systems. Academic Press, London

Elders WA, Fridleifsson GO, Albertsson A (2014) Drilling into magma and the implications of the Iceland Deep Drilling Project (IDDP) for high-temperature geothermal systems worldwide. Geothermics 49:111–118. https://doi.org/10.1016/j.geothermics.2013.05.001

Erdlac RJ Jr, Gross P, McDonald E (2008) A proposed new geothermal power classification system. GRC Transactions, vol 32. Geothermal Resources Council, Davis, California, p 322–327. https://publications.mygeoenergynow.org/grc/1028352.pdf. Accessed 17 December 2020

Faulds JE, Henry CD, Hinz NH (2005a) Kinematics of the northern Walker Lane: An incipient transform fault along the Pacific-North American plate boundary. Geology 33(6):505–508. https://doi.org/10.1130/G21274.1

Faulds JE, Henry CD, Hinz NH et al (2005b) Transect across the northern Walker Lane, northwest Nevada and northeast California: An incipient transform fault along the Pacific – North American Plate Boundary. In: Pederson J, Dehler CM (eds) Interior Western United States, Geological Society of America Field Guide 6, Geological Society of America, Boulder, Colorado, p 129–150. https://doi.org/10.1130/2005.fld006(06)

Faulds JE, Bouchot V, Moeck I et al (2009) Structural controls of geothermal systems in Western Turkey: A preliminary report. GRC Transactions, vol 33. Geothermal Resources Council, Davis, California, p 375–383. https://publications.mygeoenergynow.org/grc/1028484.pdf. Accessed 17 December 2020

Faulds JE, Moeck I, Drakos P et al (2010a) Structural assessment and 3D geological modeling of the Brady's Geothermal Area, Churchill County, (Nevada, USA). In: Proceedings, Thirty-Fifth Workshop on Geothermal Reservoir Engineering, Stanford University, Stanford, California, USA, 1–3 February 2010, SGP-TR-188. https://pangaea.stanford.edu/ERE/pdf/IGAstandard/SGW/2010/faulds.pdf. Accessed 17 December 2020

Faulds JE Coolbaugh MF Benoit D et al (2010b) Structural controls of geothermal activity in the Northern Hot Springs Mountains, Western Nevada: The Tale of three geothermal systems (Brady's, Desert Peak, and Desert Queen) GRC Transactions, vol 34. Geothermal Resources Council, Davis, California, p 675–684. https://publications.mygeoenergynow.org/grc/1028722.pdf. Accessed 17 December 2020

Faulds JE, Coolbaugh M, Bouchot V et al (2010c) Characterizing structural controls of geothermal reservoirs in the basin and range, USA, and western Turkey: developing successful exploration strategies in extended terranes. In: Proceedings World Geothermal Congress 2010, Bali, Indonesia, 25–30 April 2010, Paper Number: 01163. https://www.geothermal-energy.org/pdf/IGAstandard/WGC/2010/1163.pdf. Accessed 17 December 2020

Ferrill DA, Smart KJ, Morris AP (2020) Fault failure modes, deformation mechanisms, dilation tendency, slip tendency, and conduits v. seals. In: Ogilvie, S.R., Dee, S.J., Wilson RW, Bailey WR (eds) Integrated Fault Seal Analysis. Special Publications, 496, Geological Society, London, p 75–98. https://doi.org/10.1144/SP496-2019-7

Feuillet N, Manighetti I, Tapponnier P et al (2002) Arc parallel extension and localization of volcanic complexes in Guadeloupe, Lesser Antilles. J Geophys Res Solid Earth 107(B12):2331. https://doi.org/10.1029/2001JB000308

Flynn G, Ghusn G Jr (1983) Geologic and hydrogeologic research in the Moana geothermal system, Washoe County, Nevada. GRC Transactions, vol 7. Geothermal Resources Council, Davis, California, p 417–421. https://publications.mygeoenergynow.org/grc/1001103.pdf. Accessed 17 December 2020

Fox Hodgson S (2013) Focus on Chile. Part III. GRC Bull 42(3):28–37

Fucheng L, Suna Z, Zhanga J (2018) Influence of mid-crustal rheology on the deformation behavior of continental crust in the continental subduction zone. J Geodyn 117:88–99. https://doi.org/10.1016/j.jog.2018.04.002

Genter A, Traineau H, Bourgine B et al (2000) Over 10 years of geological investigations within the European Soultz HDR project, France. In: Iglesias E, Blackwell D, Hunt T et al (eds) Proceedings World Geothermal Congress 2000, Kyushu–Tohoku, Japan, 28 May – 10 June 2000, p 3707–3712. https://www.geothermal-energy.org/pdf/IGAstandard/WGC/2000/R0710.PDF. Accessed 17 December 2020

Gibbs AD (1984) Structural evolution of extensional basin margins. J Geol Soc London 141:609–620. https://doi.org/10.1144/gsjgs.141.4.0609

González-Partida E, Carrillo-Chávez A, Levresse G et al (2005) Hydrogeochemical and isotopic fluid evolution of the Los Azufres geothermal field, Central Mexico. Appl Geochem 20:23–39. https://doi.org/10.1016/j.apgeochem.2004.07.006

Grant MA, Bixley PF (2011) Geothermal Reservoir Engineering, 2nd edn. Academic Press, Burlington

Grant MA, Donaldson IG, Bixley PF (1982) Geothermal Reservoir Engineering. Academic Press, New York

Haenel R, Rybach L, Stegena L (1988) Fundamentals of geothermics. In: Haenel R, Rybach L, Stegena LA (eds) Handbook of Terrestial Heat Flow Density Determination. Kluwer Academic, Dordrecht

Hanano M (2004) Contribution of fractures to formation and production of geothermal resources. Renewable Sustainable Energy Rev 8:223–236. https://doi.org/10.1016/j.rser.2003.10.007

Hedenquist JW, Browne PRL (1989) The evolution of the Waiotapu geothermal system, New Zealand, based on the chemical and isotopic composition of its fluids, minerals and rocks. Geochim Cosmochim Acta 53:2235–2257. https://doi.org/10.1016/0016-7037(89)90347-5

Hedenquist JW (1990) The thermal and geochemical structure of the Broadlands-Ohaaki geothermal System, New Zealand. Geothermics 19:151–185. https://doi.org/10.1016/0375-6505(90)90014-3

Heicken G (1982) Geology of geothermal systems. In: Edwards LM, Chilingar GV, Rieke HH et al (eds) Handbook of Geothermal Energy. Gulf Publishing Company, Houston, Texas, pp 177–217

Henley RW, Ellis AJ (1983) Geothermal systems ancient and modern: A geochemical review. Earth-Sci Rev 19:1–50

Henneberger RC, Browne PRL (1988) Hydrothermal alteration and evolution of the Ohakuri hydrothermal system, Taupo Volcanic Zone, New Zealand. J Volcanol Geotherm Res 34:211–231. https://doi.org/10.1016/0377-0273(88)90034-0

Herrera R, Montalvo F, Herrera A (2010) El Salvador country update. In: Proceedings World Geothermal Congress 2010, Bali, Indonesia, 25–30 April 2010, Paper Number: 0141. https://www.geothermal-energy.org/pdf/IGAstandard/WGC/2010/0141.pdf. Accessed 17 December 2020

Hickson C, Ferraris F, Rodriguez C et al (2011) The Mariposa geothermal system, Chile. GRC Transactions, vol 35. Geothermal Resources Council, Davis, California, p 817–825. http://pubs.geothermal-library.org/lib/grc/1029336.pdf. Accessed 17 December 2020

Hickson C, Rodríguez C, Sielfeld G et al (2012) Mariposa geothermal system: A large geothermal resource in central Chile (320MWe inferred), XIII Congreso Geológico Chileno, Antofagasta, Chile, 5–9 August, 2012, p 583–585. https://biblioteca.sernageomin.cl/opac/DataFiles/14127_pp_583_585.pdf. Accessed 17 December 2020

Higley DK, Mitchell H, Roberts LNR et al (2005) 1-D/3-D Geological model of the Western Canada Sedimentary Basin. Mt Geol 42 (2):53–66

Holl HG (2015) What Did We Learn about EGS in the Cooper Basin?. Document Number: RES-FN-OT-RPT-01179, Geodynamics Limited, Milton, Australia. https://doi.org/10.13140/RG.2.2.33547.49443. Accessed 17 December 2020

Hochstein MP (1988) Assessment and modelling of geothermal reservoirs (small utilization schemes). Geothermics 17(1):15–49. https://doi.org/10.1016/0375-6505(88)90004-1

Holden L, Wallace L, Beavan J et al (2015) Contemporary ground deformation in the Taupo Rift and Okataina Volcanic Centre from 1998 to 2011, measured using GPS. Geophys J Int 202(3):2082–2105. https://doi.org/10.1093/gji/ggv243

Ingebritsen SE, Sorey ML (1988) Vapor-dominated zones within hydrothermal systems; evolution and natural state. J Geophys Res Solid Earth 93:12635–13655. https://doi.org/10.1029/JB093iB11p13635

Jolie E, Faulds J, Moeck I (2012) The development of a 3D structural geological model a part of the geothermal exploration strategy – A case study from the Brady's geothermal system, Nevada, USA. In:

Proceedings, Thirty-Seventh Workshop on Geothermal Reservoir Engineering, Stanford University, Stanford, CA, USA, 30 January – 1 February 2012, SGP-TR-194. https://pangea.stanford.edu/ERE/pdf/IGAstandard/SGW/2012/Jolie.pdf. Accessed 17 December 2020

King D, Metcalfe E (2013) Rift zones as a case study for advancing geothermal occurrence models. In: Proceedings, Thirty-Eighth Workshop on Geothermal Reservoir Engineering, Stanford University, Stanford, CA, 11–13 February 2013, SGP-TR-198. https://pangea.stanford.edu/ERE/pdf/IGAstandard/SGW/2013/King.pdf. Accessed 17 December 2020

Kious WJU, Tilling RI (1996) This dynamic Earth: the story of plate tectonics, online edition. U.S. Geological Survey, Washington, DC. https://pubs.usgs.gov/gip/dynamic/dynamic.html. Accessed 17 December 2020

Lee KC (2001) Classification of geothermal resources by exergy. Geothermics 30:431–442. https://doi.org/10.1016/S0375-6505(00)00056-0

Magni V (2019) The effects of back-arc spreading on arc magmatism. Earth Planet Sci Lett 519:141–151. https://doi.org/10.1016/j.epsl.2019.05.009

Majer EL, Peterson JE (2007) The impact of injection on seismicity at the Geysers, California geothermal field. Int J Rock Mech Min Sci Geomech 44:1079–1090. https://doi.org/10.1016/j.ijrmms.2007.07.023

Majorowicz J, Minea V (2015) Geothermal energy potential in low enthalpy areas as a future energy resource: Identifying feasible targets, Quebec, Canada, study case. Resources 4(3):524–547. https://doi.org/10.3390/resources4030524

Manning CE, Ingebritsen SE (1999) Permeability of the continental crust: Implications of geothermal data and metamorphic systems. Rev Geophys 37:127–150. https://doi.org/10.1029/1998RG900002

Mannington W, O'Sullivan M, Bullivant D (2004) Computer modelling of the Wairakei-Tauhara geothermal system, New Zealand. Geothermics 33:401–419. https://doi.org/10.1016/j.geothermics.2003.09.009

Martinez AM (2013) Case history of Los Azufres: Conceptual modelling of a Mexican geothermal field. Short Course V on "Conceptual Modelling of Geothermal Systems" organized by UNU-GTP and LaGeo, Santa Tecla, El Salvador, 24 February – 2 March, 2013, UNU-GTP-SC-16–08. http://www.os.is/gogn/unu-gtp-sc/UNU-GTP-SC-16-08.pdf. Accessed 17 December 2020

Micklethwaite S (2009) Mechanisms of faulting and permeability enhancement during epithermal mineralization: Cracow goldfield, Australia. J Struct Geol 31:288–300. https://doi.org/10.1016/j.jsg.2008.11.016

Micklethwaite S, Sheldon HA, Baker T (2010) Active fault and shear processes and their implications for mineral deposit formation and discovery. J Struct Geol 32:151–165. https://doi.org/10.1016/j.jsg.2009.10.009

Minissale A (1991) The Larderello geothermal field: a review. Earth-Sci Rev 31(2):133–151. https://doi.org/10.1016/0012-8252(91)90018-B

Minissale A, Vaselli O, Chandrasekharam D et al (2000) Origin and evolution of 'intracratonic' thermal fluids from central western peninsular India. Earth Planet Sci Lett 181(3):377–394. https://doi.org/10.1016/S0012-821X(00)00200-4

Mo X, Hou Z, Niu Y et al (2007) Mantle contributions to crustal thickening during continental collision: Evidence from Cenozoic igneous rocks in southern Tibet. Lithos 96:225–242. https://doi.org/10.1016/j.lithos.2006.10.005

Moeck I, Backers T (2011) Fault reactivation potential as critical factor during reservoir stimulation. First Break 29:67–74. https://doi.org/10.3997/1365-2397.2011014

Moeck IS (2014) Catalog of geothermal play types based on geologic controls. Renewable Sustainable Energy Rev 37:867–882. https://doi.org/10.1016/j.rser.2014.05.032

Moeck I, Kwiatek G, Zimmermann G (2009) Slip tendency, fault reactivation potential and induced seismicity in a deep geothermal reservoir. J Struct Geol 31:1174–1182. https://doi.org/10.1016/j.jsg.2009.06.012

Moeck I, Hinz N, Faulds J et al (2010) 3D geological mapping as new method in geothermal exploration: a case study from Central Nevada. GRC Transactions, vol 34. Geothermal Resources Council, Davis, California, p 807–812. http://pubs.geothermal-library.org/lib/grc/1028742.pdf. Accessed 17 December 2020

Momita M, Tokita H, Matsudo K et al (2000) Deep geothermal structure and the hydrothermal system in the Otake–Hatchobaru Geothermal Field. 22nd New Zealand Geothermal Workshop, Auckland, New Zealand, 8–10 November 2000. http://www.geothermal-energy.org/pdf/IGAstandard/NZGW/2000/Momita.pdf. Accessed 17 December 2020

Morley CK, Nelson RA, Patton TL et al (1990) Transfer zones in the East African rift system and their relevance to hydrocarbon exploration in rifts. AAPG Bull 74:1234–1253. https://doi.org/10.1306/0C9B2475-1710-11D7-8645000102C1865D

Muffler LJP (1973) Geothermal resources. In: Brobst DA, Pratt WP (eds) United States Mineral Resources. US Geol Survey Prof Paper 820:251–261

Muffler LJP (1976) Tectonic and hydrologic control of the nature and distribution of geothermal resources. In: Proceedings of the Second U.N. Symposium of the Development and Use of Geothermal Resources, San Francisco, California, vol 1, Lawrence Berkeley Laboratory, University of California, San Francisco, p 499–507.

Muffler LJP (1979) Assessment of geothermal resources of the United States – 1978. Geological Survey Circular 790, U.S. Geological Survey, Arlington, Virginia. https://doi.org/10.3133/cir790. Accessed 17 December 2020

Muffler P, Cataldi R (1977) Methods for regional assessment of geothermal resources. Larderello Workshop on Geothermal Resource Assessment and Reservoir Engineering, 12–16 September 1977, Open File Report 77–870, U.S. Geological Survey. https://pubs.usgs.gov/of/1977/0870/report.pdf. Accessed 17 December 2020

Nakakuki T, Mura E (2013) Dynamics of slab rollback and induced back-arc basin formation. Earth Planet Sci Lett 361:287–297. https://doi.org/10.1016/j.epsl.2012.10.031

Nicholson K (1993) Geothermal Fluids: Chemistry and Exploration Techniques. Springer Verlag, Berlin

Nortje GS, Rowland JV, Sporli KB et al (2006) Vein deflections and thickness variations of epithermal quartz veins as indicators of fracture coalescence. J Struct Geol 28:1396–1405. https://doi.org/10.1016/j.jsg.2006.05.006

Pambudi NA (2018) Geothermal power generation in Indonesia, a country within the ring of fire: Current status, future development and policy. Renewable Sustainable Energy Rev 81(2):2893–2901. https://doi.org/10.1016/j.rser.2017.06.096

Phillips BR, Ziagos J, Thorsteinsson H et al (2013) A roadmap for strategic development of geothermal exploration technologies. In: Proceedings, Thirty-Eighth Workshop on Geothermal Reservoir Engineering, Stanford University, Stanford, CA, USA, 11–13 February 2013, SGP-TR-198. https://www.energy.gov/sites/default/files/2014/02/f7/exploration_technical_roadmap2013.pdf. Accessed 17 December 2020

Ramos VA (2009) Anatomy and global context of the Andes: Main geologic features and the Andean orogenic cycle. In: Kay SM, Ramos VA, Dickinson, WD (eds) Backbone of the Americas: Shallow Subduction, Plateau Uplift, and Ridge and Terrane

Collision. Geological Society of America Memoirs, 204, p 31–66. https://doi.org/10.1130/2009.1204(02)

Reed MJ (1983) Assessment of low-temperature geothermal resources of the United States – 1982. Geological Survey Circular 892, U.S. Geological Survey, Alexandria, Virginia. https://doi.org/10.3133/cir892. Accessed 17 December 2020

Rogers SF (2003) Critical stress-related permeability in fractured rocks. Geological Society, London, Special Publications 209:7–16

Rosendahl BR, Reynolds DJ, Lorber PM et al (1986) Structural expressions of rifting: Lessons from Lake Tanganyika, Africa. Special Publications, 25, Geological Society, London, p 29–43. https://doi.org/10.1144/GSL.SP.1986.025.01.04

Rowland JV, Sibson RH (2001) Extensional fault kinematics within the Taupo Volcanic Zone, New Zealand: Soft-linked segmentation of a continental rift system. New Zealand J Geol Geophys 44:271–284. https://doi.org/10.1080/00288306.2001.9514938

Rowland JV, Sibson RH (2004) Structural controls on hydrothermal flow in a segmented rift system, Taupo Volcanic Zone, New Zealand. Geofluids 4:259–283. https://onlinelibrary.wiley.com/doi/full/https://doi.org/10.1111/j.1468-8123.2004.00091.x

Rowland JV, Simmons SF (2012) Hydrologic, magmatic, and tectonic controls on hydrothermal flow, Taupo Volcanic Zone, New Zealand: implications for the formation of epithermal vein deposits. Econ Geol 107(3):427–457. https://doi.org/10.2113/econgeo.107.3.427

Rowland JV, Baker E, Ebinger CJ et al (2007) Fault growth at a nascent slow-spreading ridge: 2005 Dabbahu rifting episode, Afar. Geophys J Int 171:1226–1246. https://doi.org/10.1111/j.1365-246X.2007.03584.x

Rosas JC, Currie CA, He J (2016) Three-dimensional thermal model of the Costa Rica-Nicaragua Subduction Zone. Pure Appl Geophys 173:3317–3339. https://doi.org/10.1007/s00024-015-1197-4

Ruiz OV (2013) The Miravalles Geothermal System, Costa Rica, Short Course V on "Conceptual Modelling of Geothermal Systems" organized by UNU-GTP and LaGeo, Santa Tecla, Santa Tecla, El Salvador, 24 February – 2 March 2013. http://www.os.is/gogn/unu-gtp-sc/UNU-GTP-SC-16-32.pdf. Accessed 17 December 2020

Sanyal SK (2005) Classification of Geothermal systems – a possible scheme. In: Proceedings, Thirtieth Workshop on Geothermal Reservoir Engineering, Stanford University, Stanford, CA, USA, 31 January – 2 February 2005, SGP-TR-176. https://pangea.stanford.edu/ERE/pdf/IGAstandard/SGW/2005/sanyal1.pdf. Accessed 17 December 2020

Scheck-Wenderoth M, Cacace M, Maystrenko YP et al (2014) Models of heat transport in the Central European basin system: effective mechanisms at different scales. Mar Pet Geol 55(1):315–331. https://doi.org/10.1016/j.marpetgeo.2014.03.009

Schubert G, Straus JM, Grant MA (1980) A problem posed by vapour-dominated geothermal systems. Nature 287:423–425. https://doi.org/10.1038/287423a0

Sdrolias M, Müller RD (2006) Controls on back-arc basin formation. Geochem Geophys Geosyst 7:Q04016. https://doi.org/10.1029/2005GC001090

Seebeck H, Nicol A, Villamor P et al (2014) Structure and kinematics of the Taupo Rift, New Zealand. Tectonics 33:1178–1199. https://doi.org/10.1002/2014TC003569

Shi J, Shi D, Shen Y et al (2017) Growth of the northeastern margin of the Tibetan Plateau by squeezing up of the crust at the boundaries. Sci Rep 7:10591. https://doi.org/10.1038/s41598-017-09640-0

Sibson RH (1998) Brittle failure mode plots for compressional and extensional tectonic regimes. J Struct Geol 20:655–660. https://doi.org/10.1016/S0191-8141(98)00116-3

Sibson RH (2000) Fluid involvement in normal faulting. J Geodyn 29:469–499. https://doi.org/10.1016/S0264-3707(99)00042-3

Sibson RH (2001) Seismogenic framework for hydrothermal transport and ore deposition. Rev Econ Geol 14:25–50. https://doi.org/10.5382/Rev.14.02

Sibson RH (2004) Controls on maximum fluid overpressure defining conditions for mesozonal mineralisation. J Struct Geol 26(6–7):1127–1136. https://doi.org/10.1016/j.jsg.2003.11.003

Sibson RH (2020) Preparation zones for large crustal earthquakes consequent on fault-valve action. Earth Planets Space 72:31. https://doi.org/10.1186/s40623-020-01153-x

Sielfeld G, Ruz J, Brogi A et al (2019) Oblique-slip tectonics in an active volcanic chain: a case study from the Southern Andes. Tectonophysics 770:228221. https://doi.org/10.1016/j.tecto.2019.228221

Simmons SF, Browne PRL (2000) Hydrothermal minerals and precious metals in the Broadlands-Ohaaki geothermal system: Implications for understanding low-sulfidation epithermal environments. Econ Geol 95:971–999. https://doi.org/10.2113/gsecongeo.95.5.971

Singh HK, Kumar Y, Chandrasekharam D et al (2014) High-heat producing granites of East Dharwar Craton around Gugi, Karnataka, and their possible influence on the evolution of Rajapur thermal springs, Deccan Volcanic Province. India. Geotherm Energy 2:2. https://doi.org/10.1186/s40517-014-0002-4

Singh HK, Sinha S, Alam, MA, Chandrasekharam D (2020) Tracing the evolution of thermal springs in the Hazaribagh area of Eastern Peninsular India through hydrogeochemical and isotopic analyses. Geothermics 85:101817. https://doi.org/10.1016/j.geothermics.2020.101817

Stern TA, Davey FJ (1987) A seismic investigation of the crustal and upper mantle structure within the Central Volcanic Region of New Zealand. New Zealand J Geol Geophys 30(3):217–231. https://doi.org/10.1080/00288306.1987.10552618

Stimac J, Powell TS, Golla GU (2004) Porosity and permeability of the Tiwi geothermal field, Philippines, based on continuous and spot core measurements. Geothermics 33:87–107. https://doi.org/10.1016/j.geothermics.2003.03.002

Stimac J, Nordquist G, Suminar A et al (2008) An overview of the Awibengkok geothermal system, Indonesia. Geothermics 37:300–331. https://doi.org/10.1016/j.geothermics.2008.04.004

Symithe S, Calais E, de Chabalier JB et al (2015) Current block motions and strain accumulation on active faults in the Caribbean. J Geophys Res Solid Earth 120:3748–3774. https://doi.org/10.1002/2014JB011779

Taguchi S, Nakamura M (1991) Subsurface thermal structure of the Hatchobaru geothermal system, Japan, determined by fluid inclusion study. Geochem J 25:301–314. https://doi.org/10.2343/geochemj.25.301

Taguchi, S (2001) Otake-Hatchobaru geothermal field. In: Feebrey CA, Hayashi T, Taguchi S (eds) Epithermal Gold Mineralization and Modern Analogues, Kyushu, Japan, Guidebook Prepared for the Society of Economic Geologists Field Trip, 28 October –3 November 2001, Guidebook Series of the Society of Economic Geologists, vol 34. Society of Economic Geologists, Littleton, Columbia, p 173–178. https://doi.org/10.5382/GB.34

Vanderhaeghe O, Medvedev S, Fullsack P et al (2003) Evolution of orogenic wedges and continental plateau: insights from crustal thermal-mechanical models overlying subducting mantle lithosphere. Geophys J Int 153:27–51. https://doi.org/10.1046/j.1365-246X.2003.01861.x

Walker JD, Sabin AE, Unruh JR et al (2005) Development of genetic occurrence models for geothermal prospecting. GRC Transactions, vol 29. Geothermal Resources Council, Davis, California, p 309–313. https://publications.mygeoenergynow.org/grc/1022614.pdf. Accessed 17 December 2020

Weides S, Moeck I, Majorowicz J et al (2012) Geothermal exploration of Paleozoic formations in Central Alberta. Can J Earth Sci 50 (5):519–534. https://doi.org/10.1139/cjes-2012-0137

Verma RK (1991) Geodynamics of the Indian Peninsula and the Indian Plate Margin. Oxford and IBH Publishing, New Delhi

Villamor P, Berryman KR, Ellis SM et al (2017) Rapid evolution of subduction-related continental intraarc rifts: The Taupo Rift, New Zealand. Tectonics 36:2250–2272. https://doi.org/10.1002/2017TC004715

White WM (2016) Hot Spots and Mantle Plumes. In: Harff J, Meschede M, Petersen S et al (eds) Encyclopedia of Marine Geosciences. Encyclopedia of Earth Sciences Series. Springer, Dordrecht, p 316–327. https://doi.org/10.1007/978-94-007-6238-1_14

Wilmarth M, Stimac J (2014) Worldwide power density review. In: Proceedings, Thirty-Ninth Workshop on Geothermal Reservoir Engineering, Stanford University, Stanford, CA, USA, 24–26 February 2014, SGP-TR-202. https://pangea.stanford.edu/ERE/pdf/IGAstandard/SGW/2014/Wilmarth.pdf. Accessed 17 December 2020

Wilmarth M, Stimac J (2015) Power density in geothermal fields. In: Proceedings World Geothermal Congress 2015, Melbourne, Australia, 19–24 April 2015, Paper Number: 16020. https://www.geothermal-energy.org/pdf/IGAstandard/WGC/2015/16020.pdf. Accessed 17 December 2020

Wood CP, Brathwaite RL, Rosenberg MD (2001) Basement structure, lithology and permeability at Kawerau and Ohaaki geothermal fields, New Zealand. Geothermics 30:461–481. https://doi.org/10.1016/S0375-6505(01)00003-7

A GIS-based Approach for Morphometric Analysis of Jamini Basin and Its Subwatersheds: Implication for Conservation of Soil and Water Resources

S. C. Bhatt, Rubal Singh, Rahul Singh, Mohd Saif, and M. M. Singh

Abstract

Jamini a tributary of Betwa River has Jamrar, Onri, Sajnam, Shanzad and Baragl five subwatersheds. The quantitative morphometric analysis of its subwatersheds was carried out by using advanced techniques of remote sensing and GIS. The linear, areal and relief parameters along with hypsometric values were generated for each subwatershed of Jamini basin. The remotely sensed Shuttle Radar Topographic Mission (SRTM) Digital Elevation Model (DEM) data with a capture resolution of 3.0 arc second and pixel resolution of 30 m were used to extract the drainage map, density, contour, aspect and other maps. The mean bifurcation ratio (3.79 to 5.04) of Jamini and its all subwatersheds indicate that the drainage is structurally controlled. The Rho coefficient values reveal that the higher hydrologic storage during the flood and a decrease in the erosion during elevated discharge are observed in the basin. The drainage density (1.0 to 1.04) indicates that the basin is impermeable and shows low groundwater storage capacity in hard rock granitic terrain. The form factor (0.18 to 0.49), elongation and circulatory ratios indicate that the Jamini basin and subwatersheds are significantly elongated and show low peak flow of longer duration. The values of Ruggedness and Melton ruggedness number infer that the basin is slightly rugged and less prone to erosion and sediment transport. The S-shaped hypsometric curves show that the basin is susceptible to moderate erosion and the hypsometric index values (0.49 to 0.50) imply that the basin has mature geomorphic terrain with moderately eroded landscapes.

Keywords

Morphometric analysis • Hypsometry • Correlation matrix • Jamini basin • Central India

1 Introduction

Morphometric parameters reveal the quantitative evaluation of shape and size of landforms and characteristics of the basin (Horton 1945; Melton 1958; Strahler 1964; Miller 1953; Pareta and Pareta, 2011; Prakash et al. 2016, 2019; Bhatt et al. 2017, 2020; Bhatt 2020). It can be defined as an indicator of landform structure and hydrologic processes, erosional and physical properties of soil (Horton 1932; 1945; Strahler 1964; Smith 1950). Morphometric parameters determine the lithological and structural control on drainage and characteristics of fluvial landscapes (Pati et al. 2006; 2008; Prakash et al. 2016; 2017; Pareta and Pareta 2011; Bhatt et al. 2017, 2020). The findings of morphometric analysis give information about hydrogeologic and erosion-prone area and characters of surface and groundwater potentials of a watershed (Aparna et al. 2015).

The morphotectonic and morphometric studies of tributaries of the Ganga river have been done by several researchers (Tangri 1992; Shukla et al. 2012). Contrary to this the morphometric and morphotectonic parameters of most of the tributaries of Yamuna particularly coming from Bundelkhand craton have not been studied in detail except by few workers (Prakash et al. 2016; 2017; Pareta and Pareta 2011; 2012; Bhatt et al. 2017, 2020; Bhatt, 2020; Panda et al. 2019). Therefore, in present work, the morphometric and hypsometric analysis has been carried out to explore the

S. C. Bhatt (✉) · R. Singh · M. M. Singh
Institute of Earth Sciences, Department of Geology, Bundelkhand University, Jhansi, India
e-mail: scbhatt@bujhansi.ac.in

R. Singh
Department of Geology & Geophysics, Indian Institute of Technology, Kharagpur, India

M. Saif
Department of Earth Sciences, Indian Institute of Technology Bombay, Mumbai, Maharashtra, India
e-mail: 203060026@iitb.ac.in

A. K. Shandilya et al. (eds.), *Geological and Geo-Environmental Processes on Earth*, Springer Natural Hazards,
https://doi.org/10.1007/978-981-16-4122-0_16

status of water and soil resources in the subwatersheds of Jamini basin.

The natural resources are playing a significant role in the economic and social development of all countries and their overexploitation is causing a threat to the economic development of all developing countries and are unbalancing the ecosystems (Bhatt 2020). The productivity of agricultural cropland is depending on the presence of soil water and resources. Despite rich in natural resources and having good river network Bundelkhand is facing severe problems with water and land resources. Due to a small cropped area (35 to 40%) and low rainfall, it belongs to a hot spot of dry land (Bhatt 2020). Due to drought and low agricultural productivity, it falls under privilege region and people are facing starvation and famine. Therefore, there is a need to protect these resources and the implications of GIS and ASTER/SRTM-based studies are used to monitor soil erosion. The morphometric and hypsometric data of subwatersheds of Jamini River would be useful input to plan the conservation of soil and water resources.

2 Study Area

Jamini a tributary of Betwa River occupies 4584.42 sq km area and is 201 km in length. It is originating from Vindhyan mountain in Sagar district of Madhya Pradesh, India and is flowing SW to NE direction. It joins Betwa near Orcha, Tikamgarh district of Madhya Pradesh. It has five subwatersheds viz. Shahjad, Sajnam, Baragl, Onri and Jamrar and covers 4584.42 sq km area. Its basin lies between 24°4′36″ N to 25°13′4″ N latitude and 78°17′49″ E to 78°56′23″ E longitude (Fig. 1). Jamini basin is a rugged granitic terrain showing undulating topography with average elevation. The semiarid climatic conditions and erratic rainfall are very common in this study area.

Its subwatershed Shahjad emerges from Dushai village and meets Jamni near Hazaria village of Lalitpur district (Fig. 1). The watershed covers 1111.49 sq km area and lies between 24°25′41‶N to 25°2′14′N latitude and 78°17′49‶E to 78°38′40′E longitude. Sajnam another subwatershed occupying 988.72 sq km area originates in SW of Lalitpur and meets Jamni at Chandawali (Lalitpur, U.P.). Baragl covering 320.80 sq km area is meeting to Jamini near Didora of Tikamgarh district. It lies between 24°55′15‶N to 25°9′21′N latitude and 78°38′37‶E to 78°53′21′E longitude. Onri emerges in the northeast of Madhya Pradesh and has confluence with Jamini near Tikamgarh. It is occupying 320.80 sq km area and is lying between 24°55′15‶N to 25°9′21′N latitude and 78°38′37‶E to 78°53′21′E longitude. Jamrar a subwatershed of Jamini originating in the southern part of Lalitpur is meeting to Jamini near a junction located between Sujanpura and Minora villages of Tikamgarh district, M.P. (Fig. 1).

3 Geology of the Study Area

The Jamini basin mainly consists of Bundelkhand granites, Vindhyan Supergroup, Bijawar and Deccan Traps. The larger part of the basin is covered by Bundelkhand granites in the north. The southeastern borders of the basin area are surrounded by Bijawar Group of rocks. The Vinhyans are exposed in the southeastern and southwestern margins whereas the southern boundaries of the basin are occupied by Deccan Traps (Fig. 2).

Bundelkhand craton comprises three major groups of rocks. The oldest Bundelkhand Gneissic Complex consists of Tonalite Trondhjemite granodiorite (TTG) gneisses and mafic gneisses (3.5–3.2 Ga). The Banded Iron Formation, Quartzites and metavolcanics are important rock unites of Bundelkhand Greenstone Complex. The various types of granitoids, quartz reefs and doleritic dykes belong to Bundelkhand Granitic Complex (2.6–2.0 Ga; Bhatt and Mahmood 2008, 2012; Bhatt et al. 2011, 2017; Bhatt and Gupta 2014; Bhatt 2014; Bhatt and Hussain 2008, 2012; Singh and Slabunov 2015; Singh et al. 2021). The coarse-grained porphyritic and medium-grained pink granite at places showing gneissic foliation belong to Bundelkhand Granitic Complex (Basu 1986; Bhatt and Mahmood 2012; Bhatt 2014; Pati et al. 1997, 2007).

Rewa Group belonging to upper Vindhyan Supergroup consists of Panna shale, Lower Rewa sandstone, Jhiri shale and Upper Rewa sandstone (Govindgarh and Dramnandganj sandstone). This group varying in thickness from 100 to 300 m is showing diamond-bearing conglomerate in its upper and lower boundaries. It consists of a series of shale and sandstone Formations and is containing Kimberlite-derived diamondiferous conglomerate. The medium to fine-grained iron pigmented arenaceous sandstone belonging to Upper Rewa sandstone Formation is characterized by planar, trough and herringbone cross stratification. The laminated shale, sandstone, flagstone, porcellanites and limestone overlying the Semri Group are important units of Kaimur Group of Upper Vindhyan (Fig. 2). A basal conglomerate containing pebbles of jasper is observed an important unit of Kaimur Group in Bundelkhand.

The Bijawar Group belonging to Middle Proterozoic Formations were deposited over the Bundelkhand craton along its southeastern (Hirapur and Sonarai Basins) and northwestern (Gwalior Basin) margins. The Bijawar Group consists of a succession of a basal conglomerate and quartzite overlain by hornstone breccia, limestone, phyllitic shales, red jaspers, and dioritic traps. Quartzites and sandstones, sometimes conglomeratic form the basal beds resting on gneisses and a siliceous limestone is found with the quartzites. These rocks are again well exposed along the

Fig.1 Location map of the study area

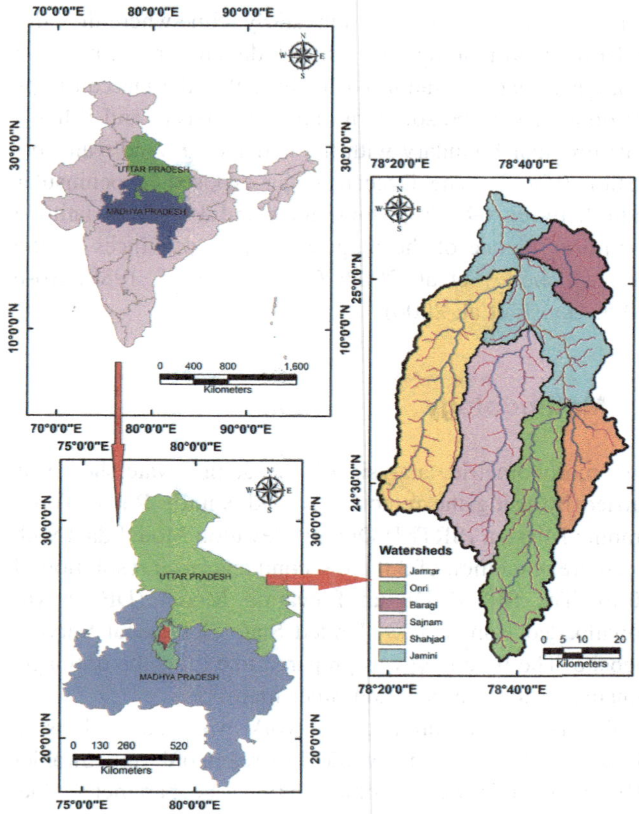

Fig. 2 Geological map of the study area

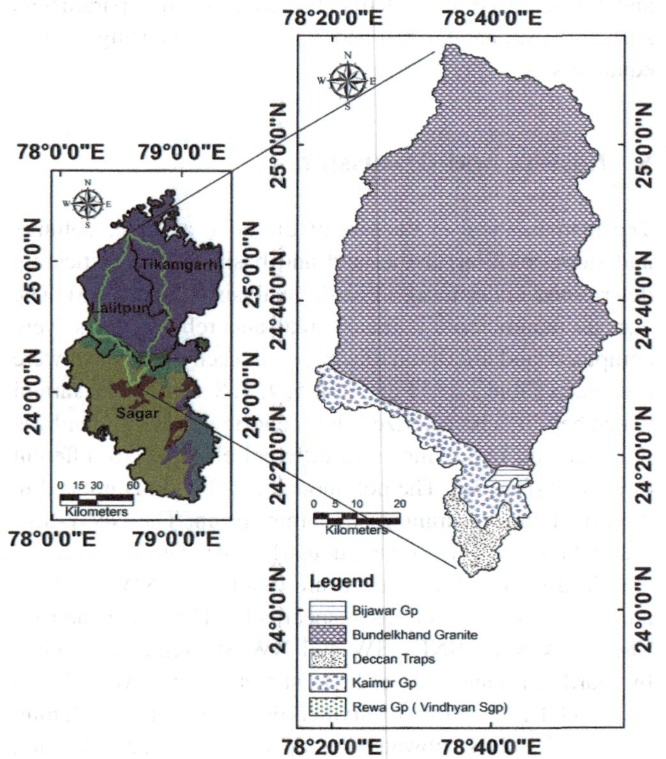

southeastern edges of the Vindhyan Syncline where they dip under the Semri Group under the Vindhyan Supergroup. The geological maps of Jamni basin show that the Deccan traps are dominant in the southern part of the basin, and it has a very irregular boundary with the underlying Vindhyan Formation (Fig. 2). The Deccan Traps exposed in peninsular India between 69°–79°E longitudes and 16°–22°N latitude comprises of one of the biggest volcanic provinces on the earth (Mahoney et al. 2002; Courtillot et al. 1988; Bose 1995; Ghosh et al. 2006).

4 Methodology

The morphometric analysis of the Jamni watershed was carried by using the remotely sensed Shuttle Radar Topographic Mission (SRTM) Digital Elevation Model data with a capture resolution of 3.0 arc second and pixel resolution of 30 m. The SRTM Digital Elevation Model (DEM) was downloaded from USGS (United State Geological Survey) website. The base map was prepared to extract the drainage, density, contour, aspect and other maps.

Watershed and drainage network was analyzed using Horton's (1945) scheme while in-stream ordering Strahler (1964) methods were adopted. The basic parameters like length of the individual stream, basin area, and perimeter of the basin were extracted from the ArcGIS-10.3 geodatabase and further used for calculating morphometric parameters such as linear, areal, relief and tectonic aspects using various equations.

5 Results and Discussion

The mathematical evaluation of the area, altitude, volume, and slope, profiles of the land and basin characteristics are determined under morphometric analysis (Clark 1966). The various parameters of linear, areal and relief aspects were computed and the thematic maps of each watershed were prepared (Table 1 and Figs. 3, 4, 5, 4, 7, 8 and 9). The Jamini watershed is characterized by dendritic to sub dendritic drainage and is flowing from different elevation to different directions (Fig. 3a). The polymodal distribution is noticed in the orientation of drainage of Jamini basin. The NNW-SSE orientation is highly conspicuous (Fig. 4). The first, second and third lower-order streams are trending in NW-SE, E-W and NNW-SSE directions respectively (Fig. 4a, b and c). The NNW-SSE, NNE-SSW and NW-SE trends are shown by fourth, fifth and sixth order of streams respectively (Fig. 4 d, e and f). The fifth to sixth order of streams in Jamini watershed and its subwatersheds (Jamrar, Shahzad, Sajnam, Onri and Buragi) were validated in the study area (Figs. 3a, 5a, 6a, 7a, 8a and 9a).

5.1 Linear Parameters

All linear, areal and relief parameters of Jamini watershed and its subwatersheds were analyzed and are shown in Table 1. The nature of surface runoff can be expressed by stream length. The streams having smaller length are characterized by steeper slope while the streams of longer length are indicative of gentle slope and flat topography. According to Horton's law (1945), the total length of the stream segment is observed maximum in the first order and it decreases as the stream order increases. The relationship between watershed area and stream length is characterized by the predominance of headward erosion for the drainage network development and expansion (Hack 1957). The maximum stream length (2339.5 km) is computed for first order and minimum (77.21 km) for sixth-order streams of Jamini watershed (Table 1). Similarly, the maximum length for the first order and minimum length for the highest order of each subwatershed (Sajnam, Shajad and Onri subwatersheds) was calculated (Table 1). The presence of a large number of smaller stream lengths in Jamini watershed and its subwatersheds implies that the terrain is consisting of impermeable rock formations.

The maximum stream length ratio observed in Onari (4.34) and Shahjad (3.51) subwatershed of Jamini basin and minimum in Baragl (2.87) and Jamrar (2.06) subwatershed. It infers that the evolution of all streams of Jamini basin is controlled by lithology and geological structures. The ratio of a number of streams of a given order (Nu) to the number of streams to the next higher order is defined as the bifurcation ratio (Strahler 1964). It varies from 2.0 in flat or rolling drainage basins to 4.0 in mountainous or highly dissected drainage basins (Horton 1945). The bifurcation ratio ranging from 2.67 (Baragl) to 8.00 (Onri) and the mean bifurcation ratio varies from 3.79 to 5.04 in subwatersheds of Jamini basin (Table 1). It was observed that the bifurcation ratio and mean bifurcation ratio is more than 3 (Table 1). It reveals that the Jamini basin and its subwatersheds were developed in a lithologically and structurally controlled terrain. The Rho coefficient (ρ) values are ranging from 0.53 (Jamrar) to 0.69 (Shahjad) in subwatersheds of Jamini basin (Table 1) and it is computed 0.59 for Jamini basin. These values are indicating that the high drainage density is intended to increase higher hydrologic storage during the flood.

5.2 Areal Parameters

The drainage density, frequency, drainage texture, form factors, circulatory ratio and other parameters were discussed under this subheading. Drainage density is defined as the sum of stream lengths per unit area and is an expression of closeness or spacing of channels (Horton 1945 and Strahler 1964). The total area of Jamini basin is 4584.42 sq

Table 1 showing Linear, areal and relief parameters of Jamini and its subwatersheds (JW-Jamni watershed; SW1-Shahjad subwatershed; SW2-Sajnam subwatershed; SW3-Baragl subwatershed; SW4-Onri subwatershed; SW5-Jamrar subwatershed)

S.No	Parameters	JW	SW1	SW2	SW3	SW4	SW5
Linear Parameters							
(1)	Stream Order (Su)	1 to 6	1 to 5	1 to 5	1 to 5	1 to 5	1 to 5
(2)	Stream Length (Lu), km	4586.41	1142.97	1006.51	327.40	913.75	358.52
(3)	Mean Stream Length Ratio (Lurm)	2.85	3.51	2.93	2.87	4.34	2.06
(4)	Weighted Mean Stream Length Ratio (Luwm)	2.45	2.61	2.49	2.33	2.89	2.16
(5)	Mean basin Width (Wb), km	53.61	15.52	15.77	12.59	12.48	10.34
(6)	Main Channel Length (Cl), km	169.69	99.16	85.24	39.16	92.51	43.93
(7)	Fitness ratio (Rf)	0.35	0.40	0.39	0.33	0.41	0.36
(8)	Wandering ratio (Rw)	1.98	1.38	1.36	1.54	1.32	1.29
(9)	Maximum Basin Length (Lb), km	85.52	71.60	62.71	25.47	70.09	33.97
(10)	Basin Perimeter (P), km	484.99	246.42	220.55	116.98	227.17	122.84
(11)	Bifurcation Ratio (Rb)	4.4–5.36	4.26–6.0	4.0–6.25	2.67–4.88	3.0–8.0	3.0–5.0
(12)	Mean Bifurcation Ratio (Rbm)	4.81	5.04	4.90	3.79	5.03	3.91
(13)	Weighted mean bifurcation ratio (Rbwm)	4.54	4.39	4.49	4.58	4.67	4.62
(14)	Rho coefficient (ρ)	0.59	0.69	0.59	0.76	0.86	0.53
(15)	Length area relation (Lar)	220.24	94.12	87.73	44.65	81.52	47.14
(16)	Lemniscate's (k)	1.59	4.61	3.98	2.02	5.62	3.29
(17)	Shape factor (Sf)	1.59	4.61	3.98	2.02	5.62	3.29
(18)	Texture ratio (Rt)	5.22	2.52	2.46	1.55	2.20	1.70
19)	Sinuosity, S (Higher Order)	1.27	1.26	1.33	1.51	1.32	1.24
Areal Parameters							
(1)	Basin Area (A), sq km	4584.42	1111.49	988.72	320.80	874.82	351.14
(2)	Relative Perimeter (P_r), km	9.45	4.51	4.48	2.74	3.85	2.86
(3)	Stream Frequency (Fs), km^{-2}	0.71	0.72	0.71	0.72	0.73	0.76
(4)	Drainage Density (Dd), km^{-1}	1.00	1.03	1.02	1.02	1.04	1.02
(5)	Drainage Intensity (Di), km^{-1}	0.71	0.69	0.69	0.71	0.70	0.74
(6)	Infiltration Number (If), km^{-3}	0.71	0.74	0.72	0.73	0.76	0.78
(7)	Elongation Ratio (Re)	0.89	0.52	0.56	0.79	0.48	0.62
(8)	Circularity Ratio (Rc)	0.24	0.23	0.26	0.29	0.21	0.29
(9)	Circularity Ration (Rcn), km	9.45	4.51	4.48	2.74	3.85	2.86
(10)	Compactness Ratio (Cc)	2.02	2.08	1.98	1.84	2.17	1.85
(11)	Form Factor (Ff)	0.63	0.22	0.25	0.49	0.18	0.30
(12)	Length of Overland Flow (Lg), km	0.5	0.48	0.49	0.49	0.48	0.49
(13)	Constant of Channel Maintenance (C), km	1.00	0.97	0.98	0.98	0.96	0.98
(14)	Drainage Texture (Rt), km^{-1}	6.69	3.27	3.17	1.98	2.81	2.17
Relief Parameters							
(1)	Minimum Height of Basin (z), m	241	278	294	274	330	333
(2)	Maximum Height of Basin (Z), m	580	495	559	396	580	421
(3)	Total Basin Relief (H), m	339	217	265	122	250	88
(4)	Slope Analysis	1.16–42.4	1.18–33.4	1.17–37.3	1.07–24.8	1.33–42.4	0.98–17.9
(5)	Contour Interval (Cin), m	30	30	30	20	30	20
(6)	Relief Ratio (Rh)	3.96	3.03	4.22	4.79	3.57	2.59

(continued)

Table 1 (continued)

S.No	Parameters	JW	SW1	SW2	SW3	SW4	SW5
(7)	Absolute Relief (Ra), m	580	495	559	396	580	421
(8)	Relative Relief Ratio (Rhp)	0.07	0.09	0.12	0.10	0.11	0.07
(9)	Dissection Ratio (Dis)	0.58	0.44	0.47	0.31	0.43	0.21
(10)	Gradient Ratio (Rg)	3.96	3.03	4.22	4.79	3.57	2.59
(11)	Watershed slope (Sw)	3.96	3.03	4.22	4.79	3.57	2.59
(12)	Ruggedness Number (Rn)	0.34	0.22	0.27	0.12	0.26	0.89
(13)	Melton's Ruggedness Number (MRn)	5.01	6.51	8.43	6.81	8.45	4.69
(14)	Hypsometric Integral (HI)	0.498	0.497	0.498	0.492	0.500	0.500

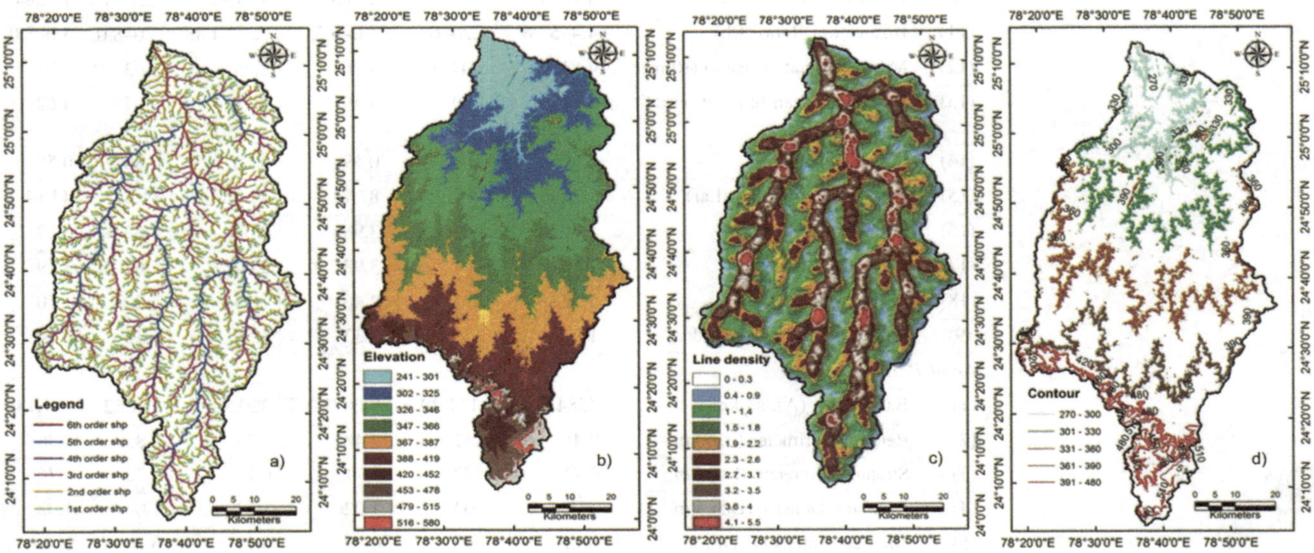

Fig. 3 **a** Drainage map; **b** Digital Elevation model (DEM); **c** Drainage density map and **d** Contour map of Jamini basin

km. The basin length is 85.52 km and the basin perimeter is 484.99 km. Among the five subwatersheds, Shahjad subwatershed has the largest perimeter (246.42 km). It is a factor determining the time travel by water (Schum 1956). Smith (1950) classified five types of density; extremely low (0 to 1), low (1 to 2), moderate (2 to 4), moderately high (4 to 6) and high (>6) density. The low density was observed in all subwatersheds of Jamini basin (Table 1; Figs. 3c, 5c, 6c, 7c, 8c and 9c). It reveals that the basin has dominantly consisted of permeable soil in a larger area with low erosion exposure. Stream frequency is expressed as the number of streams per unit area in a basin or it is a ratio of the total number of streams (Nu) in a basin (Horton 1945). The stream frequency (in km-2) is also classified as low (0 to 5), moderate (5 to 10), moderately high (10 to 15), high (15 to 20) and very high (20 to 25) by Horton (1945). The stream frequency (Fs) estimated for sub-watersheds and Jamini basin is listed in Table 1. It ranges from 0.71 to 0.76 (Table 1). The low value of frequency shown by each watershed is suggestive of occurrence of permeable soil and dense vegetation cover in the basin.

The low density shown by all subwatersheds of Jamini basin reveals that the coarse basin texture is predominant with permeable soil and thick vegetal cover. The elongation ratio is defined as the ratio of the diameter of a circle having the same area as the basin and maximum basin length (Schum 1956). The value of the elongation ratio generally varies from 0.6 to 10 over a large variety of climatic condition and geologic structures (Strahler 1964). The elongation value can be classified into three categories; elongated (0 to 0.7), oval (0.8 to 0.9) and circular (>0.9). The values of elongation ratio calculated for each subwatershed of Jamini basin ranges from 0.48 to 0.79 (Table 1). These values indicate that the all subwatersheds of Jamini basin are

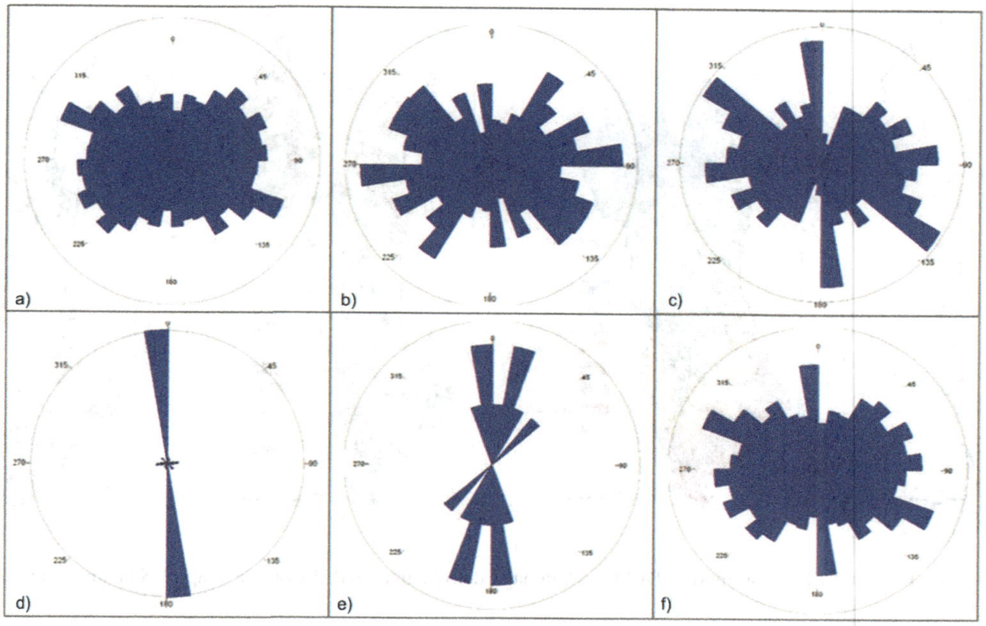

Fig. 4 **a–f** Rose diagrams showing Orientation of I, II, III, IV and V order of streams and mean orientation of all streams

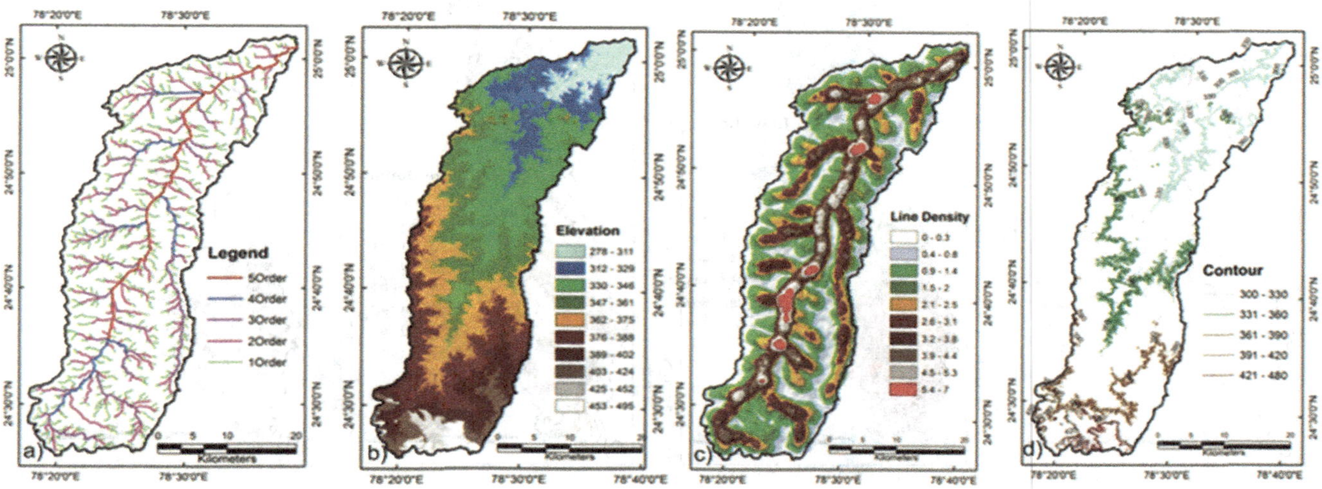

Fig. 5 **a** Drainage map; **b** Digital Elevation model (DEM map); **c** Drainage density map and **d** Contour map of Shahjad watershed

elongated (Figs. 3, 5, 6 7, and 9) except Baragl. Baragl is slightly oval (Fig. 8). Overall the Jamini basin is elongated in shape and is associated with low relief (Table to 1). The circulatory ratio is expressed as the ratio of basin area to the area of the circle (Miller, 1953; Strahler, 1964). The circulatory ratio is influenced by the length and frequency of streams, geological structure, land use land cover, climate, relief and slope of the basin (Miller, 1953). All subwatersheds of Jamini watershed are showing low values of circulatory ratio (Table 1). It implies that the Jamini basin is elongated and consists of permeable and homogenous geological terrain. All subwatersheds of Jamini basin are showing circulatory ratio (Rc) less than 0.4. It is indicative of a mature stage of watershed development.

Form factor (Ff) is represented by the shape of the watershed. It is a ratio of watershed area and square of total watershed length (Horton, 1945). The high values of form factor (>0.78) would always have a circular basin with high peak flows for shorter duration whereas the low values (<0.45). Form factor would have an elongated basin with low peak flows for a longer duration. The value of the form factor calculated for all subwatershed of Jamini basin ranges from 0.18 to 0.49 (Table 1). It indicates that all subwatersheds are elongated and exhibit low peak flow with longer duration.

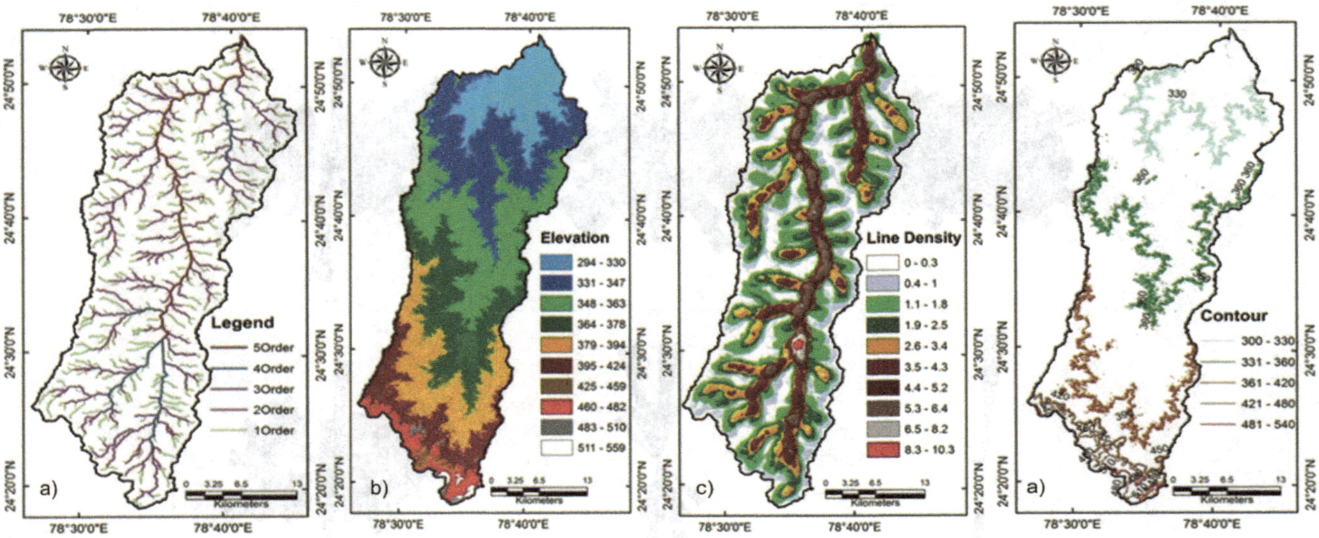

Fig. 6 **a** Drainage map; **b** Digital Elevation model (DEM); **c** Drainage density map and **d** Contour map of Sajnam watershed

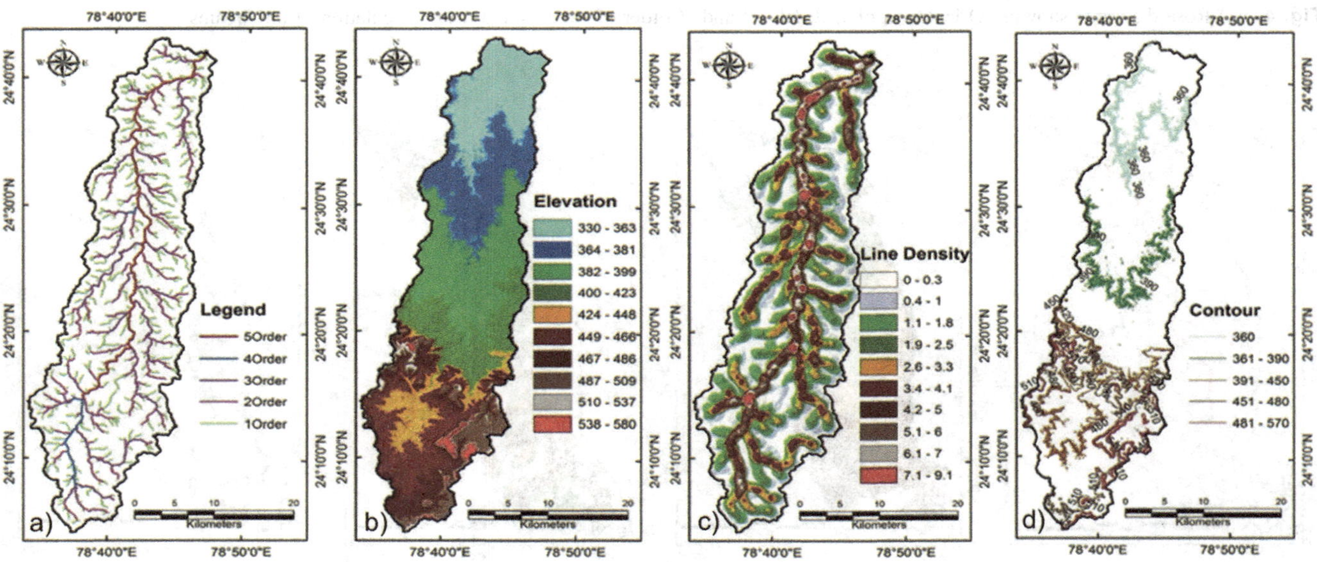

Fig. 7 **a** Drainage map; **b** Digital Elevation model (DEM); **c** Drainage density map and **d** Contour map of Onri watershed

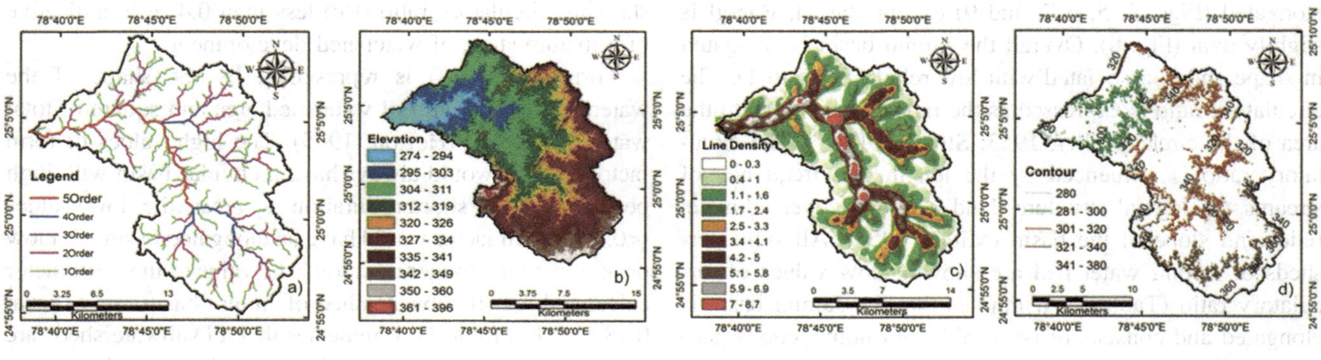

Fig. 8 **a** Drainage map; **b** Digital Elevation model (DEM); **c** Drainage density map and **d** Contour map of Bagal watershed

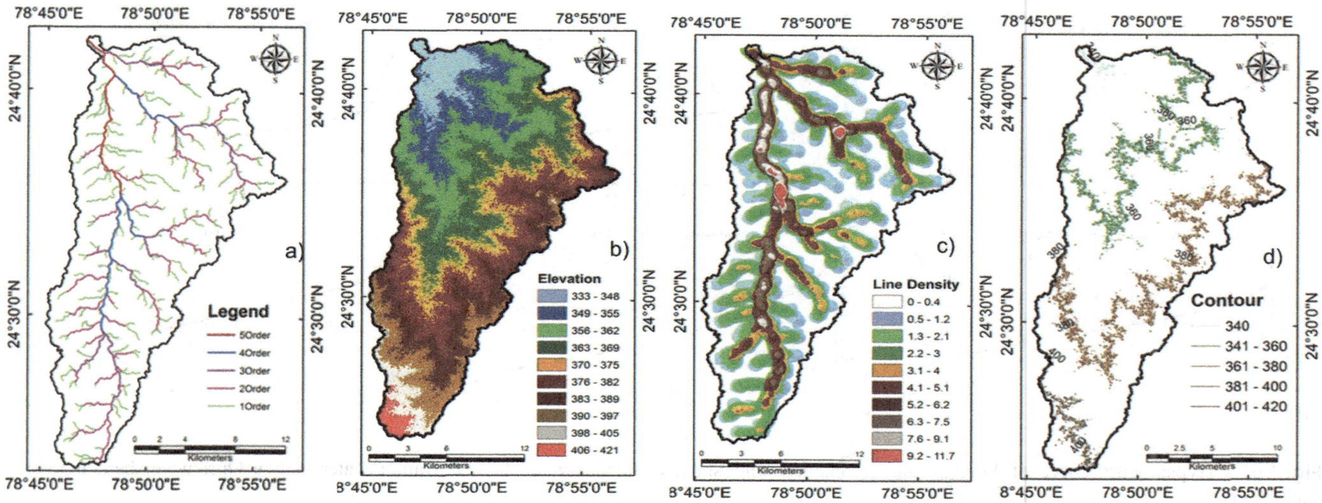

Fig. 9 **a** Drainage map; **b** Digital Elevation model (DEM); **c** Drainage density map and **d** Contour map of Jamrar watershed

5.3 Relief Parameters

The various relief parameters are given in Table 1. The highest and lowest points of a basin are referred to as the total relief of the basin. The stream gradient and influences of flood pattern and volume of sediments that transported are determined by basin relief (Hadley and Schumm 1961). The relief can be expressed as an area of a basin in three-dimensional view (Sahu et al. 2016). Basin relief measures the overall steepness of drainage and is an indicator of the intensity of erosion that occurred on the slope of a basin. The altitudinal differences in the Jamini basin (highest relief point 580 m and lowest point 241 m) and its subwatersheds from the sea level are listed in Table 1. The relief ratio is defined as the ratio of maximum basin relief to the longest horizontal distance of the basin measured parallel to the major stream (Samal et al. 2015). The relief ratio calculated for Jamini basin and its subwatersheds varies from 2.79 to 4.79. It indicates that the moderately steeper terrain is prone to erosion.

Ruggedness number (Rn) is expressed as the product of basin relief and drainage density (Strahlar, 1958; Vijith and Satheesh, 2006). It is related to the level of smoothness and roughness of the basin terrain or surface unevenness (Selvan et al. 2011). The ruggedness value is ranging from zero to one. The relatively more rugged terrain shows one ruggedness value. Contrary to this the smother terrain has ruggedness value less than one or zero. If the value of drainage density and relief is high then the value of ruggedness becomes higher. It indicates that the slope is not only steeper but is also long (Strahler1958). According to (Farhan et al. 2015), the ruggedness can be classified into four categories: subdued morphology (<0.1), slight morphology (0.1 to 0.4), moderate morphology (0.4 to 0.7) and sharp morphology (0.7 to 1.0) and high morphology (>1.0) refers to the high erosion-prone area. All subwatershed of Jamini basin except Jamrar subwatershed are showing ruggedness value from 0.12 to 0.27 (Table 1). It indicates that all subwatershed except Jamrar are less susceptible to erosion and are quite smoother. The Jamar watershed is showing rugged and sharp terrain and is more prone to erosion. Overall the value of ruggedness of Jamini basin is 0.34 (Table 1). It infers that the Jamini basin is slightly rugged and less prone to erosion and sediment transport.

5.4 Hypsometric Curve

Hypsometry is defined as the overall slope and the forms of drainage basin (Strahler 1952). These curves are related to the volume of the soil mass in the basin and the amount of erosion that had occurred in a basin against the remaining mass (Hurtrez et al. 1999). It is also used to differentiate between erosional landforms evolved at different cycles of drainage (youth, mature and old stages). Apart from erosional stages of landform, the hypsometric data can be used to determine the influence of tectonic, climatic and lithological factors on the evolution of drainage (Schumm 1956; Strahler 1956; Lifton and Chase 1992; Chen et al. 2003; Huang and Niemann 2006).

It can also explore the level of erosion of soil that has occurred in the past geological time (Bishop et al. 2002). The hypsometric curves have been classified into three types of landforms based on the shape of the curve; (1) young stage, (2) mature stage, and (3) old stage (Strahler 1952). The convex curves associated to the youth stage are indicating that the terrain is slightly eroded and undissected. The S-shaped curves matching with the mature stage are concave

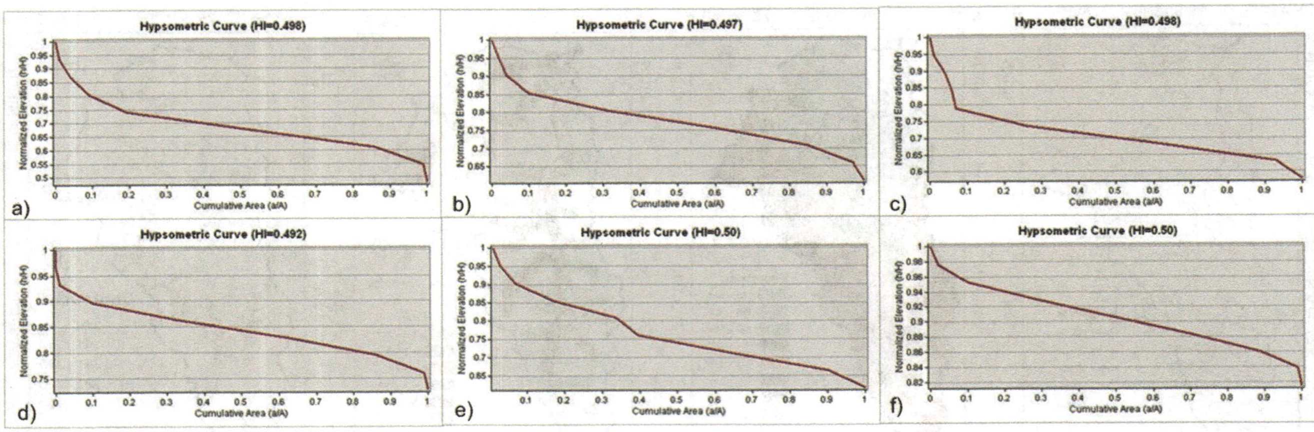

Fig. 10 **a** Hypsometric curve of Jamni basin; **b** Shahjad watershed; **c** Sajnam watershed; **d** Baragl watershed; **e** Onri watershed; **e** Jamrar watershed

upward at a higher altitude and convex downward at relief at a lower height and indicative of the occurrence of moderately eroded regions. The intensively eroded and deeply dissected landforms associated to concave curves are indicating the old stage of drainage.

In the present paper, all hypsometric curves resemble to S-shaped curve (Figs. 10a–f). It infers that all subwatershed of Jamini basin showing S-shaped curves have experienced moderate erosion activities during the development of the landscape. The hypsometric index value of all subwatersheds of Jamini basin ranging from 0.492 to 0.500 (Table 1) implies that the landscapes have reached the matured stage of evolution.

5.5 Correlation of Linear, Areal and Relief Parameters of Each Watershed of Jamini Basin

Inspection of the correlation matrix is important because it can point out the association among the variables that can show the overall coherence of the data set and indicate the participation of the individual parameter in several influencing factors. Inter parameter relationship also provide information on drainage behaviour and pathway of variables in a river system. Multiple regression analysis was carried and their results are given in Tables 2, 3 and 4 (Figs. 11, 12 and 13). Among the linear parameters shown in Table 2

Table 2 Correlation matrix of selected linear parameters

	Lurm	Luwm	Rf	Rw	Rbm	Rbwm	ρ	Lar	k	Sf	Rt	S
Lurm	**1**											
Luwm	**0.980**	**1**										
Rf	**0.692**	**0.770**	**1**									
Rw	–0.161	–0.132	–0.544	**1**								
Rbm	**0.690**	**0.803**	**0.799**	0.052	**1**							
Rbwm	0.042	0.010	–0.183	–0.097	–0.392	**1**						
ρ	**0.847**	**0.733**	0.308	–0.190	0.206	0.293	**1**					
Lar	0.058	0.166	–0.073	**0.856**	0.498	–0.223	–0.246	**1**				
k	**0.686**	**0.711**	**0.939**	–0.754	**0.564**	0.011	0.456	–0.379	**1**			
Sf	**0.686**	**0.711**	**0.939**	–0.754	**0.564**	0.011	0.456	–0.379	**1.000**	**1**		
Rt	0.003	0.114	–0.108	**0.864**	0.462	–0.221	–0.294	**0.998**	–0.414	–0.414	**1**	
S	0.068	–0.075	–0.480	0.050	–0.468	0.196	0.495	–0.379	–0.304	–0.304	–0.386	**1**

Lurm: Stream length used in the ratio; **Luwm:** weighted mean stream length ratio; **Rf:** Fitness ratio; **Rw:** wandering ratio; **Rbm:** mean bifurcation ratio; **Rbwm:** weighted mean bifurcation ratio; **ρ:** Rho coefficient; **Lar:** length area relation; **k:** lemniscate's; **Sf:** shape factor; **Rt:** texture ratio; **S:** Sinuosity (higher order)

Table 3 Correlation matrix of selected areal parameters

	Pr	Fs	Dd	Di	If	Re	Rc	Rcn	Cc	Ff	Lg	C	Rt
Pr	**1**												
Fs	–0.544	**1**											
Dd	–0.690	0.282	**1**										
Di	–0.172	**0.804**	–0.296	**1**									
If	–0.648	**0.943**	**0.577**	**0.577**	**1**								
Re	**0.552**	–0.277	–0.874	0.335	–0.539	**1**							
Rc	–0.403	0.360	–0.384	**0.614**	0.117	0.401	**1**						
Rcn	**1.000**	–0.544	–0.690	–0.172	–0.648	**0.552**	–0.403	**1**					
Cc	0.329	–0.331	0.454	–0.631	–0.071	–0.486	–0.995	0.329	**1**				
Ff	**0.612**	–0.324	–0.873	0.286	–0.573	**0.996**	0.322	**0.612**	–0.410	**1**			
Lg	**0.616**	–0.213	–0.966	0.381	–0.509	**0.873**	0.434	**0.616**	–0.493	**0.864**	**1**		
C	**0.690**	–0.282	–1.000	0.296	–0.577	**0.874**	0.384	**0.690**	–0.454	**0.873**	**0.966**	**1**	
Rt	**1.000**	–0.524	–0.687	–0.155	–0.629	**0.548**	–0.406	**1.000**	0.332	**0.608**	**0.612**	**0.687**	**1**

Pr: Relative perimeter; **Fs:** Stream frequency; **Dd:** Drainage density; **Di:** Drainage intensity; **If:** Infiltration number; **Re:** Elongation ratio;**Rc:** Circularity ratio;**Rcn:** Circularity ratio; **Cc:** Compac**Pr:** Relative perimeter; **Fs:** Stream frequency; **Dd:** Drainage density; **Di:** Drainage intensity; **If:** Infiltration number; **Re:** Elongation ratio;**Rc:** Circularity ratio;**Rcn:** Circularity ratio; **Cc:** Compactness coefficient; **Ff:** Form factor; **Lg:** Length of overland flow; **C:** Constant of channel maintenance; **Rt:** Drainage texture

Table 4 Correlation matrix of selected relief parameters

	H	Rh	Rhp	Dis	Rg	Sw	Rn	MRn	H I
H	**1**								
Rh	0.244	**1**							
Rhp	0.141	0.495	**1**						
Dis	**0.982**	0.319	0.140	**1**					
Rg	0.244	**1.000**	0.495	0.319	**1**				
Sw	0.244	**1.000**	0.495	0.319	**1.000**	**1**			
Rn	–0.430	–0.723	–0.604	–0.548	–0.723	–0.723	**1**		
MRn	0.242	0.419	**0.979**	0.226	0.419	0.419	–0.618	**1**	
HI	0.266	–0.707	–0.164	0.105	–0.707	–0.707	**0.596**	–0.048	**1**

H: Total basin relief; **Rh:** Relief ratio; **Rhp:** Relative relief ratio; **Dis:** Dissection ratio; **Rg:** Gradient ratio; **Sw:** Watershed slope; **Rn:** Ruggedness number; **MRn:** Melton ruggedness number; **HI:** Hypsometric integral

(linear parameters): Lurm, Luwm, Rf, Rbm, k and Sf bear a positive correlation among each other. P shows positive correlation with Lurm, Luwm; Lar, Rw and Rt are positively correlated among each other; while Sf shows strongest positive correlation with k. Rbwm, and S show the partial positive correlation or inverse correlation with the rest of the parameters which do not show the dependency on any other linear parameter.

In Table 3 (areal parameters): Pr, Re, Rcn, Ff, Lg, C and Rt are positively correlated among each other; Dd and Rc show positive correlation with If and Di, respectively; while Fs, Di and If are positively correlated among each other. Pr, Rcn and Rt show the strongest positive correlation among each other. Cc shows the partial positive correlation or inverse correlation with the rest of the areal parameters which do not show the dependency on any other parameter.

Relief parameters are also strongly correlated with each other which are shown in Table 4 (relief parameters), where: Dis positively correlated with H; Rg, Sw and Rh show strongest correlation among each other; MRn and HI bear a positive correlation with Rhp and Rn respectively; HI positively correlated with Rn. The relationship of linear, areal and relief parameters of Jamini basin and its subwatersheds were correlated in Figs. 13a, b and c. The uniform trends of these parameters imply that the basin was evolved in a geologically and geomorphologically homogenous terrain.

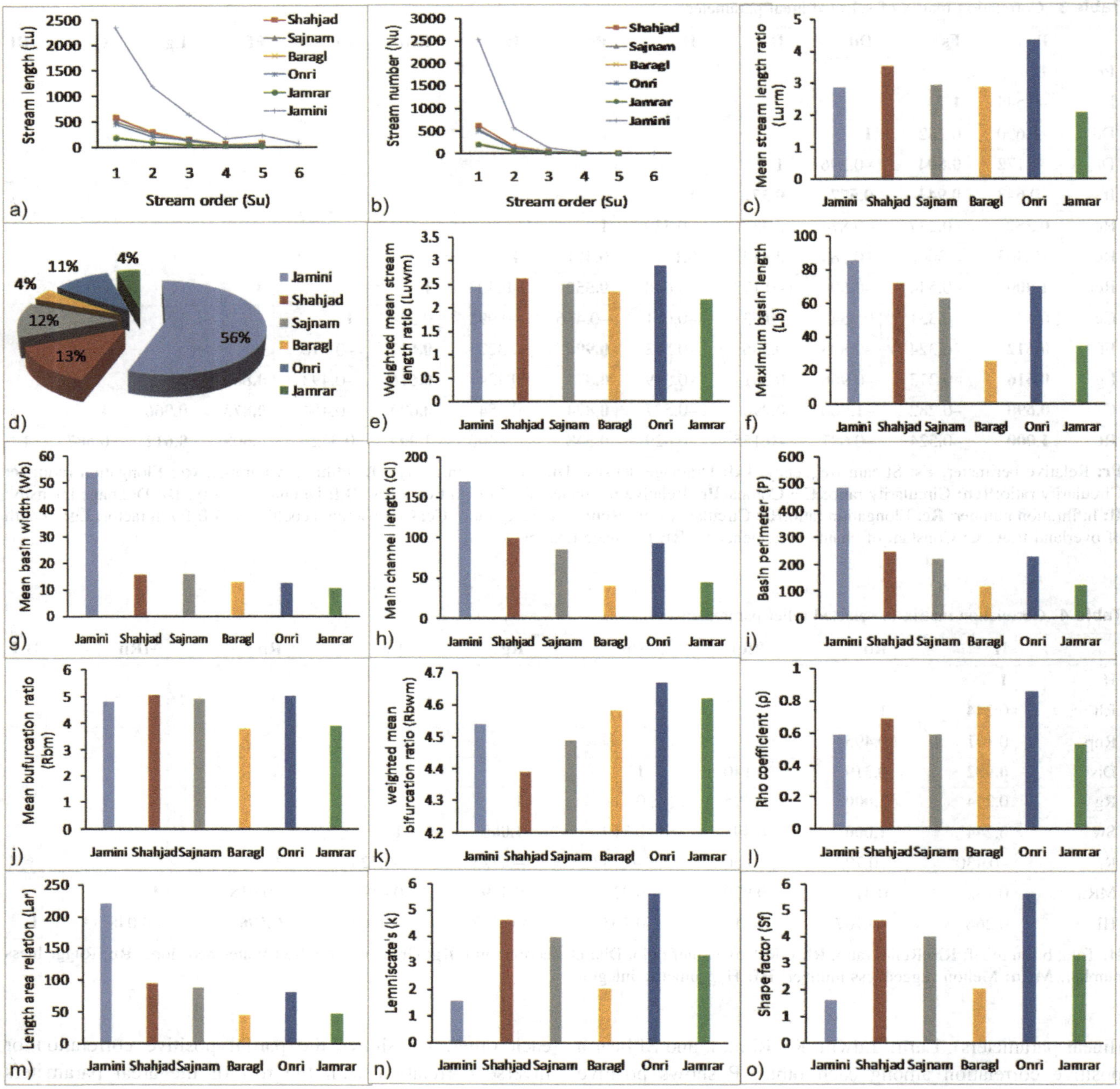

Fig. 11 shows the relationship between stream number versus stream order (**a**) and stream length versus stream order (**b**). It also shows the area percentage of Jamini river and its subwatersheds (**c–o**)

6 Conclusions

The morphometric analysis envisages that the elongated Jamini basin and its subwatersheds are showing impermeable to semi-permeable hard rock terrain in a high elevated area and permeable soil in the low lying area. The Jamini basin and its subwatersheds are significantly elongated and are more susceptible to erosion and sediment transport in a high elevated area. The S-shaped hypsometric curve indicates that the terrain has experienced moderate erosion process during landscape development. The hypsometric index value is indicative of the maturity stage of landscapes development. A statistical tool like correlation matrix analysis can be an effective method for characterization of correlation among multiple parameters, which determine the behavior of river basin. In this study, linear, areal and relief parameters are tabled in distinct correlation matrices and are

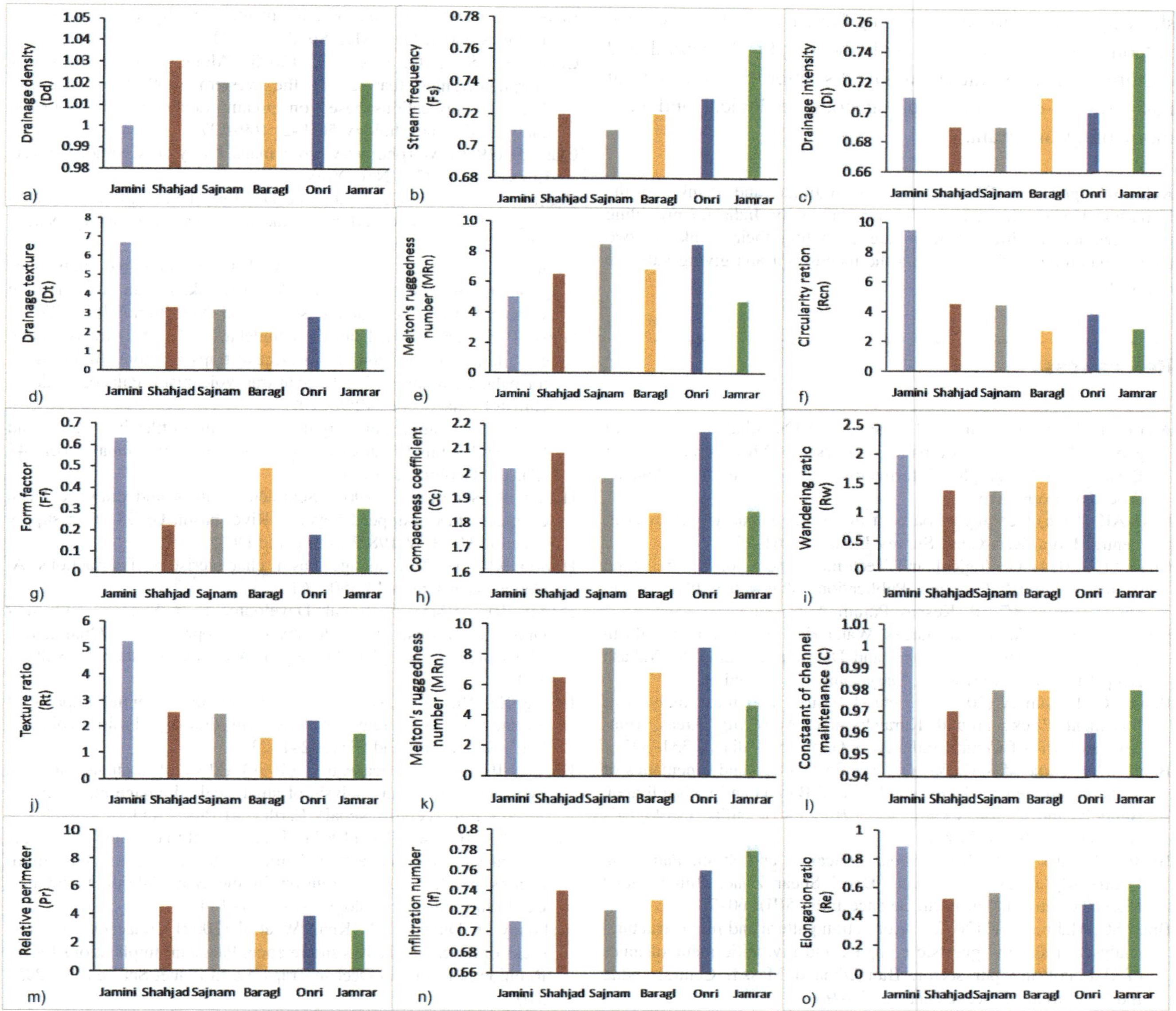

Fig. 12 The relationship of areal and relief parameters of each subwatershed of Jamini basin is represented by various trends

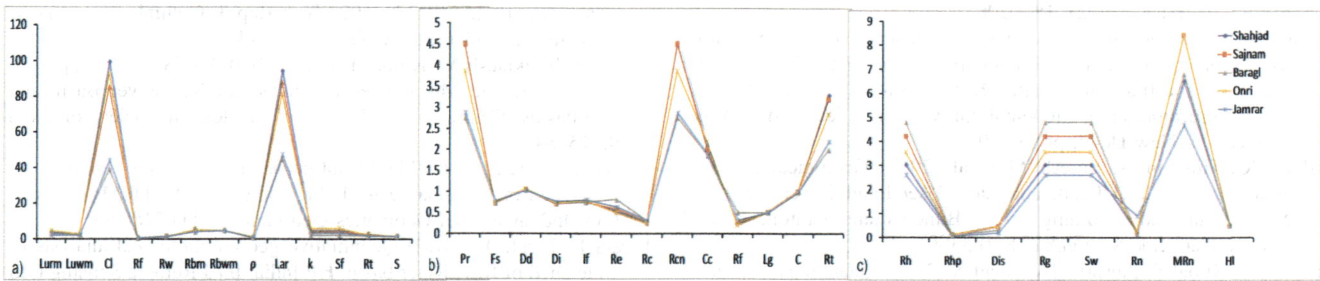

Fig. 13 shows the relationship between linear (**a**), areal (**b**), and relief parameters (**c**)

showing strong correlation among the cluster. This statistical outcome further reasserted with the graphical method and uniform trends of these parameters infer that the Jamini basin was evolved in a homogenous geological and geomorphological terrain.

Acknowledgements We express our gratitude and thanks to the Department of Geology, Bundelkhand University, India for providing the necessary facilities. Authors are extending their thanks to two anonymous referees for reviewing the manuscript and giving valuable suggestions.

References

Aparna P, Nigee K, Shimna P et al (2015) Quantitative analysis of geomorphology and flow pattern analysis of Muvattupuzha River Basin using Geographic Information system. Journal of Aquatic Procedia. 4:609–616

Basu AK (1986) Geology of parts of the Bundelkhand granite massif. Central India: Rec. Geol. Survey India. 117:61–124

Bhatt SC (2014) Geological and Tectonic Aspects of Bundelkhand Craton, Central India Angel Publication, Bhagwati publishers and Distributors C-8/77-B, Keshav Puram, New Delhi

Bhatt SC (2020) Natural Resources, Water Harvesting and Drought in Central India, Shree Publishers and Distributors 22/4735, Prakash Deep Building Ansari Road, Daryaganj, New Delhi

Bhatt SC, Husain A (2008) Structural History and fold Analysis of Basement Rocks around Kuraicha and Adjoining Areas, Bundelkhand Massif Central India Jour. Geol Soc India 72:331–347

Bhatt SC, Mahmood K (2008) Deformation pattern and kinematics of folds in basement rocks exposed around Babina and Ghisauli area. Bundelkhand Craton, Central India Bulletin of Indian Geologist's Association 41(1 & 2):1–16

Bhatt SC, Hussain A (2012) Shear Indicators and Strain Pattern in Quartz Mylonites of Chituad– Deori Shear Zone, Bundelkhand Massif, Central India, Earth Science India 5(II): 60–78

Bhatt SC, Mahmood K (2012) Deformation pattern and microstructural analysis of sheared gneissic complex and mylonitic metavolcanics of Babina-Prithvipur sector. Bundelkhand Massif, Central India: Indian Journal of Geosciences 66:79–90

Bhatt SC, Gupta MK (2014) Microstructural Analysis and Strain Pattern in Mylonites and Implications of Shear Sense Indicators in Evolution of Dinara- Garhmau Shear Zone, Bundelkhand Massif, Central India, The Indian Mineralogist, (Journal of Mineralogical Society of India), 48 (2): 186–206

Bhatt SC, Singh VK, Hussain A (2011) Implications of shear indicators for tectonic evolution of Mauranipur shear zone, Bundelkhand craton, central India. In: Singh VK, Chandra R (eds) 2nd Proceeding of Precambrian Continental Growth and Tectonism. Angel Publication, New Delhi, pp 36–49

Bhatt SC, Hussain A, Suresh M et al (2017) Geological structure control on Sukhnai Basin and Land Use/ Land Cover Pattern in Mauranipur and adjoining areas, Bundelkhand Craton, Central India. J Geol Soc Sri Lanka 18(2):53–61

Bhatt SC, Rubal S, Ansari MA, Bhatt S (2020) Quantitative morphometric and morphotectonic analysis of Pahuj catchment basin. Central India. JGSI 96(5):513–520

Bishop MP, Shorder JF, Bonk R et al (2002) Geomorphic Change in high mountains: A Western Himalaya Perspective. Global Planet Change 32(4):311–329

Bose MK (1995) Deccan basalts, mantle activity and extensional tectonics. Gond Geol Mag Sp. 2:267–281

Chen YC, Sung Q, Cheng K (2003) Along-strike variations of morphotectonics features in the western foothills of Taiwan: Tectonic implications based on stream-gradient and hypsometric analysis. Geomorphology 56(1–2):109–137

Clark JI (1966) Morphometry from maps Essay in Geomorphology. Elsevier Publ. Co., New York

Courtillot V, Feraud G, Vendamme D, Moreau MG, Besse G (1988) Deccan flood basalts and the Cretaceous/ Tertiary boundary. Nature 333:843–846

Farhan Y, Ali A, Omar E, Nisrin A (2015) Quantitative analysis of geomorphometric parameters of Wadi Kerak, Jordan, using remote sensing and GIS. J. Water Research and Protection 7:456–475

Ghosh P, Sayeed MRG, Islam R, Hundekari SM (2006) Inter-basaltic clay (bole-bed) horizons from Deccan traps of India: Implications for palaeo-climate during Deccan trap volcanism. Palaeogr Palaeoclimatol Palaeoecol 242:90–109

Hack JT (1957) Studies on longitudinal stream profiles in Virginia and Maryland. United States Geological Survey Professional paper 249 (B): US Geological Survey

Hadley R, Schumm S (1961) Sediment sources and drainage basin characteristics in upper Cheyenne River basin, USGS water supply paper 1531- B W 198.Washingaton DC

Horton RE (1932) Drainage basin characteristics Transactions A. Geophysical Union 13:350–361

Horton RE (1945) Erosional Development of Streams and their Drainage Basins: A Hydrophysical Approach to Quantitative Morphology. Geological Society of American Bulletin 56(3):275–370

Huang XJ, Niemann JD (2006) Modelling the potential impacts of groundwater hydrology on long-term drainage basin evolution. Earth Surf Proc Land 31:1802–1823

Hurtrez JE, Sol C, Lucazeau F (1999) Effect of drainage area on hypsometry from an analysis of small-scale drainage basins in the Siwalik hills (central Nepal). Earth Surf Proc Land 24:799–808

Lifton NA, Chase CG (1992) Tectonic, climatic and lithologic influence on landscape fractal dimension and hypsometry: Implications for landscape evolution in the San Gabriel Mountains. California. Geomorphology 5(1–2):77–114

Mahoney JJ, Duncan RA, Khan W et al (2002) Cretaceous volcanic rocks of the South Tethys suture zone, Pakistan: Implications for the Reunion hotspot and Deccan Traps. Earth Planet Sci Lett 203:295–310

Melton MA (1958) Correlations structure of morphometric properties of drainage systems and their controlling agents. J Geol 66:442–460

Miller VC (1953) A Quantitative Geomorphic Study of Drainage Basin Characteristics on the Clinch Mountain Area, Virginia and Tennessee, Proj., NR 389–402, Tech Rep 3, Columbia University, Department of Geology, ONR, New York

Panda B, Venkatesh M, Kumar B et al (2019) A GIS-based Approach in Drainage and Morphometric Analysis of Ken River Basin and Sub-basins. Central India, Journal of Geological Society of India 93:75–84

Pareta K, Pareta U (2011) Quantitative morphometric analysis of watershed of Yamuna basin, India using ASTER (DEM) data and GIS. Indian Journal Geomatics Geosciences 2(1):248–269

Pareta K, Pareta U (2012) Quantitative geomorphological analysis of watershed of Ravi River basin, HP India. Int J Remote Sensing GIS 1:47–62

Pati JK, Raju S, Mamgain VD et al (1997) Record of gold mineralisation in parts of Bundelkhand granitoid complex (BGC) J Geol Soc India 50: 601–606

Pati JK, MalviyaVP PK (2006) Basement re-activation and its relation to Neotectonic activity in and around Allahabad Ganga Plain. Journal of Indian Society of Remote Sensing 34:524–526

Pati JK, Patel SC, Pruseth KL et al (2007) Geology and geochemistry of giant quartz veins from the Bundelkhand craton, central India and their implications. J Earth Syst Sci 116(6):497–510

Pati JK, Lal J, Prakash K et al (2008) Spatio-temporal shift of western bank of the Ganga River, Allahabad city and its implications. J Indian Soc Rem Sens 36:289–297

Prakash K, Mohanty T, Singh S et al (2016) Drainage Morphometry of the Dhasan river basin, Bundelkhand craton, Central India using Remote Sensing and GIS techniques. Journal of Geomatics 10:121–132

Prakash K, Mohanty T, Pati JK et al (2017) Morphotectonics of Jamini River basin, Bundelkhand craton, Central India; using remote sensing and GIS Technique. Appl Water Sci 7:3767–3782

Prakash K, Rawat D, Singh S et al (2019) Morphometric analysis using SRTM and GIS in synergy with depiction: a case study of the Karmanasa River basin. North Central India Apply Water Sci 9 (1):13. https://doi.org/10.1007/s13201-018-0887-3

Sahu N, Reddy O, Kumar N et al (2016) Morphometric analysis in basaltic terrain of central India using GIS techniques: A case study. Appl Water Sci. https://doi.org/10.1007/s13201-0160442-z

Samal DR, Gedam SS, Nagarajan R (2015) GIS based drainage morphometry and its influence on hydrology in parts of Western Ghats region, Maharashtra. India. Geoc Int 30(7):755–778

Schumm SA (1956) Evolution of drainage systems and slopes in badlands at Perth Amboy, New Jersey. Geol Soc Am Bull 67:597–646

Selvan MT, Ahmad S, Rashid SM (2011) Analysis of the geomorphometric parameters in high altitude Glacierised terrain using SRTM DEM data in central Himalaya, India. ARPN Journal of Science and Technology 1(1):22–27

Shukla UK, Srivastava P, Singh IB (2012) Migration of Ganga river and development of cliffs in Varanasi Region during late Quaternary: role of active tectonics. Geomorphology 171–172:101–113

Singh VK, Slabunov A (2015) The Central Bundelkhand Archaean greenstone complex, Bundelkhand Craton, Central India: geology, composition, and geochronology of supracrustal rocks. Int Geol Rev 57(11–12):1349–1364

Singh VK, Slabunov AI, Nesterova NS et al (2021) Tectonostratigraphic terranes of the Bundelkhand Craton (Indian Shield). In: Shandilya AK, Singh VK, Bhatt SC, Dubey CS (eds) Geological and Geo-Environmental Processes on Earth, pp 155–164 (this volume)

Smith KG (1950) Standards for grading texture of erosional topography. Am J Sci 248:655–668

Strahler AN (1952) Hypsometric Analysis of Erosional Topography. Bulletin of the Geological Society of America 63:1117–1142

Strahler AN (1956) Quantitative Slope Analysis. Bulletin of the Geological Society of America 67:571–596

Strahler AN (1958) Dimensional analysis applied to fluvially eroded landforms. Geol Soc Am Bull 69:279–300

Strahler AN (1964) Quantitative Geomorphology of Drainage Basin and Channel Network. Handbook of Applied Hydrology: 39–76

Tangri AK (1992) Satellite remote sensing as a tool in deciphering the fluvial dynamics and applied aspects of Ganga Plain. In: Singh IB (ed) Proceedings of the workshop, Gangetic Plain: Terra Incognita. Department of Geology, Lucknow, p 73–84

Vijith H, Satheesh R (2006) GIS-based morphometric analysis of two major upland sub- watersheds of Meenachil river in Kerala. J Indian Soc Remote Sens 31:181–185

Morphometric and Morphotectonic Studies of Sindh Basin, Central India, Using Advance Techniques of Remote Sensing and GIS

S. C. Bhatt, Rubal Singh, Rahul Singh, V. K. Singh, and Adesh Patel

Abstract

In the present paper, morphometric and morphotectonic attributes of Sindh basin of Central India were attempted. The remotely sensed Shuttle Radar Topographic Mission (SRTM) Digital Elevation Model (DEM) data were used to extract drainage, density and contour maps. The drainage of Sindh and its subwatersheds exhibiting dendritic pattern show mean bifurcation ratio ranging from 3.85 to 5.66. It reveals that the drainage flowing in hilly and dissected area is geologically and structurally controlled. The low density (0.82) and frequency (0.4) implies that the basin is consisting of coarse permeable soil with dense vegetation. The values of elongation and form factor ratio infer that the basin is elongated and less prone to erosion. The ruggedness values infer that the Sindh basin is slightly rugged and less prone to erosion and sediment transport. The drainage orientations of Sindh River and its subwatersheds showing polymodal distribution are represented by highly conspicuous NNE-SSW and NNW-SSE trends. The four sets of major lineaments trending NE-SW, NW-SE, WNW-ESE and NNW-SSE are observed in the Sindh basin. The high density of lineaments represented by fractures, faults, foliations and bed rock structures are controlling the drainage of low order streams. The high order NE-SW trending streams show major tectonic control.

S. C. Bhatt (✉) · R. Singh · V. K. Singh · A. Patel
Department of Geology, Bundelkhand University, Jhansi, India
e-mail: scbhatt@bujhansi.ac.in

V. K. Singh
e-mail: vinodksingh@bujhansi.ac.in

R. Singh
Department of Geology and Geophysics, Indian Institute of Technology, Kharagpur, India

1 Introduction

In present scenario, the natural resources are playing significant role in socioeconomic development of a country. The excessive depletion of these resources is affecting the ecosystem and is degrading the environment globally (Bhatt 2020). Land degradation has become one of the important factor to decline agricultural productivity and soil fertility. Bundelkhand craton located in central India is facing severe problem of soil erosion, land degradation and drought. Its 11,000 square kilometer area is occupied by waste land and maximum part is affected by gulley erosion. The rocky degraded land affected by gullies and ravines is more dominant in catchment and lower basins of most of the Cratonic Rivers (Betwa, Ken, Sindh and Pahuj) of Bundelkhand. Despite having good network of rivers, ponds and wells this region is facing scarcity of water. Therefore, there is a need to conserve water, soil and other resources for development of society in such underprivileged area. The morphometric analysis is considered the best tool to understand the physical, hydrological and geological properties of a basin and for planning of water harvesting and soil protection schemes.

Morphometric analysis is a method that reveals the shape, size, hydrological, erosional and physical properties of landforms and soil of a basin (Horton 1932, 1945; Strahler 1956, 1964; Melton 1958; Miller 1953; Clark 1966; Pareta and Pareta 2012; Samal et al. 2015; Prakash et al. 2016, 2019; Bhatt et al. 2017, 2020; Smith 1950). It gives information about the characteristics of drainage system, topographic pattern, dimension of landforms, lithological and structural control on drainage pattern (Pati et al. 2006, 2008; Pareta and Pareta 2011; Prakash et al. 2016, 2017; Panda et al. 2019; Bhatt et al. 2017, 2020). The morphometric parameters also explore the status of erosion-prone areas, potential of surface and groundwater resources (Huang and Niemann 2006; Aparna et al. 2015). The geological and geomorphological studies based on remote sensing and GIS

A. K. Shandilya et al. (eds.), *Geological and Geo-Environmental Processes on Earth*, Springer Natural Hazards,
https://doi.org/10.1007/978-981-16-4122-0_17

are playing significant roles in understanding the landform processes and physical and erosional properties of soil (Horton 1945; Strahler 1952; 1964; Pareta and Pareta 2011; Prakash et al. 2016, 2019; Bhatt et al. 2017, 2020).

The morphometric analysis of Sindh basin and its subwatersheds would be more significant in planning the strategies for conservation of soil and water resources. The lineament analysis would be significant in understanding the tectonic influence and structural control on drainage.

2 Study Area

Sindh a major tributary of Yamuna River originates from Malva plateau of Vidhisha district of Madhya Pradesh in India and is 470 km in length (Figs. 1 and 2a). Its total catchment area is spreading in 26699 square km. It covers 461 km in Madhya Pradesh and 9 kilometer in Uttar Pradesh. Geologically the study area consists of Bundelkhand granitoids, sedimentaries of Gwalior basin. The detailed geology of the region studied by several workers (Bhatt and Hussain 2008, 2012; Bhatt and Mahmood 2008, 2012; Bhatt and Gupta 2014; Bhatt et al. 2011; Bhatt 2014, 2020; Singh and Slabunov 2015; Singh et al. 2021a, b). Sindh is flowing north-northeast through the districts of Guna, Ashoknagar, Shivpuri, Datia, Gwalior and Bhind in Madhya Pradesh to join the Yamuna River in Jalaun district, Uttar Pradesh, just after the confluence of the Chambal River with the Yamuna River. It is lying between 21°17′ N to 26°36′ N latitude and 74°02′ E to 82°26′ E longitude (Fig. 1). Important tributaries (subwatersheds) of Sindh are Pahuj, Parbati, Mahuar, Kunwari and Vaishali. The basin receives maximum precipitation from June to September.

Pahuj another subwatershed of Sindh is flowing through granitic terrain of Jhansi district and meets to Sindh in Jalaun district. It is also known as Puspavati and is flowing in northeast direction. It lies between 25°18′30″ and 25°31′27″ North latitude and 78°17′33″–78°33′22″ East longitude in Central India. It is 175 km in length and occupies 4258.40 sq km area. A small river having small course and dried during summer season is known as Mahuar. It is also known as Samoha river and is passing through Karera wildlife sanctuary. It is forming small subwatershed of Sindh. Kunwari, locally called Kwari a tributary of Sindh is flowing through Bhind and Morena districts of Madhya Pradesh. Kailaras, Sheopur, Morena and Bijeypur are its small tributaries.

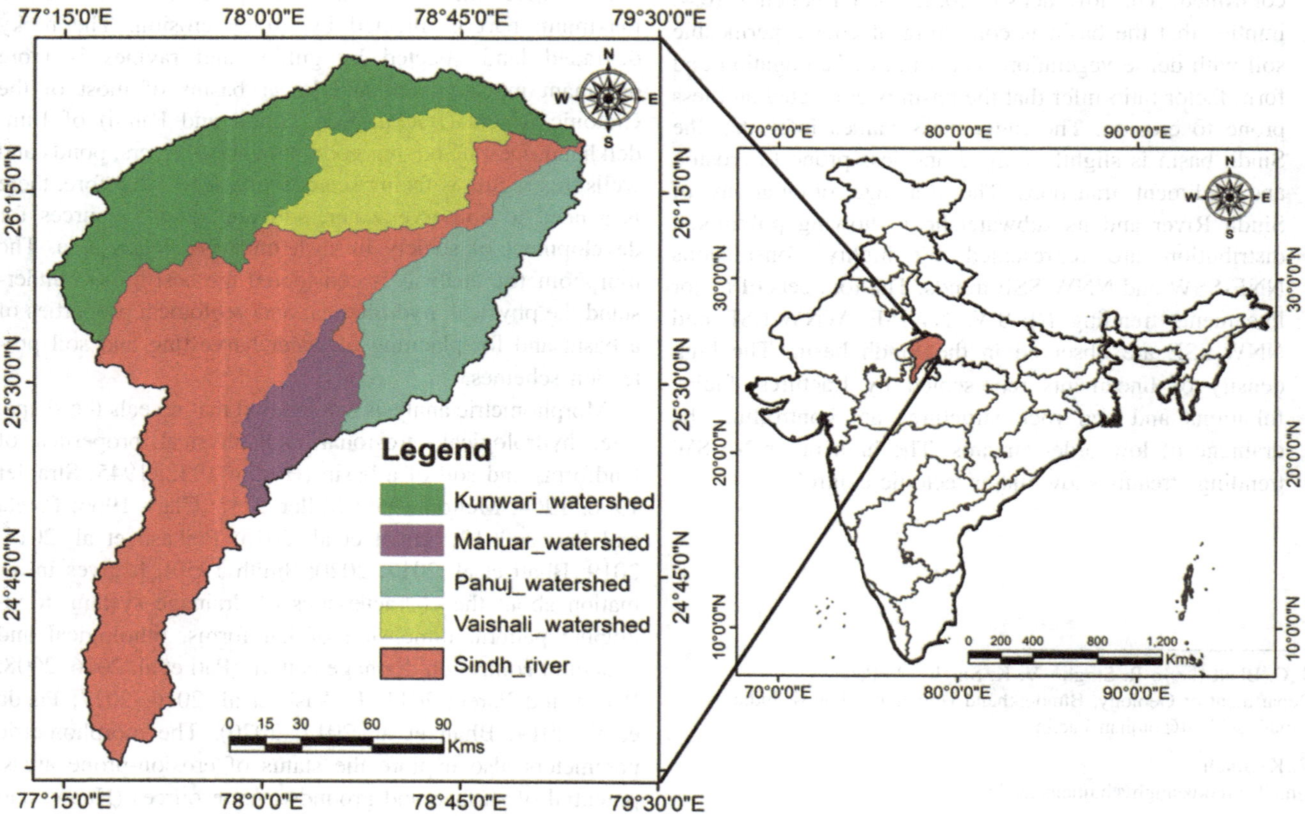

Fig. 1 Location map of Sindh and its subwatersheds

3 Methodology

In the present work, the remotely sensed SRTM and DEM data with capture solution of 3 arc second and pixel solution of 30 m were used to extract the drainage, density, contour, aspect and other maps (Figs. 2, 3, 4, 5 and 6). The DEM data were mosaic to subset the study region and the drainage map was extracted from the DEM data. The map was projected into proper coordinate system and WGS 1984 UTM Zone 44 N was used as appropriate tools in Arc Toolbox. The ArcGIS software was used to determine the length, area of the subwatershed and main watershed and stream order has been found out using the system, and the Arc Hydrotool in ArcGIS-10 software.

4 Results and Discussion

4.1 Morphometric Analysis

4.1.1 Linear Aspects

Linear parameters are one-directional character revealing the stream order, stream number, stream length, stream number ratio and collection of segments (Horton 1945; Strahler 1957). Based on Strahler's method the stream order was classified. Each rill or gulley generated first is designated as first order stream. When two first order streams meet then the second order stream is developed. Similarly, when two similar order streams confluence then the higher order streams is formed. Eighth orders of streams were identified in the Sindh basin (Fig. 2a) and seventh in Kunwari (Fig. 3a), fifth in Mahuar (Fig. 4a), sixth in Vaisali subwatershed (Fig. 5a) and sixth order of streams in Pahuj subwatersheds were validated (Fig. 6a). The drainage, density, contour, aspect and slope maps were extracted from the DEM (Figs. 2b–d, 3b–d, 4b–d, 5b–d and 6b–d). The total number of streams are found 11,003. Out of which 8493 belong to first order, 1955 to second order, 428 to third order, 93 to fourth order, 20 to fifth order, 11 to sixth order, 2 to seventh order and one to eighth order. The smaller streams are more prominent in steeper slope while the longer streams are represented by gentle and flat terrains. The number of streams decreases as the order increases and the stream length is inversely proportional to the number of stream (Horton 1945). The maximum number of streams (8493) were recorded in low order of streams of Sindh and its subwatersheds. The basin and its subwatersheds are showing dendritic to subdendritic stream patterns and were evolved in a massive hard rock terrain with numerous stream flows. The mean stream length ratio computed for Sindh and its subwatershed varies from 1.90 to 2.74 (Table 1). The bifurcation ratio is defined as the ratio of number of streams of a given order (N_u) to the number of streams to next higher order (Strahler 1964). The values of bifurcation ratio computed for Sindh basin and its subwatershed are given in Table 1. Its value varies minimum of 2 for flat or rolling drainage basins and 3–4 in mountainous or highly dissected drainage basins (Horton 1945). The bifurcation ratio computed for Sindh is ranging from 1.81 to 5.50 and for its subwatersheds varies from 2.00 to 9.00 (Table 1). Mean bifurcation ratio varies from 3.85 to 5.66 (Table 1). This infers that the basin was evolved in homogenous geological terrain and is influenced by geological structures.

4.1.2 Areal Parameters

The two-dimensional properties of a basin are discussed under areal parameters. The total basin area is spreading in 27,820 square kilometer and the perimeter of a basin is 1394.22 km. Horton (1945) and Strahler (1964) explained drainage density as the sum of stream lengths per unit area or it is an expression of closeness or spacing of channels. The density can be classified into five types (Smith 1950); extremely low (0–1), low (1–2), moderate (2–4), moderately high (4–6) and high (>6) density. The high density is related to fine texture with impermeable soil, sparse vegetation and high relief showing high run off and high potential of erosion (Horton 1945; Strahler 1964). Contrary to this the low density representing coarse texture is indicating occurrence of permeable soil with thick vegetation cover and low relief (less prone to erosion activity) in the basin. The value of density calculated for Sindh and its subwatersheds is ranging from 0.76 to 1.10 (Table1; Figs. 2c, 3c, 4c, 5c and 6c). It infers that the basin is consisting of coarse permeable soil with, thick vegetation and is showing gentle slope less susceptible to erosion. Horton (1945) defined stream frequency as the number of streams per unit area in a basin or it is a ratio of total number of streams (N_u) in a basin. It can be classified as low (0–5 km^{-2}), moderate (5–10 km^{-2}), moderately high (10–15 km^{-2}), high (15–20 km^{-2}) and very high (20–25 km^{-2}) by Horton (1945). The stream frequency (F_s) calculated for Sindh and its subwatersheds varies from 0.39 to 1.39. It is indicative of occurrence of permeable soil and dense vegetation cover in the basin.

The elongation ratio is defined as the ratio of diameter of a circle having the same area as of the basin and maximum basin length (Schumm 1956). Four categories of elongation ratio are found: elongated (0–0.7), less elongated (0.7–0.8), oval (0.8–0.9) and circular (>0.9). The value of elongation ratio for Sindh and its subwatersheds ranges from 0.45 to 0.62 (Table 1). It implies that the Sindh basin and its subwatersheds are elongated with moderate relief and are showing low vulnerability to erosion. The circulatory ratio refers to the ratio of watershed area (A) to the area of a circle having the same circumference as the perimeter of the watershed. Its value ranges from 0 to 1 it is influenced by geological structures, stream frequency, stream length,

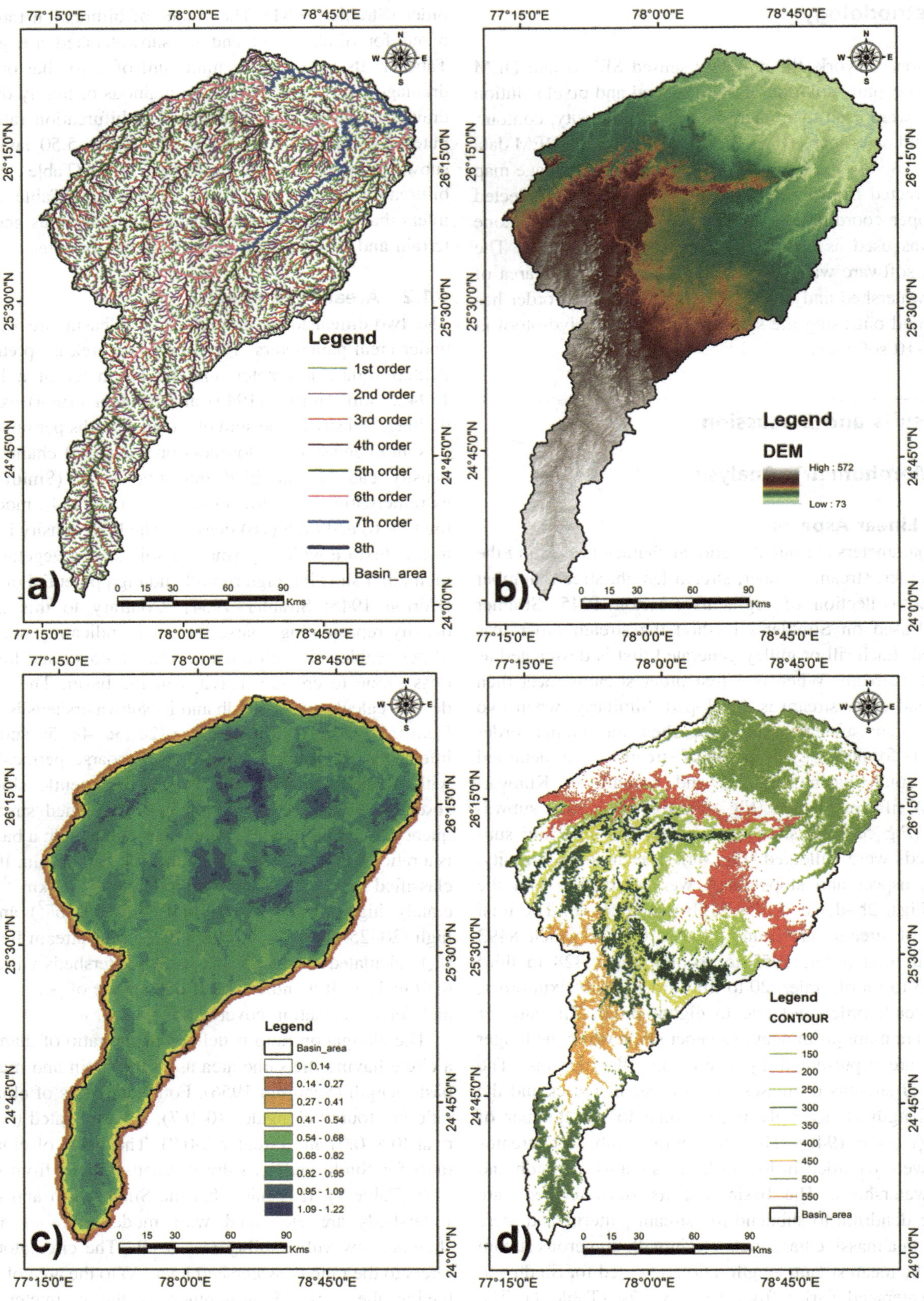

Fig. 2 **a** Drainage map, **b** DEM map, **c** Density map and **d** Contour maps of Sindh basin

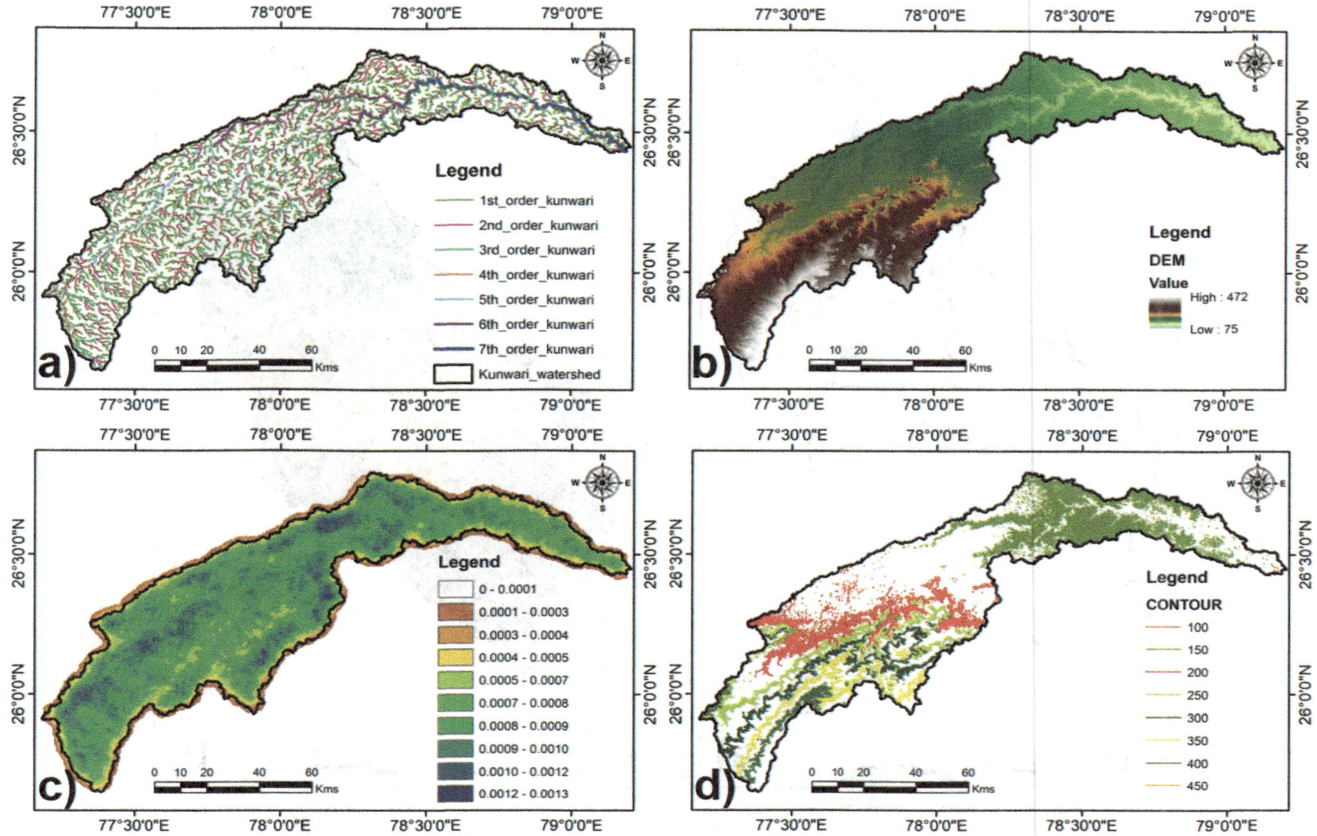

Fig. 3 **a** Drainage map, **b** DEM map, **c** Density map and **d** Contour map of Kunwari subwatershed

climate land cover, relief and slope of a watershed (Miller 1953; Strahler 1964). It also indicates the young, mature and old maturity level of the geomorphic cycle (Magesh et al. 2011). The circulatory ratio of Sindh watershed is computed 0.18. It is indicative of youth to mature stage of watershed development.

The shape of the watershed is represented by form factor (Horton 1945). It is a ratio of watershed area and square of total watershed length (Horton 1945). The high values of form factor (>0.78) would always have circular basin with high peak flows for shorter duration whereas the low values (<0.45) form factor would have elongated basin with low peak flows for longer duration. The value of form factor for Sindh and its subwatersheds varying from 0.16 to 0.30 (Table 1) indicates that the basins are elongated and show low peak flow of longer duration.

4.1.3 Relief Parameters

The relief parameters are three-dimensional characters of a basin and can be expressed in terms of area, volume and altitude of a watershed (Sahu et al. 2016). Basin relief is the difference in height between the highest and lowest points of the watershed (Hadley and Schumn 1961). It is a parameter that determines the stream gradient and influences flood pattern and volume of sediments that can be transported (Hadley and Schumm 1961). The minimum elevation in Sindh basin is 73 m and the maximum height is 572 m above sea level. The relief ratio has direct relation with erosion potential of a basin. The value of relief ratio computed for Sindh basin and its subwatersheds (Table 1) implies that the low erosion risk is in the gentle slopes of granitic and Vindhyan hills.

Ruggedness (Rs) is defined as the product of basin relief and drainage density (Strahlar 1958). It has direct relationship with structural complications and erosion potential of a basin since it joins the gradient and its length (Pareta and Pareta 2011; Bhatt et al. 2020). The ruggedness value is ranging from zero to one. The relatively more rugged terrain shows one ruggedness value. Whereas the smother terrain shows less than one or zero ruggedness value. If the value of drainage density and relief are high then the value of ruggedness becomes higher. It indicates that the slope is not only steeper, but it is long also (Strahler 1958). Farhan et al. (2015) has given the four categories of ruggedness value: subdued morphology (<0.1), slight morphology (0.1–0.4), moderate morphology (0.4–0.7) and sharp morphology (0.7–1.0) and high morphology (>1.0) refers to high erosion-prone area. The ruggedness value of Sindh basin and

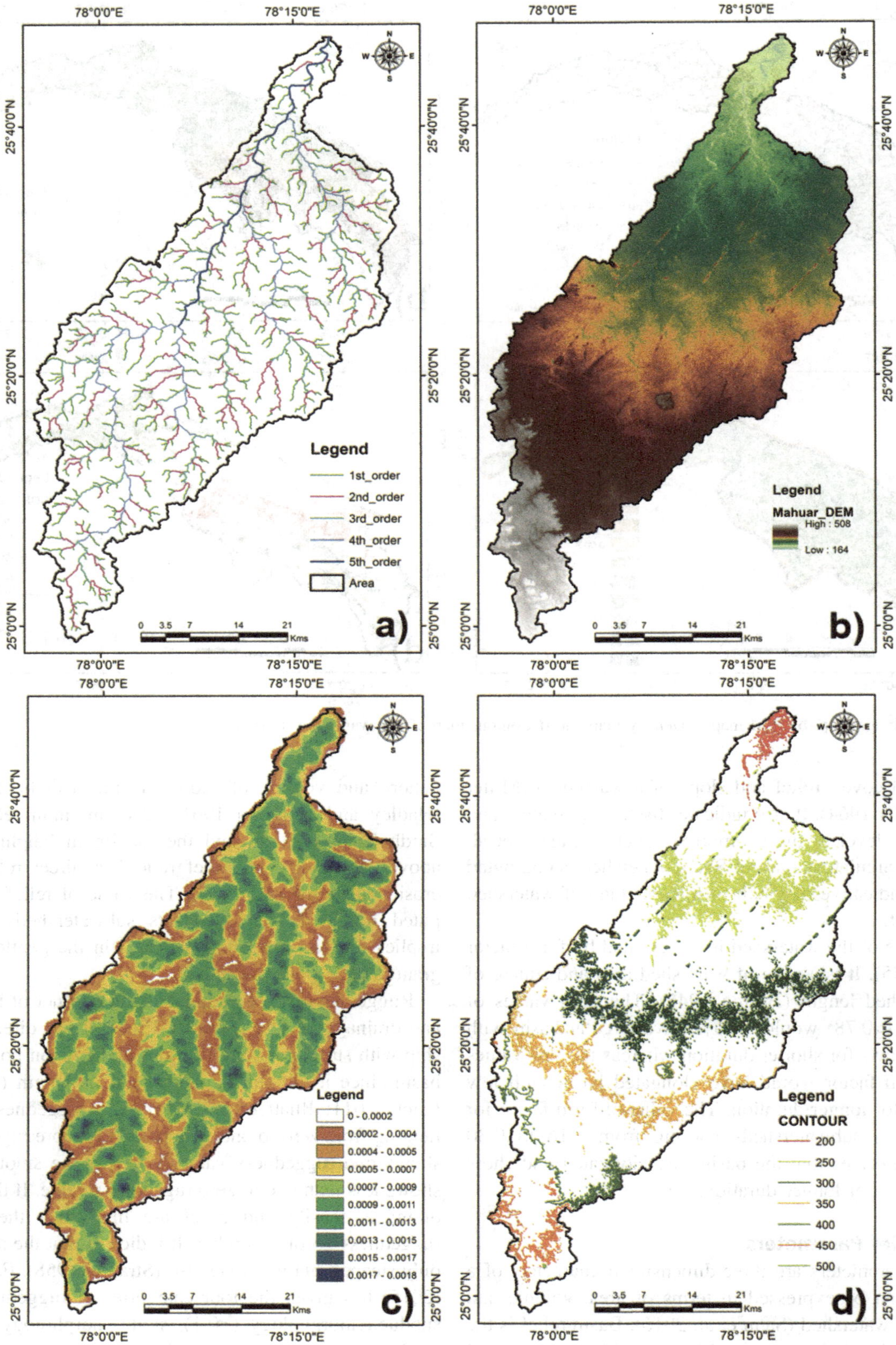

Fig. 4 **a** Drainage map, **b** DEM map, **c** Density map and **d** Contour map of Mahuar subwatershed

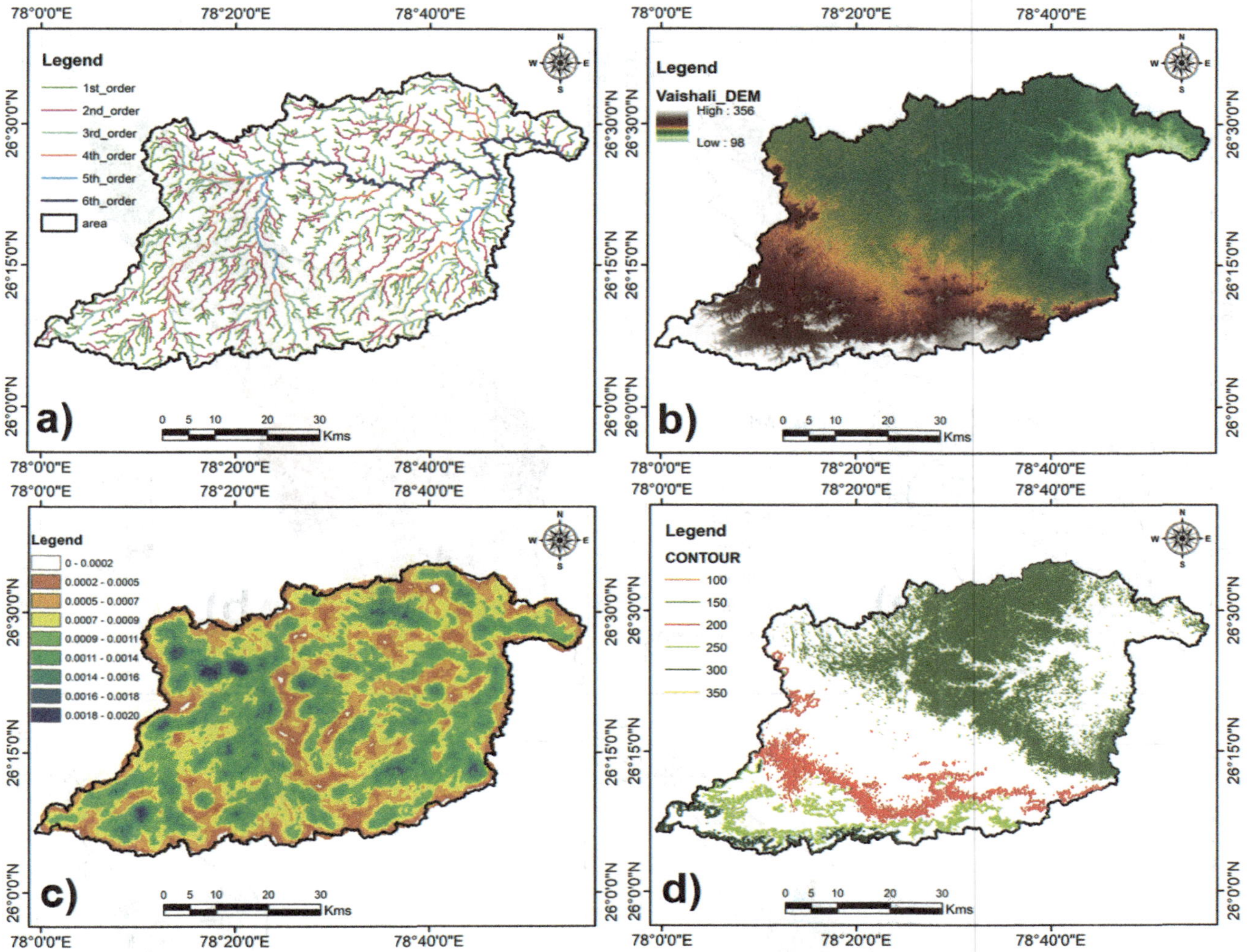

Fig. 5 **a** Drainage map, **b** DEM map, **c** Density map and **d** Contour map of Vaisali subwatershed

its subwatershed varies from 0.26 to 0.41 (Table1). It indicates that Sindh watershed has smoother terrain and has low potential of erosion. It infers that the Sindh is slightly rugged and less prone to erosion and sediment transport.

4.2 Morphotectonic Parameters

4.2.1 Drainage Pattern

The characteristic of streams can be used as indicators of underlying geological structures. Stream incisions, diversions and drainage network give strong evidence to understand geological and structural history of the area and the properties of streams are very important for the study landform processes (Strahler 2002). The dendritic to subdendritic drainage pattern is shown by the Sindh watershed and its all subwatersheds (Figs. 2a, 3a, 4a, 5a and 6a). The drainage orientations of Sindh River and its subwatersheds are shown by polymodal distribution Figs 7, 8, 9, 10 and 11. The NNE-SSW (Figs. 7b, e, 8a, d, e, 9d, e, 10c–f and 11b, c, e) and NNW-SSE trends (Figs. 7c, d, 8c, d, 9a, c, d, e, g and 11d) of streams are highly conspicuous.

The first and second order streams marked by rills and gullies are showing variable trends. The dominant WNW-ESE (Figs. 7a, 8a, g, h, 9d, 10a, g and 11a, g, i) and NW-SE (Figs. 7a, f, 8h, 9d, 10a, g, and 11a, i) trends were noticed in first order streams of Mahuar, Kunwari, Vaishali and Pahuj subwatersheds of the Sindh watershed. The maximum of rose diagrams of third and fourth order streams of all subwatersheds trends in NNW-SSE (Figs. 7c, d, 8c, d, 9a, c, d, e, g and 11d) and WNW-ESE directions (Figs. 7a, 8a, g, h, 9d, 10a, g, 11a, g, i). The azimuthal trend of higher order streams (V and VI order) showing straight and curved courses are trending in NE-SW (Figs. 7f, 9b, 10e, g and 11e), ENE-WSW (Fig. 9d) and NNE-SSW (Figs. 7b, e, 8a, d, e, 9d, e, 10c–f and 11b, c, e) directions.

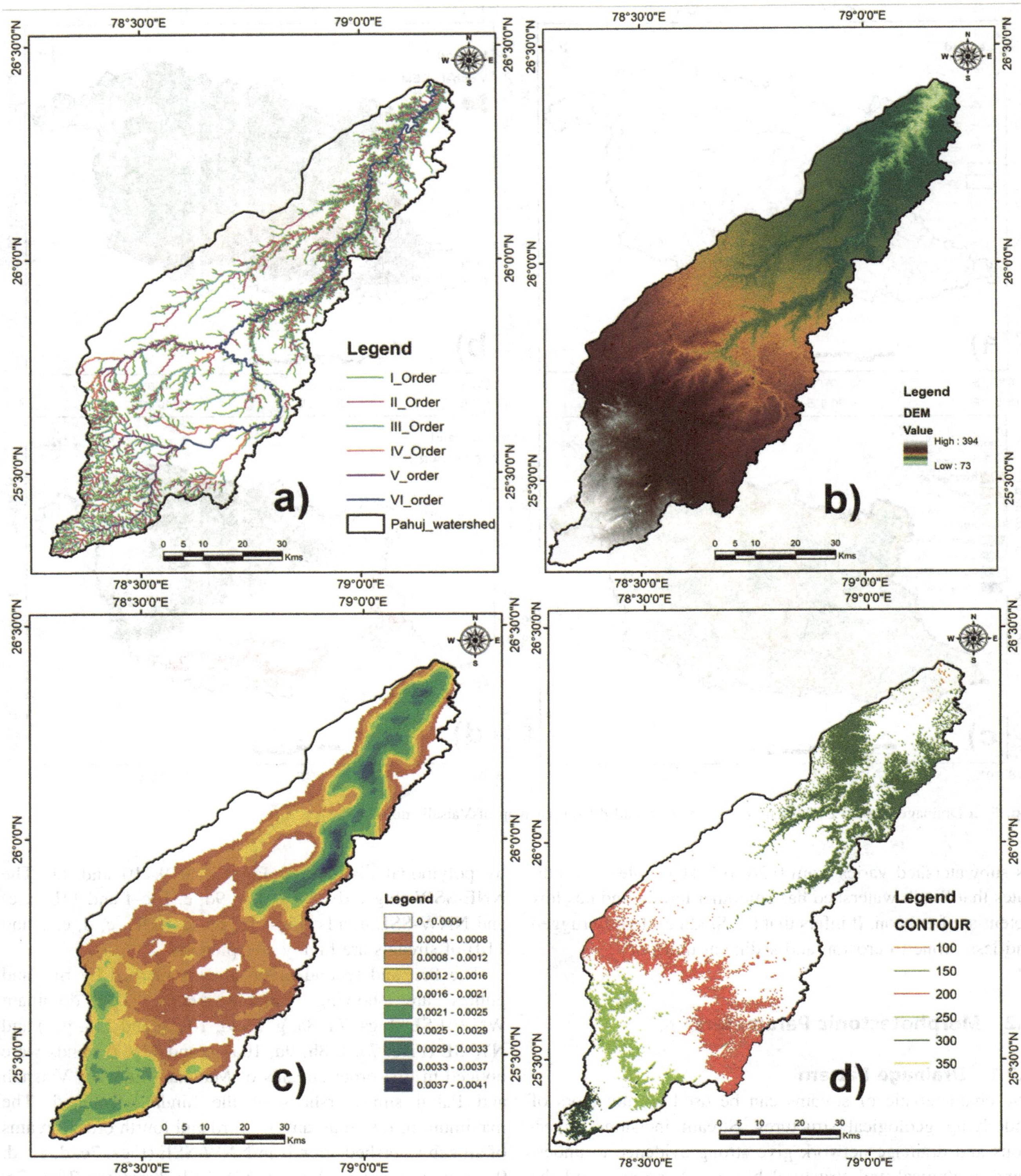

Fig. 6 **a** Drainage map, **b** DEM map, **c** Density map and **d** Contour map of Pahuj subwatershed

Table 1 Linear, areal and relief parameters of Sindh basin and its subwatersheds

S. no	Parameters	Sindh river	Mahuar river	Kunwari river	Vaisali river	Pahuj river
	Linear parameters					
(1)	Stream order (S_u)	1–8	1–5	1–7	1–6	1–6
(2)	Stream length (L_u) (Km)	22,754.9	1374.91	5614.09	2754.11	4700.78
(3)	Total number first order streams (N_1)	8586	567	2043	923	4555
(4)	Total number of streams (N_u)	11,091	735	2629	1203	5939
(5)	Mean stream length ratio (L_{urm})	1.996	2.74	2.21	2.55	1.90
(6)	Weighted mean stream length ratio (L_{uwm})	2.20	2.52	2.19	2.25	2.18
(7)	Fitness ratio (Rf)	0.28	0.20	0.16	0.3	0.37
(8)	Maximum basin length (Lb) (Km)	314.46	95.00	208.36	102.32	150.49
(9)	Basin perimeter (P) (Km)	1394.22	328.70	908.46	421.74	482.40
(10)	Bifurcation ratio (R_b)	1.81–5.5	4.12–5.44	2.0–4.49	3.00–4.98	4.00–9.00
(11)	Mean bifurcation ratio (R_{bm})	3.92	4.89	3.85	3.98	5.66
(12)	Weighted mean bifurcation ratio (R_{bwm})	4.39	4.39	4.51	4.31	4.34
(13)	Texture ratio (R_t)	6.16	1.72	2.25	2.19	9.44
	Areal parameters					
(14)	Basin area (A)	27,819.99	1816.41	6775.75	3123.61	4258.40
(15)	Relative perimeter (P_r)	19.95	5.53	7.46	7.41	8.83
(16)	Stream frequency (F_s)	0.40	0.40	0.39	0.39	1.39
(17)	Drainage density (D_d)	0.82	0.76	0.83	0.88	1.10
(18)	Drainage intensity (D_i)	0.49	0.53	0.47	0.44	1.26
(19)	Infiltration number (If)	0.33	0.31	0.32	0.34	1.54
(20)	Elongation ratio (R_e)	0.60	0.51	0.45	0.62	0.49
(21)	Circularity ratio (R_c)	0.18	0.21	0.10	0.22	0.23
(22)	Form factor (F_f)	0.28	0.20	0.16	0.30	0.19
(23)	Length of overland flow (L_g)	0.61	0.66	0.60	0.57	0.45
(24)	Constant of channel maintenance (C)	1.22	1.32	1.21	1.13	0.91
(25)	Drainage texture (D_t)	7.95	2.24	2.89	2.85	12.31
	Relief parameters					
(26)	Minimum height of basin (z) (m)	73	164	75	98	73
(27)	Maximum height of basin (Z) (m)	572	508	472	356	394
(28)	Total basin relief (H) (m)	499	344	397	258	321
(29)	Contour interval (C_{in}) (m)	50	50	50	50	50
(30)	Relief ratio (R_h)	1.59	3.62	1.91	2.52	2.13
(31)	Ruggedness number (R_n)	0.41	0.26	0.33	0.23	0.35

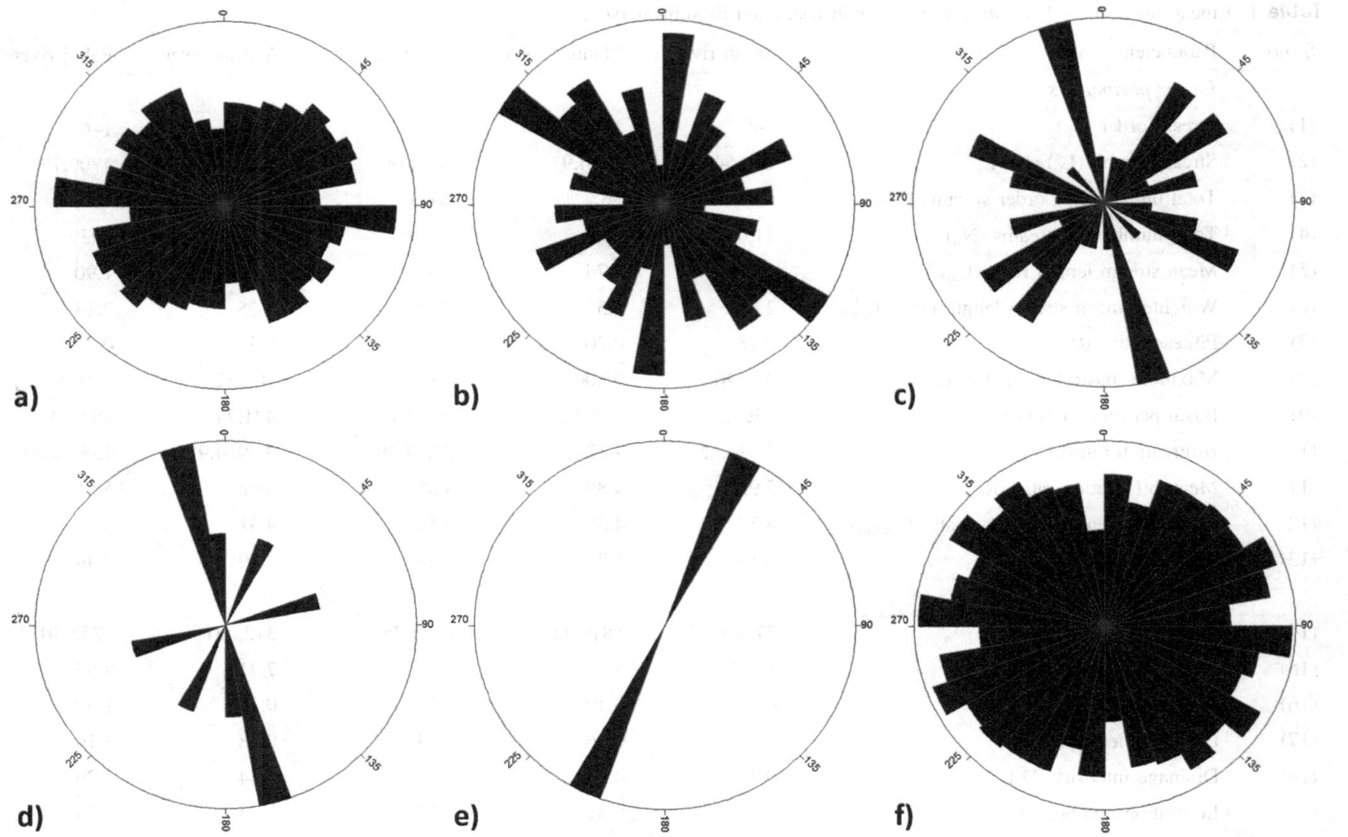

Fig. 7 Rose diagrams showing orientation of first (**a**), second (**b**) third (**c**), fourth (**d**), fifth stream order (**e**) and mean of stream orders (**f**) in Mahuar subwatershed

4.2.2 Orientation of Lineaments

The structural and morphological features and intensity of deformation on earth's surface are represented by important lineaments (Nur 1982). The lineament map of the Sindh basin and its subwatersheds was prepared by using SRTM and DEM Geomatica software (Fig. 12). The lineaments were classified based on orientations of structural and morphological features, drainage pattern, textural contrast and tonal differences. The four sets of major lineaments showing NE-SW, NW-SE, WNW-ESE and NNW-SSE were commonly observed in this study (Figs. 12 and 13). The orientations of lineaments extracted for each subwatershed of Sindh basin are represented by rose diagrams (Fig. 13). The NE-SW trending major lineaments following the trend of major tectonic unit in the study area are dominant in all subwatersheds of the basin (Fig. 13a, b, d, e). The NW–SE lineaments (other major linear features) were also observed in the basin (Fig. 13c).

4.2.3 Relationship Between Orientations of Lineaments and Drainage Pattern

The tectonic processes, climatic perturbations and their interactions at long time scales exert primary controls on the trajectory of landscape evolution, which defines the physical appearance of landscape (Bishop 2007). Each river basin in India evolved in different geological and tectonic history and developed in millions of years of time scale (Tangri 1992). The north flowing cratonic rivers are tectonically inactive and are represented by low relief granitic hills and reefs (Shukla et al. 2012; Jain et al. 2020). The Archean to Palaeoprterozoic gneissic and granitic terrain of Bundelkhand craton delineated by E-W and NE-SW shear zones are inherently controlled the drainage of these cratonic rivers (Bhatt et al. 2017, 2020; Bhatt 2020).

The NE-SW lineaments showing low frequency resemble the NE-SW trends of high order streams of subwatersheds of Sindh basin. The high frequency lineaments trending in WNW-ESE, NW-SE and NNW-ESE are following the

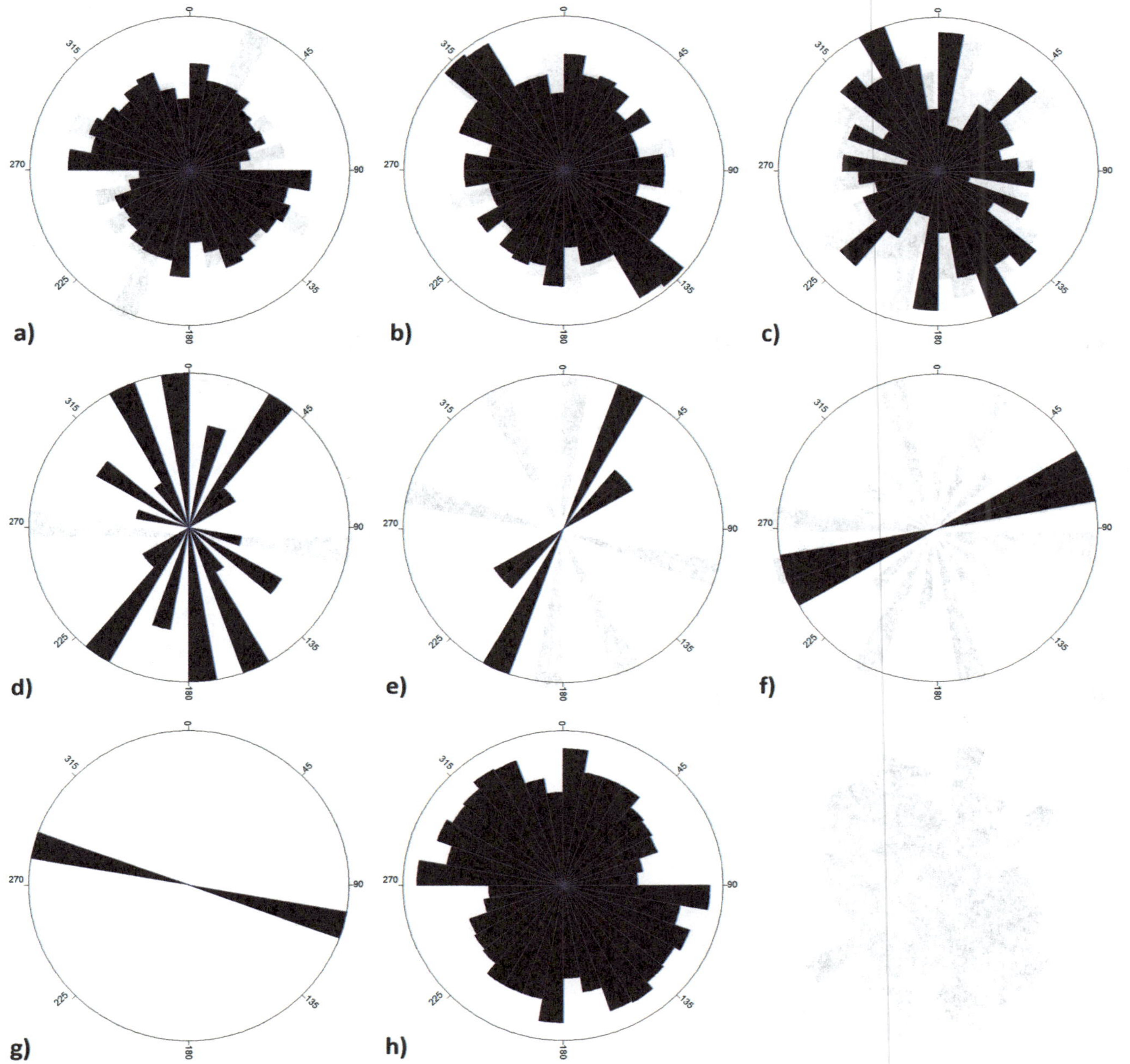

Fig. 8 Rose diagrams showing orientation of first (**a**), second (**b**) third (**c**), fourth (**d**), fifth (**e**), sixth (**f**), seventh stream order (**g**) and mean of stream orders (**h**) in Kunwari subwatershed

trends of low order streams (I, II and III order) of subwatersheds of the basin. The E-W trending lineaments are representing E-W crustal shear zones and the NE-SW and NW-SE lineaments are following the trends of quartz reefs and dykes respectively. The high density of lineaments represented by fractures, faults, foliations and bedrock structures are controlling the drainage and forming the passages to low order streams.

5 Conclusions

The morphometric analysis envisages that the elongated Sindh basin and its subwatersheds are showing impermeable to semi-permeable hard rock terrain in high elevated area and permeable soil in low lying area. The Sindh basin and its subwatersheds are significantly elongated and are more

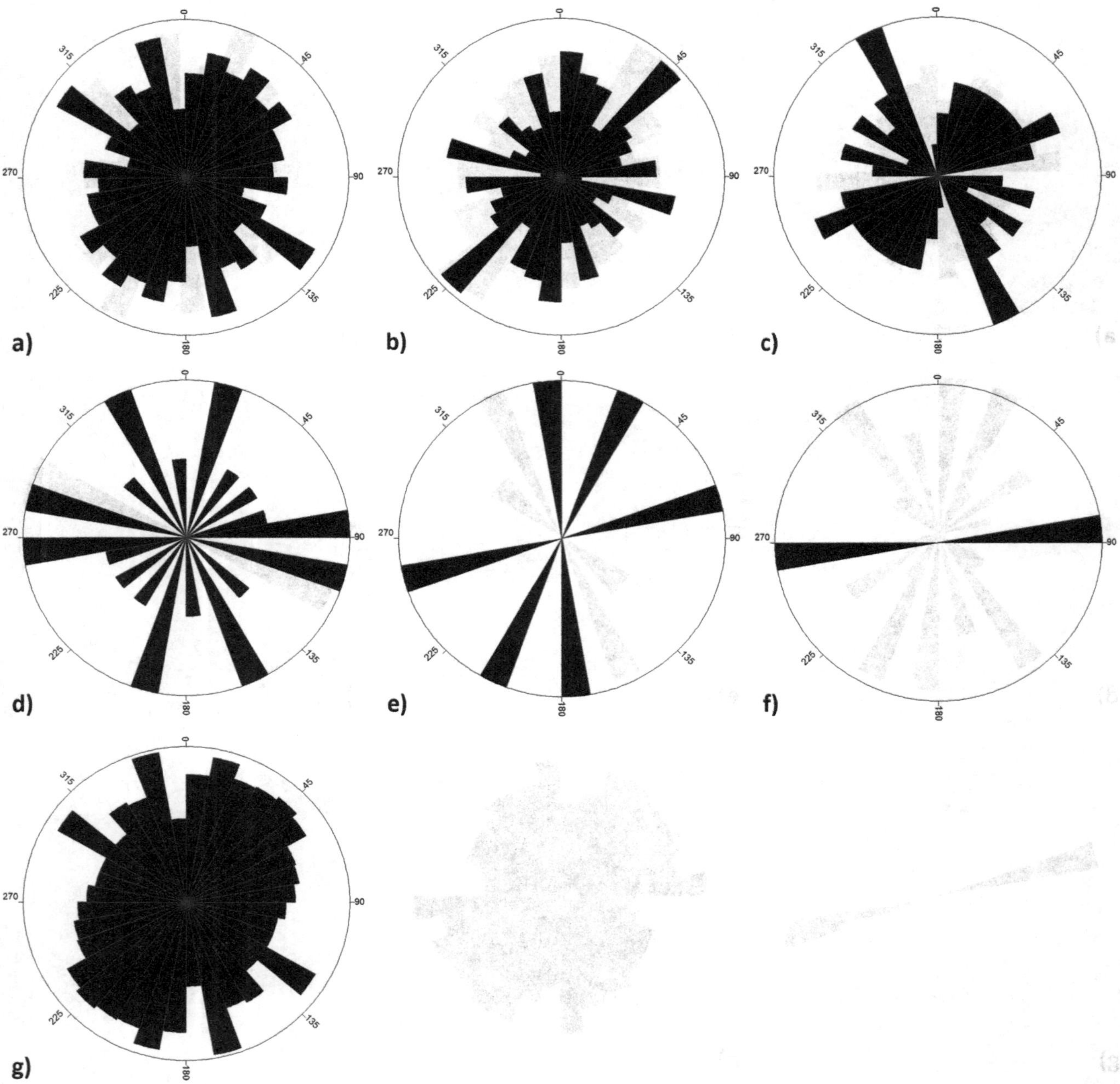

Fig. 9 Rose diagrams showing orientation of first (**a**), second (**b**) third (**c**), fourth (**d**), fifth (**e**), sixth (**f**), and mean of stream orders (**g**) in Vaishali subwatershed

susceptible to erosion and sediment transport in high elevated area. The low-density E-W, NE-SW and NW-SE lineaments are controlling the high order streams in the area. The high density of lineaments represented by fractures, faults, foliations and bedrock structures are controlling the drainage of low order streams.

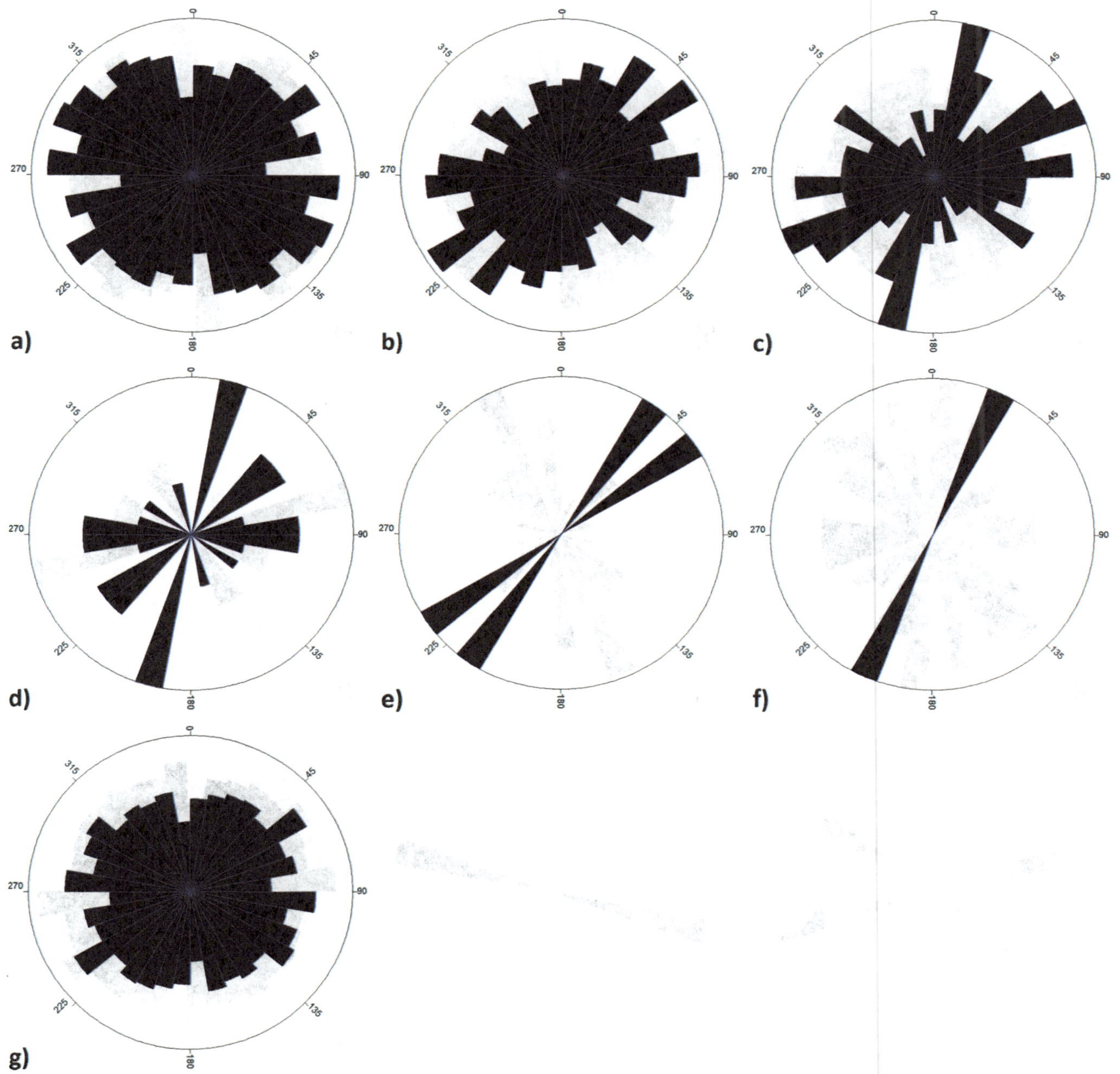

Fig. 10 Rose diagrams showing orientation of first (**a**), second (**b**) third (**c**), fourth (**d**), fifth (**e**), sixth (**f**) and mean of stream orders (**g**) in Pahuj subwatershed

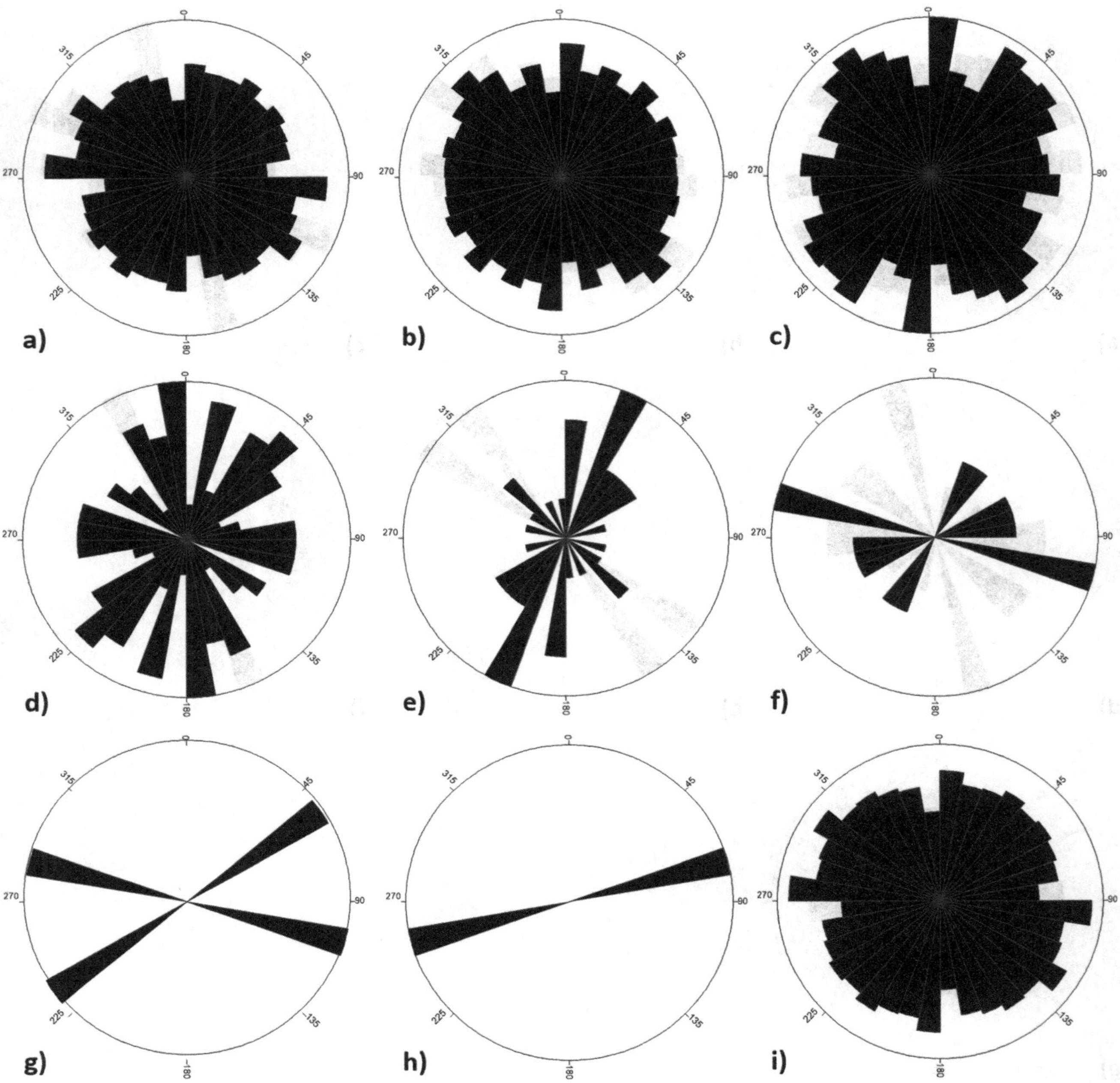

Fig. 11 Rose diagrams showing orientation of first (**a**), second (**b**) third (**c**), fourth (**d**), fifth (**e**), sixth (**f**), seventh (**g**), eighth stream order (**h**) and mean of stream orders (**i**) in Sindh basin

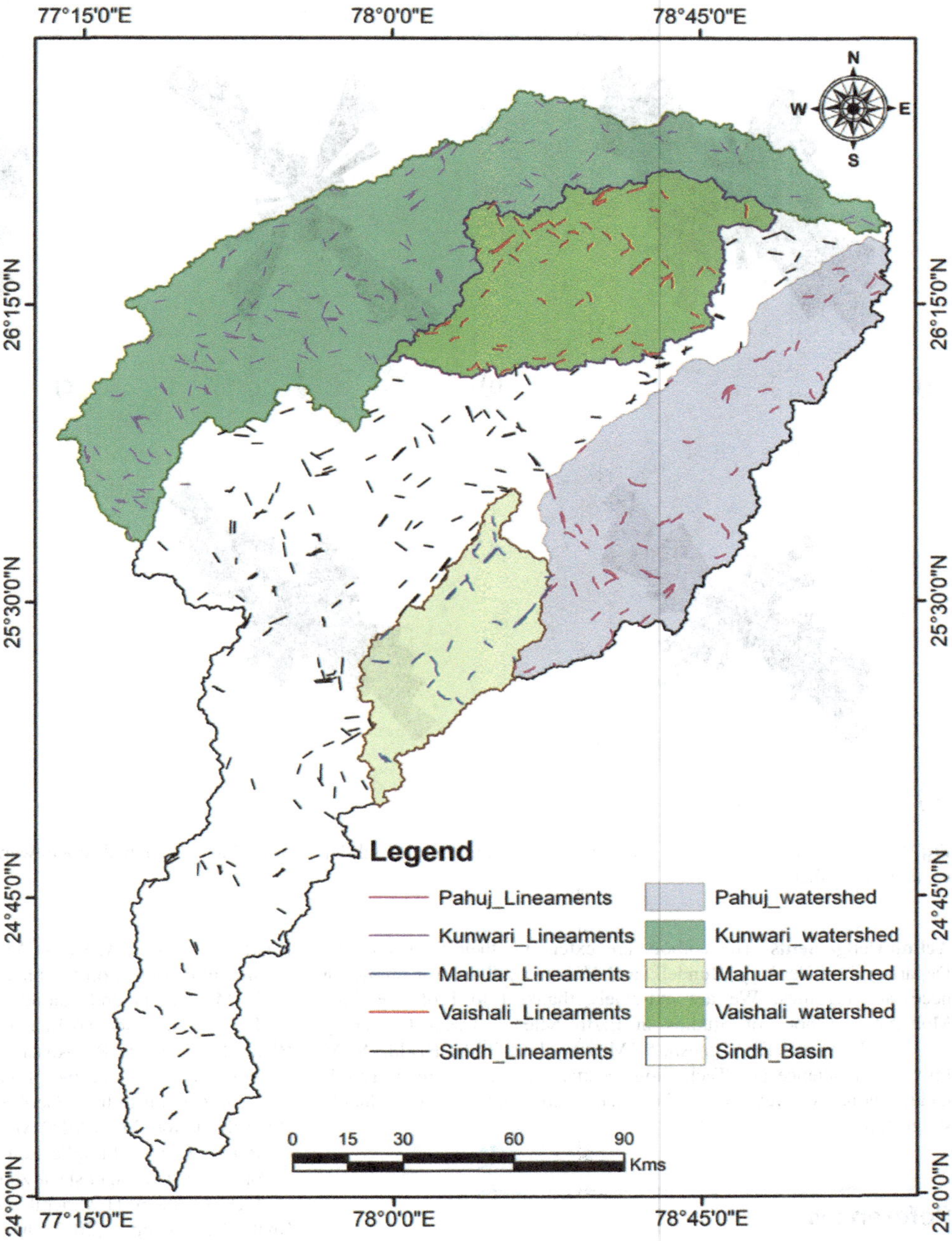

Fig. 12 Lineament map of Sindh basin and its subwatersheds

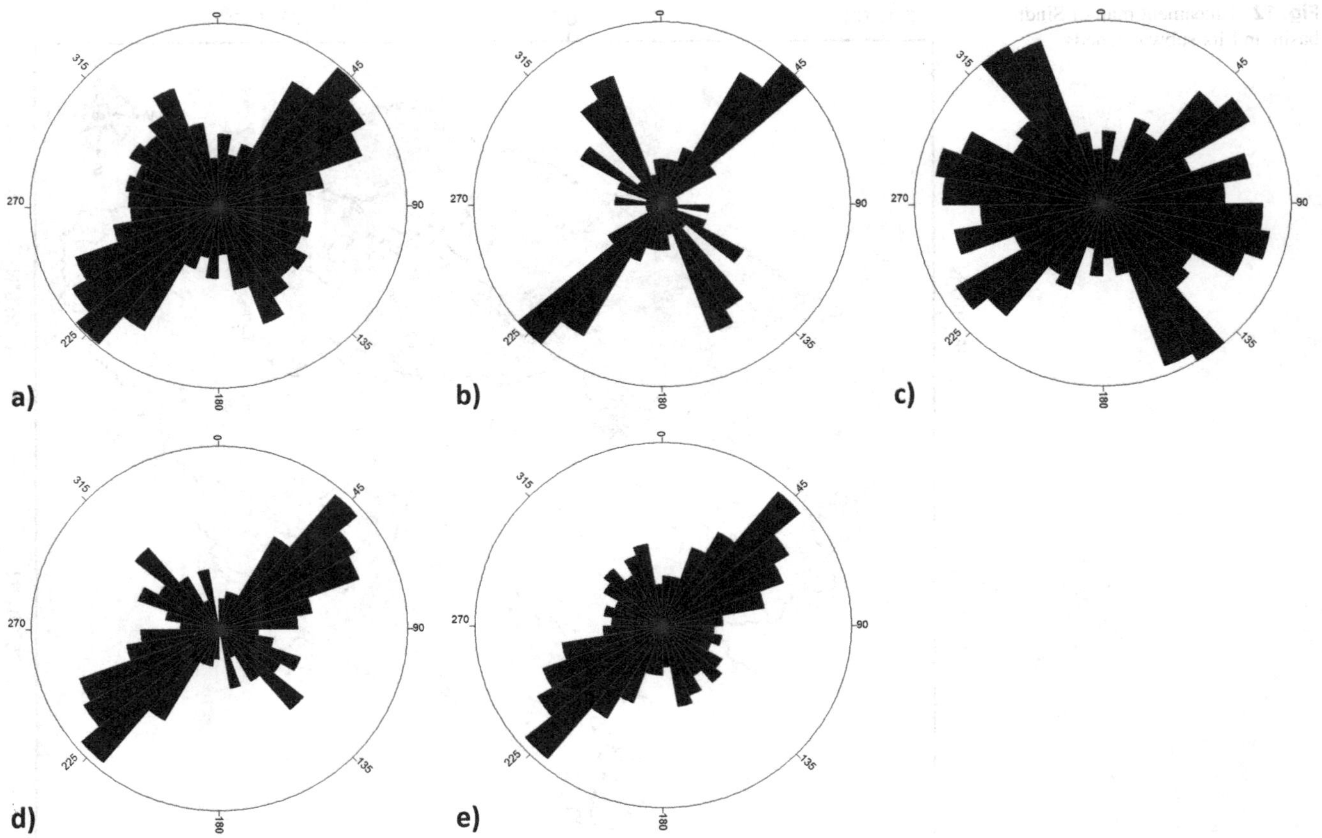

Fig. 13 Rose diagram showing orientation of lineaments; **a** Sindh basin, **b** Mahuar watershed, **c** Kunwari watershed, **d** Vaishali watershed and **e** Pahuj watershed

Acknowledgements The authors are extending their thanks to the Department of Geology, Bundelkhand University, India for providing necessary facilities. We are extremely thankful to Professor S.N. Mohapatra, School of Studies in Earth Science Jiwaji University Gwalior, India and Dr Vignesh, SRM School of Public Health, SRM Institute of Science and Technology Kattankulathur, Chennai, Tamilnadu, India for reviewing the manuscript and giving valuable suggestions.

References

Aparna P, Nigee K, Shimna P, Dissia TK (2015) Quantitative analysis of geomorphology and flow pattern analysis of Muvattupuzha River Basin using geographic information system. J Aquat Proc 4:609–616

Bhatt SC, Husain A (2008) Structural history and fold analysis of basement rocks around kuraicha and adjoining areas, Bundelkhand Massif Central India. J Geol Soc India 72:331–347

Bhatt SC, Mahmood K (2008) Deformation pattern and kinematics of folds in basement rocks exposed around Babina and Ghisauli area Bundelkhand Craton, Central India. Bull Indian Geol Assoc 41(1 & 2):1–16

Bhatt SC, Mahmood K (2012) Deformation pattern and microstructural analysis of sheared gneissic complex and mylonitic metavolcanics of Babina-Prithvipur sector Bundelkhand Massif, Central India. Indian J Geosci 66:79–90

Bhatt SC, Hussain A, Suresh M, Balasooriya NWB (2017) Geological structure control on Sukhnai Basin and land use/land cover pattern in Mauranipur and adjoining areas, Bundelkhand Craton, Central India. J Geol Soc Sri Lanka 18(2):53–61

Bhatt SC, Singh R, Ansari MA, Bhatt S (2020) Quantitative morphometric and morphotectonic analysis of Pahuj catchment basin Central India. J Geol Soc India 96(5):513–520

Bhatt SC, Gupta MK (2014) Microstructural analysis and strain pattern in mylonites and implications of shear sense indicators in evolution of Dinara-Garhmau shear zone, Bundelkhand Massif, Central India. Indian Mineral (J Mineral Soc India) 48(2):186–206

Bhatt SC, Hussain A (2012)Shear indicators and strain pattern in quartz mylonites of Chituad—Deori Shear Zone, Bundelkhand Massif, Central India. Earth Sci India 5(II):60–78

Bhatt SC, Singh VK, Hussain A (2011) Implications of shear indicators for the tectonic evolution of Mauranipur shear zone, Bundelkhand craton, Central India. In: Singh VK, Chandra R (eds) 2nd Proceeding of Precambrian continental growth and tectonism, pp 36–49. Angel Publication, New Delhi

Bhatt SC (2014) Geological and tectonic aspects of Bundelkhand Craton, Central India. Angel Publication, Bhagwati publishers and Distributors C-8/77-B, Keshav Puram, New Delhi, p190

Bhatt SC (2020) Natural resources, water harvesting and drought in Central India, p 130. Shree Publishers and Distributors 22/4735, Prakash Deep Building Ansari Road, Daryaganj, New Delhi110002

Bishop MP (2007) Long term landscape evolution: linking tectonics and surface processes, Earth, Surf Processes and. Landforms 32:329–365

Clark JI (1966) Morphometry from maps. Essay in Geomorphology. Elsevier Publ. Co., New York

Farhan Y, Ali A, Omar E, Nisrin A (2015) Quantitative analysis of geomorphometric parameters of Wadi Kerak, Jordan, using remote sensing and GIS. J Water Res Prot V 7:456–475

Hadley R, Schumm S (1961) Sediment sources and drainage basin characteristics in upper Cheyenne River basin, USGS water supply paper, 1531-B, W, 198.Washingaton DC

Horton RE (1932) Drainage basin characteristics, Transactions A. Geophysical Union 13:350–361

Horton RE (1945) Erosional development of streams and their drainage basins: a hydrophysical approach to quantitative morphology. Geol Soc Am Bull 56(3):275–370

Huang XJ, Niemann JD (2006) Modelling the potential impacts of groundwater hydrology on long-term drainage basin evolution. Earth Surf Proc Land 31:1802–1823

Jain V, Sonam Singh A, Sinha R, Tandon SK (2020) Evolution of modern river systems: an assessment of 'landscape memory' in river systems. Episode 43:535–551

Mangesh NS, Chandrasekar N, Soundranayagam JP (2011) Morphometric evaluation of Papanasam and Manimuthar watersheds, part of Western Ghats Tirunelueli District, Tamil Nadu, India: a GIS approach. Environ Earth Sci 64:373–381

Melton MA (1958) Correlations structure of morphometric properties of drainage systems and their controlling agents. J Geol 66:442–460

Miller VC (1953) A quantitative geomorphic study of drainage basin characteristics on the Clinch Mountain Area, Virginia and Tennessee, Proj, NR 389–402, Tech Rep 3, Columbia University, Department of Geology, ONR, New York

Nur A (1982) The origin of tensile fracture lineaments. Struct Geol 4:31–40

Panda B, Venkatesh M, Kumar B, Anshumali, (2019) A GIS-based approach in drainage and morphometric analysis of Ken River Basin and sub-basins Central India. J Geol Soc India 93:75–84

Pareta K, Pareta U (2011) Quantitative morphometric analysis of watershed of Yamuna basin, India using ASTER (DEM) data and GIS. Indian J Geomat Geosci 2–248

Pareta K, Pareta U (2012) Quantitative geomorphological analysis of watershed of Ravi River basin HP, India. Int J Rem Sens GIS 1: 47–62

Pati JK, Malviya VP, Prakash K (2006) Basement re-activation and its relation to neotectonic activity in and around Allahabad Ganga Plain. J Indian Soc Rem Sens 34:524–526

Pati JK, Lal J, Prakash K, Bhusan R (2008) Spatio-temporal shift of western bank of the Ganga River Allahabad City and Its Implications. J Indian Soc Rem Sens 36:289–297

Prakash K, Mohanty T, Singh S, Chaubey K, Prakash P (2016) Drainage morphometry of the Dhasan river basin, Bundelkhand craton, Central India using Remote Sensing and GIS techniques. J Geomat 10:121–132

Prakash K, Mohanty T, Pati JK, Chaubey SS, K, (2017) Morphotectonics of Jamini River basin, Bundelkhand craton, Central India; using remote sensing and GIS Technique. Appl Water Sci 7:3767–3782

Prakash K, Rawat D, Singh S, Chaubey K, Kanhaiya S, Mohanty T (2019) Morphometric analysis using SRTM and GIS in synergy with depiction: a case study of the Karmanasa River basin North Central India. Appl Water Sci 9(1):13. https://doi.org/10.1007/s13201-018-0887-3

Sahu N, Reddy O, Kumar N, Nagaraju M, Srivastava R, Singh KS (2016) Morphometric analysis in basaltic terrain of central India using GIS techniques: a case study. Appl Water Sci. https://doi.org/10.1007/s13201-0160442-z

Samal DR, Gedam SS, Nagarajan R (2015) GIS based drainage morphometry and its influence on hydrology in parts of Western Ghats region, Maharashtra India. Geoc Int 30(7):755–778

Schumm SA (1956) Evolution of drainage systems and slopes in badlands at Perth Amboy New Jersey. Geol Soc Am Bull 67:597–646

Shukla UK, Srivastava P, Singh IB (2012) Migration of Ganga river and development of cliffs in Varanasi Region during late Quaternary: role of active tectonics. Geomorphology 171–172:101–113. https://doi.org/10.1016/j.geomorph.2012.05.009

Singh PK, Verma SK, Singh VK et al (2021a) Geochronology and petrogenesis of the TTG gneisses and granitoids from the Central Bundelkhand granite-greenstone terrane, Bundelkhand Craton, India: Implications for Archean crustal evolution and cratonization. Precambr Res 359:106210

Singh VK, Slabunov AI, Nesterova NS et al (2021b) Tectonostratigraphic terranes of the Bundelkhand Craton (Indian Shield). In: Shandilya AK, Singh VK, Bhatt SC, Dubey CS (eds) Geological and Geo-environmental Processes on Earth, pp 155–164 (this volume)

Singh VK, Slabunov A (2015) The Central Bundelkhand Archaean greenstone complex, Bundelkhand Craton, Central India: geology, composition, and geochronology of supracrustal rocks. Int Geol Rev 57(11–12):1349–1364. https://doi.org/10.1080/00206814.2014.919613

Smith KG (1950) Standards for grading texture of erosional topography. Am J Sci 248:655–668

Strahler AN (1952) Hypsometric analysis of erosional topography. Bull Geol Soc Am 63:1117–1142

Strahler AN (1956) Quantitative slope analysis. Bull Geol Soc Am 67:571–596

Strahler AN (1957) Quantitative analysis of watershed geomorphology. Trans Am Geophys Union 38:913–920

Strahler AN (1958) Dimensional analysis applied to fluvially eroded landforms. Geol Soc Am Bull 69:279–300

Strahler AN (1964) Quantitative geomorphology of drainage basin and channel network. Handbook of applied hydrology, pp 39–76

StrahlerAN SAH (2002) A text book of physical geography. John Wiley and sons, New York

Tangri AK (1992) Satellite remote sensing as a tool in deciphering the fluvial dynamics and applied aspects of Ganga Plain. In: Singh IB (ed) Proceedings of the workshop, Gangetic Plain: Terra Incognita. Department of Geology, Lucknow, pp 73–84

Vijith H, Satheesh R (2006) GIS-based morphometric analysis of two major upland sub-watersheds of Meenachil river in Kerala. J Indian Soc Rem Sens 31:181–185

Morphometric Analysis of Rohni Watershed, Upper Betwa Basin, Bundelkhand Region, Central India

S. C. Bhatt, Sumit Mishra, Rubal Singh, and S. Bhatt

Abstract

Rohni is a subwatershed of Dhasan river of Betwa river system. The Morphometric analysis of Rohni watershed was carried out by using remote sensing and GIS techniques and using ASTER data. Based on hierarchical rank five subwatersheds streams are identified which are represented by the dendritic to subdendritic drainage pattern. The drainage, density, slopes and relief maps were prepared. The corresponding trends of each watershed with Precambrian structures and mean bifurcation ratio (4.61) suggest that the drainage pattern is structurally controlled. The high value of Rho Coefficient is suggestive of high hydrologic storage during a flood and shows less effect of erosion. The low value of the form factor indicates that the basin is elongated and is less prone to flood. The circulatory ratio implies that the youth and mature stages are dominant in Rohni basin. The low drainage density (1.51) is indicative of the presence of less permeable soil with a low infiltration rate and less erosion exposure in the basin. The low ruggedness and Melton ruggedness values indicate that the terrain is smoother and moderately rugged and is less susceptible to soil erosion. The drainage orientation of Rohni watershed is represented by polymodal distribution (NE-SW, E-W and NW–SE). The NNW-SSE and NE-SW trending major lineaments are prominent in the study area. The trends of low orders streams (NE-SW, NNW-SSE) coinciding the trends of major lineaments are indicative of tectonic control. The ENE-WSW trending high order streams show major tectonic control.

Keywords

Morphometric analysis • Rohni Watershed • Bundelkhand • Central India

1 Introduction

Bundelkhand craton spreading in 28,000 sq km area of the northern segment of Peninsular India is characterized by limited water and land resources. The Ken, Betwa, Dhasan, Pahuj and Sindh are important rivers of this craton. It induces low to average rainfall (300–900 mm) and is characteristics of the semi-arid condition. Morphometric analysis is a mathematical evaluation of shape, size and dimension of landforms and configuration of earth's surface (Agarwal 1998; Obi Reddy et al. 2002). The linear, areal, relief, a gradient of channel network and ground slope of the basin is measured by morphometric evaluation (Nautiyal 1994; Nag and Chakraborty 2003; Maghesh et al. 2012b; Rai et al. 2017). The information about hydrological and geomorphological processes and drainage characteristics within a watershed is studied under morphometric analysis (Strahler et al. 1964; Pareta and Pareta 2011; Aparna et al. 2015; Prakash et al. 2017; Bhatt et al. 2017, 2020; Bhatt 2020) Morphometric analysis also provides quantitative information about watershed and drainage system (Strahler 1964).

The morphometry of a basin reveals the geological and geomorphological processes undergone in geological period and their influence on drainage basin is significant in understanding the landform processes, a physical and erosional property of soil (Horton 1945; Strahler 1952; Strahler 1964; Shreve 1969). Morphometric analysis is significant to understand hydrogeological processes and landforms structures (Horton 1932; Soni 2017; Pareta and Pareta 2012;

S. C. Bhatt (✉) · R. Singh
Institute of Earth Sciences, Department of Geology, Bundelkhand University, Jhansi, UP, India
e-mail: scbhatt@bujhansi.ac.in

S. Mishra
HNB Garhwal University, Srinagar, Garhwal (UK), India

S. Bhatt
Institute of Environment and Development Studies, Bundelkhand University, Jhansi, UP, India

A. K. Shandilya et al. (eds.), *Geological and Geo-Environmental Processes on Earth*, Springer Natural Hazards,
https://doi.org/10.1007/978-981-16-4122-0_18

Prakash et al. 2017; Bhatt et al. 2017, 2020; Bhatt 2020). Present work aims to assess the surface water potential of Rohni watershed and to explore techniques for basin management and water harvesting. It also gives information about the status of erosion and floods and potential of surface and groundwater resources.

2 Study Area

The Rohni river originates from Amoda reserve forest (RF) of Vindhyan hills and meets at Dhasan near Jhankar Village of Sagar District. It is 43.29 km in length. The Rohni river is a tributary of Dhasan river of Betwa river system. Betwa is major drainage flowing from south to north and confluences to the Yamuna at Hamirpur city. The investigated area is occupying 295.8 sq km. area. It lies between $78^0 45'$E–$79^0 00'$E longitude to $24^0 30'$N–$24^0 15'$N latitude (Fig.1).

3 Geomorphology

The study area belonging to the southern part of Bundelkhand craton is represented by undulating topography and low lying erosional landforms (Bhatt 2020). These topographic features are characterized by low to moderately elevated (200–300 m) granitic hillocks and quartz reefs (Fig. 1). The linear quartz reefs emplaced NE-SW shear zones and are thickly vegetated. The erosional granitic hills, quartz reefs and low lying depositional landforms are very common in the study area (Fig. 1).

4 Geological Setup

Bundelkhand Tonalite Trondhjemite Gneissic (TTG) Complex (3.1–3.5 Ga), Archean Greenstone Complex and Bundelkhand Granitic Complex (2.5 Ga) are important major Groups identified in the study area (Bhatt and Hussain 2008, 2012; Bhatt and Mahmood 2012; Bhatt and Gupta 2014; Bhatt 2014; Bhatt et al. 2011, 2015, 2017; Singh and Slabunov 2015; Singh et al. 2021; Fig. 2). The large outcrops of TTG gneisses striking NE-SW are widely exposed in the study area. The gneissosity is defined by leucocratic quartzofeldspathic and melanocratic bands. The main constituents of these rocks are quartz, feldspar (mostly orthoclase) and biotite.

The Peridotite, Pyroxeninite and Gabbro are important rock units of greenstone complex (Fig. 2). The E-W trending lensoidal bodies of Peridotite are exposed in the eastern and western parts of the study area and are characterized by two generations of quartz veins. These massive fine-grained rocks mainly consist of altered olivine, pyroxene and opaque and mafic minerals (30–40%) along with low fraction (10%) of plagioclase. The dark grey unfoliated Pyroxenites are exposed in the southwestern part of the study area. The olivine (>25%) and few leucocratic minerals (quartz and plagioclase; 7%) are the main constituents of these rocks. Fine to medium-grained light to dark grey gabbro associated with ultramafic rocks are also mapped near the eastern part of the study area. The main mineral constituents of these rocks are altered plagioclase, pyroxenite, ferromagnessium, and hornblende and anthophyllite minerals.

Porphyritic and medium to light grey granites are identified as important granitoids of Bundelkhand Granitic Complex. The outcrops of porphyritic granite are exposed around Mandawra-Girar road. These are medium to coarse-grained pinkish-red rocks mainly constitute large phenocryst of quartz and K-feldspar. At few places, these granitoids are sheared and intruded by pegmatic and quartz veins. The light grey, coarse-grained granite consisting of phenocrysts of quartz and feldspar is exposed in the northwest and southeast of the study area.

5 Methodology

The mathematical approach for the estimation of various morphometric parameters was attempted in this study. The stream ordering was done by methods proposed by Strahler (1956) and Arc hydro tool in ArcGIS 10.3 software (Fig. 3). The linear aspects were determined by using methods of Horton (1945) and Nag and Chakraborty (2003). The areal and relief aspects were also studied by applying methods of Horton (1945); Schumn (1956); Strahler (1952, 1956, 1964); Nag and Chakraborty (2003) and Pareta (2011).

The remote sensing techniques and GIS tools of 30 m resolution and Advanced Spaceborne Thermal Emission and Reflection Radiometer (ASTER) data were used to prepare Digital Elevation Model (DEM), drainage and density maps (Figs. 1, 2, 4, 5, 6, 8).

6 Results and Discussion

The quantitative and mathematical evaluation of the basin, altitude, volume, slope, a profile of the land and drainage basin characteristics of the catchment area are incorporated in the morphometric analysis (Clark 1966). In the present work, all morphometric parameters (linear, areal and relief) were discussed. Rohni is a subwatershed of Dhasan and has been classified into five orders of streams.

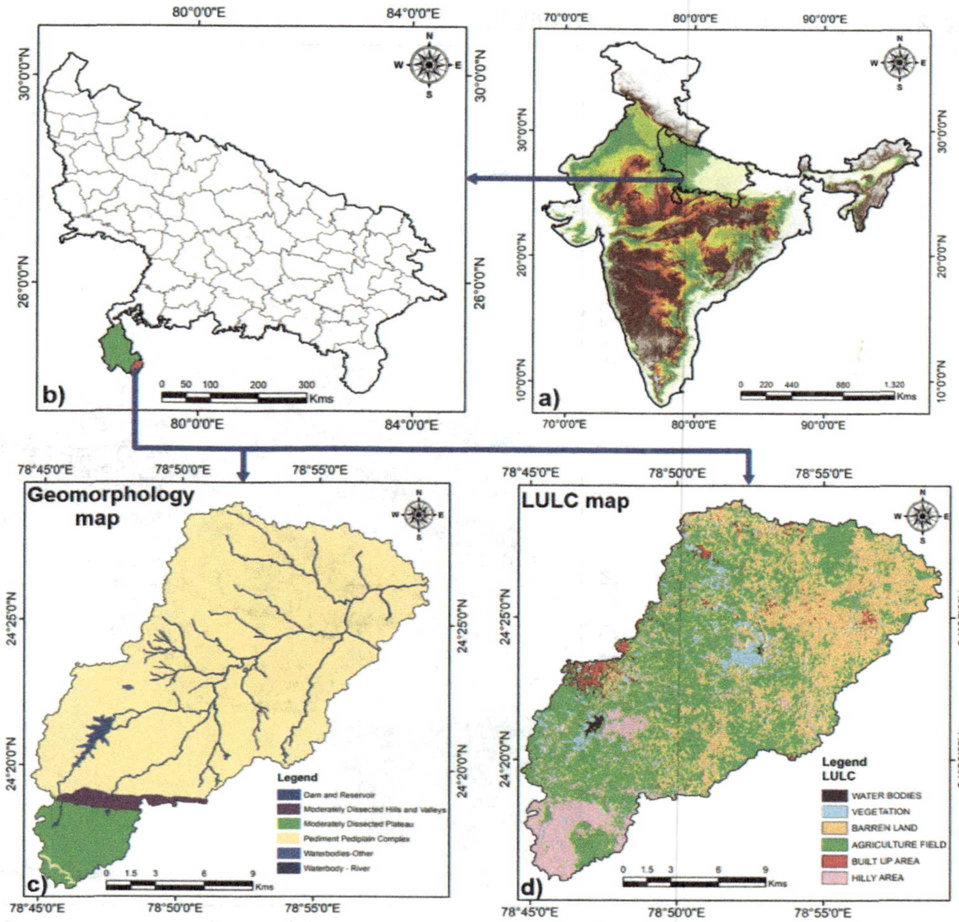

Fig. 1 Location map, Land use/Land cover map of the study area

6.1 Drainage Pattern

The drainage pattern depicting the distribution of water in a stream is significant in exploring the status of water, agricultural and forest resources in a particular river basin. The dendritic to subdendritic drainage pattern is observed and five orders of streams are identified in the study area (Fig. 5). Most of the higher-order streams (IV and V) showing curved and straight course are flowing along with major fractures and faults and show structural influence. The drainage orientation of Rohni watershed is represented by polymodal distribution (N-S, NE-SW, E-W and ENE-WNW; Fig. 7). The first and second-order streams are showing NE-SW and N-S orientations (Fig. 7a, b) while the III and IV orientations order streams are exhibiting NW–SE and ENE-WSW orientations respectively (Fig. 7a, b, c, d, f). These orientations of drainage are corresponding with the orientations of major tectonic units of the study area. It infers that the drainage pattern of Rohni river is controlled by major tectonic elements.

6.2 Morphometric Analysis

The quantitative morphometric measurement of linear, areal and slope aspects was carried out for Rohni watershed (Table 1).

6.2.1 Linear Aspects

The linear aspects including stream order, stream length, mean stream length ratio and bifurcation ratio was determined by using quantitative approach (Table 1). Stream ordering for numeric links to stream network is the first step of the stream orders and can be ranked by applying methods advocated by Strahler (1964, 1952) and Horton (1945). In the present work, the Strahler method has been used for ranking stream orders. According to this method, if the two or more similar order drainages will meet at a point then it will be converted into higher next order. Five order of stream is allocated to Rohni subwatershed. According to Horton's law, the number of streams decreases as the stream order increases. About 327 and 61 streams were traced for first

Fig. 2 Geological map of Rohni River

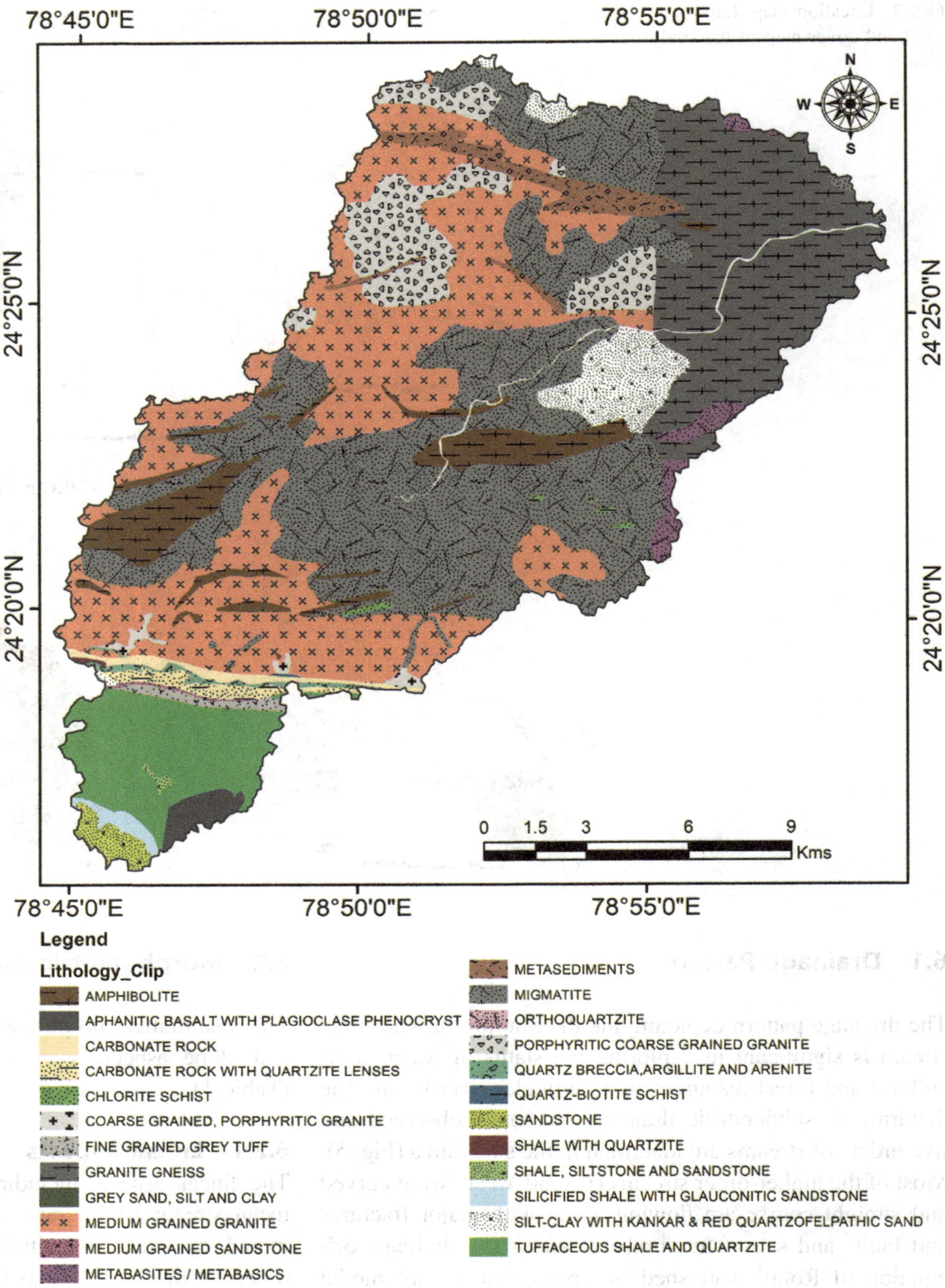

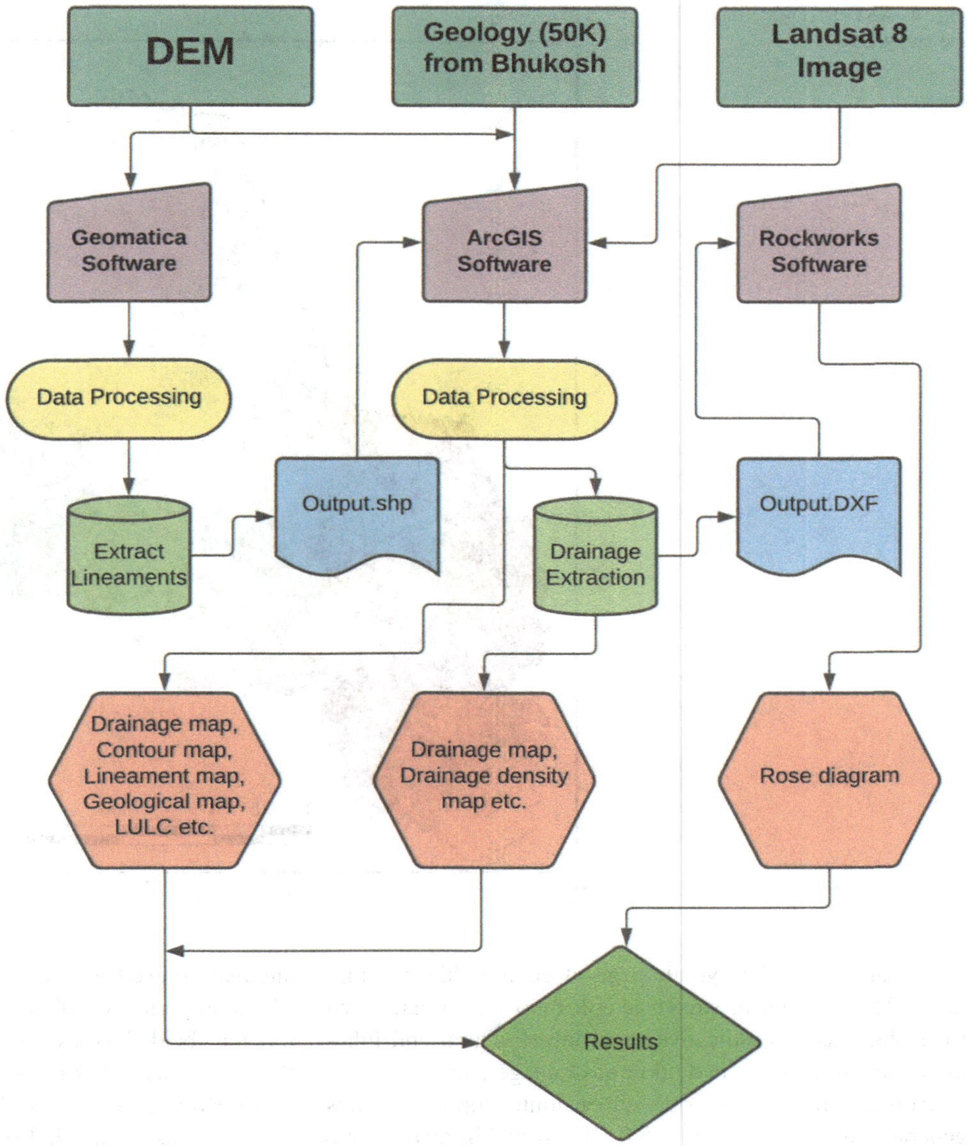

Fig. 3 Methodology shown in the flow chart

and second order of streams, respectively. The number of the stream decreases for third (12), fourth (02) and fifth (01) higher order of streams (Table 1).

Stream length of the Rohni river was computed with the help of ArcGIS software and methods advocated by Horton (1945; Table 1). Stream length has been calculated by applying Horton's method (1945) for all stream/ micro watershed of the area. The total stream length is 446.61 km (Table 1). The maximum length of 239.44 km and 108.46 km was contributed by first and second-order streams respectively (Table 1). Contrary to this the length of streams is significantly decreasing in third (65.14 km), fourth (27.54 km) and fifth (6.02 km) higher order of streams. Mean stream length (L_{um}) is a dimensional property stating the characteristic size of components of a drainage network and its contributing watershed surfaces (Strahler 1964). It can be determined by dividing the total length of stream of an order by a total number of segments in the order. (Table1). Stream Length Ratio (L_{urm}) is the ratio of the mean (L_u) of segments of order (S_u) to mean length of segments of the next lower order (L_{u-1}), which tends to be constant throughout the successive orders of a basin (Horton 1945). Horton's law of stream lengths refers that the mean stream lengths of stream segments of each of the successive orders of a watershed tend to approximate a direct geometric sequence in which the first term (stream length) is the average length of segments of the first order (Table 1). The change of stream length ratio from one order to another order

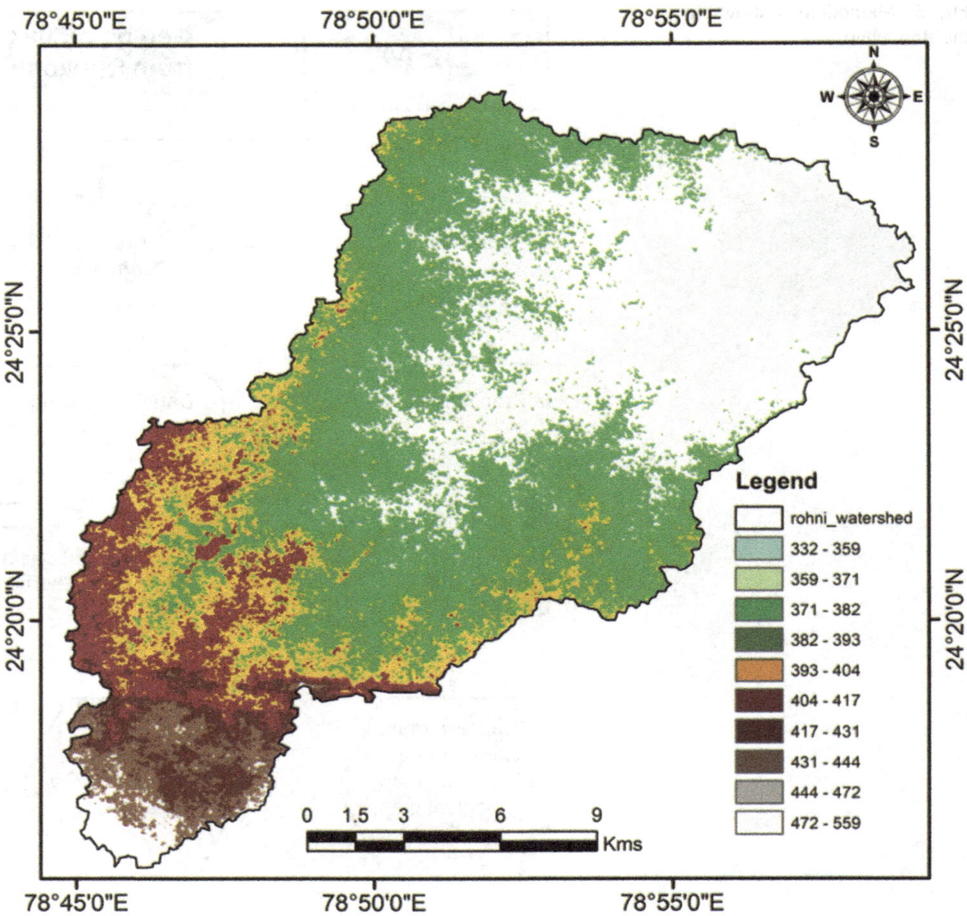

Fig. 4 ASTER (DEM) map of the study area

is indicating their late youth stage of geomorphic development. The high mean stream length ratio (2.11) calculated for Rohni basin is indicative of strong structural and lithological control on the evolution of drainage pattern.

Bifurcation ratio (R_b) is the dimensionless property and is generally ranging from 3.0 to 5.0. The mean bifurcation ratio (4.61) calculated for Rohni watershed indicates that the drainage pattern is controlled by the geological structures. The variation in bifurcation ratio (Rb) is due to the strong influence of various types of geological structures and lithological characters of rocks on drainage. Weighted Mean Bifurcation Ratio (Rbwm) determined for Rohni watershed (5.31) is prone to flooding and the higher value of Rho Coefficient ($\rho = 0.46$) is suggestive of high hydrologic storage during a flood.

6.2.2 Areal Parameters /Basin Geometry

The density, frequency, form factors, drainage texture and other factors were discussed under this heading. Gregory and Walling (1973) defined the basin length as the longest in the basin in which are end being the mouth. The length of Rohni basin is 29.98 km and the total area is 295.80 km^2 Basin perimeter is measured along the divides between watersheds and may be used as an indicator of watershed size and shape. The total perimeter of Rohni basin is 109.45 km. Form factors may be defined as the ratio of basin area to the square of the basin length and is used to predict flow intensity in the basin (Horton 1932, 145). The value of the form factor would always be less than 0.754 (for a perfectly circular watershed). Smaller the value of form factor, more elongated will be the watershed. The watershed with high form factors has high peak flows of shorter duration, whereas elongated watershed with low form factor has low flow peak of longer duration. The low value of the form factor (0.33) of Rohni basin indicates that the basin is elongated and is not prone to flood (Table 1). Schumm (1956) defined elongation ratio as the ratio of the diameter of a circle of the same area as the basin to the maximum basin length. The varying slopes of a watershed can be classified with the help of the index of elongation ratio, i.e., circular (0.9–0.10), oval (0.8–0.9), less elongated (0.7–0.8), elongated (0.5–0.7), and more elongated (<0.5). The elongation ratio (0.65) computed for Rohni basin suggests that the basin is elongated and shows low flow peaks.

Texture ratio an important factor in this analysis is depending on the underlying lithology, infiltration capacity and relief aspect of the terrain (Smith 1950). Circularity

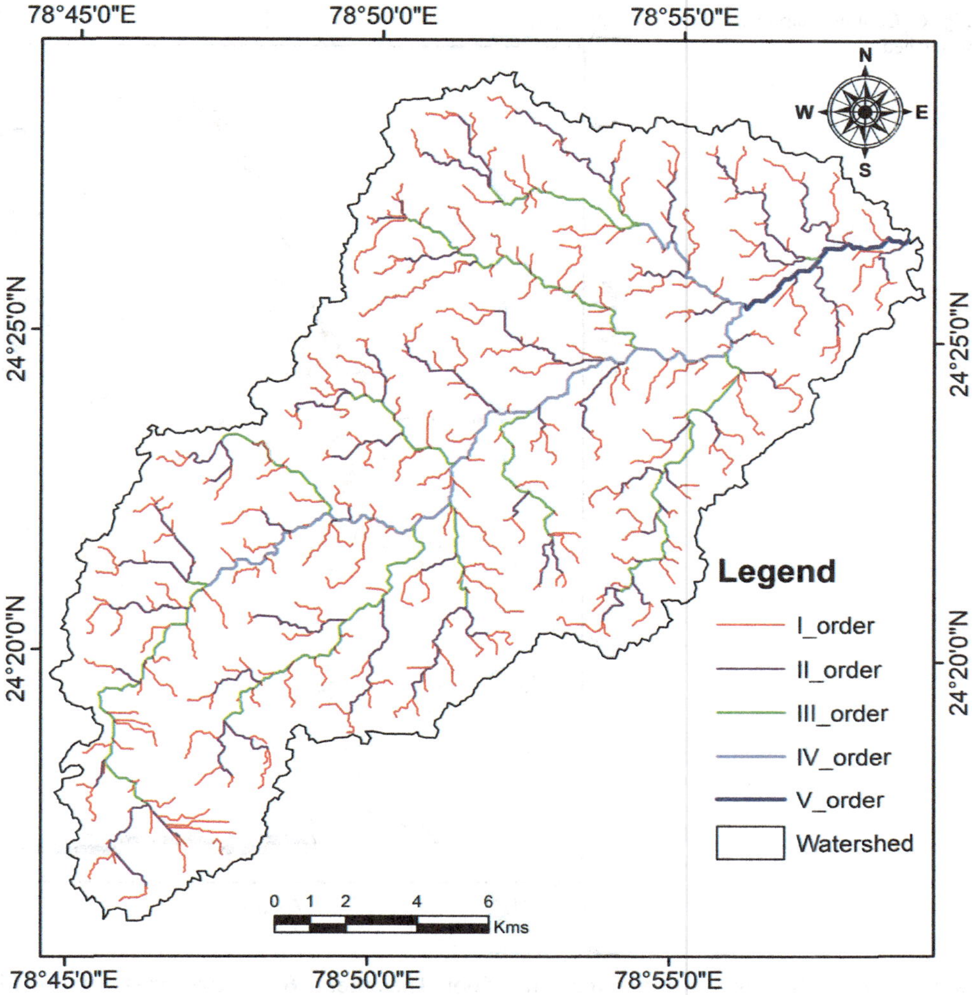

Fig. 5 Drainage map of Rohni watershed

parameter is explained as the ratio of basin area to the area of a circle having the same perimeter as the basin. The circulatory ratio 0.31 calculated for Rohni basin envisages that the youth and mature stages are dominant in Rohni basin. Smith (1950) has classified drainage texture into five different textures i.e., very coarse (<2), coarse (2 to 4), moderate (4 to 6), fine (6 to 8) and very fine (>8). The coarse texture (3.68) of Rohni basin implies that the basin mainly consists of permeable soil in low relief and shows the moderate potential of groundwater recharge.

The drainage frequency is invented by Horton (1932, 1945) means stream frequency (or channel frequency) is defined as the number of stream segments per unit area. The stream frequency ($F_s = 1.36$) calculated for Rohni watershed shows a relationship between level form and drainage and shows low eroded landform. Drainage density is defined as the stream length per unit area in the region of a watershed (Horton 1945, 1932; Strahler 1952).

Drainage density is a better quantitative expression to the dissection and analysis of landform, although a function of climate, lithology and structures and relief history of the region can finally use as an indirect indicator to explain, those variables as well as the morphogenesis of landform (Fig. 8). The moderate drainage density (1.51) is indicative of the occurrence of less permeable terrain with low infiltration rate and erosion exposure in the basin. The higher the infiltration number, the lower will be the infiltration and higher will be a run-off. Horton (1945) defined this term to refer to the length of the run of the rainwater on the ground surface before it is localized into definite channels. The higher infiltration number 2.05 computed for Rohni basin is indicative of lower infiltration and higher runoff.

6.2.3 Relief Aspects

The total relief of the river basin is explained as the difference in the elevation between the highest point of a

Fig. 6 Contour map of Rohni watershed

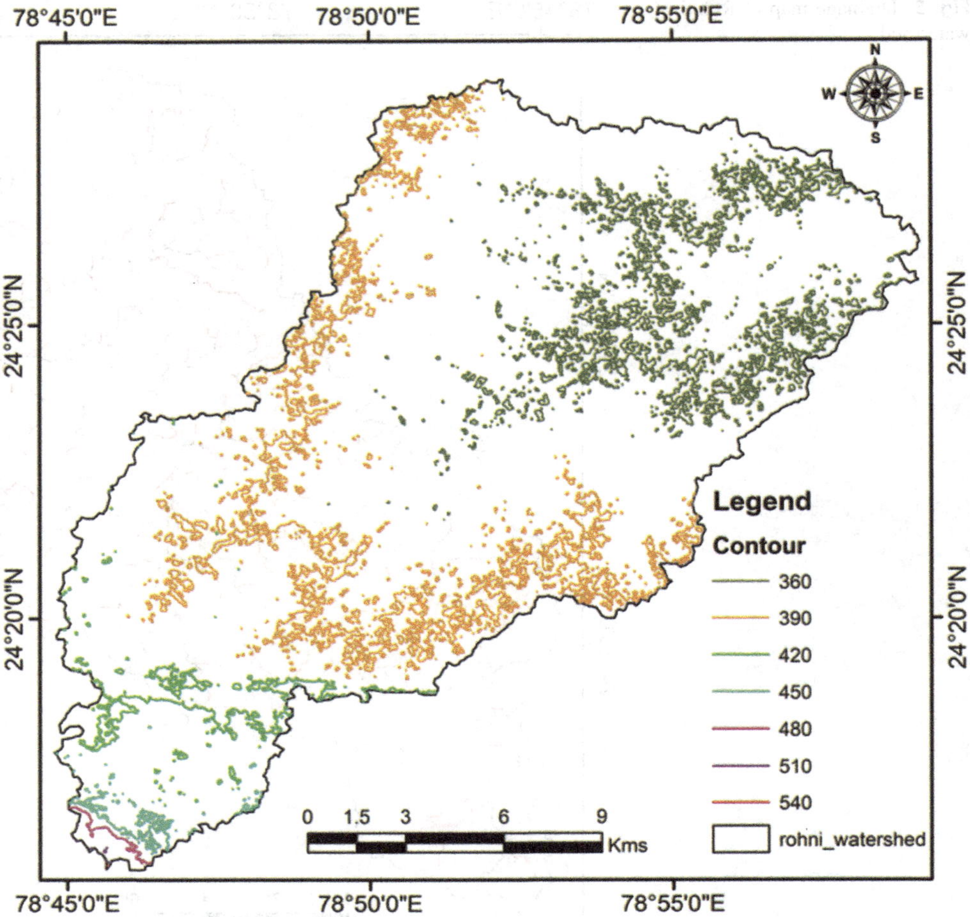

watershed and the lowest point on the valley floor. The relief ratio may be defined as the ratio between the total relief of a basin and the longest dimension of the basin parallel to the main drainage line (Schumm 1956). The possibility of a close correlation between relief ratio and hydrologic characteristics of a basin suggested by Schumm (1956) who found that sediments loose per unit area is closely correlated with relief ratios. The flood pattern and volume of transported sediment are influenced by the stream gradient. The relief ratio (0.008) indicates that the gradient in Rohni catchment basin is moderately steeper and less prone to erosion.

It is the product of the basin relief and the drainage density and usefully combines slope steepness with its length. The low ruggedness value (0.34) calculated for Rohni basin implies that the terrain is smoother and less rugged with a low gradient and less susceptible to soil erosion. The Melton ruggedness value (0.013) indicates that the Rohni watershed is a debris flood watershed.

6.3 Orientations of Lineaments and Their Relationship with Orientation of Drainage Pattern

The lineament map of the Rohni watershed was prepared by using Shuttle Radar Topographic Mission (SRTM) Digital Elevation Model (DEM) Geomatica software (Fig. 9). The NNW-SSE and NE-SW trending lineaments were identified in the Rohni watershed (Fig 9a, b). The orientations of major and minor lineaments of Rohni watershed are shown in the rose diagrams (Fig. 9b). The NNW-SSE and NE-SW trending major lineaments are corresponding with the major tectonic units of this watershed (Fig. 9a, b).

The Archean to Palaeoprterozoic gneissic and granitic terrain of Bundelkhand craton delineated by E-W and NE-SW shear zones are inherently controlled the drainage of these cratonic rivers (Bhatt et al. 2017, 2020; Bhatt 2020). The NNW-SSE and NE-SW trending low frequency lineaments are following the orientations of high order streams of

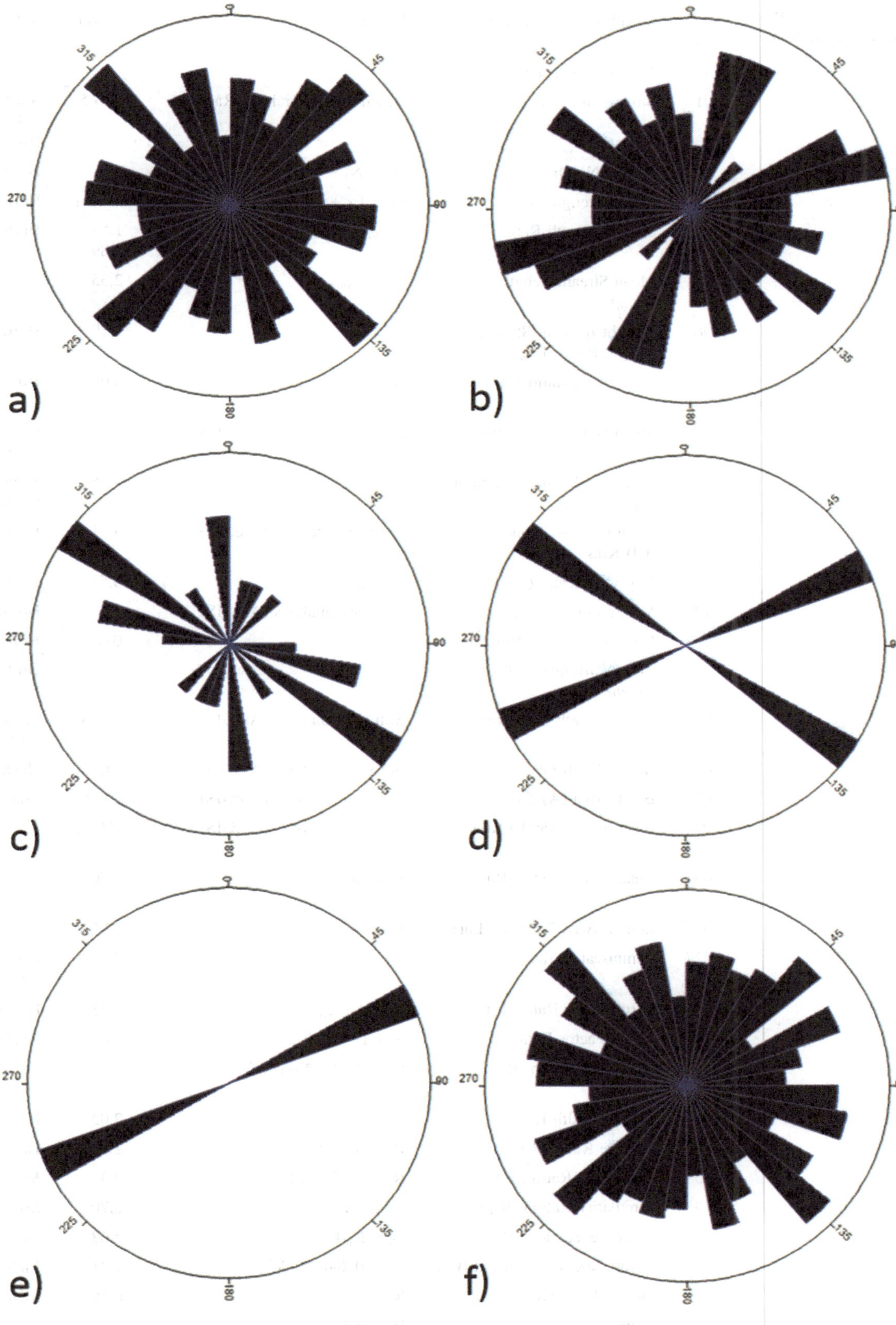

Fig. 7 **a** Orientation of first order drainage, **b** Orientation of second order drainage, **c** Orientation of third order drainage, **d** Orientation of fourth order drainage, **e** Orientation of fifth order drainage, and **f** Mean orientation of Rohni river Drainage

Table 1 Linear, Areal and Relief parameters of Rohni watershed

S. No	Morphometric parameter	Formula	Result	References
A	*Linear parameters*			
01	Stream Order (S_u)	Based on Hierarchical Rank	1 to 5	Strahler (1964), Horton (1945)
02	Stream Number (N_u)	$N_u = N_1 + N_2 + N_3 + \ldots\ldots + N_n$	403	Horton (1945)
03	Stream Length (L_u)	$L_u = L_1 + L_2 + L_3 + \ldots\ldots + L_n$	446.61	Horton (1945)
04	Stream Length Ratio (L_{ur})	L_{u+1}/ L_u	1.58–3.89	Horton (1945)
05	Mean Stream Length Ratio (L_{urm})	$L_{ur} * L_{ur\text{-}r}$	2.55	Horton (1945)
06	Weighted Mean Stream Length Ratio (L_{uwm})	L_{uwm}	2.10	Horton (1945)
07	Bifurcation Ratio (R_b)	$R_b = N_u/ N_{u+1}$	2.00–6.00	Horton (1945)
08	Mean Bifurcation Ratio (R_{bm})	$R_{bm} = (R_{b1} + R_{b2} + \ldots + R_{bn}$	4.61	Schumm (1956)
09	Weighted Mean Bifurcation Ratio (R_{bwm})		4.36	Schumm (1956)
10	Main Channel Length (Cl) Kms	Software analysis (ArcGIS)	36.10	Horton (1945)
11	Rho Coefficient (ρ)	$\rho = L_{ur}/R_b$	0.53	Horton (1945)
12	Valley Length (Vl)	Software analysis (ArcGIS)		Horton (1945)
13	Basin elongation direction		049	Horton (1945)
B	*Areal Parameters/ Basin Geometry*			Horton (1932)
01	Basin Length (L_b) Kms	Software analysis (ArcGIS)	29.98	Schumm (1956)
02	Basin Width (W_b)	Software analysis (ArcGIS)	14.75	Hack (1957)
03	Basin Area (A) SqKms	Software analysis (ArcGIS)	295.8	Srahler (1964)
04	Basin Perimeter (P) Kms	Software analysis (ArcGIS)	109.45	Schumm (1956)
05	Relative Perimeter (P_r)	$P_r = A/P$	2.70	Schumm (1956)
06	Length Area Relation (Lar)	$L_{ar} = 1.4*A^{0.6}$	42.53	Hack (1957)
07	Lemniscate's (k)	$k = L_b^2/A$	3.03	Chorley (1957)
08	Form Factor Ratio (F_f)	$F_f = A / L_b^2$	0.33	Horton (1932)
09	Shape Factor Ratio (R_s)	$S_f = L_b^2 / A$	3.03	Horton (1956)
10	Elongation Ratio (R_e)	$R_e = 2 / L_b * (A /\pi)^{0.5}$	0.64	Schumm (1956)
11	Length Width Ratio		2.02	Horton (1945)
12	Texture Ratio (R_t)	$R_t = N_1 / P$	2.99	Horton (1945)
13	Circularity Ratio (R_c)	$R_c = 12.57*A/P^2$	0.31	Miller (1953)
14	Circularity Ration (R_{cn})	$R_{cn} = A/P$	2.70	Miller (1953)
15	Drainage Texture (D_t)	$D_t = N_u/P$	3.68	Horton (1945)
16	Compactness Coefficient (C_c)	$C_c = 0.2841*P/A^{0.5}$	1.81	Melton (1957)
17	Fitness Ratio (R_f)	$R_f = Cl/P$	0.46	Melton (1957)
18	Wandering Ratio (R_w)	$R_w = Cl/L_b$	1.21	Schumm (1956)

(continued)

Table 1 (continued)

S. No	Morphometric parameter	Formula	Result	References
C	*Drainage texture analysis*			
01	Stream Frequency (F_s)	$F_s = N_u/A$	1.36	Horton 1945
02	Drainage Density (D_d)	$D_d = L_u/A$	1.51	Strahler (1952)
03	Constant of Channel Maintenance	$C = 1/D_d$	0.66	Schumm (1956)
04	Drainage Intensity (D_i)	$D_i = F_s/D_d$	0.90	Faniran (1968)
05	Infiltration Number (If)	$If = F_s*D_d$	2.05	Faniran (1968)
06	Length of Overland Flow (L_g) Kms	$L_g = A/2*L_u$	0.33	Faniran (1968)
D	*Relief Aspects/Parameters*			
01	Height of Basin Mouth (z) m	Software analysis (ArcGis)	355	–
02	Maximum Height of the Basin (Z) m	Software analysis (ArcGis)	485	–
03	Total Basin Relief (H) m	$H = Z - z$	130	Strahler (1952)
04	Relief ratio (R_{hl})	$R_{hl} = H / L_b$	0.004	Schumm (1956)
05	Relative Relief Ratio (R_{hp})	$R_{hp} = H * 100 / P$	0.17	Melton (1957)
06	Gradient Ratio (R_g)	$R_g = (Z - z) / L_b$	0.004	Sreedevi (2004)
07	Ruggedness Number (R_n)	$R_n = D_d * (H /1000)$	0.24	Singh and Dubey (1994)
08	Melton Ruggedness Number (MR_n)	$MR_n = H / A^{0.5}$	0.007	Sreedevi (2004)
09	Mean slope of the basin		0.25	–
10	Watershed slope (S_w)	$S_w = H/L_b$	0.004	Melton (1965)

Rohni basin. The orientations of high frequency lineaments resemble the trends of low order streams of the basin. It reveals that the low order streams were controlled by older bedrock structures. Contrary to this the high order streams were controlled by major tectonic units of the area.

7 Field Observations

Field observations in the study area show the sinisterly rotated porphyroclast of quartz and feldspar (Fig. 10a, b) dominantly occur in sheared granitic rocks in the downstream side of the Rohni reservoir. The strike (E-W to ESE-WNW) of sheared rocks corresponding with orientation directions of higher orders streams indicate strong structural control of geological structures on drainage pattern.

8 Conclusions

The morphometric analysis of Rohni watershed reveals that the drainage represented by dendritic to subdendritic pattern and had an influence on lithological and geological structures. The streams of low order showing smaller length, fine texture and high gradient are moderately susceptible to soil erosion in the catchment area. Contrary to this the high order streams having larger length are less prone to soil erosion in the flat plain area. The less permeable subsoil and low infiltration rate and moderate population of streams of low erosive nature are dominant in the basin. The basin is significantly elongated with a dominant mature stage and is less prone to flood. The gradient is moderately steeper, less rugged and moderately erosive.

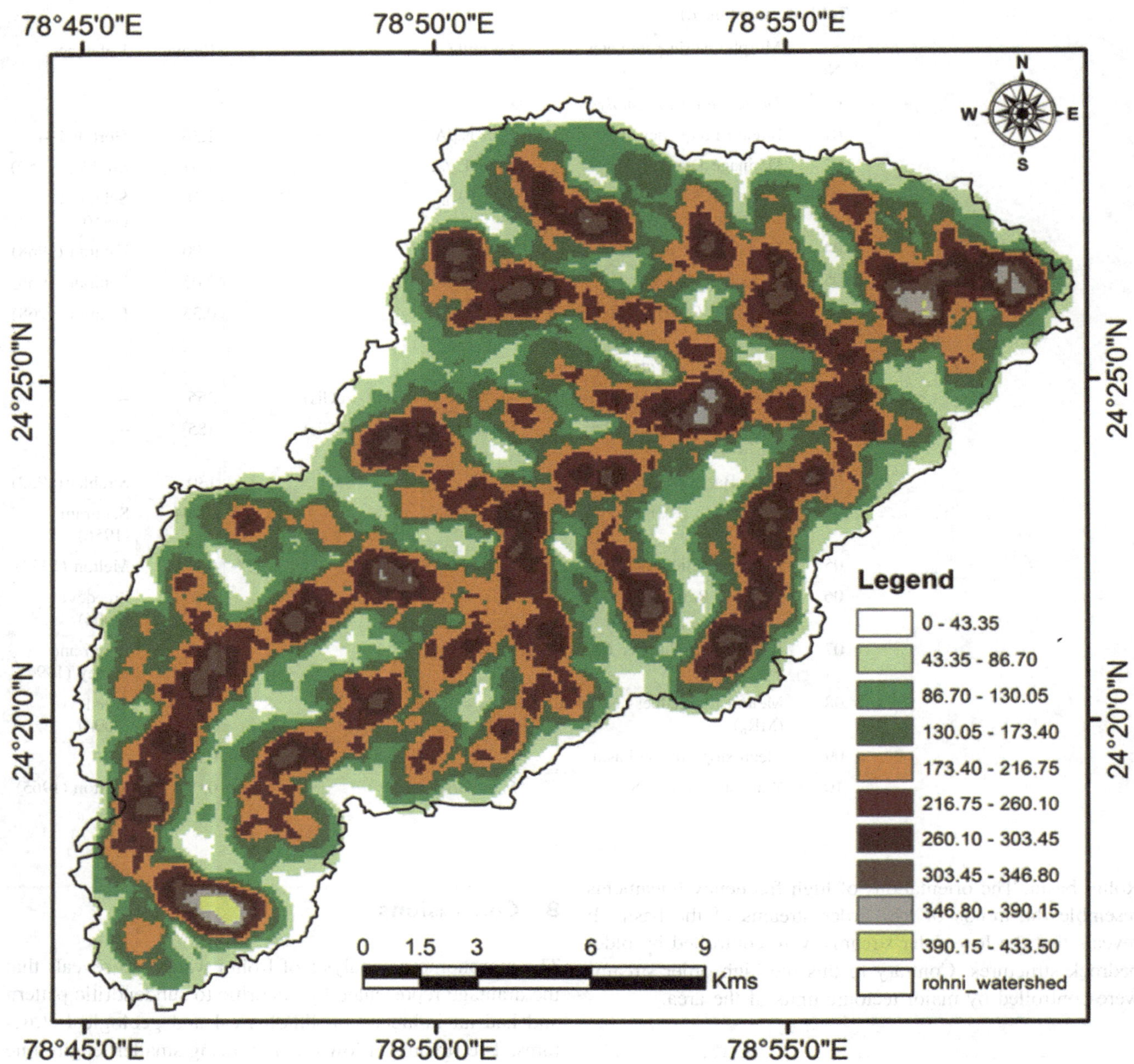

Fig. 8 Drainage density map of Rohni watershed

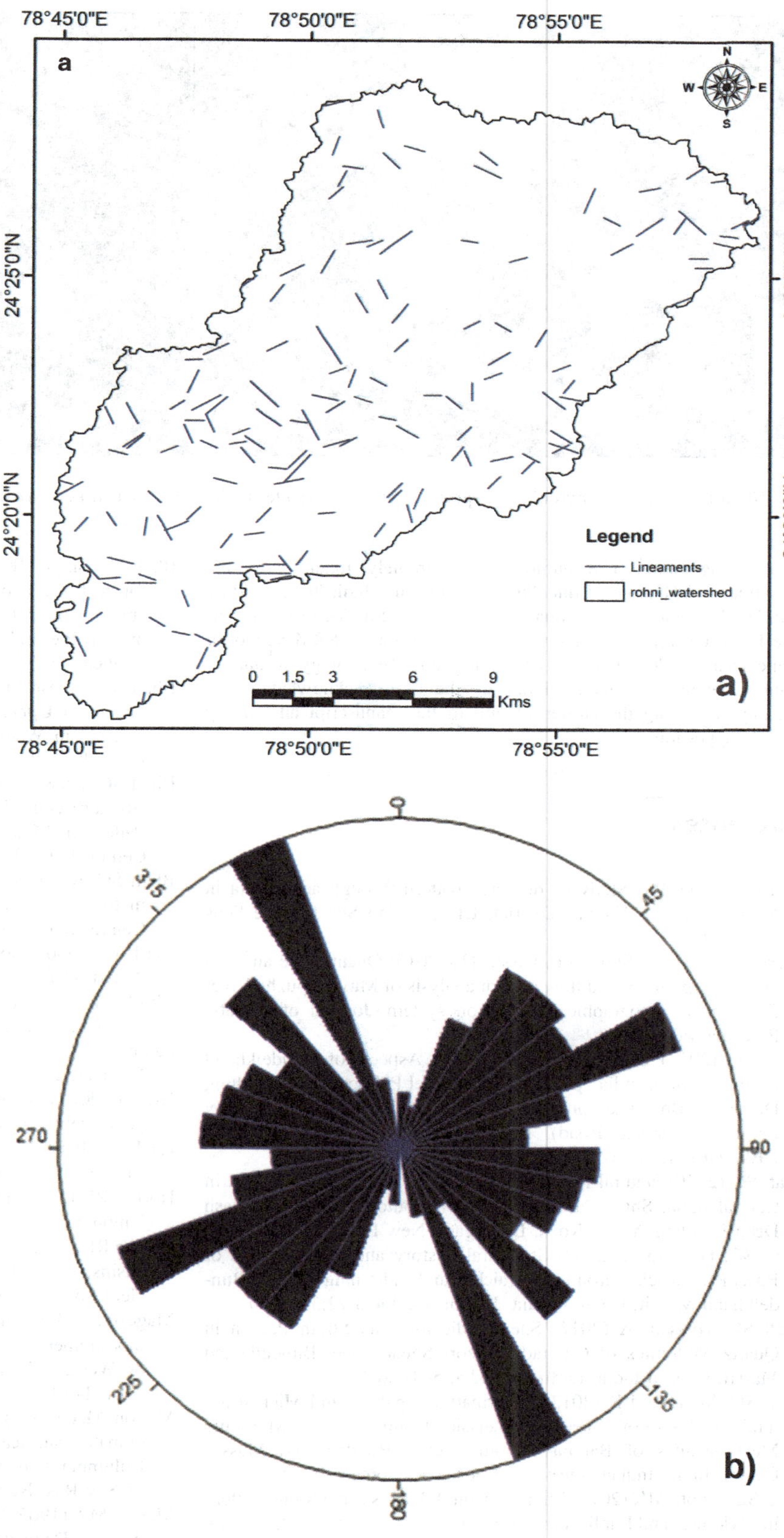

Fig. 9 a Lineament map of Rohni river watershed. b Rose plot showing orientations of lineaments

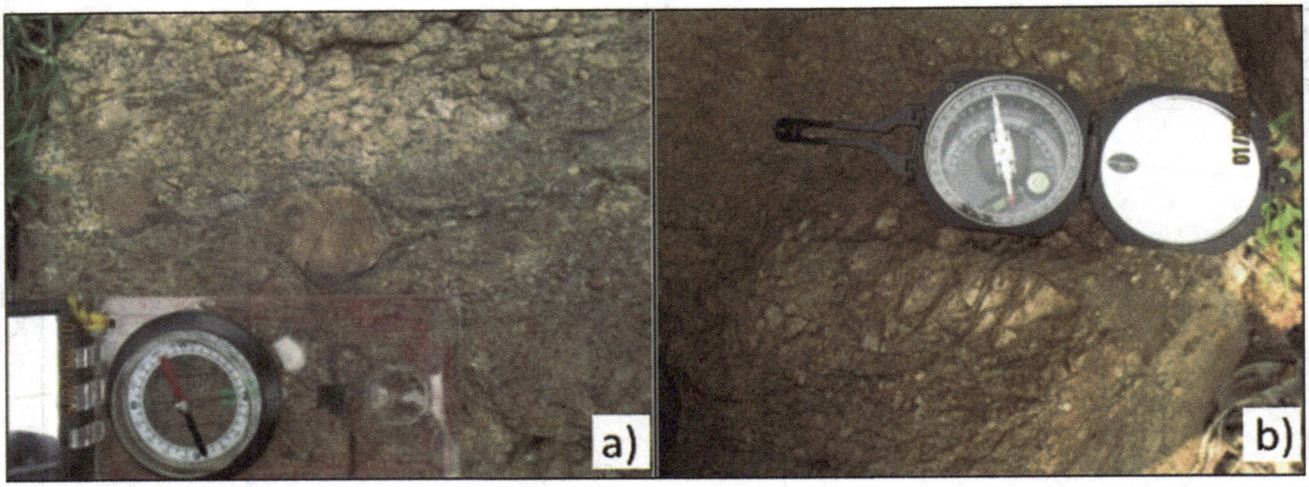

Fig. 10 a Fractured porphyroclast of quartz in sheared granite, **b** sinisterly rotated porphyroclast of feldspar in sheared granite

Acknowledgements The authors are extremely thankful to the Department of Geology, Bundelkhand University, India for providing facilities. Professor S. N. Mohapatra, School of Studies in Earth Science Jiwaji University Gwalior, India and Dr Vignesh, SRM School of Public Health, SRM Institute of Science and Technology Kattankulathur, Chennai, Tamilnadu, India are thankfully acknowledged for critically reviewing the earlier version of the manuscript and giving fruitful suggestions.

References

Agarwal CS (1998) Study of drainage pattern through aerial data in Naugarh area of Varansi District, UP. J Indian Soc Remote Sens 26:169–175

Aparna P, Nigee K, Shimna P, Dissia TK (2015) Quantitative analysis of geomorphology and flow pattern analysis of Muvattupuzha River Basin using Geographic Information system. Journal of Aquatic Procedia 4:609–616

Bhatt SC (2014) Geological and Tectonic Aspects of Bundelkhand Craton, Central India" published by, Angel Publication, Publishing Division: Bhagwati publishers Distributors C-8/77-B, Keshav Puram, (Lawrence Road), Near Keshav Puram Metro Station. NEW DELHI- 110035:190

Bhatt SC (2020) Natural Resources, Water Harvesting and Drought in Central India, Shree Publishers and Distributors 22/4735, Prakash Deep Building Ansari Road, Daryaganj. New Delhi 110002:130

Bhatt SC, Hussain A (2008) Structural History and Fold analysis of Basement Rocks around Kuraicha and adjoining areas, Bundelkhand Massif, Central India. J Geol Soc India 72:331–347

Bhatt SC, Hussain A (2012) Shear Indicators and Strain Pattern in Quartz Mylonites of Chituad – Deori Shear Zone, Bundelkhand Massif, Central India. Earth Sci India. 5(II):60–78

Bhatt SC, Mahmood K (2012) Deformational pattern and Microstructural analysis of Sheared Gneissic Complex and Mylonitic Metavolcanics of Babina-Prthipur Sector. Bundelkhand Massif, Central India, Indian Journal of Geosciences 66(1):79–90

Bhatt SC, Gupta MK (2014) Microstructural Analysis and Strain Pattern in Mylonites and Implications of Shear Sense Indicators in Evolution of Dinara- Garhmau Shear Zone, Bundelkhand Massif, Central India, The Indian Mineralogist. J Mineral Soc India 48(2):186–206

Bhatt SC, Singh VK, Hussain A (2011) Implications of shear indicators for tectonic evolution of Mauranipur shear zone, Bundelkhand craton, central India. In: Singh VK, Chandra R (eds) 2nd Proceeding of Precambrian Continental Growth and Tectonism. Angel Publication, New Delhi, pp 36–49

Bhatt SC, Suresh M, Gupta MK (2015) Structural Control on Drainage Pattern of Upper- Middle Pahuj River and Implication of Remote Sensing in Watershed Management Bundelkhand Craton. Central India, Jour of Multidisciplinary Scientific Research 2(4):9–14

Bhatt SC, Hussain A, Suresh M, Balasooriya NWB (2017) Geological structure control on Sukhnai Basin and Land Use / Land Cover Pattern in Mauranipur and adjoining areas. Bundelkhand Craton, Central India, Jour Geol Soc of Sri Lanka. 18(2):53–61

Bhatt SC, Rubal S, Ansari MA, Bhatt S (2020) Quantitative morphometric and morphotectonic analysis of Pahuj catchment basin. Central India. J Geol Soc India 96(5):513–520

Clark C (1966) Morphometry from Map, Essay in Geomorphology, American Elsvier Publishing Company Newyork, 235–274

Chorley RJ, Malm DEG, Pogorzelski HAA (1957) New Standard for Estimating Drainage Basin ShapeAmer 1 Sci 255:138–141

Faniran A (1968) The Index of Drainage Intensity - A Provisional New Drainage Factor. Aust J Sci 31:328–330

Gregory KJ, Walling DE (1973) Drainage basin form and process a geomorphological approach. Edward Arnold, London

Hack JT (1957) Studies of longitudinal stream profiles in Virginia and Maryland, USGS professional paper, US Geological Survey 94 B

Horton RE (1932) Drainage Basin Characteristics. Trans Am Geophys Union 13:350–361

Horton RE (1945) Erosional development of streams and their drainage basins; hydrophysical approach to quantitative morphology. Bull Geol Soc Am 56:275–370

Magesh NS, Jitheshlal KV, Chandrasekar N, Jini KV (2012) GIS based morphometric evaluation of Chimmini and Mupily watershed, parts of Western Ghats. Thrissur District, Kerala, Indian Earth Sci 5 (2):111–121

Melton MA (1957) An analysis of the relations among elements of climate, surface properties and geomorphology, Dept. Geol., Columbia University, Technical Report, 11, Proj.NR389–042off of Nav Res, New York

Melton MA (1965) The Geomorphic and Paleoclimatic Significance of Alluvial Deposits in Southern Arizona. J Geol 73(1):1–38

Miller VC (1953) A quantitative geomorphic study of drainage basin characteristics in the Clinch mountain area. Department of Geology, Columbia University, Virginia and Tennessee

Nag SK, Chakrborty S (2003) influence of rock types and structures in the development of drainage network in hard rock area. J India Soc Remote Sens 31(1):25–35

Nautiyal MD (1994) Morphometric analysis of a drainage basin, district Dehradun. Uttar Pradesh, J Indian Soc Remote Sens 22(4):251–261

Obi Reddy GE, Maji AK, Gajbhiye KS (2002) GIS for morphometric analysis of drainage basins. GIS India 4:9–14

Prakash K, Chaubey K, Singh S, Mohanty T, Singh CK (2017) Morpphometric Analysis of Satna River Basin, Central India. Indian J Geomorphol 22(1):41–60

Pareta K, Pareta U (2011) Quantitative morphometric analysis of a watershed of Yamuna Basin, India using ASTER (DEM) data and GIS. Int J Geomatics Geosci 2(1):248–269

Pareta K, Pareta U (2012) Quantitative geomorphological analysis of watershed of Ravi River basin, HP, India. Int J Remote Sensing GIS 1:47–62

Rai PK, Mohan K, Mishra S, Ahmad A, Mishra VN (2017) A GIS-based approach in drainage morphometric analysis of Kanhar River Basin, India". Appl Water Sci 7:217–232

Schumm SA (1956) Evolution of Drainage Systems & Slopes in Badlands at Perth Anboy. New Jersey, Bull Geol Soc America 67:597–646

Shreve RW (1969) Stream lengths and basin areas in topologically random channel networks. J Geol 77:397–414

Singh VK, Slabunov A (2015) The Central Bundelkhand Archaean greenstone complex, Bundelkhand Craton, Central India: geology, composition, and geochronology of supracrustal rocks. Int Geol Rev 57(11–12):1349–1364

Singh VK, Slabunov AI, Nesterova NS et al (2021) Tectonostratigraphic terranes of the Bundelkhand Craton (Indian Shield). In: Shandilya AK, Singh VK, Bhatt SC, Dubey CS (eds) Geological and Geo-Environmental Processes on Earth, pp 155–164 (this volume)

Smith KG (1950) Standards for grading texture of erosional topography. Am J Sci 248:655–668

Soni S (2017) Assessment of Characteristics of Chakrar watershed in Madhya Pradesh, india, using geospatial technique Applied Water Science 7:2089–2102

Strahelr AN (1952) Hypsometric (area-altitude) analysis of erosional topography, Bull Geol, Soc Am 63

Strahler AN (1956) Quantitative Slope Analysis. Bulletin of the Geological Society of America 67:571–596

Strahler AN (1957) Quantitative analysis of watershed geomorphology. Trans Am Geophys Union 38:913–920

Strahler AN (1958) Dimensional analysis applied to fluvially eroded landforms. Geol Soc Am Bull 69:279–300

Strahler AN (1964) Quantitative Geomorphology of Drainage Basin and Channel Network, Handbook of Applied Hydrology, 39–76

Quantitative Morphometric Analysis of Watershed of Upper Pahuj Basin, Bundelkhand Region, Central India

S. C. Bhatt, Rubal Singh, M. Taiyab, and S. Bhatt

Abstract

In the present study, a detail morphometric evaluation of upper Pahuj basin is attempted to delineate drainage pattern, linear, areal, relief, geological control and other aspects. The Arc GIS (10.4 version) and ASTER data were used to extract drainage, density, aspect, slope, contour and other thematic maps. The five orders of streams were identified in the study area. The relationship between geological structures and different orders of the stream was established. About fifty parameters were analysed. The higher values of the bifurcation ratio in higher orders and its mean value (4.47) indicate that the drainage pattern is inherently controlled by geological structures. The moderate drainage density (2.20) and low frequency (2.86) are suggestive of low infiltration rate and have a thin vegetal cover on less permeable rocky terrain. The low values of elongation ratio (0.62) and form factor ratio (0.30) infer that the basin catchment is elongated and is showing low peak flows of greater interval. The circulatory ratio (0.34) computed for upper Pahuj catchment implies that the youth stage was dominant in the evolution of the watershed. The ruggedness number (0.30) indicates that the terrain has a moderate slope and is showing slight morphology.

Keywords

Morphometric analysis • Upper Pahuj basin • Bundelkhand • Central India

S. C. Bhatt (✉) · R. Singh · M. Taiyab
Department of Geology, Institute of Earth Sciences, Bundelkhand University Jhansi, Jhansi, UP, India
e-mail: scbhatt@bujhansi.ac.in

S. Bhatt
Institute of Environment and Development Studies, Bundelkhand University Jhansi, Jhansi, UP, India

1 Introduction

The morphometry of a basin reveals the geological and geomorphological processes undergone in geological period and influence the drainage basin is significant in understanding the landform processes, a physical and erosional property of soil (Horton 1945; Strahler 1952, 1964). The development of quantitative physiographic methods for evolution and behaviour of surface drainage networks have been widely used in geomorphology for the last several years (Horton 1945). Morphometry is a measurement and mathematical evaluation of the earth's surface, shape, size and dimension of landforms (Agarwal 1998; Obi Reddy et al. 2002). The implications of linear, areal, relief and tectonic aspects of a drainage basin can be understood by carrying detail morphometric analysis (Singh and Singh 1997; Nag and Chakraborty 2003; Sreedevi et al. 2004; Pareta and Pareta 2011; Bhatt et al. 2017, 2020; Bhatt 2020). It will play an important role in the understanding of geo-hydrological behaviour of basin and expresses the prevailing climate, geology, geomorphology, structural antecedents of the catchment. Morphometry is also important in the assessment of groundwater potential, groundwater management, pedology and environmental assessment. The progressive development of fluvial landforms along with drainage pattern showing inherent control of lithology and tectonics is discussed under drainage basin analysis (Horton 1945; Faniran 1968; Pati et al. 2008). The proper management of watershed needs utilisation of various aspects like land, water, soil and forest resources of a particular watershed for better production and lesser hazard to natural resources (Biswas et al. 1999). The hydrological response of a river basin can be interrelated with the physiographic characteristics of the drainage basin, such as size, shape, slope, drainage density and size and length of the streams, etc. (Gregory and Walling 1973). The development of landforms and drainage network depends on the bedrock lithology and associated geological structures, hence,

A. K. Shandilya et al. (eds.), *Geological and Geo-Environmental Processes on Earth*, Springer Natural Hazards,
https://doi.org/10.1007/978-981-16-4122-0_19

information on geomorphology, hydrology, geology and land cover can be obtained by reliable information from the study of drainage pattern and texture.

The morphometric analysis of the drainage basin and channel network plays a significant role in the comprehension of the geo-hydrological nature of drainage basin and expresses the prevailing climate, geological setting, geomorphology and structural antecedents of the catchment area. A quantitative evaluation of the drainage system is a significant aspect of the drainage basin. Recently the morphometric and morphotectonic parameters of a watershed basin, geometrical characteristics of fluvial landscapes, structural anisotropy, planning of watershed and rainwater harvesting can be interpreted by carrying morphometric analysis and using Advanced Space-borne Thermal Emission and Reflection Radiometer (ASTER) and Shuttle Radar Topographic Mission (SRTM) data (Prareta 2011; Bhatt et al. 2017, 2020; Bhatt 2020).

In this work, the linear, areal and relief parameters of the watershed of Pahuj upper basin were analysed to understand the hydrological, geological, tectonic and geomorphological characters of the watershed. It gives information about the basin characteristics in terms of slope, topography, soil conditions, runoff characteristics and surface water potential, etc.

2 Study Area

The study area belongs to the upper part of the catchment of the Pahuj basin of Bundelkhand region of Central India. The toposheet G44M07 and G44M11 (1:50,000) of Survey of India were used as a base map to extract stream network and drainage map of the area and it is located in the northwestern sector of this craton. It is lying between latitude 25,018'30" to 25,027'27"N and longitude 78,017'33" to 78,031'22"E.

3 Geomorphology of the Study Area

The Study area located in the north-western part of Bundelkhand craton is represented by rugged and undulating topography with moderately elevated hills. The erosional and depositional landforms moderately elevated granitic and quartz reefs (200–300 m) and low lying alluvial plains are observed as common geomorphic features in the study area (Bhatt et al. 2015). The NE-SW trending quartz reefs are generally covered with thick vegetation. The granitic and gneissic terrains are characterised by hard and massive lithology and at few places are covered by scrub. The isolated granitic and gneissic boulders are frequently scattered at several places. The Pahuj Tal (lake) is an important water body seen in the study area. The rest of the low lying area is covered by alluvium. The fractured, weathered and sheared hard rock terrain and low lying alluvial plains are considered good potential zones for groundwater.

4 Geological Setup

Bundelkhand TTG Gneissic Complex (3.1–3.5 Ga), Greenstone Belt and Bundelkhand Granitic Complex (2.5 Ga) are identified as important major Groups in the investigated area (Bhatt and Hussain 2008; Bhatt and Gupta 2009; Bhatt and Hussain 2012; Bhatt and Mahmood 2012; Bhatt and Gupta 2014; Bhatt 2014; Singh and Slabunov 2015; Bhatt et al. 2011, 2017; Singh et al. 2021). The Archean TTG gneisses (3.4 Ga), gneisses, granite gneisses and mylonitised granite gneisses and other rocks were observed as important rock units of Bundelkhand TTG gneissic complex. Bundelkhand Greenstone belt constitutes metasedimentaries and metavolcanic of Neoarchean age. The various types of granites (grey, pink and porphyritic granites) of Precambrian age (2.5 Ga), quartz reefs, doleritic dykes and pegmatic veins belong to Bundelkhand granitic complex.

5 Data Base and Methodology

The Arc GIS (10.4) and ASTER-DEM data (Fig. 1) were used to delineate drainage map, Digital Elevation Model (DEM), watershed delineation map, density, contour, aspect and other thematic maps. The toposheets (Survey of India) were scanned, georeferenced and mosaicked in Arc GIS 10.4 platform. Stream ordering schemes first proposed by Horton (1945) then revised by Strahler (1952), were applied to order Stream (by applying a numerical value to a stream's position and size) in the basin. In the present study, morphometric analysis of basin is based on the integrated use of remote sensing and GIS technique. The remotely sensed data is geometrically rectified with respect to Survey of India (SOI) topographical maps (the year 2010) at 1:50,000. The digitisation of the drainage pattern is carried out in Arc GIS 10.4 software. (Fig. 1). For stream ordering, Horton's law is followed by designating an un-branched stream as the first-order stream, when two first-order streams join it is designated as second order. Two-second order steams join together to form third order and so on. The number of streams of each order is counted and recorded. Direct measurements of geometric characteristics (e.g., area and perimeter of the basin, length and number of streams) were automatically obtained from GIS software.

The various morphometric parameters such as area, perimeter, stream order, stream length, stream number,

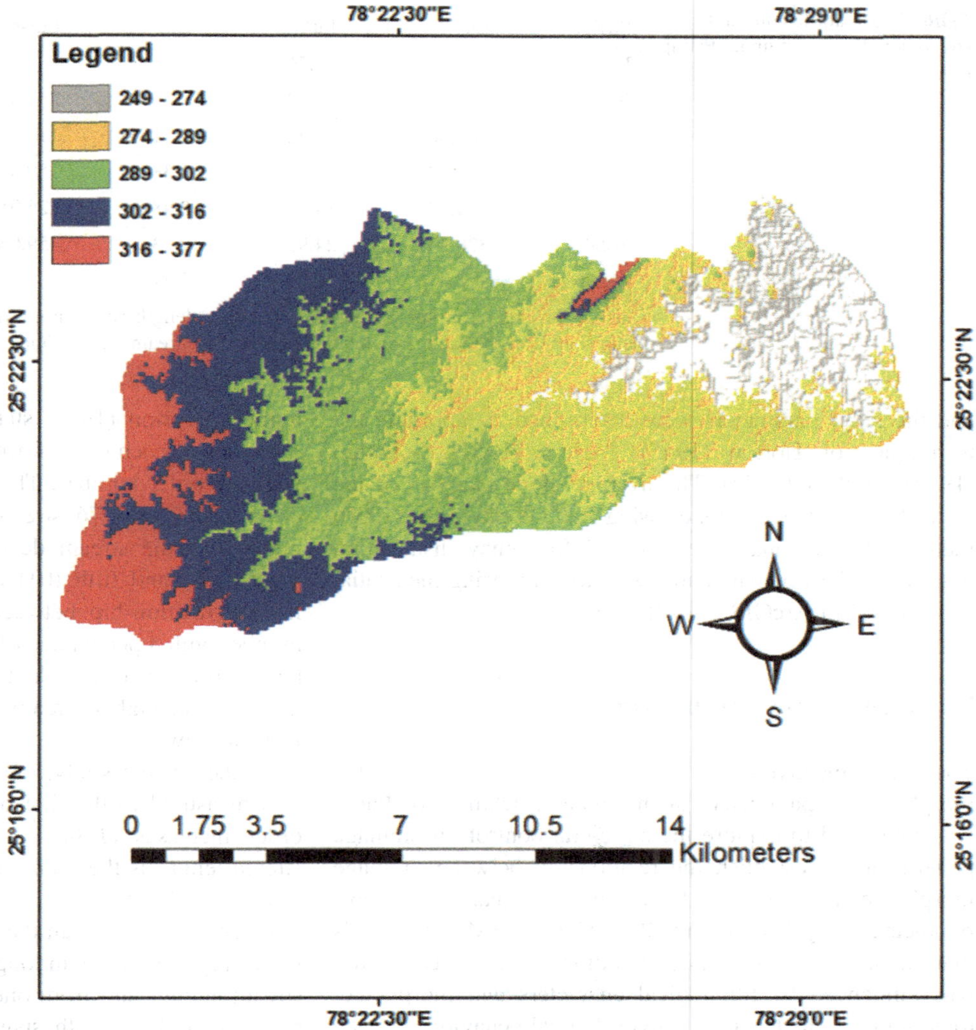

Fig. 1 Digital Elevation model (DEM) of the study area

bifurcation ratio, drainage density, stream frequency, drainage texture, length of basin, form factor, circulatory ratio, elongation ratio, length of overland flow, compactness coefficient, shape factor and texture ratio were computed (Tables 1, 2, and 3) based on the formula suggested by Horton 1945, Miller 1953, Schumn 1956 and Strahler 1964.

6 Morphometric Analysis

The linear aspects were computed by using the methods proposed by Horton (1945) and Chorley (1957). The areal aspects were quantified by using methods of Schumm (1956), Strahler (1956), Miller (1953), and Horton (1932),

Table 1 Stream order, streams number and bifurcation ratio

Su	Nu	Rb	Nu-r	Rb*Nu-r	Rbwm
I	328				4.47
II	76	4.32	404	1754.28	
III	19	4.00	95	380.00	
IV	2	9.50	21	199.5	
V	1	2.00	3	6	
Total	426	19.82	523	2339.78	
Mean		5.00			

Su: Stream order, Nu: Number of streams, Rb: Bifurcation ratio, Nu-r: Number of stream used in the ratio, Rbwm: Weighted mean bifurcation ratio

Table 2 Stream length, and stream length ratio in upper Pahuj basin

Su	Lu	Lu/Su	Lur	Lur-r	Lur*Lur-r	Luwm
I	182.67	0.56				2.07
II	70.85	0.93	1.66	253.52	420.84	
III	47.86	2.52	2.71	118.71	321.70	
IV	12.02	6.01	2.38	59.88	142.51	
V	14.93	14.93	2.48	26.95	66.84	
Total	328.33	24.95	9.23	459.06	951.89	
Mean			2.31			

Su: Stream order, Lu: Stream length, Lur: Stream length ratio, Lurm: Mean stream length ratio, Lur-r: Stream length used in the ratio, Luwm: Weighted mean stream length ratio

and the relief parameters were determined by applying the techniques of Horton (1945), Melton (1957), Schumm (1954), Strahler (1952). The average slope analysis of the watershed area was done using the Wentworth (1930) method. The Drainage density and frequency distribution analysis of the watershed area was done by using the spatial analyst tool in ArcGIS-10 software.

7 Results and Discussion

The main objective of the present study is to delineate the morphometric parameters of the upper catchment of Pahuj watershed and to explore the structural control on drainage pattern and to establish the relationship between drainage morphometry and various landforms. The quantitative morphometric analysis based on GIS ASTER DEM data reveals that the drainage pattern and evolution of each stream order was influenced by lithological characters and morphotectonic setup of the basin. The hydrological behaviour of each watershed is also documented by all computed values of linear, areal and relief data. About fifty-five parameters pertaining to linear, areal and relief aspects were quantified by using Arc GIS software and ASTER DEM data. The drainage pattern is represented by dendritic to subdendritic pattern (Fig. 2).

7.1 Linear Parameters

The linear property is a one-dimensional character used to define the linear parameters in the morphometric analysis. The stream order may be defined as a measure of the position of the stream in the hierarchy of tributaries (Strahler 1964; 1952; Horton 1945). Five order of stream were classified for the Pahuj upper catchment basin (Fig. 2). I to III orders were categorised as lower order and IV to V orders were classified as a higher order. The values of a stream number and stream length are shown in Tables 1 and 2. The total of order wise stream segments is known as stream number. Horton (1945) states that the numbers of stream segments of each order form an inverse geometric sequence with the order number. The first order has 328 and second order contributes 76 streams, respectively (Table 1). The number of the stream decreases in the higher third (19), fourth (02) and fifth (01) order of streams (Table 1). An inverse relationship between the stream order and stream number with a percentage of variables is established (Figs. 3 and 4). It indicates that the total number of streams gradually decreases in higher stream order (Fig. 4). This will follow Horton's law.

In the present study, the maximum length of the stream was measured in the first order and it decreases as stream order increases (Table 2). Horton (1945) states that the stream length is the total length of stream segment of each order. The lower order has a shorter length with a larger number of streams than the higher order. The highest order has a single stream with longer stream length. The change of stream length ratio from one order to another order is indicating their late youth stage of geomorphic development. The total length of the Pahuj catchment basin is 328 kms. The bifurcation ratio (Rb) varies from 2.00 to 9.5. The highest value (9.5) of Rb was estimated in higher order streams while the lower values were observed in lower order streams. The variable value of bifurcation ratio (ranging from 2.0 to 9.5) indicates that the drainage evolved in a homogenous geological terrain and influenced by geological structures. The value of Rho Coefficient ($\rho = 0.46$) is suggestive of high hydrologic storage during a flood.

7.2 Areal Parameters/Basin Geometry

Length of the Basin (Lb) is a maximum length of watershed measured from head to mouth (Strahler 1964) and it was measured by using Arc GIS software (Table 3). Basin Area (A) is a maximum area drained by a basin and is measured by using Arc GIS software (Table 3). The upper catchment of Pahuj basin occupies 149 km^2 area and it has a total of 22.33 km basin length. Its perimeter is 73.89 km. Form

Table 3 Morphometric analysis of upper Pahuj basin

S. No	Morphometric parameter	Formula	Result
A	*Drainage network*		
01	Stream order (Su)	Hierarchical ran	1–5
02	1st order stream (Suf)	Suf = N1	328
03	Stream number (Nu)	Nu = N1 + N2 + …Nn	426
04	Stream length (Lu) Kms	Lu = L1 + L2 …… Ln	328.33
05	Stream length ratio (Lur)	see Table 2	1.66–2.71
06	Mean stream length ratio (Lurm)	see Table 2	2.31
07	Weighted mean stream length ratio (Luwm)	see Table 2	2.07
08	Bifurcation ratio (Rb)	see Table 1	2.0–9.5
09	Mean bifurcation ratio (Rbm)	see Table 1	5.00
10	Weighted mean bifurcation ratio (Rbwm)	see Table 1	4.47
11	Main channel length (Cl) Kms	GIS Software	22.46
12	Rho coefficient (ρ)	ρ = Lur/Rb	0.46
B	*Basin geometry*		
01	Basin length (Lb) Kms	GIS Software Analysis	22.33
02	Mean basin width (Wb)	Wb = A/Lb	6.67
03	Basin area (A) SqKms	GIS Software Analysis	149.00
04	Basin perimeter (P) Kms	GIS Software Analysis	73.89
05	Relative Perimeter (Pr)	Pr = A/P	2.02
06	Length area relation (Lar)	Lar = 1.4 * $A^{0.6}$	28.19
07	Lemniscate's (k)	k = Lb^2 A	3.35
08	Form factor ratio (Ff)	Ff = A/Lb^2	0.30
09	Shape factor ratio (Sf)	Sf = Lb^2/A	3.35
10	Elongation ratio (Re)	Re = 2/Lb * (A/π) 0.5	0.62
11	Texture ratio (Rt)	Rt = N1/P	5.77
12	Circularity ratio (Rc)	Rc = 12.57 * (A/P^2)	0.34
13	Circularity ration (Rcn)	Rcn = A/P	2.02
14	Drainage texture (Dt)	Dt = Nu/P	5.77
15	Compactness coefficient (Cc)	Cc = 0.2841 * P/A 0.5	1.37
16	Fitness ratio (Rf)	Rf = Cl/P	0.30
17	Wandering ratio (Rw)	Rw = Cl/Lb	1.01
C	*Drainage texture analysis*		
01	Stream frequency (Fs)	Fs = Nu/A	2.86
02	Drainage density (Dd) Km / Kms^2	Dd = Lu/A	2.20
03	Constant of channel maintenance (Kms/Km)	C = 1/Dd	0.45
04	Drainage intensity (Di)	Di = Fs/Dd	1.30
05	Infiltration number (If)	If = Fs * Dd	6.29
06	Drainage pattern (Dp)		Dn-SubDn
07	Length of overland flow (Lg) Kms	Lg = A/2 * Lu	0.23
D	*Relief characterizes*		
01	Height of basin mouth (z) m	GIS analysis/DEM	249.00
02	Maximum height of the basin (Z) m	GIS analysis/DEM	377.00
03	Total basin relief (H) m	H = Z – z	128.00
04	Relief ratio (Rhl)	Rhl = H/Lb	0.006

(continued)

Table 3 (continued)

S. No	Morphometric parameter	Formula	Result
05	Relative relief ratio (Rhp)	Rhp = H * 100/P	0.20
06	Gradient ratio (Rg)	Rg = (Z − z)/Lb	0.006
07	Watershed slope (Sw)	Sw = H/Lb	0.006
08	Ruggedness number (Rn)	Rn = Dd * (H/1000)	0.30
09	Melton ruggedness number (MRn)	MRn = H/A$^{0.5}$	9.67

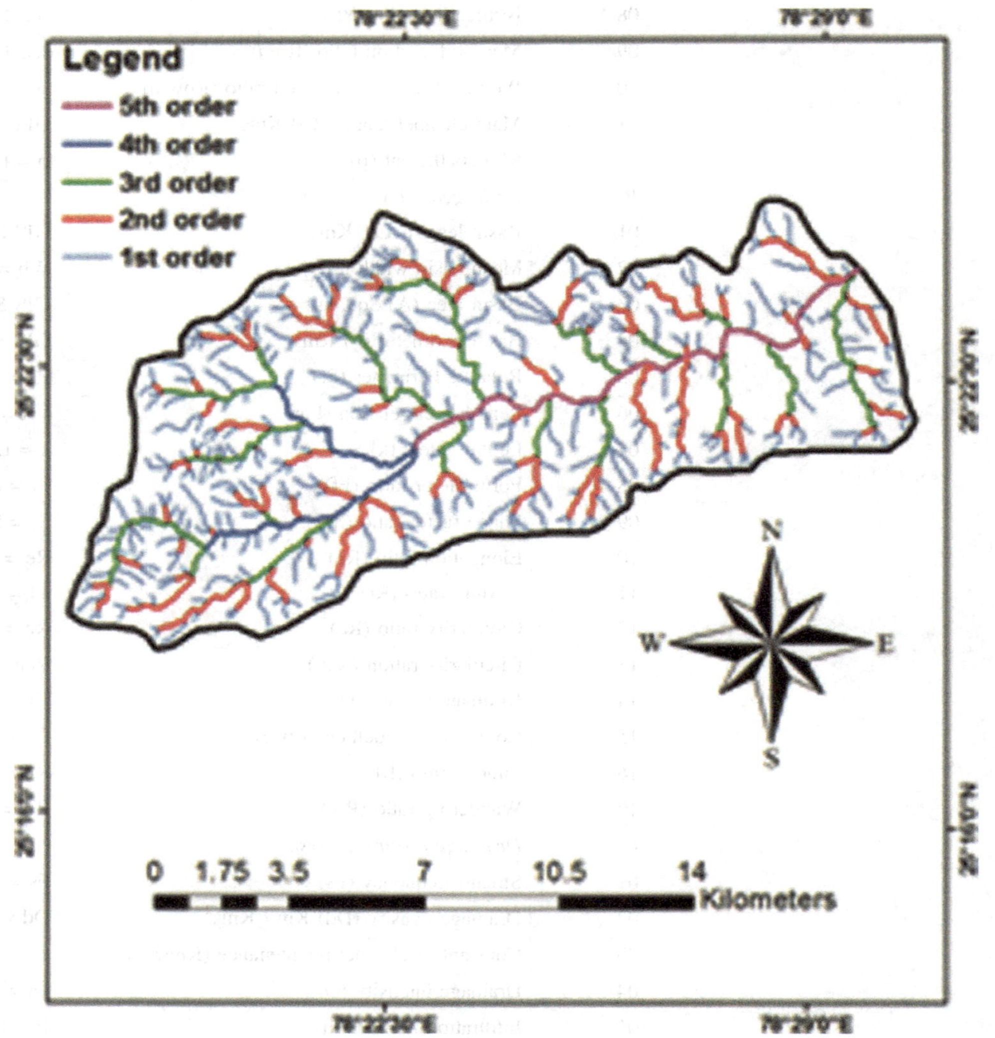

Fig. 2 Drainage map of Upper Pahuj basin

factor (Rf) is defined as the representation of the shape of the watershed (Horton 1945). The lower values of form factor are indicative of more elongated basin. The watershed with high form factors has high peak flows of shorter duration, whereas elongated watershed with low form factor has low flow peak of longer duration (Horton 1945). The value of the form factor for upper Pahuj catchment is calculated at 0.33 (Table 3). It infers that the basin is more elongated and shows lower peak flows over a longer interval of time. Elongation ratio (Re) explained as the ratio of the diameter of a circle of the same area as the watershed to the maximum basin length (Schumm 1956). The value of the elongation ratio grouped into five categories as circular (0.9–0.10), oval (0.8–0.9), less elongated (0.7–0.8), elongated (0.5–0.7), and

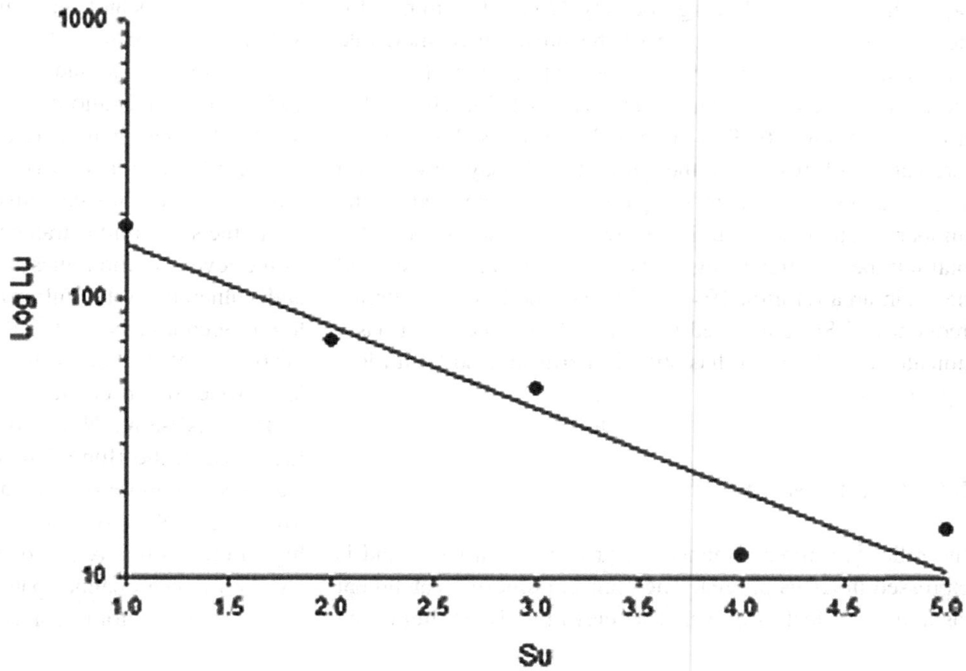

Fig. 3 Shows the relation between stream order and logarithm of stream length

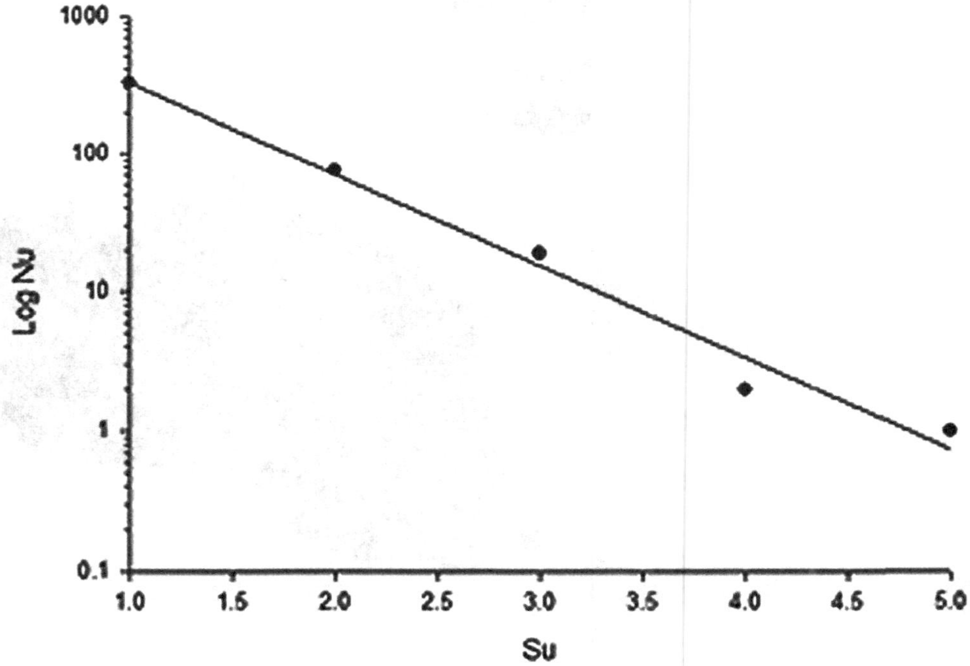

Fig. 4 The relation between stream order and logarithm of stream number

more elongated (<0.5). The elongation ratio (0.64) calculated for upper Pahuj basin implies that the basin is elongated and shows steeper slope.

Texture Ratio (Rt) is a total number of stream segments of all orders per perimeter of that area (Horton, 1945). It can be classified into five categories; very coarse (<2), coarse (2–4), moderate (4–6), fine (6–8) and very fine (>8). The upper Pahuj catchment basin has a moderate texture ratio of 5.77 (Table 3). It reveals that the basin consists of softer and more resistant harder rocks. Drainage density (Dd) is defined as the sum of stream lengths per unit area and is an expression of closeness or spacing of channels (Horton 1932). The density can be classified as extremely low (0–1), low (1–2), moderate (2–4), moderately high (4–6), and high

(>6). The moderate drainage density (2.20) determined for the area (Fig. 5) is indicative of the presence of moderate permeability with relatively good vegetative cover. Drainage texture is 5.77, infiltration number (If) is 6.29. The higher infiltration number 6.29 of upper Pahuj basin is showing low percolation of water on the ground and may have good erosion exposure. Stream Frequency (Fs) is defined as the number of streams per unit area or is the ratio between the total number of stream segments of all orders in a basin and the basin area (Horton 1945; Table 3). The low to moderate frequency (2.86) computed for upper Pahuj catchment basin indicates that the basin has slightly permeable soil with less vegetal cover.

7.3 Relief Parameters

The relief is a three-dimensional property of a basin and is expressed in terms of area. The highest point of a catchment basin is 377 feet (Fig. 6). The elevation is an important factor to understand the denudational processes of a basin. Relief is expressed as the maximum vertical distance between the lowest and highest points of a basin (Schumm, 1956). The relief ratio (0.006) indicates that the steepness of basin catchment is moderate and less susceptible to erosion. Ruggedness Number (Rn) is the product of maximum basin relief (H) and drainage density (Dd), where both parameters are in the same unit (Strahler 1956; Schumm, 1956). It refers to the level of smoothness and roughness of a basin terrain and vulnerability to soil erosion (Gutema et al. 2017). The low ruggedness value (0.30) of upper Pahuj catchment basin (Table 3) infers that the terrain is moderately steeper and is less prone to soil erosion.

E-W, NE-SW, N-S and NW–SE major lineaments are identified in the Bundelkhand craton. In the study area, the watershed is more or less in an elongated shape and is following the NE-SW lineament (Fig. 7). It reveals that the higher order streams are controlled by the major tectonic unit (NE-SW). The channel pattern and low values of sinuosity manifest less tectonic control.

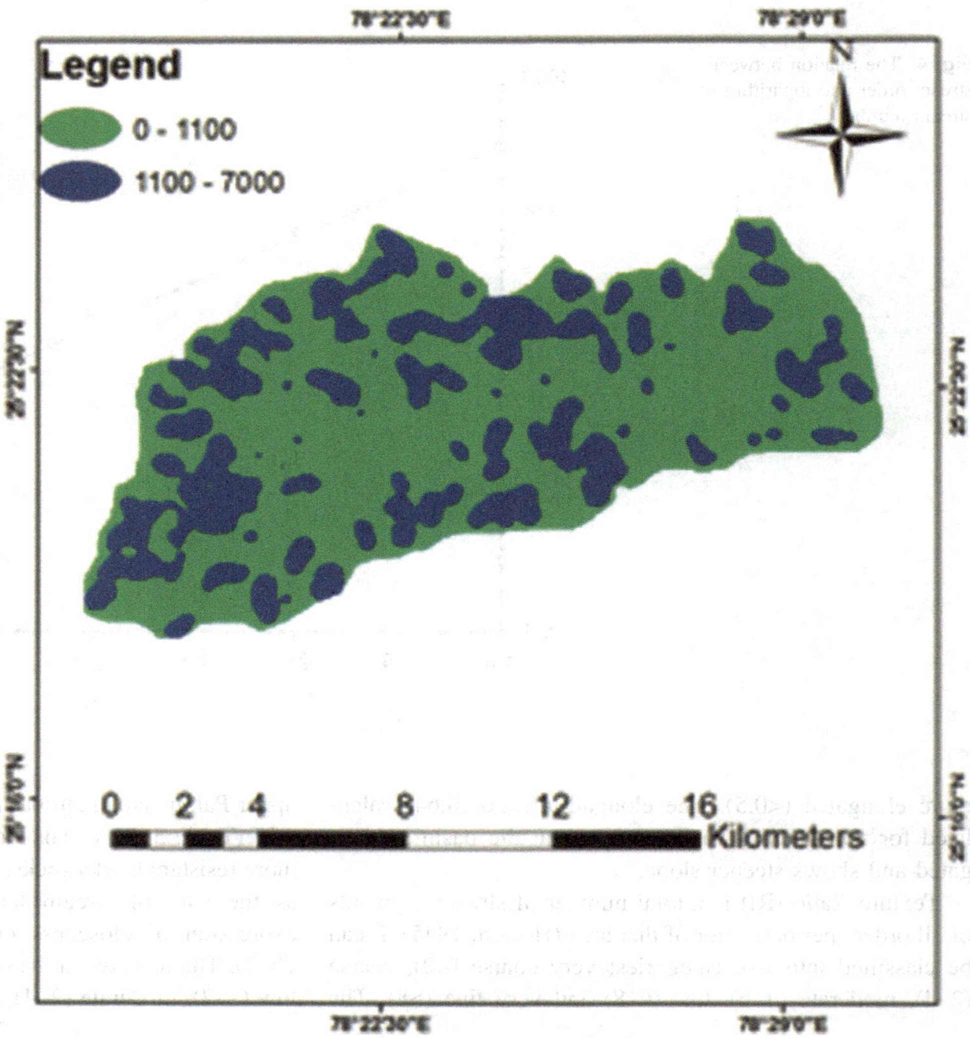

Fig. 5 Drainage density map of the watershed

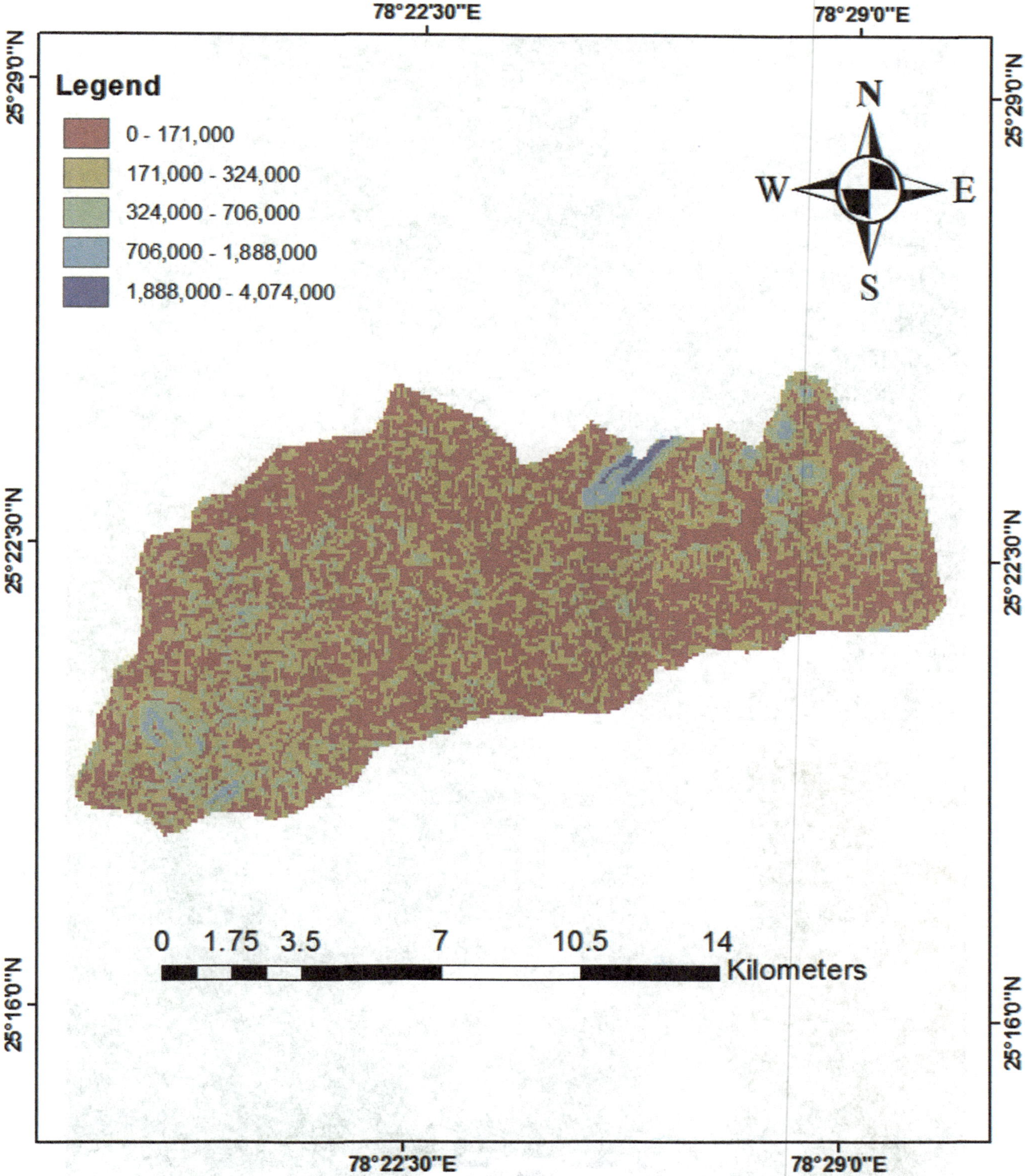

Fig. 6 Hill shade map of study area

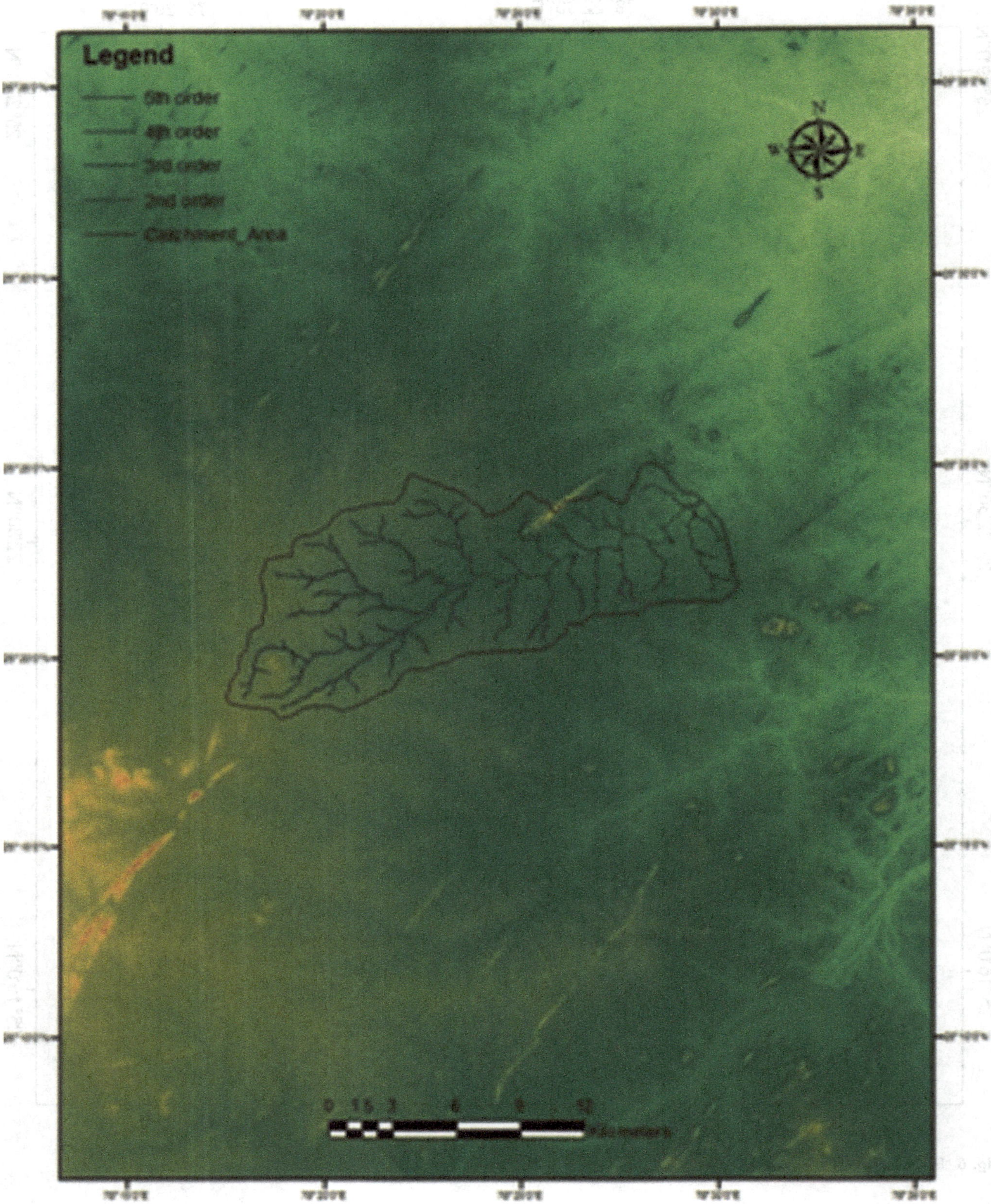

Fig. 7 The relationship between drainage network and major lineaments of the study area

8 Conclusions

Based on the above analysis the following conclusive remarks were drawn.

The morphometric analysis of Upper Pahuj basin will be helpful in depicting nature of subsurface lithology, infiltration capacity and surface runoff. The drainage evolved in a homogenous geological terrain and influenced by geological structures. The elongated basin shows moderately steeper slope and semipermeable soil with less vegetal cover. The upper Pahuj catchment basin is moderately steeper and is less prone to soil erosion. These studies would be useful in making planning for watershed management and water harvesting.

Acknowledgements The authors are extending their thanks to Department of Geology, Bundelkhand University, India for providing necessary facilities. We are extremely thankful to Professor S.N. Mohapatra, School of Studies in Earth Science Jiwaji University Gwalior, India and Dr Vignesh, SRM School of Public Health, SRM Institute of Science and Technology Kattankulathur, Chennai, Tamilnadu, India for reviewing the manuscript and giving valuable suggestions.

References

Agarwal CS (1998) Study of drainage pattern through aerial data in Naugarh area of Varanasi district, U.P. J Indian Soc Remote Sensing 26:169–175

Bhatt SC (2014) Geological and tectonic aspects of Bundelkhand Craton, Central India. Angel Publication, Bhagwati Publishers and Distributors C-8/77-B, Keshav Puram, New Delhi, 190

Bhatt SC (2020) Natural resources, water harvesting and drought in Central India. Shree Publishers and Distributors 22/4735, Prakash Deep Building Ansari Road, Daryaganj, New Delhi 110002:130

Bhatt SC, Husain A (2008) Structural history and fold analysis of basement rocks around Kuraicha and adjoining areas, Bundelkhand Massif Central India. J Soc India 2:331–347

Bhatt SC, Gupta MK (2009) Tectonic significance of shear indicator in the evolution of Dinara-Garhmau shear zone, Bundelkhand Massif, Centeral India. In: Santosh K (ed) MTM. Macmillan Publishers India Ltd., New Delhi, pp 122–132

Bhatt SC, Gupta MK (2014) Microstructural analysis and strain pattern in mylonites and implications of shear sense indicators in evolution of Dinara- Garhmau shear zone, Bundelkhand Massif, Central India. The Indian Mineralogist. J Mineral Soc India 48(2):186–206

Bhatt SC, Hussain A (2012) Shear indicators and strain pattern in quartz mylonites of Chituad – Deori shear zone, Bundelkhand Massif, Central India. Earth Sci India 5(II):60–78

Bhatt SC, Mahmood K (2012) Deformational pattern and microstructural analysis of sheared gneissic complex and mylonitic metavolcanics of Babina-Prthipur sector, Bundelkhand Massif, Central India. Indian J Geosci 66(1):79–90

Bhatt SC, Singh VK, Hussain A (2011) Implications of shear indicators for tectonic evolution of Mauranipur shear zone, Bundelkhand Craton, Central India. In: Singh VK, Chandra R (eds) 2nd Proceeding of precambrian continental growth and tectonism. Angel Publication, New Delhi, pp 36–49

Bhatt SC, Suresh M, Gupta MK (2015) Structural control on drainage pattern of upper- middle Pahuj river and implication of remote sensing in watershed management, BundelkhandCraton, Central India. J Multidiscip Sci Res 2(4):9–14

Bhatt SC, Hussain A, Suresh M, Balasooriya NWB (2017) Geological structure control on Sukhnai basin and land use/ land cover pattern in Mauranipur and adjoining areas, Bundelkhand Craton, Central India. J Geol Soc Sri Lanka 18(2):53–61

Bhatt SC, Singh R, Ansari MA, Bhatt S (2020) Quantitative morphometric and morphotectonic analysis of Pahuj catchment basin Central India. J Geol Soc India 96(5):513–520

Biswas S, Sudhakar S, Desai VR (1999) Prioritisation of sub-watersheds based on morphometric analysis of drainage basin: a remote sensing and GIS approach. J Indian Soc Remote Sens 27 (3):155–166

Chorley RJ, Maim DEG, Pogorzelski HA (1957) A new standard for estimation drainage basin shape. Am J Sci 255:138–141

Faniran A (1968) The index of drainage intensity- a provisional new drainage factor. Austrailian J Sci 31:328–330

Gregory KJ, Walling DE (1973) Drainage basin form and process a geomorphological approach. Edward Arnold, London

Gutema D, Kassa T, Sifan A, Koriche (2017) Morphometric analysis to identify erosion prone areas on the upper blue Nile using GIS: case study of Didessa and Jema sub basin, Ethopia. Int Res J Eng Technol 04(08)

Horton RE (1932) Drainage basin characteristics. Trans Am Geophys Union 13:350–361

Horton RE (1945) Erosional development of streams and their drainage basins: a hydrophysical approach to quantitative morphology. Geol Soc Am Bull 56(3):275–370

Melton MA (1957) An analysis of the relations among elements of climate, surface properties and geomorphology. Technical Report, 11: Proj NR389- 042.off of Nav Res Newyork, Department of Geology, Columbia Universirty

Melton MA (1965) The geomorphic and palaeoclimatic significance of alluvial deposits in Southern Arizona. J Geol 73(1):1–38

Miller VC (1953) A quantitative geomorphic study of drainage basin characteristics on the Clinch Mountain Area, Virgina and Tennessee. Proj., NR 389–402, Technical Report 3, Columbia University, Department of Geology, ONR, New York

Nag SK, Chakraborty S (2003) Influences of rock types and structures in the development of drainage network in hard rock area. J Indian Soc Remote Sens 31(1):25–35

Obi Reddy GE, Maji AK, Gajbhiye KS (2002) GIS for morphometric analysis of drainage basins. GIS India 4:9–14

Pareta K, Pareta U (2011) Quantitative morphometric analysis of a watershed of yamuna basin, india using ASTER(DEM) data and GIS. Int J Geo Geosci 2(1):248–269

Pati JK, Lal J, Prakash K, Bhusan R (2008) Spatio- temporal shift of western bank of Ganga river, Allahabad city and its implications. J Indian Soc Remote Sens 36:289–297

Schumm SA (1954) The relation of drainage basin relief to sediment loss. Int Assoc Sci Hydrol 36:216–219

Schumm SA (1956) Evolution of drainage systems and slopes in Badlands at Perth Anboy New Jersy. Bull Geol Soc Am 67:597–646

Singh S, Singh MC (1997) Morphometric analysis of Kanhar River basin. Natl Geogr J India 43(1):31–43

Singh VK, Slabunov A (2015) The Central Bundelkhand Archaean greenstone complex, Bundelkhand Craton, Central India: geology, composition, and geochronology of supracrustal rocks. Int Geol Rev 57(11–12):1349–1364

Singh VK, Slabunov AI, Nesterova NS et al (2021) Tectonostratigraphic terranes of the Bundelkhand Craton (Indian shield). In: Shandilya AK,

Singh VK, Bhatt SC, Dubey CS (eds) Geological and Geo-environmental Processes on Earth, pp 155–164 (this volume)
Sreedevi PD, Subrahmanyam K, Ahmad S (2004) The significance of morphometric analysis for obtaining groundwater potential zones in a structurally controlled terrain. Environ Geol 47:412–420
Strahler AN (1952) Hypsometric analysis of erosional topography. Bull Geol Soc Am 63:1117–1142
Strahler AN (1956) Quantitative slope analysis. Bull Geol Soc Am 67:571–596
Strahler AN (1964) Quantitative geomorphology of drainage basin and channel network. In: Handbook of applied hydrology, pp 39–76
Wentworth CK (1930) A simplified method of determining the average slope of land surfaces. Am J Sci 21:184–194

Morphotectonics of Kayadhu River Basin in Washim-Hingoli-Nanded Districts, Maharashtra: A Spatial Analysis

Bhagwan B. Ghute and Shaikh Md. Babar

Abstract

The morpho-tectonic and morphometric analysis of Kayadhu river basin is carried out to interpret the landscape development and stream features of the river basin with the help of quantitative analysis of geomorphic indices and field observations. This analysis is based on relief and tectonic aspects such as elongation ratio (Re), hypsometric integral (HI), basin asymmetry factor (AF), transverse topographic symmetric factor (T), stream length gradient (SL) index, longitudinal profile and lineament and microseismicity in the area. The spatial analysis has been completed with the help of remote sensing and GIS techniques. The Kayadhu river is one of the major tributaries of the Penganga river in the Deccan trap area of Maharashtra. The calculated geomorphic indices of the river, which shows positive marker of tectonic inscription on drainage network. A moderate hypsometric integral value indicate that the basin is still under mature stage of erosion. Longitudinal profile of the stream also shows the presence of Knick points. The analysis points out south westward tilting of the drainage basins with asymmetry and lineament control on smaller tributaries of the Kayadhu drainage is observed. Overall assessment of morphotectonic analysis revealed that the area is influenced by tectonics.

Keywords

Morphotectonics • Kayadhu river • Geomorphic indices • Neotectonics

B. B. Ghute (✉)
Department of Geology, Toshniwal Arts, Commerce and Science College, Sengaon, Dist. Hingoli, Maharashtra 431542, India

S. Md.Babar
Department of Geology, Dnyanopasak College, Parbhani, Maharashtra 431401, India

1 Introduction

The study area belongs to the eastern part of Deccan Basaltic Province of Maharashtra, India. The tectonics play a very important role in the evolution of the drainage basin and it reflect on structural, fluvial and morphotectonic features of the basin. Evaluation of neotectonics depends upon the geomorphic indices, which marks rock resistance, climatic changes and tectonic processes results into landscape evolution. The information about tectonic history of an area can be recovered by quantification of different morphotectonic indices, by using the techniques of remote sensing and GIS in different drainage basins by many workers including (Keller 1986; Krishnamurthy et al. 1996; Singh and Singh 1997; Obi Reddy et al. 2002; Nag and Chakraborty 2003; Narendra and Rao 2006; Pankaj and Kumar 2009; Javed et al. 2009; Babar et al. 2012; Vijith et al. 2015). It provides quantitative analysis of the river basin to understand initial ladders and the basin dynamics. Fluvial systems in the upland areas of the Deccan Trap have been studied by Rajaguru and Kale (1985). The microseismicity in Nanded and Parbhani region was studied (Babar et al. 2011; Kaplay et al. 2016). Tectonic evolution and neotectonic activity of an area can be studied by analysing the consequences of neotectonic activities over the drainage networks and basins which reflect minor and major changes in terrain morphology (Potter 1978; Salvany 2004; Garrote et al. 2008; Kothyari and Rastogi 2013; Ambili and Narayana 2014). Geomorphic indices are useful tools in evaluating tectonic activity in the area. Tectonic activities play a very important role in the morphological development of drainage basin and well reflected by structural, fluvial and morphotectonic parameters. The south eastern part of the Hingoli district of Deccan volcanic province has been experiencing micro-earthquake activity since 2016, approximately at a junction of Sirli Ghat Lineament and Wapti Lineament. In

A. K. Shandilya et al. (eds.), *Geological and Geo-Environmental Processes on Earth*, Springer Natural Hazards,
https://doi.org/10.1007/978-981-16-4122-0_20

the present study, the remote sensing and GIS approaches have been used to enlighten the morphotectonics of a Kayadhu river basin.

2 Geology of the Area

Geologically the river basin comprises simple (aa type) and compound (pahoehoe type) basalt flows (Fig. 7) of nearly horizontal lava formations. These flows have been formed due to fissured type of lava eruption during late Cretaceous to early Eocene period (Muley et al. 2012). These basalt flows belonging to the Ajanta Formation, which are stratigraphic equivalent of upper Ratangad Formation of the Wai sub-group of west Maharashtra comprising compound flows (Godbole et al. 1996; Kaplay et al. 2016).

Simple basalt flows are thick and extensive. The general characteristics of these flows are observed based on its variation from top to the bottom and degree of compaction and jointing pattern. The top surface of the basalt flows is rather undulating or weathered up to some thickness due to hydrothermal alteration, purple or greyish coloured and vesicular in nature. The middle and lower portions of the flows are free from vesicles and amygdules and they occur as thick and compact. Joints, which are contractions cracks, developed during cooling and solidification of lava always occurs in the middle and lower portion of the flow (Muley et al. 2010). The red bole bed or Tachylitic bands or paleo sole or marker bed occurs between consecutive flows (Babar et al. 2010).

The outpouring of low viscous lava through a large number of outlets result in to the compound flows (Bondre et al. 2000, 2004; Duraiswami et al. 2003, 2008). These lava flows are large, irregular and having a huge lateral extent (Duraiswami et al. 2002, 2008). The flows are marked by the presence of vesicles and amygdules. The vesicles are filled with secondary minerals like Zeolites, calcite, silica chlorophaeite, etc., to form the amygdules (Babar et al. 2010).

3 Study Area

The Kayadhu river is a tributary of the Penganga river that originates from Angarwadi village south west part of the Washim district in Ajanta hill ranges (at an elevation 580 m), and flowing over the Deccan traps. The river flows from NW-SE direction and confluences to Penganga river near Chincholi Sangam village in Nanded district. The basin is confined between N19°22″, E77°40″ in the south east and N 20°00″, E76°40″ north west (Fig. 1). The river having maximum flow length is 143.58 km and total area of 2158.38 km^2. The maximum and minimum elevations are 605 and 390 m above mean sea level. The Kayadhu river is of seventh order with a dominant dendritic drainage pattern. The basin is located in low seismic intensity zone, the study area had microseismicity in the last two years, some of them ≤ 3 with epicentre falling inside the river basin (Table 1).

4 Methodology

For the present study the analysis is based on the Advanced Spaceborne Thermal Emission and Reflection Radiometer (ASTER) Digital Elevation Model (DEM), derived from ASTER data collected from the website of Global Digital Elevation Model (GDEM). Along with the DEM, the satellite data collected from Indian Remote Sensing (IRS P6) LISS IV was also used to delineate and map the active channels and floodplain features. The digital data from IRS-P6 LISS-IV (70 km × 70 km) collected in 2016 in FCC format with high resolution (Fig. 2) and SOI toposheet map Nos. 56A/9, 56A/13, 56A/14, 56E/1, 56E/2, 56E/3, 56E/5, 56E/6, 56E/7, 56E/10, 56E/11 were used to meet the requirement of area under study. Digital satellite data were geometrically rectified and georeferenced by using Arc GIS 10.3 software for the preparation of various maps, landforms and processes of basin information is extracted from the satellite imagery, survey of India toposheet maps and extensive field visits are carried out.

5 Results and Discussion

5.1 Elongation Ratio (Re)

According to Schumm (1956) elongation ratio Re, is the ratio of diameter of a circle of the same area as the basin to the maximum basin length Table 2. The ratio ranges between 0.60 and 1.0 over a wide diversity of climate and geology. The varying slopes of the basin can be classified with the help of elongation ratio, i.e. circular (0.9–1.0), oval (0.8–0.9), less elongated (0.7–0.8), elongated (0.5–0.7) and more elongated >0.5 (Kale and Shejwalkar 2008). The elongation ratio of Kayadhu river is 0.48, which represents that the basin is more elongated in shape with low relief.

5.2 Drainage Basin Asymmetry (Af)

Drainage basin asymmetry, used to evaluate local and regional scale tectonic tilting produced asymmetry of the drainage networks and the basin, corresponds to the orientation of the basin axis (Hare and Gardner 1985; Cox 1994; Keller and Pinter 2002; Garrote et al. 2008; Giaconia et al.

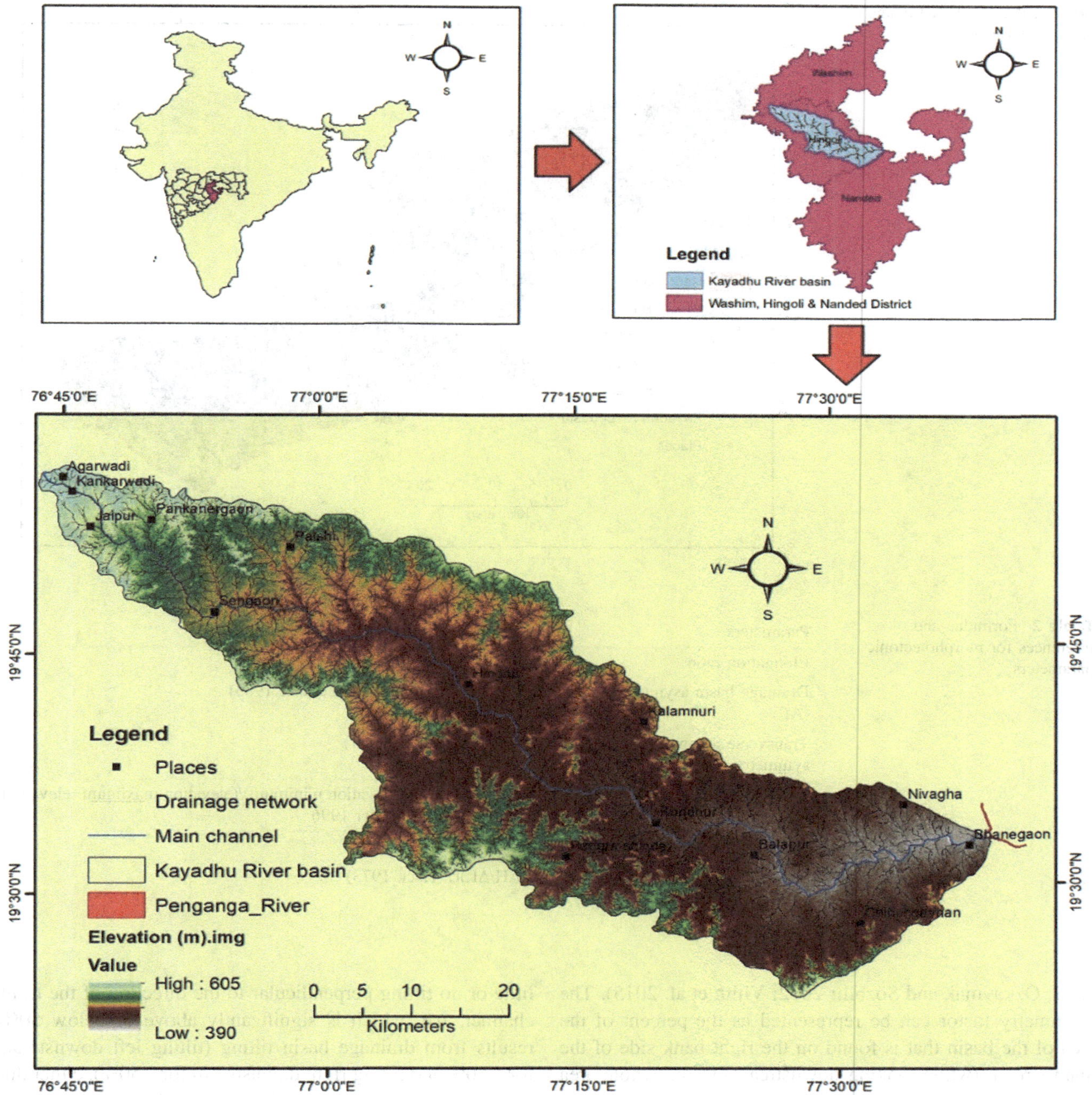

Fig. 1 Location map of study area

Table 1 Details of micro seismic events (IMD 2017; 2018; 2020)

Sr. No	YYYYMMDD	HHMMSS	Latitude	Longitude	Depth	Magnitude
1	20,181,104	185,214	19.6 N	76.9 E	15	2.5
2	20,180,913	034,902	19.7 N	76.9 E	10	2.5
3	20,170,816	063,523	19.5 N	77.2 E	10	3.0
4	20,200,602	070,900	19.5 N	77.1 E	10	3.4

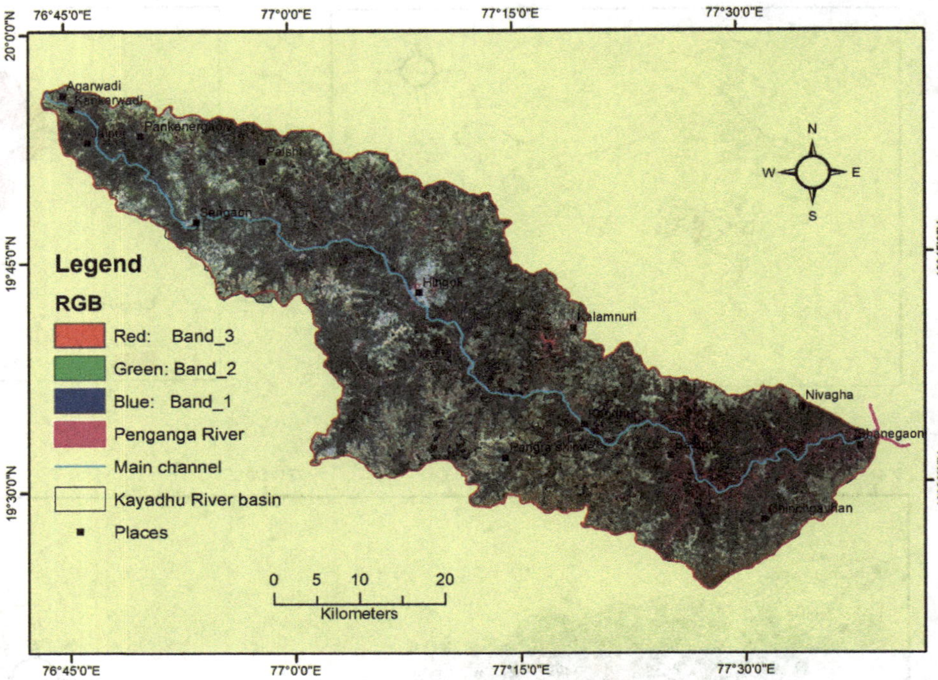

Fig. 2 LISS IV (FCC) image of Kayadhu river basin

Table 2 Formulae and references for morphotectonic parameters

Parameters	Formulae and references
Elongation ratio	$Re = 2\sqrt{Ff}/\pi$, (Schumm 1956)
Drainage basin asymmetry (AF)	AF = 100 (Ar/At), (Hare and Gardner 1985)
Transverse topographic symmetry (T)	T = Da/Dd, (Cox et al. 2001)
Hypsometric analysis	HI = (elevation mean−elevation minimum)/(elevation maximum−elevation minimum) Keller and Pinter 1996
Longitudinal profile	(Schumm 1986)
Stream length gradient (SL) index	$SL = (\Delta H/\Delta L)L$ (Hack 1973)

2012; Özkaymak and Sözbilir 2012; Vijith et al. 2015). The asymmetry factor can be represented as the percent of the area of the basin that is found on the right bank side of the main stream (while looking downstream) to the whole area of the basin and is defined as:

$$AF = 100(Ar/At) \quad (1)$$

Where Ar is the area of the basin to the right of the trunk stream and At is the total area of the drainage basin (Hare and Gardner 1985). This index identifies tectonic tilting of a drainage basin and characterises its asymmetry or symmetry, as it is sensitive to rotations normal to the axis of main stream (Tsodoulos et al. 2008; Özkaymak and Sözbilir 2012). Under stable settings of a drainage network formation, the index will be equal to 50%, AF close to 50% shows little or no tilting perpendicular to the direction of the trunk channel. AF, which is significantly above or below 50%, results from drainage basin tilting (tilting left downstream for >50% and tilting right downstream for <50%), either due to active tectonics or lithologic control. The asymmetry factor has been calculated for Kayadhu river, is found to be 47 indicating a slightly asymmetrical basin (Fig. 3).

5.3 Transverse Topographic Symmetry Factor (T)

The transverse topographic symmetry factor is used to determine possible lateral tilting and preferred stream migration in the basin (Cox et al. 2001). This index is used

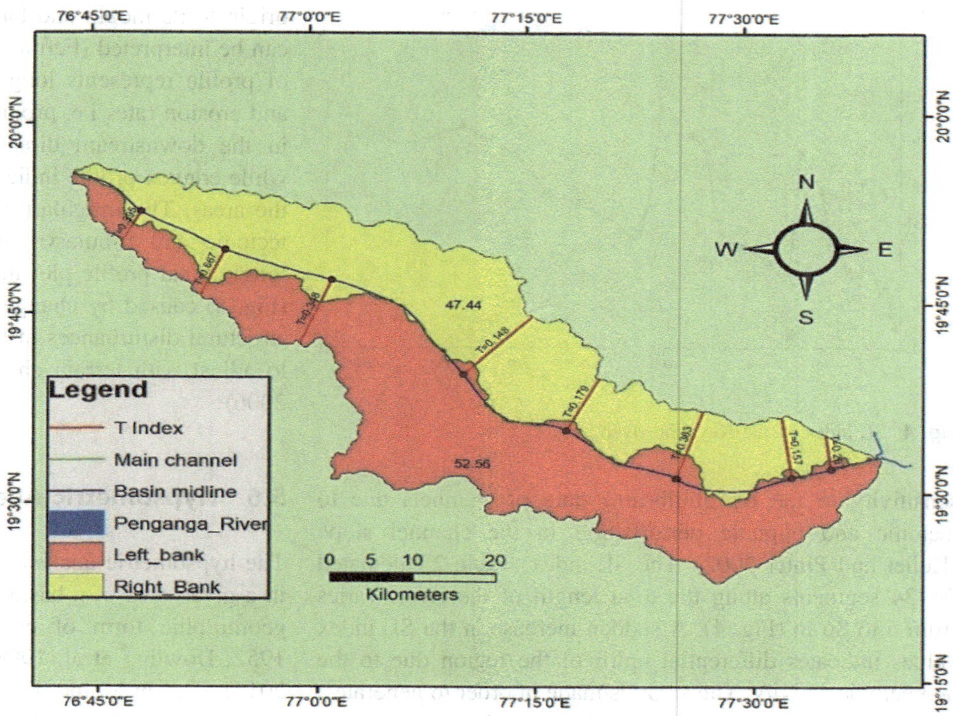

Fig. 3 Map showing transverse tographic symmetric factor (T) and drainage basin asymmetry (Af)

to provide important asymmetry information in larger drainage system and rapid uplift in the regions. The basin midline would be the location of a river that is symmetrically placed with respect to the basin divide. It also indicates the migration of the drainage system perpendicular to the principal axis of the basin due to internal fluvial processes or external forces (Keller and Pinter 2002; Salvany 2004). The T factor Table 2 can be calculated by using equation:

$$T = Da/Dd \quad (2)$$

Where Da is distance from midline of the drainage basin to the midline of the active meander belt. Dd is the distance from the basin midline to basin divide. In general, the T values range from 0 to 1 and T values close to 0 signifying symmetrical basin and stream is in the middle of the drainage basin, while values close to 1 indicate highly asymmetric basin and stream migrates laterally away from the centre of the basin margin. The T index Table 2 for the Kayadhu river (Fig. 3) shows strong asymmetry (average is 0.315). From the values of T index (Table 3), we can interpret the cause of asymmetry depends on tectonic disturbances or neotectonic structural features.

5.4 Stream Length Gradient Index (SL)

It is one of the most important index in order to know whether the area is tectonically or lithological controlled. This parameter is measured by using digital elevation map as well as contour map of the area and measuring the length of the river. By means of SL index, we can quantify the characteristics of stream gradient behaviour and its relationship with physiographical conditions, lithological control, topography and associated drainage parameters (Hack 1973; Cox 1994; Keller and Pinter 2002; Dehbozorgi et al. 2010; Giaconia et al. 2012; Vijith et al. 2015). The stream length-gradient index was first defined by Hack (1973) and computed by equation:

$$SL = (\Delta H/\Delta L)L \quad (3)$$

Where $\Delta H/\Delta L$ is the local slope of the channel segment being evaluated and L is the channel length from the divide to the midpoint of the channel reach for which index is calculated. The stream length-gradient index can be used to evaluate relative tectonic activity. The SL index is used to infer stream power and rock erodability owing to its

Table 3 Topographic transverse symmetric (T) factor calculated for the selected segments in the main stream

Segments	1	2	3	4	5	6	7	8
T	0.267	0.157	0.363	0.170	0.148	0.348	0.667	0.395
Direction of shift	WNW	WNW	ENE	ENE	NE	WSW	WSW	WSW

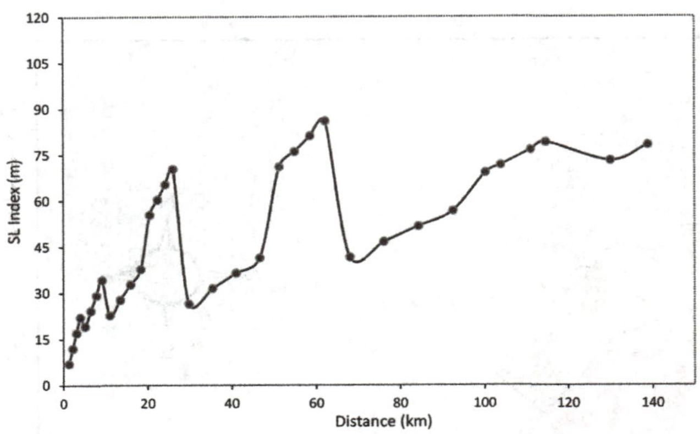

Fig. 4 SL index of the Kayadhu river

sensitivity to the disequilibrium state of channels due to tectonic and climatic perturbation in the channel slope (Keller and Pinter 2002). The SL index, Table 2 calculated for 34 segments along the total length of the river, varies from 6 to 86 m (Fig. 4). A sudden increase in the SL index values indicates differential uplift of the region due to the neotectonic activity. The study is made in order to generate a comprehensive data about the variation and pattern of neotectonic influence in the Kayadhu river basin.

5.5 Longitudinal Profile

Longitudinal profile of a river is an important geomorphic tool it provides clues to interpret the geometry of geotectonic disturbances and geomorphic evolution as well as underlying material of the area (Schumm 1986, 1993; Jain et al. 2006; Ambili and Narayana 2014; Vijith et al. 2015; Kumar et al. 2018). Longitudinal profile is a plot of river distance against elevation. It represents channel gradient of the river origin to its mouth and based on the shape of the curves it can be interpreted (Ferraris et al. 2012). The concave nature of profile represents long term equilibrium between uplift and erosion rates i.e. progressive increase in river discharge in the downstream direction (Bull and Knuepfer 1987), while convex profile indicates the upliftment is dominant in the areas. The irregularity in the profile is a result of neotectonic and tributary confluences (Schumm 1986). The longitudinal profile plot indicates a number of knick points (Fig. 5) caused by change in lithology (hard or soft) or by structural disturbances in the area therefore stream is trying to adjust with terrain characteristics (Crosby and Whipple 2006).

5.6 Hypsometric Analysis

The hypsometric analysis provides distribution of elevation in a given area of a basin and gives an idea to evaluate the geomorphic form of catchment and landforms (Strahler 1952; Dowling et al. 1998; Singh et al. 2008; Vijith et al. 2015). The hypsometric curve is related to the volume of rock in the basin that has not been eroded (Awasthi et al. 2002; Crosby and Whipple 2006). Hypsometric curves convex upward are interpreted as youthful, S-shaped curves as mature and concave upwards are old age stages of landscape evolution. In the present study percentage hypsometric integral method used (Fig. 6) by computing two ratios from contour map measurements Table 2. The first one a/A, where a is the area enclosed between a given contour within basin and basin boundary of the basin, and A is the total area of the basin. The second ratio (Keller and Pinter 2002) is h/H, where h is the height of the contour above the base level of the stream mouth and H is the total height of the basin from basin mouth. To characterise the hypsometric curve the HI

Fig. 5 Longitudinal profile of the Kayadhu river showing knick points

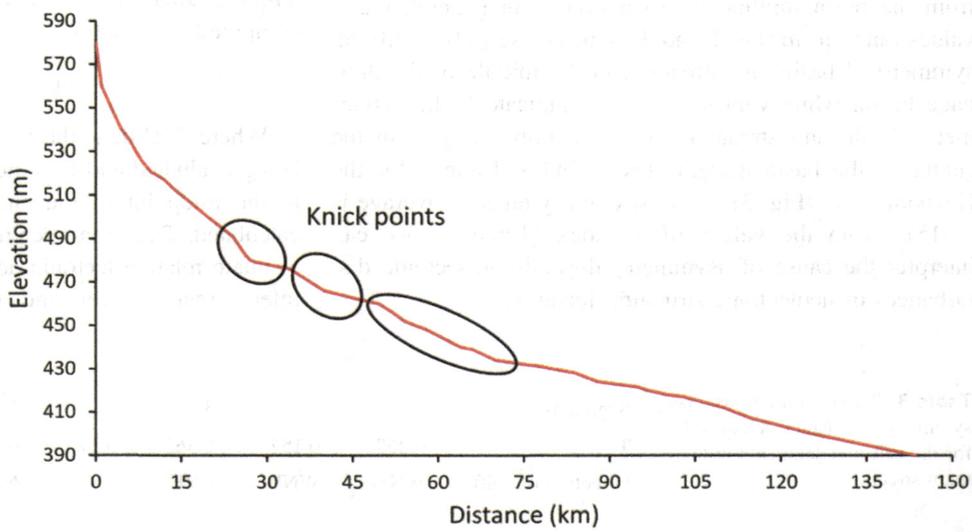

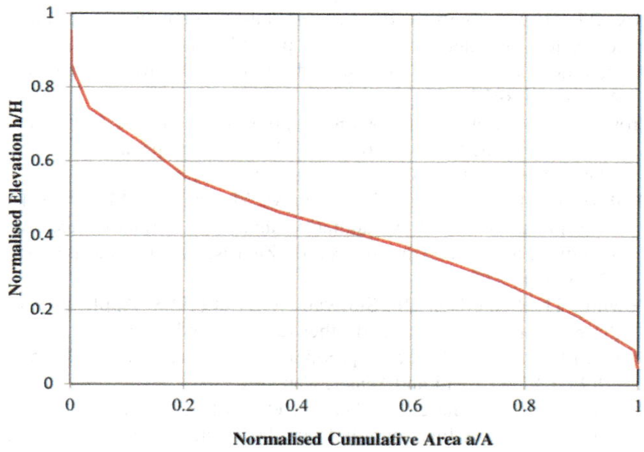

Fig. 6 Hypsometric curves with integrals for Kayadhu river basin

can be calculated as HI = (elevation mean-elevation minimum)/(elevation maximum-elevation minimum). High values of HI generally mean that the upper parts of the basin have not been eroded much and indicate a younger landscape (Vijith et al. 2015).

5.7 Lineaments and Microseismicity

Lineaments are defined as mappable linear to slightly curvilinear subsurface features, which differ distinctly from the adjacent feature and pattern, reflect subsurface information (O' Leary et al. 1976). The structural features usually related with joints, faults and lineaments may produce essentially straight rivers with minimum meandering (Twidale 2004) Orientation and extent of individual linear structure were calculated in Arc GIS 10.3 software. The orientation of various linear elements observed on the basis of ASTER, IRS P6 LISS-IV 2016 satellite data and further validated by field studies (Cantamore, et al. 1996), the preferred stream orientations are used to reconstruct neotectonics through anomalous behaviour of the streams. Lineaments are important components of the earth surface morphology and the lineament density is related to the intensity of deformation (Nur 1982). It is interesting to note that the maximum lineament frequency is found in the central part of the basin and the length of lineaments varies from 3 to 19 km (Fig. 7). Most of the drainage of the Kayadhu river follow the NW-SE and SW-NE trend. The deformation is not observed along the lineaments, these are covered by black cotton soil therefore detailed geophysical surveys along these lineaments are essential.

6 Conclusions

The morphotectonics of the Kayadhu river basin, Maharashtra, India, was studied through the various geomorphic indices, using Arc GIS10.3 software with 30 m resolution ASTER GDEM and LISS IV, IRS FCC data. Transverse topographic symmetry (T) calculated from transverse topographic profiles of 143.58 km drainage basin segments. Northern half part of the river migrates towards WSW and southern half part migrates towards south east, thus it implicates ground tilting as the driving mechanism of preferred migration. Longitudinal profile of the river indicates a number of knick points at a steep reaches in the profile caused by structural disturbance (uplift or lineaments), therefore these disturbances were caused due to tectonic

Fig. 7 Lineament and geological map of study area

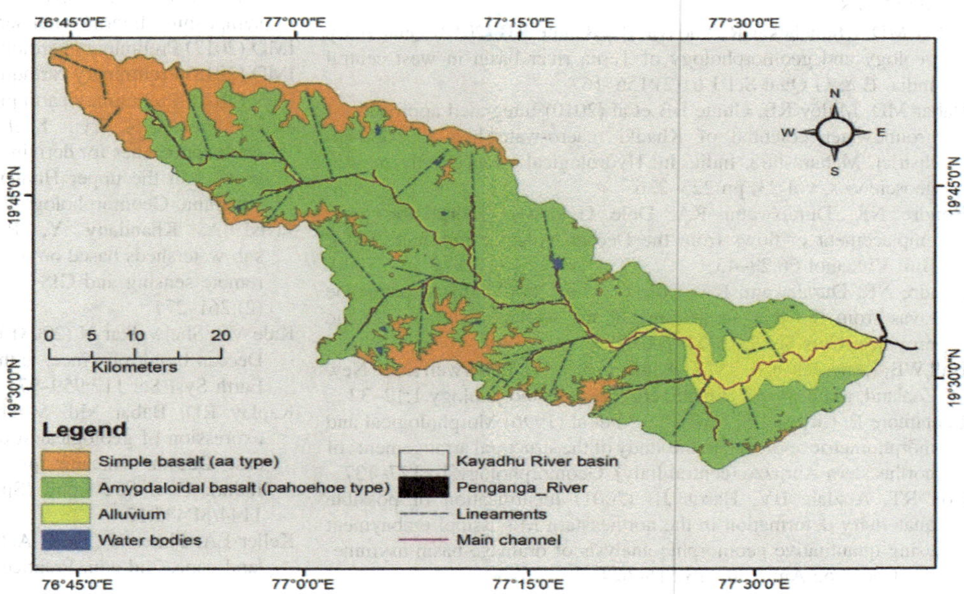

activities in the region over the development of river, which might have reflected as change in the gradient of along the longitudinal profile of Kayadhu river. The asymmetric integral values for the right bank of the basin observe (Fig. 3). The absolute difference is 2.56 indicate that the basin has slight influence tectonic activity. The calculated hypsometric integral for the Kayadhu river is 0.31 shows basin is under the mature stage of erosional development. The lineament ranges from 03 to 19 km long and show a dominant NW-SE and NE-SE trend, consisting with geological structure, and hence there is possibility of recent crustal movement in this area. The whole analysis of the basin indicates, the path of the river is tectonically influenced at some location, but most part of the basin not experienced any major disturbances.

Acknowledgements The corresponding author is thankful to Prof. Vinod Singh, Department of Geology, Bundelkhand University, Jhansi, India for his helpful comments and support and Swami Ramanand Teerth Marathwada University, Nanded for providing the financial support through research project file No. APDS/Uni.MRP/Sci.&Tech.-Geology/2017-18/2962. The authors thank two anonymous reviewers for their valuable comments and critical suggestions, which improved the manuscript significantly.

References

Ambili V, Narayana AC (2014) Tectonic effects on the longitudinal profiles of the Chaliyar river and its tributaries, southwest India. Geomorphology 217:37–47

Awasthi KD, Sitaula BK, Singh BR et al (2002) Land-use change in two Nepalese watersheds: GIS and geomorphometric analysis. Land Degrad Dev 13:495–513

Babar MD, Ghute BB, Chunchekar RV (2011) Geomorphic indicators of neotectonics from the Deccan basaltic province: a study from the upper Godavari river basin, Maharashtra, India. Int J Earth Sci Eng 4:297–308

Babar MD, Chunchekar R, Yadava MG, Ghute BB (2012) Quaternary geology and geomorphology of Terna river basin in west central India. E & G Quat Sci J 61(2):156–167

Babar MD, Muley RB, Ghute BB et al (2010) Integrated approach for groundwater potential of Khadki macro-watershed in Parbhani district, Maharashtra, India. In: Hydrological science. Advances in geosciences, vol 23, pp 223–236

Bondre NR, Duraiswami RA, Dole G (2004) Morphology and emplacement of flows from the Deccan volcanic province, India. Bull Volcanol 66:29–45

Bondre NR, Duraiswami RA, Dole G et al (2000) Inflated pahoehoe lavas from the Sangamner area of the western Deccan volcanic province. Curr Sci 78:1004–1007

Bull WB, Knuepfer PLK (1987) Adjustments by the Charwell river, New Zealand, to uplift and climatic changes. Geomorphology 1:12–32

Cantamore E, Ciccacci S, Montedel M et al (1996) Morphological and morphometric approach to the study of the structural arrangements of northeastern Abruzzo (central Italy). Geomorphology 16:127–137

Cox RT, Arsdale BV, Harris JB (2001) Identification of possible quaternary deformation in the northeastern Mississippi embayment using quantitative geomorphic analysis of drainage-basin asymmetry. Geol Soc Am Bull 113:615–624

Cox RT (1994) Analysis of drainage basin symmetry as a rapid technique to identify areas of possible quaternary tilt-block tectonics: an example from the Mississipi embayment. Geol Soc Am Bull 106:571–581

Crosby BT, Whipple KX (2006) Knickpoint initiation and distribution within fluvial networks: 236 waterfalls in the Waipaoa river, North Island, New Zealand. Geomorphology 82:16–38

Dehbozorgi M, Pourkermani M, Arian M, Matkan AA, Motamedi H, Hosseiniasl A (2010) Quantitative analysis of relative tectonic activity in the Sarvestan area, central Zagros, Iran. Geomorphology 121(3–4, 15):329–341

Dowling TI, Richardson DP, Sullivan AO et al (1998) Application of the hypsometric integral and other terrain-based metrics as indicators of catchment health: a preliminary analysis. Technical report 20/98, CSIRO Land and Water, Canberra

Duraiswami RA, Bondre NR, Dole G et al (2002) Morphology and structure of flow-lobe tumuli from the western Deccan volcanic province, India. J Geol Soc India 60:57–65

Duraiswami RA, Bondre NR, Managave S (2008) Morphology of rubbly pahoehoe (simple) flows from the Deccan volcanic province: implications for style of emplacement. J Volcanol Geoth Res 177:822–836

Duraiswami RA, Dole G, Bondre NR (2003) Slabby pahoehoe from the western Deccan volcanic province: evidence for incipient pahoehoe-aa transitions. J Volcanol Geoth Res 121:195–217

Ferraris F, Firpo M, Pazzaglia FJ (2012) DEM analyses and morphotectonic interpretation: the plio-quaternary evolution of the eastern Ligurian Alps, Italy. Geomorphology 149–150:27–40

Garrote J, Heydet GG, Cox RT (2008) Multi-stream order analyses in basin asymmetry: a tool to discriminate the influence of neotectonics in fluvial landscape development (Madrid Basin, central Spain). Geomorphology 102:130–144

Giaconia F, Booth-Rea G, Martínez-Martínez JM et al (2012) Geomorphic evidence of active tectonics in the Sierra Alhamilla eastern Betics, SE Spain. Geomorphology 145–146:90–106

Godbole SM, Rana RS, Natu SR (1996) Lava stratigraphy of Deccan basalts of Western Maharashtra. Gondwana Geol Mag (Special Publications) 2:125–34

Hack JT (1973) Stream-profile analysis and stream gradient index. US Geol Surv J Res 1:421–429

Hare PW, Gardner TW (1985) Geomorphic indicators of vertical neotectonism along converging plate margins, Nicoya Peninsula, Costa Rica. In: Tectonic geomorphology: 15th annual Binghamton symposium. International series, vol 15, pp 75–104

IMD (2017) Preliminary earthquake report. https://mausam.imd.gov.in/

IMD (2018) Preliminary earthquake report. https://mausam.imd.gov.in/

IMD (2020) Preliminary earthquake report. https://mausam.imd.gov.in/

Jain V, Preston N, Fryirs K et al (2006) Comparative assessment of three approaches for deriving stream power plots along longitudinal profiles in the upper Hunter river catchment, New South Wales, Australia. Geomorphology 74:297–317

Javed A, Khandauy Y, Rizwan A (2009) Prioritization of sub-watersheds based on morphometric and land use analysis using remote sensing and GIS technique. J Indian Soc Remote Sens 37 (2):261–271

Kale VS, Shejwalkar N (2008) Uplift along the western margin of the Deccan basalt province: is there any geomorphometric evidence? J Earth Syst Sci 117:959–971

Kaplay RD, Babar Md, Mukherji S et al (2016) Morphotectonic expression of geological structures in the eastern part of the south east Deccan volcanic province (around Nanded, Maharashtra, India). Geol Soc Lond (Special Publications). https://doi.org/10.1144/SP445.12

Keller EA, Pinter N (2002) Active tectonics. Earthquakes, uplift, and landscape, 2nd edn. Prentice Hall, Upper Saddle River, pp 362

Keller EA (1986) Investigation of active tectonics: use of surficial Earth processes. In: Wallace RE (ed) Active tectonics studies in geophysics. National Academic Press, Washington, DC, pp 136–147

Keller EA, Pinter N (1996) Active tectonics: earthquakes, uplift and landscape (Prentice Hall, Upper Saddle River, New Jersey)

Kothyari G, Rastogi B (2013) Tectonic control on drainage network evolution in the upper Narmada valley: implication to neotectonics. Geogr J 1–9

Krishnamurthy J, Srinivas G, Jayaram V et al (1996) Influence of rock type and structures in the development of drainage networks in typical hard rock terrain. ITC J 3(4):252–259

Kumar D, Singh DS, Prajapati SK et al (2018) Morphometric parameters and neotectonics of Kalyani river basin, Ganga plain: a remote sensing and GIS approach. Geol Soc India 91:679–686

Muley RB, Babar MD, Atkore SM, Ghute BB (2010) Impact of geology on groundwater and water harvesting structures in Deccan basalt area: a case study of Jhari Percolation tank in Parbhani district, Maharashtra. J Adv Sci Technol 13(1):96–101

Muley RB, Babar MD, Atkore SM, Ghute BB (2012) Hydrogeological study for identifying groundwater potential areas of Dudhgaon in Parbhani district, Maharashtra. Mem Geol Soc India 80:39–49

Nag SK, Chakraborty S (2003) Influences of rock types and structures in the development of drainage network in hard rock area. J Indian Soc Remote Sens 31(1):25–35

Narendra K, Rao KN (2006) Morphometry of the Meghadrigedda watershed, Vishakhapatnam district, Andra Pradesh using GIS and resource sate data. J Indian Soc Remote Sens 34(2):101–110

Nur A (1982) The origin of tensile fracture lineaments. J Struct Geol 4:31–40

O' Leary D, Friedman D, Poh H (1976) Lineaments, linear, lineations: some standards for old terms. Geol Soc Am Bull 87(10):1463–1469

Obi Reddy GP, Maji AK, Gajbhiye KS (2002) GIS for morphometric analysis of drainage basins. GIS India 11(4):9–14

Özkaymak C, Sözbilir H (2012) Tectonic geomorphology of the Spildağı high ranges, western Anatolia. Geomorphology 173–174:128–140

Pankaj A, Kumar P (2009) GIS based morphometric analysis of five major sub-watersheds of Song river, Dehradun district, Uttarakhand, with special reference to landslide incidence. J Indian Soc Remote Sens 37(1):157–166

Potter PE (1978) Significance and origin of big rivers. J Geol 86:13–33

Rajaguru SN, Kale VS (1985) Changes in the fluvial regime of western Maharashtra upland rivers during late quaternary. J Geol Soc India 26:16–27

Rastogi RA, Sharma TC (1976) Quantitative analysis of drainage basin characteristics. J Soil Water Conserv India 26:18–25

Salvany JM (2004) Tilting neotectonics of the Guadiamar drainage basin, SW Spain. Earth Surf Process Landf 29:145–160

Schumm SA (1956) Evolution of drainage systems and slopes in badlands at Perth Amboy, New Jersey. Geol Soc Am Bull 67:597–646

Schumm SA (1986) Alluvial river response to active tectonics. Active tectonics studies in geophysics. National Academic Press, Washington D.C., pp 80–94

Schumm SA (1993) River response to base level changes: implications for sequence stratigraphy. J Geol 101:279–294

Singh O, Sarangi A, Sharma MC (2008) Hypsometric integral estimation methods and its relevance on erosion status of north-western Lesser Himalayan watersheds. Water Resour Manag 22:1545–1560

Singh S, Singh MC (1997) Morphometric analysis of Kanhar river basin. Natl Geogr J India 43:31–43

Strahler AN (1952) Hypsometric area-altitude analysis of erosional topography. Geol Soc Am Bull 63:1117–1141

Tsodoulos IM, Koukouvelas IK, Pavlides S (2008) Tectonic geomorphology of the easternmost extension of the Gulf of Corinth Beotia, Central Greece. Tectonophysics 453:211–232

Twidale CR (2004) River patterns and their meaning. Earth-Sci Rev 67:159–218

Vijith H, Prasannakumar V, Ninu Krishnan MV et al (2015) Morphotectonics of a small river basin in the south Indian granulite terrain: an assessment through spatially derived geomorphic indices. Georisk: Assess Manag Risk Eng Syst Geohazards 9(3):187–199

Landslide Investigations Along the Road Corridor Between Nandprayag and Gopeshwar, Chamoli District, Uttarakhand Lesser Himalaya

Rahul Negi, R. A. Singh, Pooja Saini, Piyush K. Singh, and Harish Khali

Abstract

Hilly region of the Himalayas is affected from time to time by landslides due to its geological and climatic conditions. The road network is an important transportation medium in Himalayan terrain. Any kind of slope failure may lead to many problems and caused losses of life, infrastructure as well as environmental degradation. The present study of landslide lies along the road corridor between Nandprayag and Gopeshwar in Chamoli district of Uttarakhand. This is the only way to reach Badrinath and Kedarnath Dham. This route is influenced by landslide every year. The area is highly deformed and comes under lesser Himalayas. Structures like folds, fractures and joints are present. During the monsoon period heavy rainfall occurred by which debris material comes along the slope and intensity of landslides increases. Due to this, it is very sensitive for any major construction activities. Recent developmental activities particularly construction and road widening processes are still going on in the area whereby the slope geometries are disturbed that cause instability of the slopes. Six landslides (L1–L6) have been selected in the area based on their instability in which debris slide, debris fall and subsidence movement were noticed. The detailed investigations of each landslide have been done along with the kinematic analysis. The main aim of present study is to find out slope failure zone, their causes and suggest their mitigation measures.

Keywords

Landslide • Mitigation • Kinematic analysis • Climate • Lesser Himalaya

1 Introduction

Landslide is one of the major geological hazards in the Himalayan regions. It is a downward and outward movement of soil, debris and rock material under the influence of gravity and with or without the presence of water. The intensity of landslides increases during the monsoon season which directly or indirectly impacts human life, infrastructure and economy. The monsoon period is fairly long extending from June to September. Landslide is a common type of disaster originated by different causes which may be natural (geological structures, rainfall, weathering, erosion, seismicity, etc.) and manmade (major as well as minor constructional works and hill slope cutting for road widening) or combined effects of both. Recent developmental activities, particularly construction and widening of roads in hilly areas involve changes in slope geometry that makes instability of the slope.

In India, Himalayan terrain is a witness of large numbers of mass wasted incident each year due to its geological, structural and climatic conditions (Singh 2010, 2012, 2013; Singh et al. 2021). The Lesser Himalayan region is highly deformed and tectonically active zone characterised by complex structures like multiple generations of folds, fractures, joints, faults and shear planes. Due to this, it's very sensitive for any major constructions. Generally, road network is the only transportation medium in hilly terrain. The

R. Negi · P. Saini
Department of Geology, L.S.M. Government P.G. College, Pithoragarh, Uttarakhand, India

R. A. Singh (✉)
Government Degree College Gururabanj, Almora, Uttarakhand, India

P. K. Singh
Department of Earth Sciences, IIT Roorkee, Roorkee, Uttarakhand, India

H. Khali
Graphic Era University, Dehradun, Uttarakhand, India

A. K. Shandilya et al. (eds.), *Geological and Geo-Environmental Processes on Earth*, Springer Natural Hazards,
https://doi.org/10.1007/978-981-16-4122-0_21

failure of slopes may lead to obstruction in traffic. The present area was selected due to its own importance because it is a way that connects holy Badrinath and Kedarnath Dham. Many passengers as well as pilgrims have to face difficulties in traffic due to landslides. Therefore, some stability situation is necessary along the road from safety point of view. Six locations (L1–L6) have been selected based on their instability in which debris slide, debris fall and subsidence movement were found. The main aim of present study is to find out slope failure zone, their causes and suggest their mitigation measures.

Over the last few years, there are many instances when landslides resulted in great loss of human lives and property. The past major devastated landslide incidences in the Chamoli district are:

i. During the monsoon season of 2018 cloud burst incidence occurred in different part of Chamoli district. Due to which many villages, dozens of shops, vehicle, agriculture land, highway and bridges were damaged and nearly 11 human lives and 10 livestock died (Negi et al. 2018).
ii. During 16 and 17 June, 2013 cloud burst occurred in Kedarnath area and heavy rainfall in different parts of Uttarakhand. Thereby causing massive devastation to different parts of Uttarakhand. During that time in Chamoli district alone 33 human lives were lost, 31 people were injured, 1119 cattle were lost, 220 landslides were activated, 242 hectares agricultural lands and more than 3401 houses and many infrastructures were damaged. At some places National Highway-7 was partially washed out (Khanduri 2018).
iii. At midnight of 29 March, 1999 an earthquake of 6.8 magnitude occurred in Chamoli district. Due to this earthquake, more than 117 landslides had been activated/reactivated in the different parts of the Chamoli district (Kimothi et al. 2001).
iv. During 1970 heavy rainfall occurred and due to this rainfall landslides were developed. These landslides caused blockage and due to this blockage lakes viz., Belkuchi and Gohna were developed. After burst of these lakes, approximately 1000 people lost their lives, 1000 cattle had been swept away, 16 bridges & 210 houses were destroyed and 25 vehicles were washed away and the total loss was estimated at about 10 Crore rupees (Shah 2014).
v. On 5 February 2020, a huge landslide occurred at Nandprayag during widening of National Highway-7. Due to this event 5 buildings, 3 vehicles and electrical line were damaged (Negi and Singh 2020).

2 Study Area

The present study of landslides occurred in Chamoli district of Uttarakhand, India. The Chamoli district is second largest district of Uttarakhand with area about 7520 km^2. The study area bounded between Nandprayag (N30° 19′ 49″, E79° 18′ 59″) and Gopeshwar (N30° 24′ 23″, E79° 18′ 58″) along the road corridor. This area comes in Garhwal Lesser Himalaya and Survey of India toposheet number 53 N/7. The extent of the study area along the road is 24.5 km and is characterised by relatively high and rugged mountains with deeply incised valleys. Alaknanda is the major river in the area. In the study area Balkhila, Birahi Ganga, Nandakini are the important tributaries of Alaknanda River. The study of landslides is done particularly along the road section because the road network is an important way of transportation in this region. At Chamoli, NH-107A emerges with NH-7 and further going towards Kedarnath and Badrinath, respectively. Maithana, Bajpur and Kothiyalsain locations are present in the study area. During the field work, six landslides have been observed at different places in which four landslides (Fig. 1) are located along the NH-7 (Nandprayag–Chamoli) and two landslides present near the NH-107A (Chamoli–Gopeshwar).

3 Slope and Aspect

Slope is viewed as one of the major controlling factors in landslide formation. Terrain with flat and gentle slope shows either nil or very low probability for sliding. Critical slope intensifies the influence of the other factors. As slope increases the chances of formation of landslide may also increase. Accordingly, the slope map of the study area was prepared (Fig. 2a) in a GIS using contours from the toposheet. It was found that the maximum slope of the area reaches up to 66.8°. The slope amounts at landslides L1–L6 were found steep (>40°).

Aspect is analysed as slope direction which controls solar radiation receives by the surface as average sunshine per day. It is very important as it is directly related to the biogeochemical processes which operate on the slope and often cause slope stability due to the result of the aspect. Therefore, it justifies the selection of this important parameter as a determining factor in landslide formation. Generally, southern aspects are more prone to landslides because there is less moisture content in the soil and the deficiency of large trees reduces the capacity of roots to hold the ground. Aspect map of the area has been prepared (Fig. 2b). Along the road corridor the slope directions of landslides are NE (L1), almost W to WNW (L2, L3, L4 and L6) and SE (L5) directions.

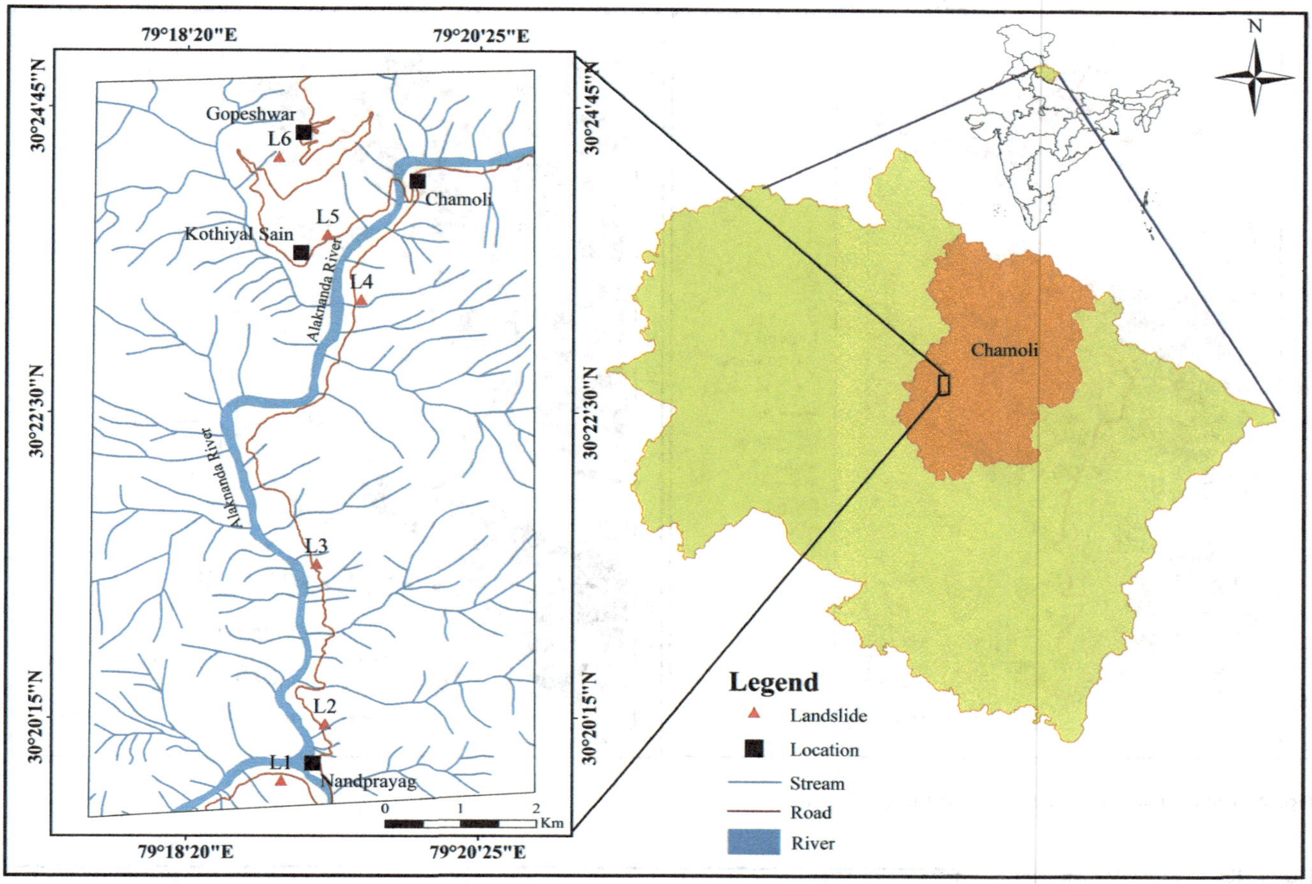

Fig. 1 Location map of landslides in the study area

4 Geology of the Area

The rocks of study area belong to Garhwal Group of Lesser Himalaya which has suffered three phases of deformation (Kumar and Agarwal 1975). The northern boundary of Garhwal Group is marked by Main Central Thrust. Saklani et al. (1991) noticed that the MCT is a composite thrust consisting of three tectonic planes, i.e. MCT–I, MCT–II and MCT–III in the Garhwal region, which was developed by the duplex mechanism of thrust tectonics. The age of Garhwal Group rocks is from Silurian to Precambrian (Mehdi et al. 1972; Agarwal 1974). These rocks are comprised of low to medium grade metamorphic and carbonate rocks that are intruded by acidic and basic igneous rocks. During the field work along the road corridor between Nandprayag and Gopeshwar, quartzite, gneiss, phyllite, schist and metabasic rocks were observed (Fig. 3), at some places river bed materials are also present. All these rocks are highly fractured, jointed, weathered and having weak cohesion along the bedding or schistosity. The Patroli, Gwanagarh and Chamoli formations come under Garhwal Group (Kumar and Agarwal 1975). The Chamoli formation characterised by quartzite rock which is coarse to fine-grained, creamy white, reddish, brownish and greenish in colour. Complex folding, faulting, shear plane, boudins, pinch and swell structures are present in the area.

5 Climate

The elevation of the study area ranges between 800 to 1600 m above the sea level. Microclimate of the district varies from place to place depending upon the altitude. There is rainy season from June to September and annual precipitation is 1231.4 mm (Kaur and Purohit 2014). Annual temperature ranges from −2 °C to 34 °C. Uttarakhand Himalaya received abnormally heavy precipitation during 2013. During June 2013 almost 220 landslides was activated in different part of the district which causes heavy loss of human lives, infrastructure and property (Khanduri 2018).

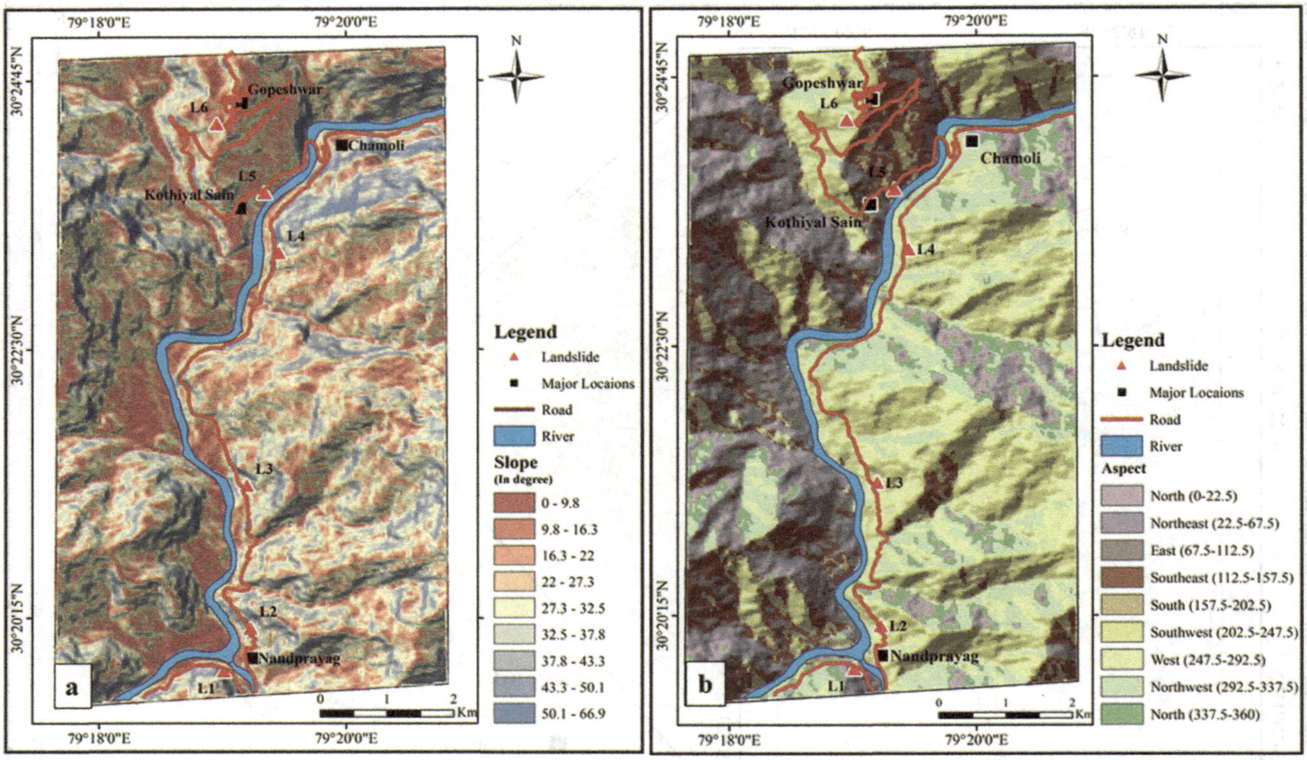

Fig. 2 **a** Slope map of the area; **b** Aspect map of the area

Fig. 3 Regional geological map of the area after Juyal et al. (2010). The red box shows boundary of the study area. MCT-Main Central Thrust

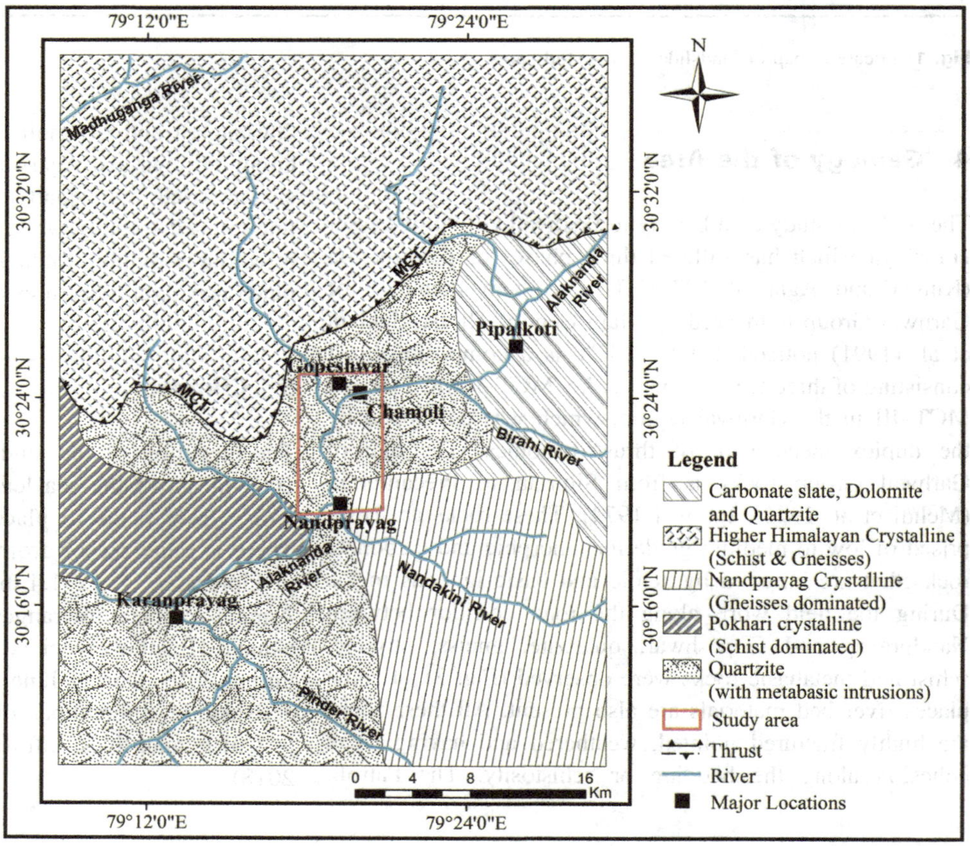

6 Seismicity

Chamoli district lies in Vth zone in the seismic map of India. A major earthquake of 6.8 magnitude occurred in March 1999 (Kimothi et al. 2001). During that time major landslides occurred in upper part of the Gopeshwar city locally named as Chada and some landslides are activated in other part of Chamoli district.

7 Landslide Investigations

In present study locations of landslides have been identified during the field work with the help of Global Position System and marked on the Survey of India toposheet no 53 N/7. Lithology, material, type of landslides and their causes have been observed. Structural data and landslide data have been collected. Slope map (Fig. 2a) and aspect map (Fig. 2b) has been prepared and used for landslides analysis. Main causative factors are water, road widening, joint, slope, aspect and weathered material which impact the road, house and environment. Kinematic analysis of one slope failure has been done at location L1 (Fig. 5f).

7.1 Morphology of the Landslides

As mentioned above during the field work six landslides have been observed at the different places and the classification of these landslides have been done on the basis of Varnes (1978).

7.1.1 Landslide-1

This landslide has occurred at 30° 19′49.98″ latitude, 79° 18′58.87″ longitude at an altitude of 901 m just after 1 km from Nandprayag towards Karnaprayag along NH-7 (Fig. 4a). This landslide is situated at the left bank of Alaknanda River. Length and width of the landslide are 45 m and 20 m, respectively. This is debris slide and rockfall type of landslide with boulders of up to approximately 1 × 0.7 m in size. This is a relatively new landslide which faces towards north-east direction. The landslide (L1) is present on steep slope of the mountain. Base rock is quartzite which is highly jointed and fractured. The bedding plane dips 50° ENE and the prominent joints dip 37° towards SW and 50° towards NNW were also observed. At this place, Kinematic analysis shows wedge failure condition (Fig. 5f). The main causes of this landslide are joints, slope and rainfall.

7.1.2 Landslide-2

This landslide is located at 30° 20′09.84″ latitude, 79° 19′ 12.12″ longitude at an altitude of 927 m just after 1 km from Nandprayag towards Chamoli along NH-7 (Fig. 4b, c). This landslide is situated at the left bank of Alaknanda River. Length and width of the landslide are 170 m and 90 m, respectively. This is a debris slide type of landslide containing largest boulders of approximately 1 × 0.5 m in size. The landslide contains debris material as well as tilted and broken trees. The slope of the landslide is steep which faces in western direction. At the landslide micaceous gneiss is present. The foliation plane is 30° due ESE and prominent joints dip 70° towards NE, 70° towards SW and 60° towards ESE. Rocks are highly jointed, fractured and weathered. Preparing of wirecated wall approximately 10 m height from the road and rock bolting process is going on. The main causes of this landslide are rainfall, groundwater present in the hill, weathered and jointed rock, slope, etc. This is the old landslide but reactivated during rainy season and affects the road.

7.1.3 Landslide-3

This landslide occurred at 30° 21′20.67″ latitude, 79° 19′10″ longitude at an altitude of 967 m near Maithana village along NH-7 (Fig. 4d, e, f). This landslide is situated at the left bank of Alaknanda River. Length and width of the landslide are 250 m and 200 m, respectively. This is the subsidence and debris slide type of landslide. Debris and river bed material is present containing boulders up to 3 × 2 m in size. This landslide is present on steep slope which faces in west direction. The landslide is present at left bank of the Alaknanda River. Highly weathered phyllite and micaceous schist are present at the landslide. Phyllite dips 30° due ESE. The causes of landslide are rainfall, slope, joints, weathered rock, loose soil, river bed material and toe cutting, etc. Scrub and grass have grown by the government on the landslide surface for stabilising the landslide. At this place tilted trees are present and approximately 150 m long stretch of NH-7 is damaged along with two houses. The foundation of houses is tilted as well as cracks developed in the building. Wirecated wall is present below and above the road. Rock bolting, netting and cementing are being done for the treatment of the landslide. During rainy season the rate of the subsidence increases.

7.1.4 Landslide-4

This landslide is present at 30° 23′18.82″ latitude, 79° 19′ 25.69″ longitude at an altitude of 991 m at Bajpur which is 1 km from Chamoli along NH-7 (Fig. 5a). Length and width

Fig. 4 **a** NH-7 damaged due to landslide (L1) present near Nandprayag town; **b** NH-7 damaged due to landslide (L2); **c** Rock bolting and gabion wall are preparing by NH workers for mitigation measure at landslide (L2); **d** Damaged house and road due to Landslide (L3) along NH-7 near Maithana village; **e** Closer view of damaged house due to landslide (L3). Up to 2 cm wide crack developed on the house just below NH-7; **f** At landslide (L3) netting, cementing and bolting have been done for mitigation measures

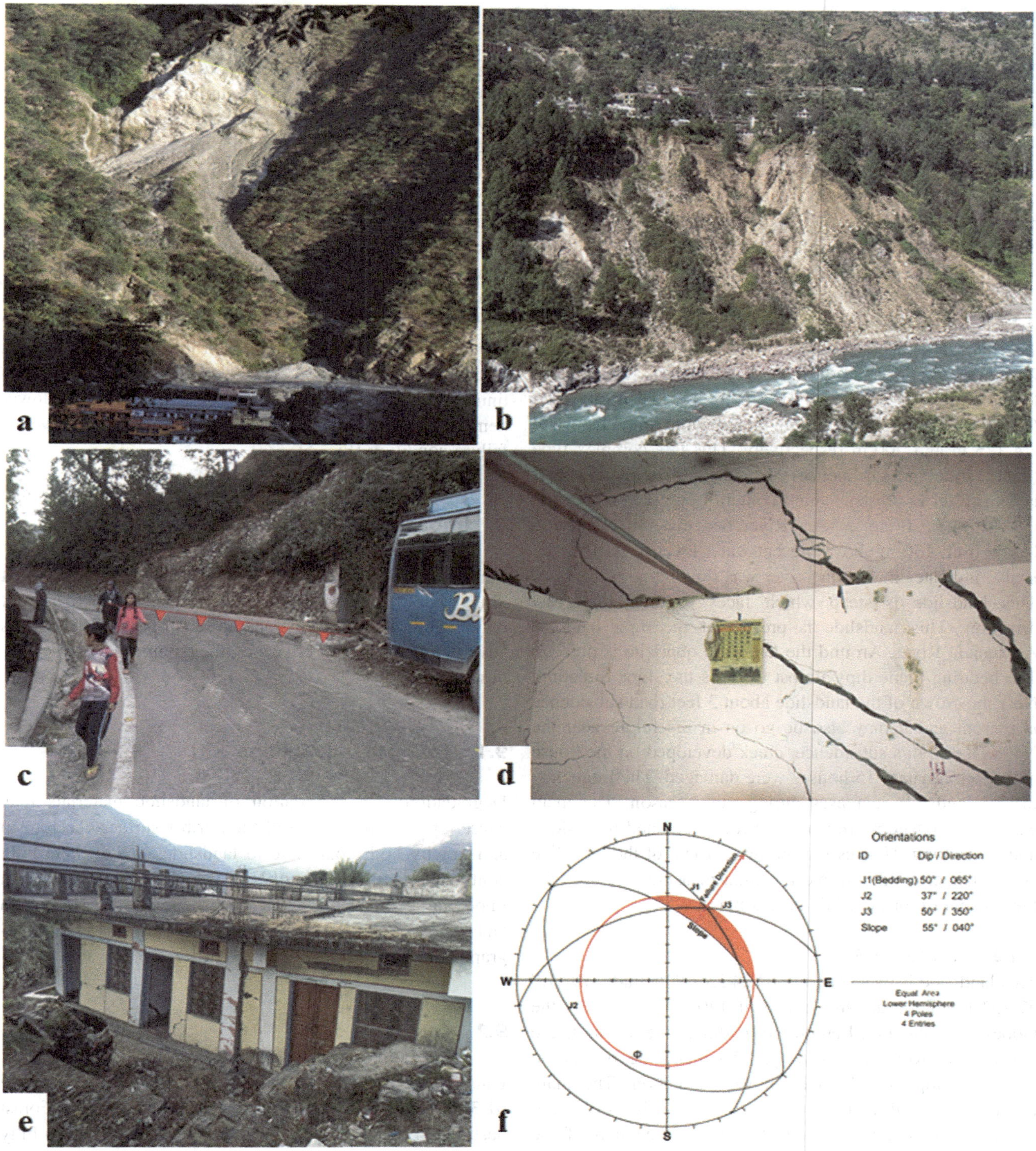

Fig. 5 **a** Debris material collected at road due to landslide (L4) near Bajpur village; **b** Landslide (L5) present below the Kothiyalsain town, Some houses are in danger condition at the crown of landslide; **c** Subsidence of road near Kothiyalsain due to L5; **d** Up to 10 cm wide cracks developed on the wall due to subsidence (L5); **e** Damaged houses due to landslide (L6) just below the Gopeshwar stadium; **f** Kinematic analysis of different discontinuities along with slope showing wedge failure condition at landslide (L1). Shaded area shows possible failure envelope and red circle denote angle of internal friction (ϕ) is 42°

of the landslide are 120 m and 50 m, respectively. This is complex type of landslide (debris slide, debris fall and debris flow). This landslide is present on left bank of the Alaknanda River along with the drainage. The slope of landslide is very steep which faces WNW direction. Quartzite and metabasic rock contact are present at this place (Fig. 5a). Rocks are highly jointed and fractured. Cemented and bulged wire-cated wall is present below the road. This landslide was reactivated during June 2016 and nearly 40 m long stretch of NH-7 had washed away. The main causes of this landslide are: rainfall, drainage, joints, weathered material and slope.

7.1.5 Landslide-5

This landslide occurred at 30° 23′49.35″ latitude 79° 19′ 20.28″ longitude at an altitude of 1025 m at Kothiyalsain town which is present on Chamoli–Gopeshwar road (NH-107A) (Fig. 5b, c, d). Length and width of the landslide are 175 m and 200 m respectively. This is a complex type (debris slide and subsidence) of landslide. Debris and river bed material is present which contain boulders of approximately 30 × 30 cm in size. River bed material is coming below from 150 m above present water level of Alaknanda. It also indicates that earlier river was flowing above 150 m. This landslide is steep which faces towards south-east direction. This landslide is present at the right bank of Alaknanda River. Around the landslide quartzite is present. The bedding plane dips almost towards the slope direction. Near the crown of the landslide about 3 feet road subsidence is present and it may also be go on in the future near the Nala. Due to this subsidence, crack developed on the house and approximately 15 houses were damaged. The frequency of this landslide increases during rainy season. The main causes of landslides are water, surface slope and rock slope and toe cutting. Houses located near crown of the landslide are in danger condition. No remedial measures are used for the prevention of landslide at this place.

7.1.6 Landslide-6

This landslide is occurring at 30° 24′23.14″ latitude, 79° 18′ 57.13″ longitude at an altitude of 1408 m just below the Gopeshwar Stadium. Length and width of the landslide are 120 m and 70 m, respectively. This landslide occurs on moderate slope which faces in west direction. The main causes are rainfall and weathered material. It is an old subsidence type of landslide due to which some part of the Gopeshwar stadium and three houses are totally damaged (Fig. 5e). The scrub and grass have grown over this landslide.

8 Kinematic Analysis

The main objective of this study is to identify possible mode of slope failure along the road cutting. Kinematic analysis (Hoek and Bray 1981) can be used for investigation or forecast the probability of slope failure (wedge failure, plane failure and toppling failure). This analysis is also the inter-relationship between the orientation of the discontinuity and the slope face to determine the condition of possible mode of failure. Kinematic analysis of slope failure has been done by using stereographic projection method. The possible mode of slope failure was determined by structural data and slope geometry. At the location (L1), three set of joints have been observed in quartzite in which joint spacing of almost 1 mm to 5 cm is noticed. The structural data of all discontinuity planes at location L1 were plotted in the lower hemisphere of the equal area projection which shows wedge failure condition (Fig. 5f).

9 Impact of Landslide on Environment

Landslides are destructive and they affect the environment directly or indirectly. Their influence on the environment may be long lasting. In the area different types of landslides have different types of effects on the environment which are described below.

9.1 Topographic Change

Degradation and aggradation of landslide materials may cause topographic change of the Earth's surface. Tectonism and gravity both contribute to landslides activity and both continually modify the Earth's surface. Hence landslide is an important geomorphic agent that shape mountainous area and redistribute sediments in gentle terrain. Large topographical changes were observed at L2, L3, L5 and L6.

9.2 Forest

Uttarakhand is a hilly region of India which covers nearly 61.32% land with forest (Sati and Bandooni 2018). Forest destruction by landslides is a common phenomenon in hilly regions. Excess rainfall and cloud burst activity can damage the forest and be wiped out by landslide. Some examples of forest degradation by landslides in the area are at locations L1, L2 and L3.

9.3 Water Quality

Landslides affect the quality of water. Mostly debris, earth and mud flow type of landslide effect the water quality. A large amount of sediments/debris material enters into river which affects the water quality. The suspended sediments and organic material in river water reduce the drinking water quality. In the area debris materials from L2, L3, L4 and L5 entered into Alaknanda River and degraded the water quality of the river. This has adversely affected the villagers who depend on the river water for their domestic and drinking purposes.

9.4 Biodiversity

When earth and debris material of landslides enter into stream/river they reduce the water quality and harm the habitat of aquatic life like fishes. Fish habitats, especially of the snow trout was greatly affected by the landslides L2, L3, L4 and L5. Landslides impact forest cover area and this directly/indirectly affects the wildlife habitat.

9.5 Agriculture Land

The material of landslide enters into agriculture land and reduces their fertility. Sometimes removal of upper humus layer of farming land has taken place at L3, L5 and L6.

10 Remedial Measures

The main dominant reasons for the slope failure are excess rainfall, permeability of soil, geological structures and cutting of mountains for road widening. The various causes of landslides that have been discussed above and according to these remedial measures are also suggested below:

i. Proper drainage should be prepared along mountain side.
ii. Debris material of the landslide L1 and L4 should be removed.
iii. The retaining wall should be constructed against the slope to give support at L1.
iv. A proper network of surface drainage should be prepared at L1 to L5. At the crown side of the landslide L2, L3 and L5 cemented surface drainage should be prepared for efficient and quick drains.
v. If possible, the direction of surface drainage should be changed at L5 and L6. The concrete check dams should be constructed near toe of the landslide L5.
vi. If possible, the nala at L4 and L5 should be cemented for proper and quick drain so that the side cutting due to nala should be stopped.
vii. The fractured zones near the road side may be sealed by using proper grouting, netting and rock bolting techniques at L1 and L4.
viii. Plantation on damaged landslide surface such as Napier grass (Pennisetum purpureum), Shrubs, Rambans, etc. protects the slope failure.

11 Discussions

The present study area is highly vulnerable for landslide hazards and seismic point of view. After the cloud burst tragedy of 2013 and 2016 many landslides have been reactivated in this area. The four landslide lies on NH-7 which connects Nandprayag and Chamoli, and two landslides present near the NH-107A which connects Chamoli and Gopeshwar. All these six landslides (L1–L6) occurred in moderate to very steep slope. Due to the toe cutting by Alaknanda River and slope cutting for road widening the area becomes more vulnerable for landslides. Heavy rainfall increases the rate of landslide in this area. The NH-7 is the only road that connects Rishikesh and Badrinath and it is a very important route also for the point of view of Badrinath and Kedarnath yatra. Every year thousands of tourists and pilgrims face many difficulties in traffic due to these landslides. In recent past few years, the changing rainfall pattern and some localised precipitation increasing landslides incidence in Uttarakhand state. During the rainy season from June to September these landslides are reactivated and at places soil and debris move downward in huge quantity and damaged the road and infrastructure from time to time. Tectonically the area is very sensitive due to some active major thrust/fault hence rocks present in the area are highly jointed, fractured and sheared. Steep slope (>40°) and aspect is also an important factor for the formation of landslides.

12 Conclusions

Road networks play an important role for transporting in hilly terrain of Himalayan region. Several remote areas connected with the main city by the road networks. During monsoon period many obstructions occurred in the transportation and many infrastructures were damaged due to landslides. By detailed analysis of these landslides, it can be concluded that the rainfall, slope, type of material, geological structures of rocks, drainage, cutting of mountain slope for road widening, toe cutting by river play an important role for developing the landslides. Some important suggestions are:

(i) The landslide is a natural disaster but we must identify the safe zone and unsafe zone.
(ii) After the identification of unsafe zone some mitigation measures should be given.
(iii) Geological and geotechnical investigations should be done before any major construction.
(iv) It is highly recommended that all the construction in the township located along the Alaknanda River should maintain respectable distance from recent high flood level.

Acknowledgements Authors are thankful to Prof. Vinod K. Singh for invitation to write this manuscript and also for fruitful discussion during preparation of the manuscript. We are also thankful to two reviewers Prof. V. Srivastava from Dept. of Geology, B.H.U., Varanasi and Prof. V. Joshi from University School of Environment Management, Guru Govind Singh Indra Prasth University, Delhi for their constructive suggestions to the manuscript.

References

Agarwal NC (1974) Discovery of bryozoan fossil in the calcareous horizon of Garhwal group, Pauri Garhwal district, UP. Himal Geol 4:600–618

Hoek E, Bray JW (1981) Rock slope engineering, 3rd edn. The Institution of Mining and Metallurgy, London

Juyal N, Sundriyal YP, Rana N et al (2010) Late quaternary fluvial aggradation and incision in the monsoon-dominated Alaknanda valley, Central Himalaya, Uttarakhand India. J Quat Sci 25 (8):1293–1304

Kaur S, Purohit MK (2014) Rainfall statistics of India. ESSO/IMD/HS R.F. report/01 (2016)/19, Lodi Road, New Delhi

Khanduri S (2018) Landslide distribution and damages during 2013 deluge: a case study of Chamoli district Uttarakhand. J Geogr Nat Disasters 8(2):1–10

Kimothi MM, Garg JK, Joshi V et al (2001) Landslide induced by Chamoli earthquake: an assessment using high resolution IRS 1C/1D data. In: Wason HR, Pachauri AK, Prakash V (eds) Proceedings of the workshop on recent earthquakes of Chamoli and Bhuj. University of Roorkee, Roorkee, pp 185–196

Kumar G, Agarwal NC (1975) Geology of the Srinagar-Nandprayag area (Alakananda valley), Chamoli, Garhwal and Tehri Garhwal districts, Kumaun Himalaya, Uttar Pradesh. Himal Geol 5:29–59

Mehdi SH, Kumar G, Prakash G (1972) Tectonic evolution of eastern Kumaun Himalaya: a new approach. Himal Geol 2:481–501

Negi R, Singh RA (2020) Landslide: a study from NH-7. Anusandhan Vigyan Sodh Patrika 8(1):29–34

Negi R, Singh RA, Singh PK et al (2018) Study of landslides in Uttarakhand during rainy season of 2018. Anusandhan Vigyan Sodh Patrika 6(1):74–82

Saklani PS, Nainwal DC, Singh VK (1991) Geometry of the composite main central thrust (MCT) in Yamuna valley, Garhwal Himalaya, India. N Jb Geol Palaont Mh 1991:364–380

Sati VP, Bandooni SK (2018) Forest of Uttarakhand: diversity, distribution, use pattern and conservation. ENVIS Bull Himal Ecol 26:21–27

Shah NC (2014) An historical Fallstruz lake, Gohna, dist Chamoli (Uttarakhand), India (1893–1970). Earth Sci India 7(3):1–14

Singh RA (2010) Manle landslides, Pithoragarh Uttarakhand. JSR 54 (1):45–49

Singh RA (2012) Dravyaman Sanchalan (Bhooskhalan): Karan evam upai. In: Singh RA (ed) Bhooskhalan evam paryavaran hraas. Gyanodaya Prakashan, Nainital, pp 67–78

Singh RA (2013) La-Jhekla landslides, Pithoragarh district, Uttarakhand, India. In: Singh RA (ed) Landslides and environmental degradation. Gyanodaya Prakashan, Nainital, pp 141–149

Singh RA, Negi R, Singh PK et al (2021) Landslide studies between Devprayag and Pali along national highway-7, Tehri district Garhwal Lesser Himalaya. JSR 65(1):32–38

Varnes DJ (1978) Slope movement, type and processes. In: Schuster RL, Krizek RJ (eds) Landslide analysis and control. Special Report, vol 176, National academy of sciences, transportation research board, Washington, DC, pp 11–33

9789811641213